SYSTEM INNOVATION FOR AN ARTIFICIAL INTELLIGENCE ERA

System Innovation for an Artificial Intelligence Era: Applied System Innovation X, contains the papers presented at the IEEE 10th International Conference on Applied System Innovation (ICASI 2024, Kyoto, Japan, 17–21 April 2024. Of the more than 600 submitted papers from 12 different countries, after review approximately a quarter was accepted for publication. The book aims to provide an integrated communication platform for researchers from a wide range of topics including information technology, communication science, applied mathematics, computer science, advanced material science, and engineering.

System Innovation for an Artificial Intelligence Era: Applied System Innovation X enhances inter-disciplinary collaborations between science and engineering, and is aimed at academics and technologists interested in the above mentioned areas.

APPLIED SYSTEM INNOVATION X. PROCEEDINGS OF THE IEEE 10TH INTERNATIONAL CONFERENCE ON APPLIED SYSTEM INNOVATION (ICASI 2024), APRIL 17–21, 2024, KYOTO, JAPAN

System Innovation for an Artificial Intelligence Era

Applied System Innovation X

Edited by

Artde Donald Kin-Tak Lam
Fujian University of Technology

Stephen D. Prior
University of Southampton

Siu-Tsen Shen
National Formosa University

Sheng-Joue Young
National United University

Liang-Wen Ji
National Formosa University

CRC Press
Taylor & Francis Group
Boca Raton London New York Leiden

CRC Press is an imprint of the
Taylor & Francis Group, an **informa** business
A BALKEMA BOOK

First published 2025
by CRC Press/Balkema
4 Park Square, Milton Park, Abingdon, Oxon, OX14 4RN

and by CRC Press/Balkema
2385 NW Executive Center Drive, Suite 320, Boca Raton FL 33431

CRC Press/Balkema is an imprint of the Taylor & Francis Group, an informa business

British Library Cataloguing-in-Publication Data
A catalogue record for this book is available from the British Library

Library of Congress Cataloging-in-Publication Data
A catalog record has been requested for this book

ISBN: 978-1-032-84452-7 (hbk)
ISBN: 978-1-032-84751-1 (pbk)
ISBN: 978-1-003-51483-1 (ebk)

DOI: 10.1201/9781003514831

Typeset in Times New Roman
by MPS Limited, Chennai, India

System Innovation for an Artificial Intelligence Era – Lam et al. (Eds)
© 2025 the Editor(s), ISBN: 978-1-032-84452-7

Table of Contents

Editor Biographies xi

An efficient algorithm for mining high average utility patterns based on the UBLN-tree 1
Ye-In Chang, Po-Chun Chuang, Yu-Zhen Yang & Kuan-Ting Lee

Exploring esports game users experience satisfaction and continuance intentions: A comparative study on desktop and mobile users 7
Li-Hsun Peng & Ming-Han Bai

Development of automatic quadratic meshing technique with unequal spreading intervals on outer and inner contours for IC CAD models 11
Jiing-Yih Lai, Pradiktio Putrayudanto, Qi Wang, Joii Huang, Pei-Pu Song & Yao-Chen Tsai

Development of a nasal swab rapid test system for quarantine hotels 19
Yu-Kun Yuan & Kuu-young Young

A reproducible fabrication process for electroactive PDMS coating 23
Anahita Amiri, Nathan van de Wouw, Ines Lopez Arteaga & Dirk J. Broer

An innovative mechanism design of maze box with puzzles as an integrated teaching aid 29
Jian-Liang Lin & Kuo-Huang Hsiao

The shift towards data-driven organizations in the Thai media industry: A framework for adaptation and transformation 34
Jackkit Sangkittiwan, Nagul Cooharojananone, Thira Chavarnakul & Tatri Taiphapoon

The effect of CEO overconfidence on foreign post-entry performance 39
Jung-Ho Lai

Developing of autonomous guided vehicle with collision avoidance system and smart object detection for dynamic collection mechanism 44
Federico David Vera Gomez, Milton Sala Pereira, Franco Lombardo Barrail, Ariel Federico Gonzalez Samaniego & Meng-Kun Liu

Design of an automated guided vehicle with obstacle avoidance and ball collection mechanism 49
Ana Cristina Ojeda Caceres, Rodrigo Ivan Arambulo Gonzalez, Gianluca Distefano Martinez, Kevin David Jaime Insfran & Meng-Kun Liu

Game-changer: AGV and robotic arm integration for automated ping pong ball collection 55
Kenzo Shirahama, Tamara von Glasenapp, Fabricio Alcaraz, Carlos Colmán & Meng-Kun Liu

A study of the performance of a kW-class Stirling engine used in a micro-combined heat and power system based on CFD simulation 59
Jin-Jia Huang, Sin-Ying Shen & Hang-Suin Yang

Intelligent car searching machine interface design for parking lots based on wayfinding behavior 64
Jia-Yang Lin, Meng-Cong Zheng, Jia-Yang Lin & Meng-Cong Zheng

A study on visitors' experience of visiting a tech art exhibition in Taiwan 70
Chih-Yu Teng & Chao-Ming Wang

Exploring the impact of virtual reality on satisfaction with using stress-relieving products 76
Ya-Chun Yang & Chao-Ming Wang

Initial investigation on more elderly-friendly gesture technology of viewing photos on smart phone 80
Ming-Hong Wang, Chih-Yuan Cheng & Chian-Fan Liou

The application of method acting in story writing instruction 85
Shih-Chieh Liao, Chian Fan, Liou Rou & Hua Zeng

Ancient chinese barbed-spring locks: From basic research to science education in museums 92
Kuo-Hung Hsiao, Yang Zhang, Xiao-Jie Zhuo & Jian-Liang Lin

Interaction force estimation for upper-limb exoskeleton 97
Yu-Hao Lai, Ming-Yang Cheng & Hsien-Hao Chiu

Contour error reduction of DMPs with time-dilation-based time scale adjustment 103
Chia-Chun Huang & Ming-Yang Cheng

Inverse study of natural convection heat transfer for four diamond-arranged tubes in a hot box 108
Han-Taw Chen, Li-Yuan Hsu, Hung-Hsiu Liu & Wei-Mon Yan

Study of 3D inverse natural convection heat transfer for single-tube plate-finned tube heat exchanger in an open cavity 114
Han-Taw Chen, Hung-Hsiu Liu, Li-Yuan Hsu, Wei-Mon Yan & Saman Rashidi

A real-time energy management control for a hybrid electric vessel using whale optimization algorithm 120
Yi-Hsuan Hung, Chien-Ming Lin & Chi Li

Effects of different micro-arc oxidation discharge parameters and $AgNO_3$ addition amount on the bioactivity of silver-containing HA coating on titanium substrate 126
Wei-Ting Lin, Yi-Ju Li, Chi-Sheng Chien, Tzu-Yi Hsieh, Jun-Wei Huang, Chieh-Tsung Wang, Tzer-Min Lee & Tsung-Yuan Kuo

Use LSTM to predict hourly effluent water quality of the pilot wastewater treatment plant 130
Yu-Hao Lin, Yu-Hao Wu & Hung-Hsiao Hsu

Effect of different air change rate on thermal comfort and energy consumption of container buildings in subtropical areas 134
Hsiu-Wen Chang, Mei-Huei Li & Chen-Yu Pan

Ventilation strategy for office — the balance between carbon dioxide concentration and air conditioner energy consumption 140
Chin-Wei Chang, Hsieh-Chih Hsu, Lee-Meng Chieh & Chen-Yu Pan

Research on air conditioning vent outlet airflow velocity, distance, and comfort 146
Yu-Ju Chen, Hsieh-Chih Hsu & Chen-Yu Pan

Exploring the usability of the system taskbars in windows 10 and windows 11 150
Meng-Ru Liu & Siu-Tsen Shen

Feasibility study of integrating immersive gastronomic experience with virtual reality 156
Liang-Yin Kuo, Jo-Hua Tseng & Chia-fang Hsu

Exploring the usability and understanding of flood potential maps: Design recommendations for essential features based on user experience evidence from different levels of disaster prevention literacy 159
Jia-Xin Guo & Meng-Cong Zheng

Numerical simulation of linear and circular ridge acoustic waveguides 165
Tai-Ho Yu, Chun-Jen Ku & Tang-Wei Guan

Leveraging the Internet of Things for sustainable waste management and improved customer engagement in Taiwanese university beverage culture 168
Jong-Chao Yeh, Kherion Jay Miranda, Wei-Cheng Chiu, Rong-Bin Lai, Kao Wei & Chen

Design and realization of mosquito trapping device using electronics components 174
Miguel Amado Gaona Armoa, Danna Gabriela López Almada, Enrique Sebastián Shrem Smith & Chen-Yang Lan

Exploring the correspondence of material influence on visual and auditory imagery in over-ear headphones 178
Tzu-Chia Huang & Chun-Heng Ho

Use landmarks as auxiliary positioning information in the presentation of the route navigation interface 183
Chuan-Hung Yuan & Meng-Cong Zheng

Research on STEAM toys with attractiveness factors and high customer loyalty 190
Tz-Ying Lin & Yu-Ling Yang

Comparison of input methods in Virtual Reality: Joystick vs. Gesture control modify 194
Liang-Yin Kuo, Yao-Yun Hsiao, Chen-Yu Chuang, Nian-Ze Hu, Jing-Chung Shen & Chia-fang Hsu

The study of virtual reality assisted teaching application—basketball assisted teaching: Enhancing flow experience 197
Kuo Liang Yin, Kitty Leung, Ng Pei Ying & Hsu Chia-fang

Sustinnovate or perish: Principles for sustainable innovation 202
Ingrid Wiederkehr, Denis Tissen, Christian Koldewey & Roman Dumitrescu

Portals across domains: A data analysis task and documentation canvas for data-driven model-based systems engineering 208
Denis Tissen, Ingrid Wiederkehr, Ruslan Bernijazov, Christian Koldewey & Roman Dumitrescu

Research on the application of generative artificial intelligence to the design process of commercial vehicle rear lamp 214
Tz-Ying Lin & Yu-Ju Lin

Greener production of lead sulfide from lead dust through sodium hydrosulfide roasting followed by water leaching 217
Li-Pang Wang & Yan-Wen Chen

An intelligent swimming recording and replay system based on computer vision and accelerometer 221
Chih-Hsien Huang, Fu-Sung Lin, Ying-Hsien Chen, Pu-Chun Liu, Hsiao-Ching Lin, Chun-Yuan Chen & Han-Wei Lian

Applying brainwave information to dynamic image form for constructing a novel interactive art 224
Hung-Ying Lee & Chao-Ming Wang

Image cropping based on photographic composition 229
Tsu-Kang Fang & Tsorng-Lin Chia

Deep learning-based yoga pose assessment system 235
Yu-Ting Tsai & Tsorng-Lin Chia

The study on how AI is used in early childhood education in Taiwan 242
Chia-fang Hsu & Liang-Yin Kuo

Research on carbon dioxide adsorption and desorption experiments of 0.5mm membrane containing 50:50 PBAT/PBS 246
Yang Yi-Ling & Tsai Jung-Chin

Precise advertising based on people flow analysis using low-power mobile devices 249
Chun-Liang Lee

Adsorption and desorption experiments using carbon dioxide as the adsorbent for membranes containing 30:70 pbat/pbs 253
Hsieh Yueh-Ling & Tsai Jung-Chin

Enhanced biodegradable polyester adsorption-desorption materials: A study of 1.0 mm membrane containing 50:50 PBAT/PBS 256
Tsai Yu-Chia & Tsai Jung-Chin

Developing micropillar array of microfluid smart sensor for Carbon Footprint on-line monitoring 259
Yun-Peng Yeh, Martin Kwek, Alan Tobias Torres Velazquez, Tais Nathaly Machado, Ana Cristina Ojeda Caceres & Tamara Raquel Von Glasenapp

Study on the graphene oxide absorbent to treat heavy metal nickel wastewater 262
Shan-Yi Shen & Yuan-Kai Chang

New style of interactive navigation system of Taipingma Museum of Fuxing Templein Xiluo 265
Shih-Mo Tseng & Chao-Ming Wang

IQLS: Framework for leveraging metadata to enable large language model based queries to complex, versatile data 270
Sami Azirar, Hossam A. Gabbar & Chaouki Regoui

Research on the packaging design of local cultural specialties in Taiwan—taking Chiayi Xingang Candies as an example 279
Huang Hsiao Chien & Hwang Shyh Huei

Control resource set and slot-based resource allocation in 5G new radio networks 287
Ya-Ju Yu & Hao-Chun Kuan

A virtual MIMO transmission scheme using network coding for B5G C-V2X communication systems 294
Cheolwoo You

Research on recycle and innovation of discarded desks and chairs 298
Li-Mei Chen & Cheng-Jia Chuang

Manufacturing supply chains for electric vehicles solutions: Identifying themes and opportunities using a systematic literature review 305
Adrian E. Coronado Mondragon

AI-empowered IoV architecture for real-time application in heterogeneous 6G 310
Linyu Zhu, Song Bao & Shoupeng Lu

Machine learning-based prediction model of thermal-induced error in mechanical press 316
Tsung-Liang Wu, Yu-Tang Cao & Jing-Yao Huang

An efficient clique-based tree approach for incremental mining of co-locations from spatial database 320
Ye-In Chang, Chen-Chang Wu & Jia-Zhen Qiu

The quantum brain hypothesis that self-entraps in the two-layer body-mind memory model dilemma—the existence of free will in conscious and unconscious psychological activities 326
Jung-Jung Wen

Examining the meaning of "Truly Attaining the Path of Vision" in Mahāyāna based on Master Xuanzang's comment about "The Inability of Confused Persons to Understand Complementarity": The Buddha Dharma of Mahāyāna was indeed expounded by the Buddha himself 332
Ping-Shi Xiao & Ling-Tze Chao

A study on psychological defense mechanisms and memory storage in multiple personality disorder: An interpretation using the three-layer psychological structure memory model 338
Wei-Hsuan Fan

Decoding the unconscious for insight into free will and memory storage – A comparative analysis of the three-layer psychological structure and the two-layer body-mind memory model 342
Li-Chen Tsai

A study about generative design of product form based on the PUR model 347
Ruoyu Wang, Yue Zong & Yi Xu

Investigating the traditional and modern symbols of Minnan Architecture: A case study of Liu Zai ancient house 354
Kai-Han Chung & Yun-Ya Huang

A service-oriented biological data storage and processing system 357
Veska Gancheva & Hristo Stoev

A Study on the predicaments and transformation in traditional tea store businesses-take the Hua-Yuan Tea Corp. in Kaohsiung as an example 362
Chien-Chun Lin & Ying-Jye Lee

Understanding the political power of water engineering: Factors of Japan, America and China in the development of water engineering in Taiwan 368
Bin-Yuan Wen & Cheh-Shyh Ting

Computer aided platform for healthcare data analytics 372
Veska Gancheva & Ivaylo Georgiev

Research on the applicability of generative artificial intelligence in image generation 378
Xin-Zhu Liu & Jiun-Ting Chen

A study on the interface usability of ticket booking process for airline websites 382
Ruan Shihan & Chien-Hsiung Chen

Characteristics of SUS304 stainless steel annealed at different temperatures 387
Sean Wu, Chien-Chih Chiang, Yu-Ting Wu, Wen-Jen Lee & Yee-Shin Chang

The cross cultural difference in University students' mobile learning between Malaysian and Taiwanese by using the A^6 metric 392
Siu-Tsen Shen & Stephen D. Prior

A study on the selection of elements and style formation of posters in sugar industry during the Japanese-Occupied Period 397
Po-Jo Lai & Shyh-Huei Hwang

The paradox of universal design's applicability in public spacefrom the church 408
Xiang-Yuan Zeng, Chen Wang, Bai-Yu Liu & Ya-Ling Yu

Author index 413

Editor Biographies

Dr. Artde D.K.T. Lam was born in Hong Kong on February 23, 1965. He received the B.S. degree in mechanical engineering from National Cheng Kung University (NCKU), Tainan, Taiwan, in 1987, and the Ph.D. degree in mechanical engineering (in field of mechanical design) from National Sun Yat-Sen University (NSYSU), Kaohsiung, Taiwan, in 1993. Currently, he is a Distinguished Professor in Department of Digital Design, Fujian University of Technology, Fuzhou, Fujian, P. R. China. His current research interests include nanotechnology, innovation design, creative design and Fractal Theory.

Dr. Stephen D. Prior has been working in the area of Field Robotics for the past 30 years. His research interest in autonomous systems relates to a shortlisted entry to the MoD Grand Challenge event in August 2008, where he led a team to design, make and test a novel drone, which consisted of a patented Y6 arrangement. On the basis of this, he founded the Autonomous Systems Lab and has been researching with a small team of staff/students working on defence-related robotic technologies. His team won the DARPA UAVForge competition in 2012, and he continues to lead teams to success in the annual IMechE UAS Challenge event. He is on the editorial board for the International Journal of Micro Air Vehicles & Unmanned Systems and has published widely on the subject. During the last few years he co-founded a Tethered UAS company developing a persistent stare capability and is the CEO of Hybrid Drones Ltd, who are developing a heavy-lift hybrid, last-mile resupply drone.

Dr. Siu-Tsen Shen has studied widely, gaining her Master degree in Industrial Design Research from the Design Academy of Eindhoven, and her PhD in Design from Goldsmiths College, University of London. She has been a visiting professor at UCL and Middlesex University, UK and is currently a Professor in Multimedia Design at the National Formosa University, Taiwan. Her research interests lie in the areas of HCI, User-Centred Design, Cross-Cultural Research, User Interface Design, and Design Team Formation using Personality Type. She has taught on a number of different programmes including Introduction to Multimedia Design, Creative Thinking & Design Methods, Cognitive Psychology and Emerging Technology. Shen is the founder and president of Multimedia Art and Design Foundation (MADF) based in London. She is an editor in chief in Journal of Multimedia Art, Design and Education (MADE) (ISSN 2634-095X; ISSN 2634-1395) and Journal of Cultural and Creative Industries Research (ISSN 2221-6170). She leads a technical committee for Innovation in Design, Communication and Engineering (ICICE) and International Conference on Applied System Innovation (ICASI).

Dr. Sheng-Joue Young was born in Tainan, Taiwan on June 22, 1981. He received his B.S. degree from Department of Physics, National Changhua University of Education, Changhua, Taiwan in 2003, M.S. and Ph.D. degree from Institute of Electro-Optical Science and Engineering and Institute of Microelectronics, National Cheng Kung University (NCKU), Taiwan in 2005 and 2008, respectively. He was a visiting research student with the Graduate School of Engineering, Chiba University, Chiba, Japan from July 2007 to September 2007. In 2010, he became an Associate Professor with the Department of Electronic Engineering, National Formosa University (NFU), Yunlin, Taiwan, and was promoted to Associate Professor on August 01, 2013. He is also a Distinguished Professor of the Institute of Creative Industries Research, Fuzhou University; period from Jan. 1, 2014 to Dec. 31, 2014, and guide the Institute to carry out academic research. His current research interests include semiconductor physics, optoelectronic devices and nanotechnology. He has published more than 100 SCI and EI papers in recent years.

Dr. Liang-Wen Ji received the B.S. degree in physics, the M.S. degree in material science, and the Ph.D. degree in electrical engineering from National Cheng Kung University (NCKU), Tainan, Taiwan. In 2005, he became an Associate Professor with the Institute of Electro-Optical and Materials Science, National Formosa University (NFU), Yunlin, Taiwan, and was promoted to Full Professor in 2007. His current research interests include semiconductor physics, optoelectronics, and nanotechnology. He is a iTIKI Fellow. He is also a member of IEEE.

System Innovation for an Artificial Intelligence Era – Lam et al. (Eds)

An efficient algorithm for mining high average utility patterns based on the UBLN-tree[1]

Ye-In Chang, Po-Chun Chuang, Yu-Zhen Yang & Kuan-Ting Lee
Department of Computer Science and Engineering, National Sun Yat-sen University, Kaohsiung, Taiwan

ABSTRACT: The average utility (AU) of the pattern is defined as the total utility of the pattern divided by the number of items of the pattern. If AU of the pattern is higher than or equal to the given threshold value, the pattern is defined as a HAUP. Among those algorithms for mining HAUP, the LMHAUP algorithm has been proposed recently. However, the LMHAUP algorithm needs to scan the database twice and sort all of transactions twice. Moreover, the LMHAUP algorithm only considers to mine patterns in the static database. Therefore, in this proposed FUP algorithm, we propose an efficient algorithm which only needs one scan of the database to mine HAUPs in the incremental database. Because for the inserted data in the incremental database, if we consider the difference between the increased threshold value and the increased AU of a pattern, we may detect some patterns which are still kept at the same passed/un-passed status. In this way, we do not need to reprocess the whole mining algorithm for the whole updated database. From our performance study, we show that our proposed approach is more efficient than the LMHAUP algorithm.

1 INTRODUCTION

The most classic algorithm for *Frequent Itemsets Mining* (FIM) is the Apriori algorithm [1]. *High Utility Pattern Mining* (HUPM) algorithms [7,11] consider both the quantity and the profit of the item. *High Average Utility Pattern Mining* (HAUP) algorithms [4-6,9,12,13] consider the length of the high utility pattern.

The utility of a pattern is defined as the quantity of the pattern multiplied by the profit of the pattern. If the utility value of the pattern is higher than or equal to the threshold value, the pattern can be considered the *High Utility Pattern* (HUP). Table 1 shows an example database DB_1. The profits of a, b, c, d, e, f in database DB_1 are $5, 8, 1, 2, 4, 7$. In Table 1, (b, 5) in transaction T_1 means the quantity of item $b$is 5. The utility of pattern X in transaction T_i is the summation of the utilities of each item in the transaction, which is contained in pattern X. For example, for pattern $\{ad\}$ in transaction T_1, we have $uti(ad, T_1) = 23$. The utility of pattern X in database DB_1 is defined as $uti_D(X)$, which is the summation of the utilities of pattern X in all of transactions. The value of TU in database DB_1 means the total utility of each item in transaction T_i, which is defined as $TU(T_i)$, where $T_i \in DB$.

The minimum utility threshold value $\min Util$is multiplied by a percentage δ and the total utility of the database. For example, we set δ to be 20%. In Table 1, we have $\min Util = 32.8$. If the utility of the pattern is higher than or equal to $\min Util$, the pattern is considered the HUP.

However, the utility of a pattern does not satisfy the *downward-closure* property (also called the *anti-monotone* property) [1]. The *downward-closure* property means that any subset of the frequent pattern must be frequent. Therefore, in [10], they have proposed the *Transaction Weighted Utilization* (TWU) [10]. The TWU of pattern X is the summation of the transaction utilities of all the transactions containing pattern X. The TWU of the superset of the pattern is lower than or equal to the TWU of the pattern. The TWU value can satisfy the *downward-closure* property.

The drawback of the HUPM algorithm is that it does not consider the length of a pattern. When the length of a pattern is increased, the utility of the pattern in a transaction is also increased. So, by using the same minimum threshold value to judge patterns with different lengths is not fair. Therefore, the *Average Utility* (AU) has been proposed by Hong *et al.*[3].

The AU is defined as the total utility of the pattern divided by its number of items within it. The average utility of pattern X is defined as $avgU_D(X)$, which is the summation of the pattern utilities in all of transactions divided by the length of the pattern. For example, for pattern $\{cd\}$ in database DB_1, we have $avgU_D(cd) = 35 \div 2 = 17.5$. If the average

[1]This research was supported in part by the National Science and Technology Council of Republic of China under Grant No. NSTC 112-2221-E-110-052.

DOI: 10.1201/9781003514831-1

utility of the pattern is higher than or equal to the threshold value min$Util$, the pattern is considered as the *high average utility pattern* (HAUP). The average utility pattern also does not satisfy with the *downward-closure* property.

Hong *et al.* [4] have proposed the Average Utility Upper Bound (*auub*) to overestimate the actual average utility of a pattern for satisfying the downward closure property. MU in Table 2 means the maximum utility in transaction T_i and it is defined as $MU(T_i)$. For example, for transaction T_1, we have $uti(a, T_1)(= 15)$. Because the maximum utility in transaction T_1 is 40, and we have $MU(T_1) = 40$. The *auub* of pattern X is defined as *auub*(X), which is the summation of MU of the transactions that contains pattern X. For example, pattern $\{c\}$ appears in transactions T_1, T_2 and T_3. We have $auub(c) = 105$. For example, pattern $\{cd\}$ appears in transactions T_1 and T_3. We have $auub(cd) = 65$. If *auub*of pattern X (*i.e.*, $\{c\}$) is lower than min$Util$, the *auub* of pattern Y (*i.e.*, $\{cd\}$), which is the superset of pattern X (*i.e.*, $\{c\}$), is also lower than min$Util$. The number of transactions containing superset Y (*i.e.*, $\{cd\}$) must be equal to or smaller than the number of transactions containing pattern X (*i.e.*, $\{c\}$).

Recently, the *Incremental High Average Pattern Mining* (IHAUPM) algorithm has been proposed. Table 3 shows the transactions which are inserted into database DB_1. For the static database, the HAUI-Miner algorithm [8] uses the *Average-Utility* (AU)-list structure to find the HAUP without generating candidates. The HAUP-Miner algorithm introduces the *Transaction-Maximum Utility Downward Closure* (TMUDC) property to reduce the search space. The TWUDC property can prune the unpromising candidates early by the *Average-Utility Upper-Bound* (*auub*). However, *auub*of the pattern is an overestimated value. The HAUI-Miner algorithm scans the database at least three times. The disadvantage of the list structure is that it needs to add the cost of the joining process. In the LMHAUP algorithm [5], it proposes two tight upper bounds. The tight upper bound can reduce more number of unpromising patterns. However, it only mines HAUPs from static databases and requires additional processing time to calculate the tight upper bounds.

The LMHAUP algorithm [5] based on the list structure has the most strict upper bound among the related algorithm [6]. The LMHAUP algorithm uses the strict upper bound to reduce the number of candidates. When we decrease the number of candidates, the search space and runtime will also be reduced. Therefore, the LMHAUP algorithm has low memory usage and better runtime performance than the other algorithm [13]. However, the LMHAUP algorithm needs to scan the database twice and sort the database twice to construct the data structure. Therefore, the LMHAUP algorithm can only be applied to the static database, but most of databases are incremental or dynamic in the real world. Therefore, they need to mine all of HAUPs again in the updated database. Therefore, to reduce the cost of mining HUPs in the incremental database, in this paper, we propose a new algorithm to efficiently discover HAUPs based on the tree structure in the incremental database. From our performance study, we show that our proposed algorithm is more efficient than the LMHAUP algorithm.

Table 1. An example database DB_1.

TID	Transaction (Item, Quantity)	TU
1	(a, 3), (b, 5), (c, 2), (d, 4)	65
2	(a, 4), (b, 5), (c, 1)	61
3	(c, 1), (d, 12)	25
4	(a, 1), (b, 1)	13

Table 2. The maximum utility (MU) of each transaction in database DB_1.

TID	1	2	3	4
MU	40	40	25	8

Table 3. An example database DB_2 for mining high average utility patterns in the dynamic database.

TID	Transaction (item, quantity)	TU	MU
5	(a, 2), (c, 1), (d, 13)	37	26
6	(b, 1), (e, 1), (f, 9)	75	63

2 A BRIEF SURVEY OF THE LMHAUP ALGORITHM

The LMHAUP algorithm [5] has proposed two tight upper-bounds. The difference between the list of the LMHAUP algorithm [5] and the list of the LIMHAUP algorithm [6] is that the list of the LMHAUP algorithm has links which point the same transaction identity in each item. The LMHAUP algorithm needs to scan the database twice and sort the database once to construct the data structure. Moreover, the LMHAUP algorithm is only applied to the static database, because it requires scanning the original database whenever the transaction is inserted.

3 THE FUP ALGORITHM

In the *Fast UPdated* (FUP) algorithm, we use the proposed $diff_pass$ value, which is used to increase AU to be a pseudo-AU to allow the pseudo-candidate to be stored temporarily. In this way, we store them in Pre_list, such that we can reduce the large disk I/O due to avoid rescanning the whole original database.

3.1 *Basic ideas*

Here, we will introduce variables and basic ideas used in our algorithm. First, we need to determine whether the pattern is HAUP or not by the given threshold value of the *average utility*. Second, to avoid rescanning the database, we will store all of transactions in UBLN-Tree in the first database scan. Third, we consider the dynamic database, we mine HAUPs in UBLN-Tree and store the difference value between the utility of the pattern and the threshold value in HAUP_list. Moreover, we will store the candidate pattern whose utility of the pattern is close to the threshold value in Pre_list. The database DB_3shown in Table 4 is used to illustrate our algorithm. Moreover, the related profits for a, b, c, d, e and f are 4, 6, 10, 2, 8 and 3, respectively. Furthermore, the related supports (*i.e.*, count) for a, b, c, d, e and f are 5, 3, 5, 2, 6 and 5, respectively. DB_4 (which contains $T_8\{(e,4),(f,6)\}$ with maximum utility = 32 and transaction utility = 50, and $T_9\{(c,3),(g,3)\}$ with maximum utility = 30 and transaction utility = 51) is for the case of data insertion.

The UBLN-Tree structure consists of three parts. First, the tree stores the item of the transaction in the node and a root node. The node stores the item, the summation of all of the maximum utilities of the transactions, and the *Count List* (*CL*) which stores the quantity of the item and the quantities of the ancestors of the item. Second, the HeaderM table stores each item and *Average Utility Upper Bound* (*auub*) of each item [9]. To quickly find the item in UBLN-Tree, we link each item to the same name of the node in the HeaderM table and in UBLN-tree. Third, tail_list stores the leaf node of the tree. In the step of sorting the tree, we can easily get every path in the tree.

$Quan(X, T_q)$is the quantity of pattern Xin transaction T_q, and $Prof(X)$ is the profit of pattern X. The AU value is defined as the total utility of the pattern divided by its number of items within it. Take pattern $\{bd\}$ as an example. We have $AU(bd) = \frac{uti(bd,T_1)}{2} + \frac{uti(bd,T_2)}{2} = \frac{6+2}{2} + \frac{6+2}{2} = 8$.

3.2 *An example to construct UBLN-tree*

Take Table 4 as an example. We insert transaction $T_1 = \{(a,2), (b,1), (d,1), (e,2), (f,4)\}$ into UBLN-Tree as shown in Figure 1. In the HeaderM table, we insert the item which appears in transaction T_1 into the HeaderM table. Because the value of $MU(T_1)$ is 16, *auub* of patterns $\{a\}$, $\{b\}$, $\{d\}$, $\{e\}$, and $\{f\}$ is 16. Items a, b, d, e and f are inserted into the HeaderM table and *auub* is stored as 16. In the tree, we start from item a. The node stores $MU(T_1)(= 16)$ and *Count List* (*CL*), which is stored with the quantity (= 2) of item a. Moreover, item a in the HeaderM table has a link to the node which stores item a and the link is shown with the dashed line.

The next node is item b. Item b is inserted into the child node of node item a. The node also stores $MU(T_1)$ (= 16) and CL of item b, with the quantity (= 1) of item b and the quantity of the parent of item b, *i.e.*, the quantity of item a (= 2). Moreover, item b in the HeaderM table has the link to the node of item b. Similarly, items d, e and f are inserted into the tree and have links to the node which stores the same item.

Table 4. The example database DB_3.

TID	Transaction (Item, Quantity)	Maximum utility	Transaction utility
1	(a, 2), (b, 1), (d, 1), (e, 2), (f, 4)	16	44
2	(b, 1), (c, 3), (d, 1), (e, 1)	30	46
3	(a, 2), (c, 3), (e, 3), (f, 4)	30	74
4	(a, 3), (c, 1), (e, 2), (f, 4)	16	50
5	(b, 1), (e, 3)	24	30
6	(a, 2), (c, 3), (f, 3)	30	47
7	(a, 2), (c, 4), (e, 1), (f, 2)	40	62

After inserting transaction $T_2 = \{(b, 1), (c,3), (d,1), (e,1)\}$, we insert item c into the HeaderM table and add $MU(T_2)$ (= 30) to *auub* of each item which has appeared in transaction T_2. Then, we insert item b into the tree. The node stores $MU(T_2)$ (= 30) and CL of the node with the quantity (= 1) of item b. Item b in the HeaderM table is linked the nodes of item b in the tree. Similarly, items c, d and e are inserted into the tree. Figure 1 shows the result of inserting transactions T_1 and T_2.

When transaction $T_3 = \{(a, 2), (c, 3), (e, 3) (f,4)\}$ is inserted, the value of $MU(T_3)$ is 24. We add 24 to *auub*of items a, c, e and f in the HeaderM table. Because transactions T_1 and T_3 have the same prefix item, item a. The node of item a is stored with the sum of $MU(T_1)$ (= 16) and $MU(T_3)$ (= 24), and CL is stored with the sum of quantities of item a in transactions T_1 and T_3. The next item is item c. Item c is inserted into the node of item a. The node stores $MU(T_3)$ (= 24), the quantity of item c (= 3) and the quantity of the parent of item c, *i.e.*, the quantity of item a(= 2). Figure 2 shows the result of inserting transactions T_1, T_2 and T_3.

After transactions $T_1 \sim T_7$ are inserted, the result is shown in Figure 3. Because we want to reduce the number of paths in the tree, we sort UBLN-Tree by values of supports in the descending order. If the supports of two items are the same, we sort its order according to the alphabetical order. The result is shown in Figure 4.

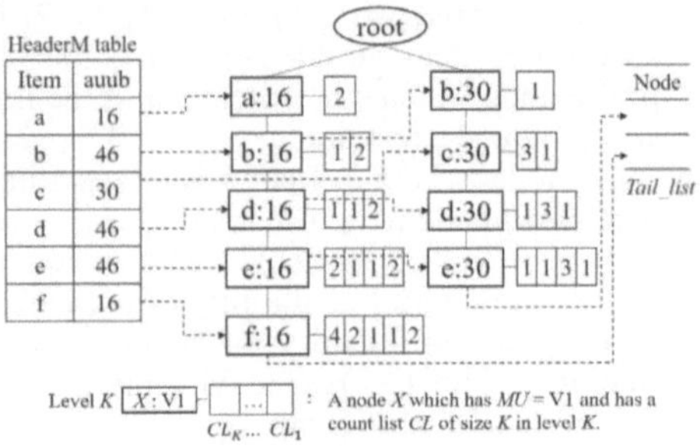

Figure 1. The example database DB_3.

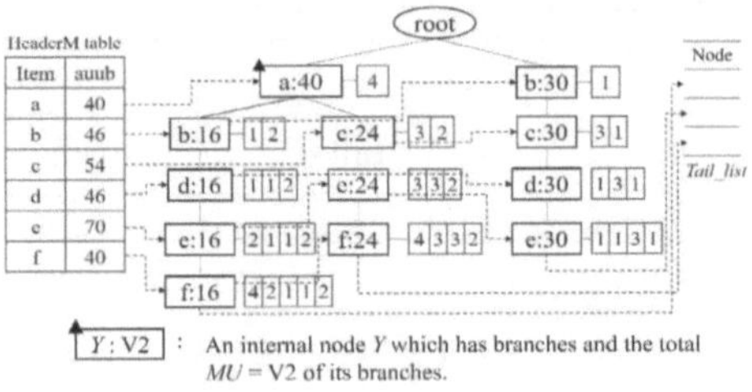

Figure 2. After inserting transactions T_1, T_2 and T_3.

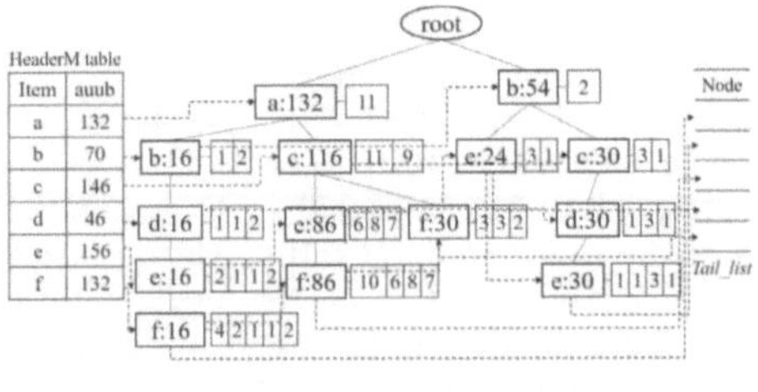

Figure 3. After transactions $T_1 \sim T_7$ are inserted.

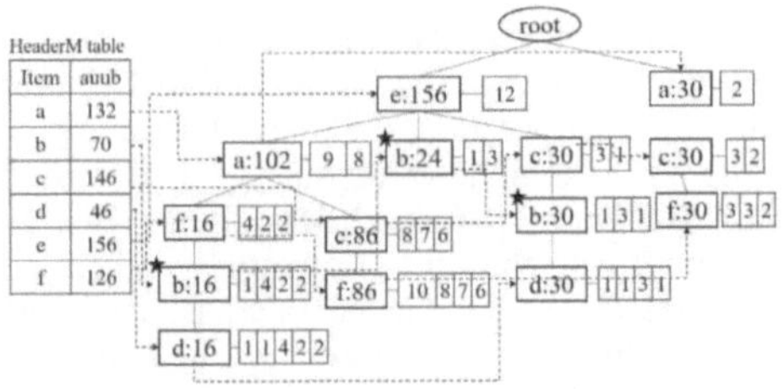

Figure 4. After transactions $T_1 \sim T_7$are inserted and sorted by supports in the descending order.

HAUP_list stores HAUPs and the difference value between the utility of the pattern and the threshold value. However, Pre_list stores the pattern that is not a HAUP. But the sum of the difference value which is between the utility of the pattern and the threshold value, and $diff_pass$ is higher than or equal to the threshold value. The difference value between the utility of the pattern and the threshold value is defined as the absolute value of the difference between $AU(P)$ and $ThreV$.

3.3 *The pruning strategy*

We use *auub*to prune the unpromising patterns early. For any pattern X, the value of $auub(X)$ is higher than or equal to $UtiD(X)$. If $auub(X)$ is lower than $ThreV$, then $UtiD(X)$ is also lower than $ThreV$. For patten Y, pattern Y is the superset of pattern X, the value of $auub(X)$ is higher than or equal to $auub(Y)$. Moreover, if $auub(X)$ is lower than $ThreV$, then $auub(Y)$ is also lower than $ThreV$. For any pattern X, if $auub(X)$ is lower than $ThreV$, then *auub* of any superset of pattern X is also lower than $ThreV$. Therefore, when *auub* of the pattern is lower than $ThreV$, the utility of the pattern and the utility of supersets of the pattern are also lower than $ThreV$. The *Auub* pruning strategy can prune the pattern and its supersets, when *auub* of the pattern is lower than $ThreV$.

3.4 *The mining process*

3.4.1 *The basic idea*

In the mining process, we use the depth first way to mine the HAUPs and items are sorted by the supports in the ascending order. The order is $\{d \to b \to f \to c \to a \to e\}$. First, we mine from the pattern which has the lowest support. We prune the pattern, if *auub* of the pattern is lower than $ThreV$. Second, if *auub* of pattern X is higher than or equal to $ThreV$, we extract all of paths that start from the leaf node of pattern X. We calculate AU of pattern X and utilities of patterns which can combine with pattern X.

The utility of pattern Y in these paths is defined as $u_path(Y)$, which is multiplied by $Quan(Y)$ in these paths and $Prof(Y)$. Third, we find the maximum utility of the pattern which can combine with pattern X, because we need to find the pattern which is the most helpful for pattern X. Pattern Y has the maximum utility in those paths. After we combine patterns X and Y, there are three possible cases. In Case 1, pattern X is a HAUP, no matter whether the utility of pattern Y is higher than AU of pattern X or not. In Case 2, pattern X is not a HAUP, and the utility of pattern Y is higher than $AU$of pattern X and $ThreV$, which means that pattern Y can help pattern X to be a HAUP. In Case 1 and Case 2, if the utility of pattern $X \cup Y$ minus the value, which is multiplied by $ThreV$ and the length of pattern X, is higher than or equal to $ThreV$, it means that pattern $X \cup Y$ will be a HAUP. However, we need to calculate the actual AU of pattern $X \cup Y$ to verify the true decision. In Case 3, pattern X is not a HAUP and the utility of pattern Y is lower than or equal to AU of pattern X, which means that pattern Y can not help pattern

X to be a HAUP. In Case 3, there is no superset of pattern *X* that can be a HAUP.

After mining pattern $X \cup Y$, we will try other patterns which are sorted according to utilities in the descending order. After mining pattern *X*, we will mine other patterns which are behind *X* according to the ascending order of *auub*. Fourth, if the pattern is a HAUP, then we store the pattern and the difference between *AU* of the pattern and *ThreV* to HAUP_list. In order to check quickly whether the pattern is still a HAUP after inserting the new transaction, we will consider the following case. If the pattern is a HAUP and the sum of the different value and new utility of the pattern is higher than or equal to the new *ThreV* of new transactions, then the pattern is also a HAUP. In order to avoid extracting the same paths again, if the sum of the utility of the pattern and $diff_pass$ is higher than *ThreV*, we store the pattern and the difference between *AU* of the pattern and *ThreV* to Pre_list.

3.4.2 *Insertion of the new data*

Because the threshold value of the updated database is increased, the different value between HAUPs and *TherV* is decreased. However, the different value between the pseudo-HAUPs and *ThreV* is increased. After we have inserted the new transactions, three possible cases happen. The database contains many transactions. We called the inserted transactions as DB_{new}, the original database as DB_{ori}, and the updated database as DB_u. There are three cases as shown in Table 5.

Table 5. The algorithm of three cases for new data insertion.

Case: Original → New	Results
Case 1-(a): HAUP → HAUP	Patterns are HAUPs.
Case 1-(b): HAUP → non-HAUP	Decided by the existing information.
Case 2-(a): non-HAUP → non-HAUP	Patterns are not HAUPs.
Case 2-(b): non-HAUP → HAUP	Decided by the existing information.
Case 3-(a): none → HAUP	Decided by the updated threshold value.
Case 3-(b): none → non-HAUP	Decided by the updated threshold value.

4 PERFORMANCE

In this section, we will present the performance study of our algorithm and the LMHAUP algorithm [5], by using the real transaction database.

4.1 *Real transaction databases*

For real transaction databases, we experiment two databases, *foodmart* and *ecommerce*. The details of two databases are shown in Table 6. Because both databases are used in the real life, we use two databases in experiments.

Table 6. Real transaction databases in the experiment.

Database Name	TotalT	ItemN	ItemMax	Density
foodmart	4141	1559	14	0.28
ecommerce	14975	3468	29	0.34

A comparison is shown in Figure 5 by using database *foodmart* under different values of *ThreP*. In Figure 5 and Table 7, the processing time and the number of candidates of our algorithm are always less than those of the LMHAUP algorithm, respectively. Because our algorithm does not need to scan the database twice and sort the database twice, the preprocessing time of our algorithm will be less than that of the LMHAUP algorithm. Moreover, in the mining step, we use three possible cases to replace the computationally complex estimated upper bounds. Therefore, the number of candidates created by our algorithm is also less than that created by the LMHAUP algorithm.

A comparison is shown in Figure 6 by using database *ecommerce* under different values of *ThreP*. In Figure 6 and Table 8, the processing time and the number of candidates of our algorithm are always less than those of the LMHAUP algorithm, respectively. The reason has been explained before.

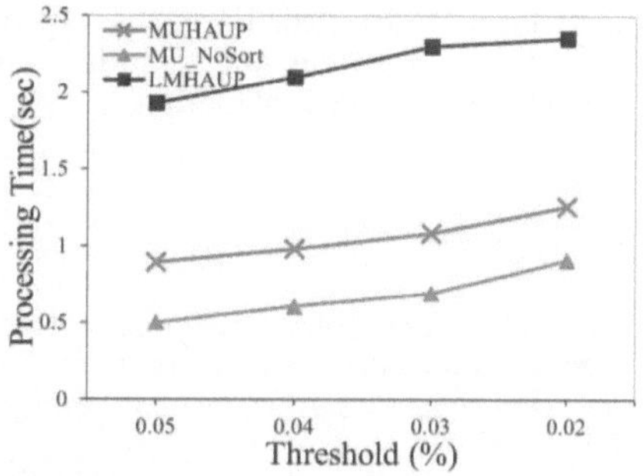

Figure 5. A comparison by using database *foodmart* under the change of *ThreP*.

Table 7. The number of candidates for database *foodmart* under the change of *ThreP*.

ThreP	LMHAUP	MUHAUP	MUHAUP_NoSort
0.05	31951	7203	10757
0.04	37111	12837	16905
0.03	39464	20513	24305
0.02	40045	30605	32507

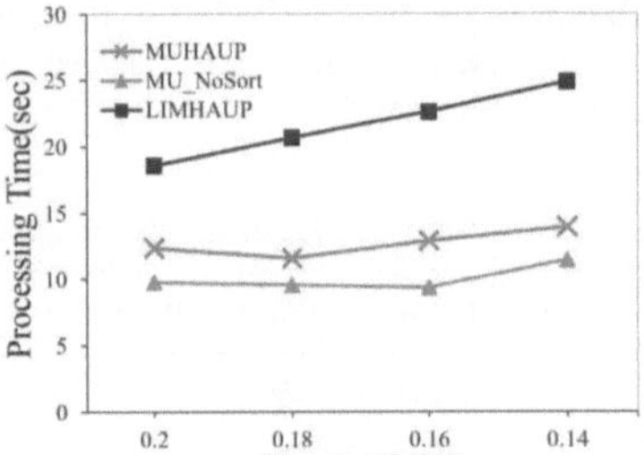

Figure 6. A comparison by using database *ecommerce* under the change of *ThreP*.

Table 8. The number of candidates for database *ecommerce* under the change of *ThreP*.

ThreP	LMHAUP	MUHAUP	MUHAUP_NoSort
0.2	9983	1216	9649
0.18	14637	1594	12764
0.16	22470	2282	15642
0.14	33271	3109	21934

5 CONCLUSION

In this paper, we have proposed the MUHAUP algorithm, which can mine high average utility patterns by only scanning the database once and sorting the tree once. Moreover, for the incremental database, we do not need to rescan the original database after new transactions are inserted. From our experiment results, we have shown that our algorithm is more efficient than the LMHAUP algorithm.

REFERENCES

[1] Agrawal R. and Srikant R. (Sept. 1994). Fast algorithms for mining association rules, *Proc. of the 1994 Int. Conf. Very Large Data Bases*, pp. 487–499.

[2] Fournier-Viger P., Lin C. W., Gomariz A., Gueniche T., Soltani Z. D. A. and Lam H. T. (2016). The Spmf open-source data mining library version 2, *Proc. 19th European Conf. on Principles of Data Mining and Knowledge Discovery (PKDD 2016) Part III*, pp. 36–40.

[3] Hong T.-P., Lee C.-H. and Wang S.-L. (2009). Mining high average-utility itemsets, *Proc. of IEEE Int. Conf. on Systems, Man and Cybernetics*, pp. 2526–2530.

[4] Hong T.-P., Lee C.-H. and Wang S. L. (July 2011). Effective utility mining with the measure of average utility, *Expert Systems with Applications*, Vol. 38, No. 7, pp. 8259–8265.

[5] Kim H., Yun U., Baek Y., Kim J., Vo B., Yoon E. and Fujita H. (Jan. 2021). Efficient list based mining of high average utility patterns with maximum average pruning strategies, *Information Sciences*, Vol. 543, pp. 85–105.

[6] Kim J., Yun U., Yoon E., Lin J. C. W. and Viger P. F. (Oct. 2020). One scan based high average-utility pattern mining in static and dynamic databases, *Future Generation Computer Systems*, Vol. 111, pp. 143–158.

[7] Lee J., Yun U., Lee G. and Yoon E. (June 2018). Efficient incremental high utility pattern mining based on pre-large concept, *Engineering Applications of Artificial Intelligence*, Vol. 72, pp. 111–123.

[8] Lin J. C. W., Li T., Viger P. F., Hong T. P., Zhan J. and Voznak M. (April 2016). An efficient algorithm to mine high average-utility itemsets, *Advanced Engineering Informatics*, Vol. 30, No. 2, pp. 233–243.

[9] Lin J. C. W., Ren S., Viger P. F., Pan J. S. and Hong T. P. (June 2018). Efficiently updating the discovered high average-utility itemsets with transaction insertion, *Engineering Applications of Artificial Intelligence*, Vol. 72, pp. 136–149.

[10] Liu Y., Keng Liao W. and Choudhary A. (2005). A two-phase algorithm for fast discovery of high utility itemsets, *Proc. of the 9th Pacific-Asia Conf. on Advances in Knowledge Discovery and Data Mining*, pp. 689–695.

[11] Nam H., Yun U., Yoon E. and Lin J. C. W. (Aug. 2020). Efficient approach of recent high utility stream pattern mining with iindexed list structure and pruning strategy considering arrival times of transactions, *Information Sciences*, Vol. 529, pp. 1–27.

[12] Sethi K. K. and Ramesh D. (Nov. 2020). High average-utility itemset mining with multiple minimum utility threshold: A generalized approach, *Engineering Applications of Artificial Intelligence*, Vol. 96, p. 103933.

[13] Wu J. M. T., Lin J. C. W., Pirouz M. and Viger P. F. (April 2018). TUB-HAUPM: Tighter upper bound for mining high-average utility patterns, *IEEE Access*, Vol. 6, pp. 18655–18669.

System Innovation for an Artificial Intelligence Era – Lam et al. (Eds)
© 2025 the Author(s), ISBN: 978-1-032-84452-7

Exploring esports game users experience satisfaction and continuance intentions: A comparative study on desktop and mobile users

Li-Hsun Peng
Department of Creative Design, National Yunlin University of Science and Technology, Yunlin, Taiwan

Ming-Han Bai
Doctoral Program, Graduate School of Design, National Yunlin University of Science and Technology, Yunlin, Taiwan

ABSTRACT: In today's prevailing video gaming culture, relevant research topics are gradually gaining attention as video games become integral to people's daily activities. However, most relevant studies have not explored the potential differences that different platforms may bring. This study focuses on the realm of MOBA esports games and aims to differentiate people's experiences and behaviors when using video games into two key aspects: game users experience satisfaction and the continuation intention. This study aims to investigate and analyze these dynamics by exploring the interaction between user experience satisfaction and continued use intention across different devices. Conduct an online survey to achieve the above goal, targeting players of the desktop version of *League of Legends* and the mobile version of *League of Legends: Wild Rift* for investigation and analysis. The mobile version of the game exhibits higher game users' experience satisfaction and continuance intentions than the computer version, and there are many significant correlations between various aspects of the game users' experience satisfaction and continuation intention. These research findings can lead to improved design directions for future MOBA esports games, and future research can conduct a progressive analysis and discussion from these perspectives.

Keywords: Esports, Video Game, User Experience, Continuance Intention

1 INTRODUCTION

Video games have become a fusion of entertainment content and mass media. As the game industry has grown, it has created a "gamer generation" and become a part of their lives (Beck and Wade 2004). In particular, in the era of the flourishing game industry, video games have been considered a part of human culture (Shaw 2010).

However, the impact of video games is broader than just the media and culture. Enhanced the popularity of video games has also been enhanced by the rise of the internet and social networking services. Electronic sports (eSports) have increasingly influenced the economy and society (Vera 2015).

The multiplayer online battle arena (MOBA) genre in esports, exemplified by games such as *League of Legends* released in 2009, has garnered a substantial fan base and significantly influenced contemporary game design. It has become a focal point of attention within the esports community (Mora-Cantallops & Sicilia 2018).

With the increasing number of games and viewers, the number of viewers watching live broadcasts has exceeded that of the National Basketball Association (NBA) in 2014. eSports games were included in the Asian Games in 2022, including *League of Legends*. eSports games have become a global mainstream phenomenon from a niche interest and even a new industry that has broken the framework of traditional sporting events (Chap *et al.* 2023).

Validates that video game and smartphone addiction have valid potential pathological concerns in the modern era (Gutierrez 2021; Shen *et al.* 2019). However, some studies argue that it is essential to consider the additional impacts these usage experiences may have on users (Snodgrass *et al.* 2017). Many research endeavors employing the concept of the Technology Acceptance Model propose that by examining the perspective of users' gaming experiences, we can understand how it influences the intention to continue using video games (Patzer *et al.* 2020; Yang and Lin 2019).

However, the technology acceptance model neglects to explore factors such as playability, narrative, and design aesthetics. Most related

DOI: 10.1201/9781003514831-2

studies have also predominantly focused on MOBA esports games specific to particular device platforms, thereby neglecting to investigate the experiential differences that different devices may bring.

This study investigates potential differences in gaming experiences and continuance intentions on different devices. Additionally, it seeks to understand how game users' experience satisfaction may influence their continuance intentions.

2 EXPERIMENTAL

2.1 *Participants and process*

Based on the recruitment methods employed in related studies (Linh Pham *et al.* 2023; Tan and Yang 2022; Tseng *et al.* 2022), this study utilizes a convenience sampling method to investigate demographic characteristics, game users' experience satisfaction, and continuance intentions. The survey period for this research is from January 2, 2024, to January 16, 2024. Using anonymous online surveys conducted on the PTT Bulletin Board System, Bahamut Gamer's Community, and social networking services like Dcard, individuals aged 20 and above who play the MOBA esports game *League of Legends* on Desktop and *League of Legends: Wild Rift* on mobile targeted for participation.

2.2 *Study tools*

This study examines the assumed relationship between user experience satisfaction and the intention to continue using the game. Investigates the study in three parts: (1) demographic and gaming experience information, (2) the Game User Experience Satisfaction Scale, and (3) the Continuance Intention Scale.

The first part of the survey examined participants' gender, age, education, occupation, annual income, gaming experience, and weekly gaming hours.

The second part consisted of a Game User Experience Satisfaction Scale to measure the experience and satisfaction with video games. It comprised eighteen items, with a total Cronbach's α value of 0.785 (Keebler *et al.* 2020). The scale included nine constructs: Usability/Playability, Narratives, Audio Aesthetics, Visual Aesthetics, Play Engrossment, Enjoyment, Creative Freedom, Personal Gratification, and Social Connectivity. The scale utilized a seven-point Likert scale, ranging from 1 (strongly disagree) to 7 (strongly agree).

The third part involved a Continuance Intention Scale to measure users' intention to continue playing. It consisted of four items and only one construct, with a total scale reliability of Cronbach's α of 0.91 (Hsiao and Chiou 2017; Patzer *et al.* 2020). This scale also employed a seven-point Likert scale, ranging from 1 (strongly disagree) to 7 (strongly agree).

2.3 *Data analysis*

The Kolmogorov-Smirnov test determined whether the samples had a normal distribution.

The results indicated that the samples for all constructs were less than .05. Therefore, this study used non-parametric statistical methods for data analysis.

2.4 *Ethics approval*

The Institutional Human Research Ethics Committee of National Chung Cheng University approved the study on January 2, 2024 (No. CCUREC112121802).

3 RESULTS AND DISCUSSION

3.1 *Demographic characteristics*

Recruited 371 participants for this study, and the demographic characteristics, including gender, age, education level, annual income, pet ownership, and prior augmented reality (AR) experience, were investigated in the demographic profile in Table 1.

Table 1. Demographic characteristics (N = 371).

Demographic		Frequency	Percentage
Gender	Male	277	74.7%
	Female	94	25.3%
Age	20-30	266	71.7%
	31-40	95	25.6%
	31-40	10	2.7%
Education	High school or below	27	7.3%
	University	262	70.6%
	Graduate Schools	82	22.1%
Occupation	Labor force	285	76.8%
	Non-labor force	50	13.5%
	Unemployed	36	9.7%
Annual income	<NT$300,000	135	36.4%
	NT$300,000-600,000	137	36.9%
	$\geq$ NT$600,000	99	26.7%
Gaming Experience	<2 Years	57	15.4%
	2-4 Years	91	24.5%
	$\geq$ 4 Years	223	60.1%
Weekly Gaming Hours	<7 Hours	147	39.6%
	7-14 Hours	133	35.8%
	14-21 Hours	44	11.9%
	$\geq$ 21 Hours	47	12.7%

3.2 *Analysis of demographic characteristics*

The chi-square test was employed to assess whether there were significant differences in the demographic characteristics of participants playing MOBA esports games. Presents the analysis results are in Table 2.

The results indicate a significant difference in gaming experience ($P < .001$) among participants playing on desktop and mobile versions. However, observe no significant differences in gender, age, education level, annual income, and weekly gaming hours.

Table 2. Chi-square test for demographic characteristics between device type (n = 371).

Demographic variable	n	Device type, n (%) Desktop (n = 274)	Mobile (n = 97)	Chi-square test $\chi2$ (df)	P value
Gender				0 (1)	.908
Male	277	205(74.8)	72(74.2)		
Female	94	69(25.2)	25(25.8)		
Age				0.3 (2)	.860
20-29	266	195(71.2)	71(73.2)		
30-39	95	72(26.3)	23(23.7)		
40-49	10	7(2.6)	3(3.1)		
Education				3.7 (2)	.161
High school or below	27	17(6.2)	10(10.3)		
University	262	191(69.7)	71(73.2)		
Graduate Schools	82	66(24.1)	16(16.5)		
Income				0.2 (2)	.915
<NT$ 300,000	135	98(35.8)	37(38.1)		
NT$ 300,000-600,000	137	102(37.2)	35(36.1)		
≥ NT$ 600,000	99	74(27)	25(25.8)		
Gaming Experience				107 (2)	<.001
<2 Years	57	41(15)	16(16.5)		
2-4 Years	91	31(11.3)	60(61.9)		
≥ 4 Years	223	202(73.7)	21(21.6)		
Weekly Gaming Hours				3.1 (3)	.383
<7 Hours	147	115(42)	32(33)		
7-14 Hours	133	92(33.6)	41(42.3)		
14-21 Hours	44	33(12)	11(11.3)		
≥ 21 Hours	47	34(12.4)	13(13.4)		

3.3 *Analysis of different device*

The Mann-Whitney U test was employed to assess whether differences exist in-game user experience satisfaction and continuance intentions between desktop and mobile MOBA esports games, as presented in Table 3.

The results indicate statistically significant differences between desktop and mobile participants regarding game user experience satisfaction related to usability/playability, visual aesthetics, and continuance intentions.

For participants using the desktop version, usability/playability (median 4.5, IQR 3.5-5.5) was lower than that of participants using the mobile version (median 5, IQR 4-6); visual aesthetics for participants using the desktop version (median 5.5, IQR 4.5-6) were lower than those using the mobile version (median 6, IQR 5-6.75); continuance intentions for participants using the desktop version (median 4.25, IQR 4.25-6) were lower than those for participants using the mobile version (median 5.5, IQR 4.75-6).

Table 3. Mann-Whitney U test of game user experience satisfaction and continuance intentions with the device type (N = 371).

Variables	Desktop Median (IQR)	Mean (SD)	Mobile Median (IQR)	Mean (SD)	P value
U/P	4.5 (3.5-5.5)	4.5 (3.5-5.5)	5 (4-6)	4.8 (1.43)	.012
N	5 (3.5-6)	5 (3.5-6)	5 (4-6)	4.84 (1.61)	.192
AA	5.5 (4.5-6)	5.5 (4.5-6)	6 (4.5-6.25)	5.49 (1.09)	.063
VA	5.5 (4.5-6)	5.5 (4.5-6)	6 (5-6.75)	5.76 (1.03)	.021
PE	4.5 (3-5.5)	4.5 (3-5.5)	4 (3.5-5.5)	4.34 (1.48)	.872
ENJ	5.5 (4.5-6)	5.5 (4.5-6)	6 (5-6.5)	5.59 (1.14)	.091
CF	4.5 (4-5.5)	4.5 (4-5.5)	5 (4-6)	4.79 (1.19)	.348
PG	6 (5.5-7)	6 (5.5-7)	6.5 (6-7)	6.24 (0.87)	.052
SC	5.5 (4-6)	5.5 (4-6)	5.5 (4-6)	5.06 (1.25)	.506
CI	5.25 (4.25-6)	5.18 (1.19)	5.5 (4.75-6)	5.46 (1.08)	.033

3.4 *Analysis of correlation*

This study employed Spearman's rank correlation analysis to examine the correlations among usability/playability, narratives, audio aesthetics, visual aesthetics, play engrossment, enjoyment, creative freedom, personal gratification, social connectivity, and continuance intentions, as shown in Table 4.

The results reveal that most nine constructs exhibit a significant positive correlation. However, play engrossment has no significant correlation between usability/playability (ρ = .091, $P <$ .001) or narrative (ρ = .078, P = .079).

Table 4. Spearman's rank correlation of game user experience satisfaction and continuance intentions (N = 371).

Variables	1	2	3	4	5	6	7	8	9	10
1. U/P	1									
2. N	.305**	1								
3. AA	.237**	.383**	1							
4. VA	.265**	.405**	.491**	1						
5. PE	.184**	.226**	.144**	.139**	1					
6. ENJ	.180**	.240**	.379**	.472**	.141**	1				
7. CF	.220**	.483**	.387**	.472**	.262**	.367**	1			
8. PG	.091	.078	.270**	.359**	.137**	.349**	.178**	1		
9. SC	.348**	.255**	.284**	.444**	.200**	.355**	.395**	.185**	1	
10. CI	.293**	.403**	.425**	.594**	.239**	.539**	.498**	.278**	.486**	1

* p < .05. ** p < .01.

3.5 *Discussion*

We observed that participants' gaming experience in *League of Legends: Wild Rift* was significantly lower than in *League of Legends*. This difference may attributed to the fact that *League of Legends* was released in 2009, while *League of Legends: Wild Rift* launched in 2020.

Additionally, it noted that usability/playability, visual aesthetics, and continuance intentions in the mobile version were significantly higher than in the desktop version. This difference could influenced by the distinct game designs tailored for different platforms.

We also observed a significant correlation between game user experience satisfaction and the intention to continue usage. Across various constructs, including usability/playability, narratives, audio aesthetics, visual aesthetics, play engrossment, enjoyment, creative freedom, personal gratification, social connectivity, and continuance intentions. Only play engrossment has no significant correlation with usability/playability and narrative.

4 CONCLUSION

This study aims to conduct an in-depth study of the differences and correlations between MOBA Esport games user experience satisfaction and continuation intention on different platform devices, focusing on players of the specific games *League of Legends* and *League of Legends: Wild Rift.*

The study revealed that MOBA Esport games have higher usability/playability, visual aesthetics, and continuance intentions in the mobile version compared to the desktop version. Furthermore, we found a significant correlation between game users' experience satisfaction and continuance Intentions.

This research confirms that electronic sports games of the same genre exhibit significant differences in game users' experience satisfaction and continuance intentions when played on different platforms. Moreover, this study identifies game users' experience satisfaction as a crucial factor influencing continuance intentions. These insights can serve as valuable directions for improving game users' experience satisfaction in future electronic sports game designs. Additionally, exploring the interactive effects among these constructs through alternative approaches would contribute to a more in-depth analysis.

FUNDING

This research is partly supported by the National Science and Technology Council under Grant No. NSTC—112-2410-H-224-023.

REFERENCES

Beck, J. C. and Wade, M. (2004). *Got game: How the gamer generation is reshaping business forever*. Harvard Business School Press.

Chap, W. K., Rao, A. C. and Pandey, P. K. (2023). A technological review on rise of esports in world economy. *Proceedings of the 4th International Conference on Information Management & Machine Intelligence*, Jaipur, India. https://doi.org/10.1145/3590837.3590916

Gutierrez, J. P. (2021). Do game transfer phenomena lead to flow? An investigation of in-game and out-game immersion among MOBA gamers. *Computers in Human Behavior Reports, 3*, 100079. https://doi.org/10.1016/j.chbr.2021.100079

Hsiao, C.-C. and Chiou, J.-S. (2017). The social influence of online collaborative community: the moderating effect of achievement. *Behaviour & Information Technology, 36*(3), 269–280. https://doi.org/10.1080/0144929X.2016.1221463

Keebler, J. R., Shelstad, W. J., Smith, D. C., Chaparro, B. S. and Phan, M. H. (2020). Validation of the GUESS-18: A short version of the game user experience satisfaction scale (GUESS). *Journal of usability studies, 16*(1), 49–62. https://doi.org/10.5555/3532747.3532752

Linh Pham, T. T., Huang, T.-L., Liao, G.-Y., Chou, Y.-J., Cheng, T. C. E. and Teng, C.-I. (2023). Cooperate to play: How in-game cooperation knowledge impacts online gamer loyalty. *Computers in Human Behavior, 143*, 107686. https://doi.org/10.1016/j.chb.2023.107686

Mora-Cantallops, M. and Sicilia, M.-Á. (2018). MOBA games: A literature review. *Entertainment Computing, 26*, 128–138. https://doi.org/10.1016/j.entcom.2018.02.005

Patzer, B., Chaparro, B. and Keebler, J. R. (2020). Developing a model of video game play: motivations, satisfactions, and continuance intentions. *Simulation & Gaming, 51*(3), 287–309. https://doi.org/10.1177/1046878120903352

Shaw, A. (2010). What is video game culture? cultural studies and game studies. *Games and culture, 5*(4), 403–424. https://doi.org/10.1177/1555412009360414

Shen, S.-T., Prior, S. D., Wang, C.-H. and Kuo, L.-Y. (2019). The negative effect of smartphone use on the happiness and well-being of Taiwan's youth population. In A. Lam, S. Prior, S.-T. Shen, S.-J. Young, & L.-W. Ji (Eds.), *Smart Science, Design & Technology* (1st Edition ed., pp. 180–182). CRC Press.

Snodgrass, J. G., Dengah, H. J. F., Lacy, M. G., Bagwell, A., Van Oostenburg, M. and Lende, D. (2017). Online gaming involvement and its positive and negative consequences: A cognitive anthropological "cultural consensus" approach to psychiatric measurement and assessment. *Computers in Human Behavior, 66*, 291–302. https://doi.org/10.1016/j.chb.2016.09.025

Tan, W.-K., & Yang, C.-Y. (2022). An exploration of MMORPG in-game virtual-item contribution to game enjoyment from the perspectives of purchase behavior and psychological ownership. *Computers in Human Behavior, 134*, 107303. https://doi.org/10.1016/j.chb.2022.107303

Tseng, F.-C., Huang, T.-L., Pham, T. T. L., Cheng, T. C. E. and Teng, C.-I. (2022). How does media richness foster online gamer loyalty? *International Journal of Information Management, 62*, 102439. https://doi.org/10.1016/j.ijinfomgt.2021.102439

Vera, J. A. C. (2015). La dimensión social de los videojuegos' online': de las comunidades de jugadores a los'e-sports'. *Index. comunicación: Revista científica en el ámbito de la Comunicación Aplicada, 5*(1), 39–51.

Yang, H.-L., & Lin, R.-X. (2019). Why do people continue to play mobile game apps? A perspective of individual motivation, social factor and gaming factor. *Journal of Internet Technology, 20*(6), 1925–1936.

System Innovation for an Artificial Intelligence Era – Lam et al. (Eds)
© 2025 the Author(s), ISBN: 978-1-032-84452-7

Development of automatic quadratic meshing technique with unequal spreading intervals on outer and inner contours for IC CAD models

Jiing-Yih Lai, Pradiktio Putrayudanto & Qi Wang
Department of Mechanical Engineering, National Central University, Taoyuan, Taiwan

Joii Huang, Pei-Pu Song & Yao-Chen Tsai
CoreTech System (Moldes3D) Co., Ltd., Hsinchu, Taiwan

ABSTRACT: The purpose of this study was to propose a method for automatic quadratic mesh generation with variable spreading intervals on outer and inner contours for IC CAD models. We have developed a similar method for fixed spreading interval on all contours. However, as the point interval is fixed, substantial number of meshes is required when a model has many tiny features on its inner regions, which frequently appears on IC CAD models. When the spreading intervals becomes different, the sizes of outer and inner meshes may become different, which results in more complex node Type distribution on a loop. Typical problems such as co-linear and mesh intersection problems may thus occur if the meshes are not connected properly. The technique used for fixed spreading interval must be revised to fit for the use for variable spreading intervals. A multi-dimensional testing module was developed under different combination of spreading intervals. The testing results were used to correct the program. The results of a complex IC CAD model under 336 combination of input dimensions were used to demonstrate the feasibility of the proposed program.

1 INTRODUCTION

Mold flow analysis has been widely used in IC packaging to help with the packaging process. In mold flow analysis, it is necessary to convert the CAD model of an IC product into solid meshes for the analysis solver. The composition of meshes can affect the accuracy and computational efficiency of the analysis. Typical types of meshes include tetrahedral, prismatic and hexahedral meshes [1]. Hexahedral mesh is better than the other two types as the elements can be larger in the non-thickness directions compared to the thickness direction. A hexahedral element is essentially composed of six faces, with all faces been quadratic. Hexahedral meshes are typically created by generating surface meshes of quadratic type first, and then creating solid meshes along the thickness direction by sweeping or warping. However, converting a surface into all quadratic meshes is still difficult for complex geometries, such as the surface of an IC product that involves many tiny circular or rectangular features.

Quadratic meshes are basically divided into two types: structured grids and unstructured grids. Structured grids are identified by regular connectivity, which is space efficient in storage, higher resolution and better convergence; while unstructured grids are identified by irregular connectivity, which requires explicit storage of neighboring entities. Most commercial software provides quadratic meshes of structured grids for simple geometries. However, generating all quadratic meshes for complex geometries is difficult. Some software provides quadratic dominant surface meshing, where triangular meshes are employed to fill in regions that cannot be described by quadratic meshes.

Generating quadratic meshes of unstructured grids is more difficult as it must be fitted for any complex geometry. Blacker *et al.* [2] proposed a method called paving algorithm to create quadratic meshes for any 2D shape. George [3] extended the paving algorithm to deal with freeform surface. Skovpen [4] applied the paving algorithm to deal with the issue of converting triangular meshes into quadratic meshes. Several studies tried to improve the paving algorithm [5,6]. In addition, initial point spreading on the contours can affect the formulation of meshes. Fixed spreading interval was commonly studied. However, variable spreading intervals on inner and outer contours may reduce total number of meshes and is more attractive for industrial applications.

2 PROBLEM STATEMENT

The purpose of this study focuses on the development of an automatic quadratic mesh generation method for IC CAD models, emphasizing on the application of variable spreading intervals on outer and inner contours. When the spreading

DOI: 10.1201/9781003514831-3

intervals on outer and inner contours are different, the deviation of adjacent mesh sizes would become large, which can increase the irregularity of node Type distribution on a loop. Typical problems such as co-linear and mesh intersection may thus occur if the meshes are not connected properly. Figure 1 depicts several examples of IC CAD models that can be meshed with different spreading intervals. One feature of these models is that the surface geometry is very complex, especially the existing of many tiny circular and/or rectangular features. Fixed spreading interval may result in increased number of meshes to maintain the shape of tiny features. However, as the space between the outer contour and other inner contours are sufficient large, it could be possible to enlarge outer meshes, while keep inner meshes small. As mesh growing continues, new meshes near the outer contour are reduced, while those near inner contours are increased. By this way, the mesh sizes may not deviate too much when two rows of meshes are finally connected.

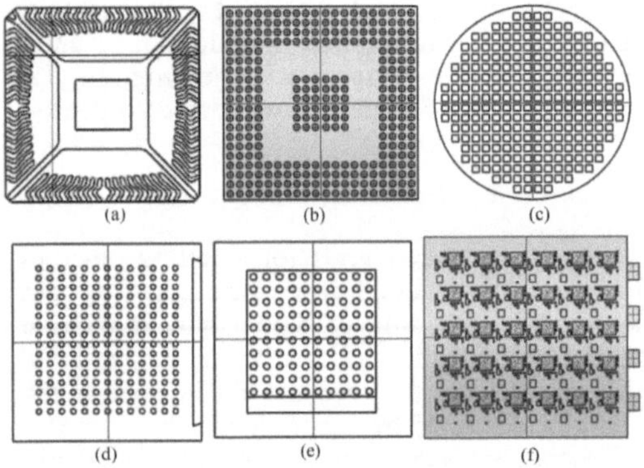

Figure 1. Example of IC CAD models.

The advantage of this concept is that it can be used for any IC CAD model of complex geometry. However, the distribution of nodes during the mesh generation process could vary significantly, which can easily deteriorate the quality of the resulting meshes if any node neighboring conditions is not considered carefully. Therefore, enormous number of testes must be implemented to discover all kinds of node neighboring conditions, and then propose appropriate rules to solve each of the conditions that might deteriorate the meshes. In addition, the computational efficiency is highly dependent on the number of inner loops on a surface because mesh intersection check and correction inside a loop or between loops must be implemented iteratively in each cycle of mesh growing. Six IC CAD models, as shown in Figure 1, with variable combination of spreading intervals, were tested to debug the program and to verify the feasibility of the proposed method.

3 METHODOLOGY

The proposed method to generate quadratic meshes of unstructured grids is based on the paving algorithm [2]. However, the original paving algorithm cannot be used to deal with complex CAD model. One of the reasons is that it must detect the occurrence of intersection and eliminate it. This process is complex and difficult to implement for complex geometries. The other reason is that the node neighboring conditions discussed in [2] were limited. The main difference of the proposed method in this study is that it avoids the occurrence of mesh intersection during the process. A loop is subdivided when it detects that mesh intersection may occur in the next cycle of mesh growing. Similarly, two neighboring loops are combined when it detects that mesh intersection may occur in the next cycle of mesh growing. Besides, the proposed method can be used for variable spreading intervals on outer and inner contours. The node Type distribution is more complex because of the variation of the mesh size throughout the model. The algorithm can easily be crashed if the node distribution patterns are not considered completely.

Figure 2 depicts the overall flowchart of the proposed method. It is mainly divided into 3 program loops L1~L3 and sixteen steps S1~S15 (S10 are repeated twice). In L1, mesh generation is implemented on all active loops recorded in *WorkRow* in sequence. Mesh intersection in the next cycle of mesh growing between the loops and inside a loop are also predicted and solved. In L2, several mesh quality improvement techniques are implemented to correct bad quality of meshes and to stitch narrow gaps, ensuring that mesh intersection will not occur in the next cycle of mesh growing. Finally, L3 is the outer loop of the entire process, where active loops in *WorkRow* are updated and checked. The process stops when *WorkRow*

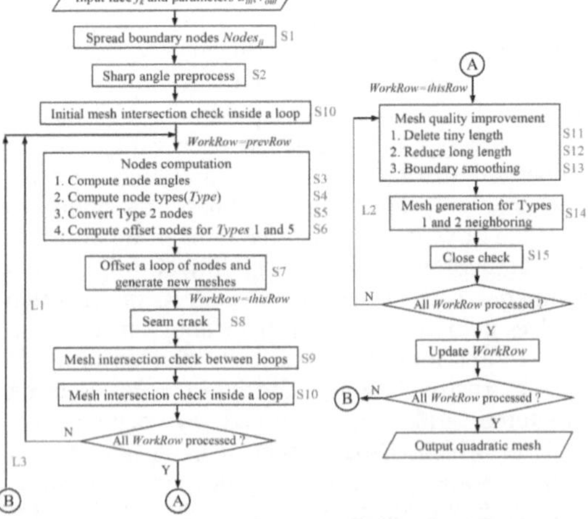

Figure 2. Flowchart of the proposed method to generate quadratic meshes of unstructured grids.

is empty. Once this process is completed, it outputs a surface mesh of full quadratic types. Mesh structure improvement and smoothing is then implemented to improve the overall quality of the meshes. Each of the steps is briefly described below:

3.1 *Spread boundary nodes (S1)*

The input is the boundary contours of a face f_k, which may contain an outer contour (called outer loops hereafter) and several inner contours (called inner loops hereafter). The parameter d_{in} represents the point interval on inner loops, while r_{out} represents the expansion ratio of the point interval between the outer loop and inner loops. Here $r_{out} \geq 1$. Let d_{out} represent the point interval on the outer loop. Then $d_{out} = d_{in}$ if $r_{out} = 1$, while $d_{out} > d_{in}$ if $r_{out} > 1$. Given a point interval, the nodes on each edge (of the outer or inner loop) can be determined. Figure 3 depicts the examples of node distribution on boundary loops for $r_{out} = 1$ and $r_{out} > 1$. The nodes are denoted $Nodes_{ji}$, where j represents the loop index and i represents the node index on the jth loop.

3.2 *Boundary loops preprocess*

When a loop has sharp angle or narrow gap, it may result in mesh intersection in the next cycle of mesh growing. Therefore, two preprocess, namely sharp angle preprocess (S2) and initial mesh intersection check inside a loop (S10), must be implemented first. Let the angle of a node i be θ_i, where θ_i represents the angle between the vectors $V_{i,\,i+1}$ and $V_{i,\,i-1}$. If $\theta_i < \theta_{tol}$, then node i is a sharp angle. Figure 4 shows the algorithm to eliminate sharp angle. First, use 7 nodes i-3 to $i + 3$ to create 5 meshes, where 2 inner nodes are added by using the iso-parametric method. Nodes among i-2 to i + 2 are then excluded from the current loop and all topological data related to nodes and meshes are updated. Second, the length l between i-3 and $i + 3$ is compared with a minimum length l_{min}. If $l < l_{min}$, then use i-4 to $i + 4$ (5 nodes) to create new meshes. The nodes on the current loop are updated accordingly. Repeat this procedure until $l > l_{min}$, which results in a wider edge that allows successful mesh growing later. As for the second preprocess S10, it is the same as the step "Mesh intersection check inside a loop", ad will be described later.

3.3 *Mesh generation on one loop (S3∼S7)*

Steps S3 to S7 represent the generation of meshes on one loop of nodes. Here *prevRow* denotes the index of the existing loop, while *thisRow* denotes the index of the new loop. First, S3 is implemented to compute all node angles on *prevRow*. Second, S4 is implemented to assign a node Type to each node. Six node Types are as follows: Type 1: $\theta_i \leq 90^0 + \alpha_1$, Type 2: $90^0 + \alpha_1 < \theta_i \leq 180^0 - \alpha_2$, Type 3: $180^0 - \alpha_2 < \theta_i \leq 180^0 + \alpha_3$, Type 4: $180^0 + \alpha_3 < \theta_i \leq 270^0 - \alpha_4$, Type 5: $270^0 - \alpha_4 < \theta_i \leq 270^0 + \alpha_5$, Type 6: $270^0 + \alpha_5 \leq \theta_i$, where α_1 to α_5 are fixed parameters. Third, S5 is implemented to convert Type 2 nodes into Type 1 or Type 3. As the node offsetting mode for Type 1 and Type 3 are different, each Type 2 must be converted to either Type 1 or Type 3. Fourth, the offsetting nodes for Type1 and Type 5 must be computed beforehand. One offsetting node is determined for a Type 1 node, while 3 offsetting nodes are determined for a Type 5 node. Finally, S7 is implemented to generate a loop of meshes for the loop *prevRow*. The new loop of nodes is denoted *thisRow*. All topological data related to the old and new loops of nodes and meshes must be evaluated also. It is noted that in S6 and S7, a parameter d_s must be evaluated to determine the offsetting distance. When $r_{out} = 1$, the edge lengths of all adjacent meshes are quite similar. Therefore, d_s can be set as d_{in}. However, when $r_{out} > 1$, adjacent edge length can be quite different. Therefore, a new algorithm is provided to determined d_s locally. Assume that l_{avg} is an averaged length of $n + 1$ nodes close to node i (i is included), n_h denotes the number of nodes with an edge length longer than l_{avg}, and n_l denotes the number of nodes with an edge length shorter than l_{avg}. Then, d_s at node i is determined as follows:

$$d_s = \begin{cases} d_{out}, \mathit{if}\, n_h > n_l \\ d_{in}, \mathit{if}\, n_h < n_l \\ 0.5 \quad (d_{in} + d_{out}), \mathit{if}\, n_h = n_l \end{cases}$$

Figure 5 gives outer and inner loops examples to show the distribution of different node Types and the types of meshes created for each Type of nodes.

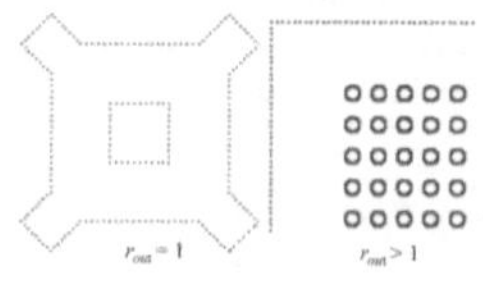

Figure 3. Spread boundary nodes.

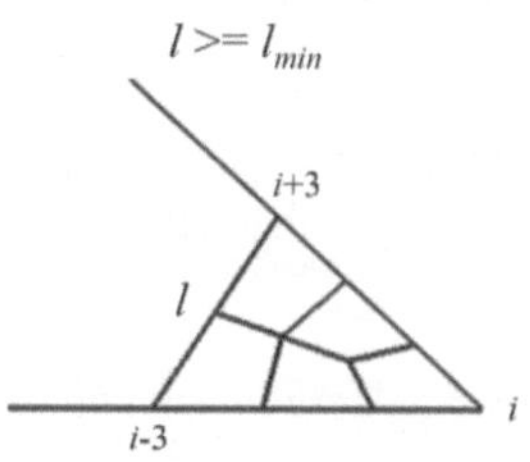

Figure 4. Sharp angle preprocess.

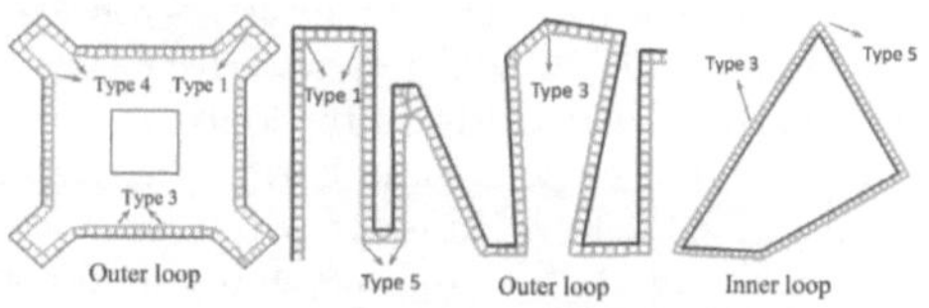

Figure 5. Mesh generation for various node types.

3.4 Seam crack (S8)

When a node angle θ_i is smaller than an allowed tolerance θ_{tol}, it may result in mesh intersection in later mesh growing. The modification of the meshes is not only related to the narrow angle, but also related to the neighboring edge length and the number of near meshes. This process is called seam crack. The conditions to perform seam crack are divided into 7 Cases: Cases 11-13, 2, 3, 41 and 42. The modification of nodes and meshes for Cases 41 and 3 is the same, and that for Cases 42 and 2 is the same.

As Figure 6 depicts, for Case 11, a new node i_{new} is set as the average of i-1 and i + 1. Nodes i, i-1 and i + 1 are moved to i_{new} and 5 meshes m_0 to m_4 are reduced to 4 meshes, where m_0 is eliminated. Cases 12 and 13 are similar to Case 11, except that the right edge length in Case 12 is much longer, while the left edge length in Case 13 is much longer. In Case 41, where the left edge length is much longer that the sum of two right edge lengths, i-1 to i + 2 are used to form a new mesh. On the contrary, in Case 42, where the right edge length is much longer that the sum of two left edge lengths, nodes i-2 to i + 1 are used to form a new mesh.

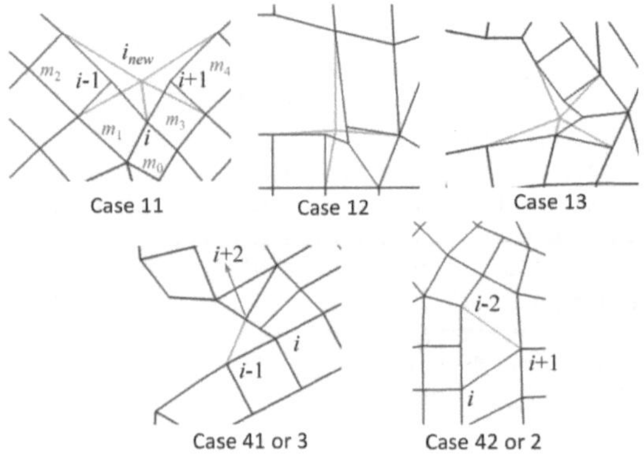

Figure 6. Seam crack.

3.5 *Mesh intersection check inside a loop (S10)*

When narrow gap exists inside a loop, it may result in mesh intersection in the next cycle of mesh growing. Therefore, in this step, the narrow gap is detected first, and then new meshes are created to bridge the gap, which divides the loop into two loops. A local mesh generation (S14) is then implemented on each new loop to remove irregular type of node distribution that might affect next cycle of mesh growing. As each loop may have multiple narrow gaps, the above process is formed as an iterative procedure until no narrower gap exists on the loop.

The algorithm to detect and correct the narrow gap is described below. Consider two nodes i and j on a loop. Compute the angles θ_i and θ_j and compute a length l_{ij} based on the angles. If $l_{ij} < c_1 \cdot l_{avg}$, where l_{avg} is an average length of the edges neighboring to i and j, and c_1 is a coefficient related to the node offsetting distance. It is noted that c_1 for Type 5 should be set larger than Type 1 as the offsetting distance for Type 5 node is longer. The detailed procedures are as follows: (1) determine a set of point pairs: compare each pair of points for the entire loop. When the above length equation is satisfied, a set of candidate point-pairs are selected. (2) determine the point pair i and j to bridge meshes: when a loop is divided into two loops, several problems must be avoided, such as insufficient number of nodes on a loop, co-linear points on a loop and self-intersection on a loop. In addition, as the indices for the first and end points are dis-continuous, additional rules must be provided to determine the final point pair correctly. (3) determine Case for bridging: the number of near meshes, the distance between i and j, and the number of nodes on each loop can divide the conditions into different Cases. (4) divide the loop and generate meshes: individual program is developed for each Case to divide the loop into two loops and create the corresponding bridge meshes. All topological data must be updated correctly. (5) implement S14 on each loop to generate meshes locally.

Figure 7 depicts the intermediate results of the step S10. A distance l_{ij} is computed for every two nodes i and j which are at least two nodes far away. When $l_{ij} < c_1 \cdot l_{avg}$, i and j are regarded as a candidate point pair. Figure 7(a) depicts all candidate point pairs for this example. Figure 7(b) shows the final point pair for bridging the meshes. The distance of them is the minimum among all candidate point pairs except those that may result in insufficient number of points on a loop or co-linear distribution of points. Figure 7(c) shows new meshes created to bridge the gap. The number of meshes along the bridging direction is related to the distance of i and j. The maximum number of meshes can be up to 4. The loop is divided into two new loops, namely L1 and L2. Figure 7(d) shows the result after S14 is implemented on each loop. Here L1 has been closed, and hence is removed from *WorkRow*; while L2 is still active for later mesh growing.

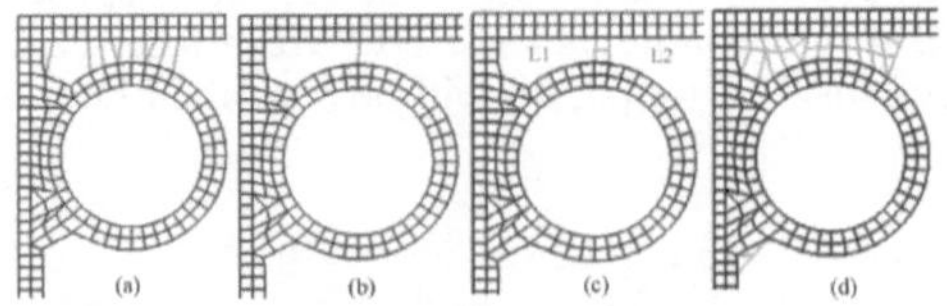

Figure 7. Intersection check inside oneloop, (a) candidate point pairs, (b) point pair for bridging, (c) bridging meshes, and (d) S14 implemented.

3.6 *Mesh intersection check between loops (S9)*

When narrow gap exists between two loops, it must also be solved by creating new meshes to bridge the gap. The algorithm to detect and bridge the narrow gap is similar to that used in S10. However, the main difference is that every pair of loops must be checked one by one. The computational time may significantly be increased when many inner loops exist. To reduce the computational time, the boundary box and orientation of each loop is computed. When the orientation of a loop is counterclockwise, the mesh is grown inside; while when the orientation is clockwise, the mesh is grown outside. Mesh intersection might exist between two loops if one of the following two conditions is satisfied: (1) one loop is counterclockwise and the other is clockwise, and (2) both loops are clockwise and the boundary boxes intersect. With the above-mentioned rules, the computational time can be reduced significantly.

Figure 8 depicts the intermediate results of the step S9. A distance l_{ij} is computed for every two nodes from different loops. A candidate point pair is determined if $l_{ij} < c_1 \cdot l_{avg}$. Figure 8(a) depicts all candidate point pairs. It is noted that insufficient number of points on a loop or co-linear problems would not occur in this step. Therefore, determining the point pair for bridging is easier than that in S10. Figure 8(b) shows the final point pair for bridging meshes. The distance of this point pair is obviously the minimum one. Figure 8(c) shows new meshes created to bridge the gap. It is noted that in S10, only one row of meshes is created as one of the new loops might have scare number of points. However, two rows of meshes are created in this step as the problem in S10 will not be occurred. Similarly, the number of meshes for bridging is between 1 to 4. Figure 8(d) shows the result after S14 is implemented on the new loop.

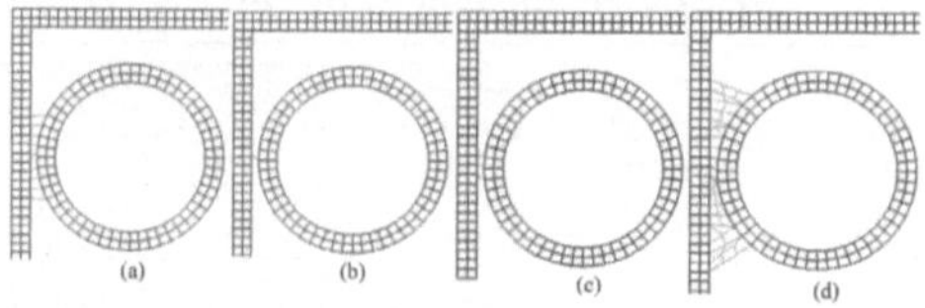

Figure 8. Intersection check between two loops, (a) candidate point pairs, (b) point pair for bridging, (c) bridging meshes, and (d) after S14 implemented.

3.7 *Delete tiny length (S11)*

When two edges neighboring to a node i is shorter than their neighboring edges, it may result in mesh intersection in later mesh growing. Based on the length ratio of neighboring edges and number of near meshes on i and $i + 1$, two types of tiny length modification, as shown in Figure 9, are described below. First, Case "ThreeTiny_Mess22": $l_1 < 0.55 l_2$, $l_3 < 0.55 l_4$ and $l_4 < 0.55 l_5$, where $l_1 = \|i - 1, i\|$, $l_2 = \|i - 2, i - 1\|$, $l_3 = \|i, i + 1\|$, $l_4 = \|i + 1, i + 2\|$ and $l_5 = \|i + 2, i + 3\|$. Also, the near meshes of both nodes i and j are 2. The modification of nodes and meshes is depicted in Figure 9(a). Two new nodes i_{new1} and i_{new2} are added in between $i - 2$ and $i + 3$. They are combined with the original 6 nodes ($i - 2 \sim i + 3$) to generate 3 new meshes. All topological data of nodes and meshes are then updated. Second, Case "OneTiny_ Mess22": $l_1 < 0.55 l_2$ and $l_3 < 0.55 l_4$, or $l_1 < 0.65 d_s$ and $l_3 < 0.65 d_s$. Also, the near meshes of nodes $i - 1$, i, $i + 1$ are 2. The modification of nodes and meshes is depicted in Figure 9(b). A new node i_{new} is put on the average point of nodes $i - 1$ and $i + 1$. Mesh m_0 is deleted, while m_1 to m_3 are modified, as shown in Figure 9(b). When the near meshes of node i, $i - 1$ or $i - 2$ are different (typically 3), the update of the topological data must be modified slightly.

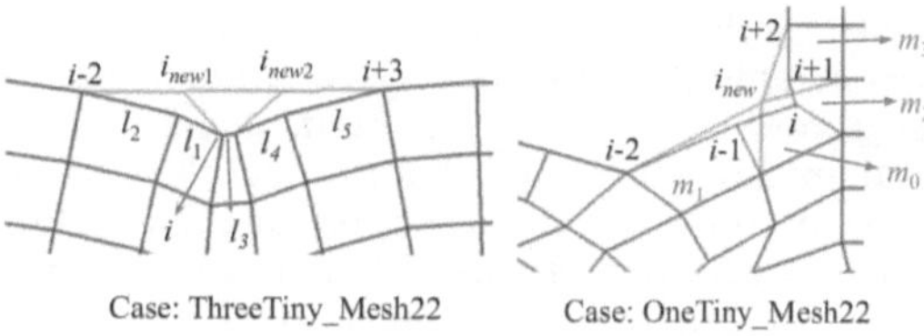

Figure 9. Delete tine length.

3.8 *Reduce long length (S12)*

When both neighboring edges of node i are too long and $\theta_i > 180^0$, new meshes at i will be expanded outside and the mesh size will become bigger and bigger. To solve this problem, two nodes can be inserted on both sides of i to divide two neighboring meshes of i into three meshes. If the following condition is satisfied, then node i is regarded as a long length: ($l_1 > c_1 \cdot l_2$ or $l_1 > c_2 \cdot d_s$) and $\theta_I > \theta_{exp}$, where $l_1 = 0.5 \cdot (\|i - 1, i\| + \|i, i + 1\|)$, $l_2 = 0.5 \cdot (\|i - 2, i - 1\| + \|i + 1, i + 2\|)$, c_1 and c_2 are coefficients, and θ_{exp} is an angle slightly larger than 180^0. Figure 10 depicts the modification of nodes and meshes when long length occurs. Two new nodes i_{new1} and i_{new1} are inserted onto the middle point of both neighboring edges of i, respectively. A new mesh m_2 can thus be generated using nodes i_p, i_{new2}, i and i_{new1}. One node each on m_0 and m_1 is also moved to one of the new nodes. All geometric and topological data of nodes and meshes are updated.

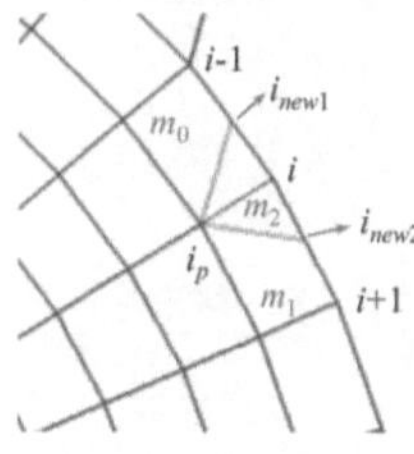

Figure 10. Reduce long length.

3.9 *Boundary smoothing (S13)*

When a new loop of nodes is generated, boundary smoothing is implemented to adjust these nodes slightly. The nodes on the new loop represent the front for later mesh growing. When the distribution of the nodes is improved, it may reduce the generation of poor quality of meshes in later cycles. The boundary smoothing algorithm is primarily based on the iso-parametric smoothing algorithm [2]. It uses the nodes on the neighboring of a node *i* to compute the new position of *i*.

3.10 *Mesh generation for Types 1 and 2 neighboring (S14)*

When Types 1 and 2 are neighboring or close to each other, it may result in narrow region. In addition, when Type 4 or 5 exists nearby, it may even complicate the shape locally. If the irregular or narrow region is not removed, it may easily result in mesh intersection in the next cycle of mesh growing. S14 is very important because during the entire process, irregular distribution of nodes and narrow regions will occur frequently. Therefore, S14 is also implemented in S9 and S10.

Figure 11 shows the flowchart of S14. For each node *i* in loop *this_Row*, its neighboring nodes and node Types are computed. A series of node distribution modes are established as rules and each mode is denoted by a parameter *this_xx*, where *xx* changes in terms of the mode. Table 1 lists the primary series of node distribution modes and each series can further be divided into many sub-modes, each of which is denoted by a *this_xx*. Table 2 gives the sub-modes of *this_B11* series. When the distribution of *i* and its neighboring nodes satisfies a rule, set *this_xx* = *i*. Once all nodes are checked, each *this_xx* is checked one by one. When a *this_xx* is not -1, it indicates that a node distribution mode is satisfied, and hence a Case index is assigned. New meshes can then be generated around *i* and its neighboring nodes for this Case. It is noted that when *this_A11* or *this_B11* occurs, additional sub-modes are distributed to ensure that all new meshes can be created appropriately. The entire process is formulated as an iterative loop to deal with all nodes that satisfy the rules. The process stops when *this_xx* = -1 after all nodes are checked.

Figure 12 gives an example to show how S14 is implemented repeatedly. The left plot on the upper row represents the loop *this_Row*. First, Case *B12_B* is detected, where *i* is Type 1 and *i* + 1 is Type 2. Two meshes are created. Then, Case *B12_A* is detected, where *i* is Type 2 and *i* + 1 is Type 1. Two meshes are created. Next, Case *A5* is detected, where both *i* and *i* + 2 are Type 1. Three meshes are created. The process continues to detect Cases *B12_B*, *A7*, *A2* and *A2* in sequence, and the corresponding meshes are created.

Table 1. Primary series of node distribution mode.

Rule Type	i-2	i-1	i	i+1	i+2	i+3
this_B11 series			1	1		
this_B12 series			1	2		
this_B21 series			2	1		
this_B22 series			2	2		
this_A11 sereis			1		1	
this_A12 series			1		2	
this_A12 series			2		1	
this_A22 series			2		2	
this_C series		5	1			
this_C series		>=4	1	>=4		
this_C series	>=4		1		>=2	
this_C series		>2	1	>=5		
this_S series			1	2	2	1
this_S series		3	1	3		
this_S series			2	3		
this_S series		4	2			2

Table 2. Node distribution mode of *this_B11* series.

<table>
<tr><th>Cases</th><th>i-5</th><th>i-4</th><th>i-3</th><th>i-2</th><th>i-1</th><th>ind</th><th>i+1</th><th>i+2</th><th>i+3</th><th>i+4</th><th>i+5</th><th>i+6</th><th>i+7</th></tr>
<tr><td>this_B11_N41P4</td><td></td><td></td><td>>=4</td><td></td><td rowspan="2">1</td><td rowspan="13">1</td><td rowspan="13">1</td><td>>=4</td><td></td><td></td><td></td><td></td><td></td></tr>
<tr><td>this_B11_N1P4</td><td></td><td></td><td></td><td></td><td></td><td></td><td>>=4</td><td></td><td></td><td></td></tr>
<tr><td>this_B11_N5</td><td></td><td></td><td></td><td></td><td>>=5</td><td rowspan="4">1</td><td></td><td></td><td></td><td></td><td></td></tr>
<tr><td>this_B11_N5P1</td><td></td><td>>=5</td><td></td><td></td><td rowspan="9">!=1</td><td></td><td></td><td></td><td></td><td></td></tr>
<tr><td>this_B11_P15</td><td></td><td></td><td></td><td></td><td>>=5</td><td></td><td></td><td></td><td></td></tr>
<tr><td>this_B11_P1</td><td></td><td></td><td></td><td></td><td></td><td></td><td></td><td></td><td></td></tr>
<tr><td>this_B11_P15_p6</td><td></td><td></td><td></td><td></td><td></td><td>1</td><td></td><td></td><td>>=5</td><td></td></tr>
<tr><td>this_B11_P15_p7</td><td></td><td></td><td></td><td></td><td></td><td>1</td><td></td><td></td><td></td><td>>=5</td></tr>
<tr><td>this_B12_P14 (=B_32)</td><td></td><td></td><td></td><td></td><td></td><td>1</td><td></td><td>>= 5</td><td></td><td></td></tr>
<tr><td>this_B11_N51</td><td></td><td>>=5</td><td></td><td>1</td><td></td><td></td><td></td><td></td><td></td><td></td></tr>
<tr><td>this_B11_P4</td><td></td><td></td><td></td><td></td><td></td><td>>=4</td><td></td><td></td><td></td><td></td></tr>
<tr><td>this_B11_N5P2</td><td></td><td></td><td></td><td>>=5</td><td></td><td><=2</td><td></td><td></td><td></td><td></td></tr>
<tr><td>this_B11</td><td></td><td></td><td></td><td></td><td>!=1, !=2</td><td></td><td></td><td></td><td></td><td></td><td></td></tr>
</table>

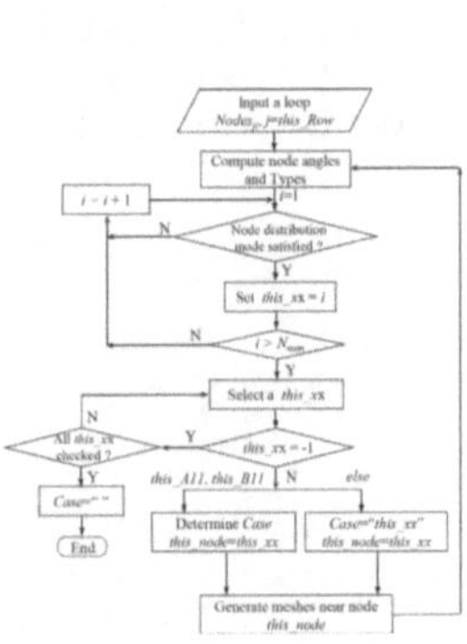

Figure 11. Flowchart of types 1 and 2 neighboring.

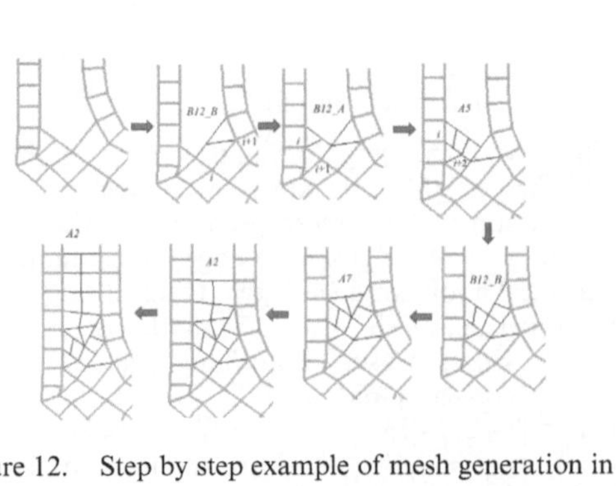

Figure 12. Step by step example of mesh generation in S14.

3.11 *Close check(S15)*

When only 8, 6 or 4 nodes left on a loop, it is necessary generate the remaining meshes in accordance with node distribution patterns. Once the loop is closed, it is removed from *Work_Row*. In S14, most loops with 6 or 8 nodes left are already dealt with. Only 4 nodes left must be dealt with. A new mesh is created by using the 4 nodes left on a loop. However, there might be some cases with 6 or 8 nodes left. They must be closed here.

3.12 *Mesh quality improvement and smoothing*

After the above process, a surface mesh for the model has been generated. However, the mesh quality of this surface mesh is still not satisfied. It is necessary to modify the mesh structure and to perform overall smoothing. The mesh smoothing is combined Laplacian smoothing and iso-parametric smoothing methods. When a node has 3 near meshes only, Laplacian smoothing is employed; while when a node has more than 3 near meshes, iso-parametric smoothing is employed. Figure 13 depicts an example that only mesh smoothing is implemented. Figure 13(a) shows the original surface mesh. It is obvious that many poor faces exist on the model. Figure 13(b) shows the result after mesh smoothing is implemented. The overall quality has been improved. However, there are still some faces with bad structure that will deteriorate the smoothing result.

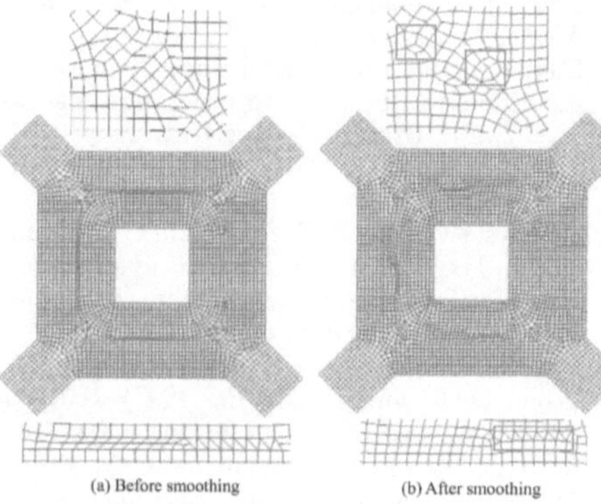

Figure 13. Mesh smoothing alone, d_{in} = 0.1 mm.

Bad structure of faces detected in this study includes (A) two diagonal nodes each with 3 near meshes, (B) two nodes surrounded by six nodes, and (C) a node with 2 near meshes only. Figure 14(a) shows the form of Type A and its correction. This case can be detected by checking the diagonal nodes i and j of a mesh m_0. If both nodes have 3 near meshes, this case occurs. Nodes i and j are moved to their midpoint. And, the original five meshes m_0~m_4 are reduced to four meshes m_1~m_4, where m_0 is deleted. Figure 14(b) shows the form of Type B and its correction. This case can be detected by checking two neighboring nodes i and j. If both nodes have 3 near meshes, this case occurs. Nodes i and j are moved to i' and j', respectively. And, m_2 and m_3 are deleted. Figure 14(c) shows the form of Type C and its correction. This case can be detected by checking the near meshes of node i. If it has 2 near meshes only, this case occurs. In this case, i is removed. Therefore, two original meshes m_0 and m_1 is reduced to m_0.

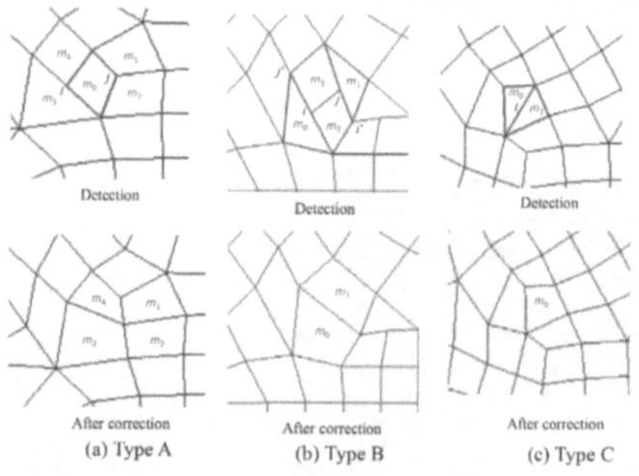

Figure 14. Mesh quality improvement.

4 RESULTS AND DISCUSSION

A program was written in C++ and was based on the Rhino CAD platform and openNURBS functions. It was implemented to test the feasibility of the proposed quadratic mesh generation algorithm for IC CAD models. A multi-dimensional testing module was developed to help the debug of the program. In the multi-testing module, the program is repeatedly tested under different combination of d_{in} and r_{out}. By this way, all kinds of node distribution conditions can be excited to discover all errors on the program. Here the CAD model in Figure 1(f) is demonstrated as it is the most complex model. For multi-dimensional testing, d_{in} varies between 0.04 to 0.06 mm, with an increment 0.001 mm, and r_{out} varies between 3.5 to 5.0, with an increment 0.1. Therefore, 336 spreading intervals were tested in this example.

Figure 15 shows the result for d_{in} = 0.05 mm and r_{out} = 1, In this case, the point intervals of both outer and inner loops are kept the same. Total number of meshes is 828,360 and the CPU time for the entire process is 69.555 sec. The sizes of all meshes are kept similar, as shown on the expansion plot in Figure 15. Figure 16 shows the result for d_{in} = 0.05 mm and r_{out} = 4, In this case, the point interval on the outer loop is 0.2 mm, while 0.05 mm for all inner loops. Total number of meshes is 362,450 and the CPU time is 26.398 sec. It is obvious that meshes near the outer loop is much larger than those near inner loops. The number of meshes is about 44% of that in the first case. In Figure 17, d_{in} and r_{out} are further changed to 0.06 mm and 5, respectively. In this case, the point interval on the outer loop is 0.3 mm, while 0.06 mm for all inner loops. Total number of

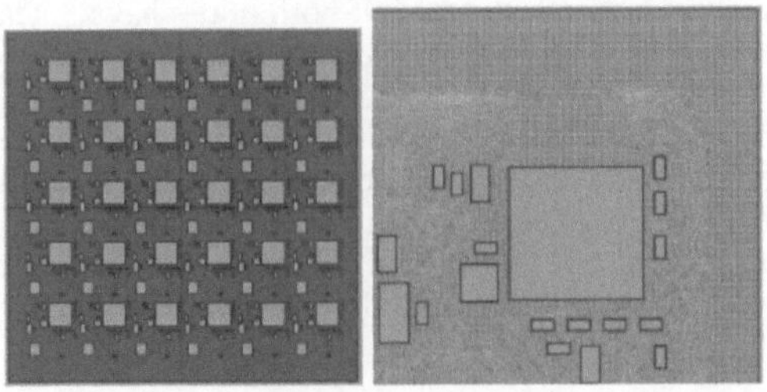

(a) d_{in}=0.05 mm, r_{out}=1

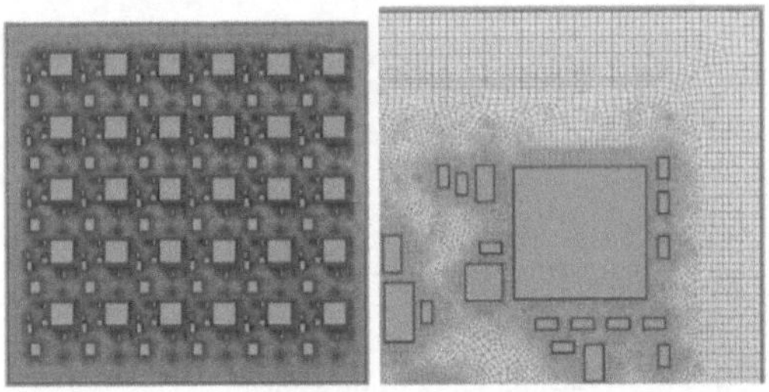

(b) d_{in}=0.05 mm, r_{out}=4

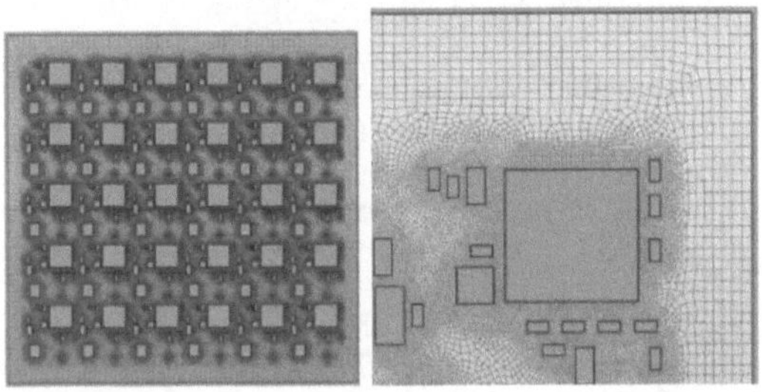

(c) d_{in}=0.06 mm, r_{out}=5

Figure 15. Several results of the IC CAD model in Figure 1f.

meshes is 264,715 and the CPU time is 21.81 sec. The number of meshes is about 32% of that in the first case. We ever run the values d_{in} = 0.06 mm and r_{out} = 8 to check the result. It results in 269,165 meshes and CPU time 17.905 sec. The number of meshes is increased slightly. It is because the space between the outer and inner loops only occupies a small part of the entire area.

5 CONCLUSION

In this study, we focus on the development of an automatic quadratic meshing technique with variable spreading intervals on outer and inner contours for IC CAD models. The proposed algorithm was mainly divided into two parts, mesh generation and quality improvement and smoothing. The process of mesh generation was divided into 16 steps. As the entire process is very complex, a multi-dimensional testing module was developed to excite all program errors. During the testing, it was found that most of the errors appeared in steps S10 and S14, especially when a larger r_{out} is applied. It is because node distribution becomes more irregular and complex when the size difference between outer and inner meshes is enlarged. The CAD model in Figure 1(f) was used to demonstrate the feasibility of the proposed algorithm. A multi-dimensional testing was implemented with d_{in} varying from 0.04 to 0.06 mm, with an increment 0.001 mm, and r_{out} varying from 2.5 to 6.0, with an increment 0.01. The program can work perfectly for all cases tested.

REFERENCES

[1] Moldex3D, Website: http://www.moldex3d.com/en/, Accessed November, 2022.

[2] Blacker T. D. and Stephenson M. B., (1991). Paving: a new approach to automated quadratic mesh generation, *International J. for Numerical Methods in Engineering*, Vol. 32, No. 4, pp. 811–847.

[3] Cass R., (1992). A general three dimensional all quadrilateral surface mesh generation algorithm, *MS. Thesis*, Brigham Young University.

[4] Skovpen A. V., (2004). Modified algorithm for unstructured quadratic meshing, *RFNC-VNITF.*

[5] Tautges T., Lober R. R., Vaughan C. and Jung J., (1995) The design of a parallel adaptive paving all-quadrilateral meshing algorithm, *Proc. The 6th International Symposium on Computational Fluid Dynamics*, Sep. 4-8, Lake Tahoe, USA.

[6] White D. R. and Kinney P., (1997) Redesign of the paving algorithm: robustness enhancements through element by element meshing, *Proc. 6th Int. Meshing Roundtable*, pp. 323–335.

System Innovation for an Artificial Intelligence Era – Lam et al. (Eds)
© 2025 the Author(s), ISBN: 978-1-032-84452-7

Development of a nasal swab rapid test system for quarantine hotels

Yu-Kun Yuan & Kuu-young Young
Graduate Degree Program of Robotics, National Yang Ming Chiao Tung University, Hsinchu, Taiwan

ABSTRACT: With the appealing non-contact nature, robots have been favored to replace human medical staffs. Even with the influence from COVID 19 has been alleviated, the society still needs to be prepared for possible pandemic occurrence in the future. In this paper, we intend to develop an autonomous nasal swab rapid test system based on using the mobile robot manipulator as the platform for quarantine hotels, which is an important facility to block pandemic spread. We first analyze the loopholes arising from antigen screening in the hotel to come up with proper system design. For robot navigation, the system is equipped with the functions of localization, path planning, and motion control. The deep learning approach combined with the image recognition method is utilized for nostril detection, and force control strategy is developed to ensure the success rate of the rapid test, in additon to the safety and comfort of the subjects. A series of experiments are conducted, with the feedbacks collected through questionnaires.

1 INTRODUCTION

One effective way to suppress the spread of pandemic, just like what we did to deal with that of COVID-19, is to reduce direct or indirect human-to-human contact, as well as to conduct inspection and quarantine at the cross-border entry. As the quarantine hotel is an important facility to block pandemic spread, For current rapid test in a quarantine hotel, the common approach is that the staff places the rapid test reagents at the door for the person(s) under quarantine to collect and test by themselves, which will then be treated as the waste. One problem is that if the quarantined person was already infected, it may invoke cross-infection due to indirect contact with the staff. Anothe concern is that different persons may not take the same way for testing, while some might even render false test results for personal reasons.

As the robot is favored to serve as a paltform with its non-contact nature [1-2], we thus propose developing an autonomous nasal swab rapid test system based on using the mobile robot manipulator. The proposed system is not only intended to meet task requirements by avoiding the loopholes arising from current antigen screening in the hotel mentioned above, but also to reduce subjects' psychological pressure. For navigation among the room and quantarined person(s), the system needs to be able to conduct localization, path planning, and motion control in the presence of both static and moving objects. For nostril detection, the commonly used image recognition method is employed with the assistance from the deep learning approach. To achieve a better success rate of the rapid test, while ensuring the safety and comfort of the subjects, a force control strategy based on impedance control is developed. We conduct a series of experiments, approved by the IRB: NYCU-REC-111-087E, Taiwan, to evaluate system performance, including nostril detection, tested nasal swab collection, robot navigation, and entire system feasibility, and also collect feedbacks from the participants through questionnaires. The results show that the proposed system is suitable for use in a quarantine hotel, and capable of reducing the risk of possible cross-infection caused by human testing.

2 PROPOSED SYSTEM

Figure 1 shows the system block diagram of the proposed autonomous nasal swab rapid test system, which includes the four main modules: nostril detection, environment detection, robot localization, and force control. Four sensors are equipped on the mobile robot manipulator, including depth camera, force sensor, laser range finder, and inertial measurement unit to provide the necessary information for these modules to execute their functions.

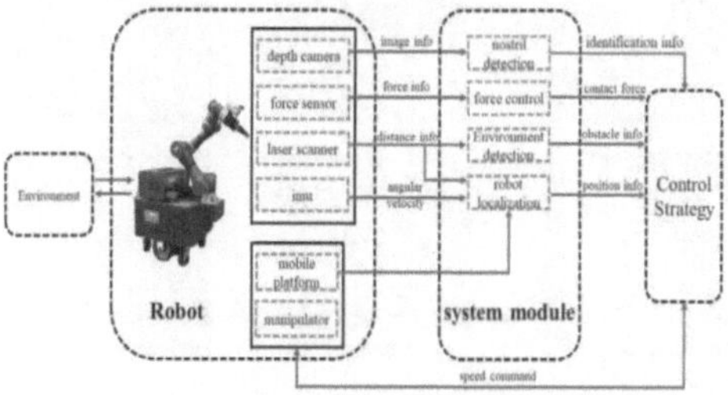

Figure 1. System block diagram for the proposed system.

DOI: 10.1201/9781003514831-4

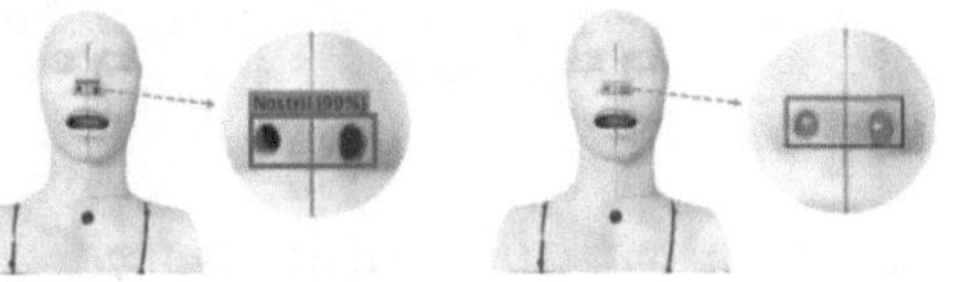

(a) nostril location (b) nostril shape and center

Figure 2. Detection results.

For nostril detection, by considering the limited computing power provided the system equipped on the mobile robot manipulator, we adopted the YOLOv4-tiny algorithm [3] along with the OpenCV library (Open Source Computer Vision Library) [4] for image processing. In this stage of the research, the mannequin is used for study. We collected 617 images of different angles, lightings, and backgrounds. Figure 2 shows the detection results, with Figure 2(a) the nostril location bounding by a box by using the YOLOv4-tiny algorithm and Figure 2(b) the shape and center of the nostril by using the Open CV.

For robot navigation, which demands environment detection and its localization, we first applied the Gmapping algorithm [5] to build the map. With the map and initial pose of the robot manipulator, the AMCL algorithm (Adaptive Monte Carlo Localization), which is effective for dynamic pose estimation, is then used to determine robot's actual position. Figure 7 shows the process by applying the AMCL for robot localization, during which the particles from a sparse distribution gradually converges, indicating the system finally recognizes the precise location of the robot within the environment.

The next step will be path planning for navigation. As there may be table, chair, bed, and others present within a quarantine hotel, along with people walking around. To deal with both static and dynamic obstacles, we propose first utilizing the A* algorithm [6] to plan a feasible path among the static obstacles, and then the DWA algorithm (Dynamic Window Approach) [7], which can adjust the moving speed to avoid the dynamic obstacles along the path. By combining both algorithms properly, the system can smoothly navigate the robot to reach the desired target without collision.

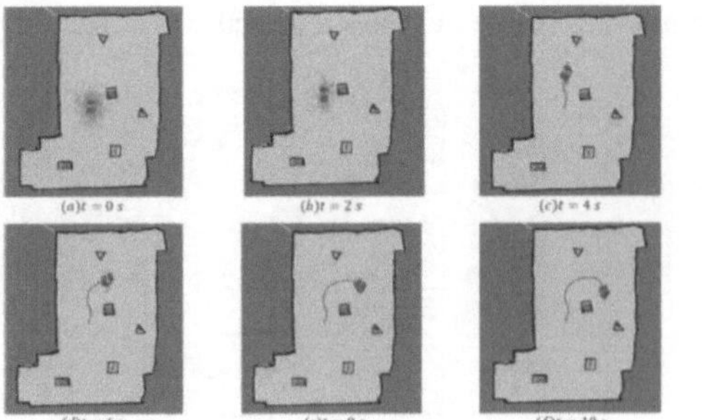

Figure 3. The process by applying AMCL for robot localization.

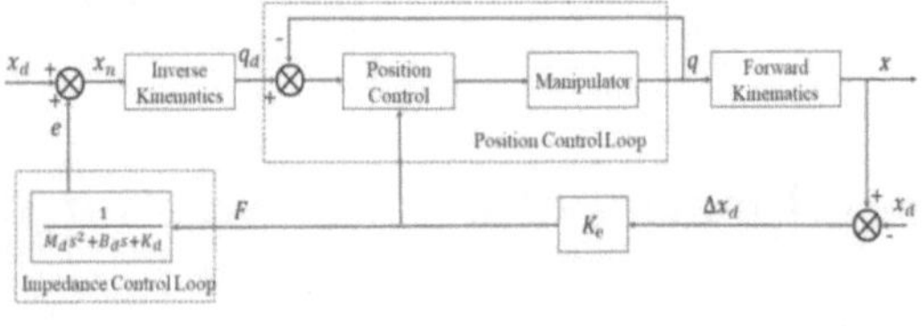

Figure 4. Position-based impedance control scheme.

The safety and comfort of the subject is of concern in our design. Therefore, we especially consider legibility, safety, and physical comfort during the execution of the nasal swab collection. It will be conducted in an easy-to-understand way to reduce the psychological pressure of the person to be inspected, along with the reminding via voice that the head should be tilted back 70 degrees and try not to move the body.

Force control is also crucial for safety and comfort insurance, which will be activated when the measured force exceeds a preset value that is considered to be allowed in the human nasal cavity. The position-based impedance control scheme, shown in Figure 4, is developed to serve for the purpose, which mainly consists of a position control in the inner loop and an impedance control at the outer. When the contact force exceeds the threshold, the system will switch from position control to that of compliance, and proceed with corresponding position adjustment, and thus render a proper contact force for the nasal swab, ensuring human safety during the process.

3 EXPERIMENT

All the experiments were conducted in our laboratory, which was arranged to be similar to an isolation room in a quarantine hotel. The 2019-nCov special nasal swab [8] manufactured by Aikang Biotechnology Co., Ltd. was selected for rapid test. Parameter settings, listed in Table 1, were determined empirically during the experiments, which satisfy the safety and comfort concern. Feedbacks from 20 participants were collected via three questionnaires: System Usability Scale (SUS) [10], an indicator of subjective usability, After-Scenario Questionnaire (ASQ) [11], for evaluation of the usability of the mobile robot manipulator, and

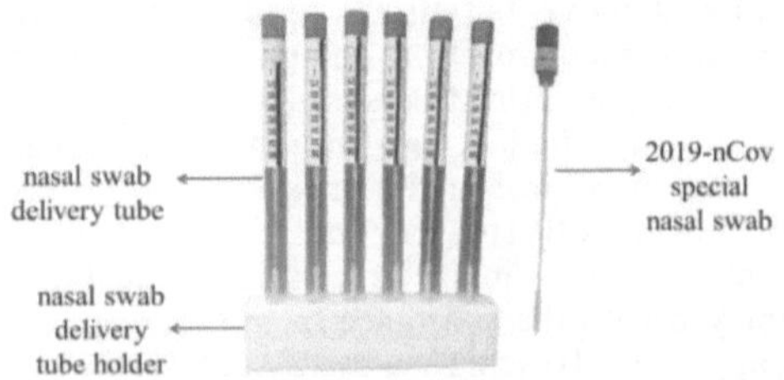

Figure 5. 2019-nCov special nasal swab.

Human-Machine Industrial Collaboration Trust Scale [12], for trust evaluation between human and robot.

Table 1. System parameter settings.

	Parameter	Value
Mobile platform	speed limit	0.4 m/s
	acceleration limit	2 m/s^2
Range of robotic arm	free space	0.85 m
	detection space	0.5 m
	collection space	0.25 m
Contact force threshold	Z direction	3 N

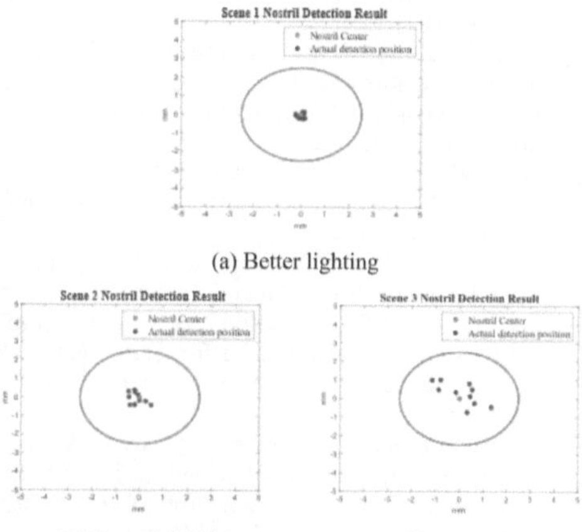

(a) Better lighting

(b) Normal lighting

(c) Poor lighting

Figure 6. Detection results for nostril center.

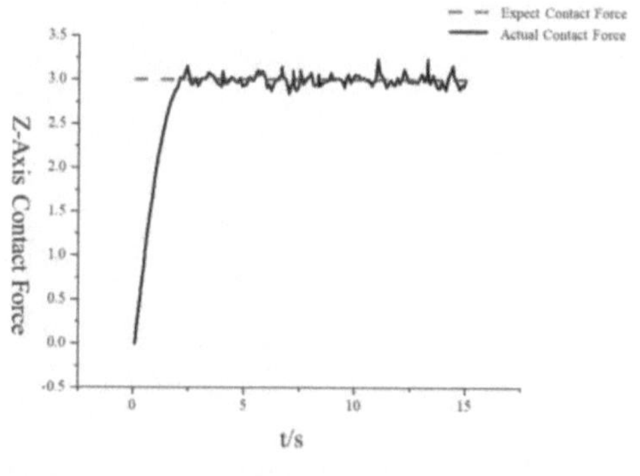

Figure 7. Force response with slight head movement.

The first set of experiments were intended to evaluate the performance of the nostril detection module. Figure 6 shows the detection results under different lightings, in which the red point indicates the center of the nostril and the blue one the identified position. From Figure 6(a)-(c), all the identified positions were very close to the center. We also conducted experiments by placing the mannequin at different locations and also with

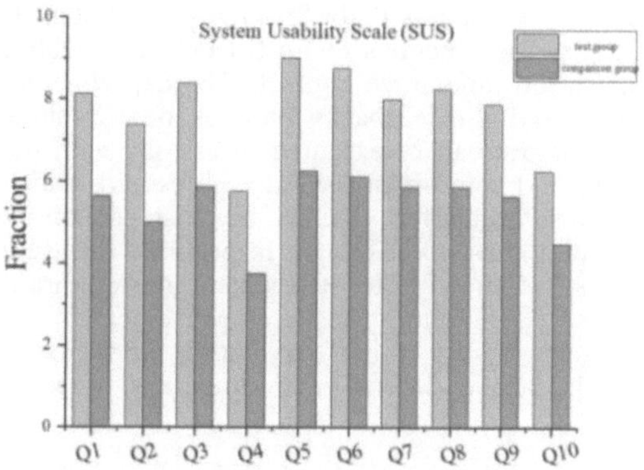

Figure 8. The result of the questionnaire for system usability scale.

different orientation. The results were also satisfactory, thus demonstrating its effectiveness.

The second set of experiments were intended to evaluate the ability of the force control scheme in handling unexpected situations, such as deviation in nostril position detection, head movement during inspection. Under such conditions, the nasal swab held by the gripper of the robot arm might be deformed, which in turn generated a contact force and consequently activated the force control scheme. Figure 7 shows the force response for a case that a slight head movement occurred during inspection. In Figure 7, the actual contact force gradually approached the preset expected force and stayed to be within a small range, implicating its ability for disturbance management.

The final set of experiments were intended to evaluate system feasibility of the entire process, including robot navigation, nostril detection, force control, and rapid test. We conducted the experiments for 15 times, and successfully completed 12 times. The 3 failures were mainly due to the imprecise navigation of the mobile robot manipulator, even though the nostril detection module indeed had identified the position of the nostril correctly, while the inspection area was out of reach. To equip the proposed system with a better navigation capability will be one of our future works. Meanwhile, the three questionnaires collected after the experiments all reveal that the experimental group yielded evidently higher score than that of the control group, as that for the questionnaire of system usability scale shown in Figure 8. It indicates that the participants favored the proposed system more. And, most of them expressed their willingness to use this system in the quarantine hotel.

4 CONCLUSION

In this paper, we have developed an autonomous nasal swab rapid test system based on using the mobile robot manipulator for quarantine hotels. It includes the functions of nostril detection, robot navigation, and force control. The experimental results and also feedbacks via questionaires show that the test can be executed sucessfully with the safety and comfort of the subject ensured, which can also effectively reduce the risk of cross-infection caused via human inspection. In future works, we intend to test the system in a quanrantine hotel, in addition to the improvement of the accuracy during robot navigation and nostril detection under differnt environmental settings.

ACKNOWLEDGMENT

This work was financially supported in part by the National Science and Technology Council, Taiwan, under grant NSTC 111-2221-E-A49-178-MY2.

REFERENCES

[1] Li S.-Q., Guo W.-L., Liu H., Wang T. and Zhou Y.-Y., (2020). Clinical application of an intelligent oropharyngeal swab robot: implication for the COVID-19 pandemic, *European Respiratory Journal*, vol. 56, no. 2, pp. 2001912.

[2] *The Autonomous Nasal Swab Collecting Robot*, Available: https://brainnavi.com/product/nasalswabrobot/, 2021.

[3] Solawetz J., (2020). *Train YOLOv4-tiny on Custom Data - Lightning Fast Object Detection*, Available: https://blog.roboflow.com/train-yolov4-tiny-on-custom-data-lighting-fast-detection/.

[4] Open Source Computer Vision Library, (2012). Available: https://opencv.org/.

[5] Grisettiyz G., Stachniss C. and Burgard W., (2005). Improving grid-based SLAM with rao-blackwellized particle filters by adaptive proposals and selective resampling, *IEEE International Conference on Robotics and Automation*, pp. 2432–2437.

[6] He S., Song T., Wang P., Ding C. and Wu X., (2023). An enhanced adaptive monte carlo localization for service robots in dynamic and featureless environments, *Journal of Intelligent & Robotic Systems*, vol. 108, no. 6.

[7] Hart P. E., Nilsson N. J. and Raphael B., (1968). A formal basis for the heuristic determination of minimum cost paths, *IEEE Transactions on Systems Science and Cybernetics*, vol. 4, no. 2, pp. 100–107.

[8] Fox D., Burgard W. and Thrun S., (1997). The dynamic window approach to collision avoidance, *IEEE Robotics & Automation Magazine*, vol. 4, no. 1, pp. 23–33.

[9] ACON® SARS-CoV-2 IgG/IgM Rapid Test, (2022). Available: https://www.aconbio.com/en/ACON/650.html.

[10] Brooke J., (1996). SUS-A quick and dirty usability scale, *Usability Evaluation in Industry*, vol. 189, no.3.

[11] Lewis J. R., (1991). Psychometric evaluation of an after-scenario questionnaire for computer usability studies: the ASQ, *ACM Sigchi Bulletin*, vol. 23, no. 1, pp. 78–81.

[12] Charalambous G., Fletcher S. and Webb P., (2016). The Development of a Scale to Evaluate Trust in Industrial Human-robot Collaboration, *International Journal of Social Robotics*, vol. 8, no. 2, pp. 193–209.

System Innovation for an Artificial Intelligence Era – Lam et al. (Eds)
© 2025 the Author(s), ISBN: 978-1-032-84452-7

A reproducible fabrication process for electroactive PDMS coating

Anahita Amiri
Institute for Complex Molecular Systems (ICMS), Eindhoven University of Technology (TU/e), The Netherlands
Department of Mechanical Engineering, Eindhoven University of Technology (TU/e), The Netherlands

Nathan van de Wouw & Ines Lopez Arteaga
Department of Mechanical Engineering, Eindhoven University of Technology (TU/e), The Netherlands

Dirk J. Broer
Department of Chemical Engineering, Eindhoven University of Technology (TU/e), The Netherlands

ABSTRACT: In this paper, we present a framework for a more controlled reproducible fabrication of PDMS coatings. Transforming the proof-of-concept of newly developed, smart surfaces into a prototype sample is a challenging process, which can bridge the gap between state-of-the-art material science and the integration of these materials in application devices. An example of such novel smart surfaces, which can show dynamic surface oscillations, is the recently developed electro-active polydimethylsiloxane (PDMS) coating that is promising for many applications, such as haptics, microfluidics and adjustable tribology. However, consistent reproducibility of the dynamic surface oscillation amplitude is highly sensitive to the PDMS fabrication process. In this study, we present a framework for fabricating a more uniform PDMS coating with a more even thickness using spin-coating. On the one hand, the developed framework enables monitoring of the key measurable sample properties during the fabrication process, such as thickness and uniformity of PDMS, which influence the dynamic surface deformation. On the other hand, we discuss the unmeasured (qualitative) sample parameters that are complex to control or hard to measure during the fabrication process and the observations on how they affect the dynamic surface deformation of PDMS. These results give a general insight into the fabrication parameters of PDMS films that can enhance their reproducibility in many PDMS-based devices. This fabrication reproducibility is a crucial feature for building a prototype sample and stepping towards practical applications of PDMS devices.

1 INTRODUCTION

Electroactive oscillating surfaces can be integrated into many application devices such as optical switching devices, surfaces with adjustable tribology, and haptic displays [1–3]. There is a surge of interest in fabricating such oscillating surfaces, which can be dynamically activated and remotely controlled. Nevertheless, until now, most of these developed surfaces are built through complex fabrication synthesis processes or with materials that possess complicated nonlinear dynamics, such as liquid crystal networks [4–7].

Recently, an oscillating electroactive polydimethylsiloxane (PDMS) coating was fabricated in [2], which shows dynamic surface oscillations when connected to an AC source. A schematic of the PDMS coating setup in [2] can be seen in Figure 1. When the set-up in Figure 1 is connected to an alternating current (AC) source, the PDMS coating can experience Maxwell stress due to the generated electrostatic attraction between the top (gold) and bottom electrodes. As a result, the coating contracts in its thickness in the presence of this electrostatic attraction force, and expands in the lateral direction. This mechanism generates a surface deformation at the location of the electrodes E_1 and E_2, in the shape of valleys and hills. Moreover, the position of these valleys and hills can be switched dynamically, thanks to the frequency of the AC source.

Since PDMS is easy to process, flexible, transparent, and has a relatively low glass transition temperature in comparison to other polymers [8,9], it is a popular candidate material for fabricating application devices [8,9]. Therefore, the coating in [2] is promising for functional smart surfaces such as haptics, self-cleaning, and particle transportation devices. However, to realize such applications based on this coating, one requires an experimental set-up, in which a prototype of PDMS coating is fabricated and its key fabrication and sample properties are measured. This is a crucial step in going from the proof-of-concept towards the fabrication of a prototype, which will ensure the reproducibility of the dynamic coating

DOI: 10.1201/9781003514831-5

behavior within a certain confidence boundary. This is also an observable trend in the literature on PDMS devices. It can be observed that in recent years, great attention has been given to the importance of reproducibility and replicability of PDMS devices, due to their popularity in many applications. This is especially observable in research fields that are mature enough to investigate the possibilities of producing functional application devices, such as microfluidic devices [10–12]. Investigating and enhancing the thickness uniformity of PDMS is among the key fabrication-related properties of PDMS coatings that are investigated for several types of PDMS devices [13–15]. As such, ensuring the reproducibility of PDMS devices, particularly in achieving thickness uniformity, is a key challenge.

Therefore, the contribution of this paper is to develop a reproducible fabrication process, for the novel electroactive PDMS coating developed in [2] but also for any other spin-coated PDMS film in other application devices. We show the fabrication challenges to develop a prototype PDMS sample, highlighting the key sample parameters that influence the PDMS surface dynamics in [2] the most.

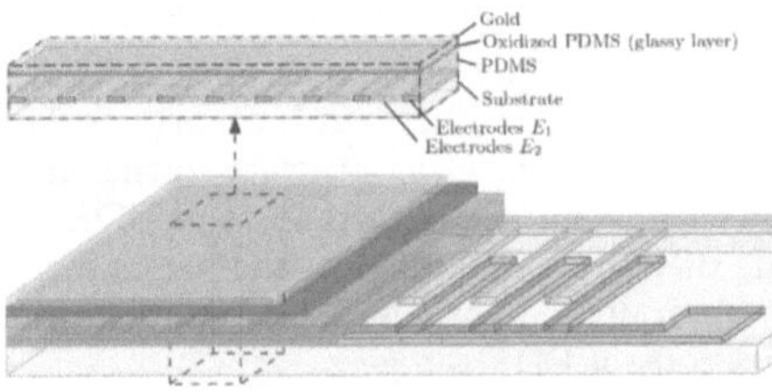

Figure 1. Schematic of PDMS coating bottom to top layers, prior to wiring.

2 MATERIALS AND FABRICATION PROCESS

The glass substrates with interdigitated Indium Tin Oxide (ITO) electrodes are about 1 cm ×1 cm, and are ordered from an external lithography company. The geometry of the electrode's interdigitated pattern is 20 μm electrode width and 10 μm gap between the electrodes. This is the optimum geometry to obtain the largest surface deformation, as was found in a simulation study in [16]. Furthermore, the PDMS precursor and the curing agent (Sylgard 184) were obtained from Dow Corning.

To fabricate a PDMS-coated device as in Figure 1, which was previously introduced in [2], a mixture of silicon precursor and 2 wt% curing agent, is prepared. After manual mixing, the mixture is degassed under reduced pressure. In the meantime, the glass substrates are cleaned through a two-step ultrasonic cleaning, where they are soaked for 20 minutes into acetone and, subsequently, 20 minutes into isopropanol. When the substrates are removed from the solvent, they are dried with a nitrogen gun. Then, the degassed mixture is spin-coated on the cleaned glass substrates to obtain a PDMS layer with a thickness of 8 μm. Next, the coating is cured at 120°C for two hours and is, subsequently, treated for 20 minutes inside the UV-Ozone oven. This treatment oxidizes and chemically modifies the top part of the PDMS layer, resulting in better compatibility with the gold layer. Moreover, this treatment is important as it can enhance the surface properties of PDMS by converting a layer on top of the coating to silica-like materials. The elastic modulus of this layer is more than 10 times higher [2] than the unoxidized PDMS layer, which can affect the dynamic surface deformation amplitude. Therefore, during the fabrication of PDMS coating, only a thin oxidized layer is desired to obtain enhanced surface properties and to not influence the amplitude of deformation. This oxidized top layer of PDMS is referred to as the "glassy layer" throughout this paper. Lastly, a thin gold layer of about 12 nm is sputter-coated on top of the coating. When the coating is ready, we connect the wires through soldering to the bottom electrodes on the glass substrate, and through conductive glue to the gold (top electrode).

3 FABRICATION OPTIMIZATION AND ANALYSIS FOR REPRODUCIBILITY

The fabrication of the PDMS coating consists of sequential steps, in which several factors can influence the surface deformation of the coating under electric stimulation. Although not all the factors can be measured easily during fabrication, some additional and feasible measurements can be defined within the fabrication steps, which enables more control over sample parameters and leads to better reproducibility.

The thickness precision and uniformity of the PDMS coating are two of the key sample parameters. To obtain the desired thickness, the parameters of the spin-coating machine, such as speed, acceleration, and spinning time can be adjusted. To ensure that the obtained thickness is the same as the desired thickness at every location, spectrophotometry can be implemented to measure the transmission of the film. One of the advantages of using PDMS material is the coating transparency, which facilitates transmission measurements and calculation of the thickness by interference. This is a non-destructive, easy, and fast measurement, which provides the necessary information for calculating the film thickness of the measured area. To proceed with this measurement, when PDMS is spin-coated on the glass substrate, we use a UV-visible spectrophotometer to measure the transmission of our coating. The transmission is measured within the incident light spectrum, over a range in which PDMS has a high transmission percentage, namely 500 nm and 800 nm [17,18]. Prior to this measurement, the background signal is obtained by measuring a bare glass substrate. This background signal is then deducted from the transmission signals of the coating. The obtained transmission signal from a uniformly coated PDMS on the glass resembles a sinusoidal signal shape related to the interference between the light reflected from the top interface and the lower interface of the PDMS. Then, we use the equation below to calculate the thickness of the coating at the measured surface area:

$$d = \frac{\Delta m}{2\sqrt{n^2 - \sin^2\theta}} \times \frac{1}{\frac{1}{\lambda_2} - \frac{1}{\lambda_1}}. \tag{1}$$

In (1), Δm is the number of minima or maxima between the chosen wavelengths of incident light λ_2 and λ_1, n is the refractive index of PDMS, which is considered to be 1.43 [19], and wavelength independent, and θ is the angle between the incident light and PDMS, which in our case is a right angle. This measurement is performed through a pin-hole with a diameter of about 5 mm, for 4 locations on the surface of each PDMS coating sample, such that, at the end, the whole coating on top of the interdigitated electrodes is measured.

Nevertheless, to obtain clear signals from spectrophotometry measurements, uniformity of the coating is a key factor. The spin-coating parameters determine the final thickness, as well as the uniformity of the coating. Initially, the spin-coating program in [2], with the parameters in Table 1 was used. However, for the PDMS mixture of our study, these settings led to splashing and non-uniformity on the coating surface, as shown in Figure 2a. This non-uniformity is also observed in the signals recorded with a UV-visible spectrophotometer, such that the signals become difficult to analyze and the minima/maxima within the signals cannot be distinguished anymore (see Figure 2c). For this reason, we aim to optimize the spin-coating program for our mixture. We conjecture that the high viscosity of PDMS, together with the high acceleration and spinning speed, plays an important role in obtaining such non-uniformity. Considering this conjecture, we increased the number of spin-coating steps from one to two, where in the first step a much lower speed and acceleration is used than in the second step. The goal is to obtain a smooth layer of the mixture over the whole glass substrate in the first step. Then, we use the second step to obtain the desired thickness of the coating by removing the excess PDMS through faster spinning. This optimized spin-coating program can be found in Table 2. By employing this program, an improved uniformity of the PDMS coating could be observed in the optical microscope image and the spectrophotometer signal, as shown in Figures 2b and **2d**, respectively.

Next, the PDMS thickness that is obtained after using the spin-coating program in Table 2 is measured and calculated. The surface of the sample is measured at different positions, where the average thickness in all measurements is about 8 μm. The measurements also show that by employing the optimized spinning program, we obtain a quite uniform thickness on our samples, where the divergence from our targeted thickness can be up to 7.5%, as indicated in Table 3.

Table 1. Initial spin coating parameters.

Parameters	Step1
Speed	7000 rpm
Acceleration	1000 rpm/s
Duration	120s

Table 2. Enhanced spin coating parameters.

Parameters	Step 1	Step 2
Speed	800 rpm	7000 rpm
Acceleration	50 rpm/s	1000 rpm/s
Duration	25 s	120 s

Table 3. PDMS coating thickness d obtained with enhanced spin coating parameters.

Number of samples	Minimum d [μm]	Maximum d [μm]	Average d [μm]
9	7.84	8.6247	8.1197

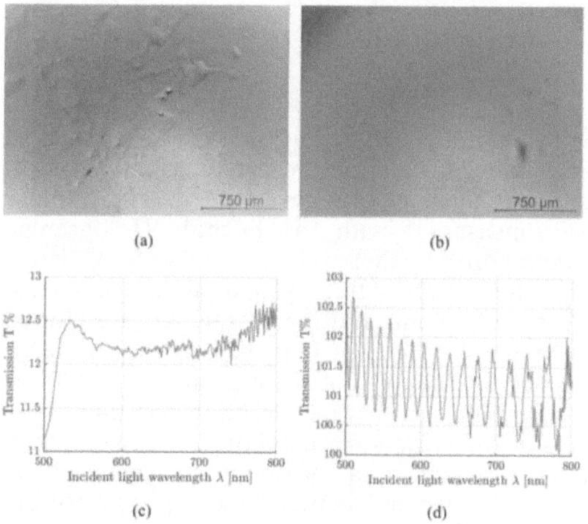

Figure 2. Influence of spinning parameters on sample uniformity. (a) and (b) are optical microscope images, and (c) and (d) are transmission signals of coating before and after enhancing the spin-coating program, respectively.

4 INFLUENCE OF PDMS THICKNESS ON SURFACE DEFORMATION AMPLITUDE

Thickness is a crucial parameter of the PDMS coating, which can be determined and measured throughout the fabrication process. To enable control over fabricating different coating thicknesses via the spin-coating machine, first, the correlation between the spinning speed and the final thickness needs to be investigated. Thereto, the spin-coating step for four samples is performed with the parameters as in Table 2 except for the spinning speed in step 2, which is varied between 4000 and 8000 rpm. The resulting thicknesses corresponding to each of the spinning speeds are illustrated in Figure 3.

Figure 1. Correlation between the spinning speed of the spin-coating machine and the obtained PDMS thickness.

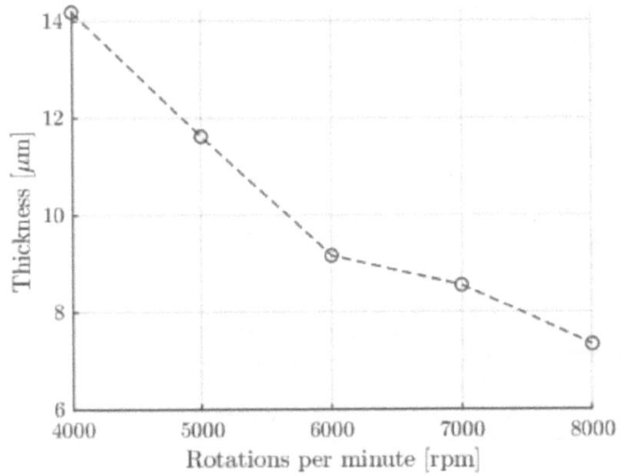

Figure 3.

Next, the effect of thickness variation on the amplitude of deformation for four samples with different thicknesses is investigated. The samples are measured with a Digital Holographic Microscope (DHM) and the amplitude of deformation is averaged through several measurements. These data points and their error bars are plotted in Figure 4. The error bars indicate the 95% confidence interval (CI) of the measured data [20]. It can be observed that increasing the coating thickness also increases the amplitude of deformation. However, there is a maximum thickness, above which the distance between the top and bottom electrodes becomes too large, which decreases the generated Maxwell force. As a result, the amplitude of deformation starts decreasing. This behavior was also found in [2] through a simulation study.

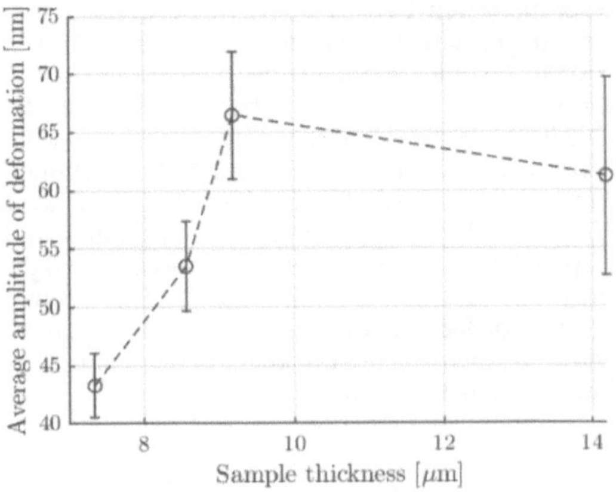

Figure 4. Influence of thickness variation on the amplitude of deformation of PDMS samples. The error bars indicate the 95% CI of the data at each thickness.

5 ELECTRODES AND ELECTRICAL CONNECTION QUALITY

An important component of the electroactive PDMS device is the electrodes, through which the Maxwell force and the surface deformation are generated. The geometry pattern of the ITO electrodes on the substrate must be precise and without defects, as it influences the quality of the electrical connection. Therefore, prior to the fabrication of each batch of samples, an optical microscope is used to check the accuracy and the presence of defects within the substrate geometry. This is done with a 10× magnification lens for five locations on the interdigitated pattern of the glass substrates, to filter out the substrates with ruptured or connected (diffused) ITO patterns between electrodes.

Similarly, the top electrode, which is the gold layer, can have an impact on the surface deformation of the PDMS device. The gold thickness is rather complex to precisely control and measure at every position on the coating. However, it is possible to ensure the uniformity of the gold layer at the macro-scale (observable) by carefully examining the samples after they are sputter-coated, by the naked eye. There is a range in thickness where the gold layer seems uniform and without any wrinkles or cracks, which can be observed by the naked eye. This range of gold thickness is investigated for the samples by varying the gold sputtering duration and keeping the current at 65 mA. A nominal thickness range (according to the machine manuals) of the gold layer between 9 nm and 17 nm is obtained. Then, by looking at the samples, we can define the upper and lower thickness threshold of the gold layer, which also marks the uniform gold layer range. For our samples, this range is between a nominal thickness of 12 nm and 14 nm, where at 11 nm gold layer cracks are initiated, and at 15 nm gold layer a wrinkling effect is observed. A picture of three samples with cracks, a uniform gold layer, and wrinkles is shown in Figure 5.

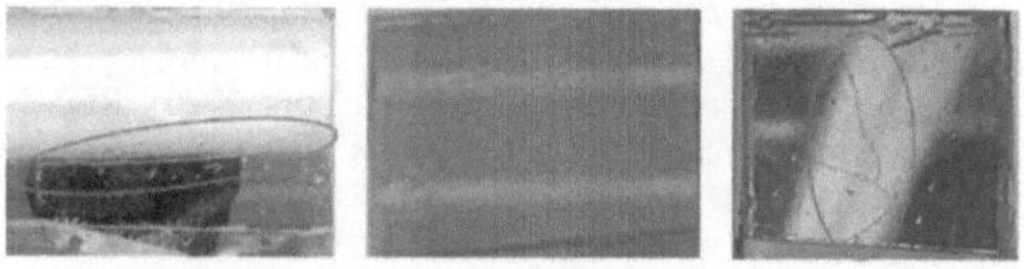

Figure 5. An image of the sample with cracks on the left, a sample with a uniform gold layer in the middle, and a sample with wrinkles on the right. The thickness of the gold layer is increased from left to right.

The last factor that influences the amplitude of deformation in the PDMS coating is the quality of the electrical connection between the source and the coating electrodes. This connection is made through wires and conductive glue, which can be enhanced during the measurements by, for example, adding more glue at the connection point between the gold and the wires. We have observed that when measuring a point on top of the coating before and after this enhancement, the coating response is improved and a more uniform response for different electrodes is obtained. This is shown in Figure 6, where the coating surface deformation in terms of height change is measured on top of six electrodes with DHM. The three red lines and blue lines correspond to the height change of the points on top of the electrode set E_1 and E_2 (see Figure 1), respectively.

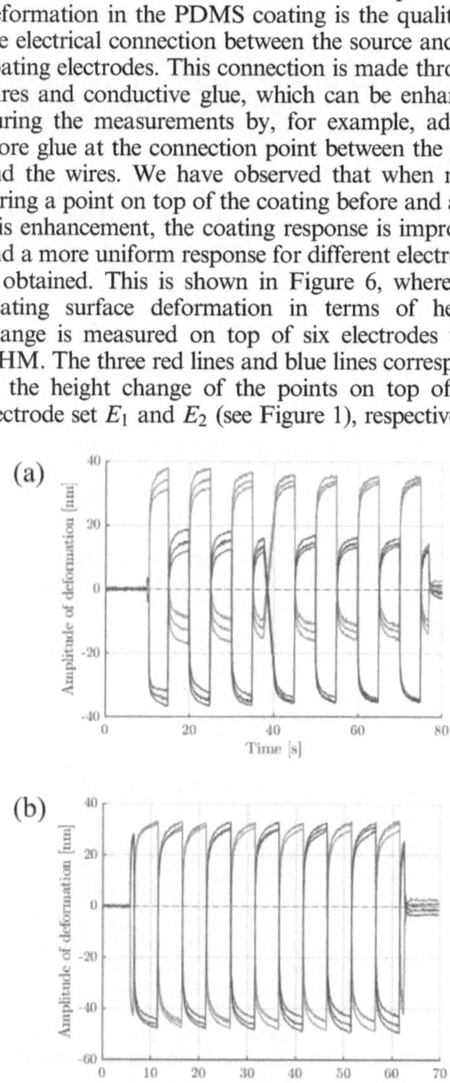

Figure 6. Height change measurements of six points on top of electrodes (a) before and (b) after enhancing the electrical connection.

6 RESULTS AND DISCUSSION

Fabricating the PDMS coating consists of consecutive steps, which implies that keeping track of each step is necessary to provide reproducibility. Concerning this, we introduced an optimized fabrication process that helps to monitor the fabrication-related properties of the PDMS coating qualitatively and quantitatively. We have shown that thickness and uniformity of the coating are among the quantitative parameters that can be measured and adjusted as desired, through a controlled fabrication process.

However, qualitative factors such as gold thickness and the quality of electrical connection are the properties of PDMS that are difficult to measure during the fabrication process.

Apart from the macro-scale uniformity of the gold layer that can be assessed with the naked eye, the gold uniformity of our sample might be affected by other factors on a smaller scale. The first factor is the sputtering method, which generates a very thin gold layer, but also a variation in the gold thickness at different locations on the coating. Furthermore, there might exist some discrepancy between the nominal value of the obtained thickness and the actual value generated by the sputtering machine.

The thickness of the glassy (oxidized) layer of PDMS plays a dominant role in the electromechanical response. This layer is formed by oxidation through the UV-ozone treatment of the PDMS layer forming a silicone oxide topcoat. However, the thickness of this layer is rather challenging to measure, as this layer is an inseparable part of the PDMS. Also, optical measurement is not accessible as the transition between the silicone oxide top to the silicone rubber goes gradually over a finite distance without a discrete refractive index step. UV-ozone treatment is commonly used in the literature to enhance the surface properties of PDMS [21,22]. Moreover, as also mentioned in [2], this layer has orders of magnitude higher stiffness and less flexibility than the unoxidized bottom part of the PDMS layer. This difference in mechanical properties can enhance the compatibility and adhesion of the gold layer with the PDMS layer. However, we conjecture that if the thickness of this oxidized layer is not negligible like the one considered in [2], which is about 0.1%, it significantly increases the stiffness of the PDMS layer. This is also a possible reason that the amplitude of surface deformation obtained with our samples is on average lower than the ones obtained in [2]. The generated thickness of this layer can be influenced by the duration of the UV-ozone treatment and the UV lamp intensity.

In the outlook of future research, it is interesting to determine the optimal balance between the gold layer uniformity, the thickness of the oxidized PDMS layer, and the surface deformation of the PDMS coating. This could be a crucial factor in adjusting the amplitude of surface deformation of the PDMS prototype used in this research when aiming at a certain deformation level for a targeted application. This can be done experimentally or by employing a physics-based model using the finite element method. Furthermore, this model can also give insights for future studies on enhancing the material parameters and response to determine the influence of the uncertain parameters that are not feasible to measure during fabrication, as discussed earlier.

7 CONCLUSIONS

In summary, in this paper, we propose a more controlled fabrication framework of PDMS coatings that includes analysis steps of fabrication-related sample parameters. Our focus is on the reproducibility of the PDMS film, highlighting fabrication challenges and the identification of the sample parameters that most influence the amplitude of deformation. Hereto, we define pre- and post-fabrication analysis and measurements, which help in controlling these highlighted parameters. The results of this study are crucial for developing a prototype sample with reproducible thickness and uniformity, which is one of the initial steps toward employing PDMS devices in practical applications.

REFERENCES

[1] Poncet P. *et al.*, (May 2017). Static and dynamic studies of electro-active polymer actuators and integration in a demonstrator, *Actuators*, vol. 6, no. 2, p. 18, doi: 10.3390/act6020018.

[2] Visschers F. L. L., Gojzewski H., Vancso G. J., Broer D. J. and Liu D., (Nov. 2019). Oscillating surfaces fueled by a continuous AC electric field, *Adv. Mater. Interfaces*, vol. 6, no. 21, p. 1901292, doi: 10.1002/admi.201901292.

[3] Visschers F. L. L., Hendrikx M., Zhan Y. and Liu D., (Jun. 2018). Liquid crystal polymers with motile surfaces, *Soft Matter*, vol. 14, no. 24, pp. 4898–4912, doi: 10.1039/C8SM00524A.

[4] Feng W., Broer D. J., Grebikova L., Padberg C., Vancso J. G. and Liu D., (Feb. 2020). Static and dynamic control of fingerprint landscapes of liquid crystal network coatings., *ACS Appl. Mater. Interfaces*, vol. 12, no. 5, pp. 5265–5273, doi: 10.1021/acsami.9b11928.

[5] Feng W., Liu D. and Broer D. J., (2018). Electric field switched surface topography of fingerprint liquid-crystal network polymer coating, in *Emerging Liquid Crystal Technologies XIII*, vol. 10555, p. 47, doi: 10.1117/12.2287180.

[6] He X. *et al.*, (2012). Synthetic homeostatic materials with chemo-mechano-chemical self-regulation, *Nature*, vol. 487, no. 7406, doi: 10.1038/nature11223.

[7] Amiri A., Caasenbrood B., van de Wouw N. and Lopez Arteaga I., (Mar. 2023). A replacement model to simulate the nonlinear dynamics of electro-responsive liquid crystal coatings, *AIP Adv.*, vol. 13, no. 3, p. 035203, doi: 10.1063/5.0138991.

[8] Eduok U., Faye O. and Szpunar J., (Oct. 2017). Recent developments and applications of protective silicone coatings: A review of PDMS functional materials, *Prog. Org. Coatings*, vol. 111, pp. 124–163, doi: 10.1016/J.PORGCOAT.2017.05.012.

[9] Raj K. M and Chakraborty S., (Jul. 2020). PDMS microfluidics: A mini review, *J. Appl. Polym. Sci.*, vol. 137, no. 27, p. 48958, doi: 10.1002/APP.48958.

[10] Cottet J., Vaillier C., Buret F., Frénéa-Robin M. and Renaud P., (Nov. 2017) A reproducible method for μ m precision alignment of PDMS microchannels with on-chip electrodes using a mask aligner, *Biomicrofluidics*, vol. 11, no. 6, p. 64111, doi: 10.1063/1.5001145/385465.

[11] Tony A. *et al.*, (Apr. 2023). The additive manufacturing approach to polydimethylsiloxane (PDMS) microfluidic devices: Review and future directions, *Polymers (Basel).*, vol. 15, no. 8, p. 1926, doi: 10.3390/polym15081926.

[12] Shakeri A., Khan S. and Didar T. F., (2021) Conventional and emerging strategies for the fabrication and functionalization of PDMS-based microfluidic devices, *Lab Chip*, vol. 21, no. 16, pp. 3053–3075, doi: 10.1039/D1LC00288K.

[13] Wang S., Kallur A. and Goshu A., (2011). Fabrication and characterization of PDMS thin film, in *SPIE OPTO, 2011*, vol. 7935, p. 79350M, doi: 10.1117/12.875247.

[14] Mata A., Fleischman A. J. and Roy S., (Dec. 2005). Characterization of polydimethylsiloxane (PDMS) properties for biomedical micro/nanosystems, *Biomed. Microdevices*, vol. 7, no. 4, pp. 281–293, doi: 10.1007/s10544-005-6070-2.

[15] Schneider F., Draheim J., Kamberger R. and Wallrabe U., (Apr. 2009). Process and material properties of polydimethylsiloxane (PDMS) for optical MEMS, *Sensors Actuators A Phys.*, vol. 151, no. 2, pp. 95–99, doi: 10.1016/j.sna.2009.01.026.

[16] Ludovicus Louis Visschers F., (2021). Bringing coating surfaces to life, Technische Universiteit Eindhoven, *Eindhoven*.

[17] Kim J., Kim K. and Pahk H. J., (Oct. 2017). Thickness measurement of a transparent thin film using phase change in white-light phase-shift interferometry, *Curr. Opt. Photonics*, vol. 1, no. 5, pp. 505–513, doi: 10.3807/COPP.2017.1.5.505.

[18] Chronis N., Liu G., Jeong K.-H. and Lee L., (Sep. 2003). Tunable liquid-filled microlens array integrated with microfluidic network, *Opt. Express*, vol. 11, no. 19, p. 2370, doi: 10.1364/OE.11.002370.

[19] Huibers P. D. T. and Shah D. O., (Oct. 1997). Multispectral determination of soap film thickness, *Langmuir*, vol. 13, no. 22, pp. 5995–5998, doi: 10.1021/la960738n.

[20] O'Brien S. F. and Yi Q. L., (Jul. 2016). How do I interpret a confidence interval?, *Transfusion*, vol. 56, no. 7, pp. 1680–1683, doi: 10.1111/trf.13635.

[21] Berdichevsky Y., Khandurina J., Guttman A. and Lo Y.-H., (Feb. 2004). UV/ozone modification of poly(dimethylsiloxane) microfluidic channels, *Sensors Actuators B Chem.*, vol. 97, no. 2–3, pp. 402–408, doi: 10.1016/j.snb.2003.09.022.

[22] Fu Y.-J. *et al.*, (Mar. 2010). Effect of UV-ozone treatment on poly(dimethylsiloxane) membranes: surface characterization and gas separation performance, *Langmuir*, vol. 26, no. 6, pp. 4392–4399, doi: 10.1021/la903445x.

System Innovation for an Artificial Intelligence Era – Lam et al. (Eds)
© 2025 the Author(s), ISBN: 978-1-032-84452-7

An innovative mechanism design of maze box with puzzles as an integrated teaching aid

Jian-Liang Lin & Kuo-Huang Hsiao
National Science and Technology Museum, Kaohsiung, Taiwan

ABSTRACT: A maze box is a trendy and perfect toy for kids or adults to solve. They can train logical thinking, the response of humans, and the control of the hand. The maze boxes have various appearances. The standard shapes of maze boxes are cubic, circular disc, and cylinder. Whatever appearances, their design concept is the same, i.e., move the obstacle (a pin or a ball) to pass through the way of the maze. Here, this paper presents a new mechanism for the maze box. The design had a maze plate, and the obstacle was a pin fixed on a sliding plate. The sliding plate slides under the maze plate. The maze plate also slides under the top face of the maze box. These two sliding motions are vertical. Thus, the maze plate can be drawn out when the sliding plate with a pin moves to the correct position. Such a design can generate a separate and staged solving process. Furthermore, the maze's path is hidden because of the maze plate under the sliding plate. The maze box can transform into a puzzle game. The operator should resolve the puzzles to get the correct codes or patterns. At the same time, the pins will be moved to the correct positions of the maze to pass through the aisles. Therefore, the innovative design of the maze box can integrate a puzzle game to deliver information. It can be a teaching aid for diverse education themes.

1 INTRODUCTION

The maze, also known as the labyrinth, dates back thousands of years, with origins in various cultures around the world. The word "Labyrinth" comes from the Greek word "Labry (πέλεκυς)," and there were several puzzle games in the ancient Greek (Doob P. R. 1992; Labyrinth Definition, Retrieved 2024). One of the earliest known examples of a maze-like structure is found in ancient Egypt, where intricate maze designs adorned the walls of tombs and temples. These mazes were often symbolic representations rather than physical constructions. The Labyrinth of Egypt, which dates from 1800 BC and is supposed as the complex structure in the reconstruction research, is near the foot of the Pyramid of Amenemhat III at Hawara, as shown in Figure 1 (Etheredge L. et al., retrieved 2024; Lloyd A.B. 1970; Lost Labyrinth of Egypt, cited 2024; Gruodis-Gimbel E., cited 2024). In Greek mythology, the most famous maze is the Labyrinth of Knossos on the island of Crete, attributed to the legendary craftsman Daedalus. According to myth, Daedalus built the labyrinth to imprison the Minotaur, a half-man, half-bull creature. This labyrinth was said to be so complex that anyone who entered it would become hopelessly lost. Therefore, throughout history, mazes have appeared in various forms and contexts, including as symbols of spiritual journeys, as architectural features in gardens and estates, and as puzzle games. Medieval European cathedrals often featured labyrinthine designs on their floors, which pilgrims would traverse as part of a spiritual ritual. In the modern era, mazes continue to be popular as recreational toys or electric games, garden attractions, and even architectural installations. Their origins are deeply rooted in human history and culture, with interpretations and applications evolving over time.

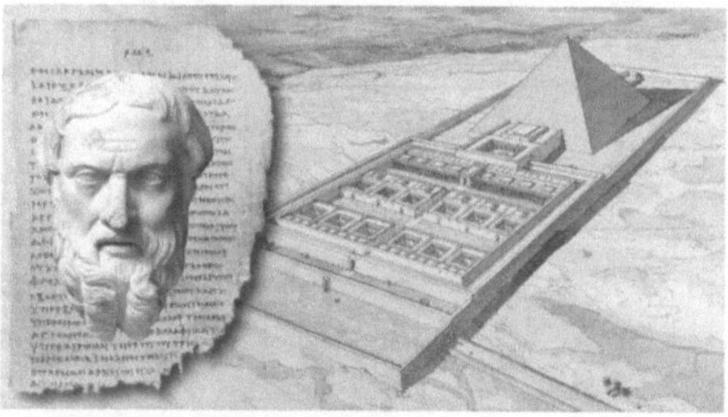

Figure 1. The reconstruction of maze near the foot of the Pyramid of Amenemhat III.

The academic study of maze boxes is rare, especially for the structure design. The maze box as a medium or tool is usually applied to biological or physical research. The earliest mazes for rodent study began in the laboratory of Dr. Edmund Sanford in the 1890s. The lab's student, Linus Kline, was interested in studying animal psychology. Since animals had been used to be as the models for human psychology, Kline applied them to study behavioral models by Darwin's work (Traetta L. 2020).

The puzzle box has a rich historical development. These puzzle boxes can be the forms of furniture and jewelry boxes with hidden compartments since the Renaissance. They were used for entertainment in England in the 19th century a(Hoffman P. 1893; Sanders W.E., Sanders 1999 C.). These earlier designs of puzzle boxes were mostly one-trick openings, and Chinese cricket boxes were involved.

One kind of puzzle box originated in Asia, particularly in Japan, with the creation of intricate wooden puzzle boxes known as "Himitsu-Bako" during the Edo period (17th to 19th centuries). This product forms a traditional skill of Japanese wood craftsmen, such as the well-known "Karakuri Box" in the Hakone Odawara region. The ingenious mechanism design is mostly made by the craftsmen's own imaginations, as shown in Figure 2 (Iwasaki S., 200; Karakuri Box, retrieved 2024) [11, 12]. These boxes were initially used to store valuable items and documents, and they featured complex locking mechanisms that required specific sequences of moves to open. Over time, the design and popularity of puzzle boxes spread beyond Japan. During the 20th century, puzzle boxes continued to evolve, with variations in materials, designs, and mechanisms. Some were made of metal instead of wood, and they often incorporated more intricate locking mechanisms and hidden compartments.

Figure 2. A Karakuri box.

Today, puzzle boxes remain popular as both puzzles and decorative items. They come in various designs and complexities, appealing to enthusiasts of all ages. Additionally, modern technologies have been integrated into some puzzle boxes, adding electronic components and augmented reality elements to enhance the puzzle-solving experience.

The maze box and the puzzle box are usually regarded the same. In fact, these two products have slightly different. The maze box emphasizes on the complicated design of maze path. But the puzzle box, also known as a trick box, is the combination of several mechanisms. Based on the previous study and the creative design of mechanisms, this paper aims to develop an innovative design that is integrated the maze box and the puzzle box.

2 CLASSIFICATIONS OF MAZE BOXES

Maze boxes can be classified based on various criteria such as complexity, material, size, and mechanism. These classifications provide a framework for understanding the diverse range of maze boxes available, catering to different skill levels, preferences, and interests. Here's a breakdown of some common classifications.

1. Complexity: The beginner level of maze boxes designed for entry-level users, featuring simple puzzles and fewer obstacles. The intermediate level of maze boxes with moderate complexity, requiring more strategic thinking and problem-solving skills. The advanced level of maze boxes for experienced users, often featuring intricate designs, multiple levels, and challenging obstacles.
2. Material: The wooden maze boxes crafted primarily from wood, often featuring traditional designs and craftsmanship. The metal maze boxes made of metal, known for their durability and sleek designs. The plastic: Maze boxes constructed from plastic, offering a wide range of designs and colors.
3. Size: The small maze boxes are compact and designed for portability and convenience. The medium Standard-sized maze boxes suitable for tabletop use. The large Maze boxes with expansive layouts or multiple compartments, offering a more immersive experience.
4. Mechanism: The mechanical maze boxes include physical components and mechanisms such as levers, gears, and sliding panels. The electronic maze boxes incorporate electronic elements such as sensors, lights, and sound effects. And, the hybrid maze boxes combine both mechanical and electronic features for a unique puzzle-solving experience.
5. Theme: The traditional maze boxes have classic designs and themes, such as geometric patterns or nature-inspired motifs. The themed maze boxes feature specific themes, such as fantasy, sci-fi, or historical settings.

Furthermore, maze boxes can be classified by shape, with various designs offering unique challenges and aesthetics. Here are some common classifications according to shape.

1. Rectangular: Rectangular maze boxes are the most common shape, featuring a box-like structure with straight edges and corners. They often have sliding panels, hidden compartments, and multiple pathways to navigate, as shown in Figure 3(a).
2. Cube: Cube-shaped maze boxes resemble a three-dimensional cube, with six square faces. Users must manipulate the sides of the cube to navigate through the maze and unlock compartments, as shown in Figure 3(b).
3. Cylinder: Cylinder-shaped maze boxes have a cylindrical structure, resembling a tube or barrel. Players must rotate the cylinder to navigate through the maze and reach the solution, as shown in Figure 3(c).
4. Sphere: Sphere-shaped maze boxes are spherical or globe-like in shape, offering a 360-degree challenge. Players must tilt and rotate the sphere to maneuver through the maze and solve the puzzle, as shown in Figure 3(d).
5. Irregular/Custom: Irregular or custom-shaped maze boxes come in various non-standard shapes, such as pyramids, hexagons, or even whimsical designs. These unique shapes add an extra layer of challenge and creativity to the puzzle-solving experience. Each shape offers its own set of challenges and gameplay dynamics, providing puzzle enthusiasts with a diverse range of options to explore and enjoy.

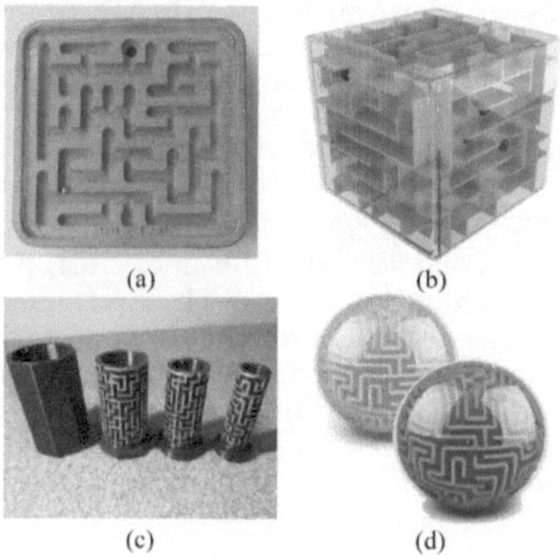

Figure 3. Various maze box by shapes.

3 THE INNOVATIVE DESIGN OF MAZE BOX WITH A PUZZLE GAME

In this paper, the maze box and the puzzle box are combined to present an innovative design of a maze box with a puzzle game. The creative product is regarded as a carrier to promote various topics or information. By integrating the knowledge of mechanisms and someone's specific research topic, this maze box with a puzzle game is an interdisciplinary product. Figure 4 is the innovative design. In what follows, the design principle and the operating process are introduced in detail.

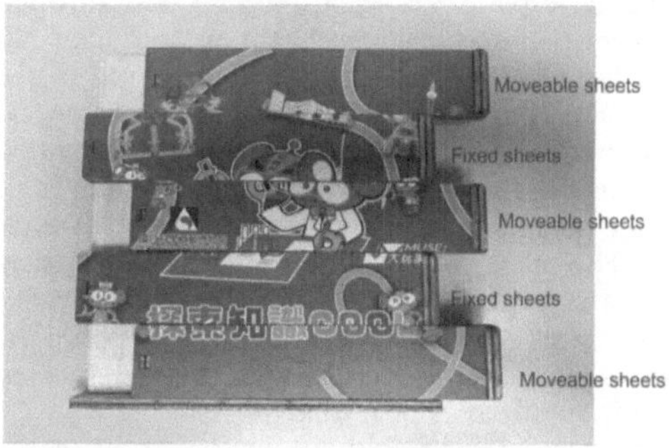

(a) front view of puzzle plate

(b) back view of puzzle plate

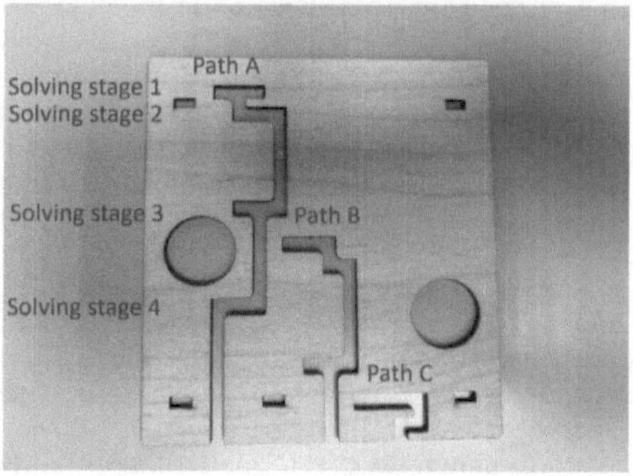

(c) maze plate

Figure 4. Mechanism introduction of the innovative maze box integrated puzzle box's function.

3.1 *Design principle of the new maze box*

The new maze box is integrated with a puzzle game. The puzzle game is the application of a research topic's knowledge, such as the climate change, the issue of net zero carbon emissions, the green energy. It can be an interdisciplinary teaching aid.

There are moveable sheets and fixed sheets on the moveable puzzle plate of the cubic box. The maze plate behind the puzzle plate is hidden to generate the choosing function for solving puzzles. The moveable sheets should be drawn right. When all moveable sheets are put on the right positions of the maze plate, i.e., each pin can pass through the corresponding maze path, the puzzle plate can be pulled. As the maze paths are increased in the maze plate, the moveable sheets have to be added to the puzzle plate. Moreover, the maze box with a puzzle game has a design with multiple stages of solving puzzles by the arrangement of the maze path. The maze path is a combination of vertical tracks and horizontal tracks. The horizontal tracks divide the path into several stages and provide the answer choices for the solving puzzle by pin locating in different positions.

Thus, based on the above introduction of the design principle, the operators can solve more puzzles by two methods. One is to increase the number of horizontal tracks in each maze path. The other is to add the number of maze paths in the maze plate. Furthermore, the maze plate and the puzzle plate are separate. They merely connect to each other through the contact of pin and track, i.e., the sliding joints. We can change the maze plates with different designs of maze paths under the same puzzle plate. The playing of the innovative maze box with a puzzle game is various.

3.2 *Mechanism analysis of the new maze box*

Based on the mechanism design of the maze box with a puzzle game, each motion in the maze box is regarded as a two-linkage mechanism, including one sliding joint, which is the most straightforward mechanism, whatever the motions of the puzzle sheets or the motion of the puzzle plate. Actually, the new product is a deformed combination lock. The password is in the form of "patterns" or "figures" instead of the "letters" or "numbers" commonly seen in general combination locks. The unlocking surface could be composed of several wooden sheets. Some sheets are movable and have pins under them. The remaining sheets are fixed and can be moved by pulling. The moveable sheets are used to match the correct pattern and find the correct password. Then, the pins can pass through the tracks of math paths on the maze plate.

3.3 *Solving process of the new maze boxes*

As shown in Figure 4, the maze plate has three maze paths, and the puzzle plate has three moveable sheets. From the left viewpoint, the maze path A and the maze path B, respectively, have four solving stages (four horizontal tracks) and three solving stages (three horizontal tracks). The final maze path C only has two solving stages (two horizontal tracks). For path A, all of the solving stages have three position choices. For path B, the 1st, the 2nd, and the 3rd have 3, 2, and 3 position choices, respectively. For path C, the 1st and the second solving stage have 4 and 2 position choices. To conclude the position choices of all maze paths, it is known that the 1st, the 2nd, the 3rd, and the 4th have 36, 12, 9, and 3 combinations. Finally, we introduce the correct operation process.

3.3.1 *The 1st step (the first solving-puzzle stage)*

To draw all moveable puzzle sheets to the correct positions, as shown in Figure 5(a). These moveable puzzle sheets align the bamboo hat, the dinosaur, and the text, respectively. Then, the puzzle plate can be pushed down.

3.3.2 *The 2nd step (the second solving-puzzle stage)*

To draw all moveable puzzle sheets to their correct positions, as shown in Figure 5(b). These moveable puzzle sheets align the Mondrian geometric kaleidoscope, the logo man, and the two museum buildings, respectively. Then, the operator can push the puzzle plate down again.

3.3.3 *The 3rd step (the third solving-puzzle stage)*

After the second step, the fifth (moveable) puzzle sheet leaves the box totally. The operator only solve the first and the third puzzle plates. To draw all remaining moveable puzzle sheets to their correct positions, as shown in Figure 5(c). These sheets align the three-dimensional red block and the dinosaur. Then, the operator can push the puzzle plate down again.

3.3.4 *The 4th step (the final solving-puzzle stage)*

After the third step, the third (moveable) puzzle sheet leaves the box totally. There is only the first plate in the top of box. To draw the first plate to the starting position. The sheet align the small man, as shown in Figure 5(d). The operator can pull out the puzzle plate totally. The opening of box is finished.

4 CONCLUSION

This paper introduced the development of maze boxes or puzzle boxes and classified these existing products by their shape and functions. Actually, the maze boxes and the puzzle boxes should be different, though they were regarded as the same in the internet information. In this paper, the maze box and the puzzle box are defined clearly. These two products had their advantages. The puzzle box often had a complicated mechanical structure but had the solving process. The structure of the maze box seems to be complicated, but the design and manufacturing are simple. Through the combination of a maze box and a puzzle box, an innovative design of a maze box with puzzles was presented. This design integrated the functions of the maze box and puzzle box and has more operation procedures for a more varied playing process. Owing to the application of puzzles, various knowledge could be added. The maze played the function of choices. The creative mechanism design could be a teaching aid for different topics. Also, it had the patent of the Republic of China.

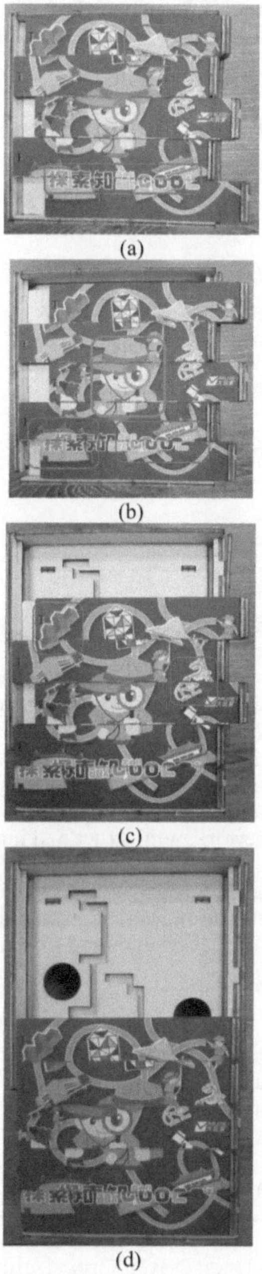

Figure 5. Solving process of the creative maze box.

REFERENCES

Doob P.R. (1992). *The Idea of the Labyrinth: from Classical Antiquity through the Middle Ages*. Ithaca: Cornell University Press.

Etheredge L. et al. Amenemhet III in: Encyclopædia Britanni. Retrieved 2024 April 6.

Gruodis-Gimbel E. The *'Lost Egyptian Labyrinth' Of Hawara: Is A 2,000-Year-Old Mystery Finally Solved? In: The archaeologist* [cited 2024 April 6]. Available at https://www.thearchaeologist.org/blog/the-lost-egyptian-labyrinth-of-hawara-is-a-2000-year-old-mystery-finally-solved

Hoffman P. (1893). *Puzzles Old and New*. London: Warne.

Iwasaki S. (2003). *Karakuri works in Hakone & Odawara*. Odawara: Karakuri Creation Group.

Karakuri Box [Internet]. Available from: https://karakuri.gr.jp/en/about-karakuri/

Labyrinth Definition [Internet]. *World History Eneyclopedia*. Available from: https://www.worldhistory.org/Labyrinth/

Lloyd A.B. (1970). The Egyptian labyrinth. *The Journal of Egyptian Archaeology*. 56:81–100. doi: 10.2307/3856044.

Lost Labyrinth of Egypt | Location, Origin & Significance in: Study.com [cited 2024 April 6]. Available from https://study.com/academy/lesson/lost-labyrinth-egypt-location-origin-significance.html.

Sanders W.E., Sanders C. (1999). *Pocket Matchsafes: Reflections of Life & Art*. Atglen: Schiffer.

Traetta L. (2020). At the beginning of learning studies there was the maze. *Open Journal of Medical Psychology*. 9, 168–183.

System Innovation for an Artificial Intelligence Era – Lam et al. (Eds)
© 2025 the Author(s), ISBN: 978-1-032-84452-7

The shift towards data-driven organizations in the Thai media industry: A framework for adaptation and transformation

Jackkit Sangkittiwan
Technopreneurship and Innovation Management Program, Graduate School, Chulalongkorn University, Thailand

Nagul Cooharojananone
Department of Mathematics and Computer Science, Faculty of Science, Chulalongkorn University, Thailand

Thira Chavarnakul
Department of Commerce, Chulalongkorn Business School, Chulalongkorn University, Thailand

Tatri Taiphapoon
Faculty of Communication Arts, Chulalongkorn University, Thailand

ABSTRACT: The media industry is undergoing a significant transformation driven by technological advancements. This presents both opportunities and challenges for media companies across all sectors. Data has become the new oil, and its value in driving successful media businesses is constantly increasing. In response to these changes, media organizations are adapting and converging with emerging forms of media, such as digital and social media. They are also transitioning towards becoming Data-Driven Organizations (DDOs). However, there is a lack of clear guidance and frameworks for this adaptation and transformation process. This qualitative research study employs a literature review and in-depth interviews with case study companies in the Thai media industry to identify a framework for adaptation and transformation towards DDOs that is suitable for the local context. To develop the DDO framework, semi-structured interviews were conducted with six executives from leading Thai media firms, encompassing both traditional and digital-native companies. Subsequently, the framework's robustness was validated through interviews with executives from emerging new media companies. The research findings present a framework that is appropriate for different types of media companies in Thailand in terms of goodness of fit, demand, and priority.

1 INTRODUCTION

Data is considered a key resource for conducting activities and business decision-making. The rise in data over the years has led many experts to say: "data is the new oil" (Parkins. 2017), and increasingly valuable. (Van der Sande *et al.* 2021) The more data there is, the more opportunities organizations must make use of such data to develop and improve their operations even better. (Sivarajah *et al.* 2017) Many organizations use data in different ways in business processes to level up and change their way of doing business even better. A study by Bokman (Bokman *et al.* 2014) found that being an organization driven by data increases its corporate return on investment by 2.6 times and has 23 times more advantages in terms of understanding and reaching customers than an organization without.

Using data as the heart of a business drive, data can be analyzed and led to enterprise decisions to run a business to generate more revenue. (Davenport *et al.* 2010)

In the media industry, technological change is considered a major driving force and poses a major challenge to the development of the media industries in all sectors (Kung. 2007). Traditional Media Company (TMC), a media organization that acts as a one-way transmitter to the recipient, has extensive effort to adapt themselves, create convergence with other emerging new media, and transform to Data-Driven Organization (DDO) to create capabilities such as the ability to deeply understand viewers' behavior individually.

The media industry in Thailand still faces several major challenges related to the use of data e.g. the challenge of systematically gathering and storing large-scale data, management of a variety of data types such as text, images, audio and video, and the integration of different data sources such as from social media. Furthermore, the challenge of reaching different segments of audiences and enhancing ability to understand, analyze and interpret accurately and quickly is one of the most difficult parts to achieve. In addition, the challenge

 DOI: 10.1201/9781003514831-7

of encouraging people in an organization to use and integrate data from analysis into business decision-making processes are utmost concerned. Moreover, media organizations also face challenges in terms of personal information and data security as this has been breached in many evidenced cases abroad.

Therefore, developing an organization to be a successful DDO requires a holistic, practical, and managerial framework to guide and drive TMC in Thailand to create efficient operations and generate maximum value and benefits in delivering media services in an environment suitable for Thailand's context.

2 EXPERIMENTAL

This research applies the concept of Diamon Framework (Leavitt. 1965, Vidgen *et al.* 2017) and in-dept interviews with case study companies in the Thai media industry to identify a framework for adaptation and transformation towards DDOs that is suitable for the local context.

Within the framework of Leavitt's Diamond Model concept, it has been applied to many researches related to enterprise development, such as Nerur's research on enterprise transition in software development (Nerur *et al.* 2005) Saeed and Wang's study of enterprise analytics (Saeed. 2013) Vidgen's studies on capacity-building in enterprise business data analysis (Vidgen *et al.* 2017) or Hupperz's (Hupperz *et al.* 2021) research on DDO and so on. The concept of Diamond Model is taken into consideration with the DDO characteristics defined by several researchers as shown in Figure 1. As a result, the draft DDO framework with five key Dimensions is proposed; Strategy and Organization, People Skill, Data Process, Data Technology and Data Value Realization.

Definition/Author	Characteristics
Mason & Patil, 2015	An organization that seeks to use data to drive decisions at all levels, decisions are made from data, and data is used to measure the success of decisions.
Kolbjørnsrud et al., 2018	Describes an information-driven organization in the form of an intelligent system that can sense, comprehend, act, learn, and explain. (SCALE Framework)
Hupperz et. al., 2021	The enterprise consists of five key components: Digital Transformation, Data Science, Data-Driven Business Model, Data Driven Innovation, and Data Analysis.
Van der Sande & Van der Meijden, 2021	An organization that integrates data analysis into the core of a business process.
AWS, 2023	Treat information as a strategic asset and build capacity to use it not only for important decisions, but also for the day-to-day operations of the organization.

Figure 1. Characteristics of DDO.

Each of the five dimensions' Sub-dimensions is determined by a literature review, and its further development is based on in-depth semi-structured interviews with six executives from top Thai media companies (Figure 2 shows the profiles of these executives and firms), including both traditional and digitally native businesses, to elucidate the particular needs for adaptation and transformation towards DDO.

Media Company	Business Type	Business Size (Annual Revenue)	Interviewee
Firm A	One of the top 3 commercial TV broadcasters	< 50m USD	Assistant Vice President - IT
Firm B	One of the top 3 state-own TV broadcasters	< 50m USD	Assistant Vice President - IT
Firm C	One of the top 10 commercial TV broadcasters	< 10m USD	Senior Manage New Media
Firm D	One of the top 3 local video streaming providers	< 10m USD	Managing Director
Firm E	One of the top 3 news website operators	< 10m USD	Managing Editor
Firm F	One of the top 3 web board operators	< 1m USD	Chief Technology Officer

Figure 2. Profiles of Thai media firms and executives.

The DDO framework for the Thai media business is then completed after the robustness of the framework is confirmed through interviews with three executives from new media companies. Likert scale 1-5 are used in the scoring matric for Goodness of Fit (GOF) and Demand while Priority is ranked by order among each Dimension.

3 RESULTS AND DISCUSSION

3.1 *Strategy and organization*

Four Sub-dimensions are proposed to form the Strategy and Organization Dimension in DDO framework i.e. New Media Strategy, Digital Leader, Data Culture and Organization Structure.

Establishing a robust New Media Strategy is crucial for TMC, and its significance cannot be overstated. Given how quickly the digital market is changing, media companies need to be extremely clear and specific about their strategies. New media has become the main channel for engagement, communication, and contact with a wide range of audiences in the modern period.

Digital Leader is essential for developing a DDO due to their ability to navigate complexity, articulate strategic vision, drive innovation, manage change, promote collaboration, and ensure ethical data practices. By providing strategic guidance and fostering Data Culture, digital leaders empower organizations to leverage data effectively, drive innovation, and maintain a competitive edge in the digital age.

Organization Structure defines roles, positions, and responsibilities for operations and decision-making within the organization (Grant, 1966) The capacity to advance DDO is increased by clearly defining of the roles, positions, and responsibilities of data stakeholders.

3.2 *People skill*

The DDO framework proposes three Sub-dimensions, namely Media and Information Literacy (MIL), Data Professional, and Data Skill, to comprise the People Skill Dimension.

MIL skills empower individuals to critically evaluate, analyze, and effectively utilize vast amounts of data and information. By enhancing MIL skills among employees, the company can ensure that its workforce is adept at navigating digital platforms, discerning credible sources, and extracting meaningful insights from data. This fosters a culture of informed decision-making, facilitates the integration of data-driven approaches into organizational processes, and ultimately enhances the company's ability to leverage data for strategic initiatives, innovation, and competitive advantage in the rapidly evolving media landscape.

A TMC would require Data Professionals, such as data analysts, data scientists, and data engineers, to support the development of a DDO because these specialists possess the expertise and technical acumen necessary to effectively manage, analyze, and interpret large volumes of data. By leveraging their specialized skills, Data Professionals can ensure the accuracy, reliability, and relevance of data insights, enabling informed decision-making and strategic planning within the organization.

To support the development of a DDO, a TMC would need its staff to have Data Skills since these abilities allow workers to efficiently gather, analyze, and use data for creative thinking and strategic decision-making. With Data Skills, employees can proficiently analyze audience behavior, market trends, and content performance, enabling the company to tailor its offerings to meet evolving consumer preferences and optimize resource allocation.

3.3 *Data process*

The Data Process Dimension in the DDO framework is recommended to be composed of four sub-dimensions: Behavioral Data Analytic, Data Governance, Data Management, and Data Privacy Protection.

A TCM would benefit from conducting Behavioral Data Analytics because such analysis provides invaluable insights into audience preferences, engagement patterns, and content consumption habits. By leveraging behavioral data, the company can tailor its content offerings, advertising strategies, and marketing campaigns to better resonate with target audiences, thereby enhancing viewer satisfaction and loyalty. Moreover, Behavioral Data Analytics enables the identification of emerging trends, forecasting of audience behavior, and optimization of content distribution channels, empowering the company to make informed decisions and stay ahead in a rapidly evolving media landscape.

Data Governance ensures the quality, integrity, and security of data assets. By establishing clear policies, standards, and procedures, the company can mitigate risks associated with data inaccuracies, inconsistencies, and unauthorized access. Data Governance also facilitates compliance with regulatory requirements and industry standards, instilling trust and confidence in data-driven decision-making processes. Furthermore, Data Governance promotes transparency, accountability, and collaboration across departments, fostering a culture of data stewardship and enabling the company to maximize the value of its data assets for strategic initiatives and operational efficiency.

Data Management ensures the availability, accessibility, and usability of data across the organization. By implementing robust Data Management practices, the TMC can organize, store, and process data efficiently, enabling timely access to accurate and relevant information. Moreover, Data Management facilitates data integration, cleansing, and enrichment, enhancing the quality and reliability of data assets. This, in turn, empowers the company to derive actionable insights, optimize operational processes, and drive innovation in content creation, distribution, and audience engagement strategies.

Safeguarding the privacy of data ensures that sensitive information, such as audience demographics and viewing behaviors, is handled securely and ethically. By implementing robust Data Privacy Protection, including encryption, access controls, and compliance with privacy regulations, the TMC can mitigate the risk of data breaches and unauthorized access. Moreover, prioritizing data privacy protection fosters transparency and accountability, enhancing the company's reputation and credibility among audiences and stakeholders. This, in turn, promotes trust and confidence, facilitating continued engagement and collaboration with consumers and partners.

3.4 *Data technology*

Content and Delivery Platform, Big Data Technology, AI & ML, Analytic & BI Tool, and Cyber Security are the five Sub-dimensions that are suggested to make up the Data Technology Dimension in the DDO architecture.

Content and Delivery Platform facilitates the seamless creation, distribution, and analysis of content across various channels. By leveraging a centralized platform, the TMC can efficiently manage content assets, track audience engagement metrics, and derive actionable insights to inform content strategy and optimization efforts. Moreover, a Content and Delivery Platform enables the personalization of content delivery, allowing the company to tailor offerings to individual preferences and behaviors, thereby enhancing viewer satisfaction and retention.

Big Data Technology enables the collection, storage, processing, and analysis of large volumes of data from diverse sources. Examples of Big Data Technologies include Hadoop for distributed data storage and processing, Apache Spark for real-time data analytics, and Apache Kafka for stream processing and data integration. By leveraging these technologies, the company can gain deeper insights into audience behaviors, content preferences, and market trends. Additionally, Big Data Technologies enable scalability and flexibility, ensuring the company can efficiently handle growing data volumes and adapt to changing business needs.

AI and ML offer powerful capabilities for analyzing vast amounts of data, predicting audience behavior, and personalizing content delivery. For example, recommendation systems powered by ML algorithms can suggest relevant articles, videos, or shows to individual users based on their past preferences and viewing habits, enhancing user engagement and satisfaction. Similarly, AI-powered content creation tools can automate tasks such as video editing, captioning, and transcription, streamlining production processes and improving efficiency. Furthermore, AI-driven analytics platforms can uncover actionable insights from data, enabling the company to optimize content strategies, target advertising more effectively, and drive revenue growth.

Analytics and BI Tools enable the TMC to extract actionable insights from data, visualize trends, and make informed decisions. Examples of Analytics and BI tools used in Thai media companies include Tableau, Power BI, and Google Analytics. These tools allow organizations to analyze audience engagement metrics, track content performance across various platforms, and identify opportunities for optimization and growth.

Cyber Security initiatives in media companies include implementing robust encryption protocols to secure data transmission, conducting regular vulnerability assessments and penetration testing to identify and mitigate potential security risks, and implementing multi-factor authentication measures to prevent unauthorized access to sensitive systems and information. By prioritizing Cyber Security, TMCs can ensure the integrity, confidentiality, and availability of their data assets, thereby enabling a safe and secure environment.

3.5 *Data value realization*

The DDO architecture proposes five sub-dimensions, notably Behavioral-based Data Driven Decision Making (BDDDM), Revenue Generation, Cost Efficiency, Customer Experience, and Innovation. to build the Data Value Realization Dimension.

BDDDM allows for a deeper understanding of audience behaviors and preferences, leading to more targeted and effective content strategies. Examples of BDDDM initiatives in media companies include analyzing viewer engagement metrics to optimize programming schedules, leveraging audience segmentation data to tailor advertising campaigns, and using viewer feedback data to inform content development decisions. By adopting BDDDM, TMCs can enhance audience engagement, increase viewer satisfaction, and drive revenue growth.

Revenue Generation from data enables monetization of data assets. Examples of Revenue Generation from data in media companies include targeted advertising based on viewer demographics and interests, subscription models tailored to individual preferences and consumption patterns, and partnerships with data providers or analytics firms to sell insights and intelligence derived from audience data.

Cost Efficiency value stems from enabling of the optimization of operational processes, resource allocation, and budget management by using data. Examples of Cost Efficiency initiatives from using data in media companies include optimization of content production schedules, elimination of inefficiencies in distribution channels and advertising campaigns, and leveraging audience insights to reduce content development costs. By harnessing data-driven Cost Efficiency, TMCs can streamline operations, minimize wastage, and maximize return on investment, thereby improving profitability.

Media companies are utilizing data to enhance their Customer Experience through various initiatives such as personalized content recommendations based on viewing history and preferences, targeted advertising campaigns based on audience demographics and interests, and interactive features like polls and surveys to collect feedback and enhance content relevance. To boost viewer happiness, loyalty, and retention, TMCs are giving utmost priority on the Customer Experience enhancement.

Data-driven Innovation fosters creativity, enhances competitiveness, and drives growth. Innovation initiatives in media companies include the development of AI-powered content recommendation engines to personalize user experiences, the implementation of data-driven content optimization tools to improve engagement metrics.

3.6 *DDO framework robustness*

Fives Dimensions and 21 Sub-Dimensions as explained in the earlier sections have been validated by three industry experts. The score matrix

Figure 3. 5 Dimensions of DDO framework.

is used to justify the robustness of DDO framework in terms of goodness of fit, demand, and priority. The findings show that the proposed model has very high score with 4.6 and 4.7 on GOF and Demand accordingly. Figure 4 below shows the average score of each parameter. The most prioritized Sub-dimensions in each Dimension are Digital Leader, MIL, Behavioral Data Analytics, Content & Delivery Platform/Big Data Technology, and BDDDM.

Dimensions	Sub-Dimensions	GOF	Demand	Priority
Strategy & Organization	New Media Strategy	4.7	5.0	2.7
	Data Culture	4.7	5.0	3.7
	Digital Leader	4.0	4.3	4.0
	Organization Structure	4.3	4.7	3.7
People Skill	MIL	5.0	5.0	5.0
	Data Professional	3.7	3.3	3.0
	Data Skill	4.0	4.0	4.0
Data Process	Behavioral Data Analytics	5.0	5.0	5.0
	Data Governance	4.3	5.0	3.7
	Data Management	4.3	5.0	2.0
	Data Privacy Protection	5.0	5.0	3.3
Data Technology	Content & Delivery Platform	5.0	5.0	4.3
	Big Data Technology	5.0	5.0	4.3
	AI & ML	4.7	4.7	1.0
	Analytics & BI Tool	5.0	5.0	3.3
	Cyber Security	5.0	5.0	2.0
Data Value Realization	BDDDM	5.0	5.0	4.7
	Revenue Generation	4.3	4.3	2.3
	Cost Efficiency	4.0	4.0	2.0
	Customer Experience	5.0	5.0	4.3
	Innovation	4.3	4.3	1.7
Total		4.6	4.7	

Figure 4. 5 dimensions.

4 CONCLUSION

In this paper, the qualitative research has been employed to identify the DDO framework for adaptation and transformation towards DDOs that is suitable for the TCM in Thailand. The DDO literatures have been studied and adapted to the local context by using in-depth interviews and measuring robustness through scoring matrix on GOF and Demand while identifying the areas where the most priority for development is needed. The results show the proposed DDO framework is very suitable for different type of media companies. It is recommended that more quantitative study be done in the future to determine the maturity and present state of various media firms.

REFERENCES

AWS. (2023). *How do you Become a Data-Driven Organization?.* https://aws.amazon.com/executive-insights/content/how-do-you-become-a-data-driven-organization/

Bokman, A., Fiedler, L., Perrey, J., & Pickersgill. A. (2014). *Five Facts: How Customer Analytics Boosts Corporate Performance.* https://www.mckinsey.com

Davenport, T. H, Harris, J. G. & Morison, R. (2010). *Analytics at Work: Smarter Decisions, Better Results.* Harvard Business Press.

Grant, R. M. (1996). Toward a knowledge-based theory of the firm, *Strategic Management Journal.*

Hupperz, M., Gür, I., Möller, F. & Otto, B. (2021). What is a Data-Driven Organization?. *Americas Conference on Information Systems*, Montreal.

Kolbjørnsrud, V., Andersen, E., Johnson, J., & Ragnvald, S. (2018). The data-driven organization: Intelligence at scale. *Forefr. Look. Ahead*, 23, 11–15.

Kung, L. (2007). Does media management matter? Establishing the scope, rationale, and future research agenda for the discipline. *Journal of Media Business Studies*, 4(1), 21–39.

Leavitt, H. J. (1965). Applied organizational change in industry, I JG March (ed.) *Handbook of Organizations*, Chicago: Rand McNally, 1144: 1170.

Mason, H., & Patil, D. J. (2015). Data driven. Creating a Data Culture.

Nerur, S., Mahapatra, R. & Mangalaraj, G. (2005). Challenges of migrating to agile methodologies. *Communications of the ACM*, 48 (5), 72–78.

Parkins, D. (2017). The world's most valuable resource is no longer oil, but data. *The Economist.* https://www.economist.com/leaders/2017/05/06/the-worlds-most-valuable-resource-is-no-longer-oil-but-data

Saeed, B. B., and Wang, W. (2013). Organisational diagnoses: a survey of the literature and proposition of a new diagnostic model. *Int. J. Information Systems and Change Management*, (6), pp. 222–238.

Sivarajah, U., Kamal, M. M., Irani, Z., & Weerakkody, V. (2017). Critical analysis of Big Data challenges and analytical methods. *Journal of Business Research*, 70, 263–286.

Van der Sande, L. & Van der Meijden, M. (2021, May 21). Why you need to capitalize on the rise of the data-driven enterprise. *Accenture.* https://www.accenture.com/nl-en/blogs/insights/data-driven-enterprise

Vidgen, R., Shaw, S., and Grant, D. (2017). Management challenges in creating value from business analytics. *European Journal of Operational Research* (261:2), pp. 626–639 (doi: 10.1016/j.ejor.2017.02.023).

System Innovation for an Artificial Intelligence Era – Lam et al. (Eds)

The effect of CEO overconfidence on foreign post-entry performance

Jung-Ho Lai
Department of Finance, National Taipei University of Business, Taipei City, Taiwan

ABSTRACT: This paper examines if CEO overconfidence can explain corporate post-entry performance when entering foreign markets. Using options-based measure of CEO overconfidence, I find that CEO overconfidence lowers foreign post-entry performance. Moreover, host country risk negatively moderates the relationships between CEO overconfidence and post-entry performance such that the negative relationship between CEO overconfidence and post-entry performance is worsen when firms enter foreign markets with higher country risk. The findings of this paper help to explain the inconclusive effect of CEO overconfidence on firm performance so far.

1 INTRODUCTION

The determinant a firm's overseas market entry performance has been a critique issue in the international business research (e.g., Brouthers and Hennart 2007; Chari and Chang 2009). For the past decades, researchers explore possible antecedents from a variety of theoretical perspectives, including transaction costs, incremental learning, bargaining, eclectic, and knowledge-based theories.. Salient determinants identified are classified into host country-specific variables, home country-specific variables, company-specific variables, and venture-specific variables (see Morschett, Schramm-Klein and Swoboda 2010 for a review). Despite that the extant literature has incorporated a variety of concerns, these theoretical approaches make the implicit assumption that entry-mode choices are based on economic considerations. However, decisions on how to enter foreign markets are characterized by complexities, incomplete information and considerable uncertainty. Under such conditions, decision-makers' psychological characteristics, such as their risk-taking propensities and belief in their ability to select superior strategies and manage successfully are likely to have an important bearing on corporate foreign-market entry performance. In this study, therefore, rather than focusing on environmental-level and firm-level characteristics as have been well researched, I relate entry performnace to personal characteristics of the top decision makers inside the firm. Building on psychological and behavior finance literatures, I argue that one important determinant is the level of CEO overconfidence, i.e., their tendency to think that they are better than they really are in terms of characteristics such as ability, judgment, or prospects for successful outcomes. Researchers show that the level of CEO overconfidence significantly affects corporate decision makings and their performances. For example, Shu, Yeh, Chiang, and Hung (2013) find that CEO overconfidence is positively correlated with the intensity of share repurchasing. However, the programs launched by overconfident CEOs were undervalued and associated with reduced post-announcement returns. Take another example, Hirshleifer, Low and Teoh (2012) find that firms with overconfident CEOs have greater return volatility, invest more in innovation, obtain more patents and patent citations, and achieve greater innovative success for given research and development expenditures. To sum up, despite of both positive and negative performance effect found, the existing empirical evidences suggest a salient impact of CEO overconfidence on firm performance.

To systematically address the issue of CEO overconfidence in the strategy context of foreign entry, I first clarify the content of CEO overconfidence and its implications for decision performance. To obtain a more complete understanding of the influence of CEO overconfidence on a firm's foreign investment actions, I go on to investigate the effect of CEO overconfidence tendency on the post-entry performance. Finally, to shed light on the inconclusive effect of CEO overconfidence on various strategic performances found so far, I investigate the moderating effect of host country risk.

2 THEORY AND HYPOTHESES DEVELOPMENT

2.1 *CEO overconfidence and its performance implications*

In the psychology literature, overconfidence is conceptualized as the tendency of individuals to

DOI: 10.1201/9781003514831-8

think of themselves as 'above average' on positive characteristics (Kruger 1999). This 'better-than-average' effect is due to a psychological attribution bias, the self-serving inclination to attribute good outcomes to one's own actions and bad outcomes to luck (Miller and Ross 1975). Kahneman and Riepe (1998, p. 54) remark that 'The combination of overconfidence and optimism is a potent brew, which causes people to overestimate their knowledge, underestimate risks, and exaggerate their ability to control events'. Since overconfident managers tend to believe that future strategic outcomes are under their control, this illusion of control drives overconfident CEO to undertake aggressive strategies and invest excessively (Malmendier and Tate 2005). Their inclinations to overestimate their own efficacy in bringing about success (Wolosin *et al.* 1973), however, could lead to inefficient strategic actions. The existing evidence supports the notion that overconfident CEOs tend to over-expand, but overestimate the probability of their success and underestimate the probability of their failure (Langer 1975; Wolosin *et al.* 1973). Such psychological bias can have important implications for the issue of corporate foreign-market entry performance, as discussed follows.

2.2 *Effect of CEO overconfidence on foreign-market entry performance*

Psychologists remark that overconfident CEO tend to overestimate the precision of their private information and overreacts to it (Daniel *et al.* 2001; Gervais and Odean 2001). Besides, they are inclined to overestimate the net discounted expected payoffs from uncertain endeavors, because they overestimate their own efficacy in bringing about success (Langer 1975; Wolosin *et al.* 1973). Empirical works show that strategies pursued by overconfident CEOs generally result in inferior outcomes. Malmendier and Tate (2008) find that returns from the strategic investments undertaken by overconfident managers are lower than the returns achieved by not overconfident managers. Doukas and Petmezas (2007) find that overconfident bidders realize lower announcement returns than rational bidders and exhibit poorer long-term performance. They further find that high-order acquisitions (five or more deals within a three-year period) are associated with lower wealth effects than low-order acquisitions (first deals). They explain this as managers' propensity to credit the initial success to their own ability and therefore become overconfident and engage in more deals with poor performance.

Considering the salience of foreign market entry in a firm's strategic profile, it thus could be biased by CEOs' hubris. I predict that overconfident CEOs may underestimate the risks underlying foreign market entries, misperceiving some entry opportunities with high risks to be low risks, and as a result are more likely to suffer from poorer post-entry performance than what is undertaken by rational managers. The above discussion leads to the following hypothesis:

Hypothesis 1: CEO overconfidence negatively impacts the performance of foreign market entry.

2.3 *Moderating effect on the relationship between CEO overconfidence and foreign-market entry performance*

Firms opting for shared ownership mode is argued subject to fewer exogenous uncertainties than opting for full acquisitions because of the target's managerial incentive. By permitting the target to share ownership, foreign entrants increase the incentives of targets' managers to share their local knowledge and relationship networks, reduce the risks associated with heterogeneous socioculture, politics, economics, and competitive dynamics in foreign markets (López-Duarte and García-Canal 2002). In contrast, a complete acquisition of a local firm creates difficulties in motivating target firm managers, as they no longer benefit from any direct ownership interests they have had in the original firm (Chi 1994). Losing high-powered incentives discourage target firm managers to transfer their critical tacit assets such as local market knowledge and relation ties with key stakeholders, exposing the foreign entrant to a higher level of information asymmetry concerning the local market and greater risk arising from environmental turbulence.

To sum up, since greater resource commitments involve increased exposure to country risks, the relationship between country risk and the share of equity sought post-acquisition by a rationale risk-averse manager is negative, which contrasts with overconfident CEO's propensity to over expand. The level of host country risk thus may negatively moderate the relationship between CEO overconfidence and post-entry firm performance.

Hypothesis 2: host country risk negatively moderates the relationships between CEO overconfidence and post-entry performance such that the negative relationship between CEO overconfidence and post-entry performance is worsen when firms entering foreign markets with higher country risk.

3 SAMPLE

A sample of U.S. firms undertaking cross-border investments including full acquisitions, partial acquisitions, and joint ventures is collected from Acquisitions and Alliances database of Securities Data Company over the period of 2010 to 2021. Securities Data Company provides the event of foreign acquisitions and joint ventures, and disclosure contains detailed information such as investment scale, scope of cooperation, mode of acquisition, and so on. To ensure that foreign investment cases that represent "large" investments by firms is included, I require that the size of the foreign investment be at least 10% of the size of the partnering firm. The initial announcement date (day 0) for each investment is confirmed by Wall Street Journal Index, where day 0 refers to the earliest date when the investment announcements have to meet the following criteria: (1) to avoid any potential confounding effect on measurement of wealth effect, the sample firms should not have other major announcements (e.g., earnings, dividends, or capital structure change announcements) three days before and after the investment announcement date; (2) daily stock return information must be available from the Center for Research in Securities Prices (CRSP) returns files; and (3) companies' financial information must be available from the Compustat files. Information of host country risk is from International Country Risk Guide (ICRG). Data on executives' options holding is drawn from Execucomp database.

4 MEASURES OF VARIABLES

4.1 *Dependent variable*

The standard event-study method is employed to estimate cumulative abnormal returns (CARs) as the measure of the dependent variable. This approach has been widely used in accounting, economics, finance, and management studies to test the influence of firm-specific events on firm value. The measure of abnormal stock returns in relation to foreign investment announcements follows Brown and Warner (1985) by using the market model to obtain estimates of the expected returns. This model captures a firm's stock price change after adjusting for general market-wide factors and the firm's systematic risk. The abnormal return for firm i on day t, ARit, is computed by:

$$AR_{it} = R_{it} - E(R_{it}/I_{t-1}), \qquad (1)$$

where Rit is firm i's actual returns on day t, and It-1 represents the information set available to the market about the firm at time t-1.

The expected return for firm i on day t is estimated by:

$$E(R_{it}/I_{t-1}) = \alpha_i + \beta_i R_{mt} \qquad (2)$$

where Rmt is the return for the market portfolio on day t, α_i is the intercept, and β_i measures the risk or sensitivity of the firm's returns relative to the market portfolio. I define Day 0 (t = 0) as the initial announcement date. I use the value-weighted CRSP Index as the proxy for the market portfolio. The parameters α_i and β_i are estimated using data for the period of 210 to 11 days before the initial announcement date (i.e., days -210 to -11). Finally, the dependent variable of two-day cumulative abnormal returns, CAR (-1, 0), is estimated by summing the daily abnormal returns over the window of days -1 and 0.

4.2 *Independent variable*

I use options-based overconfidence measure based on the premise that undiversified, risk-averse executives typically tend to exercise their own-firm stock options early if the option is sufficiently in the money (Hall and Murphy 2002). Malmendier and Tate (2008) remark that executive options give the holder the right to purchase company stock, usually at the stock price on the grant date after a four-year vesting period. Upon exercise, the holder receives shares of company stock which can be in turn immediately sold (Ofek and Yermack 2000). Since a large part of managerial compensation is equity-based and their human capital is invested in their firms, CEOs are highly exposed to company risk. As a result, risk-averse CEOs may exercise options early if the stock price is sufficiently high (Hall and Murphy 2002), the time when the marginal cost of continuing to hold the option (risk exposure) exceeds the marginal benefit (option value). Overconfident CEOs, however, persistently fail to exercise highly in-the-money vested options due to their over estimation of the firm's future returns.

Following Malmendier and Tate (2005, 2008), Confident CEO (Options) takes a value 1 if a CEO postpones the exercise of vested options that are at least 67% in the money, and 0 otherwise. While Execucomp database does not provide detailed data on a CEO's options holdings and exercise prices for each option grant, I follow Campbell *et al.* (2011) in calculating the average moneyness of the CEO's option portfolio for each year. First, for each CEO-year, I calculate the average realizable value per option by dividing the total realizable value of the options by the number of options held by the CEO. The strike price is calculated as the fiscal year-end stock price minus the average realizable value. The average moneyness of the options is then calculated as the stock price divided by the estimated strike price minus one.

4.3 *Moderator and control variables*

I measure country risk, the moderator, using the index of country risk published by the International Country Risk Guide (ICRG) for the deal year (Jung 2004). The index measure represents country risk in three exogenous areas: political, financial, and economic. The index ranges from 1 to 100, with greater numbers indicating less risk. I reverse-code the scores so that greater scores indicate greater country risk.

I control several factors that have been shown to affect entry mode choice and post-entry performance. Specifically, firm size is measured by the natural logarithm of market value of equity (as in Aulakh and Kotabe 1997). Pre-entry ROA measures a firm's return on assets (Tihanyi *et al.* 2005). Firm R&D intensity is operationalized as firm R&D expenses divided by total sales (as in Datta *et al.* 2009). I define Firm international experience by firms' foreign entry cases in the past 5 years (Aulakh and Kotabe 1997). Additional to firm level variables, I also control for the influence of environmental factors, including Host country GDP per capita, GDP growth in the past 3-year period, Size of the host market (proxied by GDP) and Cultural distance between the host and home country following prior research (e.g., Brouthers *et al.* 2008; Richards and Yang 2007; Tihanyi *et al.* 2005). I measure value of the control variables for the fiscal year prior to the focal entry.

5 EMPIRICAL RESULTS

5.1 *Regression results*

The regression results are presented in four models in Table 1. Model 1 represents the base model with only the control variables. In Models 2and 3 I include CEO overconfidence, and its interaction with host country risks. Model 4 represents the full model results including all variables. I computed VIFs (variance inflation factors) associated with independent and control variables to assess if multicollinearity is a problem in my study. The highest VIF values are 2.35 and 2.69—well below the cutoff value of 10 identified by Cohen, Cohen, West, and Aiken (2013).

Hypothesis 1 predicts that CEO overconfidence is negatively associated with post entry performance. Results presented in Model 2 of Table 1 provide strong support for this hypothesis. They indicate that CEO overconfidence is significantly negatively related to post entry firm performance ($p < .01$). Likewise, I find support for Hypothesis 2, with the interaction between CEO overconfidence and host country risk positively associated with post entry firm performance ($p < .05$). Finally, I put all variables in Model 4. The significant effects of independent variable and moderator are unchanged.

5.2 *Robustness tests*

While the inclusions of control variables help to address possible sources of unobserved heterogeneity in the data, for thoroughness I also conduct analyses using fixed-effects models to control for potential confounds related to unobserved heterogeneity (Greene 2001). None of the coefficients that are (not) significant in the original model loses (gains) significance in this alternative model, suggesting that unobserved heterogeneity is not confounding my findings. Further, to ensure that my findings are not driven by outliers, I winsorize all continuous variables at the 1st and 99th percentiles (Ahn *et al.* 2010), finding similar results. Finally, I compute the t-value using heteroskedasticity-consistent standard errors (White 1980), findings similar results. Taken together, the sensitivity analyses confirm the robustness of my findings and support their interpretation as evidence of the impact of CEO overconfidence on the outcome of foreign market entry performance.

Table 1. Regression analyses.

Variables	Model 1	Model 2	Model 3	Model 4
Intercept	3.543***	3.681***	3.068***	3.451***
Firm capitalization	-0.464***	-0.460***	-0.422***	-0.500***
Pre-entry ROA	1.668**	1.674**	1.656**	1.729**
Host country experience	0.101	0.021	0.421	0.035
Industry diversification	-0.114	-0.114	-0.163	-0.178
Host country GDP per capita	0.042***	0.042***	0.044***	0.044***
Host country GDP growth	-0.231***	-0.230**	-0.231***	-0.284***
Size of the host market	0.162***	0.126**	0.182***	0.135***
Cultural distance	-0.629***	-0.621***	-0.633***	-0.550***
Institutional distance	-0.192***	-0.124***	-0.185***	-0.173***
CEO overconfidence		-0.755***	-0.215***	-0.321***
CEO overconfidence * host country risks			0.637**	0.542**
Service industry dummy	YES	YES	YES	YES
R-squared	0.720***	0.759***	0.875***	0.546***
Number of observations	660	660	660	660

***p< 0.01, **p< 0.05, *p< 0.1.

6 CONCLUSION

Grounded in behavior decision theory, this study postulates that CEOs' exaggerated self-enhancement tendency along with information asymmetry and enormous challenges associated with going abroad can result in poorer foreign-market entry performance than rational CEOs. Since foreign entries involve considerable means-ends ambiguity with eventual outcomes extremely uncertain, the condition of bounded rationality applies, and managerial psychological bias can be expected to intervene in such choices (Mischel 2004). Supporting the predicted hypotheses, findings show that foreign-market entries led by overconficent CEOs significantly underperform those led by non-overconficent CEOs. Besides the line effect result, the examination findings of contingency provides further supports showing that the negative relationship between CEO overconfidence and post-entry performance is worsen when firms enter foreign markets with higher host country risk. Findings of this study demonstrates the value of giving attention to individual-level factors from other theoretical lense. Such multilevel analyses can provide additional insights and contrast these theories in their efficacies in predicting foreign market entry performance.

REFERENCES

Ahn, S., Jiraporn, P. and Kim, Y.S., (2010). Multiple directorships and acquirer returns. *Journal of Banking and Finance* 34(9), 2011–2026.

Aulakh, P.S. and Kotabe, M., (1997). Antecedents and performance implications of channel integration in foreign markets. *Journal of International Business Studies* 28, 145–175.

Brouthers, K.D. and Hennart, J.F., (2007). Boundaries of the firm: Insights from international entry mode research. *Journal of Management* 33(3), 395–425.

Brouthers, K.D., Brouthers, L.E. and Werner, S., (2008). Real options, international entry mode choice and performance. *Journal of Management Studies* 45, 936–960.

Campbell, T., Colin , Michael Gallmeyer, Shane Johnson, Jessica Rutherford and Brooke Stanley, (2011). CEO optimism and forced turnover, *Journal of Financial Economics* 101, 695–712.

Chari, M.D.R. and Chang, K., (2009). Determinants of the share of equity sought in cross-border acquisitions. *Journal of International Business Studies* 40 (8), 1277–1297.

Chi, T. (1994). Trading in strategic resources: Necessary conditions, transaction cost problems, and choice of exchange structure. *Strategic Management Journal*, 15 (4), 271–290.

Cohen, J., Cohen, P., West, S. G. and Aiken, L. S. (2013). *Applied Multiple Regression/Correlation Analysis for the Behavioral Sciences*. Routledge.

Daniel, K.D., Hirshleifer, D. and Subrahmanyam, A., (2001). Overconfidence, arbitrage, and equilibrium asset pricing. *Journal of Finance* 56, 921–65.

Datta, D., Musteen M. and Herrmann, P., (2009). Board Characteristics, Managerial Incentives, and the Choice Between Foreign Acquisitions and International Joint Ventures. *Journal of Management* 35, 928–953.

Doukas, J.A. and Petmezas, D., (2007). Acquisitions, overconfident managers and self-attribution bias. *European Financial Management* 13(3), 531–577.

Gervais, S. and Odean, T., (2001). Learning to be overconfident', *The Review of Financial Studies* 14, 1–27.

Greene, W. (2001). *Estimating Econometric Models with Fixed Effects*. Department of Economics, Stern School of Business, New York University.

Hall, Brian and Kevin Murphy, (2002). Stock options for undiversified executives, *Journal of Accounting and Economics* 33, 3–42.

Hirshleifer, D., Low, A. and Teoh, S., (2012). Are overconfident CEOs better innovators? *Journal of Finance* 67(4), 1457–1498.

Jung, J., (2004). Acquisitions or joint ventures: Foreign market entry strategy of US advertising agencies. *Journal of Media Economics* 17(1), 35–50.

Kahneman, D. and Riepe, M.W.,1998. Aspects of investor psychology. *Journal of Portfolio Management* 24, 52–65.

Kruger, J., 1999. Lake Wobegon be gone! the below-average effect and the egocentric nature of comparative ability judgments. *Journal of Personality and Social Psychology* 77(2), 221–32.

Langer, E.J. and Roth, J.,(1975). Heads I win, tail it's chance: the illusion of control as a function of the sequence of outcomes in a purely chance task. *Journal of Personality and Social Psychology* 32, 951–55.

López-Duarte, C. and García-Canal, E., (2002). Adverse selection and the choice between joint-ventures and acquisitions: Evidence from Spanish firms. *Journal of Institutional and Theoretical Economics* 158(2), 304–324.

Malmendier, Ulrike and Geoffrey Tate, (2005a). CEO overconfidence and corporate investment. *Journal of Finance* 60, 2661–2700.

Malmendier, Ulrike and Geoffrey Tate, (2008). Who makes acquisitions? CEO overconfidence and the market's reaction. *Journal of Financial Economics* 89, 20–43.

Miller, D.T. and Ross, M., (1975). Self-serving biases in the attribution of causality: fact or fiction? *Psychological Bulletin* 82(2), 213–25.

Mischel, W. (2004). Toward an integrative science of the person. *Annu. Rev. Psychol.*, 55, 1–22.

Morschett, D., Schramm-Klein, H. and Swoboda, B., 2010. Decades of research on market entry modes: what do we really know about external antecedents of entry mode choice? *Journal of International Management* 16(1), 60–77.

Ofek, E. and Yermack, D., (2000). Taking stock: equity-based compensation and the evolution of managerial ownership. *Journal of Finance* 55, 1367–1384.

Richards, M., Yang, Y., (2007). Determinants of foreign ownership in international R&D joint ventures: Transaction costs and national culture. *Journal of International Management* 13(2), 110–130.

Tihanyi, L., Griffith, D.A. and Russell, C., (2005). The effect of cultural distance on entry mode choice, international diversification and MNE performance: A meta-analysis. *Journal of International Business Studies* 36, 270–283.

Shu, P.G., Y.H., Yeh, T.L., Chiang and J.Y., Hung, (2013). Managerial overconfidence and share repurchase. *International Review of Finance* 13(1), 39–65.

White, H. (1980). A heteroskedasticity-consistent covariance matrix estimator and a direct test for heteroskedasticity. *Econometrica: Journal of the Econometric Society*, 817–838.

Wolosin, R.J., Sherman, S.J. and Till, A.,(1973). Effects of cooperation and competition on responsibility attribution after success and failure. *Journal of Experimental Social Psychology* 9, 220–35.

System Innovation for an Artificial Intelligence Era – Lam et al. (Eds)
© 2025 the Author(s), ISBN: 978-1-032-84452-7

Developing of autonomous guided vehicle with collision avoidance system and smart object detection for dynamic collection mechanism

Federico David Vera Gomez, Milton Sala Pereira, Franco Lombardo Barrail & Ariel Federico Gonzalez Samaniego
Department of Electromechanical Engineering, Universidad Politécnica Taiwan-Paraguay, Asunción, Paraguay

Meng-Kun Liu
Department of Mechanical Engineering, National Taiwan University of Science and Technology, Taipei, Taiwan

ABSTRACT: Reliance on human labor in an industrial environment often results in decreased productivity, and the risk of accidents also remains a prevalent concern in potentially hazardous environments. This study introduces the development of an autonomous robot car with a robot arm as a collection mechanism for efficient pick-and-place operations. The system incorporates machine learning techniques for object detection, introducing the possibility of smart collection. Microcontrollers and motor drivers are implemented to allow the robot to move and collect objects. Additionally, with infrared sensors and ultrasonic technology, the robot can navigate and avoid obstacles autonomously along its path. The robot design is compact yet customizable to enhance adaptability in different environments. A camera module and a machine learning-based recognition algorithm are implemented for smart object recognition and collection. The sensors enable the system to operate autonomously, while the camera plays a crucial role in object detection, creating essential operations for the collection mechanism. With these attributes, the system can proficiently execute tasks, even in obstructed pathways, displaying its proficiency in recognizing objects and navigating around obstacles.

1 INTRODUCTION

In industrial automation, the development of Automated Guided Vehicles (AGV) has been an essential factor in safety, improved efficiency, and versatility. This research presents the design and implementation of an innovative AGV [1] that includes a unique Collision Avoidance System and Smart Object Detection for dynamic collection tasks in complex environments. Nowadays, AGVs are gaining prominence in industries like manufacturing, warehousing, and logistics due to their ability to make operations smoother. However, while traditional AGVs are effective in structured settings, these often struggle with unexpected obstacles and changing conditions. To solve this problem, this paper presents a solution with infrared [2,3] and ultrasonic sensors [4,5] to detect and avoid obstacles in real time [6], minimizing the risk of collisions and keeping the operations running continuously.

The vehicle also features an object detection system for tasks requiring the autonomous recognition and handling of various objects. This system consists of a camera for image collection and software for image labeling [7]. Afterward, the model is trained to achieve sufficient accuracy; once achieved, the model is incorporated into an ESP32-CAM microcontroller [8], enabling effective object detection and interaction through a mechanical robot arm.

This paper details the technical specifications and operational capabilities of the developed AGV, demonstrating how the integrated Collision Avoidance System and Smart Object Detection System improve efficiency, versatility, safety, and autonomy together.

2 METHODOLOGY

2.1 *Navigation system*

The navigation system is designed to operate through DC Gearbox motors, all of which are paired with an L293D motor driver shield, which allows the control of both the speed and direction of rotation. These DC motors are connected to the wheels, making various movements possible. For optimal performance, the L293D drivers are programmed to operate at approximately 30% of the DC motors' maximum speed. This intentional speed control is necessary for maintaining stability and control. The navigation system offers a great degree of motion, allowing for up to 12 distinct movements, as shown in Figure 1. However,

 DOI: 10.1201/9781003514831-9

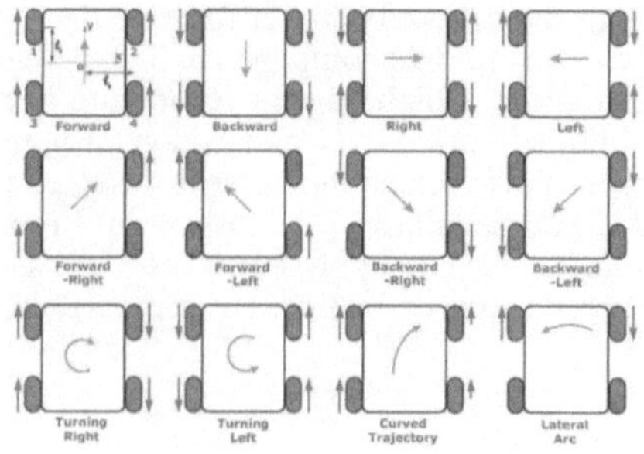

Figure 1. Overall possible movements with mecanum wheels.

practical applications often require fewer movements for optimized functionality. These movements are made possible by integrating mecanum wheels, which combined with the coordinated operation of the DC motors and L293D drivers, provide a system with a high degree of agility and responsiveness.

The navigation system is restricted to six essential movements: Forward, Backward, Right Side Way, Left Side Way, Clockwise Rotation, and Counterclockwise Rotation. This selection aligns with the robot's operational requirements. The synchronization of these components assures efficient and reliable navigation, making the system adaptable to various operational scenarios.

Two HW201 infrared sensor modules are integrated into the front of the vehicle. The sensor detects the intensity of surfaces' reflections in an environment. Upon detection, the robot starts making adjustments to avoid derailing. Complementing the infrared sensors, the robot is also integrated with HC-SR04 ultrasonic sensors [9]. It operates by emitting ultrasound at 40000 Hz that bounces back to the module if there is an obstacle; the sensor ranges from 2 cm to 400 cm, allowing the vehicle to perform more precise distance calculations and collision avoidance. Together, the HW201 and the HC-SR04 sensors ensure a collision avoidance and path-following system, making the robot adaptable to various changing environments and real-world applications. Integrating these sensors makes the project goal of being an autonomous and adaptable robot possible.

2.2 *Mechanical arm*

The 3D-printed mechanical arm employed PETG as the chosen material for its robust construction and great durability. The arm is designed with three main joints and five servo-motors: the shoulder, allowing lateral adjustments; elbow articulation, allowing the arm to bend or flex; wrist pitch, facilitating vertical adjustments of the wrist; wrist roll, enabling rotational movement of the wrist; and gripper control, essential for securely picking up and dropping objects. This mechanism allows the mechanical arm to manipulate its surroundings with high mobility [10], as shown in Figure 2.

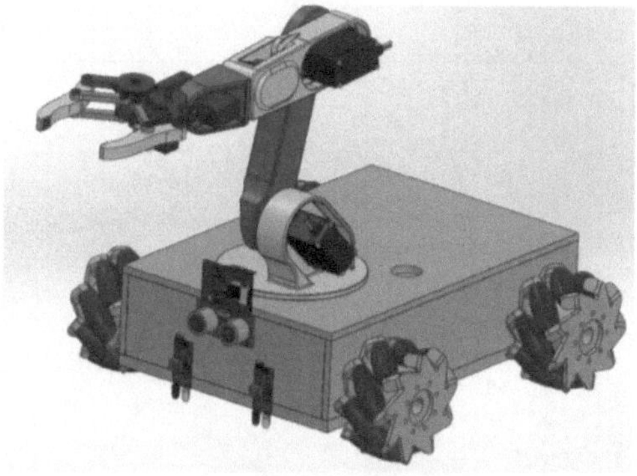

Figure 2. Robot 3D model.

To guarantee precision in these movements, the programming of the servomotors is carefully personalized; each servo motor is coded with specific delay times. In this way, distinct rotation speed configurations align with the requirements of the associated articulation; this calibration guarantees that the mechanical arm will operate with accuracy, as shown in Figure 3.

$$H_1^0 = \begin{bmatrix} \cos\theta_1 & 0 & \sin\theta_1 & 0 \\ \sin\theta_1 & 0 & -\cos\theta_1 & 0 \\ 0 & 1 & 0 & l_1 \\ 0 & 0 & 0 & 1 \end{bmatrix}$$

$$H_2^1 = \begin{bmatrix} \cos\theta_2 & -\sin\theta_2 & 0 & l_2\cos\theta_2 \\ \sin\theta_2 & \cos\theta_2 & 0 & l_2\sin\theta_2 \\ 0 & 0 & 1 & 0 \\ 0 & 0 & 0 & 1 \end{bmatrix}$$

$$H_3^2 = \begin{bmatrix} \cos\theta_3 & -\sin\theta_3 & 0 & l_3\cos\theta_3 \\ \sin\theta_3 & \cos\theta_3 & 0 & l_3\sin\theta_3 \\ 0 & 0 & 1 & 0 \\ 0 & 0 & 0 & 1 \end{bmatrix}$$

$$H_4^3 = \begin{bmatrix} \cos\theta_4 & 0 & \sin\theta_4 & 0 \\ \sin\theta_4 & 0 & -\cos\theta_4 & 0 \\ 0 & 1 & 0 & 0 \\ 0 & 0 & 0 & 1 \end{bmatrix}$$

$$H_5^4 = \begin{bmatrix} \cos\theta_5 & -\sin\theta_5 & 0 & 0 \\ \sin\theta_5 & \cos\theta_5 & 0 & 0 \\ 0 & 0 & 1 & l_4 \\ 0 & 0 & 0 & 1 \end{bmatrix}$$

Link	θ	α	r	d
1	θ_1	90°	0	l_1
2	θ_2	0	l_2	0
3	θ_3	0	l_3	0
4	θ_4	-90°	0	0
5	θ_5	0	0	l_4

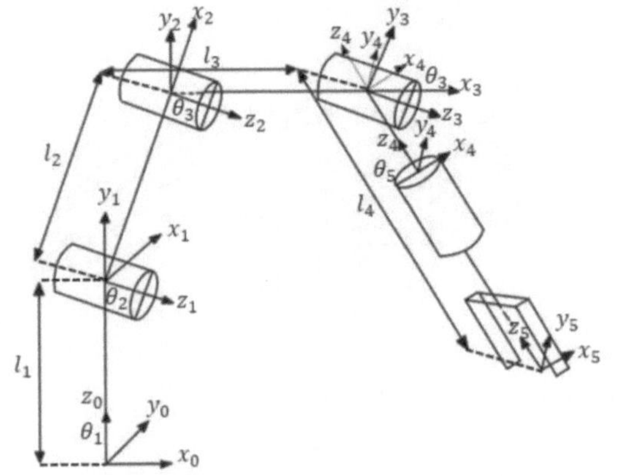

Figure 3. Mechanical arm coordinate frame.

2.3 *Curcit configuration*

The electronic circuitry lies on the Arduino Mega, the central processing unit. Widely known for its diverse capabilities, the Arduino Mega board is an important component due to its vast capacity to connect many peripherals through the 54 digital inputs/outputs, of which 15 provide PWM output and 16 analog inputs. The large number of pins on the Arduino Mega make this board the robot's perfect central control unit, responsible for the navigation system, movements of the mechanical arm, and processing of the sensors and camera module information.

The PWM pins of the Arduino allow the speed and intensity of the DC motors to be controlled. This adaptive control is of particular significance when achieving several combinations of movements in both the navigation system and the mechanical arm. In the same way, the wide variety of digital and analog pins and serial pins make possible the integration of the HW201 infrared sensor, the HC-SR04 ultrasonic sensor, and the communication with the external board ESP32-CAM, all connected to a single processing unit, as shown in Figure 4, providing a convenient and efficient autonomous system. This autonomous feature makes the robot's operation without a human operator possible. Additionally, this improves the system's adaptability.

The applied power source for the Arduino board is a 3S-Pack 18650 battery, chosen for its reliability and energy efficiency. This high-capacity battery serves as the energy reservoir, providing the necessary power to drive the motors and the other module components. Since the servo motors require a high current to operate in more challenging situations, a voltage regulator unit is necessary. This voltage regulator is designed using an LM350T regulator [11] shown in Figure 4, which guarantees a supply current of 3A, satisfying the servo's needs in more strict movements.

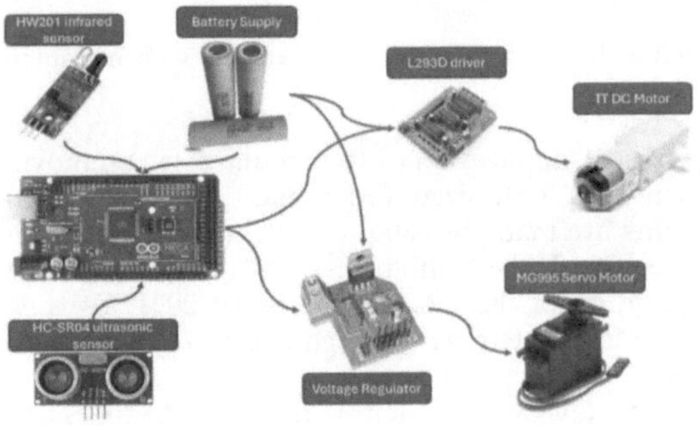

Figure 4. Hardware configuration.

3 DATA ACQUISITION AND PREPARATION

3.1 *Obstacle detection*

The car should be able to detect obstacles to avoid collisions. It should also be able to calculate the distance of the obstacle from the car; in this way, the system knows when it is proper to start an avoidance procedure. Ultrasonic sensors are used for this purpose; this sensor measures the distances of objects from 2 to 400 cm in real-time with an accuracy of 3mm. Ultrasonic sensors use the mechanism of the ultrasonic wave to check the distance between the car and an obstacle that might be close to it. The sensor consists of two ultrasonic transducers; one acts as a transmitter that converts the electrical signal into 40 KHz ultrasonic sound pulses, while the other acts as a receiver and listens for the transmitted pulses. When the receiver receives these pulses, it produces an output pulse whose width is proportional to the object's distance in front.

The width of the received pulse is used to calculate the distance from the reflected object. This can be worked out in the programming software using a simple distance-speed-time equation in which the time information is obtained from the width of the pulse, and the velocity is the sound velocity. Once the obstacle is detected to be a specific distance from the car, the processing unit starts an avoidance mechanism, and then the car returns to its path, which is detected with the help of infrared sensors.

3.2 *Track detection*

The HW201 infrared sensor consists of an infrared emitter and an infrared receiver. When the emitter emits infrared light, it hits the surface underneath the robot car. If the surface is reflective, some of this light is reflected back to the receiver. The intensity of the reflected light varies depending on the surface's reflectivity.

The infrared receiver detects this reflected light and converts it into an electrical signal. Typically, this signal's voltage level changes based on the intensity of the received infrared light. When the sensor is directly over the line, more light is reflected, resulting in a high-voltage signal. Conversely, when the sensor is off the line, less light is reflected, leading to a low-voltage signal.

As the IR sensor detects the reflected infrared light, it produces an analog voltage signal. This signal is then fed into the op-amp comparator circuit. The op-amp comparator compares this analog voltage signal with a reference voltage level, which is adjusted through a built-in potentiometer. Based on this comparison, the op-amp comparator produces a digital output signal. This digital output is then transmitted to the Arduino board. The Arduino's code then interprets this digital value to make decisions about the robot car's movement. If the value indicates that the car is off the line, the code triggers the automatic path correction mechanism to steer the car back onto the line. If the value suggests the car is on the line, the code can continue with normal operation. The flow chart of track detection is illustrated in Figure 5.

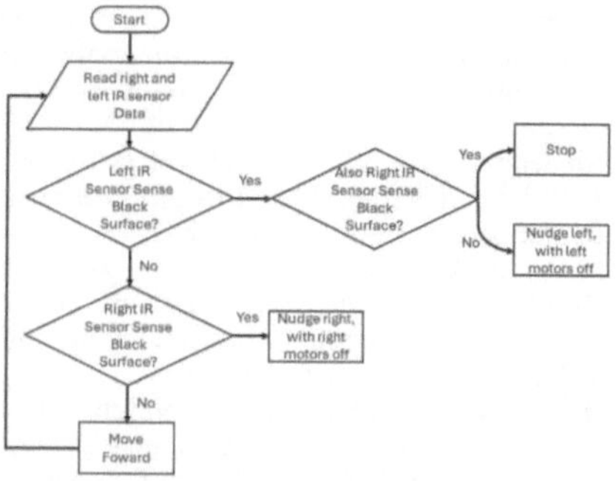

Figure 5. Flow chart of track detection.

3.3 *Object detection*

The ESP32-Cam is a camera microcontroller based on a 32-bit CPU & has a combined Wi-Fi + Bluetooth/BLE Chip. The module combines with the OV2640 Camera Module, which has the highest Camera Resolution up to 1600 × 1200. These characteristics make the module a reliable and crucial piece for the object detection feature [12].

The object detection procedure consists of several steps, as shown in Figure 6.

First, images of the desired object with an OV2640 camera are collected to improve the model's accuracy, and then the images are labeled. For this purpose, Edge Impulse is used to draw bounding boxes around the images and give them a label; after that, the model is trained using a dataset of images, as shown in Figure 7. In this step, it is important to test the model to determine how accurate it is; finally when the model is accurate enough, the model is exported into an Arduino library, which provides sample codes for the object detection, serving as a starting point of the programming and testing [7].

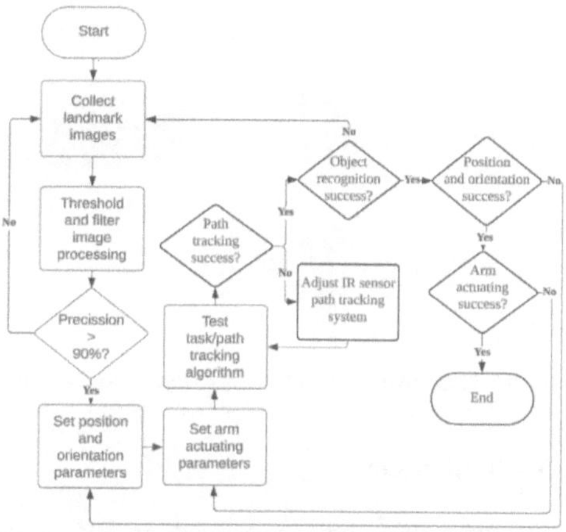

Figure 6. Flow chart of object detection.

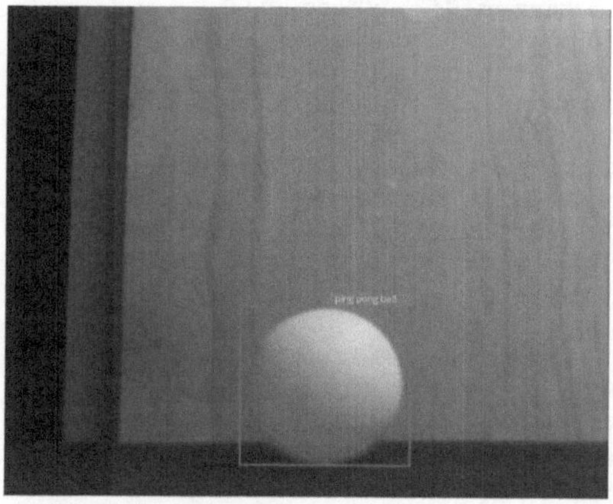

Figure 7. Object recognition.

3.4 *Object collection*

The mechanism for collecting a ping pong ball uses the ESP32-CAM module and the Arduino MEGA board as the main controllers, a mechanical arm with a gripper, involving specific steps to ensure precise coordination. Initially, the ESP32-CAM captures an image of the target area. This image is then processed locally on the camera module controller to extract position information about the object. Subsequently, this processed image data is transmitted to the Arduino MEGA board, which utilizes this information to determine the precise location of the object within its field of view and, based on this, generates specific values for the coordinate parameters corresponding to the object's position. The Arduino board sends these signals to the servo motors on the mechanical arm, moving it with precision to align the gripper accurately with the object's location to activate the gripper. The whole task of object retrieval is simplified and shown in Figure 8.

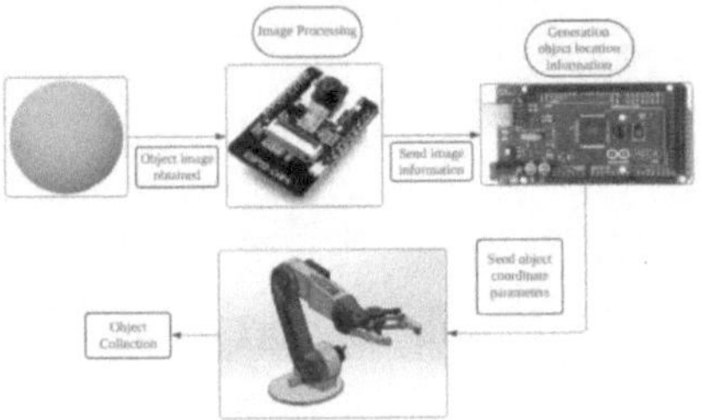

Figure 8. Hardware actuation for object collection.

4 CONCLUSION

In summary, this study proposed an innovative development of an Autonomous Guided Vehicle with a Collision Avoidance System and Smart Object Detection with dynamic collection mechanisms. The collision avoidance system, utilizing infrared and ultrasonic sensors, has proven effective in real-time obstacle detection, making possible a risk reduction for accidents. On the other hand, Smart Object Detection has demonstrated great accuracy in object recognition, making precision handling an easy task. Integrating these features resulted in a highly autonomous and efficient vehicle that fulfilled expectations. This research emphasized the importance of advanced sensor technologies, algorithms, and mechanical principles in industrial automation. The developed AGV represents a step forward in the quest for more intelligent and adaptable autonomous vehicles that can be employed in many industries, such as manufacturing, logistics, and warehousing.

ACKNOWLEDGEMENT

This work was financially supported by the Ministry of Foreign Affairs, Republic of China (Taiwan), under the Universidad Politécnica Taiwan-Paraguay (UPTP) project.

REFERENCES

[1] Kuo S.-P., Tsai K.-Y. and Leu Y.-G., (2020). Object moving system of four-wheel mobile vehicle with mechanical arm, *2020 International Automatic Control Conference (CACS)*, Hsinchu, Taiwan, pp. 1–4.
[2] ElectronicsForu. Understanding IR sensors and IR LEDs: Functions, differences, and applications. electronicsforu.com, https://www.electronicsforu.com/technology-trends/learn-electronics/ ir-led-infrared-sensor-basics.
[3] Das D. *Interfacing IR Sensor Module with Arduino*. CircuitDigest.com, https://circuitdigest.com/microcontroller-projects/interfacing-ir-sensor-module-with-arduino.
[4] Last Minute Engineers. *How HC-SR04 Ultrasonic Sensor Works & Interface It With Arduino*. lastminuteengineers.com, https://lastminuteengineers.com/arduino-sr04-ultrasonic-sensor-tutorial.
[5] Iqbal A., Ahmed S. S., Tauqeer M. D., Sultan A. and Abbas S. Y., (2017). Design of multifunctional autonomous car using ultrasonic and infrared sensors, *2017 International Symposium on Wireless Systems and Networks (ISWSN)*, Lahore, Pakistan, pp. 1–5.
[6] Liu T. and Liang Z., Design of multi-AGV intelligent collision avoidance system based on dynamic priority strategy, *2022 Asia Conference on Advanced Robotics, Automation, and Control Engineering (ARACE)*, Qingdao, China, 2022, pp. 47–51.
[7] DroneBot Workshop. *ESP32-CAM Object Detection with Edge Impulse*. dronebotworkshop.com, https://dronebotworkshop.com/esp32-object-detect.
[8] Mehendale N., (Jul. 2022). *Object Detection using ESP 32 CAM*, pp. 1–9, https://papers.ssrn.com/sol3/papers.cfm?abstract_id=4152378
[9] Nedelkovski D. *Ultrasonic Sensor HC-SR04 and Arduino – Complete Guide*. howtomechatronics.com, https://howtomechatronics.com/tutorials/arduino/ultrasonic-sensor-hc-sr04/
[10] Hanan A. R. Akkar, Ahlam Najim *Kinematics Analysis and Modeling of 6 Degree of Freedom Robotic Arm from DFROBOT on Labview*, https://www.researchgate.net/publication/310389891_Kinemics_Analysis_and_Modeling_of_6_Degree_of_Freedom_Robotic_Arm_from_DFROBOT_on_Labview
[11] Texas Instruments. LM150/LM350A/LM350 3-Amp Adjustable Regulators. https://www.ti.com/lit/ds/symlink/lm350-n.pdf
[12] Santos R. *ESP32-CAM Video Streaming and Face Recognition with Arduino IDE*. randomnerdtutorial.com https://randomnerdtutorials.com/esp32-cam-video-streaming-face-recognition-arduino-ide/

System Innovation for an Artificial Intelligence Era – Lam et al. (Eds)

Design of an automated guided vehicle with obstacle avoidance and ball collection mechanism

Ana Cristina Ojeda Caceres, Rodrigo Ivan Arambulo Gonzalez, Gianluca Distefano Martinez & Kevin David Jaime Insfran
Department of Electromechanical Engineering, Universidad Politécnica Taiwan-Paraguay, Asunción, Paraguay

Meng-Kun Liu
Department of Mechanical Engineering, National Taiwan University of Science and Technology, Taipei, Taiwan

ABSTRACT: Automated Guided Vehicles (AGVs) are robotic systems designed for efficient material handling and transportation in industries. These autonomous robots follow predetermined paths, guided by sensors, and play a crucial role in process optimization and inventory management in a dynamic environment. This project aims to design, assemble, and program a robot capable of moving, gripping, and depositing objects into a container. The vehicle's Mecanum wheels enable a wide range of movements, offering versatility in the robot's mobility. The ESP32 controller module with integrated Wi-Fi and Bluetooth is paired to an Xbox series X controller through Arduino's integrated development environment (IDE). The gripper efficiently collects ping-pong balls, places them in a temporary onboard ball container, and discharges them via the ramp-and-gate system into the final container. An ultrasonic sensor for obstacle detection and collision avoidance and an infrared line tracking sensor designed to detect a black line marked on the floor are also incorporated. This sensor integration expands the robot's adaptability and responsiveness, enhancing safety and performance across tasks and scenarios.

1 INTRODUCTION

Automated Guided Vehicles (AGVs) have revolutionized the landscape of modern industrial automation by providing efficient and flexible material handling solutions. These self-operating vehicles are designed to transport materials within manufacturing facilities, warehouses, and distribution centers without human intervention [1]. This paper focuses on developing an AGV capable of executing pick-and-place operations. The vehicle's ultrasonic sensor uses high-frequency sound waves to detect the presence, distance, and position of objects in its vicinity, and an infrared line stacking sensor is designed to detect and track lines or paths marked with an IR light source. In addition to obstacle avoidance, a mechanism gaining attention is the ball collection system. This feature allows the AGVs to collect, transport, and deposit spherical objects precisely, making them ideal for applications in sports equipment manufacturing, toy production, and waste management.

The robotic platform is characterized by its agility and robustness. The base, with a thickness of 8 mm, was meticulously fabricated using a laser engraving machine from a high-quality plastic substrate. A systematic design approach was undertaken to meticulously engineer the robot's structural components using SolidWorks, a sophisticated Computer Aided Design (CAD) software. Mecanum Omnidirectional wheels, renowned for their superior handling characteristics and multi-directional mobility capabilities, were integrated to enhance the robot's maneuverability and facilitate versatile movements. The unique design of these wheels necessitates individualized control for executing precise maneuvers [2]. This precise control is achieved by implementing L298N motor drivers, which regulate the speed and direction of the robot's movement via dual H-bridges, ensuring seamless and accurate navigation in various operational environments.

The robot's arm, comprising a base, a shoulder, an elbow, and two grippers, is a critical component. Movement in this section was orchestrated by three servo motors: SG90, MG995, and MG996R. SG90, a low-power positional motor with 180 degrees of angular motion, was complemented by MG995, a high-torque servo motor with metal gears, and MG996R, a rotational servo motor. All arm components were intricately 3D printed after being designed in the CAD program. Integration was achieved through the ESP32, serving as the main control module. The connectivity was enhanced by Bluetooth technology, facilitating seamless communication between the module and an Xbox controller for intuitive operation and control. This

DOI: 10.1201/9781003514831-10

comprehensive approach ensures a robust and efficient AGV robot capable of navigating and performing intricate pick-and-place tasks. Certain important and remarkable processes, such as setting a home position, helped greatly to facilitate and speed up the pick-and-place process. In the commitment to responsible engineering, we carefully considered the environmental impact of the design choices. Using sustainable materials and the potential for adaptability in eco-friendly practices underline our dedication to creating robotics solutions that align with global sustainability goals.

2 DESIGN AND DISTRIBUTION OF THE HARDWARE

Figures 1 and 2 show a detailed view of the robot's structural composition and operational functionality. These visual representations help better understand the robot's configuration and capabilities. A notable role in this design is the plate in which the robotic arm and the basket are integrated to optimize the space above the car. Another critical design detail is the base of the vehicle, which is configured with the shape of a "t," strategically engineered to avoid any contact of the wheels with the hardware above the vehicle, underscoring the safety and precision in the development of this robot to enhance its reliability and functionality.

Figure 1. Isometric view of the "t" base for the robot.

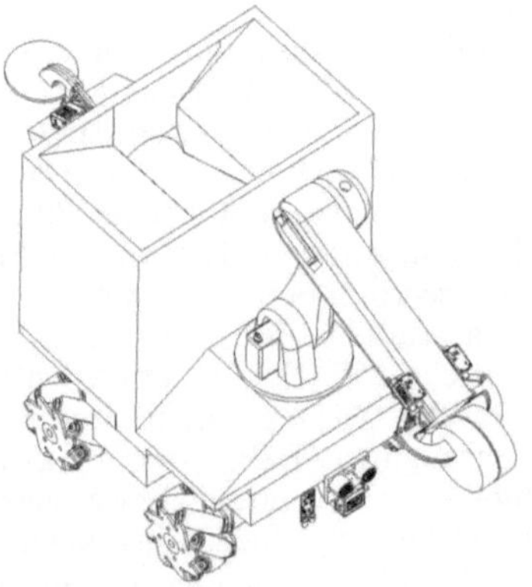

Figure 2. Isometric view of the robot.

The hardware block diagram presented in Figure 3 showcases a sophisticated robotic arm system controlled by an ESP32 microcontroller. The system includes multiple servo drivers, such as MG996 and SG90 servos, enabling precise and coordinated movements. With a rotary base, two arm links, and a circular gripper, this robotic arm demonstrates versatility in performing various tasks. Integrating different components highlights the complexity and potential applications of modern robotics. Additionally, Figure 4 provides a detailed exposition of the hardware configuration tailored specifically for the robot's arm, offering deeper insights into its structural layout and operational dynamics.

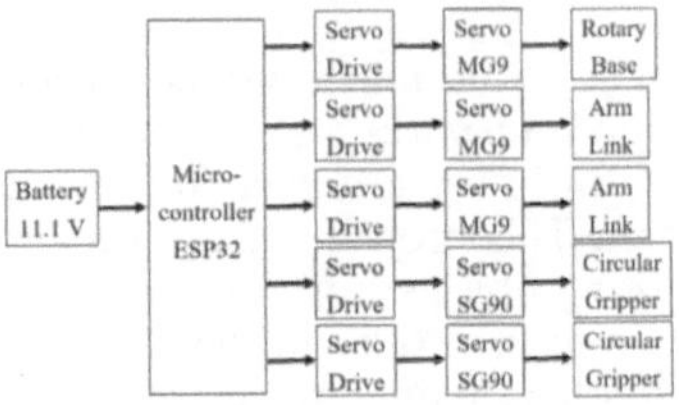

Figure 3. Block diagram of the robot's arm.

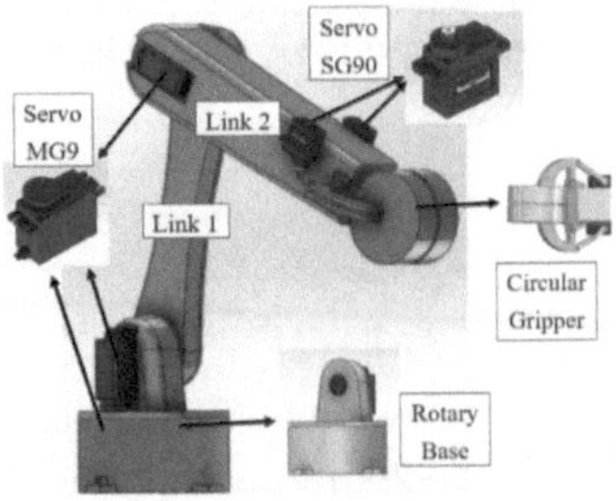

Figure 4. Hardware diagram representation of the robot's arm.

3 ANALYSIS OF THE SYSTEM

The robotic car comprises four Mecanum wheels, each connected to a motor and a motor driver. Its configuration is shown in Figure 5.

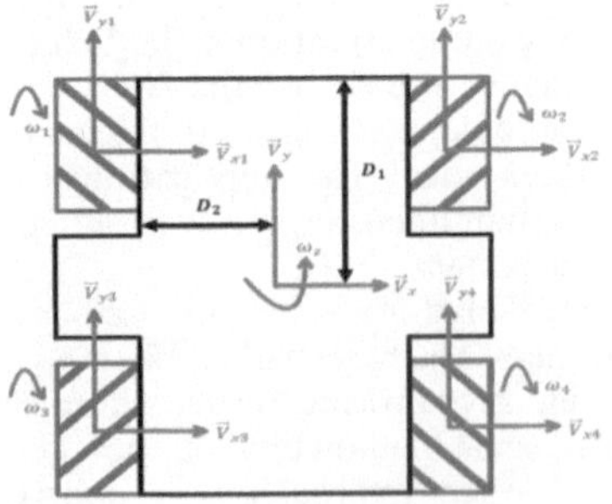

Figure 5. Mecanum wheel robotic car configuration.

Relationship of vehicle velocity and wheel velocity is given by:

$$V_{xi} = V_x + D_1\omega_z \tag{1}$$

$$V_{yi} = V_y + D_2\omega_z \tag{2}$$

Where V_x and V_y are vehicle velocity in the X and Y direction, ω_z is vehicle angular velocity around the Z axis.

This robotic car uses Mecanum wheels for movement purposes. This movement depends on the combination or coordination of movement of each mecanum wheel at that time. Table 1 shows all the combinations of these functions, with the respective linear movement as the car's output.

Table 1. Combinations for linear movements of the car.

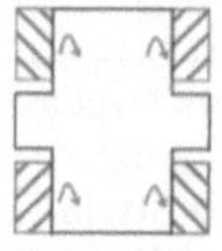	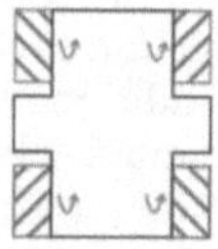	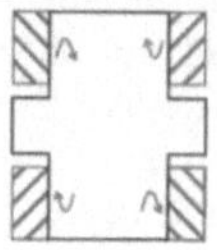
Forwards	Backwards	Left
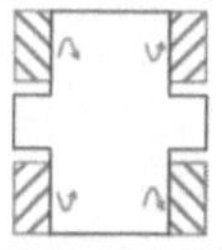		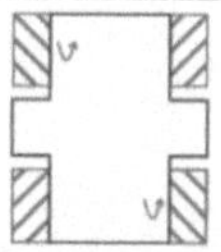
Right	Left Diagonal Forward	Left Diagonal Backwards
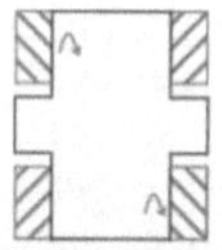	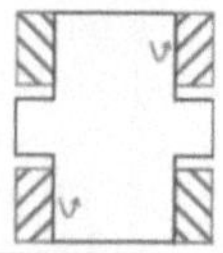	
Right Diagonal Forward	Right Diagonal Backwards	

The combinations with the respective rotational movement as an output for the vehicle are shown in Table 2.

Table 2. Combinations for rotational movements of the car.

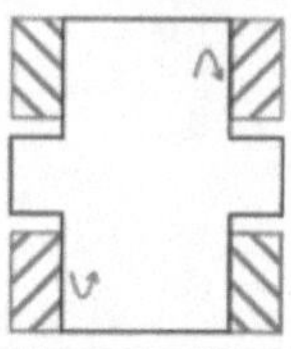	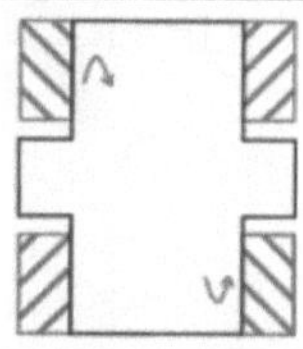
CW rotation about the mass center	CCW rotation about the mass center

The study of robotic arms involves nonlinear analysis and includes challenges related to multiple solutions, singularity points, and possible redundancies. Several studies in the previous literature consider the kinematic analysis of robotic arms [3]. For a 3-DOF manipulator with three rotating joints, the sufficient condition for a closed solution is that two adjacent joint axes intersect at one point. In the existing 3-DOF manipulator design, Axis 1 and Axis 2 are designed to intersect at one point. The coordinate systems of the manipulator are depicted in Figure 6. The DH parameters of the manipulator based on the provided coordinate systems are listed in Table 3.

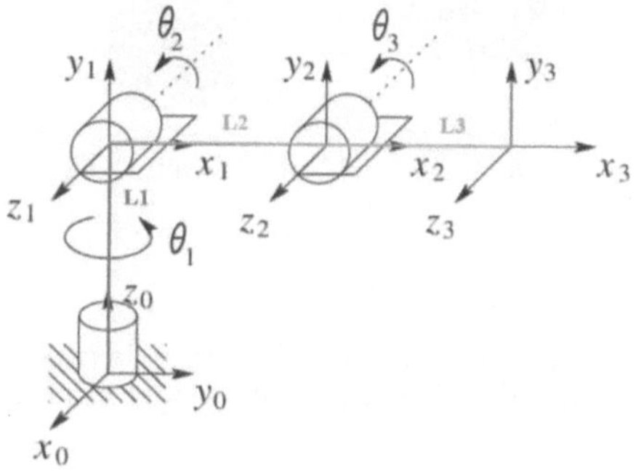

Figure 6. The coordinate system of the manipulator [3].

Table 3. Denavit-Hartenberg table for this 3-DOF manipulator.

Link	a_i	α_i	d_i	θ_i
1	0	90°	$\boldsymbol{L_1}$	θ_1
2	$\boldsymbol{L_2}$	0	0	θ_2
3	$\boldsymbol{L_3}$	0	0	θ_3

A systematic way to describe the transformation matrix for the manipulator, as a joint varies with the angle, is given.

$$T_n^0 = A_1 \ldots A_n \tag{3}$$

$$A_i = A_i q_i \tag{4}$$

$$T_3^1 = \begin{bmatrix} c\theta_3 & 0 & -s\theta_3 & L_3 c\theta_3 \\ s\theta_3 & 0 & c\theta_3 & L_3 s\theta_3 \\ 0 & 1 & 0 & 0 \\ 0 & 0 & 0 & 1 \end{bmatrix}$$

4 IMPLEMENTATION OF SENSORS

An ultrasonic sensor positioned at the front of the vehicle was utilized for object avoidance. Its function is to emit ultrasonic waves through one of its terminals called trig. These waves bounce off nearby objects and return to the module at the other terminal called echo, as depicted in Figure 7. The data outputted by the sensor is the time in microseconds it takes for the wave to return to the sensor [4].

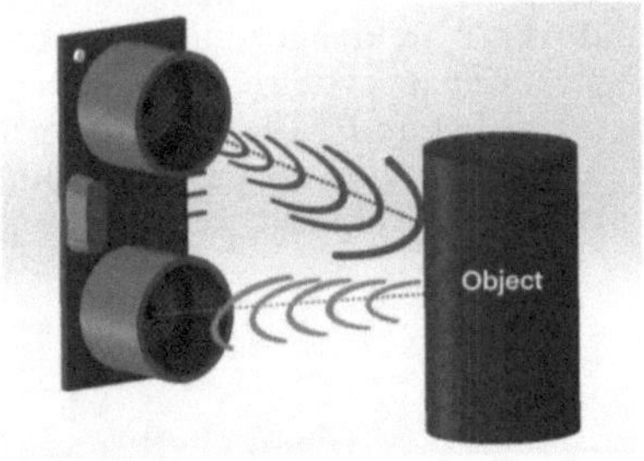

Figure 7. Representation of the ultrasonic waves bouncing.

To calculate the linear distance between the sensor and the object, the following formulas are used:

$$Speedofsound = 0.0343 cm/\mu s \quad (5)$$

$$Time = Sensortime/2 \quad (6)$$

$$Distance(D) = speed \times time \quad (7)$$

$$Distance(D) = 0.0343 \times time \quad (8)$$

Since a complete scan of what lies ahead of the vehicle is required, a servo motor is utilized to enable the ultrasonic sensor to perform a 180-degree scan. There are three important parameters: the direction in which the servo motor was turning at that moment, the angle of rotation of the servo motor, and the time recorded by the sensor. This can be observed in the following figure:

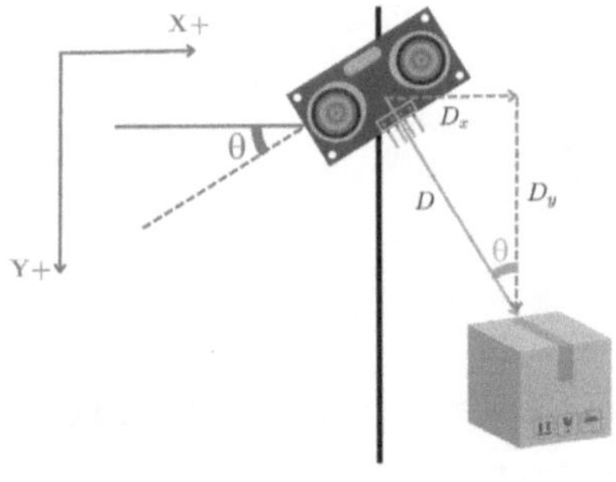

Figure 8. Diagram of distances between the sensor and the object.

The aim is to avoid collisions. Thus, the vertical and horizontal distances between the object and the sensor are calculated using the following formulas [5]:

$$D_x = D \times \sin(\theta) \quad (9)$$

$$D_y = D \times \cos(\theta) \quad (10)$$

Now, the process proceeds to maneuver the vehicle to avoid the object and return to the lane to continue its path.

$$M_x = (-1)^n \times (Width_{vehicle}/2 - D_x + d) \quad (11)$$

$$M_y = Length_{vehicle} + D_y + d \quad (12)$$

$$M_x = -(-1)^n \times (Width_{vehicle}/2 - D_x + d) \quad (13)$$

The variable d is a safety gap, with a default value of 5cm. Initially, the code is written to have the sensor change direction each time it reaches the 180-degree range. Additionally, it increments a counter variable, denoted as "n," by one unit, indicating the direction of sensor rotation. The counter's initial value is set to 0, and the sensor begins rotating clockwise; then we have:

$$\theta(ccw) \rightarrow n = odd \quad (14)$$

$$\theta(cw) \rightarrow n = even \quad (15)$$

This approach allows for efficient and safe avoidance of obstacles along the path.

The line tracking operates using infrared sensors positioned at the robot's base. These sensors emit infrared waves from a terminal and detect them using a photodiode. The infrared waves are absorbed when a black-colored surface is beneath them, causing the photodiode to receive no signals, as depicted in Figure 9. This allows the system to recognize the black line beneath the sensor [6]. Consequently, we can instruct the robot to move toward the sensor that detected the black line. This relocation ensures that the black line is positioned between the sensors.

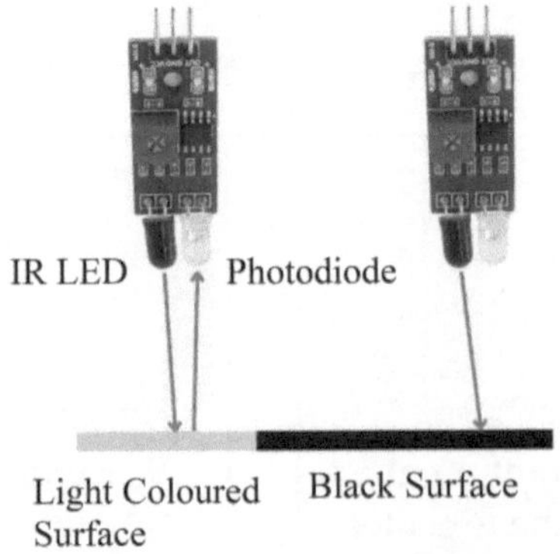

Figure 9. The representation of the wave reflection depends on the surface.

5 PROGRAMMING FLOWCHART AND CIRCUIT CONFIGURATION

The flowchart of the program procedure is illustrated in Figure 10.

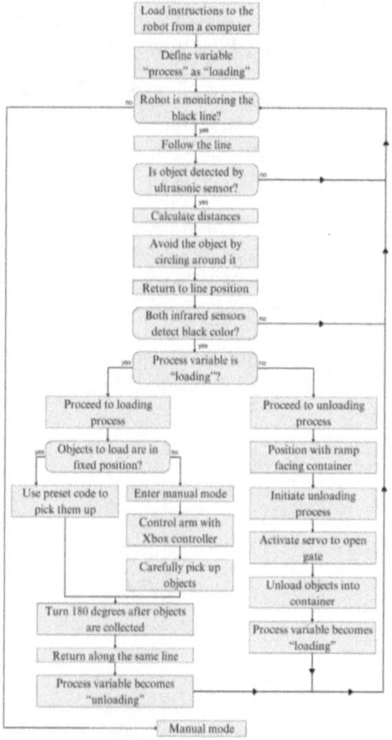

Figure 10. Programming flowchart.

Figure 11 illustrates a schematic diagram detailing the intricate connections within the electrical circuitry of the robot car, elucidating the mechanism for locomotion and arm manipulation. At the heart of this circuit lies the ESP 32, the primary controller responsible for orchestrating the vehicle's functionalities. Through pre-programmed Python scripts, the ESP 32 coordinates the motions of the motors and servo motors, enabling precise control over the car's movement and the arm's articulation. The integration of Bluetooth technology facilitates seamless communication between the controller and the ESP 32.

Subsequently, the diagram delineates the distinct components of the car, notably segregating the wheel assembly from the arm and ball-discharging mechanism. Instructions from the controller to the ESP 32 are then transmitted to the motor drivers governing the wheels. The motor control subsystem is powered by two L298N drivers strategically chosen for their compatibility with the ESP 32 microcontroller, thereby enabling nuanced control over the direction of movement.

Within the arm section of the diagram, the ESP 32 serves as the central hub, facilitating the connection of three SG90 servo motors. These servo motors are allocated specific roles, with two designated for gripper functionality and the third dedicated to manipulating the door and regulating the release of balls into the basket. Furthermore, the arm assembly incorporates three MG servo motors, comprising two MG995 units for arm and shoulder movement and a rotational MG996R Servo Motor for the base rotation.

Noteworthy is the system's reliance on power supplied by a dedicated power bank, which ensures consistent and stable motion by delivering a reliable current and maintaining a constant voltage of 5V. Additionally, the drivers responsible for controlling the robot's motion are powered by a 12V battery source, ensuring adequate energy supply for seamless operation.

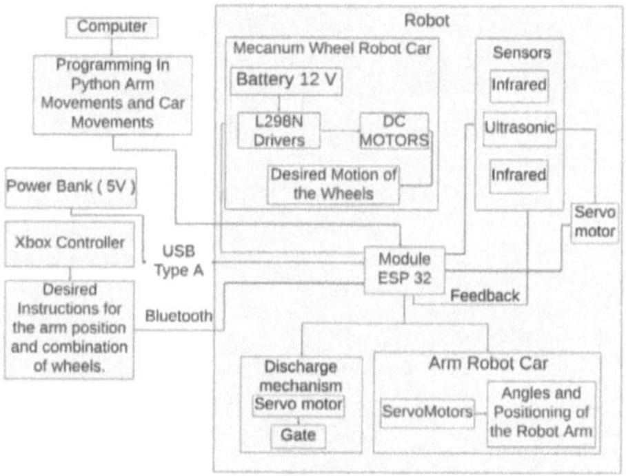

Figure 11. Block Diagram of the entire electrical circuit.

6 CONCLUSION

The project presents a comprehensive analysis and exploration of the domain of AGV robots. This paper provides a detailed explanation of the design and assembly of the vehicle robot arm. Through a detailed exploration of its mechanical design, control system, and incorporation with vehicle platforms, Developing the AGV target for executing precise and fast pick-and-place operations takes a significant step in integrating advanced robotics technologies. Across careful design and engineering, the vehicle robot arm performs a successful operation.

ACKNOWLEDGEMENT

This work was financially supported by the Ministry of Foreign Affairs, Republic of China (Taiwan), under the Universidad Politécnica Taiwan-Paraguay (UPTP) project.

REFERENCES

[1] Defensebridge. (2024, March 12). (September 2007). *A Comprehensive Guide to Autonomous Ground Vehicles (AGVs)*. https://defensebridge.com/article/a-comprehensive-guide-to-autonomous-ground-vehicles-agvs.html

[2] Doroftei I., Grosu V. and Spinu V., Omnidirectional Mobile Robot – Design and Implementation, Bioinspiration and Robotics: Walking and climbing Robots, Book edited by: Maki K. Habib, I-Tech, Vienna, Austria, EU,

[3] Shah J., Rattan S. S. and Nakra B.C. (Sept. 2013). End effector position analysis using forward kinematics for 5 DOF Pravak Robot Arm. *International Journal of Robotics and Automation* 2(3) (2013), 112–116.

[4] Kusumadjati A., Aminah N. S. and Djamal M. (April 2019). An ultrasonic sensor system for vehicle detection application. *Journal of Physics Conference Series*, 1204(1), 012017. DOI: 10.1088/1742-6596/1204/1/012017

[5] Sanmugam M. (January 19, 2023). *Distance measurement using Ultrasonic sensor and Arduino.* Retrieved from https://www.geeksforgeeks.org/distance-measurement-using-ultrasonic-sensor-and-arduino/.

[6] Pakdaman M. and Sanaatiyan M. M., (2009). Design and Implementation of Line Follower Robot, *2009 Second International Conference on Computer and Electrical Engineering*, Dubai, pp. 585–590.

[7] Mark W. Spong, Seth Hutchinson, M. (2020). *Vidyasagar - Robot Modeling and Control.* John Wiley & Sons Inc.

[8] Sure, Ramshetty K. and Savita Patil. (2014). Android Based Autonomous Coloured Line Follower Robot. *International Journel of Research and Technology,* 3.

[9] Patel G. K. S., Maitra R. K., Bhattacharya S., Moosa S., Pavan P. (2023). Robotic Car Using NodeMCU ESP8266 Wi-Fi Module. *In 2023 9th International Conference on Advanced Computing and Communication Systems (ICACCS).*

System Innovation for an Artificial Intelligence Era – Lam et al. (Eds)
© 2025 the Author(s), ISBN: 978-1-032-84452-7

Game-changer: AGV and robotic arm integration for automated ping pong ball collection

Kenzo Shirahama, Tamara von Glasenapp, Fabricio Alcaraz & Carlos Colmán
Department of Electromechanical Engineering, Universidad Politécnica Taiwan-Paraguay, Asunción, Paraguay

Meng-Kun Liu
Department of Mechanical Engineering, National Taiwan University of Science and Technology, Taipei, Taiwan

ABSTRACT: The remote-controlled wheeled vehicle is designed mainly for executing basic and repetitive operations. The purpose of this project is to create a robotic vehicle with a robotic arm and gripper able to pick up objects, collect them into an integrated storage box, and deliver them to another designated place. The design of this robotic vehicle consists of a Mecanum wheeled base that allows the vehicle to move freely in any direction. In addition, an arm and gripper are integrated, able to pick up the objects precisely, and a storage box where the collected objects will be temporarily placed. The robot uses an Arduino-based system controlled through a smartphone with an Android app connected via Bluetooth.

1 INTRODUCTION

The main purpose is to integrate the vehicle with a robotic arm to create the most efficient picking-up robot. First, it is necessary to investigate the best mobilization system and then identify an appropriate robotic arm that can be integrated with this vehicle. The problems to be solved are the variety of types of movement, the possibilities of grippers, the number of motors that can be chosen, and, lastly, the controlling method of the vehicle and robotic arm.

2 EXPERIMENT

2.1 *Design*

The Mecanum wheel is a wheel design that allows movement in any direction. As mentioned in [11], these rollers have an axis of rotation at 45° to the plane of the wheel in a plane parallel to the axis of rotation of the wheel. By alternating wheels with left and right-handed rollers so that each wheel applies force roughly at right angles to the diagonal the wheel is on, the vehicle is stable and can be made to move in any direction and turn by varying the speed and direction of rotation of each wheel. The Mecanum wheels, as opposed to the standard wheels, allow for a wider variety of movements, letting the vehicle move diagonally, to the sides, and rotate around its axis. The design for Mecanum wheels can be seen in Figure 1.

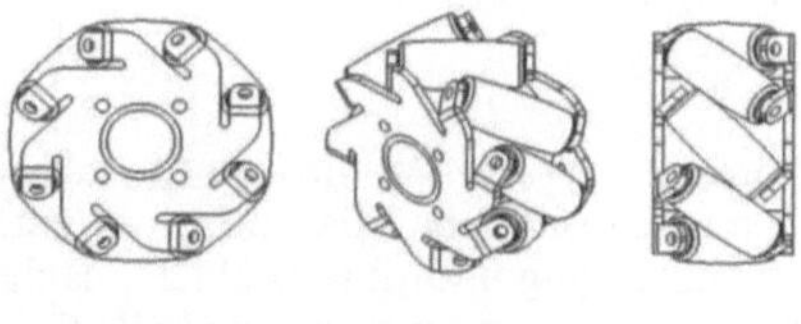

Figure 1. Mecanum wheels design.

Moving all four wheels in the same direction causes forward or backward movement, running the wheels on one side in the opposite direction to those on the other side causes rotation of the vehicle, and running the wheels on one diagonal in the opposite direction to those on the other diagonal causes sideways movement. Figure 2 shows how and where the vehicle's direction can move with the combination of rotating wheels.

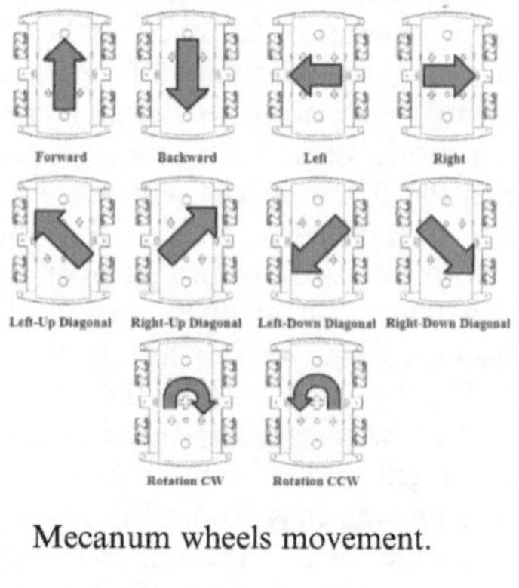

Figure 2. Mecanum wheels movement.

As stated by [2], when it comes to grippers, robot claws are one of the simplest types of them. Typically, robot claws consist of two claws that come together on opposite sides of an object to grasp it. The number of joints represents the number of Degrees of Freedom (DOF) of the robotic arm. The joints are usually actuated using servo motors, which provide the necessary torque to rotate the attached links. Microcontrollers are used to send the electrical signals required for controlling the angular motion of the servo motor shafts.

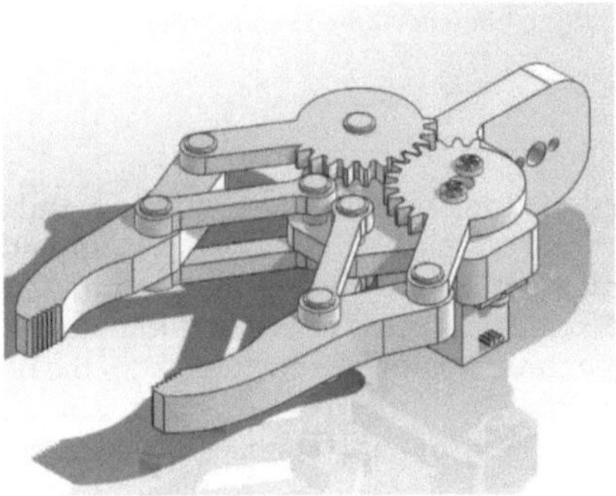

Figure 3. Gripper design.

The project requires that the robot be capable of moving to each ping-pong ball position, grabbing the ball, and storing it until the vehicle reaches the final delivery place and puts all the balls there. To select the vehicle base, first, it was necessary to design the base of the robot. A 5 mm thick acrylic base was tailored in SolidWorks and then cut with a laser machine for the final base. The material's transparency facilitates working with the wiring and the motors that will be used. Then, four walls were attached to that base, allowing the motors to be in place and allowing some space for all the wheel wiring, later to these walls; another acrylic base was cut, the same as the first one, and was put on top of the walls, as a ceiling. The robotic arm would be located in this second base. Lastly, the storage basket was also designed in SolidWorks, with an internal slope, allowing all the picked balls to go smoothly to the door for the delivery system.

For the robotic arm and gripper, a pre-made version from the Internet was selected, and then a SolidWorks design was made, as shown in Figure 3. Both parts (arm and gripper) were 3D printed, and later, the gripper was upgraded to improve friction issues and facilitate the picking-up system. During each step of the creation process, the robot was tested, first by testing the wheel's movement and then by testing the robotic arm. The design was easy to modify and optimize, so if a change was needed, making some improvements at the section of the wheel, for example, detaching the arm, could be easily done. This allowed the person optimizing the robot to get to that lower part and modify whatever part was needed. The final SolidWorks design of the robotic vehicle and arm with gripper integrated is shown in Figure 4 and Figure 5, where the main components are specified.

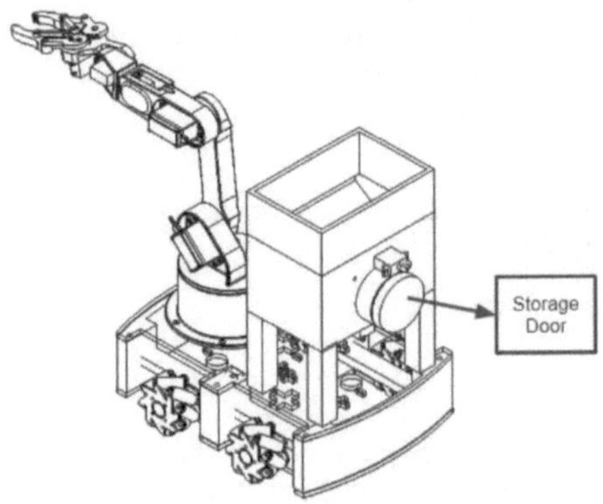

Figure 4. Final vehicle and arm assembled design.

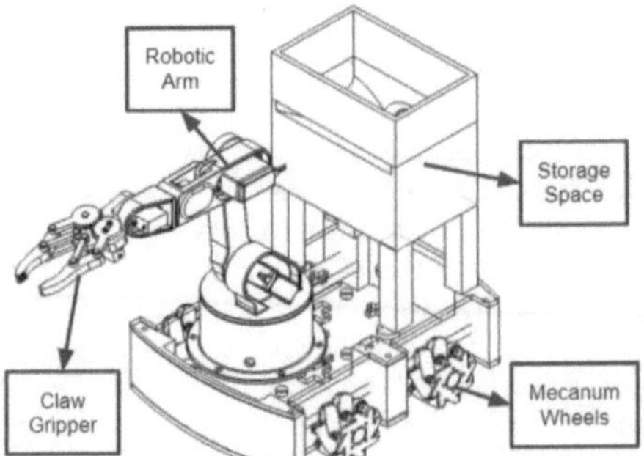

Figure 5. Final vehicle and robotic arm integrated design.

To integrate motors and Mecanum wheels, four small DC motors of 6V from 90 to 300 RPM were used to move the wheels.

2.2 *Software and control*

The robotic arm and the gripper will be controlled by a smartphone, which is different from the one controlling the wheels, meaning two pilots will be needed to control the whole robot. Two Arduino boards are used, one connected to the wheels and another to the arm's servo motors. The wheels are powered by a 12V battery of 3600 mAh. For the arm servo motors, a 20000 mAh power bank with an output of 5V was used.

Related to the software section, the vehicle and robotic arm were controlled using the Arduino IDE

(Integrated Development Environment) and the MIT App Inventor. The Arduino IDE is a software platform for programming and uploading code to Arduino boards. As stated by [10], Arduino is an open-source electronics platform that allows individuals to create interactive electronic projects. It is designed to be user-friendly, making it accessible to beginners while offering advanced features for experienced developers. It plays a crucial role in the Arduino ecosystem, enabling users to bring their electronic projects to life by writing and uploading code to Arduino boards.

As known from [5], MIT App Inventor is a visual programming environment that allows users, especially those without extensive coding experience, to create mobile applications for Android devices. It was developed by the Massachusetts Institute of Technology (MIT) and is designed to make app development accessible to many users, including educators, students, and hobbyists. In the following Figure 6, the interfaces for the applications controlling the gripper and vehicle, respectively, are shown.

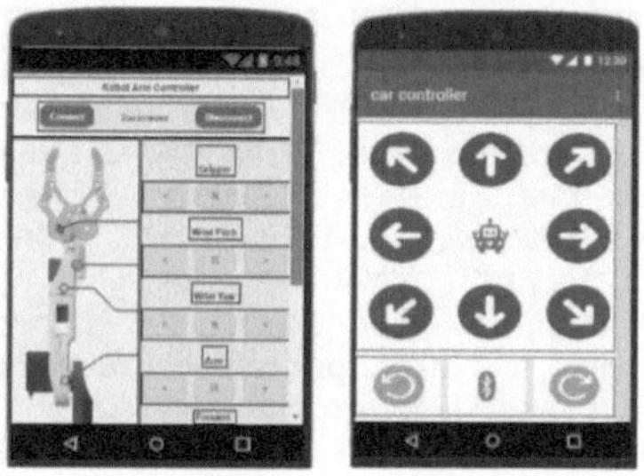

Figure 6. Robot arm app and car wheels app interface.

In addition, to supply power to the Arduino, a stable 5V power bank was used, and the 5V pin from the Arduino was also used to give power to the PCA board. This was necessary because using a power bank was more stable than using batteries. Seven servo motors were connected, five for the arm, one for the gripper, and one for the basket door to release the balls at the delivery point. Later, in Figure 7, a block diagram of the servos circuit of the arm is provided for better understanding.

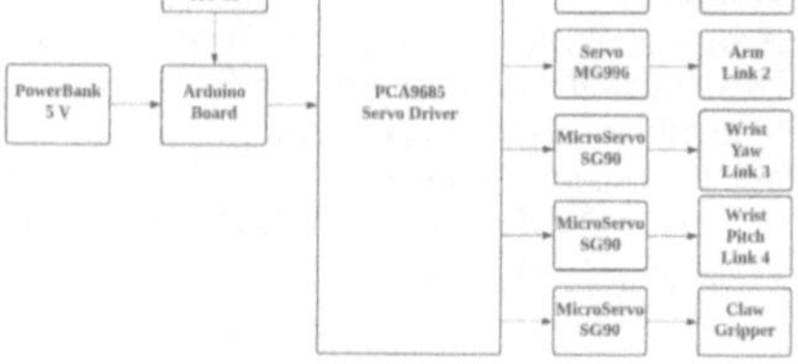

Figure 7. Block diagram of servos circuit of the arm.

3 RESULTS AND DISCUSSION

The project has been assembled, and the trials have been successful. The vehicle could successfully rotate 360 degrees in all directions without interference. For the wheels, a power-on switch was implemented, and another one was implemented for the arm, as well, so turning on and off the device easily without touching the wiring was possible. Then, a custom Android application was made through which the Mecanum wheels were controlled to move the robot in any direction. The following Figure 8 shows an image of the side view actual vehicle and arm with gripper assembly.

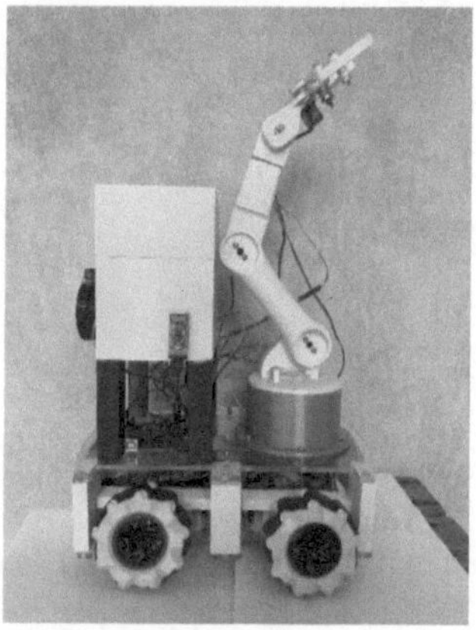

Figure 8. Side view of the robot.

Following this, one robotic arm with six degrees of freedom was added. It is used for pick-and-place operation and is also controlled by another Android mobile app. It worked efficiently, correctly picking and placing the object. The arm and gripper were operated by six servo motors. Lastly, one servo motor was assigned to the basket's door to deliver the balls to the end of the challenge. The Figure 9 shows a fclose-up of the actual gripper while Figure 10 shows the backside view of the vehicle for a clearer insight.

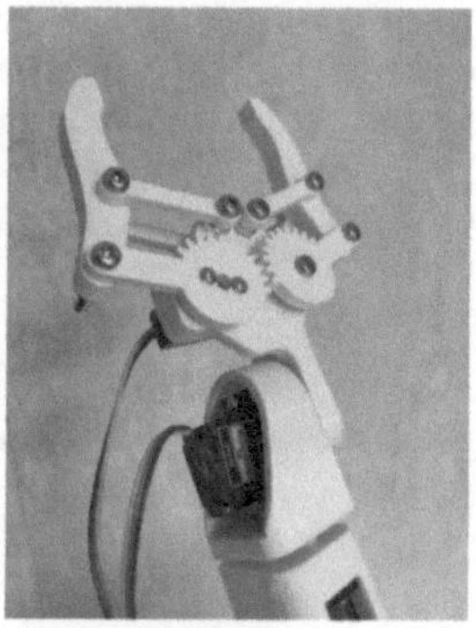

Figure 9. Gripper close-up.

Figure 10. Backside view of the robot.

4 CONCLUSION

This project provided useful insights on how to create a robot. This involves which steps were needed to take and how to solve any problem that was encountered during the process of creation of the robot. Learning how to code with the Arduino IDE was a big part of this process, and this acquired knowledge will surely represent, in the future, the possibility of making future projects using Arduino way less complicated. Learning how to make the interface and coding of an Android mobile app was an interesting approach, because in a world where everyone has a smartphone, making apps to control everything with a smartphone without the need to buy and using some external device makes everything way easier for the user, and improves the overall user experience with the product that was aspired to do. The 3D printer was a very important part of this whole process, and learning how it operates and what kind of pieces can and cannot be made for future projects will facilitate the steps of creation of the parts that are needed to fabricate. The robot could move properly and presented a capable picking-up device.

For a Future Scope, we can add the following items:

- Print arm 3D parts with a lighter configuration so the servo motors will work better.
- Choose lighter screws and nuts for the gripper for better results.
- Use stronger wheel motors to achieve better movement
- Change the gripper design to one more specific task.
- Upgrade the basket of recollection of the balls for a better final delivery system.
- Add IR and lidar sensors to avoid crashing into objects when controlling the car.

ACKNOWLEDGEMENT

This work was financially supported by the Ministry of Foreign Affairs, Republic of China (Taiwan), under the Universidad Politécnica Taiwan-Paraguay (UPTP) project.

REFERENCES

[1] CircuitBest, *Make Arduino Bluetooth controlled car with Front & Back Lights*, circuitbest.com. https://circuitbest.com/arduino-bluetooth-controlled-car-with-front-back-lights-using-arduino-uno-l293d-motor-driver-hc-05/

[2] Chandrasekar G., Gokul J, Kaviyarasan K and Rajasekar S., (2022) Mecanum grounded vehicle with robotic arm, *International Journal of Advanced Research in Science, Communication and Technology (IJARSCT)*, vol. 2, doi: https://doi.org/10.48175/568.

[3] Arduino, *Bluetooth Controlled Car*, projecthub.arduino.cc. https://projecthub.arduino.cc/samanfern/bluetooth-controlled-car-c71cd0

[4] Dejan, *DIY Arduino Robot Arm with Smartphone Control*, howtomechatronics.com. https://howtomechatronics.com/tutorials/arduino/diy-arduino-robot-arm-with-smartphone-control/

[5] About Us, *appinventor.mit.edu*. https://appinventor.mit.edu/explore/about-us.html

[6] Hackster.io, *How to Make Bluetooth Controlled car using Arduino UNO R3, H*, hackster.io. https://www.hackster.io/dhritimanhb2015/how-to-make-bluetooth-controlled-car-using-arduino-uno-r3-h-07bc57

[7] Lerda G. DIY-Tech, *KAUDA Robotic Arm*, instructables.com. https://www.instructables.com/KAUDA-Robotic-Arm/

[8] Arduino getting started, *Arduino controls Servo Motor via Bluetooth | Arduino Tutorial*, arduinogetstarted.com. https://arduinogetstarted.com/tutorials/arduino-controls-servo-motor-via-bluetooth

[9] SmartBuilds.io, *DIY Arduino Robot Arm – Controlled by Human Gestures*, smartbuilds.io. https://smartbuilds.io/diy-robot-arm-arduino-hand-gestures/

[10] Arduino, What Is Arduino?, *Arduino.cc*, Feb. 05, 2018. https://www.arduino.cc/en/guide/introduction

[11] Sanketh R. S., *Wheels*, medium.com. https://medium.com/manual-robotics/wheels-f180d4d6f2d5#:~:text=The%20Mecanum%20wheel%20is%20one

System Innovation for an Artificial Intelligence Era – Lam et al. (Eds)

A study of the performance of a kW-class Stirling engine used in a micro-combined heat and power system based on CFD simulation

Jin-Jia Huang, Sin-Ying Shen & Hang-Suin Yang
Department of Mechanical Engineering, National Chung Cheng University, Minxiong, Chiayi, Taiwan

ABSTRACT: The combined heat and power system (mCHP) uses natural gas or propane as a fuel source to generate electric power and thermal energy for home use. It can reduce the losses of electricity due to transmission. The total efficiency of energy usage is always higher than 90 %. In this study, a kW-class beta-type Stirling engine with a rhombic drive mechanism was developed for a mCHP. The performance and thermal-fluid field of the engine were simulated using the commercial software ANSYS Fluent. In the CFD model, the k-ε turbulence model was used to solve the thermal-fluid field characteristics in the engine. The simulation results were validated with experiment results and theoretical models. The results show that the proposed CFD model agrees with the experiment results and theoretical model well. The performance of the engine under different charged pressures, heating temperatures, and engine speed were studied by the proposed CFD model. The results indicate that the indicated work generated by the engine increases with charged pressure and heating temperature, but the indicated work and thermal efficiency decreases with engine speed. The maximum indicated power and the thermal efficiency of the engine can reach 3656.8 W and 7.5 %, respectively, with a heating temperature of 1073 K at 2000 rpm. The maximum thermal efficiency of the engine can reach 30 % as the engine speed decreases to 500 rpm. The proposed CFD model can be applied in the design of the Stirling engines in the future.

1 INTRODUCTION

Due to human activity causing ecological harm, scientists are committed to seeking new renewable energy sources or improving the utility of energy. Cogeneration is one of the important directions in the development. In this area, combined heat and power systems (CHP) play a significant role. The CHP can be operated by natural gas or propane to generate electrical power and apply the waste heat to produce hot water and warm gas in parallel. This allows for an increase in energy utility to 80%. The micro-combined heat and power systems (mCHP) generated a smaller wattage of electrical power. Integrating mCHP with buildings enhances the energy utility of the building, while also reducing reliance on traditional energy sources. The mCHPs helps decrease the building's carbon footprint, contributes to environmental conservation, and assists in lowering operational costs and improving sustainability.

Each part of the mCHP system is crucial, especially the power generation part. In this study, the Stirling engines are adopted in this system. Stirling engine is an external combustion engine that can be operated by arbitrary heat sources. Therefore, Stirling engines can be used as a power source in the mCHP system. Moreover, the Stirling engines exhibit a higher thermal efficiency because their theoretical efficiency is equal to Carnot efficiency. Stirling engines work with an enclosed cycle and use helium, hydrogen, air, or nitrogen as working fluid. So, there is no gas emission problem in Stirling engines. Due to these reasons, Stirling engines are always employed in the field of green technology.

The Stirling engine is composed of a displacer, a piston, an expansion space, heating tubes, a regenerator, cooling fins, and a compression space. The working fluid absorbs heat from the external heat sources by the heating tubes, and then the working fluid in the expansion space expands and drives the piston. Then the displacer drives the working fluid to the compression space through the regenerator and cooler. The heat is exhausted in the cooler and the working fluid in the compression space is compressed. In these processes, the regenerator is used to recycle the heat during a thermal cycle which is composed of a porous metal. That is why Stirling engines have higher thermal efficiency than other types of engines.

In this study, a beta-type Stirling engine with a rhombic drive mechanism applied in a mCHP is proposed and simulated [1]. The proposed model is validated with experimental results [2]. The parametric study of the engine based on the proposed model is performed. These researching results can be further used to find the optimal design in the future.

DOI: 10.1201/9781003514831-12

2 CFD MODEL & EXPERIMENTAL RESULTS

In this study, a beta-type Stirling engine with a rhombic drive mechanism is simulated using the commercial software ANSYS Fluent for computational fluid dynamics (CFD). The schematic and photograph of the proposed engine are shown in Figures 1 and 2, respectively. The basic specifications are listed in Table 1.

Table 1. The basic specifications of the engine.

Working fluid	Helium
Charged pressure	4~10 bar
Heating temperature	673~1073 K
Cooling temperature	300 K
Heater type	Tubular heater
Cooler type	Fined cooler
Piston swept volume	542 cm^3
Compression ratio	1.89

The Stirling engine is composed of several components. The tubes on the top are heating tubes which are used to observe heat. The chamber connected with the heating tubes is expansion space. The bottom chamber is compression space which is connected with cooling fins. The fins are used to release waste heat by cooling water. The component located between the heating tubes and cooling fins is the regenerator which consists of porous metal. The regenerator is used to recover the heat without use during the thermal cycle. The displacer is installed between the expansion space and compression space which is used to move working fluid. The piston is under the compression space which is used to compress and expand working fluid.

The ANSYS Fluent solves the following governing equations for thermal-fluid fields [3].

Continuity equation

$$\frac{\partial \rho}{\partial t}+\frac{\partial \rho u}{\partial x}+\frac{\partial \rho v}{\partial y}+\frac{\partial \rho w}{\partial z}=0 \quad (1)$$

Navier-Stokes equation

$$\begin{aligned}&\frac{\partial \rho u}{\partial t}+\frac{\partial \rho u^2}{\partial x}+\frac{\partial \rho uv}{\partial y}+\frac{\partial \rho uw}{\partial z}=\\&-\frac{\partial p}{\partial x}+\frac{\partial \sigma_{xx}}{\partial x}+\frac{\partial \tau_{yx}}{\partial y}+\frac{\partial \tau_{zx}}{\partial z}\\&+\frac{\partial(-\rho\overline{u'u'})}{\partial x}+\frac{\partial(-\rho\overline{u'v'})}{\partial y}+\frac{\partial(-\rho\overline{u'w'})}{\partial z}\end{aligned} \quad (2)$$

$$\begin{aligned}&\frac{\partial \rho v}{\partial t}+\frac{\partial \rho vu}{\partial x}+\frac{\partial \rho v^2}{\partial y}+\frac{\partial \rho vw}{\partial z}=\\&-\frac{\partial p}{\partial y}+\frac{\partial \tau_{xy}}{\partial x}+\frac{\partial \sigma_{yy}}{\partial y}+\frac{\partial \tau_{zy}}{\partial z}\\&+\frac{\partial(-\rho\overline{v'u'})}{\partial x}+\frac{\partial(-\rho\overline{v'v'})}{\partial y}+\frac{\partial(-\rho\overline{v'w'})}{\partial z}\end{aligned} \quad (3)$$

$$\begin{aligned}&\frac{\partial \rho w}{\partial t}+\frac{\partial \rho wu}{\partial x}+\frac{\partial \rho wv}{\partial y}+\frac{\partial \rho w^2}{\partial z}=\\&-\frac{\partial p}{\partial z}+\frac{\partial \tau_{xz}}{\partial x}+\frac{\partial \tau_{yz}}{\partial y}+\frac{\partial \sigma_{zz}}{\partial z}\\&+\frac{\partial(-\rho\overline{w'u'})}{\partial x}+\frac{\partial(-\rho\overline{w'v'})}{\partial y}+\frac{\partial(-\rho\overline{w'w'})}{\partial z}\end{aligned} \quad (4)$$

Energy equation

$$\begin{aligned}&\frac{\partial \rho h}{\partial t}+\frac{\partial \rho uh}{\partial x}+\frac{\partial \rho vh}{\partial y}+\frac{\partial \rho wh}{\partial z}=\\&\frac{\partial p}{\partial t}+u\frac{\partial p}{\partial x}+v\frac{\partial p}{\partial y}+w\frac{\partial p}{\partial z}\\&+\frac{\partial}{\partial x}\left(k\frac{\partial T}{\partial x}\right)+\frac{\partial}{\partial y}\left(k\frac{\partial T}{\partial y}\right)+\frac{\partial}{\partial z}\left(k\frac{\partial T}{\partial z}\right)\end{aligned} \quad (5)$$

In addition, the turbulence model employed in this study is the k-ε model. The proposed model was solved using PISO algorithm in this study.

As displayed in Figure 1, the volume of working space is varied with the motion of the piston and displacer. The moving boundary is controlled by the user-defined functions (UDF) in the ANSYS Fluent. The velocity of the moving boundary of the piston and displacer are:

$$v_p = R[\omega_{OA}\cos\theta + \omega_{AB}(\varepsilon_p + \cos\theta)] \quad (6)$$

$$v_d = R[\omega_{OA}\cos\theta + \omega_{AC}(\varepsilon_d + \cos\theta)] \quad (7)$$

where

$$\omega_{AB,AC} = -\frac{\omega_{OA}\sin\theta}{\sqrt{e_{AB,AC}-(\varepsilon_{p,d}+\cos\theta)}} \quad (8)$$

$$e_{AB,AC} = \frac{L_{AB,AC}}{R}, \varepsilon_{d,p} = \frac{R_f - L_{dw,pw}}{R} \quad (9)$$

where $e_{AB} = e_{AC} = 2.5$ and $\varepsilon_p = \varepsilon_d = 1.1$.

The orthogonal grids were adopted in this study. The grid sensitivity test was conducted with grid sizes ranging from 10^5 to 2×10^6 grids. The simulation results converged as the number of grids approached 10^6 grids. So the grid number used in this study is approximately 10^6. The absolute pressure and volume of the engine are obtained from ANSYS Fluent. The area surrounded by absolute pressure and volume is indicated power. The shaft power is the indicated power deducts mechanical losses.

The proposed CFD model was validated by comparing the experimental data and simulation results. Base on these results, a series of parametric studies were conducted. The effects of heating temperature, charged pressure, and engine speed are studied.

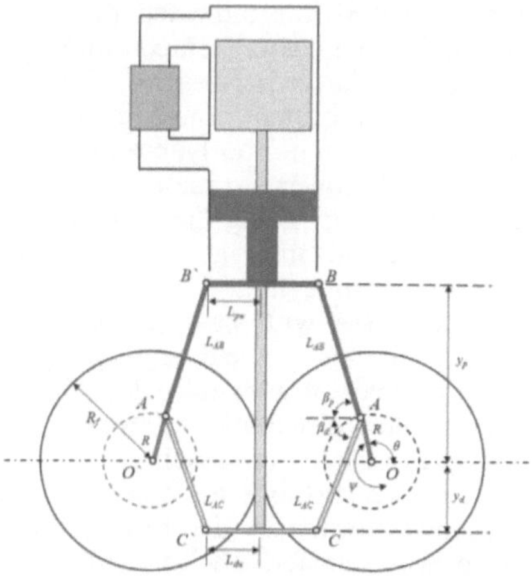

Figure 1. The schematic of the proposed engine.

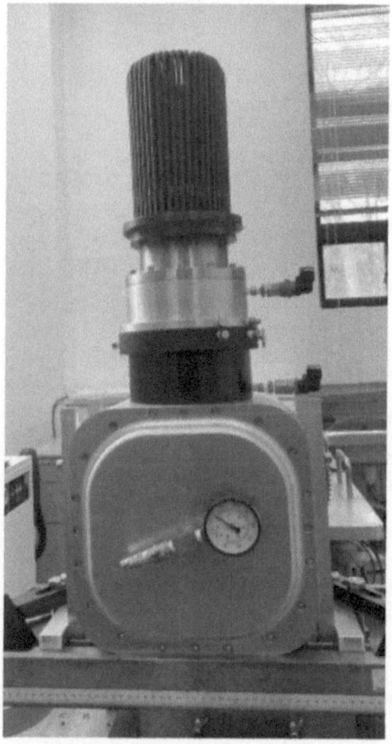

Figure 2. The photograph of the proposed engine.

3 RESULTS AND DISCUSSION

To validate the proposed CFD model, the simulation results were compared to the experimental data and numerical model. The shaft power of the engine 4 bar of helium and 673 K of heating temperature are shown in Figure 3. The results reveal that the tendency of the shaft power is close and the difference

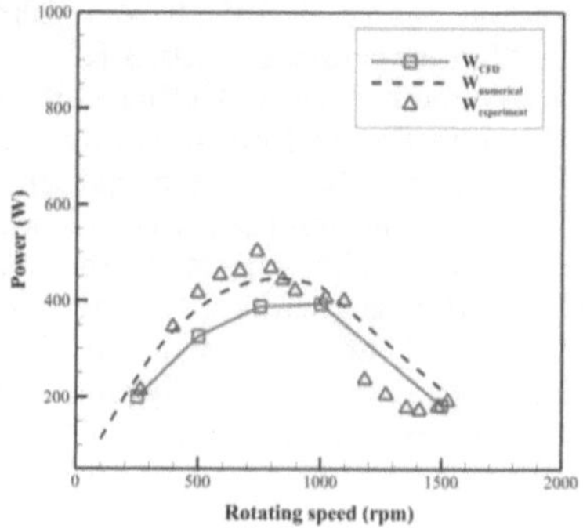

Figure 3. Comparison of experimental data, numerical, and simulation results.

between them is lower than 17.54 %. The maximum shaft power reaches about 500 W at 750 rpm.

Based on the proposed CFD model, the internal thermal-fluid field of the working fluid in the working space can be obtained. Figure 4 displays the variation in the temperature distribution of working fluid during a thermal cycle at an engine speed of 1000 rpm and a heating temperature of 873 K. It is obvious that there are two large vortices generated as the working fluid flow into the compression space and expansion space at crank angle $\theta = 0°$ and $\theta = 180°$, respectively. These vortices enhance the heat transfer between the working fluid and the wall boundary inside the compression space and expansion space.

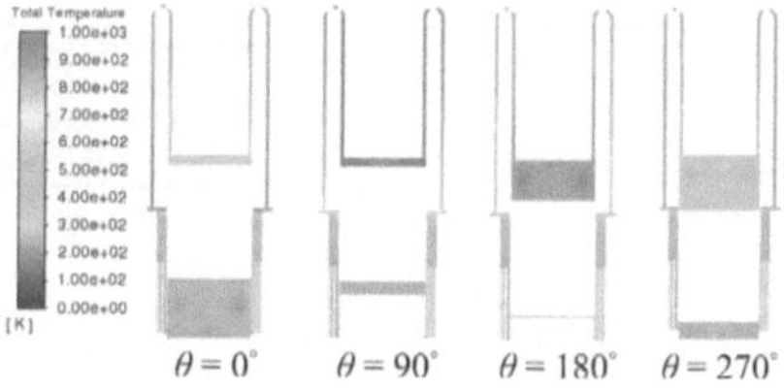

Figure 4. The temperature distribution of working fluid at different crank angles.

The effects of heating temperature, charged pressure, and engine speed on indicated power are studied here. The indicated power of the proposed engine with 8 bar of charged pressure under 673 K, 873 K, and 1073 K of the heating temperature is displayed in Figure 5. Compare with the results indicated that the shaft power decreases with engine speed as the speed higher than 1000 rpm which is caused by the increase in the imperfect heat transfer in the engine. The results reveal that the indicated power increases with charged pressure and engine speed linearly. The maximum indicated power can reach 3656.8 W at 2000 rpm under a heating temperature of 1073 K.

The corresponding thermal efficiencies of the cases displayed in Figure 5 are shown in Figure 6. It can be observed that the thermal efficiency increases with heating temperature and decreases with engine speed. The maximum thermal efficiency of the proposed engine can reach approximately 30 % at a heating temperature of 1073 K and an engine speed of 500 rpm. Then the thermal efficiency decreases to approximately 23.5 % as the

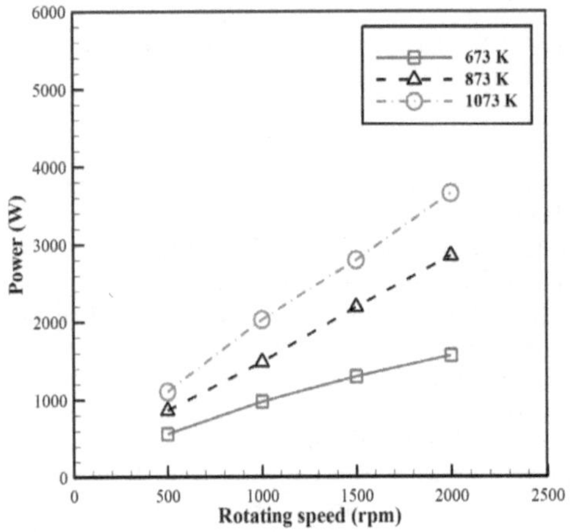

Figure 5. The indicated power under different heating temperatures and different engine speeds.

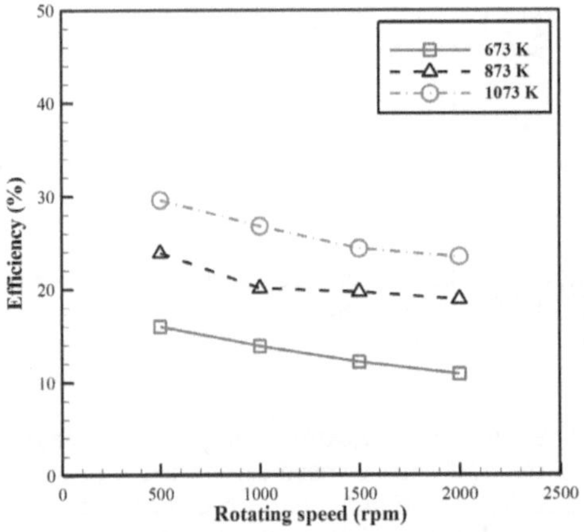

Figure 6. The thermal efficiency under different heating temperatures and different engine speeds.

engine speed is increased to 2000 rpm. Similarly, the thermal efficiency decreases from approximately 16 % to 10 % as the engine speed increases from 500 rpm to 2000 rpm at a heating temperature of 673 K.

Figure 7 displays the indicated power of the proposed engine at a heating temperature of 873 K under different charged pressures. The results reveal that the indicated power increases with charged pressure monotonously, but the increment in the indicated power decreases with charged pressure. The corresponding thermal efficiency of the cases displayed in Figure 7 is shown in Figure 8. The results indicate that the thermal efficiency decreases with charged pressure and engine speed. The higher the charged pressure, the greater the number of gas molecules. Due to the same energy being distributed to more molecules, heat transfer decreases. As a result, the variation in thermal efficiency exhibits an opposite trend to indicated power, primarily due to the increase in incomplete heat transfer. The maximum thermal efficiency reaches 26% with a charged pressure of 4 bar and an indicated power of 481.24 W. The indicated power and thermal efficiency can reach 1053.88 W and 23 %, respectively, under the same operating condition.

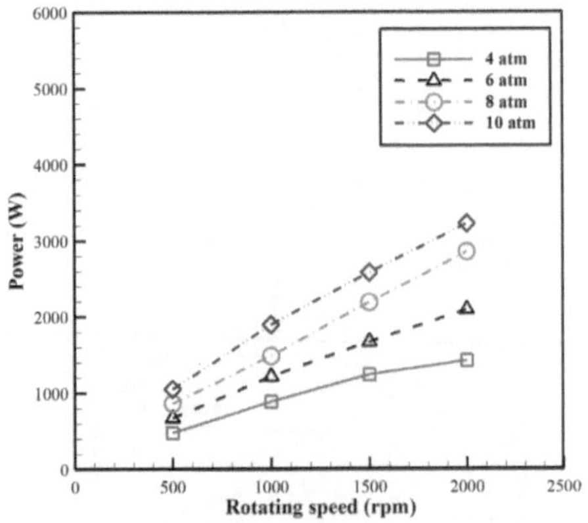

Figure 7. The indicated power under different charged pressures and different engine speeds.

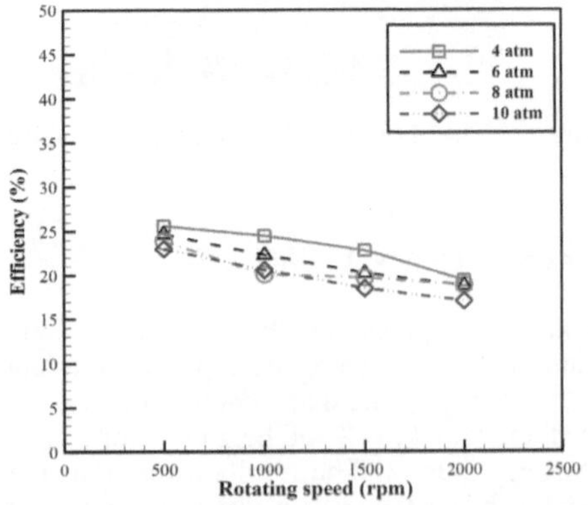

Figure 8. The thermal efficiency under different charged pressures and different engine speeds.

4 CONCLUSION

In this study, a CFD model is proposed for analyzing the performance of a kW-class beta-type Stirling engine with a rhombic drive mechanism. The thermal-fluid field was simulated using ANSYS Fluent and the results were compared with the experimental data and numerical model. The effects of charged pressure and heating temperature on indicated power and thermal efficiency are studied. The findings of this investigation are as follows:

(1) The results obtained by the CFD simulation are close to experimental data and the results obtained by the numerical method. The difference between experimental data and CFD results is lower than 17.54%.
(2) The indicated power increases with heating temperature linearly. The maximum value reaches 3656.8 W at 2000 rpm as the heating temperature is 1073 K.
(3) As the engine speed exceeds 1000 rpm, the shaft power decreases due to the imperfect heat transfer. The thermal efficiency of the proposed engine decreases from 30% to 25.5% as the engine speed increases from 500 rpm to 2000 rpm at a heating temperature of 1073 K.
(4) The indicated power increases with charged pressure and engine speed due to the increase in engine speed. However, the thermal efficiency decreases with charged pressure and engine speed. The enhancement in charged pressure leads to imperfect heat transfer in the engine. The lowest thermal efficiency is 23% at a charged pressure of 10 bar and an engine speed of 2000 rpm.

ACKNOWLEDGEMENT

The authors would like to thank the National Science and Technology Council, Taiwan, for the financial support under grant NSTC 112-2221-E-194-010-MY3.

REFERENCES

[1] H.S. Yang, MA Ali, KVR Teja, and YF Yen Parametric study and design optimization of a kW-class beta-type Stirling engine. *Applied Thermal Engineering*, 2022. 215.
[2] H.S. Yang, H.Q. Zhu, and X.Z. Xiao, Comparison of the dynamic characteristics and performance of beta-type Stirling engines operating with different driving mechanisms. *Energy*, 2023. 275.
[3] ANSYS, I., *ANSYS Fluent user's Guide*

System Innovation for an Artificial Intelligence Era – Lam et al. (Eds)
© 2025 the Author(s), ISBN: 978-1-032-84452-7

Intelligent car searching machine interface design for parking lots based on wayfinding behavior

Jia-Yang Lin
Department of Industrial Design, Master Program of Innovation and Design, Taipei Tech, Taipei, Taiwan

Meng-Cong Zheng
Department of Industrial Design, National Taipei University of Technology, Taipei, Taiwan

ABSTRACT: In Taiwan, the total number of automobiles has been climbing annually, leading to an increasing demand for parking. Currently, parking lots have gradually implemented services such as license plate recognition, occupancy detection, and intelligent vehicle locating. Among these, the intelligent vehicle locating service enables users to inquire about the parking location of their vehicle and provides a guided path between the user and the vehicle. This study focuses on the Daan Forest Park underground parking lot in Taipei, Taiwan, as the experimental site. Users were asked to use the intelligent vehicle locating kiosk and then carry out a task to find their vehicle. Through think-aloud protocols, questionnaire surveys, and semi-structured interviews, we gathered user feedback on the interface's usability and issues encountered while operating the vehicle locating kiosk. The findings indicate that the "homepage of the kiosk" suffers from an overload of search function options, making it difficult for users to discern the features. Furthermore, the display method of the "kiosk's guided path screen" confuses users about the direction they should take from their current location to their vehicle. We suggest that future iterations of the kiosk should prioritize retaining the highly used license plate search function and improve the display method of the guided paths on the map, such as adding directional cues. These improvements could enhance user operation and efficiency in locating their vehicles.

Keywords: Wayfinding behavior, Parking lots, Intelligent car searching machine, User interface design

1 INTRODUCTION

According to the research findings (Andrew Chesterton 2018), the car ownership rate in the United States is approximately 90%, making it the country with the highest car ownership rate in the world. In the Asia-Pacific region, Taiwan's car ownership rate reaches 68%. The global car ownership rate is increasing year by year, and currently, the average number of cars per household in Taiwan is 1.5, with nearly 60% of households owning one car, and nearly 30% of households owning two cars.

However, according to the statistics from the Taiwan Ministry of Transportation and Communications in 2021, the national average time spent by each driver searching for a parking space is 10.4 minutes, with Taipei City requiring 12.8 minutes, ranking first in the country in terms of parking search time. In recent years, the government has been promoting smart parking systems for parking lots, gradually implementing services such as license plate recognition, occupancy detection, and smart car locating, to improve parking turnover rates, enhance the quality of parking lot operation services, and meet the parking needs of users.

In the past, some studies have focused on indoor wayfinding behavior, such as signage design in hospitals (Samah *et al.* 2012; Wang 2023), as well as map design for underground parking lots (Zheng 2014). There have also been studies exploring individual differences in wayfinding cognition (Bishop 2012; Hund and Nazarczuk 2009). In the past, both domestic and international academic research has mainly focused on the development of smart car locating systems for parking lots (Gao *et al.* 2018; Hanzl 2020; Hao *et al.* 2022; Wang 2018). Currently, there has been no research specifically addressing the interface design of car locating query machines. Therefore, this study aims to identify issues with the operation interface of car locating query machines through the process of users operating these machines and their car-locating behaviors. Suggestions for the design of the query machine interface will be proposed to enhance users' car locating performance and reduce their valuable time spent returning to the parking lot to search for their vehicles.

DOI: 10.1201/9781003514831-13

2 EXPERIMENTAL

This study selected the "Underground Parking Lot of Daan Forest Park in Taipei, Taiwan" as the experimental site. This parking facility is divided into B1 and B2 levels, totaling two underground floors with approximately 1352 parking spaces. The main reasons for choosing this location are as follows:

(1) The underground parking lot provides a closed-off environment, isolating external visual cues. Unlike outdoor parking lots, where surrounding buildings can aid in determining the location of parking spaces, the underground setting eliminates such external references.
(2) The larger spatial scale and higher number of parking spaces in the underground facility increase the complexity of the space, thus challenging users' car locating behaviors.

Investigation revealed the presence of four smart car locating query machines within the facility, each positioned in pedestrian stairwells. The parking space numbering in sections B1 and B2 is independently coded, and the query machines are only installed in the B1 section. Therefore, this study focuses solely on conducting experiments for car locating tasks in the B1 section.

The car locating query machine at the research site offers the following search functions: car number search, time search, parking location number search, and full car number search (as shown in Figure 1). In addition, the search results of the car locating query machine will display the navigation path between the user and the vehicle (as shown in Figure 2).

Figure 1. Homepage screen of the car locating query machine.

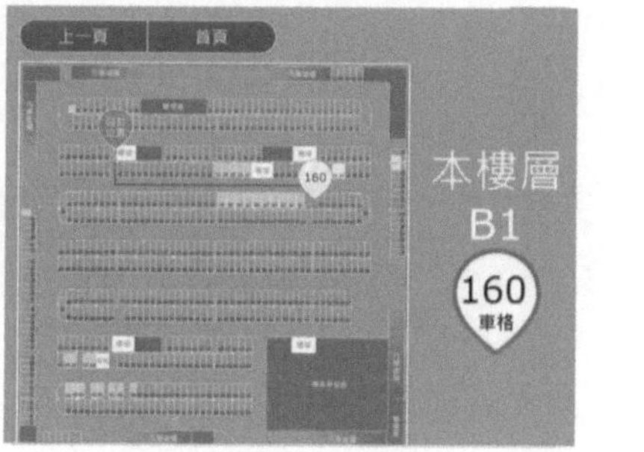

Figure 2. Car searching machine navigation path screen.

This study plans to conduct car locating task experiments with 30 participants, including 15 males and 15 females, aged between 18 and 60 years old. Participants have no prior experience operating car locating query machines and are unfamiliar with the experimental parking lot environment. Additionally, all participants in this study must possess basic navigation skills, communication abilities, and clear expression capabilities.

Before the experiment commences, participants will complete a basic demographic questionnaire to gather information about their parking experiences and previous usage of car locating query machines. The car search task is divided into missions A, B, and C, with the difficulty of each mission determined by the execution steps and walking distances (as shown in Table 1), as well as the car search task steps (as shown in Figure 3). During the experiment, the participants' processes will be recorded using the thinking-aloud method.

Before the experiment, participants will be provided with a photo of the designated parking space, which serves as the destination for the car search task

Table 1. Vehicle search task.

TaskNo.	Starting Point	End Point	Steps	Distance (m)
A	Machine #1	Parking #160	5	85
B	Machine #2	Parking #552	7	135
C	Machine #3	Parking #542	9	205

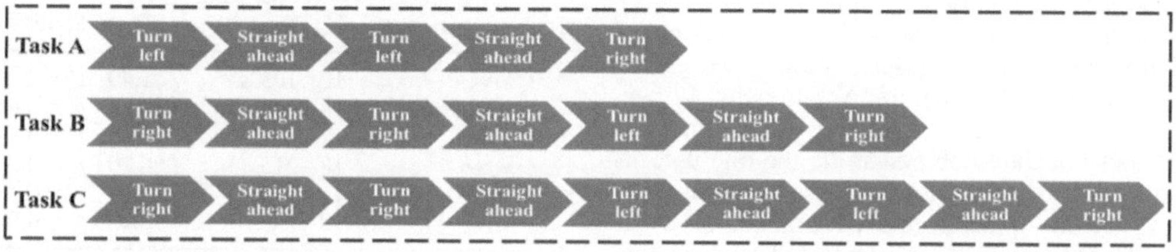

Figure 3. Task Steps

Figure 4. Vehicle search mission photos.

(as shown in Figure 4). The photo information includes the license plate number, the time when the vehicle was parked, and the parking space number, among other details. Before the car search task, participants will be allowed to choose their preferred method of searching for the vehicle. This approach aims to understand the participants' utilization rates of various car search functions provided by the search machine and to gather suggestions for improving the interface design of the search machine.

After completing the task experiments, participants will undergo interface evaluation using questionnaire scales. This study adopts the System Usability Scale (SUS), a five-point scale developed by (Brooke 1996), and the Questionnaire for User Interaction Satisfaction (QUIS), a seven-point scale developed by (Chin *et al.* 1988). Semi-structured interviews will also be conducted to gather feedback on the display and operation of the car locating query machine interface, aiming to assess users' perceptions of interface usability and identify any issues encountered while operating the query machine.

3 RESULTS AND DISCUSSION

3.1 *Basic information of subjects*

In this experiment, there are a total of 30 participants, comprising 15 males and 15 females, aged between 21 and 50 years old. According to the results of the basic questionnaire survey, regarding parking experience, out of the 30 participants, 24 have parking experience, while 6 do not.

Regarding the experience with car locating query machines, 18 participants have never used a car locating query machine before. Additionally, 12 participants have used a car locating query machine before, but not the same system as the one used in this study. Furthermore, none of these 12 participants have visited the underground parking lot of Daan Forest Park for car locating purposes. Therefore, this study considers these 12 participants as having no prior experience with car locating query machines, thus treating them as part of the valid sample.

3.2 *Task operations performance assessment*

The performance evaluation of tasks in this study refers to assessing the overall time taken, distance traveled, and interview content while participants operate the car locating query machine and reach the designated task vehicle location. This evaluation aims to explore any issues encountered by participants while operating the car locating query machine and analyze experimental findings accordingly.

3.2.1 *Gender and operational performance*

The results of independent samples t-tests on the relationship between gender and operational performance in terms of operation time show no significant differences for tasks A, B, and C. Similarly, the results of independent samples t-tests on the relationship between gender and operational performance in terms of walking distance show no significant differences for tasks A, B, and C.

3.2.2 *Relationship between parking experience and operational performance*

The results of independent samples t-tests on the relationship between parking experience and operational performance in terms of operation time show no significant differences for tasks A, B, and C. Similarly, the results of independent samples t-tests on the relationship between parking experience and operational performance in terms of walking distance show no significant differences for tasks A, B, and C.

3.2.3 *Relationship between experience with car locating query machine and operational performance*

The results of independent sample t-tests regarding the relationship between experience with the car locating machine and operational performance, specifically focusing on the operation time, are presented in Table 2. The analysis indicates that

Table 2. Operation time relationship between experience with car locating query machine and operational performance.

Task	Car locating query machine experience	N	Average operating time (s)	SD	Significance
A	Yes	12	103.83	21.40	t = -0.455,
	No	18	107.78	24.39	p = 0.653
B	Yes	12	184.00	71.85	t = -1.040,
	No	18	215.83	88.10	p = 0.307
C	Yes	12	178.25	25.67	t = -3.066,
	No	18	248.61	92.15	p = 0.006**

* $p < .05$. ** $p < .01$. *** $p < .001$.

there are no significant differences in operational performance between Task A and Task B. However, for Task C, there is a significant difference ($p = 0.006^{**}$), indicating that experience with the machine influences operational performance.

The independent sample t-test results regarding the relationship between experience with the car locating machine and operational performance, specifically focusing on the walking distance, show no significant differences for Task A, Task B, and Task C.

3.3 *Scale assessment analysis*

After completing the car locating tasks, participants in this study will sequentially fill out the System Usability Scale (SUS) and the Questionnaire for User Interaction Satisfaction (QUIS). This will enable us to assess participants' evaluations of the usability and satisfaction with the car locating query machine, and subsequently conduct post-use evaluation analysis.

3.3.1 *SUS scale evaluation analysis*

The System Usability Scale (SUS) is considered to have above-average usability if the average score is higher than 68, and below-average usability if the score is lower than 68 (Lewis and Sauro 2018). In this experiment, a total of 30 participants completed the survey, with an overall average score of 78.08. The overall usability rating is classified as Grade C, falling within an acceptable range.

3.3.1.1 *Relationship between gender and SUS scale scores* The independent samples t-test results indicate that there is no significant difference in System Usability Scale (SUS) scores between males (M = 79.00, SD = 13.15) and females (M = 77.17, SD = 14.26).

3.3.1.2 *Relationship between parking experience and SUS scale scores* The independent samples t-test results indicate that there is no significant difference in System Usability Scale (SUS) scores between participants with parking experience (M = 75.94, SD = 13.71) and those without parking experience (M = 86.67, SD = 9.17).

Table 3. Mean scores of the four dimensions of the QUIS satisfaction scale.

	Overall Response (6 questions)		Interface Presentation (4 questions)		Interface Information (4 questions)		Learning Level (6 questions)	
	Mean	Total Score	Mean	Total Score	Mean	Total Score	Mean	Total Score
Min	3.50	21.00	4.00	16.00	2.50	10.00	3.83	23.00
Max	7.00	42.00	7.00	28.00	7.00	28.00	7.00	42.00
Mean	5.46	32.73	5.78	23.10	5.73	22.93	6.04	36.27

3.3.1.3 *Relationship between experience with Ccr locating query machine and SUS scale scores* The independent samples t-test results reveal no significant difference in System Usability Scale (SUS) scores between participants with experience using the query machine (M = 80.63, SD = 13.02) and those without experience (M = 76.39, SD = 13.94).

3.3.2 *QUIS satisfaction scale evaluation analysis*

The study utilized a (Likert Scale) with seven levels and assessed user satisfaction based on four major dimensions: overall response of the interface, interface presentation, interface information, and learning level. Each dimension was scored on a scale of 1 to 7, with a minimum score of 1 and a maximum score of 7 for each question. A total of 30 participants completed the questionnaire, yielding an average score of 4. Participants rated their satisfaction higher than the average score for all four dimensions (as shown in Table 3).

3.3.2.1 *Relationship between gender and QUIS scores* The independent samples t-test results showed no significant difference between gender and QUIS satisfaction scores, with males (M = 115.67, SD = 18.68) and females (M = 114.40, SD = 15.40) exhibiting similar levels of satisfaction.

3.3.2.2 *Relationship between parking experience and QUIS scores* The independent samples t-test results revealed no significant difference between parking experience and QUIS satisfaction scores. Participants with parking experience (M = 115.21, SD = 18.48) and those without parking experience (M = 114.33, SD = 8.43) demonstrated similar levels of satisfaction.

3.3.2.3 *Relationship between experience with car locating query machine and QUIS scores* The independent samples t-test results indicated no significant difference between experience with the query machine and QUIS satisfaction scores. Participants with query machine experience (M = 119.75, SD = 12.59) and those without query machine experience (M = 111.89, SD=18.83) showed similar levels of satisfaction.

3.4 *Semi-structured interviews*

Although the experimental results of this study have quantitatively indicated that participants' usability and satisfaction with the current car locating query machine have met the standard scores, qualitative interviews conducted after the experiment revealed that 26 participants (86.7%) still suggested optimization and improvement of the machine interface. Feedback on the presentation of the interface was collected and summarized, highlighting the following key points:

(1) Regarding the "search function options on the query machine homepage," there are too many options, requiring users to think and choose

the desired search function. This makes it difficult for users to make an immediate decision and proceed with the vehicle search action.

(2) Concerning the "guidance display time on the query machine's navigation path screen," the guidance display time is too short. Users do not have enough time to memorize the navigation path and need to repeat the search action to view the navigation path again.

(3) Regarding the "display method of the navigation path screen on the query machine," users find it challenging to discern the direction they should proceed from their current location.

"Homepage screen of the car locating query machine."

Based on the experimental and interview results, it was found that 6 participants (20%) spent more than 2 seconds on the homepage of the query machine before selecting the desired search function during the car locating task.

"Car searching machine navigation path screen."

According to the experimental and interview results, it was found that 11 participants (36.7%) repeated the operation on the query machine's guidance path screen more than 2 times during the car locating task. However, among these 11 participants, 8 repeated the operation twice in Task A, 5 in Task B, and 3 in Task C.

Furthermore, based on the interview results, it was discovered that the higher number of repetitions in Task A was mainly due to the participants gradually becoming familiar with interpreting the guidance path screen of the query machine during subsequent car locating tasks.

Additionally, 20 participants (66.7%) stated that "the map display format is not easy to determine the direction of travel" when interpreting the guidance path screens for each task. In Task A, where the correct direction was left, 10 participants (33.3%) misjudged the direction of the car exit. In Task B, where the correct direction was right, 16 participants (53.3%) misjudged the direction of the car exit. In Task C, where the correct direction was also right, 3 participants (10%) misjudged the direction of the car exit.

In Task C, the number of participants who misjudged the direction was fewer compared to other tasks. Among them, 15 participants (50%) experienced detours due to misjudging the direction of the car exit in Tasks A and B, gradually becoming familiar with the experimental environment. However, in Task C, the number of participants who misjudged the direction decreased accordingly.

The participants expressed that "the navigation path indicators cannot discern the actual direction of the vehicle exit." From this, it can be inferred that the execution steps and walking distance of the tasks are not correlated with the participant's judgment of the navigation path screen. Instead, the main reason for the difficulty lies in the direction shown on the map display.

4 CONCLUSION

This study utilized a one-sample t-test to analyze the usage of query functions by participants in each car locating task. In Task A, 24 participants (80%) utilized the license plate number search function, while 6 participants (20%) used the parking space number search function. In Task B, 20 participants (66.7%) used the license plate number search function, 9 participants (30%) used the parking space number search function, and only 1 participant (3.3%) used the complete license plate number search function. In Task C, 23 participants (76.7%) used the license plate number search function, and 7 participants (23.3%) used the parking space number search function. However, the time search function was not used in any of the car locating tasks. There was a significant difference in the usage rates of various query functions among participants in the car locating tasks (p<0.001***), as shown in Table 4.

Table 4. Results of license plate and parking space number searches (N = 30).

Task	License plate number search	Parking space number search	Complete license plate number search	Significance
A	24	6	0	$t = 16.155$, $p < .001$***
B	20	9	1	$t = 11.366$, $p < .001$***
C	23	7	0	$t = 15.703$, $p < .001$***

* $p < .05$. ** $p < .01$. *** $p < .001$.

The experimental findings of this study revealed that the car number search function had the highest usage rate on the "Search Machine Homepage," followed by the parking location number search. Additionally, the time search function was not utilized by any participants across all tasks. Based on these results, the following key proposals are suggested for the future design of search machine interfaces:

(1) Interface design should adopt a simple and concise presentation approach.

(2) Prioritize retaining the car number search function, which has the highest usage rate, on the "Search Machine Homepage."

(3) Extend the display time of the navigation path map on the "Search Machine Navigation Path Screen" and incorporate directional cues to help users quickly understand the orientation of the starting point for car locating.

These proposals aim to enhance user operation and car locating performance, ultimately reducing the valuable time spent searching for vehicles.

REFERENCES

Andrew Chesterton. (2018). Retrieved from: https://www.carsguide.com.au/car-advice/how-many-cars-are-there-in-the-world-70629.

Bishop K. (2012). The role of art in a paediatric healthcare environment from children's and young people's perspectives. *Procedia - Social and Behavioral Sciences*, 38, pp.81–88.

Brooke J. (1986). *System usability scale (SUS): a quick-and-dirty method of system evaluation user information. Reading, UK: Digital equipment co ltd*, 43, 1–7.

Brooke J. (1996). Sus: a quick and dirty usability. *Usability evaluation in industry*, 189(3).

Chin J. P., Diehl, V. A. and Norman K. L. (1988). Development of an instrument measuring user satisfaction of the human-computer interface. *In Proceedings of the SIGCHI conference on Human factors in computing systems* (pp. 213–218).

Gao R., He, F. and Li, T. (2018). VeLoc: Finding your car in indoor parking structures. *Sensors*, 18(5), 1403.

Hanzl J. (2020). Parking information guidance systems and smart technologies application used in urban areas and multi-storey car parks. *Transportation Research Procedia*, 44, 361–368.

Hao R., Li X. and Liu, J. (2022). Intelligent parking and AR vehicle searching system in parking lot. *International Core Journal of Engineering*, 8(3), 209–213.

Hund A. M. and Nazarczuk S. N. (2009). The effects of sense of direction and training experience on wayfinding efficiency. *Journal of Environmental Psychology*, 29(1), pp.151–159.

Samah Z. A., Ibrahim N., Othman S. and Wahab M H A. (2012). Assessing quality design of interiors: A case study of a hospital outpatient unit in Malaysia. *Procedia – Social and Behavioral Sciences*, 35, pp. 245–252.

Wang J. Y. (2023). *Exploring wayfinding behavior in virtual environments: Designing signage systems for medical institutions.* Retrieved from https://hdl.handle.net/11296/d83k6p.

Wang S.-S. (2018). A BLE-based pedestrian navigation system for car searching in indoor parking garages. *Sensors*, 18(5), 1442.

Zheng M. C. (2014). Design of a map for an underground parking space using wayfinding and route description. *Journal of Asian Architecture and Building Engineering*, 13(2), 325–332.

System Innovation for an Artificial Intelligence Era – Lam et al. (Eds)

A study on visitors' experience of visiting a tech art exhibition in Taiwan

Chih-Yu Teng & Chao-Ming Wang
Department of Digital Media Design, Nation Yunlin University of Science and Technology, Yunlin, Taiwan

ABSTRACT: The tech art exhibition provides visitors with a unique sensory and intellectual experience, blending distinctive artistic presentations with advanced technological applications. This study aims to explore visitors' experiences at tech art exhibitions. Through this research, we seek to gain a comprehensive understanding of the variations in visitors' experiences across different domains, enabling improvements in curation and design to better cater to audience needs in future exhibitions.

The study concentrates on scrutinizing variances in visitors' experiences at tech art exhibitions across various domains. It examines visitors' experiential values throughout the exhibition, the usability of interactive interfaces of exhibits, and overall perceptual effects. Additionally, the research explores the influence of interactive devices in exhibitions, such as virtual reality, wearable devices, and various types of interactive installations, to evaluate whether these elements augment visitors' usability and acceptance of tech art.

The research utilizes questionnaire surveys and observational methods. The surveys aim to explore visitors' domain backgrounds, emotional responses to exhibits, and experiential perceptions of exhibition works. Concurrently, observations will document visitors' dwell times, interactive behaviors, and attention allocation to exhibits, providing insights into the degree to which tech art exhibitions stimulate visitors' contemplation and exploration.

Feedback and observational data collected from visitors across diverse domains will assist curators in gaining a deeper understanding of the requirements of various groups. These insights will be invaluable for informing future curatorial decisions and ensuring that exhibitions effectively cater to a broad range of audience needs.

Keywords: Tech Art, Human-Computer Interaction, Interactive Technology, Exhibition Design

1 INTRODUCTION

Tech art exhibitions represent a significant manifestation of contemporary art, seamlessly integrating technology and artistic expression to showcase unparalleled innovation and imagination. This fusion not only captivates art enthusiasts but also appeals to audiences from diverse fields, including scientists, engineers, designers, artists, and other professionals. However, our understanding of how the backgrounds of these audiences from different domains affect their experiences at tech art exhibitions is still limited.

This study utilizes the "Diffusion Couple" exhibition held at the Kaohsiung Museum of Fine Arts in 2023 as a case study (see Figure 1). motivated by curiosity about the diversity of experiences among visitors to tech art exhibitions. It is observed that visitors from different domains may have varying understandings of tech art due to their professional backgrounds, levels of knowledge, and perspectives, necessitating an in-depth investigation into the impact of these differences. Tech art exhibitions involve not only the artistic aspect but also various aspects of technological applications. This diversity allows audiences to have different focuses and interpretations. For example, a scientist may be more interested in the scientific principles and technological innovations in the works, while an artist may focus more on the form and artistic conception of the artwork. This diversity provides rich research material for this study while also posing a series of unanswered questions. The emergence of tech art exhibitions has brought about cultural integration, allowing professionals from various fields to better understand and appreciate each other's work through this exchange platform.

This study aims to understand how technology applications guide interactions between visitors from different domains and artworks, thereby influencing their experiential values, the usability of interactive interfaces of exhibits, and overall perceptual effects. It is hoped that this understanding will contribute to enhancing the cultural adaptability of exhibitions, making them more inclusive and attracting a broader audience.

Tech art exhibitions emphasize active participation by the audience and encourage them to become

DOI: 10.1201/9781003514831-14

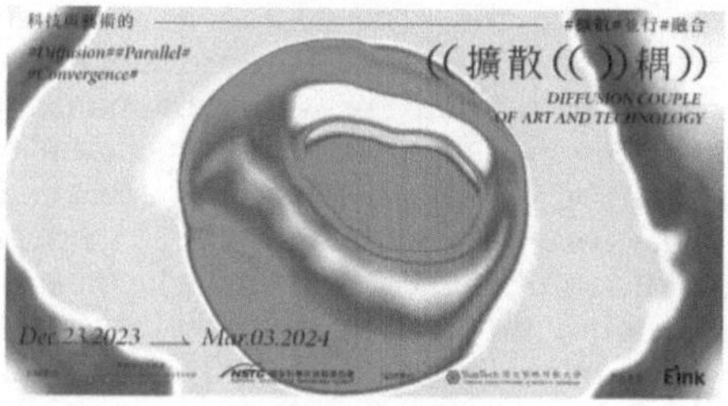

Figure 1. Main visual of the "diffusion couple" exhibition at the Kaohsiung Museum of fine arts (Source: https://youtu.be/oJrZVBGtyD0?si=oVKtyuVkDE716-oM)

part of the artwork. For curators and exhibitors, understanding the level of audience engagement in interactions is particularly crucial. This not only relates to the attractiveness of the exhibition and the satisfaction of visitors but also involves the direction of artistic creation. Through this study, it is hoped to provide curators with more effective exhibition design recommendations by gaining deeper insights into visitor participation patterns, while also promoting deeper engagement from the audience.

2 LITERATURE REVIEW

2.1 *Tech art exhibitions*

Since the Digital revolution in the 20th century, technology has influenced human life and provided more diverse materials and inspirations for artistic creation. In new media art and tech art creation, the combination of "technology" and "art" has promoted the integration of cross-media art due to the characteristics of technology such as interaction, virtuality, and digitization. Since the 1980s, there has been a global trend of interdisciplinary collaboration, especially flourishing in the art world, with more and more artists and groups in Taiwan collaborating across fields to create artworks and exhibitions using diverse media. Tech art can generally be divided into two main aspects: one explores current social issues through technological interaction, while the other explores the application of new technological techniques, broadening the perspective of new technologies.

In recent years, tech art exhibitions in Taiwan have become increasingly popular. For example, in 2014, the "Traveler of Light and Shadow - Chen Cheng-Po Interactive Exhibition" was showcased at the Chiayi Municipal Museum (Chiayi City Government Cultural Affairs Bureau/Administrative Office. January 15, 2014. "Traveler of Light and Shadow Chen Cheng-Po Interactive Exhibition" - The unprecedented digital interactive exhibition of Taiwan's first indigenous artist), marking Taiwan's first large-scale interactive art exhibition featuring a local artist, Chen Cheng-Po. The exhibition divided Chen Cheng-Po's life history and works into five main themes: "Southern Sun," "Time Travel," "Exploration Theater," "Restoration Code," and "Snow Accumulation on Yushan." In 2019, the National Taiwan Museum of Fine Arts launched the "Post-Digital Human Era - International Tech Art Exhibition" (National Taiwan Museum of Fine Arts. March 8, 2019. "Post-Digital Human Era - International Tech Art Exhibition" - An in-depth observation of human consciousness in the digital generation through the relationship between "human-nature-technology"), inviting twelve artists to explore interactive installation artworks reflecting external environments and living conditions from their own perspectives. In 2021, the National Taiwan Museum of Fine Arts hosted the "Silver Lining" International Light Art Festival, featuring eight sets of light and shadow artworks (ARTCOKIDS Art Collection. March 3, 2021. "2021 Taiwan International Light Art Festival" at the National Museum of Fine Arts - Observing the transformation between darkness and light to re-examine the connection between oneself and the environment), exploring how humans observe changes in all things during the unknown night and how they connect with natural ecosystems.

2.2 *Interactivity*

The level of "interactivity" primarily depends on the degree of interaction between the communicator and the recipient, with higher levels indicating more interaction. Borsook & Higginbotham (1991) proposed seven elements that constitute interaction: Immediacy of response, Non-sequential access of information, Adaptability, Feedback, Options, Bi-directional communication, and Grain-size. Interactivity in tech art exhibitions has become one of the key elements, conveying the author's creative concepts through user experience. Immediate feedback from interactive artworks and bidirectional communication between participants in interactive modes become elements of user experience.

Interactive artworks provide audiences with a more active and participatory artistic experience, enriching the content of exhibitions and changing the way audiences interact with artworks, thereby affecting their perception and understanding of art. The interactive modes in tech art exhibitions have become a notable research area in the art world in recent years. These interactive modes encompass various technological elements, including virtual reality, augmented reality, sensor technology, interactive art installations, digital interaction, and cross-media interaction, to provide audiences with deeper participation and experiences.

2.3 *Experiential value*

In the study by Holbrook and Hirschman (1982), it was found that consumers value the experiential aspects of pleasure, enjoyment, aesthetics, and entertainment during shopping, termed as experiential value. Subsequently, scholar Holbrook (2006) categorized the values consumers experience during shopping into external, internal, self-oriented, other-oriented, active, and passive values, including efficiency, excellence, status, respect, amusement, aesthetics, ethics, and spirituality, believing that customer value is an interactive, relative, and preference-based experience.

User experience in tech art exhibitions includes interactivity, participation, interface design, usability, sensory experience, emotional connection, social experience, learning effects, feedback, and evaluation. Interactivity is crucial for participation in user experience. Well-designed interaction can stimulate active participation from the audience, enhance their connection with the artworks, and thereby improve the overall visiting experience

2.4 *Usability of interactive interfaces*

In the studies by Davis (1989) and Venkatesh & Davis (1996), perceived usefulness and perceived ease of use were found to be significantly related to user satisfaction with information systems. Mawhinney and Lederer (1990) pointed out that user satisfaction is closely related to the perceived usefulness of information systems. Usability refers to the extent to which users can easily understand, operate, and complete tasks efficiently and satisfactorily when interacting with an application, website, product, or system interface. Improving usability aims to enable users to achieve their goals easily, reduce errors, increase satisfaction, and experience less cognitive burden during interface use.

2.5 *Perceptual effects*

Perceptual effects are crucial in tech art exhibitions, involving audiences' emotional responses, perceptual reactions to artworks, and overall satisfaction with the exhibition experience. Perceptual effects include the challenge posed by artworks to viewers' concepts, perception of technological applications, and understanding of contemporary art forms. Audience perceptual effects can reflect the depth and impact of artworks.

Tech art often resonates with audiences emotionally through technological elements, leading them to engage more deeply with the exhibited content. Depending on their professional backgrounds, audiences may have different perspectives, experiences, and value interpretations of interactive artworks.

By exploring the combination of technology and art in exhibition trends and examining the interactive relationship between communicators and recipients, we discover that elements of user experience encompass perception and understanding. Perception includes the experiential aspects and emotional connections valued by users during consumption, along with the usability of interactive artworks during the experience. These factors combined influence users' perception of the overall exhibition experience.

3 RESEARCH METHOD

The present study employed both observational and questionnaire survey methods. On January 6th and 20th, 2024, the researchers visited the exhibition area of the "Diffusion Couple" at the Kaohsiung Museum of Fine Arts. Observations were conducted on the interaction between visitors and artworks in the nine main exhibition zones. Additionally, an online questionnaire was distributed on January 24th, 2024, at the exit of the exhibition.

The research focused on the differences in the perception of artworks among visitors from different professional backgrounds and proposed the following three research hypotheses:

(1) Visitors from different professional backgrounds experience different levels of experiential value in the exhibition. (2) Visitors from different professional backgrounds perceive differences in the usability of interactive interfaces in the exhibition. (3) Visitors from different professional backgrounds perceive different perceptual effects in the exhibition.

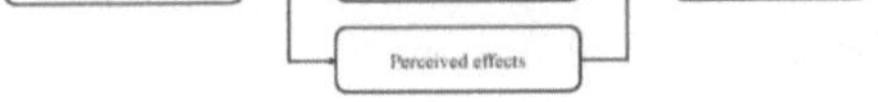

Figure 2. Research framework diagram. Developed for this study.

3.1 *On-site observation*

Systematic observations were conducted at the tech art exhibition venue, recording visitors' interactive behaviors, perceptual responses, and flow paths. Through direct observation of visitors' behaviors, reactions, and interactions at the tech art exhibition, on-site data was obtained as depicted in Figures 3 through 12. This aided in understanding the behavioral patterns and interaction modes of visitors in the exhibition space.

Figure 3. Audience interactive wall at the exhibition entrance.

Figure 4. "Unnatural" Earth Echoes.

Figure 5. Virtual Homecoming Project.

Figure 6. Future Puppetry Theater.

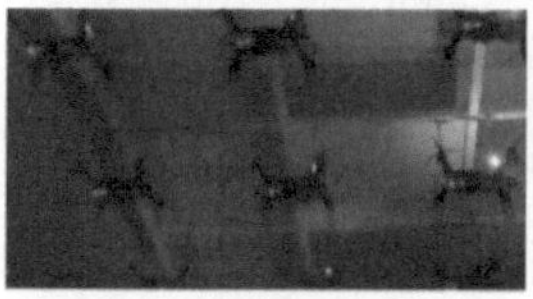

Figure 7. Weighted Beams

Figure 8. EAT Da Vinci 2.0.

Figure 9. Erhu. AI. Impressions.

Figure 10. ASOMROF.

Figure 11. Island Interpretation "Traces".

Figure 12. Embracing.

3.2 *Questionnaire survey*

Questionnaires were distributed at the exhibition exit, designed to cover aspects such as visitors' gender, age, professional field, visiting experience and experiential value, usability of interactive interfaces, and overall evaluation of perceptual effects, as shown in Table 1. The questionnaire included items coded as Q1-Q10 for experiential value, S1-S10 for usability of interactive interfaces (with S1 designed as a reverse question), and T1-T10 for perceptual effects. Responses were measured using a Likert five-point scale ranging from "Strongly Disagree" to "Strongly Agree", with scores ranging from 1 to 5 accordingly.

Table 1. Questionnaire design phase II: (Likert Scale). developed for this study.

Code	Dimension	Question
Q1	Experiential Value	I actively interacted with the exhibits during the visit.
Q2		I emotionally connected with the exhibits during the visit.
Q3		I felt the combination of contemporary technology and art during the visit.
Q4		I successfully received the design concepts and messages conveyed by the exhibitors during the visit.
Q5		I felt the artistic and creative aspects of the exhibition were very good.
Q6		The way I experienced the exhibits was pleasant.
Q7		The exhibition gave me a new sense of participation.
Q8		I believe that visiting the exhibition has impacted my personal growth.
Q9		After visiting the exhibition, I am more willing to engage in this type of exhibition.
Q10		I would recommend this exhibition to my friends or family.
S1	Usability	I find it difficult to understand the operation of the exhibits.
S2		I find the interactive nature of the exhibits addictive.
S3		I believe that the interactive experience in the exhibition enhanced my understanding of the exhibits.
S4		I found it easy when experiencing the exhibits.
S5		The interactive interfaces of the exhibits were smooth during the experience.
S6		I felt there were enough operation prompts when experiencing the exhibits.
S7		I believe the interactive interfaces of the exhibition match my usage habits.
S8		I believe the interactive exhibits provide a new way of participation.
S9		The interactivity of the exhibits added to my enjoyment of the exhibition.
S10		I received enough information and explanations in the exhibition to better understand the exhibits.
T1	Perception Effect	I felt relaxed and happy during the exhibition visit.
T2		I had profound thoughts during the exhibition visit.
T3		I felt challenged and stimulated during the exhibition visit.

(*continued*)

Table 1. Continued

Code	Dimension	Question
T4		I felt amazed or surprised during the exhibition visit.
T5		The interactive nature of the exhibition stimulated my curiosity about art.
T6		The interactive nature of the exhibition stimulated my curiosity about technology.
T7		I felt a resonance with the exhibits in the exhibition.
T8		The exhibition left me with pleasant memories.
T9		I felt excited about the development of contemporary technology and art during the exhibition visit.
T10		I believe the interactive nature of the exhibition has a significant impact on enhancing its attractiveness.

4 EXPERIMENTAL RESULT

4.1 *Descriptive statistics*

After excluding invalid questionnaires, the experimental results were analyzed based on 34 actual visitors. Among the 34 participants, there were 16 males and 18 females. Regarding the distribution of ages among participants, there were 2 individuals below 18 years old, 16 aged between 18-25 years old, 7 aged between 26-35 years old, 1 aged between 36-45 years old, and 3 aged between 46-55 years old. Additionally, there were 5 participants from natural sciences and engineering, 2 from social sciences, 3 from humanities, 1 from medical and health sciences, 5 from business and management, and 18 from arts and design. Regarding participants' experience in visiting tech art exhibitions, 23 had prior experience while 11 did not. Specific descriptive statistics are shown in Table 2.

Table 2. Descriptive statistics.

Dimension	Quantity	Percentage
Males	16	47.1%
Females	18	52.9%
Below 18	2	5.9%
18-25 years	16	47.1%
26-35 years	7	20.6%
36-45 years	1	2.9%
46-55 years	3	8.8%
56-65 years	3	8.8%
Above 65	2	5.9%
Natural Sciences and Engineering	5	14.7%
Social Sciences	2	5.9%
Humanities	3	8.8%
Medical and Health Sciences	1	2.9%
Business and Management	5	14.7%
Arts and Design	18	52.9%
Experienced Visitors	23	67.6%
Non-experienced Visitors	11	32.4%

4.2 *Questionnaire reliability analysis*

The reliability analysis results for each dimension are shown in Table 3, with reliability values ranging from .850 to .926. Therefore, the reliability of this questionnaire can be accepted.

Table 3. Reliability analysis results.

Dimension	Number of Items	Reliability
Experiential Value	10	.881
Usability of Interactive Interfaces	10	.850
Perceptual Effects	10	.926

4.3 *One-way ANOVA analysis on three major hypotheses*

This study explored three major hypotheses across different professional fields including natural sciences and engineering, social sciences, humanities, medical and health sciences, business and management, and arts and design. Through investigating whether interactive installations in the exhibition hall caused differences in experiential value among different professional fields, whether experiencing interactive exhibits caused differences in usability of interactive interfaces among different professional fields, and whether visitors from different professional fields perceived differences in the overall interactive exhibits, the one-way ANOVA results, as shown in Table 4, indicate that different professional fields did not achieve significance (p<0.05) in terms of experiential value, usability of interactive interfaces, and perceptual effects.

Table 4. One-way ANOVA analysis on three major hypotheses.

Dimension		Sum of Squares	Degrees of Freedom	Mean Square	*F*	Significance
Experiential Value Q1-Q10	Between Groups	.745	4	.186	.767	.556
	Within Groups	6.798	28	.243		
	Total	7.542	32			
Usability of Interactive Interfaces S2-S10	Between Groups	1.775	4	.444	1.243	.316
	Within Groups	9.997	28	.357		
	Total	11.772	32			
Perceptual Effects T1-T10	Between Groups	2.563	4	.641	1.917	.135
Within Groups		9.360	28	.334		
Total		11.922	32			

***p<0.05.

5 RESEARCH FINDINGS AND DISCUSSION

From the literature review, it is evident that art and technology exhibitions can be categorized into two main aspects. One aspect involves using technology to explore and address current societal issues, while the other focuses on the application of new technological advancements. Exhibitions such as "Silver Lining," "Post-digital Human Chronicles - International Art and Technology Exhibition," and "Diffusion Couple" predominantly express social issues through technological interaction, falling into the former category. On the other hand, "Travelers of Light and Shadows - Interactive Exhibition of Chen Cheng-Po's 102nd Birthday" presents new artistic forms through the incorporation of new technologies, fitting into the latter category. The "Diffusion Couple" exhibition combines both aspects, exploring how new technologies and art intersect and effectively convey social issues to visitors through interdisciplinary creations.

Observations made on-site during the "Diffusion Couple" exhibition revealed that artworks such as the interactive wall at the entrance (Figure 3), "Unnatural" Earth Echoes (Figure 4), "Future Puppetry Theater" (Figure 6), "EAT Da Vinci 2.0" (Figure 8), "ASOMROF" (Figure 10), and "Embracing" (Figure 12) actively engage users by sensing their interactions and providing immediate feedback. Visitors tended to spend 4 to 6 minutes interacting with these pieces, indicating a heightened willingness to engage. On the other hand, artworks like "Virtual Hometown Project" (Figure 5), "Weighted Beams of Light" (Figure 7), "Erhu. AI. Impressions" (Figure 9), and Island Interpretation "Traces" (Figure 11) featured performance art elements, where interaction through technology was used to narrate scenarios and discovered issues. Visitors' engagement with these pieces depended on the duration of the performances, typically lasting around 10 minutes.

Analysis of the questionnaires revealed an equal distribution of male and female visitors, with the majority falling within the 18-25 age group and predominantly from the field of arts and design. Results showed that visitors from different professional backgrounds did not significantly differ in their perceptions of experiential value, usability of interactive interfaces, and perceived effects. However, there was a slight approach to significance ($p = 0.135$) in the perceived effects dimension, although it did not meet the standard $p < 0.05$.

6 CONCLUSIONS

From a consumer perspective, analyzing experiential value and usability of interactive interfaces reveals that overall experiential perceptions do not vary significantly based on different professional backgrounds. However, from the perspective of visitor experience, there may be slight differences in preferences and levels of perception towards different interactive elements depending on the professional field.

The "Diffusion Couple" exhibition involved nine teams, resulting in nine different social issue scenarios. Visitors' resonance with these scenarios may vary depending on their understanding of the social issues, leading to different interpretations and feelings. Since this study primarily focused on overall exhibition interactions and did not conduct detailed analyses of individual artworks' interactions, visitors may experience a mixture of active and passive engagement with the artworks, experiencing them in an intertwined manner. It is recommended that future research first conducts studies comparing the interactive experiences of individual artworks and then extends the investigation to the overall exhibition experience.

REFERENCES

Borsook T. K. and Higginbotham-Wheat N. (1991). *Interactivity: What is It and What Can It Do for Computer-Based Instruction? Educational Technology,* 31: 10, pp. 11–17.

Calisir F. and Calisir F. (2004). The relation of interface usability characteristics, perceived usefulness, and perceived ease of use to end-user satisfaction with enterprise resource planning (ERP) systems. *Computers in Human Behavior*, 20(4), 505–515.

Chiayi City Government, Cultural Affairs Bureau/ Administrative Office. (2014, January 15). *"The Traveler of Light and Shadow: Chen Cheng-Po's 102 Interactive Exhibition"* - Unprecedented! Taiwan's first senior painter's digital interactive exhibition. Retrieved from https://www.chiayi.gov.tw/News_Content.aspx?n=455&s=353761

Davis F. D. (1989). Perceived usefulness, perceived ease of use, and user acceptance of information technology. *MIS Quarterly*, 319–340.

Davis F. D., Bagozzi R. P. and Warshaw P. R. (1989). User acceptance of computer technology: A comparison of two theoretical models. *Management Science*, 35(8), 982–1003.

Holbrook M.B. (2006). "Consumption experience, customer value, and subjective personalintrospection: an illustrative photographic essay", *Journal of Business Research* (59:6), pp. 714–725.

Holbrook M.B. and Hirschman E.C. (1982). "The experiential aspects of consumption:consumer fantasies, feelings, and fun", *Journal of Consumer Research* (9:2), pp. 132–140.

Hsu L., Hsu C., Tang C. and Liang C. (2010). A study on the effect of blog experiential value on users' continuous intention to use. *Journal of Information Management*, 17(4), 89–117.

Mawhinney C. H. and Lederer A. L. (1990). A study of personal computer utilization by managers. *Information & Management*, 18(5), 243–253.

National Taiwan Museum of Fine Arts. (2019, March 8). *"Post-Humanist Record: International Science and Technology Art Exhibition"* - A profound observation of human consciousness in the digital generation through the relationship of "people-nature-technology." Retrieved from https://www.ntmofa.gov.tw/News_Content.aspx?n=1384&s=66659

System Innovation for an Artificial Intelligence Era – Lam et al. (Eds)
© 2025 the Author(s), ISBN: 978-1-032-84452-7

Exploring the impact of virtual reality on satisfaction with using stress-relieving products

Ya-Chun Yang
Graduate School of Design, National Yunlin University of Science and Technology, Douliu, Taiwan
Department of Innovative Living Design, Overseas Chinese University, Taichung, Taiwan

Chao-Ming Wang
Department of Digital Media Design, National Yunlin University of Science and Technology, Douliu, Taiwan

ABSTRACT: With the impact of the global outbreak of the new coronavirus (COVID-19) and the rapid development of technology media, various stress-relieving products combined with high-tech Virtual Reality to present interactive situations have become an important topic today. The use of VR technology to lead users in Experiencing products in Virtual Reality can produce richer experience effects. This study aims to explore the application of virtual reality (VR) technology in stress relief products and analyze user experience satisfaction.

In the research process, first summarizes descriptive vocabulary and representative product samples through literature data, and extracts the 8 most representative measurement indicators and 16 evaluation criteria through expert interviews and questionnaire surveys. Then, through 8 pillow products with a stress-relieving concept, participants were allowed to use VR head-mounted displays to conduct stress-relieving product experience experiments in two virtual reality environments: nature (forest) and living room.

The research results show that compared with the past way of simply experiencing products through vision and touch, allowing users to immerse themselves in the field atmosphere constructed by VR technology to experience products can increase their concentration and sense of identity, and reduce negative emotions, especially with specific combinations. The experience is better when you are in a situation.

Keywords: Virtual Reality, Stress-Relieving Products, Situational Space, Satisfaction, Head-Mounted Display

1 INTRODUCTION

In contemporary society, individuals encounter intricate and ever-changing life pressures. According to relevant data, over half of office workers experience stress that impacts their mood. Each person adopts a unique approach to managing negative emotions and stress. Some individuals utilize relaxation techniques such as hypnosis, meditation, deep breathing, and yoga, while others seek solace and emotional relief through indulgence in sweets or by purchasing and using stress-relieving products (Chen Youzhen, 2012). When a product interacts with a user, it initially evokes perceptions and feelings through its appearance and form, harnessing positive emotions to fulfill both physical and mental needs (Mano & Oliver, 1993), thereby alleviating physical and mental stress. Numerous studies have indicated that exposure to nature can also aid in stress reduction during times of heightened stress.

Pressure permeates every aspect of life, particularly exacerbated by the global epidemic outbreak in 2020, which accelerated the proliferation of innovative applications like various stress-relief way, stress-relief products, and interactive virtual reality technology. As outlined by Burdea (1993), virtual reality comprises three key elements: interaction, immersion, and imagination, which collectively guide users towards a fully sensory immersive experience within virtual environments, facilitating interactions with products or services. Furthermore, reports from Ericsson's Consumer Behavior Research Laboratory and Industry Research Laboratory indicate that consumers perceive experiences facilitated through AR (Augmented Reality) and VR (Virtual Reality) technologies as instrumental in meeting their needs.

Building on this foundation, this study is dedicated to conducting in-depth research to investigate whether users can enhance concentration and satisfaction when utilizing stress-relieving products

DOI: 10.1201/9781003514831-15

within an immersive environment constructed using virtual reality (VR) technology.

2 OBJECTIVES

The emergence of virtual reality (VR) technology has redefined the boundaries of human interaction with space, time, and sensory perception. It enables interactive communication and immersive sensory experiences among individuals, objects, and the environment, creating a sense of immersion in the scene and encouraging physiological relaxation. For this reason, the purpose of this study is:

(a) Research to investigate whether the introduction and implementation of virtual reality technology can enhance concentration.
(b) Research to investigate whether experiencing the product in a specific environment and atmosphere will positively influence the mood and satisfaction of users of stress-relieving products.
(c) Clarify the primary factors influencing immer sion in virtual reality.

3 METHODS

This study devised the research methodology and structure in alignment with its objectives, aiming to investigate whether experiencing stress-relieving in a virtual reality environment can enhance mood and alleviate stress. Initially, a comprehensive collection of image-shaping vocabulary and representative product samples was conducted from literature, design resources, and online sources. Two professional designers were invited to conduct preliminary screening and extraction of the gathered information, which was subsequently compiled into a questionnaire. The Expert Delphi Method was employed to a questionnaire survey among, resulting in the extraction of the 8 most representative measurement indicators and 16 evaluation criteria. Subsequently, based on the data results, samples were produced, and participants were provided with VR head-mounted displays to engage in stress-reduction product experience experiments with in two virtual stress-relief environments: outdoor (natural environment) and indoor (living room).

3.1 *Screening and extraction of design elements from images*

At this stage, image-shaping vocabulary were primarily gathered through relevant literature research, resulting in the extraction of a total of 96 descriptive vocabulary "stress relief". Due to the initial abundance of image-shaping vocabulary collected, two experts with design backgrounds were invited to form a research team (as Table 1). Their task involved screening out representative pictures that matched the adjective vocabulary while eliminating inappropriate words and images, ultimately yielding 32 sets of descriptive vocabulary collected and representative product samples pictures.

Table 1. Background information of the expert panel.

Professional	professional field	job title	Gender	Working experience
A	visual communication design	brand designer	Female	Twenty
B	Product design planning	assistant researcher	Female	twenty two

3.2 *Extraction and evaluation indices of image-shaping factors*

At this stage, questionnaires and The Expert Delphi Methods were employed to assess the significance of descriptive vocabulary and representative product samples pictures derived from the aforementioned data. This process aimed to extract the most image-shaping factors and evaluation indicators, serving as the foundational framework for the subsequent stage of questionnaire design and research verification. A total of 6 experts participated in the evaluation, comprising administrative personnel, professional designers, individuals with university-level design lecturer credentials or higher, among others.

The questionnaire comprises a semi-open-ended format. descriptive vocabulary and representative product samples pictures of Stress-Relieving traits identified during initial screening were integrated into a seven-level evaluation scale. These levels range from "very different" to "strongly agree." Descriptive statistics are primarily utilized to organize and calculate the extensive data gathered from the survey sample. Following the distribution of the questionnaire, experts are invited to provide reasoned judgments and responses regarding the image-shaping factors. Upon collection of the questionnaires, SPSS software is employed to perform descriptive statistical analysis, facilitating the deletion and extraction of the acquired data (as Figure 1).

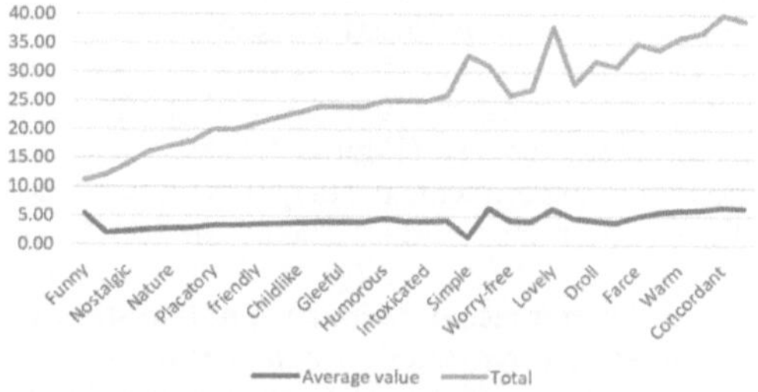

Figure 1. Results of expert evaluation of image-shaping factors.

Based on the expert evaluation results, 8 representative shaping factors indicators were identified after excluding image-shaping factors with evaluation weights lower than 5.5 (as Table 2).

Table 2. Representative shaping factors.

Lovely 6.33	Warm 6.00	Funny5.50	Concordant 6.67
Softly 5.50	悠閒的 6.50	Besotted 6.17	Comfortable 5.67

3.3 *Research verification and testing*

This chapter will make experimental sample products based on the above eight representative key indicators. Throughout the experiment, participants will utilize VR head-mounted displays with in a virtual reality environment to experience stress-relieving products in two distinct virtual scenes: "outdoor" nature and "indoor" living room (as Figure 2).

"outdoor" nature

"indoor" living room

Figure 2. Virtual scene environment image.

The questionnaire is primarily divided into two sections. The first section comprises basic information, encompassing gender, age, education, and other variables. The second section employs a Likert 7-point rating scale to administer a satisfaction perception questionnaire. The measurement indicators of the questionnaire are formulated with reference to the previous eight Representative shaping Factors indicators. The scale ranges from "1", indicating strong disagreement, to "7", indicating strong agreement. Participants in the study select their satisfaction level based on their personal perceptions.

During the experiment, a stress-relief product experience test was conducted utilizing the Meta head-mounted display (Figure 3). Prior to commencing the experiment, participants were briefed on the questionnaire and instructed on how to use the equipment. Additionally, to obtain objective and neutral reference data, participants responded to a semi-structured questionnaire. Subsequently, the questionnaire data was analyzed using independent sample t-tests with SPSS statistical software.

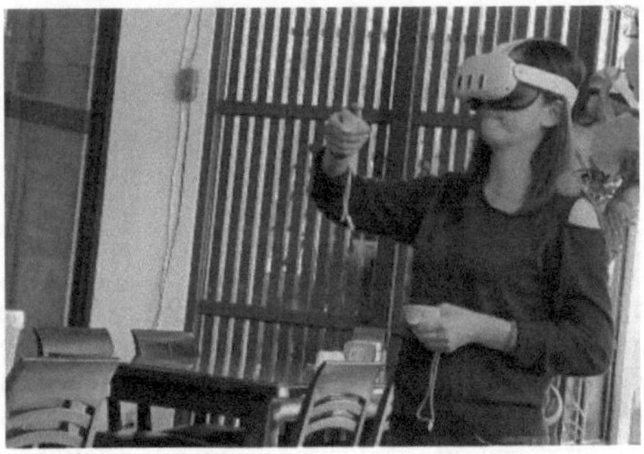

Figure 3. Participants wearing VR equipment for testing.

4 TEST RESULTS

Due to time constraints, this study employed purposive sampling to select 20 participants (13 women, 7 men) with varied backgrounds, including both design and non-design disciplines, and spanning an age range from approximately 20 to 60 years old. Following their participation in the measurements, the collected questionnaires underwent independent sample T-tests using SPSS statistical software for comparative analysis of the results.

The statistical findings indicate that when both male and female subjects engage with stress-relieving products in outdoor "nature" settings, shapes characterized by "freedom" and "leisure" are more likely to evoke positive emotions and alleviate emotional stress. Conversely, when experiencing stress-relieving products in urban "living room" environments, shapes associated with "comfort" tend to facilitate positive emotions and emotional stress relief among users (as Table 3, 4).

Table 3. "outdoor" nature.

Male	Lovely 5.86	Warm 5.71	Humorous 5.86	Concordant 6.29	Softly 6.29	Leisurely 6.86	Besotted 6.29	Comfortable 6.86
Female	Lovely 6.00	Warm 6.23	Humorous 5.54	Concordant 6.00	Softly 6.62	Leisurely 6.46	Besotted 6.00	Comfortable 6.77

Table 4. "indoor" living room.

Male	Lovely 6.14	Warm 6.43	Humorous 5.71	Concordant 6.14	Softly 6.14	Leisurely 6.29	Besotted 5.86	Comfortable 6.71
Female	Lovely 5.77	Warm 6.38	Humorous 5.46	Concordant 5.85	Softly 5.85	Leisurely 6.38	Besotted 5.77	Comfortable 6.62

In summary, the aim of this study is to investigate whether the introduction and utilization of virtual reality (VR) technology can enhance positive emotions and satisfaction, and to identify the influencing factors.

Based on the aforementioned statistical findings, it is evident that during experimental trials, whether in an "outdoor" natural setting or an "indoor" living room environment, individuals tend to experience greater comfort and stress-relieving when utilizing products with "comfortable" features. Furthermore, when exposed to stress-relieving products, the integration of VR technology facilitates users in immersing themselves within the spatial atmosphere, thus enhancing their overall experience. Wu Peifen (2016) also emphasized that immersive experiences entail specific technologies and activities, fostering heightened engagement and enjoyment. This enables users to amplify their focus during product experiences, enrich their sensory perceptions, and induce a state of physiological relaxation.

Additionally, some test subjects found the overall experience refreshing and engaging, especially since they were new to virtual reality, and expressed great appreciation for the novel experience of interacting with the product in this way.

5 CONCLUSIONS

From the perspective of the experience economy, integrating virtual reality experiences into stress-relieving products can introduce novel sensory encounters to users, thus enhancing their satisfaction with product usage.

REFERENCES

Bielinis, E., Takayama, N., Boiko, S.,Omelan, A., & Bielinis, L. (2018). The effect of winter forest bathing on psychological relaxation of young Polish adults. *Urban Forestry & Urban Greening*, 29, 276–283. https://doi. org/10.1016/j.ufug.2017.12.006

Brenton, H., Hernandez, J., Bello, F., Strutton, P., Purkayastha, S., & Firth, T. (2007). Using multimedia and Web3D to enhance anatomy teaching. *Computers & Education*. 49(1), 32–53.

Burdea, G., (1993). Virtual reality systems and applications. *Paper presented at the Electro'93 Intemational Conference*.

Chen Youzhen (2012, December 16). Muppets and chocolate are selling well despite the trend, and healing business opportunities are booming. *China Times*, page A6.

Hartig, T., Evans, G. W., Jamner, L. D., Davis, D. S., & Garling, T. (2003). Trackingrestoration in nature and urban field settings. *Journal of Environmental Psychology*.

Kaplan, S. (1995). The restorative benefits of nature: Toward an integrative framework. *Journal of Environmental Psychology*, 15(3), 169–182.

Maas, J., Verheij, R. A., Groenewegen, P. P., De Vries, S., & Spreeuwenberg, P. (2006). Green space, urbanity and health: how strong is the relation? *Journal of Epidemiology and Community Heatth*, 60(7), 587–592.

Mano, H., & Oliver, R. L. (1993). Assessing the dimensionality and structure of the consumption experience: Evaluation, feeling, and satisfaction. *Journal of Consumer Research*, 20(3), 451–466.

Nielsen, T. S., & Hansen, K. B. (2007). Do green areas affect health? Results from a Danish survey on the use of green areas and health indicators. *Health & Place*, 13(4), 839–850.

Sonntag-Öström, E., Nordin, M., Lundell, Y., Dolling, A., Wiklund, U., Karlsson, M.,Carlberg, B., & Järvholm, L. S. (2014). Restorative effects of visits to urban and forest environments in patients with exhaustion disorder. *Urban Forestry Urban Greening*, 13(2), 344–354. https://doi.org/10.1016/j.ufug.2013.12.007

Staats, H., & Hartig, T. (2004). Alone or with a friend: A social context for psychological restoration and environmental preferences. *Journal of Environmental Psychology*, 24(2), 199–211.

Tsunetsugu, Y., Lee, J., Park, B. J., Tyrväinen, L ., Kagawa, T., & Miyazaki, Y. (2013). Physiological and psychological eff ects of viewing urban forest landscapes assessed by multiple measurements. *Landscape and Urban Planning*, 113, 90–93. https://doi.org/10.1016/j.landurbplan.2013.01.014

Van den Berg, A. E., Koole, S. L., & van der Wulp, N. Y. (2003). Environmental preference and restoration. *Journal of Environmental Psychology*, 23(2), 135–146.

Villeneuve, P. J., Jerrett, M., Su, J. G., Burnett, R. T., Chen, H., Wheeler, A. J., & Goldberg, M. S. (2012). A cohort study relating urban green space with mortality in Ontario, Canada. *Environmental Research*, 115, 51–58.

White, M. P., Yeo, N. L., Vassiljev, P., Lundstedt, R., Wallergård, M. ,Albin, M., & Lõhmus, M. (2018). A prescription for "nature"—The potential of using virtual nature in therapeutics. *Neuropsychiatric Disease and Treatment*, 14, 3001–3013. https://doi.org/10.2147/ NDT.S179038

Wu, P. F. (2016). Research on user emotions and immersive experience of characteristic neighborhood navigation system.

Wu, Y. J. (2012). *Understanding is the First Step to Happiness*. Tianxia Magazine, 489, 26–28.

Yu, C. P., Lee, H. Y., & Luo, X. Y. (2018). The effect of virtual reality forest and urban environments on physiological and psychological responses. *Urban forestry & urban greening, 35*, 106–114. https://doi.org/10.1016/j.ufug.2018. 08.013

Zhang Chunxing (1983). *Psychology*. Taipei: Tung Hwa Book Company.

Zhou Wenqin, Sun Minhua, and Zhang Decong (2010). *Stress and Life*. Taipei City: Psychology.

System Innovation for an Artificial Intelligence Era – Lam et al. (Eds)
© 2025 the Author(s), ISBN: 978-1-032-84452-7

Initial investigation on more elderly-friendly gesture technology of viewing photos on smart phone

Ming-Hong Wang
Department of Visual Communication Design, Southern Taiwan University of Science and Technology, Tainan, Taiwan

Chih-Yuan Cheng
Institute of Financial & Economic Law, Southern Taiwan University of Science and Technology, Tainan, Taiwan

Chian-Fan Liou
Department of Visual Communication Design, Southern Taiwan University of Science and Technology, Tainan, Taiwan

ABSTRACT: As the population of aged people grows, the use of smart phone device is also growing in the elderly age group. Through study and research, we try to discuss a more elderly-friendly operation way of smart phone to reduce their frustration and improve their willingness of using a smart phone. This study starts from the gestures applied on the smart phone for viewing photos. Through the comparison between new and old gesture design, we can verify the design performance. Lastly, a conclusion is decided through the discussion with the product design professionals. In terms of new gesture development, the data shows that Gesture B is more user-friendly to the elderly than the traditional gesture. Through final interviews with the professionals and documentary comparison, 5 points indicate that Gesture B has better advantage for the elderly users than the traditional setting: 1. The interface is designed based on higher instinct, 2. Less fingers involved operating the phone, 3. The control bar is more obvious, 4. The reminder in images or texts is bigger, 5. Minimizes inadequacy of cognitive difficulty. On the other side, the user-unfriendly design elements are as follows: 1. Lacks past experience to refer to for operation. 2. A big change from traditional device setting. 3. Lacks visional clues or functions are hidden, 4. Unable to tell symbols and operation goals, 5. Two finger operation at the same time, 6. Confusion between the body reaction and cognitive operation.

1 INTRODUCTION

As the wireless internet is more popular and common, the development of smart phone is more advanced compared to the simple function of the traditional phones. The development of smart phone development has been more mature and diverse in functions, including photography, internet, GPS, social media and streaming media etc. It is even designed to control TV, washing machine, monitors and all types of home appliances. As the public relies on smart phone for longer usage, it has become a necessity on everyone's daily life. Information has developed and transmitted quickly, compared to the elderly, the teenagers and children usually adapts and learns quicker. Therefore, the elderly people are called "digital immigrants", which means they are the disadvantaged group who face more challenges to become digital users (H.-J. Lai 2016), such computers and smart phones in their real life. There are more and more aged people who show willingness or are forced to learn to use smart phones. Although many users show that they have difficulties or have no ideas how to use them, the number and rate of smart phone users and usage is growing fast every year. From the research in 2017, we know that the users in the group over age of 60 has a highest growth rate (Y.-C. Chen and C.-H. Kuo 2019).

According to the statistics of national development committee, about 89.9% of the people at the age between 60 and 64 use smart phones, whereas 63.2% of the age above 65. From the situation the daily internet communication and social media usage is close to 80% of their time and still fast growing (H.-J. Lai 2022). Many researches indicated that using a new technology is a challenge for the aged people. Compared to traditional computers, smart device has a touch interface for easier and intuitive operation. It is a very popular device with diverse APP available, which makes it more attractive for the elderly to learn and use (H.-J. Lai 2016; H.-J. Lai 2018). Undeniably as the

 DOI: 10.1201/9781003514831-16

technology evolves, smart phones are designed to be smaller, lighter with more and more additional functions and hidden interfaces and buttons. For most aged people, it is a big challenge for them to learn all about these functions (S.-W. Chang and K.-K. Lee 2002). As they age, their vision, hearing, physical and mental conditions all go down hill with a dramatic change in physical function, especially bad in vision, body control and cognitive operation. (P.-C. Yeh *et al.* 2020). If they encounter frustrations, fading memories and reduced attentions, the overloading of new knowledge, lack of guidance and teaching, anxiety and insufficient confidence will lead the aged people to give up early learning a new technology. Therefore, designing an easier and user-friendly method to reduce the elderly's frustrations is an important direction for helping them mentally and physically. (Cristiano, A. *et al.* 2022; Park, K.-H. *et al.* 2008; C.-Y. Hu, and L.-C. Chen 2013).

Until the end of 2050, the WHO estimates the population of the aged people is going to exceed 2 billion. The Population Reference Bureau in the US also pointed out in 2011 that the population of the people over 65 years old is going to be 2.5 times of that of young children under 4 years old in 2050 (H.-H. Ho *et al.* 2020; T.-H. Chiu *et al.* 2014). As the aged people's population grows, plus the smart phone touch interface is more user-friendly than traditional computers, the documents quoted in this article also pointed out that the population and usage time of aged people using smart phones will continue to grow in the future. After they learn to use smart phones, their satisfaction for life, social life and family connections will improve better. How to make it easier for aged people to learn to use smart phones, reduce their frustration and loading of learning a new technology is an important direction for efforts. In the past, most documentaries undertook the studies towards the setting, styles, colors, texts and font sizes of digital interface, with very little information about the improvement on gestures of operating a smart phone for aged people. Gestures are a powerful and natural tool for communications between people.

However, if it requires a complicate gesture or operation on the interface, due to their degenerated physical condition, it is going to be a difficult challenge for the elderly people. The scholar T.-L. Wong (2010) brought up the idea that the difficulty to use a touch interface must be reduced in order to facilitate the use for the aged people, and yet she didn't mention which operation is most suitable (T.-L. Wong 2010). Trewin et al. pointed out that although it is easy to operate physically, it is rather challenging and difficult for the elderly people to visually and cognitively operate it. Therefore how to study on gestures and interfaces to facilitate the elderly people to use is an important and primary design mission that has been rarely studied. The documents mentioned above talking about gestures and operations also stated that a clear control bar without repetitive clicking is more intuitive to the elderly people. However, the people who can use two fingers to control the screen are only 50% of total. There is nearly none of them who can accomplish 3 finger operation on the touch screen (S Trewin, *et al.* 2013; N.-T. Wang, and C.-M. Cheng 2015). Hence, from the document mentioned above, we can find the possibility and direction of design that reduces operational difficulty and provides the elderly with a more user friendly smart phone. This is the purpose of this research and test.

The key part of this research is on the gesture, which is through Focus Group to find out the improvement on difficult gesture. At the end of research, we found out that the regular functions, such as photo zoom in and zoom out is rather difficult for the elderly to operate, which is also a high frequency of use. This responds to the research content of C.-Y. Hu, & L.-C. Chen: the multiple touch using two fingers and the expanding and closing of two fingers to view pictures done by the people who did the tests. The middle aged or older people might have to spend more time to complete the image viewing action. Those basic operation actions that seem not difficult for regular users could be difficult for the aged people. This also causes the frustrations and retreat during the use (C.-Y. Hu, and L.-C. Chen 2013).

From the past research, we can see the biggest purpose for aged people to use a smart phone is to have a social life and to communicate with their family. Therefore there is a high usage rate for them on social media websites, and communication software. The social media tool usage rate in Taiwan is as high as 97%, which means only 3 people out of 100 have never used social media. The social function, like sharing photos and activities on Facebook is also very popular to the middle aged or older people. The age group over 65 years old also are a growing user group (T.-H. Tsai 2014). This is also a healthy support and connection to the world for the elderly living alone in the remote areas or with disabilities. In Taiwan, the research in the past also indicated that the elderly people not only had interests in learning Facebook, but also Line, especially using the stickers on Line. They love using stickers, taking photos and they hope they can make their own " Morning Greeting" stickers based on the smart phone practice lessons for the elderly in the past (H.-J. Lai 2016; H.-J. Lai 2022). It is common to see that the elderly people use a lot of pictures, photo sharing and viewing just as much as typing texts. If we can make this operation more user-friendly or simple, it will help the elderly view images and transmit them more effortlessly on the internet.

2 METHOD

This research is divided in 4 parts: Stage 1. We use Focus Group to find out the difficult gesture that is not easy for the elderly. Stage 2: We develop the new gestures and do some adjustments, focusing on the gesture of only using 1 finger. Stage 3 is going to do a Usability Test to compare the difference between traditional gestures and new gestures. Stage 4 is going to analyze and discuss the comparison between the interviews with the designer and the documents.

2.1 *Discuss the gestures that are difficult for the elderly*

This study took 5 best selling smart phone brands in Taiwan, HTC, Samsung, Sony, Apple, ASUS for samples to locate 5 people who received the test. The people who are over 60 years old and have had experience using over 1 year are the testers. They will engage into the basic operation of the 5 brands of smart phones for about 30 minutes. Through interviews and observations, we can find the rather difficult gestures for the elderly. We found out that the rather difficult gestures for the elderly operating smart phones on the market are zoom in and zoom out the photos using two fingers at the same time, as shown in Figure 1 (Gesture A). This stage also corresponds to the mentioning in the past documents: The higher number of fingers involved at the same time, the more difficult it is for the elderly regardless if they are cognitively or physically coordinate.

Figure 1. Gesture A: traditional gesture using 2 fingers to zoom in and zoom out (two fingers holding the screen at the same time to drag the image around).

2.2 *Development and design of gestures*

Figure 2. Gesture B: The figure on the right shows the control bar on the scale, which can be done by 1 single finger. When the finger swipe up, the image zooms in, if down, it zooms out.

Figure 3. Gesture C: Through spinning of image, one can use one single finger to spin the image clockwise to zoon in the photo or spin anti-clockwise to zoom out the photo.

The research team and product designers discuss then develop gestures. Based on the two finger operation on traditional smart phones for zoom in and zoom out, this can be improved for more user-friendly to the elderly. Through different collections and creations of ideas, we invited 10 elderly people for test. In the end, we concluded two operation methods as below, and they are both one single finger touch operation, which improves the traditional two finger operation. They are shown respectively as Figure 2 (Gesture B) and Figure 3 (Gesture C):

2.3 *Assessment on the usability of new and old gestures*

Through 5 indexes of usability assessments in the research: Learnability, Efficiency, Memorability, Errors, Satisfaction, we collect data feedback using 5-Point Likert Scale over 30 people at 55 years old or above who received the test. When the test starts, the researcher will explain different operations of gestures that have an effect on zooming in and out the image. We asked the testers to undertake this operation and answer the questions after completing the action. The data assessed through 5 indexes will be collected. The highest score is 5 and the lowest is 1.

(a) Learnability: Easy, simple, convenient and fast for learning.
(b) Efficiency: After learning and practice, it is easy to understand with efficiency.
(c) Memorability: After a period of time, it is not difficult to forget.
(d) Errors: Minimize the operation mistakes that users make.
(e) Satisfaction: Satisfaction level towards the whole operation experience.

3 ANALYSES AND DISCUSSION

There are three different groups for gesture testing on viewing images on smart phones. They all used fingers to zoom in and zoom out photos. The gestures are as follows:

- Gesture A: Traditional Gesture (Two fingers pressed the monitor at the same time and drag the picture around).
- Gesture B: New Gesture (Single finger to press on the right control bar to move up and down).
- Gesture C: New Gesture (Single finger to press on the central block to clockwise or anticlockwise).

At this stage, we used Independent Sample T-Test to compare between two new gestures B, C and traditional gesture A. We tried to find out a better solution for traditional gesture from the two newly developed gestures. After the analyses, the result is shown as Table 1 (A and B) and Table 2 (A and C):

Table 1. Usability of gesture A and B using T-test (*P-value <0.05).

	a. Learnability Mean (S.D)	b. Efficiency Mean (S.D)	c. Memorability Mean (S.D)	d. Errors Mean (S.D)	e. Satisfaction Mean (S.D)
A	3.80 (±0.41)	3.43 (±0.63)	3.40 (±0.50)	3.37 (±0.67)	3.13 (±0.73)
B	4.40 (±0.67)	4.30 (±0.65)	4.00 (±0.64)	4.30 (±0.47)	4.70 (±0.47)
P	0.000*	0.000*	0.000*	0.000*	0.000*

Table 1. Through the comparison between Gesture A and Gesture B using T-test for usability over 5 items, Gesture B in the new design all have a higher score than Gesture A in the traditional design in all the 5 items. It means Gesture B has better learnability, Efficiency, memorability, less error, and higher satisfaction for the elderly people to view images by zoom in and zoom out. Overall, it has better values.

Table 2. Usability for Gesture A and Gesture C using T-test (*P-value <0.05).

	a. Learnability Mean (S.D)	b. Efficiency Mean (S.D)	c. Memorability Mean (S.D)	d. Errors Mean (S.D)	e. Satisfaction Mean (S.D)
A	3.80 (±0.41)	3.43 (±0.63)	3.40 (±0.50)	3.37 (±0.67)	3.13 (±0.73)
C	3.30 (±0.65)	3.50 (±0.51)	3.80 (±0.76)	3.20 (±0.41)	3.80 (±0.76)
P	0.001*	0.652	0.020*	0.249	0.001*

Table 2. Through the comparison between Gesture A and Gesture C in Table using T-test over 5 items in usability, Gesture C in the new design is more difficult to learn than Gesture A in the traditional design with a obvious difference. However, in the aspect memorability, Gesture C has better performance than Gesture A with a significant difference. In the aspects Efficiency and error, there is no significant difference between Gesture A and C. Therefore, there is no need to discuss them in these two aspects.

To compare and analyze the comparison results of usability above, new Gesture B (Single finger pressing on the right control bar to move up and down to zoom in and out the image) is a more successful design. Lastly, this research assessed the result by consulting with the product designers for the data. Through interviews, it discussed why there is a difference in usability between Gesture B and Gesture C for the elderly. To summarize the opinions of product designers, they are as follows: New Gesture C conducts clockwise and anticlockwise movement to zoom in and out images. Although this removes the difficulty for 2 finger operation simulatiously, it is still insufficient for instinct action. Also it is not acknowledged as a reflex action. Normally "spinning" is recognized as a skill to move or spin the picture, not to zoom in and zoom out the picture. Although the instructions and explanations have been given to the users before operation, due to a longer time to initiate the cognition, this design is not considered better than traditional gesture A by significance. As for the success of Gesture B, the designer also points out that the control bar on the side might block the image on the screen. This operation elements are similar to the volume control on the panel bar in traditional TV setting. Although they have different purposes and functions, the control bar movement on the scale is a instinctive and direct reminder for the elderly. This enables the elderly to pull the bar naturally. On the other hand, it also has a more positive feedback on the result. In the future design, this traditional habitual recognition might be transplanted to the touch panel devices as a way for the elderly to learn and adapt to.

4 CONCLUSION AND ADVICE

This article has studied the more difficult action" two fingers pressing the screen to view the image by zooming in and out "with a further approach to improve it to single finger touch. As last, two user operation modes are selected for usability test. Through T-test, the comparison between traditional and new gestures can be done. Finally we can get a conclusion from the designer through the interviews. The whole process includes problem finding, design solving, quantitative tests, qualitative discussions and conclusions.

The less finger operation is used to lower the difficulty of operation. Among all the newly developed gestures, Gesture B is more successful. The main reason is. "the control bar for operation is intuitive and similar to the traditional TV cognitive operation. The failure reason for gesture C is "The operation lacks instinctive action and it doesn't have an obvious control panel." There are things in common between the study results and past study documents. The user-friendly features for the elderly in the control panel are: (1) the more intuitive touch panel design, (2) less finger operation involved, (3) easy to see control bar, (4) the image or text reminder are bigger in fonts, (5) reduce the inadequacy due to cognition change (T.-L. Wong 2010; S Trewin, *et al.* 2013; N.-T.

Wang, and C.-M. Cheng 2015; P.-C. Yeh *et al.* 2020). On the contrary, for the elderly to operate the smart phones, the unfriendly operation gestures include: (1) new operation without experience to refer to, (2) a significant different arrangement from the traditional control panel, (3) lacks visional clues or the functions are hidden, (4) unable to identify symbols and operation goal, (5) two finger operation at the same time, (6) difference between body reaction and cognition (C.-Y. Hu, and L.-C. Chen 2013; T.-H. Chiu *et al.* 2014; P.-C. Yeh *et al.* 2020).

Lastly, as the population of aged people grow and the smart devices develop swiftly, this design area has a high economic value and thus is worth more studies. Plenty of similar area-related researches have been done to discuss the smart device use for the elderly, which keeps them active in the medical platform, social platform, communication software. This lifestyle will have a positive impact on their mental health. As long as the threshold bar is lowered for the elderly to enter the digital world, it will have a positive help to the overall social development. This research only discussed gesture design and assessment on image viewing for the start. It's advised to use more interviews and experiments in the future to improve the design for more diverse and accurate content.

REFERENCES

Chang S.-W., and Lee K.-K., (2002). Studies of design considerations for human needs in high tech products using mobile handsets for senior citizens as an example. *Journal of Plastic Arts*, 357–368.

Chen Y.-C., and Kuo C.-H., (2019). Blending culture between generations: effectiveness of learning smart phones through museum artifacts and courses by elders–take "Telecom for Seniors" *Project as an Example. Technology Museum Review*, 23(4), 71–91.

Chiu T.-H., Chiang I-J., Huang H.-B., and Wang C.-Y., (2014). Prototype development and technology acceptance survey on glasses-attached mobile reading aids for senior citizens. *Journal of Library & Information Science*, 40(2), 77–92.

Cristiano, A., Musteata, S., De Silvestri, S., Bellandi, V., Ceravolo, P., Cesari, M., Azzolino, D., Sanna, A., and Trojaniello, D., (2022). Older adults' and clinicians' perspectives on a smart health platform for the aging population: design and evaluation study. *JMIR Aging*, 5(1), 1–16.

Ho H.-H., Wang L.-T., and Chang S.-H., (2020). Use of a technology acceptance model to explore the smart phone exercise application needs of middle-aged and older adults. *Journal of Gerontechnology and Service Management*, 8(2), 137–147.

Hu C.-Y., and Chen L.-C., (2013). Research on the frustration experiences of middle-aged and senior people while interacting with small touch screens-using overhead iphone as example. *Journal of Taipei University of Marine Technology*, 6(1), 112–124.

Lai H.-J., (2016). E-learning course design for older adults: Analysis of instructors' perspectives. *Journal of Gerontechnology and Service Management*, 4(3), 367–375.

Lai H.-J., (2018). Exploring the relationship between factors affecting internet behavior and life satisfaction of older adult learners with use of smartphone. *International Journal on Digital Learning Technology*, 10(4), 85–105.

Lai H.-J., (2022). Integrating digital media literacy into courses for older adults: an analysis of instructors' perspective. *Journal of Gerontechnology and Service Management*, 10(3), 125–138.

Park, K.-H., Lee, H.-E., Kim, Y., & Bien, Z. Z., (2008). A steward robot for human-friendly human-machine interaction in a smart house environment. *IEEE Transactions on Automation Science and Engineering*, 5(1), 21–25.

Trewin S., Swart C., and Pettick D., (2013). *Physical Accessibility Of Touchscreen Smartphones.* Proceedings of the 15th International ACM SIGACCESS Conference on Computers and Accessibility, 1–8.

Tsai T.-H., (2014). Design and development of an online platform and innovative social networking services for older people. *Journal of Gerontechnology and Service Management*, 2(3), 281–296.

Wang N.-T., and Cheng C.-M., (2015). Usability of paging on smartphones for the middle age and elderly. *Journal of Gerontechnology and Service Management*, 3(1), 5–14.

Wong T.-L., (2010). *A Study on the Use of Multi-touch Mobile phone for the Elderly-Using iPhone 3GS And HTC Hero for example.* Graduate dissertation, Department of Arts and Design, National Taipei University of Education.

Yeh P.-C., Ku C.-J., Lan W.-S., and Lian J.-L., (2020). Research on the usability of traditional and responsive web pages in mobile devices operated by the older adults. *Journal of Gerontechnology and Service Management*, 8(3), 219–234.

System Innovation for an Artificial Intelligence Era – Lam et al. (Eds)

The application of method acting in story writing instruction

Shih-Chieh Liao, Chian Fan, Liou Rou & Hua Zeng
Southern Taiwan University of Science and Technology, Tainan, Taiwan

ABSTRACT: This study integrates the concept of method acting into the teaching content of story writing. Method acting refers to acting techniques applied within the performance system, which, through training, assist actors in mastering roles and flexibly interpreting them.

Through observational methods and drawing from life experiences, students practice transforming their experiences into stories. They learn to deepen character development, enhance emotional expression, enrich scene description, and improve the fluidity of plot arrangement, cultivating writing abilities and creativity.

The objectives of this study are: (1) to explore whether incorporating the concept of method acting can enhance students' ability to develop characters, and (2) to investigate whether students find writing easier after learning method acting.

Over the course of eighteen weeks, combined with assignment exercises, students are guided through creative writing. Finally, cognitive load scales, intrinsic motivation scales, and learning mode satisfaction are used to understand students' satisfaction and willingness to use this teaching and learning approach in the future.

Keywords: Method acting, storytelling, teaching

1 INTRODUCTION

This study aims to integrate the concept of method acting into the classroom, allowing students to observe themselves and the people and things around them. This approach helps students explore various aspects of life, which serve as ingredients for storytelling. While understanding the world from their own perspectives, students are encouraged to consider viewpoints and perspectives of others. Through perspective shifts and repeated reflection, students practice narrating events from different angles. This not only enhances character development but also reflects the multifaceted nature of characters and society, thereby creating rich storylines. Just as the world is not simply black or white, characters are not purely good or evil. By delving into the inner world of characters, students gain a better understanding of their backgrounds, motivations, and emotions, thereby creating characters with depth and authenticity, imbuing them with a soul.

In these exercises, students draw inspiration from life rather than relying solely on imagination to create characters, scenes, and plots. In addition to effectively training narrative skills, students also practice crafting story structures, deepening character development, enhancing emotional expression, enriching scene descriptions, and improving plot fluency, thereby fostering writing skills and creativity.

2 LITERATURE DISCUSSION

In theatrical performance, to create vivid character portrayals for the audience, it's essential to navigate the balance between reality and fiction, resolving the contradictions between the two (Li, Xiang, 2021). Many actors employ method acting to bring characters to life. Beyond understanding the character's personality and life as outlined in the script, they must also be adept at observing behaviors and reactions of various types of people in real life and have a basic understanding of psychology. This concept can similarly be applied to writing, where, unlike actors passively interpreting roles, writers can actively select and create characters. Thus, aside from studying psychology, observation is the most rapid and direct method. By observing people and things nearby, writers can draw inspiration from life and understand how different individuals behave and react in various personalities and situations. With this as a foundation, students can consider how the characters they create should behave in the story to advance the plot.

This method of performance was established by the Russian theater artist Konstantin Stanislavski and has been widely applied in the theatrical world, profoundly influencing modern drama and performance. Within the method, concepts such as emotional memory, emotional transfer, emotional expression, and inner motivation are core

DOI: 10.1201/9781003514831-17

principles, serving as cornerstones for drama and performance education.

Deci, Cascio, and Krusell's cognitive evaluation theory also emphasize that an individual's behavior is influenced by different reasons and goals within the context of the environment, distinguishing various motivational patterns (Deci, E. L., Cascio, W. F., & Krusell, J, 1975). Whether in learning or creative endeavors, it's essential to create environments that are conducive to enhancing students' motivation.

3 RESEARCH METHODS

This study applies this concept to the storytelling class, targeting second-year undergraduate students in the Design College. The teaching objective is to cultivate students' text writing abilities. The instructional content includes story text structure, character development, scene description, and dialogue techniques, culminating in the production of a complete story by the end of the semester. Initially, the story will be deconstructed using story cubes to swiftly grasp the key elements composing the story and utilize mind maps to practice extending abstract theme concepts into concrete objectives, gradually constructing a complete narrative.

Character design will involve analyzing common character archetypes and personality traits using character cards. Through observational methods, students will observe individuals with similar traits in their surroundings, recording observed behaviors and situations as reference. Detailed character traits and behavioral features will be constructed for the characters based on this data. Subsequently, using association methods within a time limit of five to ten minutes, students will swiftly jot down subconscious associative words for the characters. Based on these impressions, students will write the character's autobiography, using first-person self-disclosure to not only clarify the character's external features but also delve into the internal causes, motivations, and purposes behind these external features. Through this writing process, students will gain a deeper understanding of the characters, and the autobiography will serve as a record and reference for subsequent creations to prevent issues such as character behavior deviating from the established setting.

In the story, dialogue will be a vital element in advancing the plot. In class, dialogue technique analysis will be conducted using sample works to understand the impact of different dialogue styles on the plot and reading experience. Through recording everyday dialogue styles, students will comprehend the differences between life and story dialogues. Dialogue also serves as a crucial means to showcase character personalities. Therefore, students will practice simulating dialogues in different scenarios with various character personalities, enhancing the application of character impressions and dialogue techniques for appropriate use in embedding plot twists or advancing the storyline.

Plot development will focus on analyzing and practicing conflicts and turning points in the story. Many stories tend to become predictable due to the excessive use of existing patterns, often leading readers to anticipate the ending from the beginning. In such cases, unless the writing is exceptionally skilled, readers are likely to lose interest in continuing. Creating appropriate conflicts within existing patterns to propel the plot to climax and arranging turning points to write an unexpected ending can create suspense, arousing readers' curiosity about the subsequent storyline.

This course will employ a threefold approach encompassing online, offline, and physical methods. In physical teaching, students will be encouraged to share their creative experiences and ideas. Teachers will incorporate narrative methods based on life experiences to make theoretical knowledge practical, enhancing students' understanding of classroom content. Simultaneously, offline and online discussions will be conducted to address students' creative doubts. Online platforms, such as establishing class communities, will provide spaces for sharing works, facilitating post-class discussions with teaching assistants, saving waiting time, and swiftly resolving dilemmas.

For assessing effectiveness, student learning outcomes will be evaluated through grades, anonymous questionnaires, and student interviews. Grades will primarily be based on assignments, discussions, and reports, evaluating student participation in discussions, assignment completion, and report content. Students will be asked to fill out questionnaires based on the teaching method of the semester, followed by interviews to understand their actual thoughts and learning experiences regarding the course.

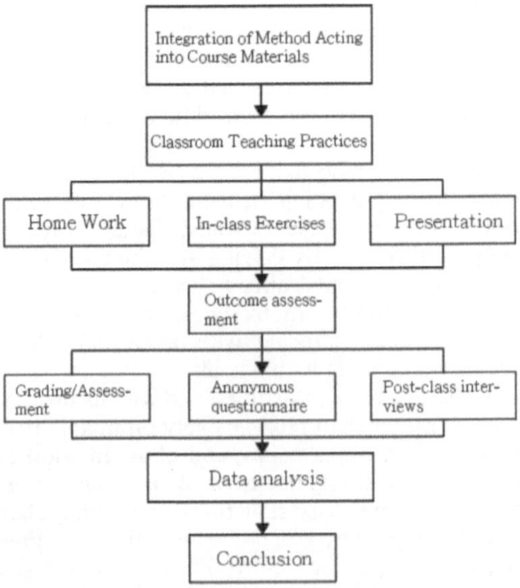

Figure 1. Research framework diagram.

The research participants for this study were students enrolled in the current academic year, divided into three classes, with approximately 108 students in total after excluding those on leave or discontinuing. The gender ratio was approximately 4:6, with an average age of 19 to 20 years old. The study evaluated its effectiveness through observations of classroom situations, assignment practices, in-class and post-class discussions, and the comprehensive assessment of end-of-term outcomes. Classroom situations were primarily assessed based on the students' focus, interaction frequency, and attendance over eighteen weeks, including student inquiries, responses, and sharing. There were a total of eleven assignments, evaluated based on submission and accuracy. Two in-class discussions and approximately five post-class discussions were held, with teaching assistants meticulously recording student participation, frequency, and discussion content to gauge improvement and corrective measures taken. End-of-term outcomes were evaluated based on the content, cover, back cover design, and layout of the books created by students. Post-semester interviews and questionnaire surveys were conducted. Three assessment methods (grades, interviews, and questionnaires) were employed to measure student learning outcomes.

Random sampling was used for the questionnaires among enrolled students, employing the Cognitive Load Scale to assess their thoughts on the teaching methods used during the academic year, aiming to understand their acceptance and learning effectiveness of these methods. Additionally, the Internal Motivation Scale and Learning Mode Satisfaction Scale were utilized to gauge students' satisfaction and willingness to use these teaching and learning methods in the future.

Table 1. Cognitive load reliability statistics.

Reliability Statistics	
Cronbach's Alpha	N of Items
.711	8

Table 2. Assignment grading criteria.

Item-Total Statistics

	Scale Mean if Item Deleted	Scale Variance if Item Deleted	Corrected Item-Total Correlation	Cronbach's Alpha if Item Deleted
In order to understand the content of this semester's textbook, many things need to be considered at the same time.	29.2143	38.660	.606	.634
The content of these learning materials is very complex.	30.2143	38.709	.658	.624
It's hard to find important information in the process of studying these materials.	30.0238	37.146	.641	.623
The design of this teaching material makes learning/creating inconvenient.	30.7143	40.502	.483	.664
It is challenging to identify and link important information while learning from these teaching materials.	30.5238	38.743	.595	.637
In the process of learning/creating, I not only try to understand the details of the content, but also try to master its structure.	28.5952	49.954	.152	.727
I am able to correctly understand all the content through the practice process of learning materials.	28.6190	53.217	-.016	.753
Using method acting and observation techniques allowed me to better understand the themes of the story writing course.	28.5952	51.661	.055	.744

Table 3. Descriptive statistics of cognitive load.

Descriptive Statistics					
	N	Minimum	Maximum	Mean	Std. Deviation
In order to understand the content of this semester's textbook, many things need to be considered at the same time.	42	1.00	7.00	4.5714	1.72701
The content of these learning materials is very complex.	42	1.00	7.00	3.5714	1.62515
It's hard to find important information in the process of studying these materials.	42	1.00	7.00	3.7619	1.81873
The design of this teaching material makes learning/ creating inconvenient.	42	1.00	7.00	3.0714	1.78603
It is challenging to identify and link important information while learning from these teaching materials.	42	1.00	7.00	3.2619	1.73991
In the process of learning/ creating, I not only try to understand the details of the content, but also try to master its structure.	42	2.00	7.00	5.1905	1.34777
I am able to correctly understand all the content through the practice process of learning materials.	42	1.00	7.00	5.1667	1.32364
Using method acting and observation techniques allowed me to better understand the themes of the story writing course.	42	2.00	7.00	5.1905	1.38349
Valid N (listwise)	42				

Table 4. Reliability statistics of intrinsic motivation.

Reliability Statistics	
Cronbach's Alpha	N of Items
.918	19

Table 5. Overall statistics of intrinsic motivation.

Item-Total Statistics				
	Scale Mean if Item Deleted	Scale Variance if Item Deleted	Corrected Item-Total Correlation	Cronbach's Alpha if Item Deleted
I think this learning/ creative approach very enjoyable.	85.1842	246.317	.827	.908
I really enjoy this learning/ creative approach.	85.1842	249.019	.824	.908
I think this to be a fun learning/ creative approach.	84.9211	252.237	.766	.910
This learning/ creative approach catches my attention.	86.4474	263.876	.764	.913
I feel anxious learning/ creating in this way.	87.0789	285.696	-.043	.929
I don't feel very relaxed during this learning/ creative process.	86.4474	277.767	.089	.928
I feel a lot of pressure learning/ creating in this way.	87.2632	285.659	-.045	.930
I think I will actively use this creative method.	85.3421	254.501	.681	.912
I believe this learning/ creative approach gives me a sense of agency.	85.1053	252.529	.732	.911
I feel like I can choose the learning/ creative approach.	84.9474	247.835	.823	.908
I feel that	85.0789	247.264	.823	.908

(*continued*)

Table 5. Continued

Item-Total Statistics				
	Scale Mean if Item Deleted	Scale Variance if Item Deleted	Corrected Item-Total Correlation	Cronbach's Alpha if Item Deleted
learning/ creating in this way is my choice.				
I think I am very good at this learning/ creating method.	86.2105	253.684	.619	.913
Compared to other classmates, I think I perform fairly well in this learning/ creating method.	86.4737	248.688	.620	.913
I am satisfied with my performance in this learning/ creating method.	85.8947	250.962	.801	.909
I am very proficient in this learning/ creating method.	86.6579	270.285	.261	.922
After learning/ creating in this way for some time, I feel that my abilities have improved.	85.6579	253.907	.819	.909
I think this learning/ creative approach helps to increase my interest.	85.5000	247.392	.836	.908
I believe this learning/ creative approach can help me acquire knowledge.	85.0263	253.161	.753	.910
I am willing to use this learning/ creative approach again because it has value for me.	85.0526	253.565	.756	.910

Table 6. Descriptive statistics of intrinsic motivation.

Descriptive Statistics					
	N	Minimum	Maximum	Mean	Std. Deviation
I think this learning/ creative approach very enjoyable.	38	2.00	7.00	5.3421	1.43840
I really enjoy this learning/ creative approach.	38	2.00	7.00	5.3421	1.34116
I think this to be a fun learning/ creative approach.	38	2.00	7.00	5.6053	1.30569
This learning/ creative approach catches my attention.	38	2.00	5.00	4.0789	.85049
I feel anxious learning/ creating in this way.	38	1.00	7.00	3.4474	1.46486
I don't feel very relaxed during this learning/ creative process.	38	1.00	7.00	4.0789	1.69867
I feel a lot of pressure learning/ creating in this way.	38	1.00	7.00	3.2632	1.55414
I think I will actively use this creative method.	38	3.00	7.00	5.1842	1.35278
I believe this learning/ creative approach gives me a sense of agency.	38	3.00	7.00	5.4211	1.34830
I feel like I can choose the learning/ creative approach.	38	3.00	7.00	5.5789	1.38782
I feel that learning/ creating in this way is my choice.	38	2.00	7.00	5.4474	1.40842
I think I am very good at this learning/ creating method.	38	1.00	7.00	4.3158	1.50863
Compared to other classmates, I think I perform fairly well in this learning/ creating	38	1.00	7.00	4.0526	1.73902

(*continued*)

Table 7. Reliability statistics for learning satisfaction.

Reliability Statistics	
Cronbach's Alpha	N of Items
.958	9

Table 8. Overall statistics for learning satisfaction.

Item-Total Statistics

	Scale Mean if Item Deleted	Scale Variance if Item Deleted	Corrected Item-Total Correlation	Cronbach's Alpha if Item Deleted
This learning task has helped me better understand the learning content.	32.9231	35.126	.858	.952
In this learning task, I made an effort to observe differences in things.	32.9744	35.657	.795	.955
Although this learning task was not easy, this learning/ creative approach was not difficult to understand.	32.8974	35.726	.812	.954
I find this method more challenging and fun than the previous method.	32.9487	35.682	.777	.956
Using this method for learning/ creating, I can gain some new discoveries or new knowledge.	32.8205	34.941	.861	.952
Using this method for learning/ creating allows me to approach learning/ creating with new methods or thought patterns.	32.8718	33.904	.863	.952
Using this method for learning/ creating helps me learn to discern the characteristics of things.	32.8462	35.134	.817	.954

(*continued*)

Table 8. Continued

Item-Total Statistics

	Scale Mean if Item Deleted	Scale Variance if Item Deleted	Corrected Item-Total Correlation	Cronbach's Alpha if Item Deleted
Using this method for learning helps me observe the differences in things.	32.8718	34.536	.895	.950
Using this method for learning helps me observe things from new perspectives.	32.8462	36.449	.777	.956

Table 9. Descriptive statistics for learning satisfaction.

Descriptive Statistics

	N	Minimum	Maximum	Mean	Std. Deviation
This learning task has helped me better understand the learning content.	39	2.00	5.00	4.0769	.83932
In this learning task, I made an effort to observe differences in things.	39	2.00	5.00	4.0256	.84253
Although this learning task was not easy, this learning/ creative approach was not difficult to understand.	39	2.00	5.00	4.1026	.82062
I find this method more challenging and fun than the previous method.	39	2.00	5.00	4.0513	.85682
Using this method for learning/ creating, I can gain some new discoveries or new knowledge.	39	2.00	5.00	4.1795	.85446
Using this method for learning/ creating allows me to approach	39	2.00	5.00	4.1282	.95089

(*continued*)

Table 9. Continued

Descriptive Statistics					
	N	Minimum	Maximum	Mean	Std. Deviation
learning/ creating with new methods or thought patterns.					
Using this method for learning/ creating helps me learn to discern the characteristics of things.	39	2.00	5.00	4.1538	.87475
Using this method for learning helps me observe the differences in things.	39	2.00	5.00	4.1282	.86388
Valid N (listwise)	39				

4 CONCLUSION

According to the questionnaire results, as shown in Table 3, students generally believe that the teaching method incorporating method acting and observational techniques in this semester is more helpful for creativity. They can correctly understand the course content through the teaching materials and practice process. Regarding the difficulty of the teaching materials, most students did not have significant concerns. Although some found it difficult to recognize and understand the content, the majority of students were still able to adapt to and utilize it well.

Based on Table 6, most students consider this teaching method to be interesting and valuable, and they feel pleasure during the process. It effectively enhances their interest in learning and improves their abilities to some extent, although they feel they still lack proficiency. Regarding satisfaction with the learning mode, Table 9 shows that students mostly provide positive feedback, agreeing with the teaching and practice methods. They also acknowledge gaining new thinking patterns from it, which helps them observe things from new perspectives. In terms of teaching effectiveness, the feedback obtained from this study will contribute to future course direction and updates, while preserving student works to serve as references for future students' end-of-term goals.

REFERENCES

Deci, E. L., Cascio, W. F., and Krusell, J. (1975). Cognitive evaluation theory and some comments on the Calder and Staw critique. *Journal of Personality and Social Psychology*, 31(1), 81–85.

Li, X. (2021). The Importance of belief and authenticity in performance. *Western Broadcasting and Television*, 3, 54–56.

System Innovation for an Artificial Intelligence Era – Lam et al. (Eds)
© 2025 the Author(s), ISBN: 978-1-032-84452-7

Ancient chinese barbed-spring locks: From basic research to science education in museums

Kuo-Hung Hsiao
Exhibition Division, National Science and Technology Museum, Kaohsiung, Taiwan

Yang Zhang
Department of Art and Design, Beijing University of Chemical Technology, Beijing, China

Xiao-Jie Zhuo
College of Art and Design, Fuzhou Technology and Business University, Fuzhou, China

Jian-Liang Lin
Science Education Division, National Science and Technology Museum, Kaohsiung, Taiwan

ABSTRACT: Barbed-spring locks, regarded as quintessential examples of locks used in ancient China and as an ingenious mechanical invention of ancient Chinese artisans. It primarily comprises the case, bolt, barbed spring, and key. Puzzle locks, which are created on the basis of barbed-spring lock design, incorporate additional components and clever features to enable locking mechanisms that cannot be overcome with only the correct key, thereby preventing theft and serving as a mental challenge. The authors of this paper reviewed the literature on Chinese barbed-spring locks and classified puzzle locks on the basis of whether they had open keyholes or hidden keyholes. This paper introduces the history of barbed-spring locks, explains the structural principles of puzzle locks, and describes the development of do-it-yourself puzzle lock kits as teaching tools. The findings of this study can be incorporated into STEM education courses and museum display education.

1 INTRODUCTION

Approximately 2,000 years ago, barbed-spring locks were already in use in ancient China. These locks are mechanical devices developed to achieve a sense of security, and they are widely used to safeguard people's safety and privacy [1]. The earliest complete barbed-spring lock, discovered in the Mausoleum of the First Qin Emperor, can be traced back to circa 220 BCE (Figure 1(a)). The widespread use of metal locks began circa 200 AD [2]; these locks were mainly crafted from bronze and iron and aesthetically enhanced through the addition of decorations or patterns on their bodies. The bronze lock depicted in Figure 1(b) is dated to approximately the 3rd century. During the Tang dynasty (618–907 AD), lock-making techniques had already reached an advanced stage, and the use of barbed-spring locks was increasingly common. Most locks were crafted with bronze, and some were made from brass, silver, or gold; these locks featured an increasingly diverse range of types, shapes, and carvings. The introduction of silver and gold gilding greatly enhanced the artistic value of these locks. Figure 1(c) depicts an 8th century gilt lock [3]. As late as the Ming dynasty

(a) Metal lock discovered in the Mausoleum of the First Qin Emperor [2]

(b) Bronze lock [3]

(c) Gilded lock [3]

(d) Hidden-keyhole lock [4]

Figure 1. Ancient Chinese barbed-spring locks.

DOI: 10.1201/9781003514831-18

(1368–1644 AD), puzzle locks that can only be opened by performing specific steps were widely manufactured and used; in contrast to a normal lock, puzzle locks cannot be quickly opened even with the correct key. On the basis of how they are opened and their keyhole design, barbed-spring puzzle locks can be distinguished into open-keyhole locks and hidden-keyhole locks. The diversity of lock types and ingenious designs showcase the extraordinary craftsmanship and intelligence of ancient Chinese artisans. Figure 1(d) is a hidden-keyhole lock from late Qing dynasty and early Republican China (1880–1930 AD) [4].

2 DESIGN PRINCIPLES AND LOCK TYPES

The diverse range of ancient Chinese puzzle locks can generally be distinguished into open-keyhole locks and hidden-keyhole locks. In particular, puzzle locks with open keyholes—locks with visible keyholes that cannot be easily opened even with the correct key—are highly distinctive. The open-keyhole lock depicted in Figure 2(a) is also known as a maze lock; to unlock it, the key can only be inserted by aligning a specific part of the key-head with a specific position in the keyhole while ensuring a specific orientation and moving the key in a specific manner. This unlocking process resembles the act of navigating a maze, hence the name. This design complicates the process of inserting the key into the keyhole even when the correct key is present and the keyhole is visible, thereby obstructing attempts by authorized personnel to open the lock [5].

Open-keyhole locks are designed with the keyhole being located at the side of the case and extending to the front, creating a three-dimensional keyhole with two symmetrical springs. The key-head comprises four protrusions of varying sizes. To open the lock, the key-head must be positioned horizontally to align with the keyhole along the side of the lock. The first step of the unlocking process involves twisting the key 90° around the z-axis. In the second step, the key is rotated around the y-axis and then shifted along the z-axis and x-axis to align the key-head with the keyhole, which should be slightly protruding out the front of the case. The third step entails moving the key along the z-axis and then along the x-axis to insert the key-head fully into the keyhole. Subsequently, the key is moved to compress the springs and release the bolt. In total, seven steps are required to open the lock illustrated in Figure 2(b) [6].

A hidden-keyhole lock integrates multiple mechanisms involving the use of holes, plates, turntables, and springs to cleverly disguise the keyhole. The actual keyhole must first be identified before the key can be inserted to open the lock.

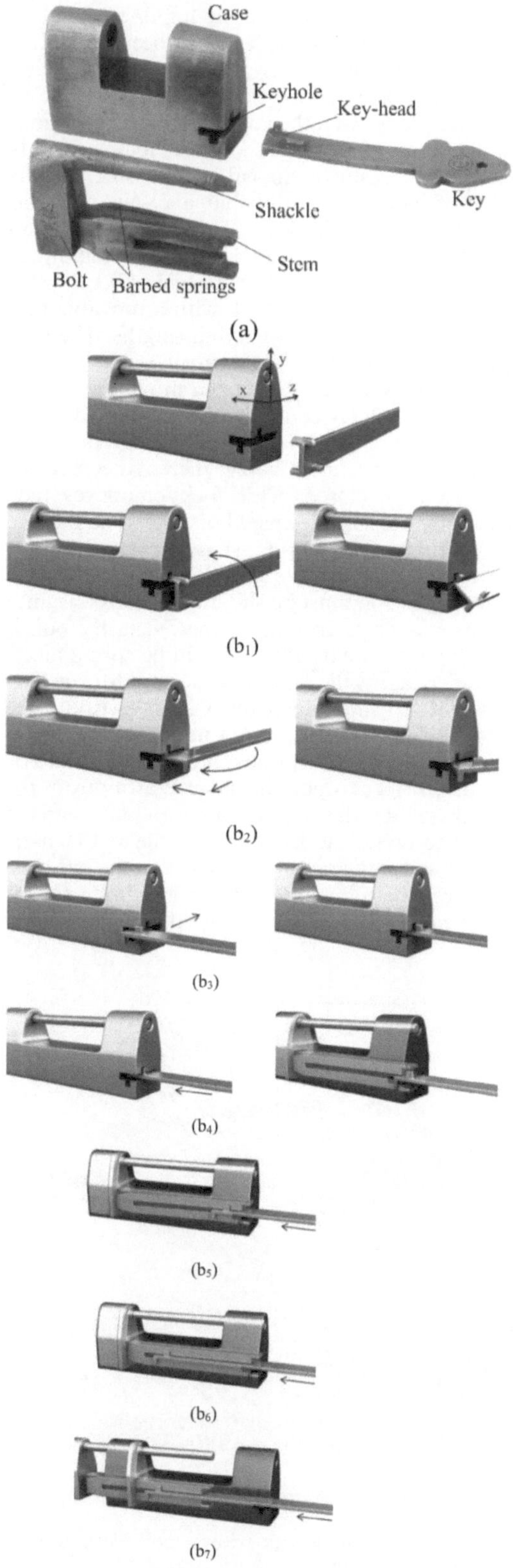

Figure 2. Opening a maze lock [6].

With this distinctive design, finding the keyhole is a challenge; after the keyhole has been identified, how to insert the key is another challenge, and even then, opening the lock requires knowledge on how to slide or twist the key. Notably, decorations on the case of a hidden-keyhole lock not only enhance the aesthetic appeal of its design and the value of its craftsmanship but may also serve as clues to opening the lock. This feature was prevalent in lock designs in ancient China, resulting in the creation of various types of locks [7].

Figure 3(a) depicts a lock with a movable button and three springs of varying lengths. The long spring, which is in contact with the inner wall of the case, prevents the bolt from moving. The end and bottom plates conceal the keyholes. Because of the length of the shackle, both the end plate and bottom plate are immovable, thereby keeping the lock securely closed. This lock comprises nine members, namely the case, bolt, movable button, long spring, middle spring, short spring, end plate, bottom plate, and key [8]. To open the lock, the movable button must be slid along the y-axis until it presses against the long spring. Next, the bolt is slid along the x-axis until the middle spring meets the inner case wall. This creates room for the bottom plate to slide along the x-axis, revealing the hole. The key pin is inserted into this keyhole to compress the middle spring, allowing the bolt to be slid out further to pull the shackle away from the end plate. At this point, the end plate can be turned to reveal the keyhole, and the key is inserted and slid forward until the short spring has been compressed, allowing the entire bolt to be removed. Figure 3(b) depicts the lock-opening process. The use of springs of varying lengths and multiple mechanisms substantially increases the difficulty of opening the lock and also enhances its educational and entertainment value.

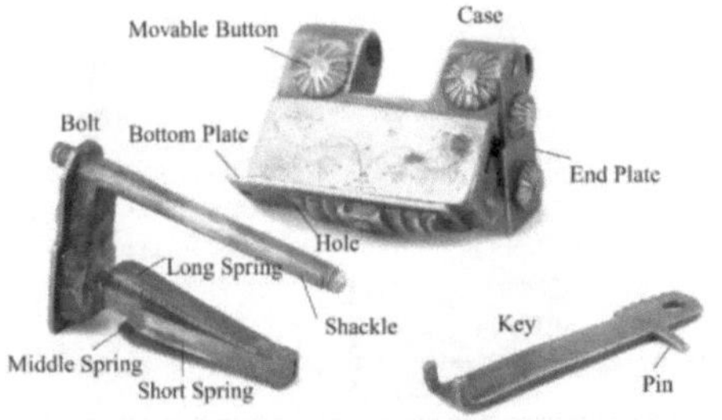

(a)

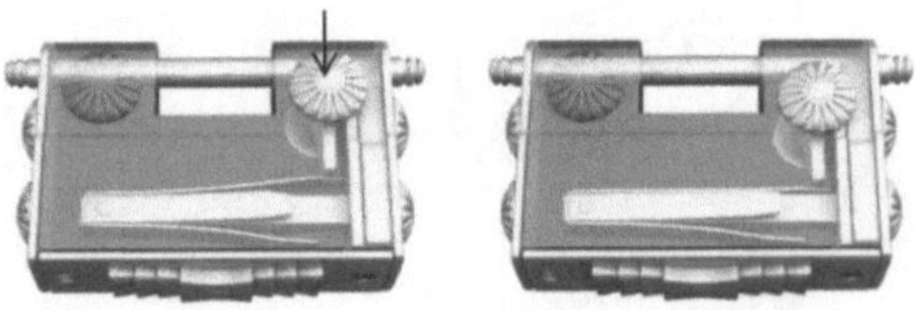

(b1) Slide the movable button

Figure 3. Opening a lock with a movable button [8] (Yizhitang Collection).

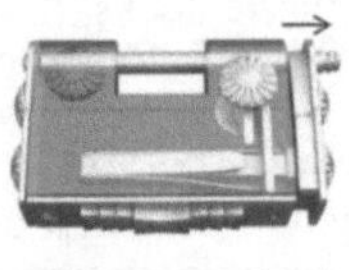

(b_2) Slide out part of the bolt

(b_3) Slide out the bottom plate

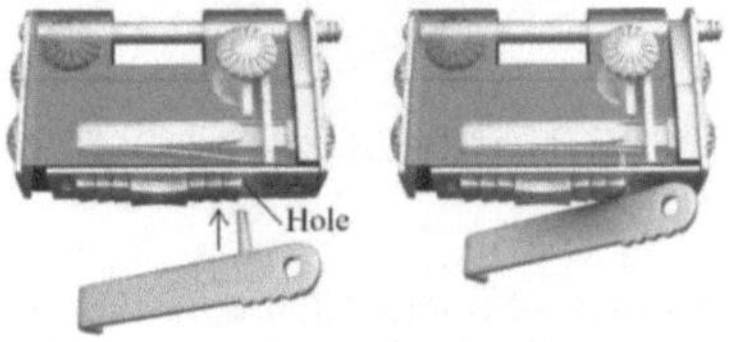

(b_4) Insert the key pin into the hole to compress the middle spring

(b_5) Slide the bolt out further

(b_6) Turn the end plate

(b_7) Insert the key into the keyhole and compress the short spring

(b_8) Remove the whole bolt

Figure 3. (Continued)

3 DEVELOPING TEACHING MATERIALS

The National Science and Technology Museum is the largest applied sciences museum in Taiwan. Its main functions are the research, collection, and exhibition of various scientific themes and artifacts and the promotion of social science and technology education; these functions are part of its mission to educate the public on major scientific and technological developments and their effects on human life [9]. Barbed-spring locks have a rich cultural history and feature a myriad of clever designs. Through the conversion of research findings on barbed-spring locks into popular science information and a collaboration with Arnos Design Company to take advantage of their production and design skills, the museum has developed a series of elegant do-it-yourself (DIY) barbed-spring lock products, which are intended to promote popular science education in museums and schools.

The optimal approach to learning a topic and sparking learners' interest is engaging in hands-on activities. The process of assembling a lock is an effective way to learn and comprehend the principles underlying the design of barbed-spring locks. Providing information on the cultural background of various barbed-spring locks, the materials used to craft these locks, various lock designs and forms, and the steps to open a lock can further enhance public awareness of the history and structural principles of these locks. In addition to providing physical replicas of traditional locks that people can interact with, a more effective educational method is teaching people to make their own locks. Thus, the DIY barbed-spring lock kits were designed to be low-cost, made from readily available materials, straightforward to assemble, and practical. By contrast, creating a metal or plastic using through the use of molds or 3D printing is costly. In consideration of production costs and quantity control, laser-cut wooden boards were determined to be the optimal choice. Furthermore, achieving sufficient elasticity is a key challenge for the production of wooden barbed springs. The solution is to integrate the lock stem and springs into a single piece; using thin barbed springs and establishing the connection point between the stem and the spring as the fulcrum can generate the elasticity required to open and lock a barbed-spring lock.

To enhance the enrichment and educational factors of the product and ensure that the public comprehends the creativity and craftsmanship of traditional artisans, creative DIY puzzle locks based on basic barbed-spring lock designs were developed as teaching materials. The DIY puzzle locks are expected to expand the market appeal of DIY locks. The unlocking mechanism of the DIY puzzle lock depicted in Figure 4(a) is similar to that of the maze lock in Figure 2. The DIY puzzle lock in Figure 4(b) is a comprehension puzzle lock that expands on the maze lock design; it comprises two barbed springs of different lengths, a sliding plate, and an additional keyhole [10].

(a) Maze lock

(b) Comprehensive puzzle lock

Figure 4. DIY barbed-spring lock kits.

4 CONCLUSIONS

Barbed-spring locks are the most quintessential examples of locks used in ancient China. Advancements over time and creativity have led to the invention of puzzle locks with clever designs, and the ingenuity inherent in these puzzle locks warrant in-depth research and promotion through museum initiatives centered on education. On the basis of a structural analysis of numerous traditional locks, puzzle locks have been distinguished into open-keyhole locks and hidden-keyhole locks. This paper presents an introduction to the history of barbed-spring locks, an explanation of the structural principles of puzzle locks, and descriptions of the development of DIY puzzle lock kits as teaching tools, assisted by the use of computer animations. The relevant research results are already presented in museums and promoted in popular science education to enhance awareness of such mechanical devices, which are based on ingenious designs and carry historical significance.

ACKNOWLEDGEMENTS

The authors are grateful to the National Science and Technology Council (Taipei, Taiwan) under grant NSTC 111-2410-H-359-001- and Humanities and Social Science Fund of Ministry of Education (China) under grant 21YJC760110 for the financial support of this work.

REFERENCES

[1] Yan, Hong-Sen, (2019). *The Beauty of Ancient Chinese Locks* (ISBN 978-7554389211). Hainan Publishing House, Haikou.

[2] Terra Cotta Archeology Team, (1981). *Archaeology and Cultural Relics*, 1 39–43.

[3] Chinese Lockset Cultural Research Association, (2014). *Atlas of Chinese Ancient Locksets.* Liaoning University Press, Shenyang.

[4] Hsiao, Kuo-Hung, (2022). *Exploring Machinations—Chinese Traditional Puzzle Locks* (ISBN 978-7-5701-2403-9). Shandong Education Press, Jinan.

[5] Kuo-Hung Hsiao, (2017). *Mechanism and Machine Theory*. 118 168–179.

[6] Kan Shi, Yang Zhang, Jian-Liang Lin, Kuo-Hung Hsiao, (2017). *The Transactions of the Canadian Society for Mechanical Engineering*. 41 433–441.

[7] Kuo-Hung Hsiao, (2018). *Mechanical Sciences*. 9 189–199.

[8] Kuo-Hung Hsiao, Yang Zhang, Kan Shi, Yu-Hsun Chen, Hai An, (2019). *Ancient Chinese Puzzle Locks. Robotics and Mechatronics* (ISBN: 978-3-030-30035-7). Switzerland: Springer. 494–501.

[9] https://www.nstm.gov.tw/English/.

[10] Kan Shi, Yang Zhang, Ming-Jie Wang, Kuo-Hung Hsiao, Yan-An Yao, (2022). *Scientific Reports*. 1211237.

System Innovation for an Artificial Intelligence Era – Lam et al. (Eds)
© 2025 the Author(s), ISBN: 978-1-032-84452-7

Interaction force estimation for upper-limb exoskeleton

Yu-Hao Lai & Ming-Yang Cheng
Department of Electrical Engineering, National Cheng Kung University, East Dist., Tainan, Taiwan

Hsien-Hao Chiu
Department of Mechanical Engineering, National Cheng Kung University, East Dist., Tainan, Taiwan

ABSTRACT: Upper-limb exoskeletons can be used in physical therapy/rehabilitation of patients and for providing assistive force to elderly people or to factory workers who need to lift/carry heavy loads. However, if the upper-limb exoskeleton that provides the assistive force does not exhibit the "compliance" property, the users could feel discomfort or may even get injured. A common approach to coping with this problem is to design a compliance control scheme in which the interaction force between the user and the upper-limb exoskeleton is exploited to adjust the desired trajectory command for the upper-limb exoskeleton to enhance "compliance". Generally, the interaction force can be either measured by a force sensor or estimated by a force-sensorless approach. In light of concerns over cost, this paper adopts the force-sensorless approach. In this paper, to assess interaction force estimation performance, several existing force-sensorless approaches such as "Generalized Momentum" and "Modified Generalized Momentum" are used in an upper-limb exoskeleton prototype developed in this paper to estimate the interaction force. Experimental results for different force-sensorless approaches are compared and analyzed.

Keywords: upper-limb exoskeleton, physical therapy/rehabilitation, compliance control, force-sensorless approach.

1 INTRODUCTION

Recently, in benefitting from the technological advancement, various kinds of exoskeletons have been developed to satisfy different needs. As the problems facing our aging society intensify, the need of power assisting devices for elderly people cannot be overlooked. As a result, many existing studies have focused on the use of exoskeletons in medical applications [1,2]. To ensure that these exoskeletons can properly execute instructions and comply with human motion, exactly how to effectively estimate human motion intention is crucial. For example, Chen et al. exploited the idea of variable admittance control and extended-state observers to perform gait prediction for lower limb exoskeletons with measurement delay [3]. In [4], an adaptive sliding mode variable admittance control scheme was employed to control the motion of a lower-limb rehabilitation exoskeleton robot [4]. Recently, exoskeletons have been used in the rehabilitation of upper limbs for patients recovering from strokes [5]. In general, the exoskeleton will guide post-stroke patients to perform resistive force training during the rehabilitation process. As a result, accurate estimation of the interaction force between the patient and the exoskeleton is crucial. In previous research [6,7], a force sensor was used to measure the interaction force between the patient and the exoskeleton. However, force sensor-based approaches are not cost effective. In addition, for certain application scenarios with limited space, it is very difficult if not impossible to install force sensors. To cope with the aforementioned difficulties, this paper uses two force sensorless observer-based approaches to estimate the interaction force between the user and the exoskeleton. One is the generalized momentum (GM) approach [8,9] and the other is the modified generalized momentum (MGM) approach [10]. One of the most attractive features of the GM approach is that the acceleration information is not required in force/torque estimation. The drawback of the GM approach is that an accurate system model is essential. Similar to the GM approach, the MGM approach also exploits the concept of generalized momentum. Building upon the GM approach, several modifications have been applied to the MGM approach so as to improve force estimation accuracy.

This paper details the construction of an upper-limb exoskeleton prototype, with three different experiments having been conducted. Since both the GM and MGM approaches are model-based, the parameter values of the exoskeleton must be identified in advance. The purpose of the first experiment was to identify the parameter values of the exoskeleton. The second experiment aimed at comparing the performance of the GM approach

DOI: 10.1201/9781003514831-19

and the MGM approach. In the third experiment, the MGM approach was combined with an admittance control scheme to control the exoskeleton.

The remainder of the paper is organized as follows. Section 2 provides a brief review of system model identification and admittance control. Section 3 introduces both the GM approach and the proposed MGM approach employed to estimate the interaction force between the exoskeleton and the user. Experimental results and conclusions are given in Section 4 and Section 5, respectively.

2 BRIEF REVIEW OF SYSTEM MODEL IDENTIFICATION AND ADMITTANCE CONTROL

2.1 *System model identification*

The exoskeleton prototype constructed in this paper can be regarded as a two-link rigid body. The general form of its dynamic model can be obtained using the Lagrange method and the classic friction model as described by Eq. (1):

$$M(\theta)\ddot{\theta} + C(\theta,\dot{\theta})\dot{\theta} + G(\theta) + F(\dot{\theta}) = \tau_c + \tau_{ext} \quad (1)$$

where $\theta \in R^2$ is the joint angle vector, $M(\theta)$ is the inertia matrix, $C(\theta,\dot{\theta})$ is the Coriolis and centrifugal matrix, $G(\theta)$ is the gravitational force vector, $F(\theta)$ is the friction vector, τ_c is the torque command, and τ_{ext} is the external torque.

To implement model-based control schemes or model-based force estimation approaches, the parameter values of the system model need to be known. If they are not known in advance, then system parameter identification needs to be performed. In this paper, parameter values such as link length, link mass, link inertia and the distance between the joint to the mass center of the link need to be identified.

To perform system identification, based on the approach proposed in [11], one can rewrite Eq. (1) into the form of regression as described by Eq. (2):

$$\tau_c + \tau_{ext} = \phi\psi \quad (2)$$

The vector of barycentric parameters is described by

$$\psi = [\psi_1\ \psi_2\ \psi_3\ \psi_4\ \psi_5\ \psi_6\ \psi_7\ \psi_8\ \psi_9] \quad (3)$$

$$\begin{cases} \psi_1 = M_1k_1^2 + M_2d_1^2 + I_1 \\ \psi_2 = M_2k_2^2 + I_2 \\ \psi_3 = M_2k_2d_1 \\ \psi_4 = M_1k_1g + M_2gd_1 \\ \psi_5 = M_2k_2g \\ \psi_6 = F_{c1} \\ \psi_7 = F_{c2} \\ \psi_8 = F_{v1} \\ \psi_9 = F_{v2} \end{cases} \quad (4)$$

The regression matrix is described by

$$\phi = [\phi_1\ \phi_2\ \phi_3\ \phi_4\ \phi_5\ \phi_6\ \phi_7\ \phi_8\ \phi_9] \quad (5)$$

$$\begin{cases} \phi_1 = \left[\ddot{\theta}_1\ 0\right]^T \\ \phi_2 = \left[\ddot{\theta}_1 + \ddot{\theta}_2\ \ddot{\theta}_1 + \ddot{\theta}_2\right]^T \\ \phi_3 = [(2\ddot{\theta}_1 + \ddot{\theta}_2)\cos(\theta_2) - (2\dot{\theta}_1\dot{\theta}_2 + \dot{\theta}_2{}^2)\sin(\theta_2) \\ \quad \ddot{\theta}\cos(\theta_2) + \dot{\theta}_1{}^2\sin(\theta_2)]^T \\ \phi_4 = [\sin(\theta_1)\ 0]^T \\ \phi_5 = [\sin(\theta_1+\theta_2)\ \sin(\theta_1+\theta_2)]^T \\ \phi_6 = \left[sign(\dot{\theta}_1)\ 0\right]^T \\ \phi_7 = \left[0\ sign(\dot{\theta}_2)\right]^T \\ \phi_8 = \left[\dot{\theta}_1\ 0\right]^T \\ \phi_9 = \left[0\ \dot{\theta}_2\right]^T \end{cases} \quad (6)$$

When performing system identification, the external torque τ_{ext} in Eq. (1) is ignored. In addition, the exciting reference trajectory described by Eq. (7) is used as a position command in Eq. (1) to excite the dynamics of the shoulder joint and elbow joint so as to obtain better identification accuracy:

$$\begin{cases} \theta_i(t) = \sum_{k=1}^{N_i} a_k^i\left[\frac{\sin(\omega_f kt)}{\omega_f k}\right] + b_k^i\left[-\frac{\cos(\omega_f kt)}{\omega_f k}\right] + \theta_{i0} \\ \dot{\theta}_i(t) = \sum_{k=1}^{N_i} a_k^i[\cos(\omega_f kt)] + b_k^i[\sin(\omega_f kt)] \\ \ddot{\theta}_i(t) = \sum_{k=1}^{N_i} a_k^i[-\omega_f kt\sin(\omega_f kt)] + b_k^i[\omega_f kt\cos(\omega_f kt)] \end{cases}, i = 1,2 \quad (7)$$

where ω_f is the fundamental frequency of the finite Fourier series; N_i is the number of harmonics of the finite Fourier series; a_k^i, b_k^i are amplitudes of the cosine and sine functions, and the finite Fourier series described by Eq. (7) has $2N_i + 1$ parameters in a_k^i, b_k^i.

By taking the constraints such as joint position/velocity/acceleration and rated torque of the motor into consideration, one can compute the condition number for each candidate exciting reference trajectory. The one with the smallest condition number is chosen as the exciting reference trajectory used in the system identification process. In this paper, a simple PD-like control law is adopted to control the exoskeleton to track the exciting reference trajectory. The trajectory tracking experiments were performed many times and joint position/torque data were measured/recorded for later use. Based on the recorded joint position data, the central difference method was used to compute the joint velocity/acceleration in an offline manner. These joint position/velocity/ acceleration data were then substituted into regression matrix ϕ in Eq. (2). By using Eq. (2) and the weighted least squares method described by Eq. (8) and (9), one can estimate the barycentric parameters:

$$\sigma = \sqrt{\|\tau_e - \phi\psi\|^2} \tag{8}$$

$$\psi_{WLS} = (\phi^T\sigma^{-1}\phi)^{-1}\phi^T\sigma^{-1}\tau_E \tag{9}$$

where ψ_{WLS} is the vector of barycentric parameters obtained using the weighted least squares method and σ is the variance.

2.2 *Admittance control*

The admittance control law [6] in the joint space can be described by Eq. (10)

$$M(\ddot{\theta}_{cmd} - \ddot{\theta}_r) + B(\dot{\theta}_{cmd} - \theta_r) + K(\theta_{cmd} - \theta_r) = \tau_{ext} \tag{10}$$

where $\ddot{\theta}_{cmd}, \dot{\theta}_{cmd}, \theta_{cmd}$ are the joint acceleration command, joint velocity command and joint position command, respectively; $\ddot{\theta}_r, \dot{\theta}_r, \theta_r$ are the adjusted joint acceleration command, adjusted joint velocity command and adjusted joint position command, respectively; τ_{ext} is the interaction torque; M is the mass parameter; B is the damping parameter; and K is the spring parameter.

With the estimated τ_{ext} and given parameter values of M, B and K, as well as $\ddot{\theta}_{cmd}, \dot{\theta}_{cmd}, \theta_{cmd}$, one can obtain $\ddot{\theta}_r, \dot{\theta}_r, \theta_r$ based on Eq. (10). In this paper, the admittance control scheme is used as the force control outer loop, while the inner position loop is used to perform trajectory tracking.

3 DISTURBANCE OBSERVER BASED INTERACTION FORCE

When performing physical therapy/rehabilitation using an exoskeleton, the information of the interaction force between the patient/human user and the exoskeleton is not only crucial in the control design of the exoskeleton but can also provide a useful assessment of rehabilitation effectiveness. Using a force sensor to retrieve the interaction force information is not cost effective. Moreover, in some application scenarios, the space is so limited that it is impossible to install a force sensor to measure the interaction force. To cope with this difficulty, this paper adopts an observer-based approach to estimate the interaction force (i.e. torque) between the human user and the exoskeleton.

3.1 *Generalized momentum approach for interaction force estimation*

The generalized momentum is defined as Differentiating the generalized momentum with respect to time t will yield

$$\dot{p} = \dot{M}(\theta)\dot{\theta} + M(\theta)\ddot{\theta} \tag{11}$$

Substituting Eq. (1) into Eq. (11) leads to

$$\dot{p} = \dot{M}(\theta)\dot{\theta} + \tau_{ext} + \tau_c - C(\theta,\dot{\theta})\dot{\theta} - G(\theta) - F(\dot{\theta}) \tag{12}$$

It is well known in robotics that matrix $\dot{M}(\theta) - 2C(\theta,\dot{\theta})$ is skew symmetric and matrix $\widehat{M}(\theta)$ is symmetric. Considering these two facts, one can obtain Eq. (13) and rewrite Eq. (12) into Eq. (14):

$$\dot{M} = C + C^T \tag{13}$$

$$\dot{p} = \tau_c + \tau_{ext} + C^T(\theta,\dot{\theta})\dot{\theta} - G(\theta) - F(\dot{\theta}) \tag{14}$$

Define residual vector r as

$$r = K\left[p - \int(\tau_c + C^T(\theta,\dot{\theta})\dot{\theta} - G(\theta) - F(\dot{\theta}) + r)dt\right] \tag{15}$$

where $r(0) = 0$ and K is a positive diagonal matrix.

Differentiating Eq. (15) with respect to time t will yield

$$\begin{aligned} \dot{r} &= K[\dot{p}(t) - (\tau + \widehat{H}^T\dot{\theta} - \widehat{G} - \widehat{F} + r)] \\ &= K[\tau^*_{ext} - r] \end{aligned} \tag{16}$$

Taking the Laplace transform of Eq. (16) will result in

$$\frac{r_i(s)}{\tau^*_{ext,i}(s)} = \frac{K_i}{s + K_i} \tag{17}$$

As suggested by Eq. (17), the residual vector can be regarded as the output of the interaction force through low-pass filtering. As the value of gain K_i increases, the difference between the residual vector and the interaction torque decreases. Namely, the estimation accuracy improves. However, the robustness to noise becomes worse as the values of K_i increase. Consequently, there is tradeoff between estimation accuracy and robustness. The generalized momentum (GM) approach is a popular method for estimating the interaction torque.

3.2 *Modified generalized momentum approach for interaction force estimation*

Compared with the generalized momentum (GM) approach [10], the modified generalized momentum (MGM) approach has several modifications. One is to introduce a feedforward term in the low pass filter. As a result, the 1st order low pass filter in the GM approach is modified into a 2nd order filter described by Eq. (18):

$$\frac{\widehat{\tau}_{ext}}{\tau^*_{ext}} = \frac{k_3 s + 1}{\left(\frac{1}{k_1}\right)s^2 + k_2 s + 1} = \frac{k_1 k_3 s + k_1}{s^2 + k_1 k_2 s + k_1} \tag{18}$$

where k_1, k_2, k_3 are gain constants.

Converting Eq. (18) into the time domain and solving for τ_{ext} will give

$$\widehat{\tau}_{ext} = \int_0^t \begin{matrix} (-k_1 k_2 \widehat{\tau}_{ext} + k_1 p + k_1 k_3 \tau^*_{ext} \\ -k_1 \int_0^t (\tau + \widehat{H}^T \theta - \widehat{G} - \widehat{F} + \widehat{\tau}_{ext}) dt) dt \end{matrix} \tag{19}$$

where $h = \int_0^t (k_1 k_3 \tau^*_{ext}) dt$ is the feedforward term.

Substituting Eq. (14) into h will yield

$$h = k_1 k_3 (p - \int_0^t (\tau + \widehat{H}^T \dot{\theta} - \widehat{G} - \widehat{F}) dt) \tag{20}$$

The 2nd order filter used in the MGM approach has a damping term to stabilize the system and also reduce the system overshoot. In addition, the use of a feedforward term in the MGM approach increases a zero for the system so as to improve the system's transient performance.

4 EXPERIMENTAL SETUP AND RESULT

4.1 *Experimental setup*

Figure 1 shows the experimental platform a 2 DOF upper-limb exoskeleton—used in this paper. Two AC servomotors are installed: one at the elbow joint and another at the shoulder joint of the upper-limb exoskeleton. Three different experiments have been conducted.

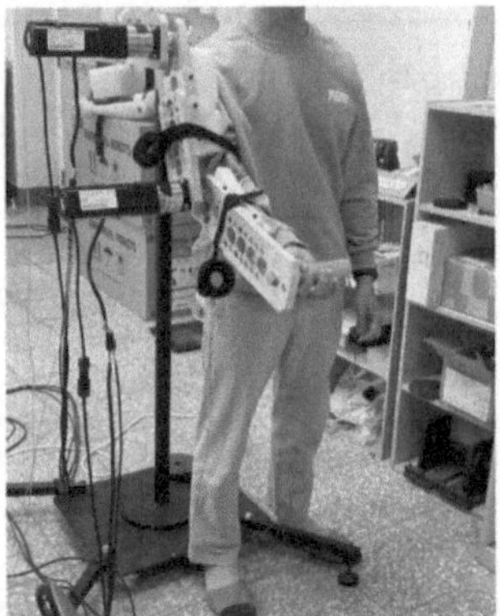

Figure 1. Experimental platform: 2 DOF upper-limb exoskeleton.

4.2 *Experiment #1: System parameter identification*

The GM approach and the MGM approach are model-based approaches. The system identification method introduced in Section 2.2 is used in this paper to identify the parameter values of the exoskeleton prototype. In the experiment, a simple PD-like controller was used to track the exciting reference trajectory described by Eq. (7). Based on the recorded joint position data and torque command, barycentric parameters are estimated using Eq. (9).

To assess the accuracy of system identification results, another trajectory tracking experiment was conducted. The identified system parameters were used to reconstruct the torque command. The relative torque command error percentage defined by Eq. (21) is used as the performance index of identification accuracy. In Eq. (21), τ_{real} is the actual torque command, while τ_{fit} is the reconstructed torque command. Figure 2 shows the torque command and the reconstructed torque command for both the elbow joint and shoulder joint of the upper-limb exoskeleton. The relative torque command error listed in Table 1 indicates that the identification accuracy is satisfactory:

$$\text{Relative Error} = \frac{\tau_{real} - \tau_{fit}}{\tau_{real}} \times 100\% \tag{21}$$

Table 1. Relative torque command error.

	Shoulder Joint	Elbow Joint
Norm_Error	7.72%	12.44%

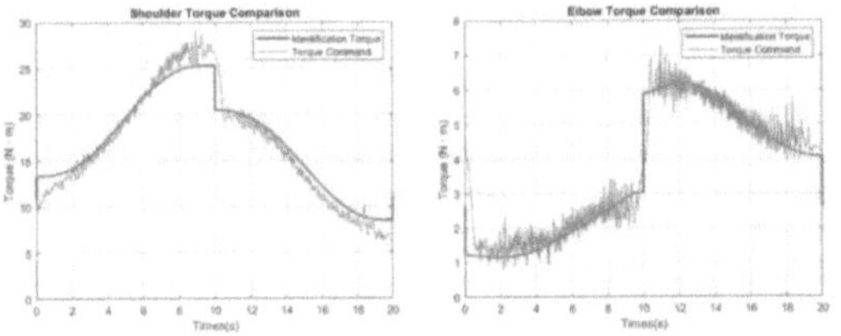

Figure 2. Torque command and reconstructed torque command for both elbow joint and shoulder joint of upper-limb exoskeleton.

4.3 *Experiment #2: Performance comparison between different interaction force estimation approaches*

In this experiment, a user is asked to hold the endpoint of the 2nd link of the upper-limb exoskeleton and moves along with the upper-limb exoskeleton. Both the GM approach and the MGM approach are used to estimate the interaction torque between the user and the upper-limb exoskeleton. Experimental results are shown in Figure 3 (shoulder joint) and Figure 4 (elbow joint). The results shown in Figures 3 and 4 indicate that the MGM approach yields smoother interaction torque estimation results than the GM approach. In addition, we designed an experiment to verify the effectiveness of the interaction torque estimation approaches. First, the human user does not wear the upper-limb exoskeleton. The upper-limb exoskeleton performs a trajectory following motion and the torque command is recorded. Then, the human user is asked to wear the upper-limb exoskeleton and moves along with it. Torque command and the interaction torque estimated using both the GM approach and the MGM approach are recorded. Figure 5 shows the interaction torques estimated using both the GM approach and the MGM approach, and the difference in torque command recorded in two experiments. By observing the results shown in Figure 5, one can find that the interaction torques estimated using both the GM approach and the MGM approach are in good accordance with the difference in torque command recorded in the two experiments. The results shown in Figure 5 and Table 2 indicate that both the GM approach and the MGM approach exhibit satisfactory performance in estimating the interaction torque.

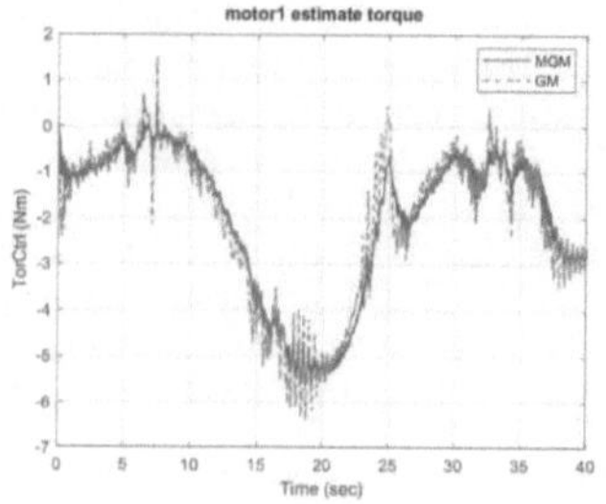

Figure 3. Interaction Torque at shoulder torque joint estimated using GM and MGM.

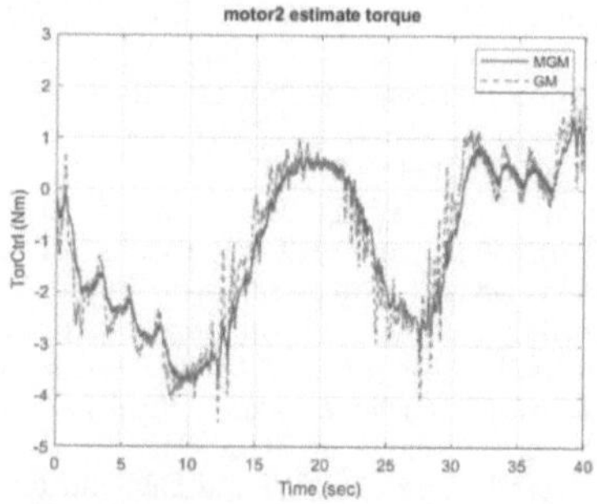

Figure 4. Interaction Torque at elbow torque joint estimated using GM and MGM.

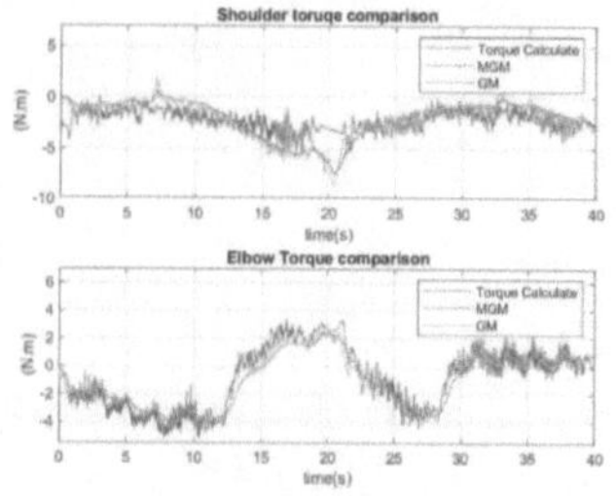

Figure 5. Interaction forces estimated using both GM and MGM with different in torque command recorded in two experiments.

Table 2. Estimation error of different approaches.

RMS Error(N)	GM	MGM
Shoulder Joint	1.8604	1.8526
Elbow Joint	0.6456	0.6779

4.4 *Experiment #3: admittance control*

In order to perform physical therapy/rehabilitation, the human user is often asked to wear an exoskeleton and move along a specific trajectory. In this experiment, the user will exert force on the upper-limb exoskeleton. Based on the interaction forces estimated using the MGM approach, an admittance control scheme is employed to adjust the trajectory command so that the compliance between the user and the exoskeleton is increased. Experimental results shown in Figure 6 indicate that the trajectory command has been adjusted to comply with the user so as to ensure the user's safety. Figure 7 shows the torque command and the interaction torque estimated using the MGM approach.

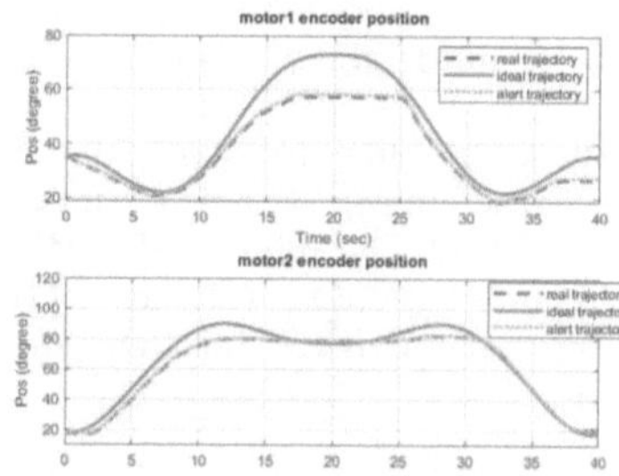

Figure 6. Original trajectory command, adjusted trajectory command, and actual trajectory.

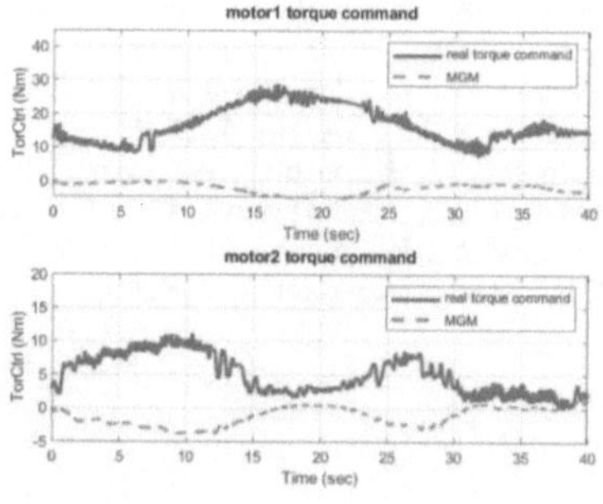

Figure 7. Torque command and estimated interaction torque.

5 CONCLUSION

For safety concerns, information on the interaction force is crucial when using an exoskeleton in physical therapy/rehabilitation. Most existing exoskeleton systems use force sensors, EMG or

EGG to measure the interaction force between the patient and the exoskeleton. However, in light of concerns over cost, this paper adopts the force-sensorless approach. In particular, two force-sensorless approaches—GM and MGM—are adopted in this paper to estimate the interaction torque. The estimated interaction torque is used in an admittance control scheme to ensure the user's safety as well as the increase in compliance. In addition, different degrees of strength of rehabilitation can be achieved by adjusting the parameters of the admittance control scheme. Several experiments have been conducted. Experimental results indicate that both the GM approach and the MGM approach exhibit satisfactory performance in estimating the interaction torque, while the MGM approach yields smoother interaction torque estimation results than the GM approach. Experimental results also show that, based on the interaction torques estimated using the MGM approach, the admittance control scheme is able to adjust the trajectory command to comply with the user so as to ensure the user's safety.

ACKNOWLEDGMENT

The authors would like to thank the National Science and Technology Council (under Grant No.: NSTC 112-2221-E-006 -135 -MY2) and Ministry of Education (Higher Education SPROUT Project), Taiwan, for their support of this research.

REFERENCES

[1] Strausser K. A. and Kazerooni H., (2011). The development and testing of a human machine interface for a mobile medical exoskeleton, in *Proc. of the 2011 IEEE/RSJ International Conference on Intelligent Robots and Systems*, San Francisco, CA, USA, pp. 4911–4916.

[2] Kagirov, I., Kapustin, A., Kipyatkova, I., Klyuzhev, K., Kudryavcev, A., Kudryavcev, I., Loskutov, Y., Ryumin, A. and Karpov, A., (Feb. 2021). Medical exoskeleton 'Remotion' with an intelligent control system: Modeling, implementation, and testing, *Simulation Modelling Practice and Theory*, vol. 107, p. 102200.

[3] Chen Z., Guo Q., Li T., Yan Y. and Jiang D., (Nov. 2023). Gait prediction and variable admittance control for lower limb exoskeleton with measurement delay and extended-state-observer, *IEEE Transactions on Neural Networks and Learning Systems*, vol. 34, no. 11, pp. 8693–8706.

[4] Tu, Y., Zhu, A., Song, J., Shen, H., Shen, Z., Zhang, X., and Cao, G., (Apr. 2020). An adaptive sliding mode variable admittance control method for lower limb rehabilitation exoskeleton robot, *Applied Sciences*, vol. 10, no. 7, p. 2536.

[5] He C., Xiong C.-H., Chen Z.-J., Fan W., Huang X.-L. and Fu C., (2021). Preliminary assessment of a postural synergy-based exoskeleton for post-stroke upper limb rehabilitation, *IEEE Transactions on Neural Systems and Rehabilitation Engineering*, vol. 29, pp. 1795–1805.

[6] Li Z., Huang B., Ye Z., Deng M. and Yang C., (Dec. 2018). physical human–robot interaction of a robotic exoskeleton by admittance control, *IEEE Transactions on Industrial Electronics*, vol. 65, no. 12, pp. 9614–9624.

[7] Wang W., Qin L., Yuan X., Ming X., Sun T., and Liu Y., (May 2019). Bionic control of exoskeleton robot based on motion intention for rehabilitation training, *Advanced Robotics*, vol. 33, no. 12, pp. 590–601.

[8] De Luca A., Albu-Schaffer A., Haddadin S. and Hirzinger G., (2006). Collision detection and safe reaction with the DLR-III lightweight manipulator arm, in *Proc. of the 2006 IEEE/RSJ International Conference on Intelligent Robots and Systems*, Beijing, China, pp. 1623–1630.

[9] Cong Z., Honglei A., Wu C., Lang L., Wei Q. and Hongxu M., (2020). Contact Force Estimation Method of Legged-Robot and Its Application in Impedance Control, *IEEE Access*, vol. 8, pp. 161175–161187.

[10] Chien S.-H., Wang J.-H. and Cheng M.-Y., (2020). Performance comparisons of different observer-based force-sensorless approaches for impedance control of collaborative robot manipulators, in *Proc. of the 2020 IEEE Conference on Industrial Cyberphysical Systems (ICPS)*, Tampere, Finland, pp. 326–331.

[11] Swevers J., Verdonck W. and De Schutter J., (Oct. 2007). Dynamic model identification for industrial robots, *IEEE Control Systems Magazine*, vol. 27, no. 5, pp. 58–71.

System Innovation for an Artificial Intelligence Era – Lam et al. (Eds)
© 2025 the Author(s), ISBN: 978-1-032-84452-7

Contour error reduction of DMPs with time-dilation-based time scale adjustment

Chia-Chun Huang & Ming-Yang Cheng
Department of Electrical Engineering, National Cheng Kung University, East Dist., Tainan, Taiwan

ABSTRACT: Recently, Dynamic Movement Primitives (DMPs) have been used to cope with the trajectory planning problems of robotic systems. The DMP can encode a given demonstration trajectory into a function that is independent of time. The shape of the DMP can be modified by this time-independent function. In particular, the temporal scaling term of the DMP can be used to adjust the velocity along the demonstration trajectory. Several existing trajectory planning studies have pointed out that the portion of the trajectory with a larger curvature is likely to result in a larger contour error. In addition, one way to reduce contour error is to lower the feedrate along the trajectory. As a result, feedrate scheduling along the trajectory should take the curvature information into consideration. In the theory of general relativity, a gravitational field will affect the curvature of its nearby space. This paper exploits this concept to propose a DMP-based trajectory planning approach for robot manipulators. In particular, the curvature of the desired trajectory is used to adjust the temporal scaling term of the DMP so that the contour error can be reduced. In this paper, the demonstration trajectory is represented in a NURBS form, while the DMP is used to generate the trajectory command for a contour following experiment conducted on a 2-DOF planar robot arm. Experimental results indicate that the proposed approach can indeed reduce contour error.

1 INTRODUCTION

Compared with conventional CNC machine tools, industrial robot manipulators recently have exhibited better flexibility when dealing with highly complex tasks in Cartesian space. As a result, more and more factories have adopted industrial robot manipulators as their machining equipment. However, due to the fact that the industrial robot manipulator has highly complex dynamics and kinematics, the number of factors that will affect the contouring error for industrial robot manipulators is much larger than that for CNC machine tools [1]. Therefore, proper motion planning is crucial for industrial robot manipulators, especially in machining tasks.

Dynamic Movement Primitives (DMP) was proposed by S. Schaal *et al.* in 2003 [2], in which a stable second order system is designed so that its system state converges to the desire goal state (i.e. equilibrium point; attractor). Moreover, the demonstration trajectory is encoded by a function which is independent of time. This time-independent function is exploited to modify the system trajectory, while a temporal scaling term is introduced to control the convergence rate of the system state. However, this kind of approach has potential problems. As a result, a modified approach was proposed in [3] to cope with the scaling problem encountered in the original DMP approach. In addition, DMP plus was proposed in [4] to improve trajectory fitting accuracy. Although these modified methods can enhance the performance of the DMP-based approach, they fail to consider the scenario of a contour following task in real-world machining. Namely, contour error exists between the trajectory command and the real trajectory. As pointed out in [5,6], contour error depends on the curvature of the trajectory command. In addition, the reduction of contour error can be achieved by adjusting the feedrate command [7,8]. To overcome the drawbacks of the DMP-based methods, the proposed approach exploits the idea that the temporal scaling term of DMP can affect the convergence rate of the system state (i.e. feedrate command), in which the curvature of the trajectory command is used to adjust the value of the temporal scaling term of DMP. According to the theory of general relativity, a gravitational field will affect the curvature of its nearby space. By exploiting this concept, the curvature is regarded as the distance between the current position and the center of a virtual gravitational field in the proposed approach, which can be used to adjust the temporal scaling term of DMP. To verify the effectiveness of the proposed approach, several contour following experiments have been conducted on a 2 DOF robot arm. Experimental results indicate that the proposed approach indeed is feasible.

DOI: 10.1201/9781003514831-20

The remainder of the paper is organized as follows. Section 2 gives a brief review on DMP. Section 3 introduces the proposed time-dilation-based time scale adjustment approach for DMP. Experimental setup and results are provided in Section 4. Conclusions are addressed in Section 5.

2 BRIEF REVIEW ON DYNAMIC MOVEMENT PRIMITIVES

As mentioned in [2], for Dynamic Movement Primitives (DMP), a stable second order system such as the well-known spring-damper-mass system described by Eq. (1), is designed so that its system state converges to the desire goal state.

$$\ddot{y} = \alpha_y(\beta_y(g - y) - \dot{y}) \tag{1}$$

where$y$, $\dot{y}$, $\ddot{y} \in \mathbb{R}^n$ are the system state, 1st derivative of the system state, and 2nd derivative of the system state, respectively; g is the goal state; and α_y, β_y are system parameters.

If the values of the system parameters are chosen properly, the system state will converge to the goal state g. To control how the system state converges to goal state g, nonlinear function $f(x) : \mathbb{R} \to \mathbb{R}^n$ called the forcing term is used to learn demonstration trajectory y_{demo} so as to describe the convergence process. In addition, to decouple the forcing term and the time variable, the forcing term can be constructed using a canonical system described by Eq. (2).

$$\dot{x} = -\alpha_x x \tag{2}$$

where α_x is positive constant.

For any initial x_0, the system state will converge to 0. In this paper, x_0 is set to 1. When $x = 0$, it represents that system state y for Eq. (1) converges to goal state g.

Eq. (3) shows a common form for the forcing term.

$$f(x) = \frac{\sum_{i=0}^{N-1} \omega_i \psi_i(x)}{\sum_{i=0}^{N-1} \psi_i(x)} x \tag{3}$$

where ψ_i is the basis function, and ω_i is the weight of the basis function.

Adding temporal scaling term τ to Eq. (2) and substituting Eq. (2) and (3) into Eq. (1) will yield

$$\begin{aligned} \tau^2\ddot{y} &= \alpha_y(\beta_y(g - y) - \tau\dot{y}) + f(x)(g - y_0) \\ \tau\dot{x} &= -\alpha_x x \end{aligned} \tag{4}$$

In Eq. (4), the temporal scaling term is used to adjust the time scale of the original trajectory command. Note that the reason why the forcing term is multiplied with $(g - y_0)$ is to ensure that the demonstration trajectory can be scaled up/down spatially. In addition, if there is a slight change in the goal state or the sign of the goal state changes, the trajectory generated by the original DMP method may experience a significant deviation from the demonstration trajectory. To deal with this problem, the MDMP (Modified DMP) method developed in [4] decouples the forcing term and $(g - y_0)$, which is described by

$$\begin{aligned} \tau^2\ddot{y} &= \alpha_y(\beta_y(g - y) - \tau\dot{y}) - \alpha_y\beta_y(g - y_0)x + \alpha_y\beta_y f(x) \\ \tau\dot{x} &= -\alpha_x x \end{aligned} \tag{5}$$

The MDMI method is adopted in this paper. In particular, as described by Eq. (6), the basis function in the forcing term adopted in this paper is a Gaussian function.

$$\begin{aligned} &\psi_i(x) = exp(-h_i(x - c_i)^2) \\ &\text{where} \quad c_i = exp\left(-\alpha_x i \frac{1}{N}\right) \\ &h_i = \frac{1}{(c_{i+1} - c_i)^2}, \quad \text{for } i = 0, 1, ..., N-2 \\ &h_{N-1} = h_{N-2} \end{aligned} \tag{6}$$

Based on Eq. (5), one can derive the forcing function f_{target} to be learnt. When performing learning, τ is set to 1 and f_{target} is described by Eq. (7).

$$f_{target} = \frac{\ddot{y}_{demo} - \alpha_y(\beta_y(g - y_{demo}) - \dot{y}_{demo}) + \alpha_y\beta_y(g - y_0)x}{\alpha_y\beta_y} \tag{7}$$

where y_{demo}, $\dot{y}_{demo}$, $\ddot{y}_{demo}$ are the demonstration trajectory, 1st derivative of the demonstration trajectory, and 2nd derivative of the demonstration trajectory, respectively.

Weightings ω_i can be calculated using LWR, in which

$$\omega_i = \frac{\sum_{j=1}^{L} x_j \psi_j(x_j) f_{target,j}}{\sum_{j=1}^{L} x_j^2 \psi_j(x_j)}, L = \text{length}(y_{demo}) \tag{8}$$

3 THE PROPOSED TIME-DILATION-BASED TIME SCALE ADJUSTMENT APPROACH FOR DMP

In contour following tasks, the contour error for a point on the trajectory with a larger curvature is likely greater than that for a point with a smaller curvature. In addition, a smaller feedrate often results in a smaller contour error [9]. As a result, one strategy to reduce contour error for a contour following task is to reduce the feedrate where the curvature is large. Since the temporal scaling term τ in DMP can be used to adjust the convergence rate of the system state (i.e. feedrate along the trajectory), this paper proposes an approach that adjusts the value of the temporal scaling term in DMP so as to reduce contour error. According to the theory of general relativity, the gravitational field will result in time-space distortion. Moreover, the larger the curvature caused by the gravitational field, the slower time becomes. The concept of gravitational time dilation in the theory of general relativity is exploited in the proposed approach to adjust the temporal scaling term. In particular, the reciprocal of the curvature is regarded as the distance r between a point and the center of a non-rotating sphere. Consequently, one can extend the concept of gravitational time dilation in the theory of general relativity to adjust the temporal scaling term τ in DMP using Eq. (9).

$$\tau = \frac{\tau_0}{\sqrt{1-\frac{r_0}{r}}} \tag{9}$$

where r_0 is the radius of the non-rotating sphere that is smaller than r (i.e. the reciprocal of the curvature), and τ_0 is the initial temporal scaling term.

r becomes smaller when the curvature becomes larger. According to Eq. (9), temporal scaling term τ becomes larger when the curvature becomes larger. Figure 1 is the plot of τ with respect to r.

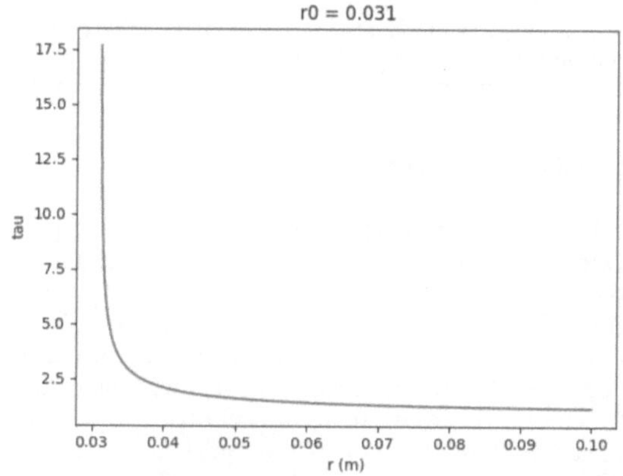

Figure 1. Plot of τ with respect to r ($r_0 = 0.031$).

4 EXPERIMENTAL SETUP AND RESULTS

4.1 *Generation of demonstration trajectory*

The demonstration trajectory $\vec{C}(u)$used in this paper is generated by a NURBS curve interpolator [10] with desired feedrate profile $f(t)$. One can use Eq. (10) to calculate parameter value $u(kT)$ at the kth sampling time instant so as to compute position command $\vec{C}(u)$ at the kth sampling time instant. Table 1 lists the parameter settings for $\vec{C}(u)$, and the plot of $\vec{C}(u)$ is shown in Figure 2.

$$\begin{aligned} u(kT+T) &\approx u(kT) + \frac{du(kT)}{dt}T + \frac{1}{2!}\frac{d^2u(kT)}{dt^2}T^2 \quad T \\ &\approx u(kT) + f(kT)\left\|\frac{d\vec{C}(u)}{du}\right\|^{-1} \\ &+ \frac{1}{2}\left[\frac{a(t)\left\|\frac{d\vec{C}(u)}{du}\right\|^{-1} - f^2(kT)\left\|\frac{d\vec{C}(u)}{du}\right\|^{-4}\left(\frac{d\vec{C}(u)}{du}\cdot\frac{d^2\vec{C}(u)}{du^2}\right)}{}\right]T^2 \end{aligned} \tag{10}$$

where T is the sampling time and $a(\cdot)$ is the derivative of the feedrate.

Table 1. Parameter settings for NURBS curve $\vec{C}(u)$.

Degree	3	
Control Points	X (m)	Y (m)
1	0.250	-0.300
2	0.204	-0.122
3	0.372	-0.020
4	0.448	0.101
5	0.250	0.300
Knots	[0, 0, 0, 0, 0.5, 1, 1, 1, 1]	

Figure 3 shows the curvature profile for the demonstration trajectory in NURBS form.

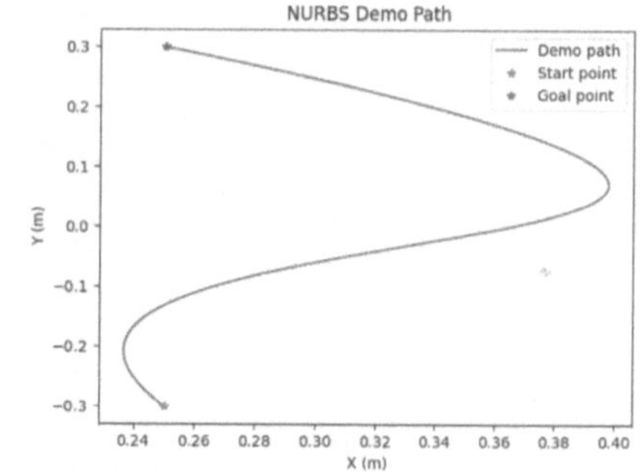

Figure 2. Demonstration trajectory generated by a NURBS curve interpolator.

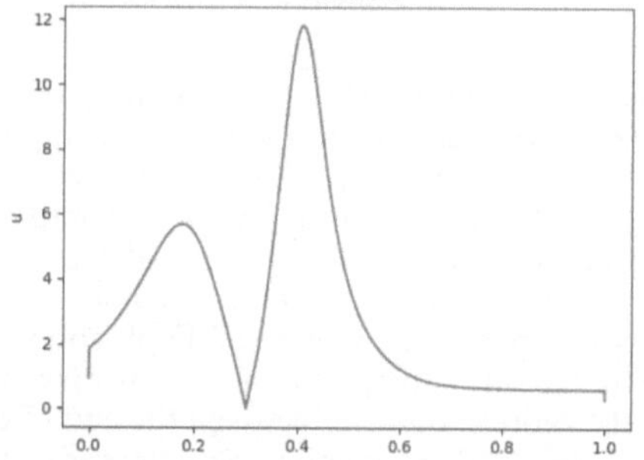

Figure 3. Curvature profile for the demonstration trajectory in NURBS form.

4.2 *Contour following experiment conducted on a 2 DOF planar robot*

To evaluate the performance of the proposed approach, a contour following experiment is conducted on an experimental platform consisting of a 2 DOF Planar robot. At first, the demonstration trajectory for DMP is generated using the parameter settings listed in Table 1 and Eq. (10). Figure 4(a) shows the demonstration trajectory and the modified trajectory generated by the original DMP method, while Figure 4(b) shows the demonstration trajectory and the modified trajectory generated by the proposed time-dilation-based time scale adjustment approach. Clearly, there exists some deviations between the demonstration trajectory command and the modified trajectory command generated by the original DMP method.

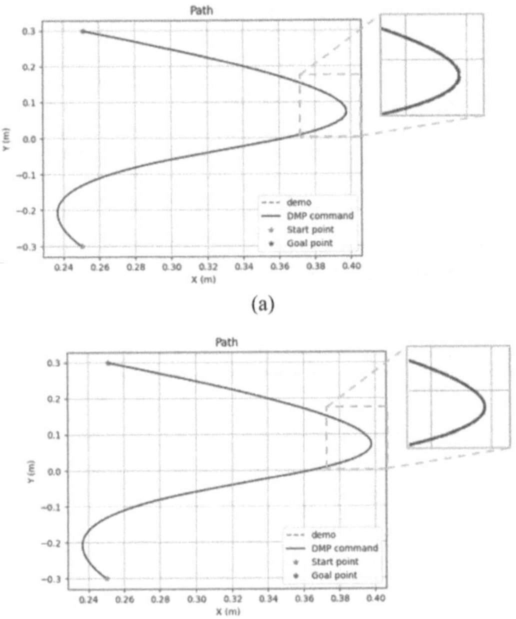

Figure 4. Trajectory command generated by DMP based approaches (a) modified trajectory command generated by the original DMP method (b) modified trajectory command generated by the proposed time-dilation-based time scale adjustment approach.

The Cartesian trajectories generated by both the original DMP method and the proposed approach need to be converted into the joint space commands for the robot. Figure 5 shows a comparison of the contour error between the trajectory command from the original DMP method and the proposed approach on a planar 2 DOF robot arm. Note that in this paper, the contour error is defined to be the shortest distance between the end-effector of the robot and the demonstration trajectory. In addition, the horizontal axis of the plot in Figure 5 is parameter u for the NURBS curve. Table 2 lists a comparison of the performance index for the original DMP method and the proposed approach. Experimental results shown in Figure 5 and Table 2 show that the proposed approach yields smaller contour error than the original DMP method.

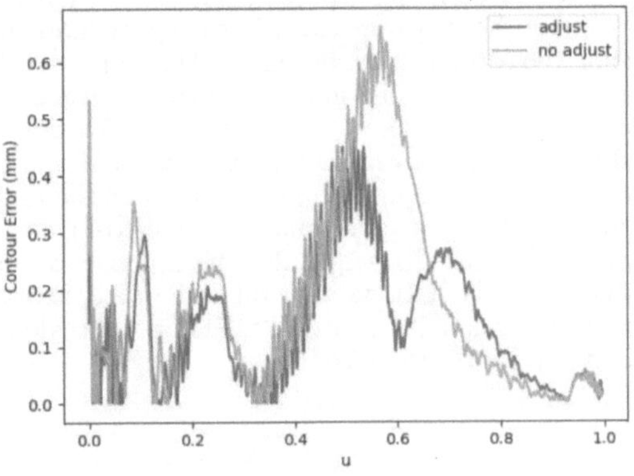

Figure 5. Comparison of contour error.

Table 2. Comparison of performance indices.

	Original DMP	Proposed approach	Percentage of contour error reduction
Average contourerror	0.1590 mm	0.1286 mm	19.12 %
Maximum contourerror	0.6615 mm	0.4652 mm	29.67 %

5 CONCLUSIONS

When applying DMP to trajectory planning problems for robot manipulators, the curvature for the demonstration trajectory is often not a constant. In addition, the contour error resulting from a contour following task for a robot manipulator depends on the curvature of the demonstration trajectory. Thereby, to reduce contour error, this paper exploits the concept of gravitational time dilation to adjust the temporal scaling term in DMP. In particular, by adjusting the temporal scaling term in DMP, one can reduce the contour error for a contour following task. Results obtained from a contour following task conducted on a planar 2 DOF robot arm verify the effectiveness of the proposed approach.

ACKNOWLEDGMENT

The authors would like to thank the National Science and Technology Council, Taiwan, for their support of this research under Grant No.: NSTC 112-2221-E-006 -135 -MY2.

REFERENCES

[1] Z.R. Zhu, X.W. Tang, C. Chen, F.Y. Peng, R. Yan, L. Zhou, Z.P. Li, J.W. Wu, "High precision and efficiency robotic milling of complex parts: Challenges, approaches and trends," *Chinese Journal of Aeronautics*, vol. 35, issue 2, pp. 22–46, Feb 2022.

[2] S. Schaal, A. Ijspeert, A. Billard, "Computational approaches to motor learning by imitation," *Philosophical Transactions of the Royal Society of London. Series B: Biological Sciences"*, vol. 358, issue. 1431, pp. 537–547, Mar 2003.

[3] M. Ginesi, N. Sansonetto, P. Fiorini, "Overcoming some drawbacks of dynamic movement primitives," *Robotics and Autonomous Systems*, vol. 144, 103844, Oct 2021.

[4] R. Wang, Y. Wu, W.L. Chan, K.P. Tee, "Dynamic movement primitives plus: for enhanced reproduction quality and efficient trajectory modification using truncated kernels and local biases," in *Proc. of the IEEE/RSJ Int. Conf. Intell. Robots Syst*, 2016, pp. 3765–3771.

[5] D. Zhang, Z. Wang, M. Tomizuka, "A variable-parameter-model-based feedforward compensation method for tracking control," *IEEE/CAA J. Automatica Sinica*, vol. 7, no. 3, pp. 693–701, May 2020.

[6] N. Huang, L. Hua, X. Huang, Y. Zhang, L. Zhu, D. Biermann, "B-spline-based corner smoothing method to decrease the maximum curvature of the transition curve," *Journal of Manufacturing Science and Engineering*, vol. 144, issue 144, 054503, May 2022.

[7] J. Yang, X. Yin, Y. Sun, "PSO-based feedrate optimization algorithm for five-axis machining with constraint of contour error," *Machines*, vol. 11, issue 144, 501, Apr 2023.

[8] Y. Guo, W. Niu, J. Zhou, H. Liu, "Near-time optimal feedrate planning for the NURBS curve considering interpolation error constraints," *Robotics and Computer-Integrated Manufacturing*, vol. 86, 102679, Apr 2024.

[9] J.L. Wong, M.-Y. Cheng, T.-Y. Huang, J.-C. Kuo, J.-M. Wu, "On the relationship between the desired feedrate and contour error constraints for 2-DOF robot Manipulators performing parametric curve following," in *proc. of the 2023 International Automatic Control Conference (CACS)*, 2023, pp. 1–8.

[10] M.-Y. Cheng, M.-C. Tsai, J.-C. Kuo, "Real-time NURBS command generators for CNC servo controllers," *International Journal of Machine Tools and Manufacture*, vol. 42, issue 7, pp. 801–813, May 2002.

System Innovation for an Artificial Intelligence Era – Lam et al. (Eds)
© 2025 the Author(s), ISBN: 978-1-032-84452-7

Inverse study of natural convection heat transfer for four diamond-arranged tubes in a hot box

Han-Taw Chen, Li-Yuan Hsu & Hung-Hsiu Liu
Department of Mechanical Engineering, National Cheng Kung University, Tainan, Taiwan

Wei-Mon Yan
Department of Energy and Refrigerating Air-Conditioning Engineering, National Taipei University of Technology, Taipei, Taiwan

ABSTRACT: The unknown heat transfer rates of the heater in the top and bottom tubes and an appropriate flow model using an inverse three-dimensional (3D) computational fluid dynamics (CFD) method combined with the least squares method and excessive experimental temperature data are predicted for natural convection heat transfer in a hot box with four diamond-arranged tubes. The present estimates are obtained under the constant heat transfer rate tube assumption. In addition, the heat dissipation from the kth tube, the unknown heat transfer rates of the remaining two heaters, the velocity pattern and the air temperature contours are also determined. The 3D heat conduction equation for tubes and the conservation equations of air are solved by coupling. Through testing of various turbulence models, it is found that the root mean square error between the numerical results of temperature obtained by the zero-equation model and the experimental data is smaller than that of other flow models. To verify the accuracy of the selected flow model, the obtained Nusselt number on the kth tube and the velocity pattern are also compared with existing correlation and interferometer photographs, respectively. The estimates obtained are compared with those obtained from four square-arranged tubes. A review of the literature revealed that few researchers have performed this study.

1 INTRODUCTION

Mun *et al.* [1] used the immersed boundary method based on the second-order finite volume method to investigate two-dimensional (2D) laminar natural convection in a square enclosure with four isothermal horizontal circular cylinders located at different rhombus positions. The cylinder temperature is higher than the wall temperature of the enclosure. Chen *et al.* [2] applied the 3D CFD method combined with the isothermal wall assumption, experimental temperature data to select an appropriate flow model and predict the 3D natural convection of the plate heat exchanger in a hot box. Chen *et al.* [3] applied a 3D inverse CFD method along with the constant heat transfer rate tube assumption (*CHTRTA*), experimental data and the least squares method (*LSM*) to predict the unknown heat transfer rate Q_k generated by the heating coil in the kth tube and the appropriate flow model for square-arranged tubes. The results show that the RNG k-ε model and enhanced wall treatment (EWT) are selected as an appropriate flow model and near-wall treatment. However, the Q_k value of each tube is assumed to be the same. The heat conduction equation of the tube is also not considered. It is known that too many thermocouples may affect air velocity field and temperature field. Thus, this study applies this inverse CFD method combined with *LSM*, *CHTRTA* and excessive experimental data.

T_a^m, T_w^m and T_{tk}^m to predict the appropriate flow model and Q_k value of each tube. However, the Q_k value of each tube is not necessarily the same. To verify the accuracy of the obtained estimates, the root mean square error (*RMSE*) between the numerical results of T_a, T_w and T_{tk} obtained from the appropriate flow model and experimental data T_a^m, T_w^m and T_{tk}^m needs to be smaller than that of other flow models. Furthermore, the obtained velocity patterns and the average Nusselt number on each tube are also compared with interferometer photographs [4,5], existing results [5] and correlation [6]. This study is significantly different from Refs. [1,3].

2 INVERSE CFD SIMULATION

A rhombus-shaped arrangement of four tubes in a cuboid hot box passes from its left wall to its right wall. The value of Q_k is assumed to be unknown and will be predicted. The effect of radiation is considered. Figure 1 shows the schematic representation of 2D physical geometry. Various flow models and wall functions are selected from ANSYS fluent 18.0 [7] for validation testing. The results show that the

DOI: 10.1201/9781003514831-21

zero-equation model can be selected as an appropriate flow model. Under the assumption of constant thermal properties except for the density in the buoyancy term of the momentum equation, incompressible flow, inviscid dissipation and the Boussinesq approximation, the 3D steady-state conservation equations of the zero-equation model can be expressed as follows. [7]

$$\frac{\partial u_i}{\partial x_i} = 0 \tag{1}$$

$$u_j \frac{\partial u_i}{\partial x_j} = -\frac{1}{\rho}\frac{\partial p}{\partial x_i} + \nu_{eff}\frac{\partial^2 u_i}{\partial x_j^2} + g_i\beta\delta_{i2}(T_a - T_\infty) \tag{2}$$

$$c_p u_j \frac{\partial T_a}{\partial x_j} = \frac{k_{eff}}{\rho}\frac{\partial^2 T_a}{\partial x_j^2} + \frac{S_h}{\rho} \tag{3}$$

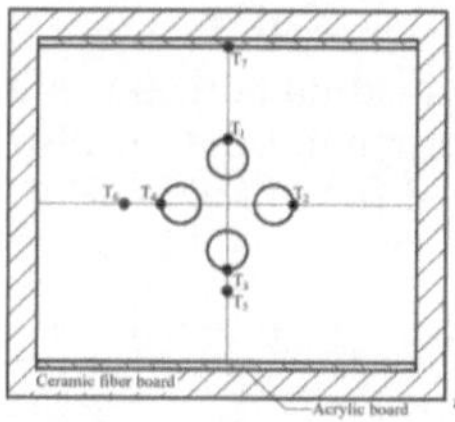

Figure 1. Schematic diagram of physical geometry on y-z plane with measurement locations.

where p, T_a, ρ, β, c_p, ν_{eff} and k_{eff} are the pressure, temperature, density, volumetric thermal expansion coefficient, specific heat, effective kinematic viscosity, and effective thermal conductivity of the air, respectively. g_i represents the component of gravitational acceleration. ν_{eff} and k_{eff} are defined as $\nu_{eff} = \nu + \nu_t$ and $k_{eff} = k_a + \rho\nu_t c_p/Pr_t$. ν_t is $\nu_t = \sqrt{2}l_m^{\,2}S_{ij}$. ν and k_a are the kinematic viscosity and thermal conductivity, respectively. The mixing length l_m is $l_m = \min(0.419d, 0.09d_{\max})$. d and $d_{\max}$ respectively represent the distance and maximum distance from the boundary of the computational domain. The mean strain rate S_{ij} is defined as $S_{ij} = (\partial u_i/\partial x_j + \partial u_j/\partial x_i)/2$. Pr_t is 0.9. S_h is the volumetric heat source. T_∞ is the ambient temperature.

The 3D steady-state heat conduction equations of the tube wall, the hot box wall and the heating coil are as follows.

$$\nabla^2 T_{tk} = 0 \tag{4}$$

$$\nabla^2 T_w = 0 \tag{5}$$

and

$$k_{e,h}\nabla^2 T_{hc} + \frac{Q_k}{V_h} = 0 \text{ for } 0 \le z \le W, \quad r \le D_i/2 \tag{6}$$

where T_{tk}, T_w and $T_{hc,k}$ are the temperature of the kth tube, the wall temperature of the hot box and the heating coil temperature inside the kth tube, respectively. $k_{e,h}$ and V_h are the effective thermal conductivity and volume containing the heating coil, respectively. V_h is $V_h = \pi D_i^2 L_t/4$. The unknown heat transfer rate Q_k produced by the heating coil in the kth tube will be predicted. The Q_1 and Q_3 values in the top and bottom tubes are assumed to be different. Due to symmetry, the Q_2 and Q_4 value in the two intermediate tubes is assumed to be the average of Q_1 and Q_3.

2.1 *Boundary conditions*

The computational domain is the same as that in Ref. [8]. The no-slip boundary conditions are assumed on the surfaces of each tube and hot box and on the boundaries of the computational domain. The temperature gradient is set to zero at boundaries other than the top boundary. The matching conditions of heat flux and temperature at the interfaces between the air and the hot box wall, the air and the kth tube wall, and the heating coil and the kth tube wall are written as follows.

$$T_w = T_a \text{ and } k_w\frac{\partial T_w}{\partial n} = k_a\frac{\partial T_a}{\partial n} \tag{7}$$

$$T_{tk} = T_a \text{ and } k_t\frac{\partial T_{tk}}{\partial n} = k_a\frac{\partial T_a}{\partial n} \tag{8}$$

$$T_{hc} = T_{tk} \text{ and } k_{e,h}\frac{\partial T_{hc}}{\partial n} = k_t\frac{\partial T_{tk}}{\partial n} \tag{9}$$

where n is the outward normal direction on the boundary.

The heat transfer coefficients h_{wi} and h_{wo} on the inner and outer surfaces of the hot box wall are determined from the following equations.

$$-k_w\frac{\partial T_w}{\partial n} = h_{wo}(T_w - T_\infty) \tag{10}$$

$$k_w\frac{\partial T_w}{\partial n} = h_{wi}(T_w - T_\infty) \tag{11}$$

where h_{wo} and h_{wo} both include convective and radiative heat transfer coefficient.

The boundary conditions at the inlet and outlet of each tube and the space containing the heating coil are as follows.

$$\partial T_t/\partial x = 0 \text{ and } \partial T_{hc}/\partial x = 0 \quad at\ x = -L_t/2 \text{ and } L_t/2 \tag{12}$$

The boundary conditions on the outer surface of the kth tube can be expressed as follows.

$$q_{kk}(\varphi, x) = -k_a\frac{\partial T_a}{\partial n} = h_{k\varphi}(T_{tk} - T_\infty) \tag{13}$$

$$Q_{hk} = \int_{-L/2}^{L/2}\int_0^{2\pi} q_{hk}(\varphi, x)d\varphi dx = A_h\bar{h}_k(\overline{T}_{tk} - T_\infty) \tag{14}$$

where A_h is $A_h = \pi D L_t$. $h_{k\varphi}$ and $\overline{h}_k$ are the local and average heat transfer coefficients on the outer surface of the kth tube, respectively.

If the heat conduction equation of the tube wall is not considered, Eq. (14) can be expressed as follow [8].

$$Q_k = A_h \overline{h}_k (\overline{T}_{tk} - T_\infty) \tag{15}$$

The local and average Nusselt numbers $Nu_{k\varphi}$ and Nu_k for the kth tube are respectively defined as follows.

$$Nu_{k\varphi} = h_{k\varphi} D / k_a \tag{16}$$

$$Nu_k = \overline{h}_k D / k_a \tag{17}$$

Many common correlations between the Nusselt number and the Rayleigh number for the natural convection heat transfer characteristics around an horizontally heated cylinder in free air have been proposed [3,4]. The correlation for square arrangements in Ref. [3] is cited to investigate its suitability for this study. This correlation between the Nusselt number Nu_{tk} and the Rayleigh number Ra_{tk} for the k-th tube is introduced as follows.

$$Nu_{tk} = c Ra_{kd}^{1/3} \tag{18}$$

where Ra_{tk} is defined as $Ra_{tk} = \frac{g\beta(\overline{T}_{tk} - T_\infty)D^3}{\nu\alpha}$. c is 0.1347 for the 1st tube, 0.1758 for the 3rd tube and 0.1553 for the 2nd and 4th tubes. $\overline{T}_{tk}$ is the average temperature of the k-th tube.

The correlation between Nu_{tk} and Ra_{tk} for $2.5 \leq S_t/D \leq 4.25$ in Ref. [6] is as follows.

$$Nu_{tk} = 0.63 Ra_{kd}{}^{0.254} [0.138(S_t/D) + 0.62454] \tag{19}$$

2.2 *Least squares method*

Q_k is predicted by the least squares method combined with four experimental temperature data T_{tk}^m. As shown in Ref. [3], the sum of squares of the total error E is as follows.

$$E(Q_1, Q_3) = \sum_{k=1,3} \left(T_{tk}^m - T_{tk} - \frac{\partial T_{tk}}{\partial Q_k} \Delta Q_k \right)^2 \tag{20}$$

To minimize E, the first derivative of E with respect to Q_1 and Q_3 is taken to be zero. Then, ΔQ_1 and ΔQ_3 can be obtained as follows.

$$\Delta Q_1 = \frac{CD_1 - BD_2}{AC - B^2} \tag{21}$$

$$\Delta Q_3 = \frac{AD_2 - BD_1}{AC - B^2} \tag{22}$$

where $A = \sum_{k=1}^4 \left(\frac{\partial T_k}{\partial Q_1}\right)^2$, $B = \sum_{k=1}^4 \left(\frac{\partial T_k}{\partial Q_1}\frac{\partial T_k}{\partial Q_2}\right)$, $c = \sum_{k=1}^4 \left(\frac{\partial T_k}{\partial Q_2}\right)^2$, $D_1 = \sum_{k=1}^4 \left(T_k - T_k^m\right)\frac{\partial T_k}{\partial Q_1}$ and $D_2 = \sum_{k=1}^4 \left(T_k - T_k^{mea}\right)\frac{\partial T_k}{\partial Q_2}$.

Q_1^k and Q_3^k are assumed to be the kth guess value, then Q_1^{k+1} and Q_3^{k+1} can be expressed as:

$$Q_1^{k+1} = Q_1^k + \Delta Q_1 \tag{23}$$

$$Q_3^{k+1} = Q_3^k + \Delta Q_3 \tag{24}$$

The initial guesses Q_1^0 and Q_3^0 for Q_1 and Q_3 can be given arbitrarily. It is found from Ref. [6] that the experimental value of Q_k is 6.46 W. Q_1^0 and Q_3^0 are both set to 6.00 W. The convergence can be achieved when $\left|Q_1^{k+1} - Q_1^k\right| / Q_1^k < 0.01$ and $\left|Q_3^{k+1} - Q_3^k\right| / Q_3^k < 0.01$. The total number of iterations N_{it} for each case is about five.

The conditions for selecting an appropriate flow model are as follows. The estimate of Nu_{tk} obtained by the appropriate flow model is the closest to the existing correlation compared to other selected flow models. In addition, the $RMSE$ between the obtained numerical results for T_{tk}, T_w and T_a and experimental data is also the smallest among all selected flow models. $RMSE$ is expressed as follows.

$$RMSE = \sqrt{\left[\left(\frac{T_w^m - T_w}{T_w^m}\right)^2 + \sum_{(k=1)}^4 \left(\frac{T_{tk}^m - T_{tk}}{T_{tk}^m}\right)^2 + \sum_{j=1}^2 \left(\frac{T_{aj}^m - T_{aj}}{T_{aj}^m}\right)^2\right] / 7} \tag{25}$$

3 EXPERIMENTAL METHODS AND PROCEDURES

The experiments of this study have been completed in Ref. [6]. The detailed experimental methods and procedures can be found in Refs. [3,6]. The length, height and width of the rectangular enclosure are 0.06 m, 0.06 m, and 0.09 m, respectively. The 3D schematic diagram of the experimental setup is shown in Figure 2. Figure 3 shows the schematic diagram of physical geometry in 2D. The width W_b of the hot box is 0.26 m. Its height and length L_b are both 0.22 m. The W_b and L_b values of the ceramic fiber board and acrylic board are the same as those of the hot box. The thickness of acrylic plate and ceramic fiber board are 0.005 m and 0.02 m, respectively. A heating coil with an input power of 200 W is wound around an aluminum rod with a tube diameter of 0.012 m and a length of 0.3 m. The aluminum rod is then placed in the hollow aluminum tubes to heat the air. The distance S_{tw} between the top wall and the center of the top tube is $S_{tw} = (L - \sqrt{2}S_t)/2$. The outer diameter, thickness and length of each hollow tube are respectively 0.027 m, 0.002 m and 0.230 m. The emissivity of aluminum tube and acrylic is 0.06 and 0.9, respectively. The distance S_t between the centers of the top tube and the middle tube is 0.045 m. The thermal conductivity of the aluminum tube at 350 K is 239 W/mK. [3,6]

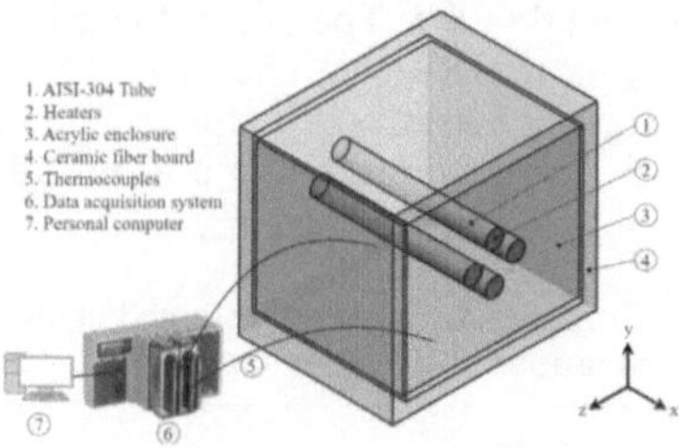

Figure 2. Schematic diagram of physical geometry of the experimental device.

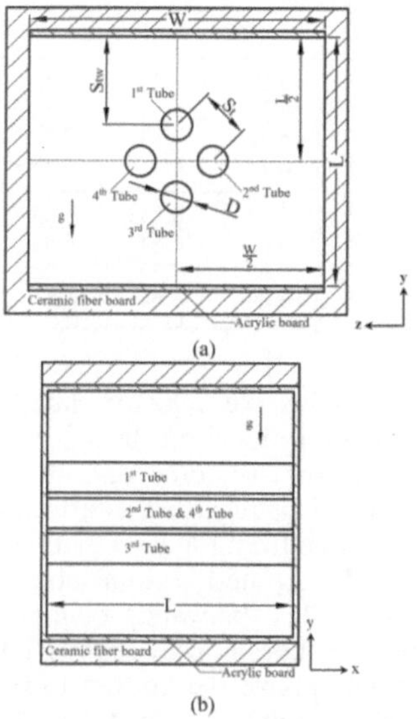

Figure 3. Schematic diagram of physical model (a) y-z plane, (b) x-y plane.

The hot box is heated by the heating coil for about 3 hours until all the measured temperatures reach a steady state. Seven T-type thermocouples are applied to measure T_{tk}^{m}, T_{w}^{m} and T_{aj}^{m}. A thermocouple is placed outside the computational domain to measure T_{∞}. The maximum accuracy of T-type thermocouple is within 0.4%. All thermocouple readings are recorded from t=0 to steady state. In general, the experiment needs to be repeated more than two times.

4 RESULTS AND DISCUSSION

The thermophysical properties are evaluated at $(T_{tk}+T_{\infty})/2$. They are determined by $[(T_{t1}+T_{t2}+T_{t3}+T_{t4})/4+T_{\infty}]/2$ in Ref. [3]. S_t is 0.045 m. Table 1 shows a comparison between the estimates of Q_k and $RMSE$ obtained from various flow models and experimental data. It is found that the $RMSE$ obtained from the zero-equation model is the smallest compared to the RNG and REAL k-ε models combined with EWT. The obtained estimate of Q_k is also closest to the experimental value than that obtained by the two k-ε models. Thus, the zero-equation model is chosen as an appropriate flow model. This result is different from laminar flow in Ref. [2] and the RNG k-ε model combined with EWT in Ref. [3]. This means that the choice of an appropriate flow model may be related to tube arrangement and boundary conditions. It is worth mentioning that the isothermal tubes and constant heat transfer rate tubes in Refs. [2,3,6] are assumed respectively. The Q_k value of each tube is assumed to be the same in Refs. [3,6]. Due to heat loss and the influence of the tube wall, the heat transfer rate Q_{hk} dissipated from the tube is lower than the estimated and experimental values of Q_k. The estimates of Q_{hk} are found to be 30% and 0.31% lower than the estimated and experimental values of Q_k, respectively. The estimate of T_k is close to the average temperature of the tube.

Table 1. Comparison between estimates obtained from various flow models.

	Diamond arrangement, T_{∞}=296.25 K			
	Present estimates, $Q_1 \neq Q_3$			
	Zero-eq.	RNG, EWT	REAL, EWT	Exp. data Q_1=Q_3 [6]
T_{t1} (K)	351.71	354.28	353.86	353.12
T_{t2} (K)	346.51	348.14	345.33	350.47
T_{t3} (K)	341.66	345.78	341.70	343.07
T_{t4} (K)	346.56	348.69	345.68	350.86
T_{a1} (K)	309.27	305.89	321.69	310.69
T_{a2} (K)	325.71	326.76	322.47	321.95
T_{w1} (K)	325.18	328.93	320.18	321.25
Q_{h1} (W)	4.27	3.78	9.33	–
Q_{h2} (W)	4.45	4.09	7.94	–
Q_{h3} (W)	4.64	4.86	6.40	–
Q_{h4} (W)	4.45	4.22	7.93	–
Q_{t1} (W)	6.28	5.69	11.59	6.46
Q_{t2} (W)	6.31	5.89	9.86	6.46
Q_{t3} (W)	6.33	6.09	8.12	6.46
Q_{t4} (W)	6.31	5.89	9.86	6.46
$\bar{T}_{t1}$ (K)	351.50	354.04	353.77	–
$\bar{T}_{t2}$ (K)	346.65	348.24	345.78	–
$\bar{T}_{t3}$ (K)	341.53	345.63	341.64	–
$\bar{T}_{t4}$ (K)	346.64	348.74	345.93	–
$RMSE$ (%)	0.94	1.30	1.57	–

Table 2 shows a comparison of estimates obtained from various flow models for Q_k =6.46 W in Ref. [6]. It can be seen from Tables 1 and 2 that due to the buoyancy and blocking effects of the four tubes, the temperature T_{t1} of the uppermost tube is the highest and the temperature T_{t3} of the bottom tube is the lowest. Due to symmetry, the tube temperatures T_{t2} and T_{t4} are almost the same. The estimates of Q_k obtained by the zero-equation model and the RNG k-ε and REAL k-ε models combined with SWF is higher than the experimental values of Q_k. The estimates of Q_k obtained by the RNG k-ε and REAL k-ε models combined with EWT are the same and lower than the experimental values of Q_k. The present results are different from those in Ref. [6]. It is also not easy to select an appropriate flow model under the assumption that the heat transfer rates of the four tubes are the same. However, the appropriate flow model can be easily selected through this inverse CFD method. The above results show that the influence of the tube wall and the assumptions of boundary conditions cannot be ignored.

Table 2. Comparison of estimates obtained from various flow models for T_∞ = 296.25 K and Q_k = 6.46 W [8].

		RNG k-ε		REAL k-ε	
	Zero-eq.	SWF	EWT	SWF	EWT
T_{t1} (K)	353.52	353.30	354.17	353.60	354.24
T_{t2} (K)	348.30	348.62	348.54	348.58	348.59
T_{t3} (K)	344.60	344.21	343.89	343.91	343.79
T_{t4} (K)	348.30	348.64	348.54	348.70	348.51
T_{a1} (K)	314.89	312.85	311.47	312.49	311.10
T_{a2} (K)	324.26	323.82	323.81	323.21	323.96
T_{w1} (K)	321.51	323.41	323.49	324.10	323.55
$\bar{h}_1$ (W/m^{2}K)	6.10	6.17	5.58	5.77	5.50
$\bar{h}_2$ (W/m^{2}K)	6.71	6.72	6.18	6.74	6.17
$\bar{h}_3$ (W/m^{2}K)	7.23	7.34	6.78	7.35	6.79
$\bar{h}_4$ (W/m^{2}K)	6.71	6.32	6.18	6.72	6.18
RMSE (%)	0.70	0.54	0.50	0.54	0.51
Q_k (W)	6.82	6.87	6.30	6.88	6.30

Table 3 shows a comparison between the estimates obtained by $\overline{T}_{tk}$ and T^m_{tk} and existing correlations (18) and (19). The obtained estimates of Nu_{tk} and $\bar{h}_k$ are found to be in good agreement with the correlations (18) and. Therefore, the present estimates have good accuracy. The estimates of Nu_{tk} and $\bar{h}_k$ obtained by $\overline{T}_{tk}$ are slightly different from those obtained by T^m_{tk}. The estimates of Nu_{tk} and $\bar{h}_k$ obtained by Q_{hk} are smaller than those obtained by Q_k because the heat conduction equation of the tube wall is taken into account. The estimate of Q_{hk} is approximately 0.7 times the estimate of Q_k. This means that the heat conduction equation for the tube wall may need to be considered. The $\bar{h}_1$ estimate is 25% lower than the $\bar{h}_3$ estimate. It is found in Ref. [6] that for triangular-arranged tubes, the thermal plumes from the two bottom cylindrical tubes significantly affect the wall heat transfer rate of the top cylindrical tube for the triangular-arranged array. For the triangular configuration of three isothermal tubes with the same temperature, the average Nusselt number of the top tube is 37% smaller than that of the single tube. Therefore, the present estimate of $\bar{h}_k$ is reasonable. The heat transfer rate from the top tube is strongly affected by the thermal plumes rising from the two bottom tubes.

Table 3. Comparison between obtained estimates and existing correlations.

	Estimates		Eq. (19)	Eq. (18)	
	Eq. (14)	Eq. (15)	$\overline{T}_{tk}$	$\overline{T}_{tk}$	T^m_{tk}
Ra_{t1}	75211				76575
Ra_{t2}	70856				74325
Ra_{t3}	65786				70487
Ra_{t4}	70846				74662
Nu_{t1}	3.80	5.59	5.43	5.43	5.72
Nu_{t2}	4.38	6.21	6.43	6.43	6.53
Nu_{t3}	5.12	6.96	7.13	7.10	6.88
Nu_{t4}	4.38	6.43	6.43	6.43	6.54
$\bar{h}_1$(W/m^{2}K)	3.94	5.80	5.83	5.83	5.96
$\bar{h}_2$(W/m^{2}K)	4.53	6.42	6.64	6.64	6.78
$\bar{h}_3$(W/m^{2}K)	5.26	7.15	7.31	7.28	7.07
$\bar{h}_4$(W/m^{2}K)	4.53	6.64	6.64	6.64	6.88

Figure 4 presents the velocity pattern obtained from zero-equation model on the x = 0 plane. The arrangement of the top tube and the two intermediate tubes can be regarded as a triangular configuration. It is seen from Figure 4 that the buoyancy plumes above the top and bottom tubes rise almost steadily upward. The buoyancy plumes above the two intermediate tubes appear slightly inward. The buoyancy plume above the bottom tube rises steadily and flows around the top tube, as shown in Figure 10 of Ref. [6]. These results also coincide with interferometer photograph [4,5]. The buoyant plume above the top tube induces that above the other tubes and rise to hit the top wall of the hot box. Two symmetrical natural circulations are formed at the upper left and right corners of the hot box. The air then flows downward along the side walls. It is induced by plumes from the top tube and the two intermediate tubes before reaching the bottom of the hot box. Then, it is heated again by these three tubes. The cold air stationary at the bottom of the hot box is also induced by plumes from four tubes. This cold air flows upward along the bottom tube into the area between the two intermediate tubes. It is heated and continues to travel along the path of the plume above the top tube.

The air temperature contours on the x = 0 plane is shown in Figure 5. It is seen from Figure 5 that the thermal plumes above the 2nd and 4th tubes are slightly inclined inward, as shown in Figure 11 of Ref. [6]. The thermal plumes above the top and bottom tubes rise steadily in the vertical direction. The bottom half of the top tube is significantly affected by the heated thermal plumes arising from the two intermediate tubes. This means that the air temperature field between

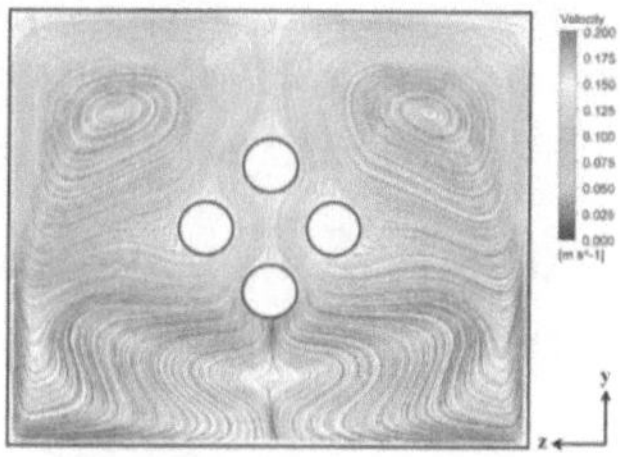

Figure 4. Velocity pattern obtained from zero-equation model on $x = 0$ plane.

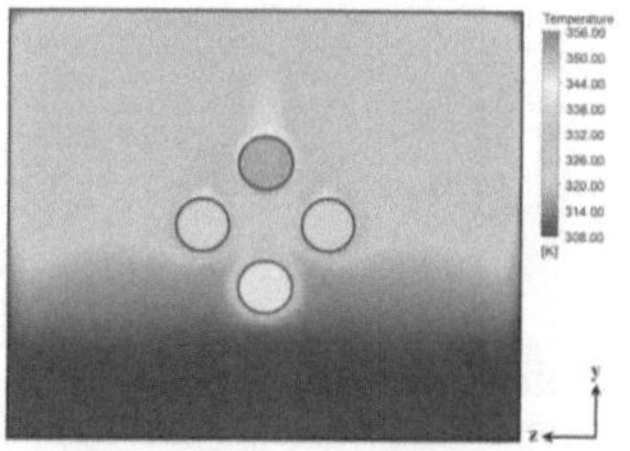

Figure 5. Air temperature contours obtained from zero-equation model on $x = 0$ plane.

the two intermediate tubes exhibits interaction. The cold air flows upward from the bottom of the bottom tube, so that its temperature is lower than that of the other three tubes. Therefore, the lengths of the buoyancy plume and thermal plume above the top tube are longer than those above the other three tubes.

The hot thermal plume further engulfs the top tube. This causes the air temperature around it to rise. The temperature gradient intensity decreases significantly near the top tube. Therefore, an overall deterioration in the heat transfer rate at the top tube surface is expected.

The values of S_t/L and Ra_{dk} in this study are closer to those in Figure 5a of Ref. [2]. However, their velocity patterns and air temperature contours are significantly different. This deviation may be due to the fact that the temperatures of the four tubes are assumed to be the same in Ref. [2] under the 2D laminar flow assumption.

5 CONCLUSIONS

The main goal of this study is to apply the CFD method combined with the least squares method and excessive experimental temperature data to predict Q_k, Nu_k, and 3D turbulent natural convection heat transfer and fluid flow characteristics for four diamond-arranged tubes. The results show that the temperatures of the four tubes are not kept the same. The zero-equation model is more suitable for this study. This result is different from the RNG k-ε model combined with EWT for square-arranged tubes [3]. This means that the appropriate flow model and wall function chosen may be related to the geometry of the tube arrangement. The obtained velocity pattern agrees with the interferometric image in Refs. [4,5]. The buoyancy and thermal plumes above the two intermediate tubes are slightly tilted inward. The buoyancy and thermal plumes above the uppermost tube are longer than those above the other tubes. The obtained estimates of Nu_k are in good agreement with the correlations (18) and (19). The estimate of Nu_1 is the largest and the estimate of Nu_3 is the smallest. Since there is no interference from the 2nd and 4th tubes, the estimates Nu_1 and Nu_3 are also close to those for four square-arranged tubes in Ref. [3].

ACKNOWLEDGEMENTS

The authors acknowledge the financial support from the National Science and Technology Council of the Republic of China in Taiwan (Grant No. NSTC 112-2221-E-006-054).

REFERENCES

[1] Mun G. S., Park Y. G., Yoon H. S., Kim M., Ha M. Y., (2017). Natural convection in a cold enclosure with four hot inner cylinders based on diamond arrays (Part-I: Effect of horizontal and vertical equal distance of inner cylinders), *Int. J. Heat Mass Transf.* 111, 755–770.

[2] Chen H. T., Chang S. C., Hsu M. H., You C. H., (2121). Experimental and numerical study of innovative plate heat exchanger design in simplified hot box of SOFC, *Int. J. Heat Mass Transf.* 181, 121880.

[3] Chen H. T., Hsu M. H., Yang K. C., Chang K.H., Liu K.C., (2022). Study of inverse natural convection-conduction heat transfer for in-line tube heat exchanger in a hot box with experimental data, *J. Taiwan Inst. Chem. Eng.* 141, 104600.

[4] Narayan S., Singh A. K., Srivastava A., (2017). Interferometric study of natural convection heat transfer phenomena around array of heated cylinders, *Int. J. Heat Mass Transf.* 109, 278–292.

[5] Holman J. P., (1989). *Heat Transfer*, McGraw-Hill, Inc., Singapore, 1st printing, pp. 330–331.

[6] Hsu M. H., (2021). *Study on 3D natural convection heat transfer characteristics of four heating tubes in a closed cavity (MS thesis)*, National Cheng Kung University, Taiwan.

[7] *Fluid Dynamics Software*, Fluent 18, Lebanon, NH-USA, 2018.

System Innovation for an Artificial Intelligence Era – Lam et al. (Eds)
© 2025 the Author(s), ISBN: 978-1-032-84452-7

Study of 3D inverse natural convection heat transfer for single-tube plate-finned tube heat exchanger in an open cavity

Han-Taw Chen, Hung-Hsiu Liu & Li-Yuan Hsu
Department of Mechanical Engineering, National Cheng Kung University, Tainan, Taiwan

Wei-Mon Yan
Department of Energy and Refrigerating Air-Conditioning Engineering, National Taipei University of Technology, Taipei, Taiwan

Saman Rashidi
Department of Mechanical Engineering, Ferdowsi University of Mashhad, Mashhad, Iran

ABSTRACT: An inverse computational fluid dynamic together with the least squares technique and three fin experimental data is used for prediction of the unknown heat transfer rate of the heater Q, an appropriate flow model, and the heat dissipation from the fins Q_{fn} and the tube Q_t. The potentials of a number of turbulence models for the simulation of this problem are evaluated. The present estimates are obtained under the constant heat transfer rate. Through testing of various turbulence models, the RNG k-ε model provides the closest results to existing results and correlations, and its root mean square error is smaller than that of other models. The obtained estimates of h_{f2} and h_t are also compared with existing results and correlations to benchmark the accuracy of the selected model. The results show that the obtained estimates of h_{f2} and ht are close to existing results and the proposed correlations. Compared to the isothermal tube assumption, this inverse technique not only needs less experimental temperature data but also provides estimates of Q, Q_{fn} and Q_t.

1 INTRODUCTION

The inverse heat conduction method (*IHCM*) combined with T_k^{mea} is used to predict the heat transfer coefficient on the middle fin in this study [1]. It is found in Ref. [1] that the fin is divided into eight sub-fin regions. The average of the experimental temperature data measured at four different locations is regarded as the fin base temperature. Therefore, 12 thermocouples are needed to measure the required experimental temperature data. However, too many thermocouples can affect the flow field and temperature field. For this reason, this study applies the inverse CFD method along with three experimental fin temperature data and the least squares method (LSM) to predict the unknown heat transfer rate Q, the dissipated heat Q_{fn} from the nth fin and an appropriate flow model. The heat transfer coefficients $\bar{h}_t$ and $\bar{h}_{f2}$ on the tube and middle fin are also predicted. The root mean square error (*RMSE*) obtained from an appropriate flow model needs to be smaller than that of other flow models. To validate the accuracy of the chosen appropriate flow model, the estimates of $\bar{h}_t$, $\bar{h}_{f2}$ and Q_{f2} are compared with those in Ref. [2] and the correlation in Refs. [3,4].

2 3D INVERSE CFD METHOD

The schematic physical geometry of this study is shown in Figure 1. The positions of the eight thermocouples placed on the middle fin are shown in Figure 2. The laminar flow and various turbulent models are selected from Ansys Fluent 18 [5] for validation testing. The effect of radiation is considered. As shown in Ref. [2], the RNG k-ε turbulence model is more suitable for this study compared to other flow models. Under the assumption of constant thermal properties except for the density in the buoyancy term of the momentum equation, incompressible flow, inviscid dissipation and the Boussinesq approximation, the 3D steady-state conservation equations of the RNG k-ε model can be expressed as follows.

$$\frac{\partial u_i}{\partial x_i} = 0 \tag{1}$$

$$u_j \frac{\partial u_i}{\partial x_j} = 2\nu_{eff} \frac{\partial S_{ij}}{\partial x_j} - \frac{1}{\rho}\frac{\partial p}{\partial x_i} + g_i \beta \delta_{i2}(T_a - T_\infty) - \frac{2}{3}\frac{\partial (k\delta_{ij})}{\partial x_i} \tag{2}$$

$$c_p u_j \frac{\partial T_a}{\partial x_j} = \frac{k_{eff}}{\rho}\frac{\partial^2 T_a}{\partial x_j^2} + 2\nu_{eff}\frac{\partial u_i S_{ij}}{\partial x_j} - \frac{S_h}{\rho} \tag{3}$$

DOI: 10.1201/9781003514831-22

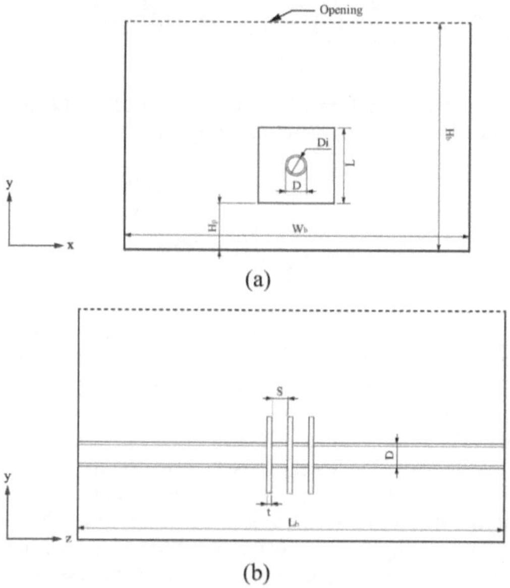

Figure 1. Schematic diagram of physical model and computational domain in dashed lines. (a) x-y plane, (b) y-z plane.

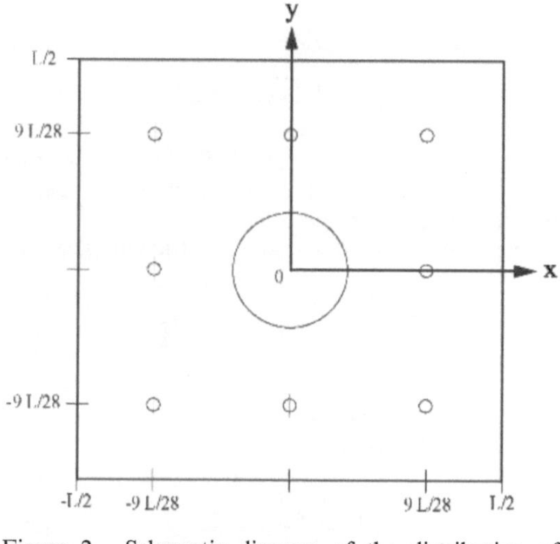

Figure 2. Schematic diagram of the distribution of measurement locations on the middle fin.

where u_i, T_a, ρ, β and c_p represent the velocity component, temperature, density, volumetric thermal expansion coefficient and specific heat of air, respectively. p stands for pressure. g_i represents the component of gravitational acceleration. x_1, x_2 and x_3 respectively represent x, y and z.

The turbulent kinematic viscosity ν_t is $\nu_t = C_\mu k^2/\varepsilon$. The effective kinematic viscosity ν_{eff} and effective thermal conductivity k_{eff} are defined as $\nu_{eff} = \nu_t + \nu$ and $k_{eff} = k_a + \rho c_p \nu_t / Pr_t$. δ_{i2} and δ_{ij} are Kronecker deltas. The mean strain rate S_{ij} is defined as $S_{ij} = (\partial u_i/\partial x_j + \partial u_j/\partial x_i)/2$. The turbulent Prandtl number Pr_t is 0.85. S_h is the volumetric heat source. k is the turbulent kinetic energy.

$$u_j \frac{\partial k}{\partial x_j} = \frac{\partial}{\partial x_j}\left(\alpha_k \nu_{eff} \frac{\partial k}{\partial x_j}\right) + \frac{G_k + G_b}{\rho} - \varepsilon \quad (4)$$

$$u_j \frac{\partial \varepsilon}{\partial x_j} = \frac{\partial}{\partial x_j}\left(\alpha_\varepsilon \nu_{eff} \frac{\partial \varepsilon}{\partial x_j}\right) + C_{1\varepsilon}\frac{\varepsilon}{\rho k}(G_k + C_{3\varepsilon} G_b) - C_{2\varepsilon}\frac{\varepsilon^2}{k} - \frac{C_\mu \eta^3 (1-\eta/\eta_0)}{1+\beta\eta^3}\frac{\varepsilon^2}{k} \quad (5)$$

where ε is the turbulent dissipation rate. α_k, α_ε, C_μ, $C_{1\varepsilon}$, $C_{2\varepsilon}$, η_0, and β are given as $\alpha_k = \alpha_\varepsilon = 1.393$, $C_\mu = 0.0845$, $C_{1\varepsilon} = 1.42$, $C_{2\varepsilon} ==1.68$, $\eta_0 = 4.38$ and $\beta = 0.012$. $C_{3\varepsilon}$ is set to zero when the air velocity is perpendicular to gravity. G_k and G_b respectively represent the turbulent kinetic energy due to the average velocity gradient and buoyancy. η represents the ratio of the characteristic time scale of turbulence to the mean flow field. The definitions of G_k, G_b and η can be found in Ref. [5].

The 3D steady-state heat conduction equations for fins, tube and heater are as follows.

$$\nabla^2 T = 0 \text{ at } z = 0 \text{ and } \frac{\partial T_t}{\partial z} = 0 \text{ at } z = W_b \quad (6)$$

$$\nabla^2 T = 0 \text{ for } 0 \le z \le W_b, \quad D_i/2 \le r \le D/2 \quad (7)$$

$$\mathrm{k_{e,h}} \nabla^2 T_{hc} + \frac{Q}{V_h} = 0 \text{ for } 0 \le z \le W_b, \quad r \le D_i/2 \quad (8)$$

where T, T_t and T_{hc} are the fin temperature, tube temperature and the temperature of the thermofoil heater, respectively. D_i and D are the inner and outer diameters of the hollow tube, respectively. $\mathrm{k_{e,h}}$ and V_h are respectively the effective thermal conductivity of the thermofoil heater and the volume of the inner tube containing it. V_h is $V_h = \pi D_i^2 L_t/4$. The unknown heat source Q generated by the thermofoil heater will be predicted.

2.1 *Boundary conditions*

At the top opening of the box, the air temperature and air velocity are assumed to be T_∞ and zero, respectively. The pressure is considered to be atmospheric pressure. The adiabatic boundary condition is assumed at the edges of the fins and at the side and bottom walls of the box. The no-slip boundary conditions are assumed on all solid surfaces. The matching conditions for heat flux and temperature are as follows.

- At k-th fin-air interface:

$$T = T_a \text{ and } k_f \frac{\partial T}{\partial n^*} = k_a \frac{\partial T_a}{\partial n^*} = h_{fk}(T_\infty - T) \quad (9)$$

- At tube-air interface:

$$T_t = T_a \text{ and } k_t \frac{\partial T_t}{\partial n^*} = k_t \frac{\partial T_a}{\partial n^*} = h_t(T_\infty - T_t) \quad (10)$$

- At fin-tube interface:

$$T = T_t \text{ and } k_f \frac{\partial T}{\partial n^*} = k_t \frac{\partial T_t}{\partial n^*} \quad (11)$$

where k_f and k_t are the thermal conductivity of the fin and tube, respectively. h_{fk} and h_t are the heat transfer coefficients on the k-th fin and tube, respectively. Both h_{fk} and h_t include the convective and radiative heat transfer coefficients. n^* is the outward normal to the boundary. The boundary conditions are as follows.

- The tube and heater at the inlet and outlet:

$$\frac{\partial T_t}{\partial z} = 0 \text{ at } z = 0 \text{ and } \frac{\partial T_t}{\partial z} = 0 \text{ at } z = W_b \quad (12)$$

- On the inner and outer surfaces of the tube:

$$q_t(\varphi, x) = -k_t \frac{\partial T_t}{\partial r} = h_t(T_t - T_\infty) \text{ at } r = D/2 \quad (13)$$

$$Q_h = -k_t A_t \frac{\partial T_t}{\partial r} \text{ at } r = D_i/2 \quad (14)$$

where q_t and h_t are the heat flux dissipated from the tube and the local heat transfer coefficient on the tube, respectively. Q_h is the heat transfer rate applied to the tube.

2.2 *Grid point system*

The computational domain is the same as Ref. [2]. A hexahedral unstructured grid system with non-uniform distribution of grid points is used to obtain numerical results. The relative convergence criteria for velocity and air temperature are 10^{-6} and 10^{-8}, respectively. The number of iterations is set to 10,000 times. Generally, the numerical results can be converged after about 6000 iterations.

3 LEAST SQUARES METHOD (LSM)

The residual sum of squares (RSS) between the numerical results of the fin temperature obtained at the 1st, 4th and 6th measurement positions and the experimental data T_1^{mea}, T_4^{mea} and T_6^{mea} is shown below.

$$RSS = \sum_{k=1,4,6} \left(T_k^m - T_k - \partial T_k/\partial QQ\right)^2 \quad (15)$$

The minimization of RSS with respect to Q can determine the correction value Q as:

$$Q = \frac{\sum_{k=1,4,6} \left(T_k^m - T_k - \partial T_k/\partial Q\right)}{\sum_{k=1,4,6} \left(\partial T_k/\partial Q\right)^2} \quad (16)$$

The initial guess Q_0 is set to 15 W in this study. The new estimate Q_m is denoted as $Q_m = Q_{m-1} + \Delta Q$ for $m = 1, 2, ..., N_{it}$. This calculation process is repeated until $|Q/Q|$ is less than 0.01. The total number of iterations N_{it} for each case is about five. The experimental value obtained from $Q = I_{Ac}V_{Ac}$, where I_{Ac} and V_{Ac} are the current and voltage through the thermofoil heater, respectively.

The average heat transfer coefficient $\bar{h}_{fk}$ and the heat dissipation Q_{fk} on the kth fin are expressed as follows.

$$\bar{h}_{fk} = \frac{\int_{A_f} h_{fk}(x,y)(T - T_\infty)dA}{\int_{A_f}(T - T_\infty)dA} \quad (17)$$

$$\begin{aligned} Q_{fk} &= 2\int_{A_f} h_{fk}(x,y)(T - T_\infty)dA \\ &= 2\bar{h}_{fk}A_f(\bar{T}_{fk} - T_\infty) \end{aligned} \quad (18)$$

where A_f is $A_f = L^2 - \pi D^2/4$. h_{fk} and T_t are the local heat transfer coefficient on the k-th fin and tube temperature, respectively.

The average heat transfer coefficient $\bar{h}_t$ is as follows.

$$\bar{h}_t = \frac{\int_0^{2\pi} h_t(T_t - T_\infty)d\varphi}{\int_0^{2\pi}(T_t - T_\infty)d\varphi} \quad (19)$$

If T_t is independent of φ, Eq. (19) can be rewritten as follows.

$$\bar{h}_t = \frac{\int_0^{2\pi} h_{t,\varphi}d\varphi}{2\pi} \quad (20)$$

The heat conduction equation of the tube is considered. The dissipated heat Q_t from the tube can be obtained from Eq. (13) as follows.

$$\begin{aligned} Q_t &= \int_{-L_t/2}^{L_t/2}\int_0^{2\pi} q_t(\varphi, z)d\varphi dz \\ &= \int_{-L_t/2}^{L_t/2}\int_0^{2\pi} h_t(T_t - T_\infty)d\varphi dz \\ &= A_t\bar{h}_t(\bar{T}_t - T_\infty) \end{aligned} \quad (21)$$

where A_t is $A_t = \pi D L_t$, φ is the counterclockwise angle from the horizontal axis on the right side of the tube. $\overline{T}_t$ is the average tube temperature.The local and average Nusselt numbers $Nu_{t,\varphi}$ and Nu_t are defined as follows.

$$Nu_{t,\varphi} = h_{t,\varphi} D / k_a \quad (22)$$

$$Nu_t = \overline{h}_t D / k_a \quad (23)$$

Three common correlations between Nu_t and the Rayleigh number Ra_d for the natural convection heat transfer characteristics around an unrestricted horizontally heated cylinder in free air are as follows. [31,32]

$$Nu_t = 0.48 Ra_d^{1/4} \text{ for } 10^4 \leq Ra_d \leq 10^7 \quad (24)$$

$$Nu_t = 0.36 + \frac{0.51 Ra_d^{1/4}}{\left[1 + (0.559/Pr)^{9/16}\right]^{4/9}} \quad (25)$$

$$Nu_t = \left[0.6 + 0.387\left(\frac{Ra_d^{1/6}}{\left[1 + (0.559/Pr)^{9/16}\right]^{8/27}}\right)\right]^2 \quad (26)$$

where Ra_d is defined as $Ra_d = \frac{g\beta(\overline{T}_t - T_\infty)D^3}{\nu a}$. The Prandtl number Pr is set to 0.7.

The modified correlation proposed in this study is as follows.

$$Nu_t = 0.51 Ra_d^{1/4} \text{ for } 10^4 \leq Ra_d \leq 10^7 \quad (27)$$

The root mean square error (*RMSE*) obtained numerical results of T_0 and T_k and the experimental data T_0^{mea} and T_k^{mea} is as follows.

$$RMSE = \sqrt{\frac{1}{9}\sum_{k=0}^{8}\left(\frac{T_k^{mea} - T_k}{T_k^{mea} - 273.15}\right)^2} \quad (28)$$

4 EXPERIMENTAL METHODS AND PROCEDURES

To avoid repetitions, detailed diagrams of the experimental setup and experimental methods can be found in Ref. [2]. The fin thickness t, fin length L and k of the fin are 0.001 m, 0.126 m and 14.9 W/mK. The emissivity of acrylic and fins is 0.9 and 0.17, respectively. The outer diameter D, thickness and length of the hollow tube are 0.0273 m, 0.002 m and 0.55 m, respectively. The outer surface area A_t of the tube and the side surface area A_f of the fin are 0.04717 m^2 and 0.009415 m^2, respectively. A cylindrical rod with a length of 0.1 m and an outer diameter of 0.02 m is wrapped around a thermofoil heater with an outer diameter of 1.5 mm. Power input of 200 W is supplied to the heater [2]. Then, this rod is inserted into the hollow tube to heat the tube and fins. All measured temperatures reach a steady state if the tube and fins are heated for about 2 hours by the rod with heating coils. Figure 2 shows a schematic diagram of the distribution of measurement locations on the middle fin. 12 T-type thermocouples are used to measure T_0 and fin temperature T_k^{mea}, as shown in Refs. [1,2]. The fin base temperature T_0 is the average temperature measured at four different interface locations between the fin base and the tube. The accuracy of T-type thermocouple is ±0.4%.

5 RESULTS AND DISCUSSION

The thermophysical properties required to obtain numerical results for fins and tube are determined by $(\overline{T}_f + T_\infty)/2$ and $(\overline{T}_t + T_\infty)/2$, respectively. The fin spacing S is 0.015 m. It is worth noting that three experimental fin temperature measurements T_1^{mea}, T_4^{mea} and T_6^{mea} are required. Table 1 shows a comparison between the estimates obtained from two different flow models, the inverse results [1] and those in Refs. [2,6]. It is seen in Table 1 that the *RMSE* obtained from the RNG k-ε model is smaller than that obtained the zero-equation model. The difference between T_0 obtained by the RNG k-ε model and the experimental data [1,2,6] is 0.8%. It is found from Ref. [1] that four thermocouples placed at the interface between the middle fin and the tube are fixed at four different positions at the base of the fin using satlon cyanoacrylate adhesive. The gap between the middle fin and the tube is filled with satlon cyanoacrylate adhesive to reduce the contact thermal resistance between them. The effect of contact thermal resistance on the estimates under the isothermal tube assumption is smaller than that under the assumption of a constant heat transfer rate within the tube. The hybrid method combined with the isothermal tube assumption in Ref. [1] can only obtain the results of $\overline{h}_{f2}$. The estimate of $\overline{h}_t$ and the heat dissipated from 1st and 3rd fins are not determined. The $\overline{h}_{f2}$ estimates obtained by the zero-equation model and the RNG k-ε model are also close to the inverse result [1] and those in Ref. [2].

It is seen from Table 1 that the estimate of Q obtained by the RNG k-ε model is 24.61 W. This estimate of Q is smaller than the experimental value of 200 W. This difference may be caused by the heat loss dissipated from the inner tube to the surroundings and contact thermal resistance. Therefore, the heat loss in this study can be as high as 88%.

Table 2 presents a comparison between the obtained estimates of Nu_t and $\overline{h}_t$ and the existing correlations (24)-(27). It is found from Tables 1 amd 2 that the estimate of $\overline{h}_t$ obtained from the RNG k-ε model is closer to the proposed modified correlation (27) than the zero-equation model. The estimates of Nu_t and $\overline{h}_t$ obtained from the RNG k-ε model are closer to the correlations (24) and (27) than the correlations (25) and (26). Thus, the RNG k-ε model is selected as the appropriate flow model. This result is the same as that of Ref. [2]. The above results show that this inverse CFD method has good accuracy and is more practical than the hybrid method combined with the isothermal tube assumption [2]. Thus, the RNG k-ε model are chosen to obtain the estimates for this study.

Table 1. Comparison between estimates obtained from various flow models and existing results for T_∞ = 299.0 K.

	Present estimates, Constant heat transfer rate		Isothermal [2]	
	RNG k-ε	Zero-eq.	RNG k-ε	Exp. data
T_0 (K)	339.65	340.22	336.86	336.86
T_1 (K)	310.69	310.53	312.45	314.24
T_2 (K)	318.89	318.85	320.64	322.25
T_3 (K)	310.69	310.54	312.47	314.08
T_4 (K)	314.27	314.29	312.82	312.21
T_5 (K)	314.27	314.30	312.89	311.94
T_6 (K)	308.10	308.19	307.24	307.14
T_7 (K)	313.27	313.53	311.41	309.65
T_8 (K)	308.10	308.19	307.26	305.04
$\overline{T}_f$ (K)	312.29	312.30	312.50	312.07
$\overline{T}_t$ (K)	342.08	342.72	–	–
$\overline{h}_{f1}$ (W/m^{2}K)	6.06	6.45		–
$\overline{h}_{f2}$ (W/m^{2}K)	12.75	12.98	13.06	14.05
$\overline{h}_t$ (W/m^{2}K)	7.48	8.15	–	
Q_{f1} (W)	2.27	2.37	–	–
Q_{f2} (W)	3.19	3.25	3.35	3.51
Q_t (W)	16.59	18.34	–	–
Q (W)	24.61	26.65	–	–
$RMSE$(%)	7.24	7.58	3.96	–

Table 2. Comparison between the obtained estimates of Nu_t and existing correlations.

	Estimates					
	Zero	RNG	(24)	(25)	(26)	(27)
Ra_d	64029			63346		
Nu_t	8.72	8.01	7.615	6.56	6.90	8.09
$\overline{h}_t$	8.89	8.16	7.76	6.68	7.03	8.25

The numerical results in Figures 3 and 4 are obtained from the RNG k-ε model. Figure 3 presents the velocity pattern at $z = S^*/2$, where S^* is defined as $S^* = (t + S)/2$. Figure 4 shows the air temperature contours at $z = S^*/2$. It is seen from Figures 3 and 4 that due to the presence of the side walls of the box and the different assumptions of the tube, the velocity patterns and air temperature contours obtained are slightly different.

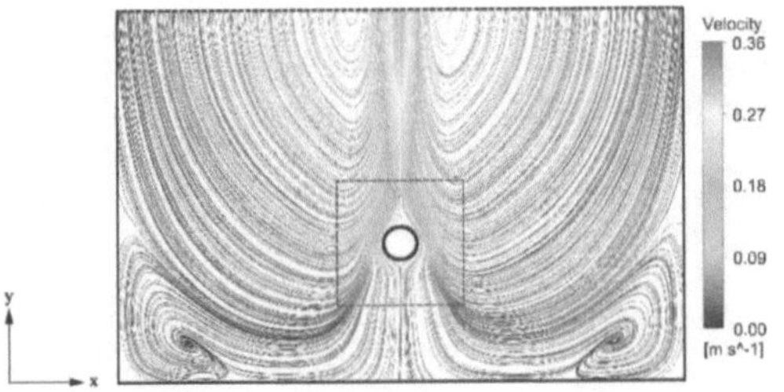

Figure 3. Velocity patterns on $z = S^*/2$ plane.

The air in the box with the upward opening is heated by the circular tube and fins and flows upward due to the buoyancy plume generated. The upward-moving natural convective boundary layer flow occurs between two adjacent fins. Then, the cold air from the surroundings flows into the box through the top opening. This air flows downward along the side walls toward the bottom and is induced by low-velocity buoyancy plume from the tube. Another cold air stagnant at the bottom of the box is also driven upward through the tube and the space between two adjacent fins. It enters between the two fins and flows through the horizontal tube. The cold air from two different areas is heated and continues to travel along the plume path above the tube. They are heated again by the hot tube and fins. The above flow phenomena lead to the appearance of two symmetrical circulations to appear at the two bottom corners. The buoyancy plume above the tube rises steadily. The faster air velocity occurs in the narrow area above the top opening of the tube.

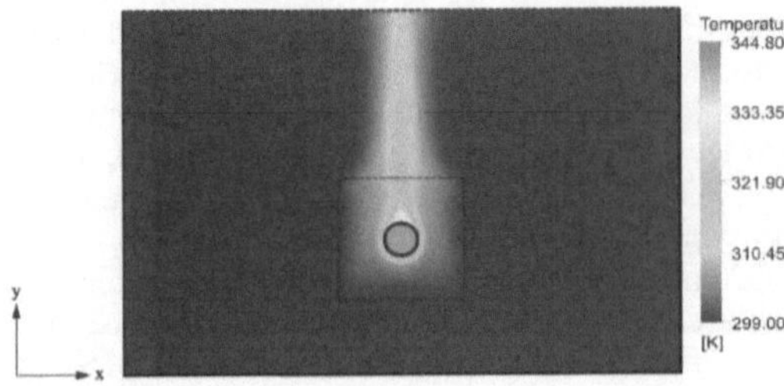

Figure 4. Air temperature contours on $z = S^*/2$ plane.

6 CONCLUSION

The inverse 3D CFD method combined with constant heat transfer rate assumption, least squares method and three fin experimental data is used to predict the fluid flow and heat transfer characteristics, Q, appropriate flow model, Q_{fn}, and Q_t. It is found that the *RMSE* obtained by the RNG k-ε model is closer to the experimental data than that obtained by the zero-equation model. The estimates of and Q_{f2} obtained from the RNG k-ε model are close to existing inverse and numerical results [1,2,6]. However, the estimates of Nu_t obtained from the RNG k-ε model are in better agreement with correlations (34) and (37) than those obtained from the zero-equation model. This demonstrates that the selected RNG k-ε model chosen is accurate. The present estimates have good accuracy and reliability. This inverse CFD method can be used to verify the accuracy of existing correlations. It is worth mentioning that the estimates obtained from the constant heat transfer rate tube assumption requires fewer experimental temperature measurements compared to the isothermal tube assumption. The estimates of $\bar{h}_t$, Q and Q_{f1} were not obtained in Refs. [1,2] with the isothermal tube assumption. The thermal plume occurs above the tube between the two fins, so the air velocity is the fastest in the area above the tube compared to other areas. The impact of contact thermal resistance on the obtained estimates and heat loss cannot be ignored. Its effect under the isothermal tube assumption is less than that under the constant heat transfer rate tube assumption. The temperature and velocity fields of the air between the two fins are affected by the tubes and the fins. A review of the literature revealed that this study has not yet appeared in the public literature.

REFERENCES

[1] Chen H.T. and Chou J.C., (2006) Investigation of natural-convection heat transfer coefficient on a vertical square fin of finned-tube heat exchangers, *Int. J. Heat Mass Transf.* 49 3034–3044.

[2] Chen H.T., Lin Y.S., Chen P. C. and Chang J.R., (2016) Numerical and experimental study of natural convection heat transfer characteristics for vertical plate fin and tube heat exchangers with various tube diameters, *Int. J. Heat Mass Transf.* 100 320–331.

[3] Arpaci V.S., Kao S.H. and Selamet A., (1999) in: *Introduction to Heat Transfer*, Prentice-Hall, NJ, pp. 571–574.

[4] Churchill S.W. and Chu H., (1975) Correlating equations for laminar and turbulent free convection from a horizontal cylinder, *Int. J. Heat Mass Transf.* 18 1049–1053.

[5] *Fluid Dynamics Software*, Fluent 18, Lebanon, NH-USA, 2018.

[6] Lin Y.S., (2021) *Study on 3D Natural Convection Heat Transfer Characteristics of Four Tubes in a Closed Cavity (MS thesis)*, National Cheng Kung University, Taiwan.

System Innovation for an Artificial Intelligence Era – Lam et al. (Eds)

A real-time energy management control for a hybrid electric vessel using whale optimization algorithm

Yi-Hsuan Hung
Undergraduate Program of Vehicle and Energy Engineering, National Taiwan Normal University, Taipei, Taiwan

Chien-Ming Lin
Department of Industrial Education, National Taiwan Normal University, Taipei, Taiwan

Chi Li
Department of Mechatronic Engineering, National Taiwan Normal University, Taipei, Taiwan

ABSTRACT: This study focuses on the optimal energy management of a hybrid electric vessel with PEM fuel cell/lithium batteries hybrid energy sources. Firstly, MATLAB/Simulink was utilized to build a dynamics model incorporating energy sources, the transmissions, the traction motors, the propellers and the nonlinear vessel body. In order to control the hybrid energy sources, we built the DCDC converter model in MATLAB/ Simulink , the DCDC converter model was in the rear part of PEM fuel cells to manage the output power of PEM fuel cell. Through the semi-active structure, the control system will be more consistent with the real situation and more effectively. A Whale Optimization Algorithm (WOA) was employed for calculating the Energy Distribution Ratio (EDR) between two electric energy sources. There are 10 search agents and 50 iterations were set in this algorithm, to optimize the objective function, which is the total energy consumption. To validate the reliability, two rapid-prototyping controllers simulated the energy management system and the system dynamics with real-time computation. A multi-mode rule-based control strategy (RBCS) was developed to verify the benefit of WOA. The RBCS has four modes, including battery mode, fuel cell mode, fuel cell charge mode and hybrid mode. The results show that WOA improved 15%+ energy consumption compared with RBCS. In conclusion, this research significantly enhances the energy efficiency of hybrid electric vessels. The real vehicle implementation of the control unit will be conducted in the future.

1 INTRODUCTION

With the international community committed to achieving net-zero emissions and sustainable development goals by 2050, countries are inclined to use electric vehicles as a means of transportation [1]. Not only are vehicles transitioning towards electrification with changes in energy policies, but vessels are also no exception. The International Maritime Organization (IMO), established the International Convention for the Prevention of Pollution from Ships (MARPOL) in the latter half of the last century to regulate pollution caused by ships at sea [2]. Appendix VI specifically regulates air pollution caused by vessels, and as regulations become increasingly stringent, the transition of vessels' energy sources toward green energy is becoming prominent in the future [3]. Therefore, this study aims to optimize the energy management system of harbor tugboats through the Whale Optimization Algorithm (WOA). It is anticipated that this optimization will effectively reduce energy consumption, thus achieving the goal of energy conservation and carbon reduction.

2 LITERATURE REVIEW

Energy management strategies can be slightly divided into the following two major categories: the most easily used is Rule Base Control (RBC), which is an energy management system that is controlled by "if-else" syntax with the knowledge and experience of experts. For example, the power of the fuel cell is divided into a number of equal parts and the power of the fuel cell is determined by the state of charge (SOC) of the lithium battery [5]. The fuzzy control is also another kind of rule base control, and there is also literature that uses fuzzy control as an energy management strategy for electric vehicles [8]. Some optimization methods are to observe natural phenomena or animal behavior through Metaheuristic algorithms, also known as bionic algorithms, such as Gray Wolf Optimization Algorithm (GWO), African Vulture Optimization Algorithm (AVOA), etc. [6,7]. From above, the algorithm used in this study is the Whale Optimization Algorithm (WOA), which emulates the behavior of whales during the feeding process. This algorithm simulates the whale's

DOI: 10.1201/9781003514831-23

behaviors during the prey process to find the best food source in the space (for the optimal control variables). In this paper, the WOA will optimize the power distribution between lithium batteries and fuel cells to minimize the ship's energy consumption, so that it can effectively reduce power consumption and extend the traveling mileage.

3 VESSEL DYNAMICS MODELING

This study takes DAMEN STAN TUG 1205 T7 as an example to retreive vessel-related physical parameters [8], and combines it with lithium battery and fuel cell modules, high-power motor modules and DC/DC converter modules to form a dual electric ship model. The vessel parameters are shown in Table 1.

Table 1. The vessel parameters.

Vessel parameter	Value
Mass (include added mass)	57.75 t
Thrust deduction coefficient	0.145
Wetted area	341.5 m^2
Gearbox reduction ratio	3.82
Air resistance coefficient	0.83
Water resistance coefficients	0.0045
Water density	1025 kg/m^3
Diameter of propeller	1.05 m

3.1 *Sailing driving cycle*

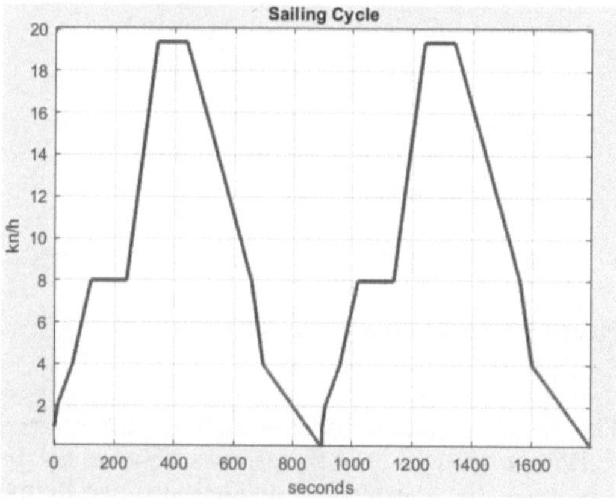

Figure 1. Sailing driving cycle.

Because there is no international standardized operation applied in the field of vessels, this study designed a sailing driving cycle according to the working characteristics of the tugboat as the test conditions of this study. It simulates the tugboat in the working process of the speed state, the time of a cycle is 900 seconds, and the maximum speed is 19.4 knots / h.

3.2 *Driving model*

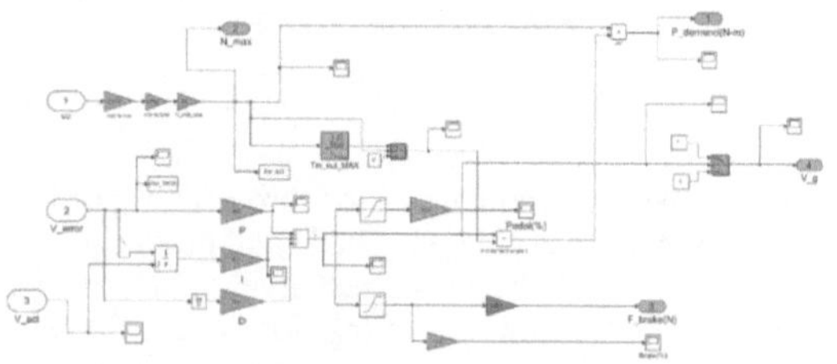

Figure 2. Driving model.

In this study, the PID controller is used to simulate the acceleration and deceleration behaviors of a ship. It can make the experimental results closed to the real situation.

The module calculates the velocity gain by scaling, integrating and differentiating the vehicle error values.

The velocity gain is calculated by equation (1). Note that if the velocity gain is less than zero, it means the braking is in progress, and the braking torque can be calculated by equation (2).

$$V_{gain}(t) = \left[K_p(\Delta V) + K_I \int_0^t (\Delta V)dt + K_d \frac{d}{dt}(\Delta V)\right] \tag{1}$$

$$F_b(t) = G_b\left[K_p(\Delta V) + K_I \int_0^t (\Delta V)dt + K_d \frac{d}{dt}(\Delta V)\right] \tag{2}$$

3.3 *Traction motor model*

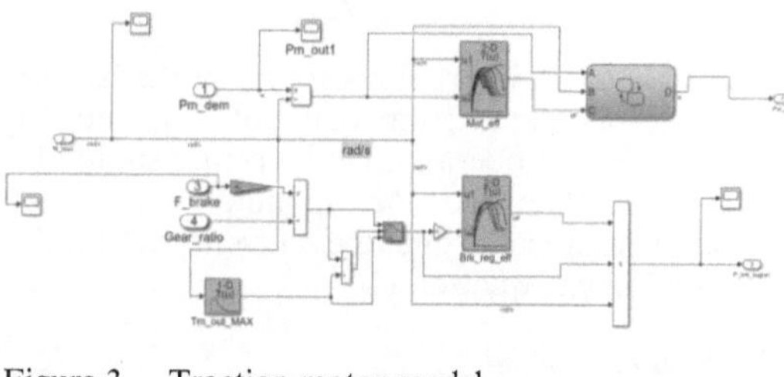

Figure 3. Traction motor model.

The energy management system calculates the power required at the next time step, and the output efficiency and brake recharge efficiency can be obtained using a look-up table:

$$T_m = \frac{P_m}{N_m} \tag{3}$$

$$\eta_m(t) = f(T_m(t), N_m(t)) \tag{4}$$

$$\eta_{m,dis} = \eta_m(T_{drv}, N_m) \tag{5}$$

$$\eta_{m,reg} = \eta_m(T_{reg}, N_m) \quad (6)$$

$$if\, P_m(t) > 0 \; P_{in} = T_{drv} \times N_m \div \eta_{m,dis} \quad (7)$$

$$elseif\, P_m(t) < 0 \; P_{reg} = T_{reg} \times N_m \div \eta_{m,reg} \quad (8)$$

Using the above equations, the torque and speed of the motor can be calculated. The actual output power of the motor and the brake recharge power can be calculated. If the motor is in traction mode, the output power can be calculated by using Equation (7), and if the braking is in progress, the braking power can be calculated by using Equation (8).

3.4 *Lithium battery model*

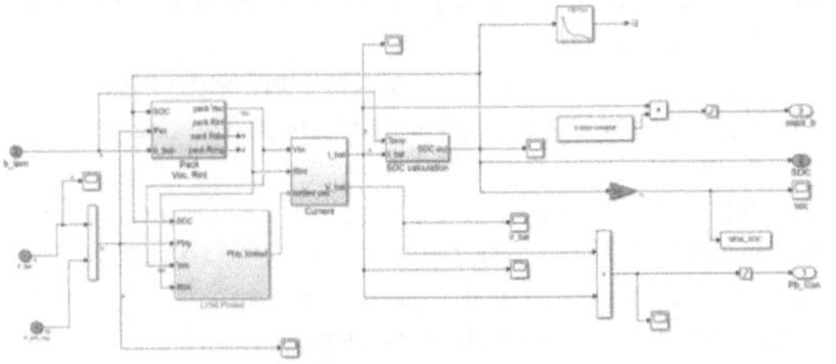

Figure 4. Lithium battery model.

The lithium battery model designed in this study was operated at 25°C and the charging and discharging currents were determined from the internal resistance of the battery and the open-circuit voltage. The State of Charge (SOC) of the battery is estimated as follows:

$$R_{bat} = f(SOC_{bat}, T_{bat}) \quad (9)$$

$$I_{bat} = \frac{V_{oc} - \sqrt{{V_{oc}}^2 - 4 \times P_d \times R_d}}{2 \times R_{bat}} \quad (10)$$

The equivalent internal resistance of the battery should be calculated from equation (9), and the current SOC and operating temperature should be used to calculate the corresponding equivalent internal resistance, while the current should be calculated from equation (10).

3.5 *Fuel cell model*

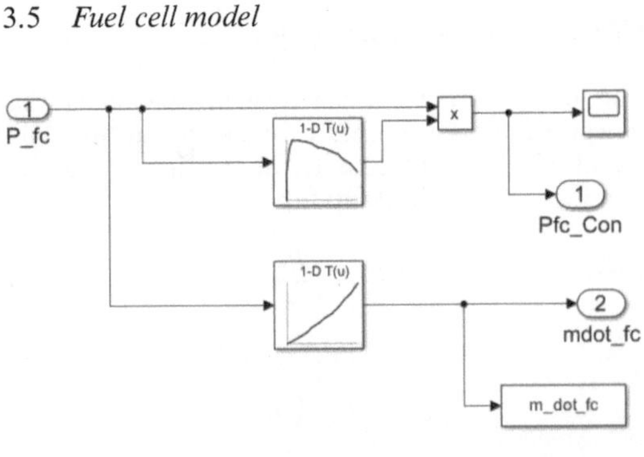

Figure 5. Fuel cell model.

The fuel cell efficiency and hydrogen consumption tables can be used to calculate output efficiency and hydrogen consumption. Equations (11, 12) show the relationship for fuel cell efficiency and hydrogen consumption, respectively:

$$\eta_{fc}(t) = \eta_{fc}(P_{fc}(t)) \quad (11)$$

$$Consumption_{H2}(t) = Consumption_{H2}(P_{fc}(t)) \quad (12)$$

3.6 *DC converter model*

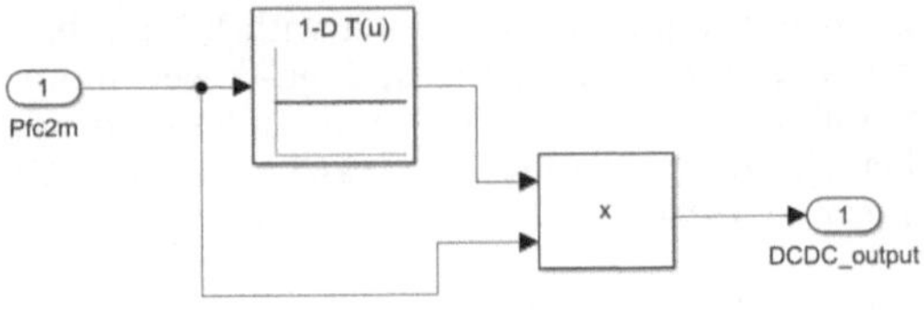

Figure 6. DC converter model.

In this study, a DC converter model was built at the back-end of the fuel cell, which can be more consistent with the real situation through the semi-active architecture, and it can make the controller more efficient.

The relationship between the output of the DC converter back-end and the fuel cell output is shown in Equation (13).

$$P_{DCDC} = P_{fc} \cdot \eta_{DCDC}(P_m) \quad (13)$$

3.7 *Vessel dynamic model*

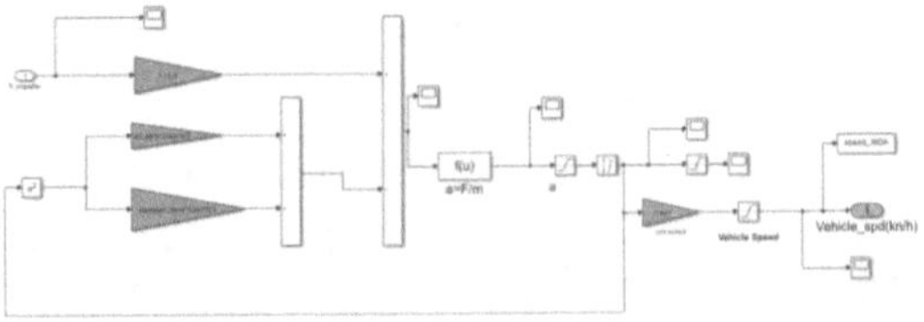

Figure 7. Vessel dynamic model.

The vessel dynamic model is used to estimate the resistance of the ship in the process of traveling. In this study, the resistance is divided into two items, one is the air resistance and the other is the hydrodynamic resistance, as shown in Equation (14, 15). The latter contains two items of the water friction resistance and the wave making resistance.

$$F_{wind} = \frac{1}{2} \times 0.83 \times \rho_{air} \times A_{vessel}(V_{vessel})^2 \quad (14)$$

$$F_{hydro} = \frac{1}{2} \times 0.0045 \times \rho_{sea} \times S_{vessel}(V_{vessel})^2 \quad (15)$$

The actual propulsion force can be found by subtracting the total resistance from the ship's thrust, as shown in equation (16),

$$T_{act} = T(1 - t_d) - \left(F_{wind} + F_{hydro}\right) \quad (16)$$

There are many different calculation methods for ship thrust T. In this study, equation (17) is used to calculate it. DHP is delivery power of motor and η_o is propeller output efficiency, the efficiency of the tugboat is between 0.3~0.6, in this study, we assume the efficiency of propeller is 0.6.

$$T = DHP \times \eta_o \times 74.5/V_{vessel} \quad (17)$$

3.8 *Transmission model*

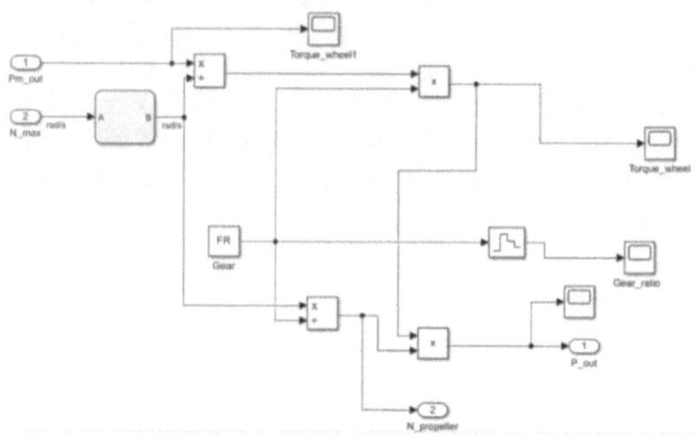

Figure 8. Transmission model.

Equation (18) divides the output power by the motor speed and multiplies it by the final deceleration ratio to obtain the output torque of the motor, which is designed with reference to the final deceleration ratio of DAMEN STAN TUG 1205 T7.

$$T_m = P_m \div N_m \times FR \quad (18)$$

4 ENERGY MANAGEMENT STRATEGY

4.1 *Rules base control (RBC)*

Considering the operating conditions and working situations of harbor-side tugboats, this study designs four different operating modes to determine the power of lithium batteries and fuel cells by lithium battery SOC and vessel demand speed, as shown in Table 2, and the corresponding power distribution ratios of these four modes are shown in Table 3.

Table 2. Mode type.

Mode	Condition
Fuel Cell Mode	Velocity over than 15 knot/h
Hybrid Mode	Velocity between 10 knot/h–15 knot/h
Battery Mode	Velocity less than 10 knot/h
Recharging Mode	SOC less than 40%

Table 3. Mode type.

Mode	Power Distribution Ratio
Fuel Cell Mode	$P_{fc2m} = 0.7P_d$ $P_{bat} = 0.3P_d$ $P_{fc2b} = 0$
Hybrid Mode	$P_{fc2m} = 0.3P_d$ $P_{bat} = 0.7P_d$ $P_{fc2b} = 0$
Battery Mode	$P_{fc2m} = 0$ $P_{bat} = P_d$ $P_{fc2b} = 0$
Recharging Mode	$P_{fc2m} = P_d$ $P_{bat} = 0$ $P_{fc2b} = P_{fc,\max} - P_d$

4.2 *Whale optimization algorithm (WOA)*

The Whale Optimization Algorithm is a bionic optimization algorithm developed by Seyedali Mirjalili and Andrew Lewis in 2016 by observing the feeding mode of humpback whales [9]. The whales encircle their prey with a circular spiral path during the feeding process, and the optimization problem can be solved by simulating the whale's movement path during the feeding process. The optimization problem is solved by simulating the movement path of the whale during the foraging process.

In the optimization calculation of WOA, there are two different calculation methods, namely exploration and exploitation, which are determined by the pace coefficient $|\vec{A}|$, $|\vec{A}|>1$ for exploration and $|\vec{A}|<1$ for exploitation.

There are two important equations in the exploration stage. They are equation (19) and equation (20). Equation (19) is used to calculate the distance between the whale and the prey, in which $X^*(t)$ is the location of the best solution, and by substituting the distance from this equation into equation (20), the space of the next search can be found, in which $|\vec{A}|$ and $\vec{C}$ are coefficient vectors, and $|\vec{A}|$ is calculated from equation (21), in which $\vec{a}$ is a function that decreases linearly from 2 to 0, as shown in equation (22). and $\vec{r}$ denotes a random number between 0 and 1, *iter* is the current of iterations and $I_{\max}$ is the maximum of iterations.

$$|\vec{D}| = |\overline{C}\cdot \vec{X^*}(t) - \vec{X}(t)\,| \quad (19)$$

$$\vec{X}(\mathrm{t}+1) = \vec{X^*}(t) - |\vec{A}| \times |\vec{D}| \quad (20)$$

$$|\vec{A}| = 2 \quad \vec{a}\cdot\vec{r} - \vec{a} \quad (21)$$

$$\vec{a} = 2 - iter\cdot\left(\frac{2}{I_{\max}}\right) \quad (22)$$

$$\overline{C} = 2\cdot\vec{r} \quad (23)$$

When $|\vec{A}|$ is less than 1, it will enter the exploitation stage, the search space starts to shrink inward and gradually converge to the optimal solution. There are two ways to find the next search range in this stage, they are Shrinking Encircling Mechanism and Spiral Updating Position, as shown in Equation (24), each of them has a 50% chance of being chosen.

$$\vec{X}(t+1) = \begin{cases} \vec{X^*}(t) - |\vec{A}|\cdot|\vec{D}| \\ |\vec{D'}|\cdot e^{bl}\cdot\cos(2\pi l) + \vec{X^*}(t) \end{cases} \tag{24}$$

$$|\vec{D'}| = |\vec{X^*}(t) - \vec{X}(t)| \tag{25}$$

5 REAL-TIME SIMULATION

To validate the practical application of the designed controllers, a real-time platform is constructed using two rapid-prototyping controllers (Micro-box, Terasoft, Taiwan). These controllers enable real-time simulation of both the energy management system and the ship's dynamic model. Figure 9 illustrates the experimental setup, with one Micro-box simulating the Energy Management System (EMS) and the other emulating the vessel's dynamic model. The Micro-box on the right provides essential outputs to the left Micro-box for real-time simulation, including speed gain, power demand, and SOC. Post-real-time simulation, the left Micro-box forwards the output power of the fuel cell and lithium battery to the right Micro-box. Conversion of the Simulink model into a real-time model is necessary before simulation, as shown in Figures 10 and 11.

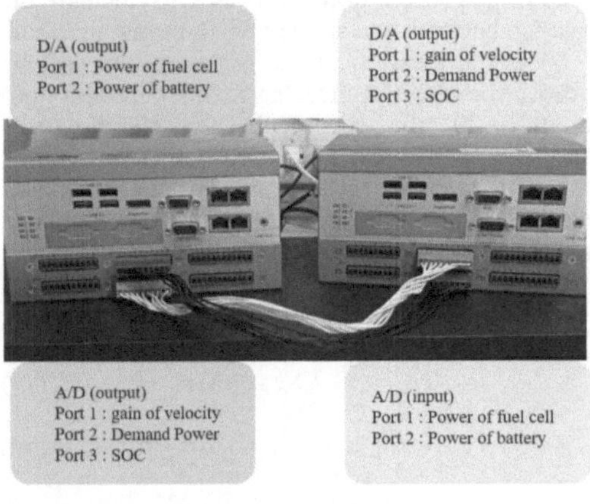

Figure 9. Real-time simulation platform.

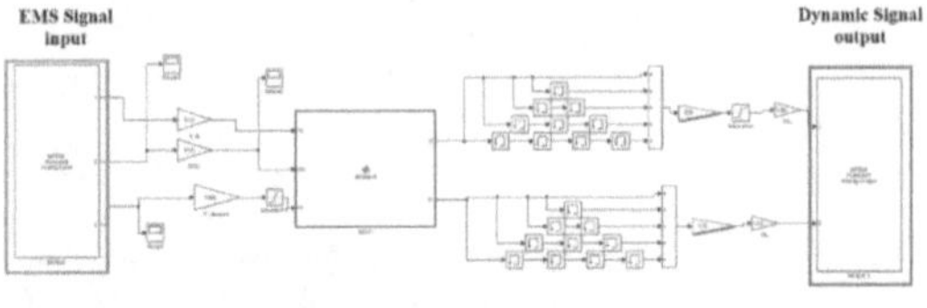

Figure 10. EMS real-time model.

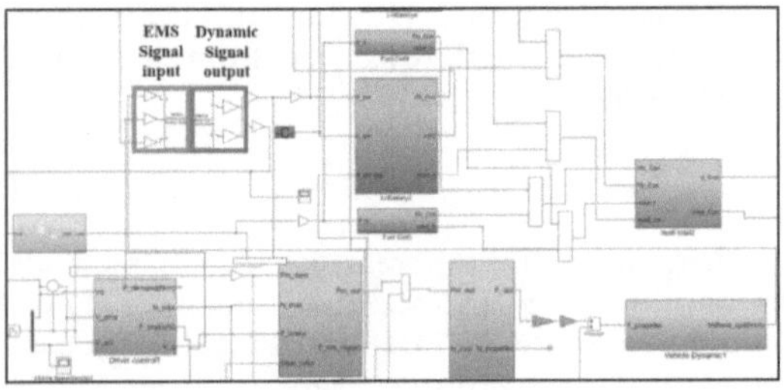

Figure 11. Dynamic real-time model.

6 RESULT AND DISCUSSION

This study uses WOA as an optimization method to conduct simulation tests within a customized driving cycle to find the power allocation ratio with the smallest energy consumption to improve power consumption. By comparing the power consumption and SOC changes of WOA and RB, it can be confirmed whether WOA can optimize power allocation when applied to electric ship energy management systems.

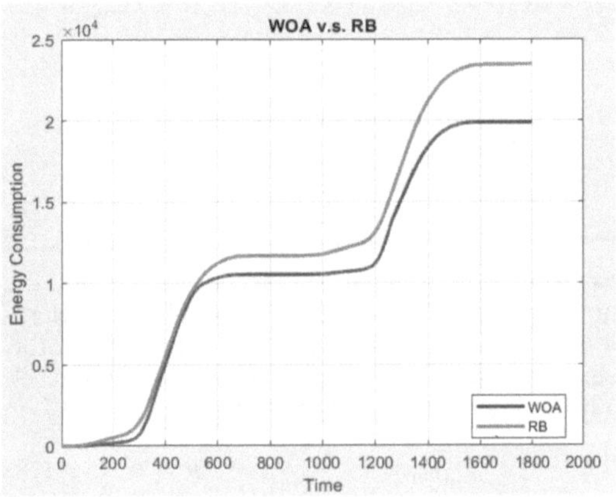

Figure 12. Energy consumption.

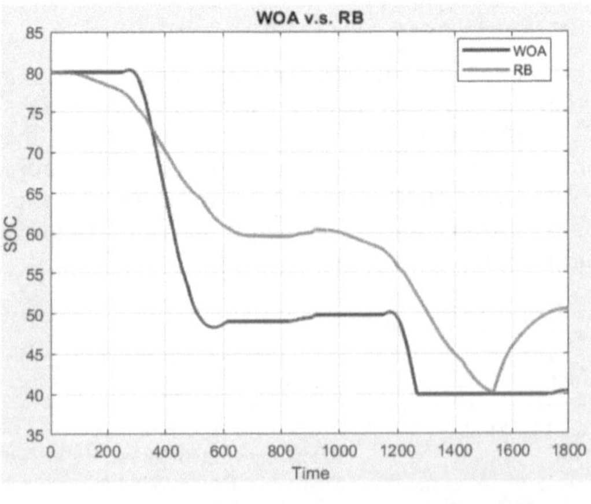

Figure 13. SOC comparison.

Figures 12 and 13 show the overall energy consumption and SOC changes of WOA and RB respectively. Since RB uses both fuel cells and lithium batteries for energy supply at 500 second, the energy consumption is slightly higher, while WOA only uses fuel cells for power supply, so the overall energy consumption is slightly lower. In the RB control strategy, the fuel cell charges the lithium battery at 1500 second, so the final overall energy consumption is higher.

7 CONCLUSION AND FUTURE WORK

In this study, by applying the Whale Optimization Algorithm to the system architecture of the dual power source electric boat and simulating it under specific test conditions, it can be found that the power consumption is better than that of the RB, and the experimental results show that the energy consumption is improved by 15.2%, as shown in Table 4.

Table 4. Comparison of energy consumption.

	RB	WOA	Improvement
Energy consumption (kJ)	23.452	19.883	15.2%

In this study, two control strategies, WOA and RB, are used for comparison. Due to the simpler structure of RB, it is more difficult to obtain the optimal solution and less accurate in practical experience, and the following items describe the future works.

7.1 *Comparison of multiple algorithms*

Since only the whale optimization algorithm is used as the optimization method in this study, other algorithms can be used as a comparison to find out more suitable algorithms for electric vessels in future studies.

7.2 *Consideration of state of health*

In this study, we only consider the effect of SOC on the output power of the battery in the objective function, but the State of Health (SOH) of the battery also affects the output power of the battery. Hence, the SOH can be added to the objective function to calculate the output power of the battery in order to obtain more accurate results.

7.3 *Ship dynamic model improvement*

Since the dynamic model of the ship used in this study is relatively simple, it may lead to the discrepancy between the experimental results and the reality. A more complicated high-ordered physical model can be used to make the experimental results closer to the real situation.

REFERENCES

[1] Veza I., Asy'ari M.Z., Idris M., Epin V., Rizwanul Fattah I.M. and Spraggon M. (2023). Electric vehicle (EV) and driving towards sustainability: Comparison between EV, HEV, PHEV, and ICE vehicles to achieve net zero emissions by 2050 from EV, *Alexandria Engineering Journal*, vol. 82, pp. 459–467

[2] Roy W. V., Scheldeman K., Nieuwenhove A. V., Merveille J. B., R. Schallier and F. Maes. (2023). Current progress in developing a MARPOL Annex VI enforcement strategy in the Bonn Agreement through remote measurements, *Marine Policy*, vol.158, 105882

[3] Van Roy W., Van Roozendael B., Van Nieuwenhove A., Scheldeman K., Merveille J. B., Weigelt A., ... and Maes F. (2023). *Assessment of the international maritime regulations' impact in the North & Baltic Sea*, Nature Portfolio

[4] Wu J., Zhang Y. and Ruan J. (2023). Rule and optimization combined real-time energy management strategy for minimizing cost of fuel cell hybrid electric vehicles, *Energy*, In Press, Journal Pre-proof, 129442.

[5] Boumediene S., Nasri A. and Hamza T. (2023). Fuzzy logic-based Energy Management System (EMS) of hybrid power sources: Battery/Super capacitor for electric scooter supply, *Journal of Engineering Research*, In Press, Corrected Proof.

[6] Mirjalili S., Mirjalili S. M. and Lewis A. (2014). Grey Wolf Optimizer, *Advances in Engineering Software*, Vol. 69, pp. 46–61.

[7] Abdollahzadeh B., Gharehchopogh F. S. and Mirjalili S. (2021). African vultures optimization algorithm: A new nature-inspired metaheuristic algorithm for global optimization problems, *Computers & Industrial Engineering*, Vol. 158.

[8] Chen L., Gao D. and Xue Q. (2023). Energy management strategy for hybrid power ships based on nonlinear model predictive control, *International Journal of Electrical Power & Energy Systems*, vol. 153, 109319

[9] Mirjalili S. and Lewis A. (2016). The Whale Optimization Algorithm, *Advances in Engineering Software*, Vol 95, pp. 51–67.

System Innovation for an Artificial Intelligence Era – Lam et al. (Eds)
© 2025 the Author(s), ISBN: 978-1-032-84452-7

Effects of different micro-arc oxidation discharge parameters and $AgNO_3$ addition amount on the bioactivity of silver-containing HA coating on titanium substrate

Wei-Ting Lin, Yi-Ju Li & Chi-Sheng Chien
Department of Orthopaedics, Chi Mei Medical Center, Tainan, Taiwan

Tzu-Yi Hsieh, Jun-Wei Huang & Chieh-Tsung Wang
Department of Mechanical Engineering, Southern Taiwan University of Science and Technology, Tainan, Taiwan

Tzer-Min Lee
Institute of Oral Medicine, National Cheng Kung University, Tainan, Taiwan

Tsung-Yuan Kuo
Department of Mechanical Engineering, Southern Taiwan University of Science and Technology, Tainan, Taiwan

ABSTRACT: Hydroxyapatite (HA) has excellent bioactivity and biocompatibility, and has been widely used clinically as a surface coating for biometallic materials. However, HA has poor antibacterial ability. Ag is recognized as the best antibacterial agent, but its addition may have adverse effects on bioactivity, and adding too much may also cause cytotoxicity. The purpose of this study is to use micro-arc oxidation (MAO) to prepare silver-containing HA coatings on the surface of titanium substrates under different applied voltages (300, 400V), with a fixed applied current (0.6A) and duration time (5min). The electrolyte uses 0.2 mol/L $C_4H_8CaO_5{\cdot}H_2O$ and 0.1 mol/L $NaH_2PO_4{\cdot}2H_2O$ as basic solutions, with 0, 0.001, and 0.002 mol/L $AgNO_3$ added respectively. It was found that the coating surface with or without the addition of Ag can form flower-like structures with excellent biomedical properties. A trace amount (0.11~0.66 at%) of Ag element appears in the EDS spectrum of the silver-added sample. However, there is no Ag or its compounds in the XRD patterns, which corresponds to the peaks of Ti, TiO_2-A (anatase), TiO_2-R (rutile), HA, and DCPD, and is super hydrophilic (contact angle $\approx 0°$). In bioactivity tests, a large amount of apatite was produced and roughly covered the surface after immersion in artificial simulated body fluid (SBF) for 7 days. It shows that excellent bioactivity can still be maintained when the addition amount of $AgNO_3$ is 0, 0.001, and 0.002 mol/L.

1 INTRODUCTION

With today's rapid advances in medical technology, the replacement technology and service life of artificial implants have been greatly improved. However, bacterial infections may occur during the implant surgery and on the implant itself, which may have a serious impact on the patient's health and treatment outcomes and microbial resistance to antibiotics is increasing over time [1], making it more difficult to treat the infection in the affected area.

Titanium and its alloys have been widely used as artificial implants due to their excellent biocompatibility, corrosion resistance, and biomechanical capabilities [2]. However, Ti is a bioinert material, resulting in poor osseointegration. Surface treatment is required to form a bioactive layer on the titanium substrate surface. Currently, the most common method is to coat the surface with a layer of bioactive ceramic materials, such as hydroxyapatite (HA, $Ca_{10}(PO_4)_6(OH)_2$) [3]. The micro-arc oxidation (MAO) process can obtain a coating of substrate oxide. At the same time, coatings with desired properties can be obtained by adjusting the electrolyte, and the effects of thermal decomposition caused by high-temperature processes can be reduced [4]. To reduce bacterial infection during the implantation process, antibacterial agents can be added to the surface of the implant. Currently applicable inorganic antibacterial agents include silver (Ag), carbon, zinc, copper, etc. Among them, silver can inhibit the growth and reproduction of microorganisms in the early postoperative period and minimize the risk of inflammation after implant placement [5]. In addition, Ag also has good biocompatibility and biodegradability. It has become the most widely used antibacterial material.

In this study, the MAO process was performed on a Ti substrate, using calcium acetate

 DOI: 10.1201/9781003514831-24

($C_4H_8CaO_5{\cdot}H_2O$) and sodium dihydrogen phosphate ($NaH_2PO_4{\cdot}2H_2O$), which can form HA, as the basic electrolyte solution, and adding different concentrations of silver nitrate ($AgNO_3$). The MAO process is used to drive the surface of the Ti substrate to form an Ag-containing HA coating to achieve the purpose of improving bioactivity and antibacterial properties.

2 EXPERIMENTAL

In this study, a commercially available pure titanium disc (Grade 2, Golden Win Steel Industry Co., Ltd., Taiwan) with dimensions of 14 mm (diameter) × 3 mm (thickness) was selected as the base material, and was sequentially ground with 1000-grit SiC sandpaper, cleaned in acetone and distilled water, and then air-dried. The parameters of MAO are set as two different applied voltages (300, 400V), fixed discharge current (0.6A), and discharge time (5min). The electrolyte uses 0.2 mol/L calcium acetate ($C_4H_8CaO_5{\cdot}H_2O$) and 0.1 mol/L sodium dihydrogen phosphate ($NaH_2PO_4{\cdot}2H_2O$) as the basic solution, and three different concentrations of $AgNO_3$ (0, 0.001, 0.002 mol/L) are added. For all coating processes, the working distance (discharge distance) between the anode (substrate) and cathode of the MAO was fixed at 15 mm, and the electrolyte temperature was maintained at approximately 20°C via a cooling system. The artificial simulated body fluid (SBF) test used in this study was developed based on the study by Kokubo *et al.* [6]. The ion concentrations of SBF and human blood plasma are shown in Table 1. The prepared samples were soaked in SBF at a constant temperature of 37°C for 1 and 7 days respectively. The soaked samples were then washed with deionized water, dried at room temperature, and then subjected to microstructural observation. The morphology of apatite formation on each sample coating was evaluated to assess its bioactivity.

Table 1. Inorganic ion concentrations of SBF and human blood plasma.

(mM)	Na^+	K^+	Ca^{2+}	Mg^{2+}	Cl^-	HCO_3^-	HPO_4^{2-}
SBF	142.0	5.0	2.5	1.5	148.8	4.2	1.0
Blood	142.0	5.0	2.5	1.5	103.0	27.0	1.0

3 RESULTS AND DISCUSSION

3.1 *Surface morphology of MAO coatings*

Figure 1 shows the SEM surface morphology of MAO coatings with different $AgNO_3$ concentrations (0, 0.001, and 0.002mol/L) added at applied voltages of 300V and 400V, fixed applied current (0.6A), and duration time (5min). The results show that the 300V samples have a plate-like structure and a small amount of flower-like structure, and there are also a small number of micropores of about 5–10 μm on the coating surface. However, the coating morphology of the 400V samples are quite different from that of the 300V samples, showing a flower-like structure and a small number of micropores. Under the same applied voltage, the coating morphology does not change significantly with different $AgNO_3$ concentrations. According to relevant studies, the flower-like structure of the coating can greatly increase the hydrophilicity [7] and can also improve the early adhesion effect of cell culture [8].

Figure 1. SEM surface morphology of MAO coatings with various $AgNO_3$ concentrations added under different applied voltages (0.6A; 5min).

3.2 *EDS analysis results for MAO coating surfaces*

Figure 2 shows the EDS analysis of MAO coatings with different $AgNO_3$ concentrations under an applied voltage of 300V. The results show that the coating without $AgNO_3$ in the electrolyte is mainly composed of Ca, P, O, and Ti elements; while the coating with $AgNO_3$ added has more Ag elements, and as the $AgNO_3$ concentration increases, the Ag content increases slightly (0.11 ~0.52 at-%). This is because the charged silver ions help promote the entry of silver into the coating under the influence of electric field electrophoresis [9]. In the coating of the 300V sample, the Ca/P ratio of the coating without $AgNO_3$ is about 1.16, while the (Ca+Ag)/P ratio of the coatings with $AgNO_3$ is about 1.19~1.39. Under the 400V samples, the Ca/P ratio of the coating without adding $AgNO_3$ is approximately 1.45. However, with the addition of $AgNO_3$, the (Ca+Ag)/P ratio increases to about 1.77~1.98, which is very close to the Ca/P ratio of HA (1.67). Ciobanu *et al.* [10] showed that during the MAO process, Ag^+ will replace Ca^{2+} in HA. Therefore, the addition of Ag will increase the (Ca+Ag)/P ratio of the coating, and the higher the applied voltage, the more obvious the effect.

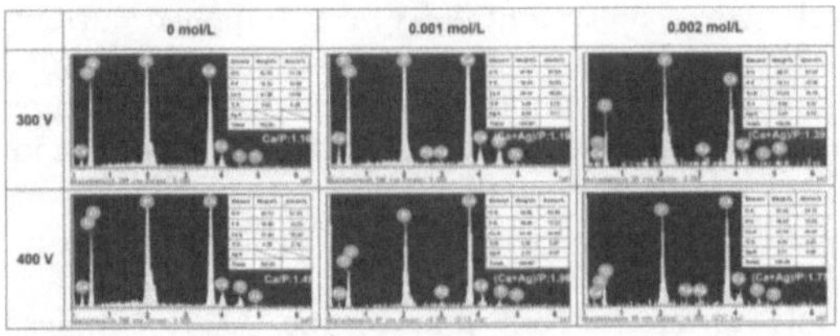

Figure 2. EDS analysis of MAO coatings with various $AgNO_3$ concentrations added at different applied voltages (0.6A; 5min).

3.3 *XRD analysis results for MAO coating surfaces*

Figure 3 shows the XRD patterns of MAO coatings with different $AgNO_3$ concentrations under an applied voltage of 300V, showing that the main compounds of all coatings are Ti, anatase (TiO_2-A), HA ($Ca_{10}(PO_4)_6(OH)_2$), and DCPD ($CaHPO_4 \cdot 2H_2O$). When the applied voltage is 400V (Figure 4), the compound types are similar, except that there is an additional rutile phase (TiO_2-R), and the diffraction intensity of DCPD is significantly lower than the 300V sample. Among them, the more peculiar DCPD is the precursor of HA [11] and is easily formed in phosphate solution. Although in this study, only the flower-like structure (HA) appeared when the applied voltage was 400V, this should be because there is still a small amount of DCPD (plate-like structure) remaining inside the coating [12]. In addition, no peaks of silver or its compounds appear in the X-ray diffraction pattern. According to previous research [13], this may be because Ag is located on the surface of the coating or the size of the Ag particles is too small, resulting in the inability to measure accurately. It may also be that the Ag content is too small and no diffraction peak appears.

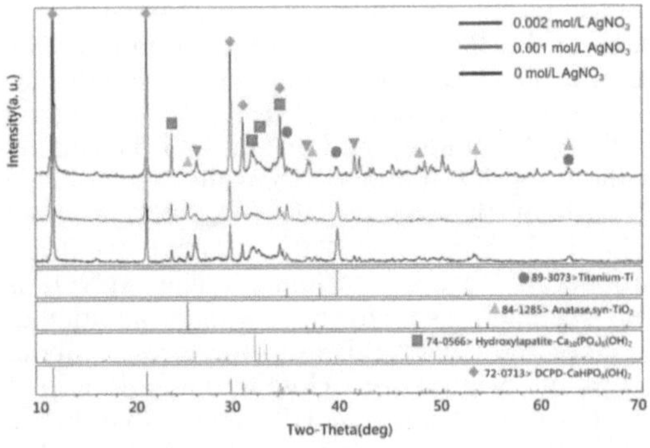

Figure 3. XRD patterns of MAO coatings prepared with different $AgNO_3$ contents added at an applied voltage of 300V (0.6A; 5min).

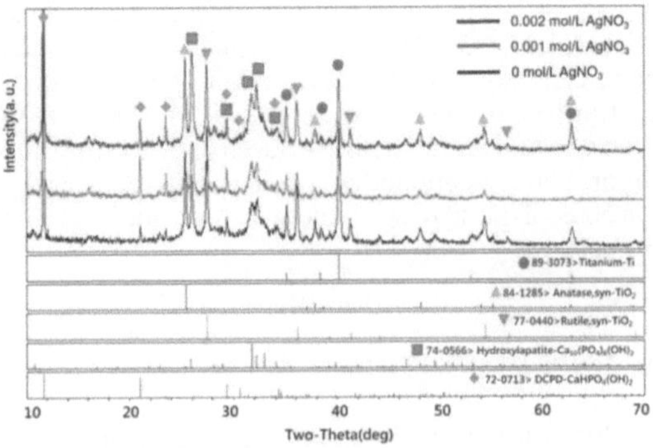

Figure 4. XRD patterns of MAO coatings prepared with different $AgNO_3$ contents added at an applied voltage of 400V (0.6A; 5min).

3.4 *Wettability (hydrophilicity) of MAO coatings*

Figure 5 shows the hydrophilicity (contact angle) measurement results of the MAO coating and pure Ti substrate under an applied voltage of 400V. A lower contact angle value indicates an improved hydrophilicity and thus a better wettability. The results show that the contact angle of the original ground titanium substrate surface is about 47° and only shows slight hydrophilicity. However, the MAO coating with flower-like structure obtained a super-hydrophilic surface with a contact angle of approximately 0°. The hydrophilicity of solid substrates mainly depends on surface morphology (especially surface roughness) and surface chemistry [14]. In terms of surface morphology, the surface of the flower-like structure coating is rougher than that of the Ti substrate without MAO treatment. In terms of surface chemistry, HA contains OH^-, while DCPD usually exists in the form of hydrate. Both OH^- and hydrate are beneficial to improving hydrophilicity. Therefore, the contact angle of MAO coatings containing HA and DCPC is greatly reduced and approaches 0° [15].

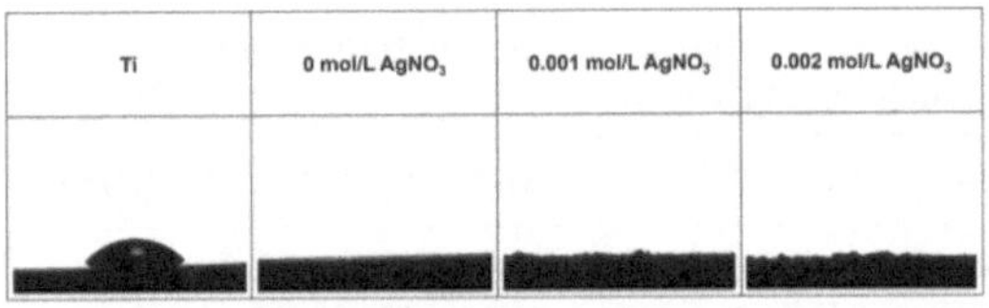

Figure 5. Analysis of contact angle of MAO coating and Ti substrate with different $AgNO_3$ concentrations (400V; 0.6A; 5min).

3.5 *In vitro bioactivity of MAO coatings*

Figure 6 shows the apatite growth of the MAO coatings and substrate when soaked in SBF with an applied voltage of 400V for 1–7 days. After immersion for 1 day, apatite nucleation occurred

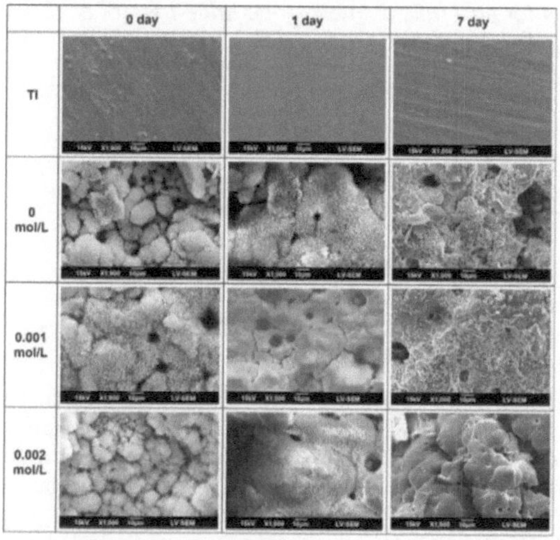

Figure 6. SEM micrographs of the growth of apatite after MAO coatings with different $AgNO_3$ concentrations were soaked in SBF for 0, 1, and 7 days (400V; 0.6A; 5min).

in all samples except for the Ti substrate. After 7 days, the surfaces of all coating samples with $AgNO_3$ added were completely covered with apatites, but there was still no apatite nucleation on the Ti surface.

According to relevant literature [16], the release of Ag^+ ions may cause the simultaneous adsorption of Ca^{2+} and PO_4^{3-}, driving the accelerated nucleation of apatite. Therefore, coatings containing Ag and higher HA content have better bioactivity.

4 CONCLUSION

1. After the MAO process, the coatings with an applied voltage of 300V all showed a plate-like structure; the 400V coatings had a flower-like structure. The addition of $AgNO_3$ will not have a significant impact on the coating structure.
2. For 300V coating, the Ca/P ratio without adding $AgNO_3$ is about 1.16, and the (Ca+Ag)/P value with $AgNO_3$ is about 1.19∼1.39; for 400V coating, the Ca/P ratio without adding $AgNO_3$ is about 1.45, the (Ca+Ag)/P value when adding $AgNO_3$ is about 1.77–1.98. Coatings with added $AgNO_3$ will contain trace amounts of silver (0.11–0.66 at-%).
3. The compounds of the MAO coating consist mainly of Ti, anatase (TiO_2-A), HA, and DCPD, while the specimen with a discharge voltage of 400V has more rutile phases (TiO_2-R), and its DCPD is significantly lower than that of 300V.
4. MAO coatings are all super-hydrophilic (contact angle 0°) due to the presence of HA and DCPD phases. In the SBF test, after 1 day of immersion, apatite nucleation occurred in all MAO coatings; after 7 days, the surface of the coating added with $AgNO_3$ was roughly covered with apatite, showing excellent bioactivity.

ACKNOWLEDGMENTS

The authors gratefully acknowledge the financial support provided to this research by the Ministry of Science and Technology of the Republic of China (Taiwan) under Contract No. NSTC-110-2221-E-218-010- and NSTC 112-2221-E-218-021-.

REFERENCES

[1] Brown E.D. and Wright G.D. (2016). *Nature*, 529 (7586), 336–343.
[2] Stich T., Alagboso F., Křenek T., Kovářík T., Alt V. and Docheva D. (2021). *Bioeng. Transl. Med.* 7, e10239.
[3] Szcześ A., Hołysz L. and Chibowski E. (2017). *Adv. Colloid Interface Sci.* 249, 321–330.
[4] Yerokhin A.L., Nie X., Leyland A., Matthews A. and Dowey S.J. (1999). *Surf. Coat. Technol.*, 3144, 73–93.
[5] Dunn D.S., Raghavan S. and Volz R.G. (1993). *J Biomed Mater Res.*, 27, 895–900.
[6] Kokubo T. and Takadama H. (2006). *Biomaterials*, 27, 2907–2915.
[7] wok D.Y. and Neumann A.W. (1999). *Adv. Colloid Interface Sci.*, 81, 167–249.
[8] Huang Q. and T.A. Elkhooly, X. Liu, R. Zhang, X. Yang, Z. Shen, Q. Feng, (2016), *Colloids Surf. B.*, 145, 37–45.
[9] Zhang L., Gao Q. and Han Y. (2016). *J. Mater. Sci. Technol.*, 32, 919–924.
[10] Ciobanu C. S., Iconaru S. L., Pasuk I., Vasile B. S., Lupu A. R., Hermenean A., Dinischiotu A. and Predoi D., (2013). *Mater. Sci. Eng. C.*, 33, 1395–1402.
[11] Ohta K., Kikuchi M., Tanaka J. and Eda H. (2002). *Chem. Lett.*, 31, 894–895.
[12] Chen K.T., Huang J.W., Lin W.T., Kuo T.Y., Chien C.S., Chang C.P. and Lin Y.D. (2022). *Materials*, 43, 1–18.
[13] He X., Zhang X., Wang X. and Qin L. (2017), *Coatings*, 7, 45–67.
[14] Lin S.H., Chien C.S., Kuo T.Y., Hsieh Y.H., Huang J.W., Chin W.H., Chang C.P., Lee T.M. and Lee H. T. (2020), *Surf. Coat. Technol.*, 40, 126400.
[15] Liu S., Yang X., Cui Z., Zhu S. and Wei Q. (2011). *Mater. Lett.*, 65 1041–1044.
[16] Rajendran A., Pattanayak D. K. (2020). *Mater. Sci. Eng. C.*, 109, 110558.

System Innovation for an Artificial Intelligence Era – Lam et al. (Eds)
© 2025 the Author(s), ISBN: 978-1-032-84452-7

Use LSTM to predict hourly effluent water quality of the pilot wastewater treatment plant

Yu-Hao Lin, Yu-Hao Wu & Hung-Hsiao Hsu
Department of Environmental Engineering, National Chung Hsing University, Taichung, Taiwan

ABSTRACT: The effectively real-time prediction of the effluent wastewater in a Wastewater Treatment Plant (WWTP) could provide decision reference for low pollution. Long Short-Term Memory neuro network (LSTM) performed better on handling the prediction of time series. However, the real-time monitoring data of a WWTP were insufficient in Taiwan, and the problem associated with the discharge of nitrogen into water has received great attention. Therefore, this study monitored and collected the hourly water treatment data from the lab scale pilot WWTP, and the monitoring features included the oxidation reduction potential in the anaerobic tank, the dissolved oxygen in the oxic tank, the mixed liquor recirculation ratio, and so on. These monitoring features and operational conditions at the last 5 hours were fed into LSTM to predict effluents such as chemical oxygen demand (COD) and nitrate-nitrogen ($NO_3^- - N$), at the next hour. The results show that the LSTM could predict the real-time effluent COD in the pilot plant. Future studies into this should devote effects to optimize the hyper-parameters of the LSTM to improve the prediction of effluent $NO_3^- - N$.

1 INTRODUCTION

The optimization of how to better operate Wastewater Treatment Plants (WWTPs) is the first step through which the discharge of pollutants into the environment the environment can be minimized (Ly *et al.* 2022). The accuracy of water quality forecasting models can not only respond to the influence of variable wastewater quality but also improve both the operational efficiency and the energy saving of WWTPs (Wu *et al.* 2016; Zhang *et al.* 2023). But the forecasting models based on physical and biochemical mechanisms, such as Activated Sludge Model, are a not only a time-consuming task but also a non-real-time one due to heavy computational burden (Wongburi and Park 2023). In contrast to the mechanism-based models, data-driven models for improving the operation of WWTPs are considered as promising approaches for use in the future (Xu *et al.* 2021), for this heavily relies on the data quality of the water parameter generated from WWTPs (Bahramian *et al.* 2023). An artificial Neural Network (ANN) is one of the classical data-driven models, and the concept of DL is used to construct deep Neural Network (NN) architectures based on ANN in order to efficiently extract features associated with highly non-linear data (Midhun *et al.* 2014). So, DL models have been employed to identify the environmental and traffic images (Chen *et al.* 2022, 2023; Liang *et al.* 2021; Lin *et al.* 2024b; Mao *et al.* 2022) and make the prediction of time series pollutant (Lin *et al.* 2024a). Long Short-Term Memory neural network (LSTM) based on the architecture of a recurrent neural network (RNN) has helped improve the vanishing gradient and long-distance dependency problems in RNN (Hochreiter and Schmidhuber 1997). So, for example, Wongburi *et. al.* (2023) used LSTM to predict the effluent biochemical oxygen demand of WWTP in which the inputs of the LSTM were the influent total suspended solids (TSSs), Total Phosphorus, total Kjeldahl nitrogen, and Ammonia Nitrogen ($NO_3^- - N$) (Wongburi and Park 2023).

However, these studies predicted the effluents from the actual WWTPs based on monthly mean values or daily mean values of water quality data. That is, this lack of data obtained in-between shorter time intervals, such as hourly data, to capture the dynamic variations of the treatment process variables would limit the online control of WWTPs and the improvement made in the operation of WWTPs (Xie *et al.* 2022). Therefore, online sensors should be used to conduct real-time monitoring and obtain big data of water parameters in different tanks of WWTPs to precisely predict effluent water quality and optimize the operation of WWTP. In addition, because of the more stringent limits in the discharge of nitrogen wastewater in Taiwan, the optimization of the AO-based wastewater treatment process has received great attention.

Hence, this study aimed to use LSTM to predict the quality of the hourly effluent wastewater of pilot WWTPs based on the real-time water treatment data. First, this study collected the

DOI: 10.1201/9781003514831-25

monitoring data of water treatment from our lab scale pilot of AO WWTP based on the experiments of different influents and control variables, and the effluent water quality includes Chemical Oxygen Demand (COD), Nitrate-Nitrogen ($NO_3^- - N$). The input features of the LSTM were the Oxidation Reduction Potential (ORP) in the anaerobic tank, the, DO in the oxic tank, the effluent of COD, and so on. The size of the input features was that generated within the last 4 hours before the start of the next operation. The output features of the LSTM were the effluent water quality at the next hour.

2 METHODOLOGY

As shown in Figure 1, this study built a lab scale pilot WWTP composed of an anaerobic tank, an oxic tank, and a settler, together with a monitoring tank of effluent water. Before experiments were conducted, microbial communities were inoculated with activated sludge and cultured in the AO pilot for more than 2 weeks. The pilot plant was equipped with several online sensors to conduct real-time monitoring and collect the hourly data of the water quality, as shown in Figure 3, such as pH, ORP and EC in the anaerobic tank, DO, SS and ORP in the oxic tank, and so on. Prior data preprocessing included data cleaning and data normalization of the raw motioning data. As to the process of data cleaning, abnormal values were excluded from the data using the three-sigma rule (Pukelsheim 1994), and then KNN was employed to reconstruct the missing data. Finally, the Min-Max normalization method was adopted to eliminate the effect of variation in the scale and units and then map the data values between [0, 1]. These data after the preprocess procedures were divided on 90:10 split for training and testing the LSTM.

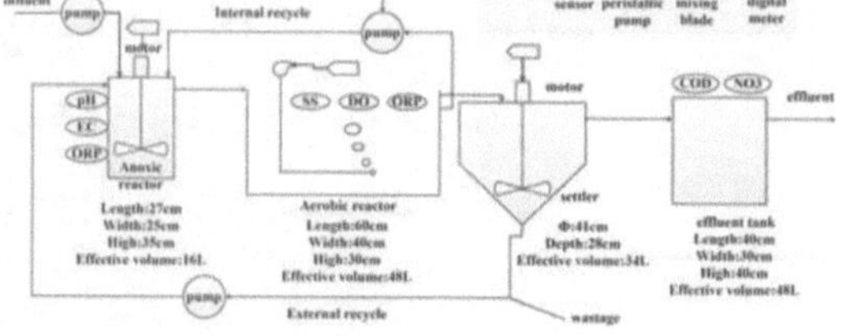

Figure 1. Architecture of the pilot plant.

A LSTM consists of feature input layers, output layers, and a series of LSTM cells and long-term memory blocks. Each LSTM cell consists of short self-recurrent memory cell (C_t), hidden state (h_t), input gate (i_t), output gate (o_t), and forget gate (f_t), and so the performance of several cells at several consecutive time intervals could link several consecutive time intervals of short-term memories into long-term memory blocks (Greff *et al.* 2016). The cell takes appropriate monitoring items at this time t (x_t), the hidden state at last time t-1 (h_{t-1}), and the cell state at last time *t-1* (C_{t-1}) as input features. The outputs of the cell were the C_t and h_t at this time *t*. Eqs. (1)-(8) mathematically described the operating process of a LSTM cell.

$$i_t = \sigma(w_i \cdot [h_{t-1}, x_t] + b_i) \tag{1}$$

$$o_t = \sigma(w_o[h_{t-1}, x_t] + b_o) \tag{2}$$

$$f_t = \sigma(w_f \cdot [h_{t-1}, x_t] + b_f) \tag{3}$$

$$C_t = f_t \times C_{t-1} + i_t \times A_t \tag{4}$$

$$A_t = \tanh(w_a \cdot [h_{t-1}, x_t] + b_a) \tag{5}$$

$$h_t = o_t \times \tanh(C_t) \tag{6}$$

$$\sigma(x) = \frac{1}{1 + e^{-x}} \tag{7}$$

$$\tanh(x) = \frac{e^x - e^{-x}}{e^x + e^{-x}} \tag{8}$$

$$MSE = \frac{1}{n}\sum_{i=1}^{n}(y_i - \hat{y}_i)^2 \tag{9}$$

where w_i, w_o, w_f, and w_a indicate the weights of the i_t, o_t, f_t, and the stored memory (A_t) at time t, respectively; b_i, b_o, b_f, and b_a indicate the bias of the i_t, o_t, f_t and A_t, respectively. $\sigma(\ )$ and tanh( ) refer to the sigmoid and hyperbolic tangent activation functions, respectively (Lin *et al.* 2024a). Detail architecture and formulate of LSTM could refer to the studies of (Hochreiter and Schmidhuber 1997). As to the hyperparameters of LSTM in this study, the slide window size, the number of neurons in layer 1, the number of neurons in layer 2, the learning rate, the batch size, the optimizer, activation functions, were past 6 hr, 50, 30, 0.1, 128, Adam, and relu function. Mean Square Error (*MSE*), as shown in Eqs. (9), was the loss function used to train the LSTM, where *n* is the ground-truth number, and y_i, $\hat{y}_i$, and $\bar{y}$ are the *i*th prediction, the *i*th ground-truth and the average of $\hat{y}$ over the n data, respectively.

3 RESULTS AND DISCUSSION

Figures 2 (a)-(b) show the real-time 1481 and 2000 monitoring data collected hourly from the pilot plant for effluent COD and $NO_3^- - N$, respectively, under different inputs and operational conditions. The time series of the effluents of COD in Figure 2(a) shows the effluent COD concentration changed from about 32 ppm to about 8 ppm, and these data (orange line) after the data preprocess indicated that KNN could impute the missing values in the data. Similarly, as shown in Figure 2(b), the effluent $NO_3^- - N$ concentration changed from about 55 ppm to about 150 ppm, and the orange line indicated that KNN could also fill the missing data of effluent $NO_3^- - N$.

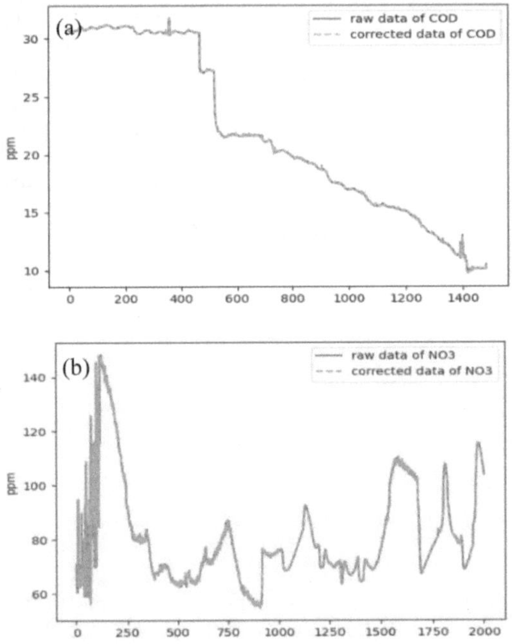

Figure 2. Effluent concentration of (a) COD and (b) $NO_3^- - N$ in the AO pilot plant.

Figures 3(a)-(b) represent the train and train and validation of the LSTM in predicting the time series of effluent COD at next hour, where the blue curves are the actual monitoring values, and the organ curves are the prediction of the LSTM. Compared with the performance of LSTM during the train, the LSTM performed relatively low accuracy during the validation because possibly there were fewer low-concentration COD data to train the LSTM. These fewer low-concentration COD data could be observed in Figure 2(a). Similarly. Figures 4(a)-(b) illustrate the train and train and validation of the LSTM used to predict the time series of effluent $NO_3^- - N$ at next hour. In the train of the LSTM on the effluent prediction of $NO_3^- - N$, this LSTM performance relatively low accuracy compared to the train of the LSTM on the effluent prediction of COD. This is possibly because the LSTM had unoptimized structure, and so the LSTM also performed lower accuracy during the validation of the LSTM, as show in Figure 4(b). These results indicated that the proposed LSTM could perform better on the prediction of effluent COD than the prediction of effluent $NO_3^- - N$ at next hour. In the future, future studies could optimize the hyperparameters of the LSTM to improve the predictions.

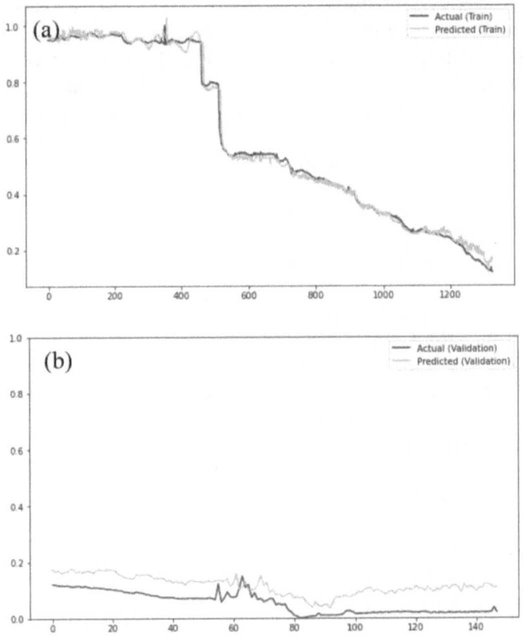

Figure 3. (a) Train and (b) validation of the LSTM on the effluent prediction of COD.

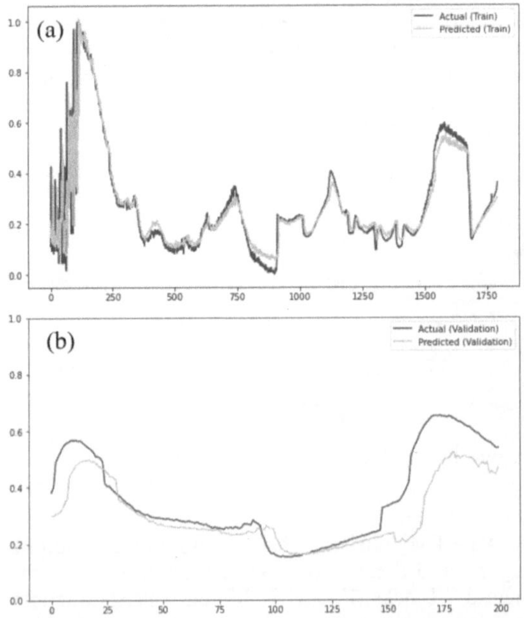

Figure 4. (a) Train and (b) validation of the LSTM on the effluent prediction of $NO_3^- - N$.

4 CONCLUSION

The accurate prediction of the effluents of WWTPs based on the influents and operational conditions would improve the performance of WWTPs, especially for short-term or real-time prediction. The real-time monitoring data of a WWTP were insufficient in Taiwan and nitrogen discharge in water became a concern. Hence, in this study, the monitoring data of water treatment were collected hourly from our lab scale pilot WWTP to train and validate LSTM. These monitoring features, such as ORP and EC in the anaerobic tank, DO and ORP in the oxic tank, and so on, at the last 5 hours were fed into LSTM to predict the effluents of COD and $NO_3^- - N$ at the next hour. The results demonstrated that the optimized LSTM could predict the real-time effluent of COD but perform relatively lower accuracy on the effluent of $NO_3^- - N$. This is possibly because the LSTM had unoptimized structure. Future study could optimize the hyperparameters of the LSTM to improve the predictions of pilot WWTP.

REFERENCES

Bahramian M., Dereli R.K., Zhao W., Giberti M. and Casey E. (2023). Data to intelligence: The role of data-driven models in wastewater treatment. *Expert Systems with Applications* 217, 119453.

Chen W.-C., Deng M.-J., Liu P.-Y., Lai C.-C. and Lin Y.-H. (2023). A framework for real-time vehicle counting and velocity estimation using deep learning. *Sustainable Computing: Informatics and Systems* 40, 100927.

Chen W.-C., Liu P.-Y., Lai C.-C. and Lin Y.-H. (2022). Identification of environmental microorganism using optimally fine-tuned convolutional neural network. *Environmental Research* 206, 112610.

Greff K., Srivastava R.K., Koutník J., Steunebrink B.R. and Schmidhuber J. (2016). LSTM: A search space odyssey. *IEEE Transactions on Neural Networks and Learning Systems* 28(10), 2222–2232.

Hochreiter S. and Schmidhuber J. (1997). Long short-term memory. *Neural Computation* 9(8), 1735–1780.

Liang C.-M., Lai C.-C., Wang S.-H. and Lin Y.-H. (2021). Environmental microorganism classification using optimized deep learning model. *Environmental Science and Pollution Research* 28, 31920–31932.

Lin M.-D., Liu P.-Y., Huang C.-W. and Lin Y.-H. (2024a). The application of strategy based on LSTM for the short-term prediction of PM2.5 in city. *Science of the Total Environment* 906, 167892.

Lin Y.-H., Mao W.-L. and Fathurrahman H.I.K. (2024b). Development of intelligent municipal solid waste sorter for recyclables. *Waste Management* 174, 597–604.

Ly Q.V., Truong V.H., Ji B., Nguyen X.C., Cho K.H., Ngo H.H. and Zhang Z. (2022). Exploring potential machine learning application based on big data for prediction of wastewater quality from different full-scale wastewater treatment plants. *Science of the Total Environment* 832, 154930.

Mao W.-L., Chen W.-C., Fathurrahman H.I.K. and Lin Y.-H. (2022). Deep learning networks for real-time regional domestic waste detection. *Journal of Cleaner Production* 344, 131096.

Midhun M., Nair S.R., Prabhakar V.N. and Kumar S.S. (2014). *Deep Model for Classification of Hyperspectral Image Using Restricted Boltzmann Machine*, pp. 1–7.

Pukelsheim F. (1994). The three sigma rule. *The American Statistician* 48(2), 88–91.

Wongburi P. and Park J.K. (2023). Prediction of wastewater treatment plant effluent water quality using recurrent neural network (RNN) models. *Water* 15(19), 3325.

Wu X., Yang Y., Wu G., Mao J. and Zhou T. (2016). Simulation and optimization of a coking wastewater biological treatment process by activated sludge models (ASM). *Journal of Environmental Management* 165, 235–242.

Xie Y., Chen Y., Lian Q., Yin H., Peng J., Sheng M. and Wang Y. (2022). Enhancing real-time prediction of effluent water quality of wastewater treatment plant based on improved feedforward neural network coupled with optimization algorithm. *Water* 14(7), 1053.

Xu R., Cao J., Fang F., Feng Q., Yang E. and Luo J. (2021). Integrated data-driven strategy to optimize the processes configuration for full-scale wastewater treatment plant predesign. *Science of the Total Environment* 785, 147356.

Zhang W., Zhao J., Quan P., Wang J., Meng X. and Li Q. (2023). Prediction of influent wastewater quality based on wavelet transform and residual LSTM. *Applied Soft Computing* 148, 110858.

System Innovation for an Artificial Intelligence Era – Lam et al. (Eds)
© 2025 the Author(s), ISBN: 978-1-032-84452-7

Effect of different air change rate on thermal comfort and energy consumption of container buildings in subtropical areas

Hsiu-Wen Chang
Department of Architecture, National Cheng Kong University, Tainan, Taiwan

Mei-Huei Li
Department of Architecture, Tamkang University, Taipei, Taiwan

Chen-Yu Pan
Department of Architecture, National Cheng Kong University, Tainan, Taiwan

ABSTRACT: Recent years, Container building has gradually become a common choice for temporary buildings. Its usage characteristics, including short time periods and frequent personnel entry and exit, exacerbate the situation of keeping doors and windows tightly closed when air conditioning is in use, making improving ventilation conditions and indoor air quality a major challenge. This study aims to investigate the impact of different air change rate (0 to 10 ACH) on temperature maintenance, indoor thermal comfort, and energy consumption in container buildings under the hot and humid weather conditions of summer. From the experimental results, it is evident that air change rate of 6 ACH or more make it difficult to maintain the temperature at the set temperature and result in an increase in energy consumption ranging from 5% to 19%. Compared to the non-ventilated groups, the groups with ventilation experienced an average decrease of 42% in thermal comfort, a decrease of 14% and 31% in thermal acceptance and preference respectively; while an increase of 25% in airflow sensation, but a decrease of 14% in acceptance. Additionally, the thermal sensation results obtained from the questionnaire survey of the participants indicate a cooler perception compared to predictions by the predicted mean vote (PMV) formula.

Keywords: Air change rate, thermal comfort, energy consumption, container building, subtropical area

1 INTRODUCTION

In recent years, container buildings have gradually become a preferred option for commercial or office spaces in Taiwan, particularly for temporary buildings, due to its convenience, mobility, and relatively lower construction time and costs compared to traditional buildings. Container buildings typically have smaller interior spaces and air conditioning systems, usually equipped with split-type air conditioning units. This leads to the lack of return air, which make the amount of outdoor air directly affect indoor thermal environment and energy consumption. As container buildings gain popularity, the efficiency of its air conditioning environment has become a topic of discussion.

In everyday use of air conditioning, indoor doors and windows are often kept closed to avoid the influence of outdoor air conditions on cooling effectiveness, leading to the accumulation of pollutants. Meanwhile, container buildings are often utilized in short and concentrated time periods, such as at construction sites, short-term events, or in spaces with frequent personnel entry and exit that makes occupants pursue a faster cooling effect. Moreover, the opening parts of container buildings are usually smaller then normal buildings due to structural factors, in some cases, there're even no opening parts in the buildings, which exacerbate these conditions. Currently, carbon dioxide (CO_2) concentration is commonly used as a standard to judge whether ventilation is adequate. The CO_2 concentration standard set by the Environmental Protection Administration in Taiwan is below 1000 ppm, and the American Society of Heating, Refrigerating and Air-Conditioning Engineers (ASHRAE) also recommends that CO_2 concentration in general indoor spaces should not exceed 1000 ppm. Exposure to high CO_2 levels can lead to drowsiness, reduced work efficiency, and similar conditions may happen when the room is in a state of closed, while increasing the air change rate can effectively improve this situation. [1–3]

Taiwan, located on subtropical area with monsoon climate, is characterized by high temperatures and humidity in summer. Tainan, the south of Taiwan, is situated south of the Tropic of Cancer,

DOI: 10.1201/9781003514831-26

experiences temperatures exceeding 30°C on more than 90% of days from June to September. In such conditions, the higher air change rate in indoor spaces, the more hot, humid outdoor air is entered, resulting in an increase in indoor temperature and the energy consumption of air conditioning operation indirectly. [4] Some ventilation regulations for office spaces are shown in Table 1, and based on Taiwan's regulations, the air change rate for 20-foot container buildings is estimated to be between 3 to 5 ACH (air change rate per hour). This study aims to examine changes in indoor thermal comfort of personnel and energy consumption under the same air conditioning set temperature but with different air change rates, while also assessing whether existing regulations balance comfort and energy efficiency benefits.

Table 1. Regulations of ventilation volume.

Regulations	Type of Space	Standard	Unit	Air Change Rate in Container Building
ASHRAE 55-2010	Office	5 CFM	per person	1.6 ACH
Building Technical Regulations (Taiwan)	Office	8 CMH	per m^2	3.7 ACH
	Meeting Room	12 CMH	per m^2	5.5 ACH
Occupational Safety and Health Facilities Rules (Taiwan)	Workplace	0.6 CMM	per person (less than $5.7m^3$ occupied by each person)	N/A
		0.4 CMM	per person ($5.7–14.2m^3$ occupied by each person)	4.6 ACH
		0.3 CMM	per person ($14.2–28.3m^3$ occupied by each person)	N/A
		0.14 CMM	per person (more than $28.3m^3$ occupied by each person)	N/A

2 EXPERIMENTAL

This study focuses on a container building space converted from a 20-foot shipping container. The internal dimensions of the container unit are approximately 2.7*5.9*2.4m (H*L*W), with an interior area of about 14.16 m^2 and a volume of approximately 38.23 m^3, while the total area of opening parts is approximately $6.44m^2$. The primary construction materials for the walls and roof are mineral wool panels, providing slightly lower airtightness compared to traditional container processing methods but facilitating equipment installation.

The interior is equipped with split-type air conditioning units and exhaust fans. Additionally, an automatic monitoring system is installed to monitor indoor temperature, humidity, CO_2 concentration, energy consumption (power, current, voltage), and other information, recording these values during experiments. During the experiments, the air conditioning set temperature is maintained at 24°C. Measurement instruments include a WBGT thermometer (measuring dry-bulb temperature, black-bulb temperature, and relative humidity) and seven small-scale temperature and humidity measuring devices. Additionally, a hot-wire anemometer is used to measure airflow velocity at several points, from which the average indoor airflow velocity is calculated.

Based on preliminary experiments, in a 20-foot container space with six occupants and the exhaust fans for mechanical ventilation were closed (i.e., air change rate of 0 ACH, with only ventilation through door gaps), the CO_2 concentration can approach 2000 ppm within one hour. When the air change rate is set to 2 ACH, the concentration exceeds the standard of 1000 ppm at around 20 minutes, with a stable concentration of approximately 1100 ppm

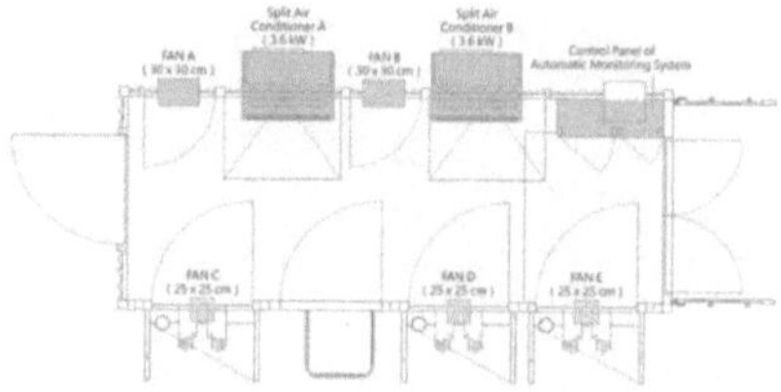

Figure 1a. Equipment in the experimental field.

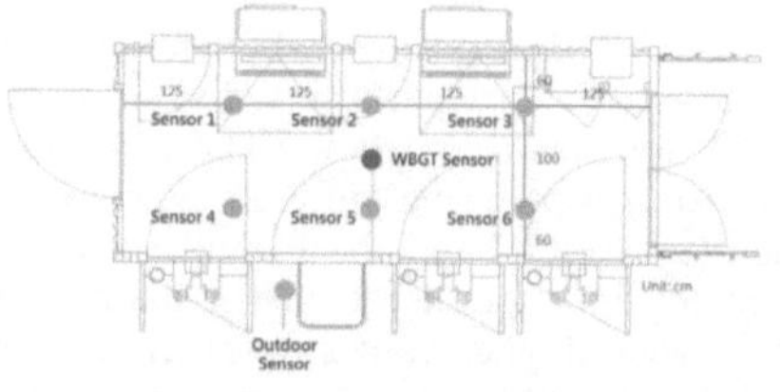

Figure 1b. Locations of sensors in the experimental field.

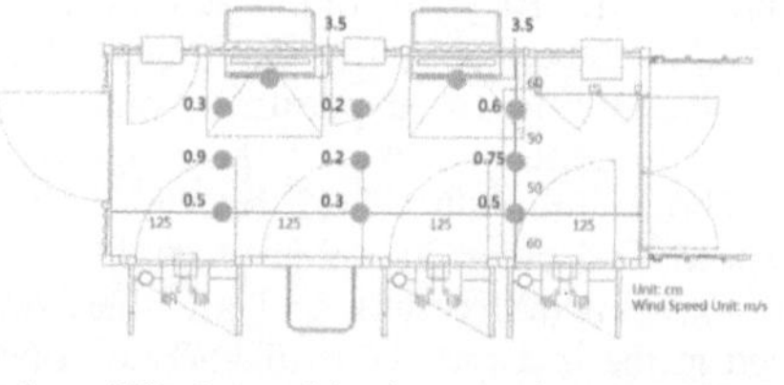

Figure 1c. Wind speed in the experimental field.

Table 2. Content and options of the questionnaire.

Content	Options						
	−3	−2	−1	0	1	2	3
Thermal Sensation Vote (TSV)	Cold	Cool	Slightly cool	Neutral	Slightly warm	Warm	Hot
Thermal Comfort (TC)	Very uncomfortable	Uncomfortable	Slightly uncomfortable	Neutral	Slightly comfortable	Comfortable	Very comfortable
Thermal Acceptance (TA)	N/A	Totally unacceptable	Barely unacceptable	N/A	Barely acceptable	Totally acceptable	N/A
Thermal Preference (TP)	N/A	Much warmer	A bit warmer	No change	A bit cooler	Much cooler	N/A
Air Movement Vote (AMV)	Very stuffy	Stuffy	Slightly stuffy	Neutral	Slightly windy	Windy	Very windy
Air Movement Acceptance (AMA)	N/A	Totally unacceptable	Barely unacceptable	N/A	Barely acceptable	Totally acceptable	N/A
Discomfort Reactions	Add clothes, headache, trembling, cold hands or feet, runny or stuffy nose, sneezing, dry mouth, drowsiness, others						

(±100 ppm). Continuing to increase the rate, the CO_2 concentration exceeds 1000 ppm around 30 minutes, and then stabilizes at 1000 ppm (±100 ppm), unable to decrease below the standard level. Therefore, the air change rate range in this experiment is set from 0 to 10 ACH, with each interval of 2 ACH forming one group, totaling six groups.

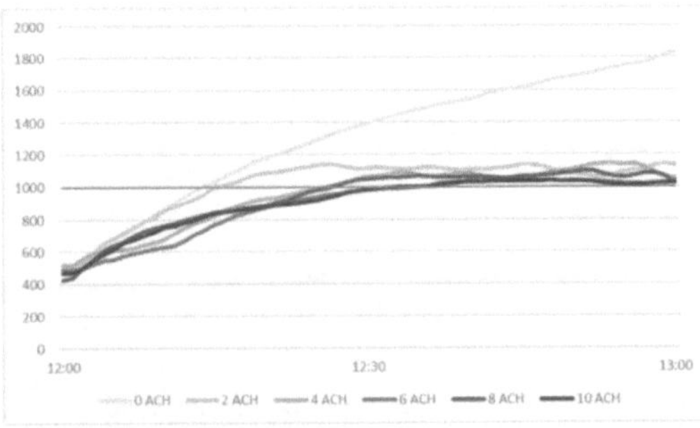

Figure 2. CO_2 concentration in different air change rate.

The comfort assessment method for this experiment consists of two parts: questionnaire surveys and PMV formula calculations. The PMV formula is used to compare the differences between the questionnaire results and the formula calculations. The content and options of questionnaire listed in Table 2, where higher values for thermal comfort, thermal acceptability, and air movement acceptability indicate better results, while the thermal preference value is optimal when closer to 0. Participants are sourced from graduate students at the Department of Architecture, NCKU, aged between 20 and 30 years old. Each experimental group includes six participants distributed across six test points. During the experiment, outdoor temperatures are controlled within the range of 30 to 32°C; relative humidity, CO2 background levels, wind speed, and other factors have minimal impact on the experimental results and are therefore not included in the necessary controlled conditions. The clothing insulation level required for PMV formula

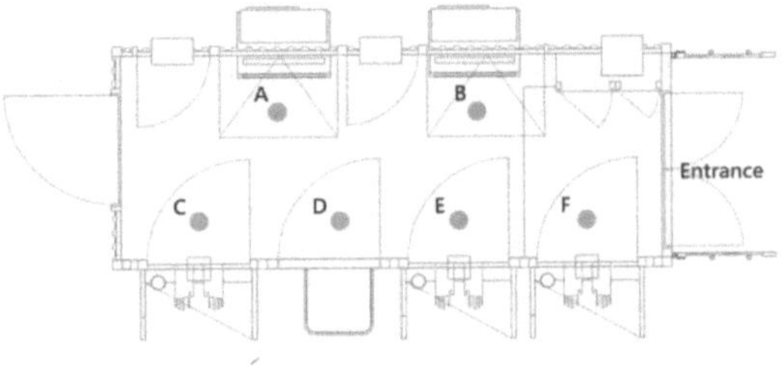

Figure 3. Locations of participants during the experiment.

calculations is set at 0.6 clo (e.g., short-sleeved shirts, shorts, or lightweight trousers for summer attire), consistent with the attire worn by participants during the experiment. Metabolic rate is estimated at 1.2 met for seated, office-type work activities.

3 RESULTS AND DISCUSSION

In terms of maintaining indoor temperature, the performance is similar for air change rate of 4 ACH or less. However, with 6 ACH or more, the stable temperature is approximately 0.8°C higher compared to the no ventilation group. Additionally, the cooling time is longer, especially with 10 ACH, which is 33 minutes longer (an increase of 80.49%) compared to the no ventilation group.

Table 3. Effect of indoor temperature maintenance under different air change rate.

Air Change Rate	0 ACH	2 ACH	4 ACH	6 ACH	8 ACH	10 ACH
Stable temperature	23°C	23.4°C	23.2°C	23.8°C	23.7°C	23.8°C
Cool down duration	41 min	54 min	53 min	55 min	58 min	74 min
Increase percentage	±0%	+31.71%	+29.27%	+34.15%	+41.46%	+80.49%

The energy consumption curve is shown in Figure 5 as a line chart. Air change rate of 0 to 4 ACH show variable frequency operation starting approximately one hour after initiation, with intervals between shutdowns lasting about 15 to

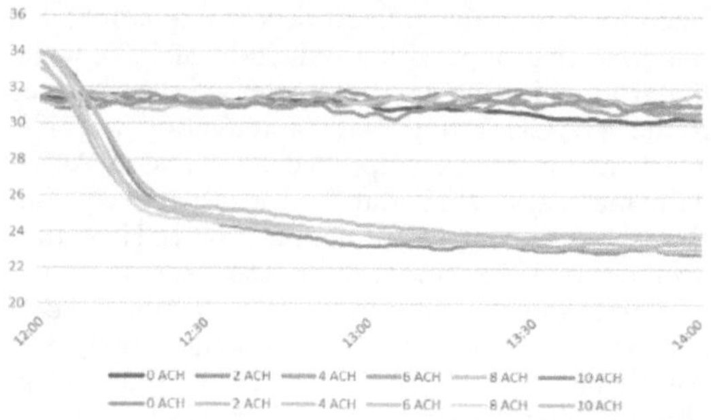

Figure 4. The line chart of indoor temperature maintenance under different air change rate (Unit:°C).

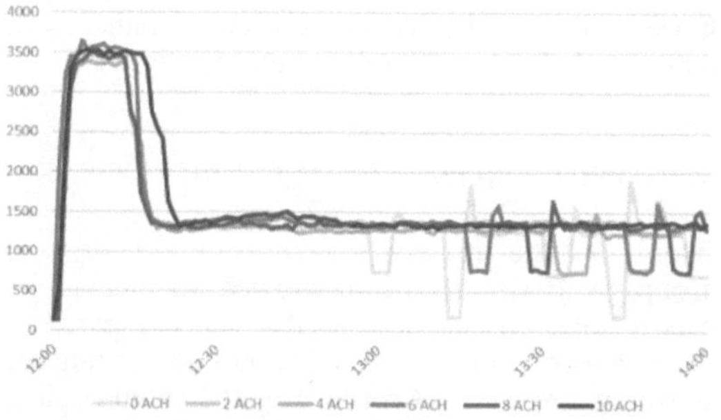

Figure 5. The line chart of energy consumption under different air change rate (Unit:W).

Table 4. Comparison of energy consumption under different air change rate (Unit:kWh).

Air Change Rate	0 ACH	2 ACH	4 ACH	6 ACH	8 ACH	10 ACH
Average energy consumption per hour (includes startup phase)	1.409	1.565	1.478	1.580	1.519	1.670
Increase percentage	±0%	+11.07%	+4.94%	+12.13%	+7.80%	+18.51%
Average energy consumption per hour (without startup phase)	1.160	1.363	1.260	1.315	1.313	1.388
Increase percentage	±0%	+17.48%	+8.65%	+13.38%	+13.15%	+19.67%

20 minutes. The groups of 6 to 8 ACH exhibit variable frequency operation starting around one and a half hours into the experiment, with shutdown intervals lasting about 30 minutes. However, the 10 ACH group shows no occurrence of variable frequency operation throughout the two-hour testing period.

While there is some margin of deviation in overall energy consumption, it still exhibits a high correlation (R = 0.754). The total energy consumption percentage relative to the baseline is higher when excluding the startup phase. This indicates that air change rate primarily affects the subsequent usage phase, with minimal impact on energy consumption during the initial cooling period. The group with the least increase in energy consumption is the 4 ACH group, showing increases of 4.94% (with startup phase included) and 8.65% (without startup phase) relative to the baseline. On the other hand, the group with the highest increase in energy consumption is the 10 ACH group, showing increases of 18.51% and 19.67% relative to the baseline, respectively.

The PMV calculation formula, used by ASHRAE to predict human thermal sensation, is complex. Therefore, in this experiment, we measured dry-bulb temperature, black-bulb temperature, relative humidity, air velocity, and controlled clothing and metabolic rates, to utilize the calculation tool named CBE Thermal Comfort Tool, which is provided by Center for the Built Environment for PMV calculation.

Results from the questionnaire survey indicate that PMV calculation results for all six experimental groups fall within the defined comfort range (−0.5 to 0.5). Only the results for the 10 ACH ventilation group exceed 0, indicating that the majority of thermal sensation predictions are categorized as "slightly cool." Additionally, the thermal sensation values obtained from the questionnaire fall within this range for all experimental groups except for the 2 ACH group, which has a slight deviation of −0.06 below the comfort range, still categorized as "slightly cool".

The results of the questionnaire survey are shown in Table 5. Compared to the baseline air change rate of 0 ACH, the thermal comfort and thermal acceptance of all other groups decreased.

Table 5. Comparison of thermal comfort under different air change rate.

Air Change Rate	0 ACH	2 ACH	4 ACH	6 ACH	8 ACH	10 ACH
Predicted Mean Vote (PMV)	±0%	+79%	−53%	+80%	−59%	−148%
Thermal Sensation Vote (TSV)	±0%	+58%	−37%	−11%	−65%	−53%
Thermal Comfort (TC)	±0%	−72%	−47%	−37%	−16%	−38%
Thermal Acceptance (TA)	±0%	−20%	−21%	−5%	−7%	−18%
Thermal Preference (TP)	±0%	23%	−17%	62%	−187%	−62%
Air Movement Vote (AMV)	±0%	−5%	−40%	−9%	+142%	+34%
Air Movement Acceptance (AMA)	±0%	−18%	−19%	−22%	−1%	−9%

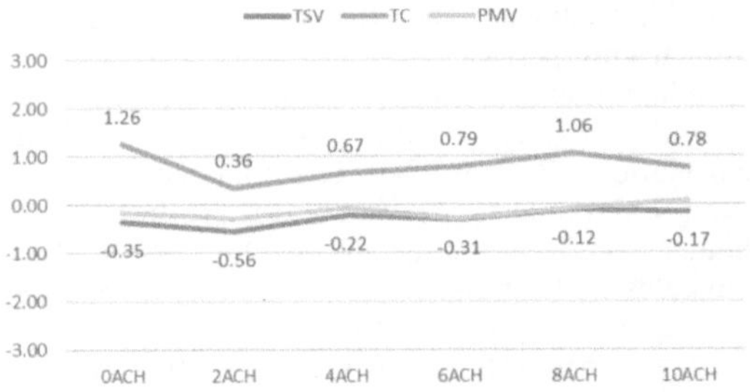

Figure 6a. Results of thermal comfort from questionnaire.

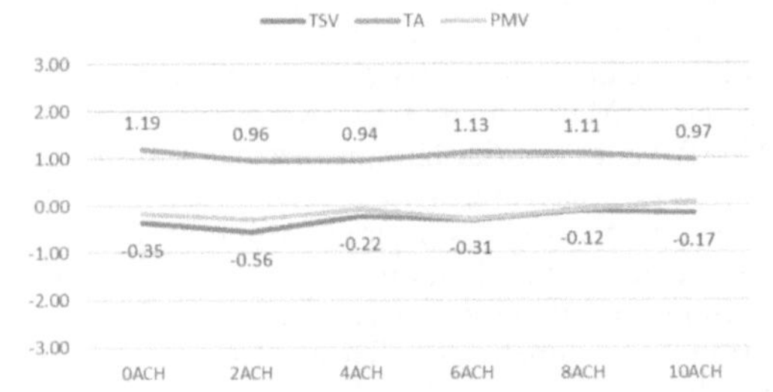

Figure 6b. Results of thermal acceptance from questionnaire.

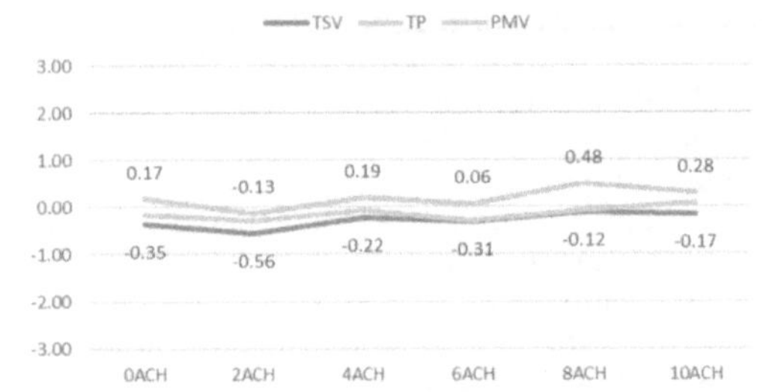

Figure 6c. Results of thermal preference from questionnaire.

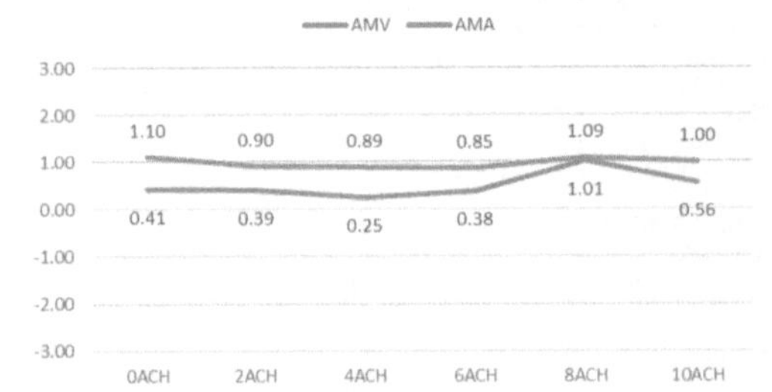

Figure 6d. Results of air movement acceptance from questionnaire.

The group with the greatest decrease in comfort is the 2 ACH group, with a reduction of 72%, while the group with the greatest decrease in acceptance is the 4 ACH group, with a reduction of 21%. Thermal preference improved at 2 and 6 ACH, with increases of 23% and 62% respectively, but showed the worst performance at 8 ACH, with a decrease of 187% compared to the baseline.

Higher values of air movement perception indicate a more "ventilated" sensation among participants. Air change rate of 8 ACH and above showed significant differences in perceived ventilation effects, with values increasing by 142% and 34% respectively. However, air movement perception slightly decreased by 5–40% for air change rate of 2 to 6 ACH. In terms of acceptance, all groups showed a decrease compared to the baseline, with a maximum decrease of 22%. However, the decrease in acceptance for the 8 and 10 ACH groups was smaller, with reductions of only 1% and 9% respectively.

4 CONCLUSION

Based on the current experimental results, the groups with ventilation all showed an increase in energy consumption of at least 4.94% compared to the non-ventilation group. When not accounting for startup phase (i.e., not considering the initial cooling phase), the average energy consumption increased by at least 8.65%. It can be inferred that if the air conditioning operating time is extended, the differences between the groups will become more pronounced. Additionally, although the outdoor air conditions were controlled within a range of ±1°C, various factors such as differences in heat generation among participants and slight variations in enthalpy, which made the results a little deviation among groups, but still showing a high correlation.

In terms of thermal comfort, both the PMV calculations and the questionnaire results fell within the defined comfort range, with only a 12% deviation observed in the questionnaire results for the 2 ACH group. Additionally, the PMV-predicted thermal sensation results were slightly higher than the questionnaire results in this series of experiments, indicating that participants perceived the environment to be cooler than predicted by the PMV model.

Regarding other aspects of thermal comfort evaluation, the groups with ventilation showed an average decrease of 42% in thermal comfort compared to the non-ventilation group. Thermal acceptance and preference also decreased by 14% and 31% respectively. Air movement perception increased by 25%, but acceptance decreased by 14%. Above 8 ACH of the air change rate, air movement perception increased, while acceptance decreased by an average of 14%. It may be influenced by the rise in indoor temperature, but still remaining within the "acceptable" range.

Maintaining ventilation at 4 ACH achieved the best balance in maintaining indoor temperature, low energy consumption, and CO_2 concentration. Although thermal comfort decreased by −47%, it still fell within the "slightly comfortable" range, consistent with existing regulations in Taiwan. Notably, the 6 ACH group showed the best performance, with only a 5% decrease in thermal acceptance and a 62% increase in thermal preference, although the decrease in thermal comfort was larger at 37%. The 8 ACH group, despite having the worst thermal preference, exhibited relatively better performance in thermal comfort and acceptance, with only a 16% and 7% decrease respectively, and better control over the rise in CO_2 concentration. If energy consumption is not a consideration, adjusting the air change rate to 6 to 8 ACH could be an option to improve indoor comfort.

Future plans include increasing the number of experiments to obtain more precise statistical values and incorporating Computational Fluid Dynamics (CFD) numerical simulations to examine the distribution of CO_2 concentration, airflow direction, and temperature uniformity under different air change rate in detail.

FUNDING

This work was supported by the National Science and Technology Commission of Taiwan, National Cheng Kung University Epidemic Prevention Center (Grant No.: NSTC 113-2321-B-006-007-), and the Stratified Green Foundation National Cheng Kung University Research Center Energy Materials Research Center (Hi-GEM) and the MOE Teaching Practice Research Project (Grant No.: H112-A165).

REFERENCES

[1] Fusuo Xu, Zhi Gao. (2022). Study on indoor air quality and fresh air energy consumption under different ventilation modes in 24-hour occupied bedrooms in Nanjing, using Modelica-based simulation. *Energy and Buildings, vol. 275.*

[2] Seppänen O. A., Fisk W. J., Mendell M. J. (1999). Association of Ventilation Rates and CO2 Concentrations with Health and Other Responses in Commercial and Institutional Buildings. *Indoor Air*, 9(4), 226–252.

[3] Runa T. Hellwig, Gearoid P. Lydon, Asit Kumar Mishra (2020). Assessment of thermal comfort and building-related symptoms in air-conditioned offices in tropical regions: a case study in Singapore and Thailand. *Frontiers in Built Environment*, vol. 6

[4] Wargocki P., Wyon D. P., Sundell, Clausen G., Fanger P. O. (2000). The effects of outdoor air supply rate in an office on perceived air quality, sick building syndrome (SBS) symptoms and productivity. *Indoor Air*, 10, 222–236.

System Innovation for an Artificial Intelligence Era – Lam et al. (Eds)
© 2025 the Author(s), ISBN: 978-1-032-84452-7

Ventilation strategy for office — the balance between carbon dioxide concentration and air conditioner energy consumption

Chin-Wei Chang & Hsieh-Chih Hsu
Department of Architecture, National Cheng Kung University, Tainan, Taiwan

Lee-Meng Chieh
National Taichung University of Science and Technology, Taichung, Taiwan

Chen-Yu Pan
Department of Architecture, National Cheng Kung University, Tainan, Taiwan

ABSTRACT: In modern times, people often choose to close doors and windows and turn on air conditioning to quickly lower indoor temperatures. However, the importance of ventilation is frequently overlooked. Striking a balance between indoor thermal comfort, ventilation and air conditioner energy consumption can be challenging. This study aims to provide users with recommendations on air conditioning and ventilation strategies by measuring the power consumption of air conditioners, ACH, and indoor temperature and CO2 concentrations. The conclusions are as follows: While higher ACH can accommodate more indoor occupants, they may lead to higher indoor temperatures and increased energy consumption by air conditioning equipment. The energy consumption of outdoor air conditioning units is positively correlated with the ACH. This suggests that, although ASHRAE recommends a minimum ACH of 4 air exchange per hour for office spaces, achieving exactly 4 air exchange per hour may not be necessary in practice due to the fluctuating density of office personnel. This study recommends that users of air conditioning in office spaces should flexibly adjust ventilation rates based on the actual number of indoor occupants to achieve energy-efficient results while maintaining a comfortable indoor environment.

Keywords: Air exchange rate (ACH), thermal comfort, energy consumption, office building

1 INTRODUCTION

In the hot summer season, modern individuals often choose to close doors and windows and turn on air conditioning to rapidly reduce indoor temperatures. However, the importance of ventilation is often overlooked. Prolonged absence of fresh outdoor air intake for ventilation in enclosed spaces can lead to discomfort due to poor Indoor Air Quality (IAQ), known as Sick Building Syndrome (SBS) (Dusan Licina *et al.*, 2021) [1].

In order to balance indoor thermal comfort and ventilation, or to avoid excessive energy consumption in maintaining overall indoor comfort, this study aims to measure air conditioning power consumption, ventilation rates, and indoor temperature and CO2 concentration using a power meter. The goal is to find combinations that satisfy a balanced benefit and provide users of air conditioning systems with recommendations for air conditioning and ventilation strategies.

Most air conditioning users nowadays focus solely on indoor comfort, often keeping office spaces tightly closed to maintain a stable temperature. While this practice ensures a consistently comfortable indoor environment, it neglects the exchange of air between the indoors and outdoors, leading to an increase in indoor carbon dioxide (CO2) levels due to continuous human respiration (Marques *et al.* 2019) [2]. As CO2 concentration gradually rises, the indoor air quality (IAQ) worsens. Prolonged absence of fresh outdoor air intake for ventilation can result in Sick Building Syndrome (SBS), causing symptoms such as allergies, headaches, eye, nose, or throat infections, drowsiness, nausea, and decreased concentration (Qiao *et al.* 2022) [3].

In 1858, Pettenkofor proposed 0.1% CO2 concentration as a standard, a benchmark adopted by many countries. It has been confirmed that exceeding 1000 ppm of CO2 concentration can cause dizziness, fatigue, and decreased concentration in individuals (Azuma *et al.* 2018; Persily 2015; Shriram *et al.* 2019) [4] [5] [6]. Many indoor pollutants are positively correlated with CO2 concentration, making changes in CO2 concentration an indicator for controlling indoor air quality (Chatzidiakou *et al.* 2015; Ramalho *et al.* 2015) [7] [8].

DOI: 10.1201/9781003514831-27

To meet IAQ standards, there are three common ventilation methods: natural ventilation, mechanical ventilation, and hybrid ventilation. While natural ventilation requires no mechanical facilities or management and can improve indoor thermal comfort (Shaeri *et al.* 2023) [9], reduce CO2 concentration (Muelas *et al.* 2022) [10], and mitigate virus transmission (Fawwaz Alrebei *et al.* 2022) [11], its ventilation effects are challenging to predict and control. In practice, this ventilation method may not achieve acceptable IAQ in many situations. Mechanical ventilation, on the other hand, provides a more predictable ventilation frequency, duration, and air changes per hour (ACH) (Wei Su *et al.* 2022) [12].

Regarding ventilation frequency regulations for office spaces, the American Society of Heating, Refrigerating and Air-Conditioning Engineers (ASHRAE) recommends a minimum of 4 to 10 air changes per hour (ACH). However, increasing outdoor air intake will also increase energy consumption. Therefore, balancing indoor hygiene requirements and reducing outdoor air intake can significantly save energy, especially when indoor air quality is otherwise satisfactory. This is particularly relevant in buildings with significant fluctuations in the number of occupants (Chien 1992) [13].

2 EXPERIMENTAL

This study focuses on the indoor comfort (temperature, CO2 concentration) and air conditioning energy consumption in office spaces during the summer season with different ventilation frequencies. The following explanations will address (1) the experimental field, (2) the experimental methods and equipment, and (3) the experimental combinations and condition settings.

2.1 *Experimental field*

In this study, the focus was on an old office space with a raised floor and ceiling renovations of 15 centimeters. The air conditioning system consists of six ceiling vents. The actual indoor height is 240 centimeters. With users sitting in the office, the distance from the head to the vent is approximately 115 centimeters. The office is equipped with a split air conditioning system (Mitsubishi FDUM medium and low static pressure concealed VRF inverter split air conditioner) with indoor cooling capacities of 5.6 kW and 7.1 kW. During the experiment, a single 5.6 kW indoor unit was used. Ventilation is achieved by introducing fresh air through a variable-frequency fan and air valve. The fresh air is guided to the return air of the air conditioning indoor unit through a custom-made pipeline, accompanied by window opening for exhaust air (EA).

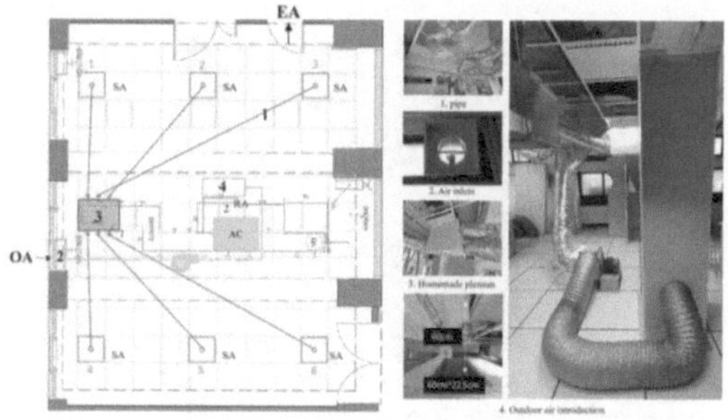

Figure 1. Experimental Site Conditions: Experimental Laboratory Equipment Layout Plan and Space Photos.

2.2 *Experimental methods and equipments*

This study used CO2 as a tracer gas for the concentration decay method to estimate ventilation rates. The experimental procedure involved releasing CO2 from a CO2 cylinder into the experimental area until it reached a stable concentration (3000 ppm). Subsequently, the air conditioning and ventilation systems were activated. Using the T&D TR-76Ui instrument (refer to Figure 2), the temperature and CO2 concentration changes at various measurement points inside were recorded. The time required for the indoor CO2 concentration to decay from the initial concentration of 3000 ppm to the target concentration of 600 ppm was calculated. This time, in turn, allowed the estimation of the corresponding ventilation rate. The calculation formula is expressed by the following equation: (1):

$$N = \frac{\left(\sum_{j=1}^{n} t_j\right) \bullet \sum_{j=1}^{n} \ln\left[C(t_j) - C(t_j)_{out}\right] - n \bullet \sum_{j=1}^{n} t_j \bullet \ln\left[C(t_j) - C(t_j)_{out}\right]}{n \bullet \sum_{i=1}^{n} t_j^2 - \left(\sum_{j=1}^{n} t_j\right)^2} \quad (1)$$

N: Air Changes per Hour (h^{-1}); n: the total number of measured time points;

$C(t_j)_{out}$: outdoor CO_2 concentration.

The power consumption of the air conditioning system was measured using instrument (c) 87786 AZ WBGT, which recorded the current values for both the indoor and outdoor units.

The instrument (b) T&D TR-76Ui records data every 30 seconds and can measure temperature, humidity, and carbon dioxide concentration. The measurement ranges are 0 – 55°C for temperature, 10 – 95%RH for humidity, and 0 – 9,999ppm for carbon dioxide. The accuracy is ±0.5°C, 5%RH, and ±(50 ppm + 5% of reading). The instrument (c) HOKI PW3360-20RU4 records data every 10 seconds with a recording frequency of 50Hz, and it measures current in the range of 50A.

2.3 *Experimental combinations and condition settings*

This study investigates the use of air conditioning and ventilation during the summer in Tainan. The

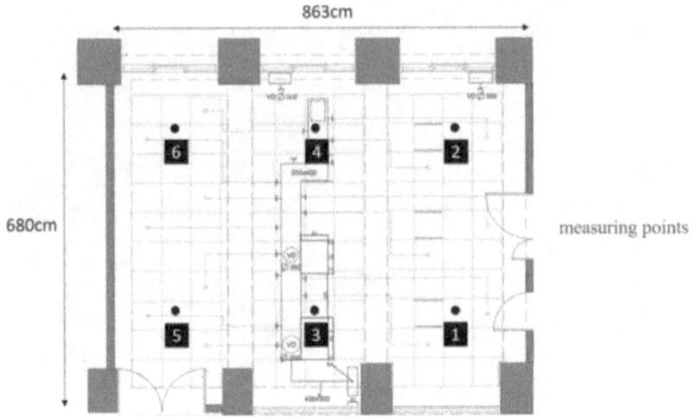

Figure 2. Instrument measuring point.

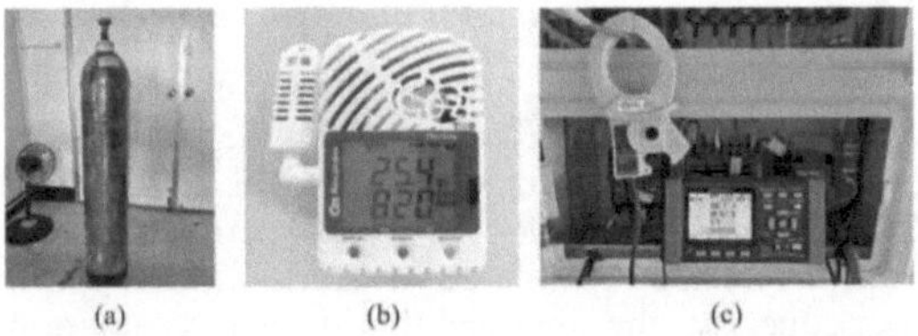

Figure 3. Instruments Used in the Experiment: (a) CO2 cylinder; (b) T&D TR-76Ui (c) HOKI PW3360-20.

following explanations will be provided for (1) indoor and outdoor conditions during the summer, along with air conditioning temperature settings, (2) cooling ventilation frequency settings, and (3) personnel measurement settings.

2.3.1 *Summer indoor and outdoor conditions and air conditioning temperature settings*

(1) Outdoor Temperature: >29°C
(2) Initial Indoor Experiment Temperature: 27°C~28°C
(3) Air Conditioning Set Temperature: 25°C

Table 1. The experimental site corresponds to the required ventilation volume and ideal fan frequency for different ventilation frequencies.

ACH	Ventilation volume(CMH)	Required Wind Velocity (m/s)	Fan frequency (Hz)
1	140	0.29	140
2	280	0.58	280
3	420	0.86	420
4	560	1.15	560
5	700	1.44	700
6	840	1.73	840
7	980	2.02	980
8	1120	2.30	1120

Experimental field space volume: 6.8m × 8.63m × 2.4m (H) = $140m^3$

2.3.2 *Cooling ventilation frequency settings*

This study aims to explore the differences under different ventilation frequencies by adjusting the frequency of the experimental fan to control the ventilation frequency. Table 2 shows the ideal fan frequency corresponding to the wind speed and ventilation volume. In practice, the experiment verifies the final ventilation frequency through the tracer gas concentration decay method. A total of 11 sets of data were collected under conditions ranging from 0 to 8 ventilation frequencies in mixed ventilation (as shown in Table 3).

Table 2. Ventilation frequency experiment combinations.

group	Tem set point	ACH	OA	EA (window open)
1	25	0	fan	○
2	25	1.4	fan	○
3	25	2.3	fan	○
4	25	2.7	fan	○
5	25	3.4	fan	○
6	25	4	fan	○
7	25	4.4	fan	○
8	25	5	fan	○
9	25	6	fan	○
10	25	7	fan	○
11	25	8	fan	○

Table 3. Experimental combinations for different occupancy densities.

group	Tem set point	personnel density	Number of people indoors	ACH
1	25	normal	8	2
2	25	high	25	5

2.3.3 *Personnel measurement settings*

In this study, to validate the impact of air conditioning ventilation strategies and actual occupancy on indoor CO2 concentration and energy savings, two ventilation demand scenarios were set based on different office personnel densities. Under the premise of a reasonable office setting, common references for personnel density (people/square meter) are 0.1 (Europe), 0.5 (USA), 0.2 (Japan), and 0.14-0.05 (Taiwan, Ministry of Economic Affairs Office Building Energy-Saving Technology Handbook). Using these references, two experimental conditions were established with indoor personnel numbers of 25 (high density) and 8 (low density), and the corresponding ventilation requirements were calculated as follows.

2.3.3.1 *High personnel density* $60m^2$ × 0.4 person/m^2 = 24~25 people; 25 people × $30m^3$/hr = $750m^3$/hr (4~5 ACH)

2.3.3.2 *Low personnel density* $60m^2$ × 0.14 person/m^2 = 8 people; 8 people × $30m^3$/hr = $240m^3$/hr (2 ACH)

3 RESULTS AND DISCUSSION

This section discusses the summer conditions in Tainan, Taiwan (outdoor temperature above 29°C, indoor temperature ranging from 27°C to 28°C). With the air conditioning set at 25°C, a hybrid ventilation system (mechanical supply, natural exhaust) is employed to meet the indoor ventilation needs. The study involves practical measurements and analysis of different ventilation frequencies, focusing on (1) indoor temperature and humidity and (2) actual power consumption and frequency conversion of the indoor and outdoor units of the air conditioner. Based on these results, optimal air conditioning ventilation frequencies and strategies are comprehensively compared and recommended.

3.1 *Air exchange rate and comfort*

Regarding indoor thermal comfort, the study references the ASHRAE 55-1992 recommendation of a summer comfort operative temperature of 24.5°C as the baseline. The comfort range is set from 23°C to 26°C. The temperature results are depicted in the following figure. From the figure, it can be observed that except for the condition with 8 air changes, other lower ventilation frequency combinations result in indoor temperatures within the comfort range. The fluctuation in the temperature curve for ventilation frequencies of 1.4, 2.3, 2.7, 3.4, and 4 changes is attributed to the variable frequency generated by the air conditioner's outdoor unit.

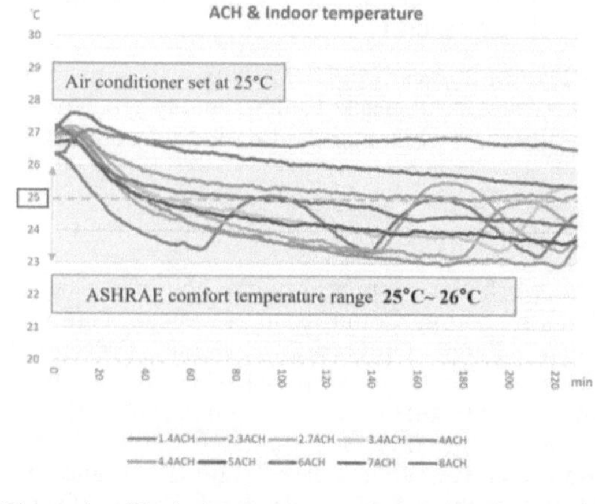

Figure 4. The indoor temperature variations over time at different ventilation frequencies with the air conditioner set to 25°C.

3.2 *Air exchange rate and power consumption*

3.2.1 *Air exchange per hour for different personnel densities*

The hourly variations in the electrical current of the indoor and outdoor air conditioner units for the two scenarios of office occupancy, 24 people (ACH: 5) and 8 people (ACH: 2), are depicted in the following figure. For both scenarios, the indoor CO_2 concentration can be maintained within 1000ppm (Figures 5 and 6), and the average current for the indoor unit is 0.58 (A), while the average current for the outdoor unit is 10.71 (A) for the 8-people scenario and 11.94 (A) for the 25-people scenario (Figures 7 and 8). With fewer ACH, the energy consumption of the air conditioning system is reduced due to a lower outdoor air load.

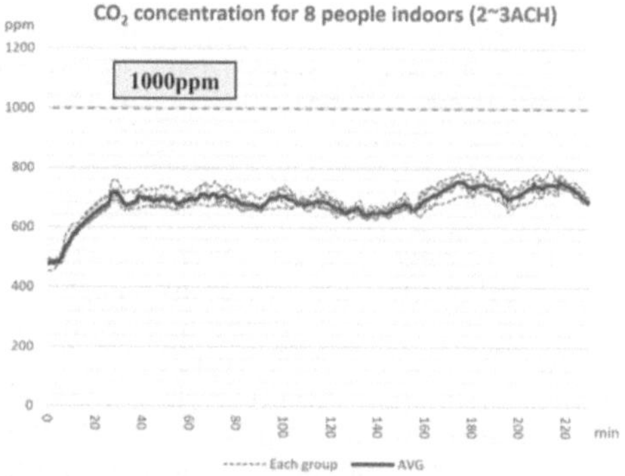

Figure 5. CO_2 concentration for 8 people indoors.

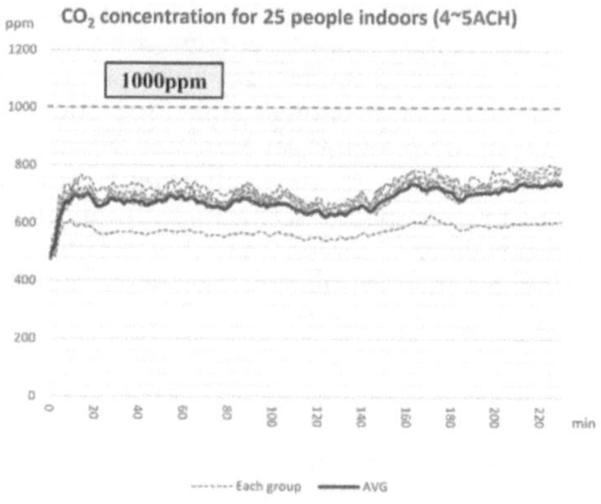

Figure 6. CO_2 concentration for 25 people indoors.

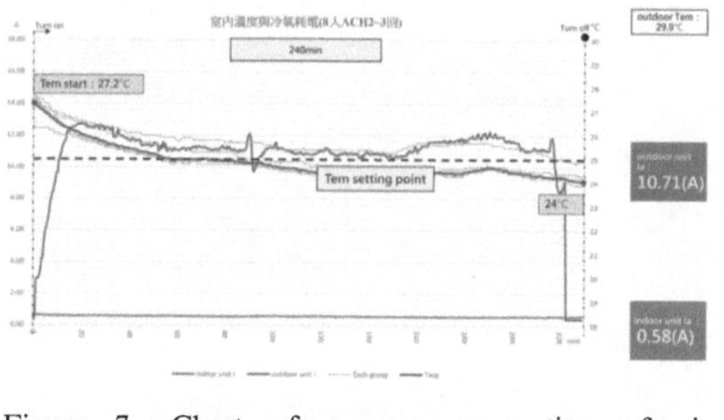

Figure 7. Chart of power consumption of air conditioning equipment and indoor temperature changes over time for 8 people indoors (ACH: 2).

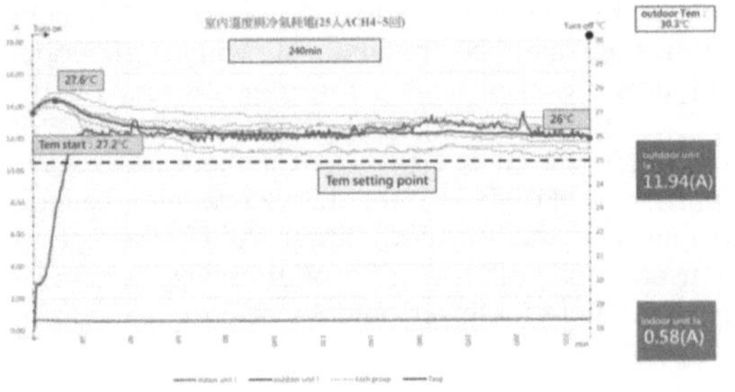

Figure 8. Chart of power consumption of air conditioning equipment and indoor temperature changes over time with 25 people indoors (ACH: 5).

3.2.2 *Air exchange rate and power consumption*

The figure below illustrates the positive correlation between the outdoor unit current and the number of ACH. The outdoor unit current is 5.25 (A) when there is no ventilation (0 cycles), and it gradually increases with the number of ACH, reaching the maximum equipment power consumption of 13.5 (A) at 5 ACH. The reason behind this is that with fewer ACH, the air conditioning system has a lower outdoor air load, allowing the indoor temperature to reach the set target of 25°C more easily. This prompts the system to adjust its frequency, resulting in lower energy consumption. However, around 5 ACH, the outdoor air load becomes significant enough that the air conditioning system cannot generate the benefits of variable frequency. Additionally, comparing the results of the occupancy scenarios with and without people, it is observed that the current consumption is quite similar at the same number of ACH.

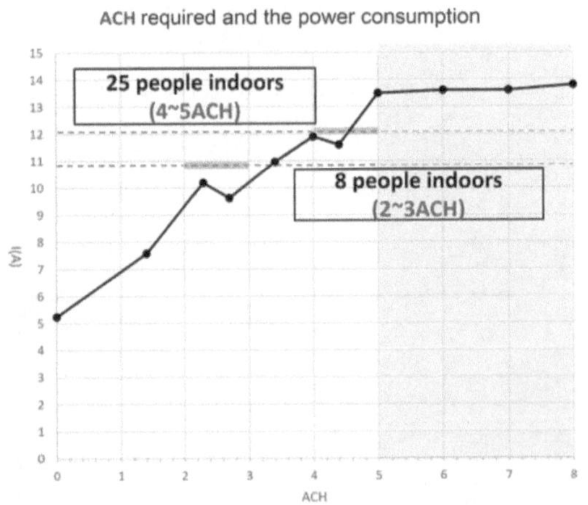

Figure 9. Relationship diagram between air exchange rate and air conditioner outdoor unit current flow.

4 CONCLUSION

The study aimed to explore the impact of different ACH on air conditioning energy consumption and the indoor comfort of cooling spaces, considering temperature and CO2 concentration. The conclusions and recommendations are as follows:

(1) Higher ACH, while accommodating more occupants, may lead to higher indoor temperatures (as observed in this study, up to 8 cycles) and increased energy consumption of the air conditioning equipment.
(2) There is a positive correlation between the energy consumption of the outdoor unit of the air conditioning system and the number of ACH. The study found that the equipment's energy consumption reaches its maximum value around 5 ACH.
(3) The actual occupancy scenarios with two different groups of occupants and their corresponding ACH show similar performance in terms of air conditioning equipment energy consumption. The results suggest that fewer occupants require fewer ACH, leading to more energy-efficient operation of the air conditioning system.
(4) While ASHRAE recommends a minimum of four ACH for office spaces, the study suggests that office space air conditioning users should flexibly adjust the ACH based on the actual number of occupants to achieve a balance between indoor comfort and energy savings, especially considering the variable density of office personnel.

These findings provide insights into optimizing ventilation strategies to achieve energy-efficient air conditioning while maintaining indoor comfort.

FUNDING

This work was supported by the National Science and Technology Commission of Taiwan, National Cheng Kung University Epidemic Prevention Center (Grant No.: NSTC 113-2321-B-006-007-), and the Stratified Green Foundation National Cheng Kung University Research Center Energy Materials Research Center (Hi-GEM) and the MOE Teaching Practice Research Project (Grant No.: H112-A165).

REFERENCES

[1] Dusan L. and Serra Y., (2021). Occupant satisfaction with indoor environmental quality, sick building syndrome (SBS) symptoms and self-reported productivity before and after relocation into WELL-certified office buildings, *Building and Environment*, 204, 108183.

[2] Marques G., Ferreira C. R. and Pitarma R., (2019). Indoor Air Quality Assessment Using a CO2 Monitoring System Based on Internet of Things, *Journal of Medical Systems*, p.67.

[3] Qiao R., Lou X., Sun Y., and Liu Y., (2022). Effects of occupant behaviors on perceived dormitory air quality and sick building syndrome symptoms among female college students, *Indoor Air*, 32(11), e13153.

[4] Azuma K., Kagi N., Yanagi U., and Osawa H., Effects of low-level inhalation exposure to carbon dioxide in indoor environments: A short review on human health and psychomotor performance, *Environment International*, 2018, pp. 51–56.

[5] Persily. A. (2015). Challenges in developing ventilation and indoor air quality standards: The story of ASHRAE Standard 62. *Building and Environment*, 2015, 91, 61–69.

[6] Shriram. S, Ramamurthy. K, and Ramakrishnan. S. (2019) Effect of occupant-induced indoor CO2 concentration and bioeffluents on human physiology using a spirometric test, *Building and Environment*, 149, 58–67.

[7] Chatzidiakou. L, Mumovic. D and Summerfield. A. (2015). Is CO2 a good proxy for indoor air quality in classrooms? Part 1: The interrelationships between thermal conditions, CO2 levels, ventilation rates and selected indoor pollutants. *Building Services Engineering Research and Technology*, 36(2), 129–161.

[8] Ramalho. O, Wyart. G, Mandin. C, Blondeau. P, Cabanes. P.-A, Leclerc. N, Mullot. J.-U, Boulanger. G, and Redaelli. M. (2015). Association of carbon dioxide with indoor air pollutants and exceedance of health guideline values, *Building and Environment*, 93, 115–124.

[9] Shaeri. J, Mahdavinejad. M, Vakilinejad. R, Bazazzadeh. H and Monfared. M. Effects of sea-breeze natural ventilation on thermal comfort in low-rise buildings with diverse atrium roof shapes in BWh regions, *Case Studies in Thermal Engineering*, 2023, 41, 102638.

[10] Muelas. Á, Remacha. P, Pina. A, Tizné. E, El-Kadmiri. S, Ruiz. A, Aranda. D and Ballester. J. (2022). Analysis of different ventilation strategies and CO2 distribution in a naturally ventilated classroom, *Atmospheric Environment*, 283, 119176.

[11] Fawwaz Alrebei. O, Obeidat. L. M, Ma'bdeh. S. N, Kaouri. K, Al-Radaideh. T and Amhamed. A. I. (2022). Window-Windcatcher for Enhanced Thermal Comfort, Natural Ventilation and Reduced COVID-19 Transmission, *Buildings*, 12(6).

[12] Wei Su, Zhengtao Ai, Jing Liu, Bin Yang, Faming Wang (2023). Maintaining an acceptable indoor air quality of spaces by intentional natural ventilation or intermittent mechanical ventilation with minimum energy use. Applied, *Energy*, 348, 121504

[13] Qian Yiming, (1990). Air Conditioning and Energy Saving in High-rise Buildings, pp. 271–350.

System Innovation for an Artificial Intelligence Era – Lam et al. (Eds)

Research on air conditioning vent outlet airflow velocity, distance, and comfort

Yu-Ju Chen, Hsieh-Chih Hsu & Chen-Yu Pan
Department of Architecture, National Cheng Kung University, Tainan, Taiwan

ABSTRACT: Older office spaces were often built at an earlier stage and their air-conditioning systems may not be modern enough, causing many problems. In addition, it is common that the distance from the ceiling to the top of the head of the users in old offices is about 115 cm, which leads to the phenomenon of windshock caused by air-conditioning blowing directly onto the human body. Therefore, this study aims to develop a new type of air vent for the common air conditioning outlets in Taiwan, and takes the old office building space as the research target, and finally presents the research results by means of actual measurement and software simulation. The experimental design of this study adjusts the direction of the outlet connector, the height of the static pressure box, and the factors of the outlet style, and actually measures the changes in wind speed and airflow pattern caused by different combinations of designs, in order to design air conditioning outlets that are comfortable for people. In the first stage, the air velocity of these 40 types of air vents was measured when the air velocity inside the duct was 5m/s. In the second stage, the visualisation experiment was used to test the airflow pattern, and in the third stage, a combination of structures was selected for comparison with simulation software. For meeting space, it is recommended to use side connector with 20cm static pressure box and TYPE2 16 holes outlet, for office, it is recommended to use two types of outlets, side connector with 20cm static pressure box and TYPE1 9 holes outlet, and side connector with 10cm static pressure box and TYPE1 9 holes outlet.

1 INTRODUCTION

Air conditioning vents can reduce energy consumption by increasing indoor air circulation. By improving the efficiency of air circulation, not only can it effectively regulate the indoor temperature, but also help to improve the quality of the indoor environment and reduce energy consumption, which is of substantial help to the goal of energy saving and carbon reduction. In addition, if the comfort level in the office decreases and the air circulation is uneven, resulting in a temperature gap, some areas will be overheated while others will be too cold, which will have a direct impact on the comfort and work efficiency of the staff. Therefore, it is important to ensure good indoor air quality and even air circulation to improve the comfort of the office environment and work efficiency. In this study, we investigated the office space on a campus of a comprehensive university in southern China, which was constructed between 1970 and 100 years ago, and found that the percentage of office space with an indoor height of 2.4 metres was 22%, and considering the fact that commercially available air vents usually lack static pressure boxes, the airflow is blown directly at the human body, which results in uncomfortable wind impacts. In addition, many older offices have limited space for in-ceiling fittings, limiting the location and size of air outlets, and these constraints affect the feasibility and effectiveness of appropriate air outlet installations. Therefore, the purpose of this study is to develop an air-conditioning outlet with uniform air velocity for steady flow in old office spaces with limited height of in-ceiling equipment, to balance the air volume and direction of wind, to reduce the wind shock effect, to disperse air in a more uniform way, to avoid sharp airflow directly affecting the human body, to enable people to enjoy good air quality at work, to improve work efficiency and concentration, and to effectively Improve indoor air circulation, thus creating a more comfortable and pleasant working environment. The American Refrigeration and Air Conditioning Association ASHRAE 55-2010 suggests that the upper limit of air velocity in the comfort zone defined by PMV is 0.20 m/s for those who are lightly dressed (clothing 0.5 clo to 0.7 clo) and in the case of prolonged static sitting (metabolic rate of 1.0 to 1.3), and the upper limit of air velocity in office environments where the working air temperature is higher than 25.5° should be 0.8 m/s in the case of those whose working air temperature is lower than 25.5°, and the upper limit of air velocity in the case of those whose working air temperature is lower than 25.5° should be 0.8 m/s in the case of those whose working air temperature is higher than 25.5°.

DOI: 10.1201/9781003514831-28

However, many experiments have found that wind speeds between 0.3 m/s and 0.5 m/s are the most comfortable.

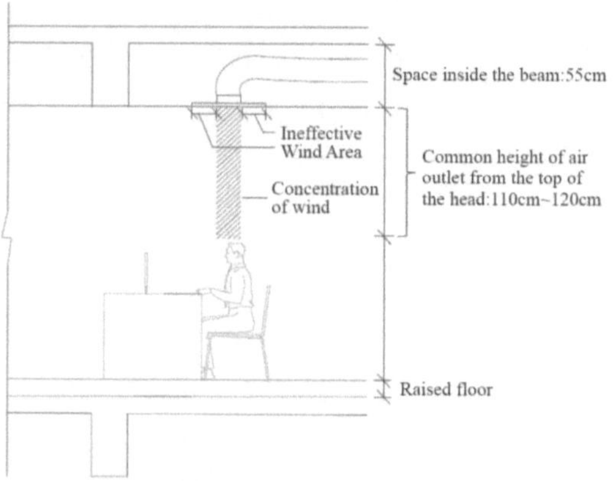

Figure 1. Common conditions of air-conditioning vents and users in old offices.

2 EXPERIMENTAL

This study adopts a practical measurement method, using a hot-wire anemometer to measure the wind speed and a smoke machine to visualise the air flow. These two measurement methods complement each other, providing concrete data support on the one hand, and demonstrating the air flow status intuitively through visualisation on the other. The focus of the evaluation is on the airflow efficiency (i.e. the average air velocity at the outlet) and uniformity (the uniformity of the air velocity at the outlet), which are compared to each other for the purpose of analysing and evaluating the airflow conditions.

There are three criteria for assessment 1. The smaller the standard deviation 2. Outlet air efficiency between 2.5m/s ~ 3.5m/3. Overhead wind speed of 0.3m/s–0.5m/s.Uniformity in accordance with the standard deviation formula, the formula is

$$s = \sqrt{\frac{\sum (x_i - \bar{x})^2}{n-1}} \tag{1}$$

s = sample standard deviation
xi = individual values of the data, i = 1~n
n = number of data

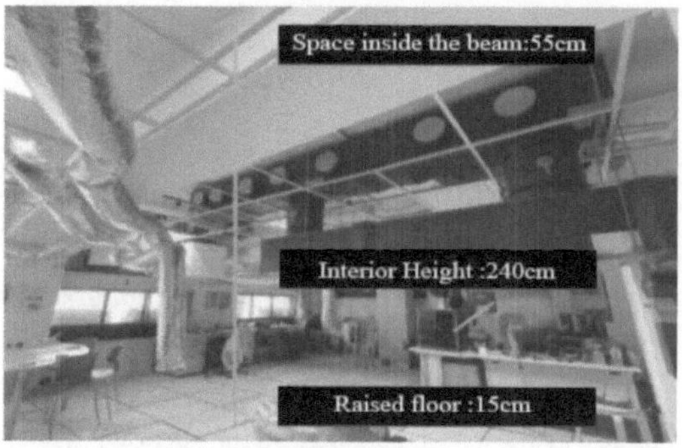

Figure 2. Experimental Space.

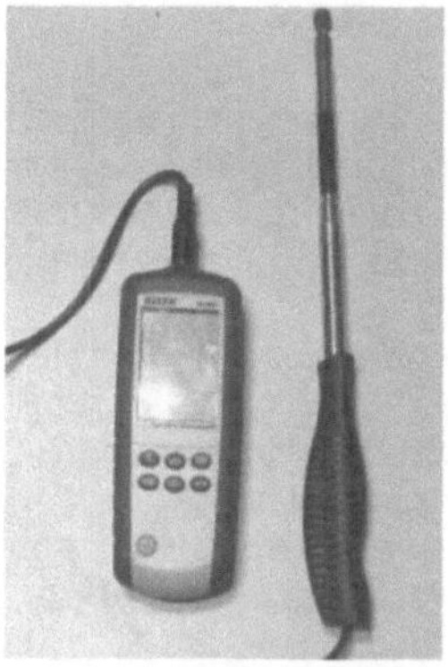

Figure 3. Thermal anemometer.

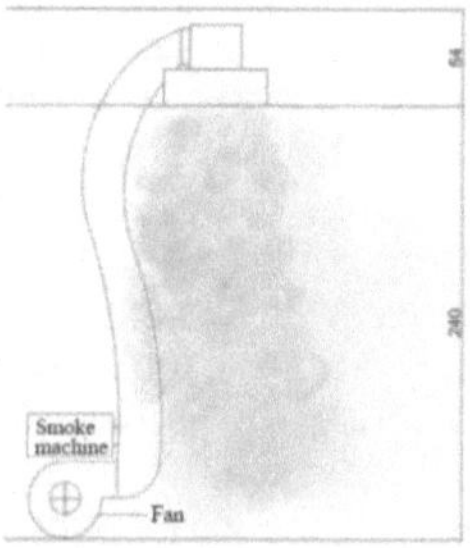

Figure 4. Fans are used to send smoke to the air outlet.

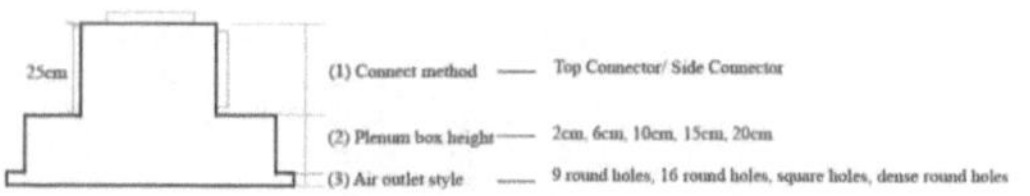

Figure 5. Air outlet design.

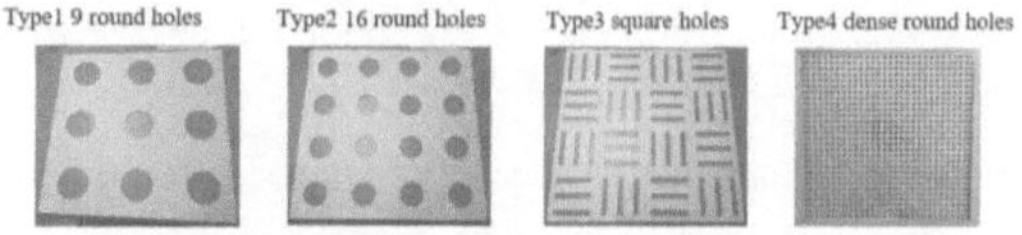

	Hole shape	Hole area(cm^2)	Number of holes	Total opening area(m^2)	Porosity
Type1	circle	75.9261	9	0.0683	0.19
Type2	circle	42.709	16	0.0683	0.19
Type3	square	14.0102	48	0.0672	0.19
Type4	circle	0.7854	870	0.0683	0.19

Figure 6. Four types of air outlet design.

In this study, the design of the downdraft air outlet is modified to involve three main variables. Firstly, there is the type of receiver, which is divided into two types: top receiver and side receiver, which affects the receiver state and thus the buffering and mixing of airflow. The second variable is the height of the plenum box. In this study, the height is divided into five heights, such as 2cm, 6cm, 10cm, 15cm and 20cm, which affects the effect of the secondary homogeneous mixing of the airflow. The third variable is the type of air outlet, which includes nine round holes, sixteen round holes, dense round holes, and square holes, which is predicted to have an effect on the airflow pattern.

These four types of outlets, the same opening rate, changes in the shape of the hole, hole area and number of holes, etc., it is predicted that these changes will have an impact on the airflow pattern of the outlet, perhaps at the same rate of opening, the changes in the shape of the hole and the number of holes will have a different impact on the effect of the outlet and the distribution of airflow, and this subtle difference may be able to bring about unexpected changes in the characteristics of the airflow, which will in turn affect the overall quality of the air flow.

3 RESULTS AND DISCUSSION

The results of the experiment are presented in three comparisons, the first is to compare the way of connectors, the second is to compare the height of the plenum box, and the third is to compare the style of the air outlet.

Figure 7 shows The side connector is better than the top connector in terms of air distribution uniformity, which is a significant difference under the same static pressure box height and style.

Table 1 shows the experimental results do not show any significant regularity in terms of the height of the variable bi-static box.

Figure 8 shows the Changes in outlet style have little or no significant effect on air uniformity.

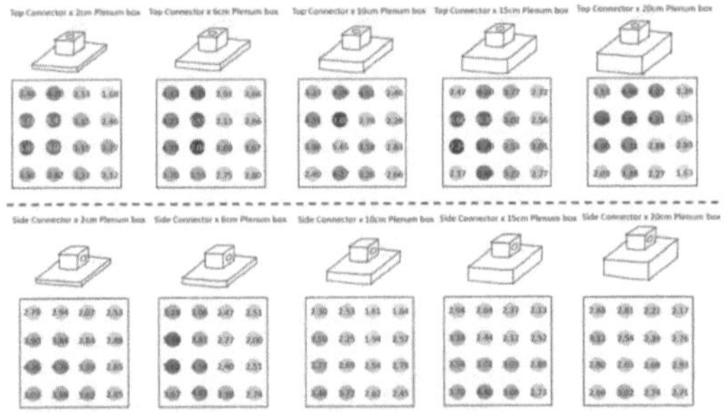

Figure 7. Compare the effect of top and side connectors.

Table 1. Compare the effect of static Plenum box height.

	Side Connector Overhead Air Velocity	Top Connector Overhead Air Velocity
9 round holes	6cm <2cm <10cm <15cm <20cm	15cm <10cm <2cm <20cm <6cm
16 round holes	6cm <2cm <10cm <15cm <20cm	10cm <2cm <15cm <6cm <20cm
dense round holes	2cm <20cm <10cm <6cm<15cm	6cm <10cm <20cm <2cm <15cm
square holes	6cm <2cm <15cm <10cm <20cm	2cm <10cm <15cm <6cm <20cm

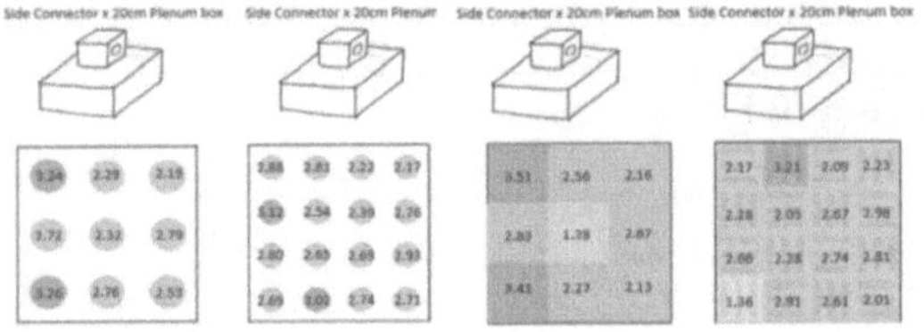

Figure 8. Compare the effect of outlet style.

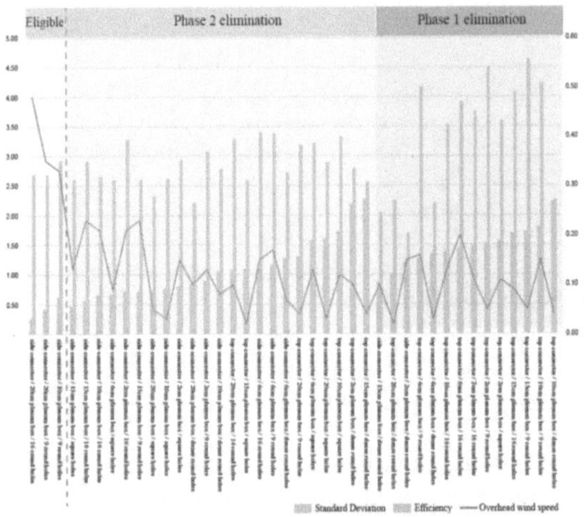

Figure 9. Comparison of the uniformity, efficiency and head wind speed of 40 types of outlets.

4 CONCLUSION

Comparing the uniformity, efficiency and head wind speed of 40 types of outlets, the uniformity was ranked from high to low, the first stage eliminated outlets with efficiency less than 2.5m/s and more than 3.5m/s, and the second stage eliminated outlets with a head wind speed of less than 0.3m/s. In the end, there were only three outlets that met the filtering criteria, which were the 20cm static pressure box for the side connector and the 16-hole outlets for the TYPE2. In the end, only three outlets met the screening criteria, namely, the side connector with 20cm static pressure box and TYPE2 16-hole outlet, the side connector with 20cm static pressure box and TYPE1 9-hole outlet, and the side connector with 10cm static pressure box and TYPE1 9-hole outlet.

FUNDING

This work was supported by the National Science and Technology Commission of Taiwan, National Cheng Kung University Epidemic Prevention Center (Grant No.: NSTC 113-2321-B-006-007-), and the Stratified Green Foundation National Cheng Kung University Research Center Energy Materials Research ·Center (Hi -GEM) and the MOE Teaching Practice Research Project (Grant No.: H112-A165).

REFERENCES

ASHRAE, A. ANSI/ASHRAE 55-(2010). *Thermal Environmental Conditions for Human Occupancy*, American Society of Heating, Refrigerating and Air-Conditioning Engineers, Inc., Atlanta, Georgia, 2010.

Hu Yan, Hidenori Okuzaki, (2009). *Synyh.Met.* 159 2225–2228.

Jönsson S.K.M., Birgerson J., Crispin X., Greczynski G., Osikowicz W., Denier van der Gon A.W., Salaneck W.R., Fahlman M., (2003). *Synyh. Met.* 139 1–10.

Kim J.Y., Jung J.H., Lee D.E., Joo J., (2002). *Synyh. Met.* 126 311–316.

Sungeun Park, Sung Ju Tark, Donghwan Kim, Curr., (2011). *Appl. Phys.* 11 1299–1301.

Szabolcs Szekeres, Attila Kostyák, Ferenc Szodrai, Imre Csáky, (2022). investigation of ventilation systems to improve air quality in the occupied zone in office buildings, *Buildings*, 12(4), 49

Timpanaro S., Kemerink M., Touwslager F.J., De Kok M. M., Schrader S., (2004) *Chem, Phys. Lett.* 394 339–343.

Wang Y., J. Phys, (2009). *Conf. Ser.* 152 012023.

Yijie Xia, Kuan Sun, and Jianyong Ouyang, (2012). *Adv. Mater.* 24 2436–2440.

System Innovation for an Artificial Intelligence Era – Lam et al. (Eds)

Exploring the usability of the system taskbars in windows 10 and windows 11

Meng-Ru Liu & Siu-Tsen Shen
Department of Department of Multimedia Design, National Formosa University, Hu-Wei, Yunlin, Taiwan

ABSTRACT: In the modern era of advanced computer technology, computers have become the main tool for our daily efficient operations, and the Windows operating system is one of the most common operating systems in the country. Many software programs rely on features on the system taskbar, such as the ability to rotate and preview active application windows and switch between them. This study aims to explore the usability of the Windows 10 system taskbar and the Windows 11 system taskbar. Although the market share of Windows 10 is much higher than that of Windows 11, the layout of the Windows 11 system taskbar has changed significantly, changing to a center-aligned design. The research process includes three steps: 1. Explore the functions and comparisons of system taskbars in literature, 2. Use the SUS usability questionnaire scale and modify it to fit the questions of system taskbars, 3. Analyze the questionnaire to understand users' opinions on Windows 10 and common preferences for the Windows 11 system taskbar. Questionnaire results: A total of 319 respondents were surveyed, of which 317 questionnaires were valid. Research results show that average users generally believe that the Windows 11 system taskbar provides a better user experience, and have a higher awareness of the taskbars of both systems. In addition, the views between the two system work columns are relatively consistent. However, compared to Windows 10, the Windows 11 system taskbar may be slightly more complex and may require more technical support. Therefore, it is recommended to reduce the complexity of the system taskbar, for example: integrating the search function with Copilot to minimize user feature choices, while ensuring users can achieve operational objectives more effectively. Additionally, merging widgets with notifications, allowing users to customize the order of viewing the notification area, further simplifying the information load of the system taskbar.

Keywords: System Taskabar, Information Architecture, Usability

1 INTRODUCTION

According to the National Digital Development Research Report (2023), 69% of users in our country use computers for office tasks, indicating the prevalence of computers in our lives. Whether it's multitasking between efficient office work, multimedia entertainment, or gaming, computers have brought about significantly high levels of productivity. Statcounter (2024 a), Statcounter (2024 b) reported that Microsoft Windows operating system accounted for 72.79% of the total computer operating systems in Taiwan from December 2022 to December 2023, compared to other operating systems, with Windows 10 at 67.46% and Windows 11 at 26.52%. This indicates the widespread use of Windows operating systems in Taiwan, hence the choice to focus on Windows operating systems for this study.

In computer operation, the system taskbar is an essential element of the Windows operating system. It not only allows for switching between application windows, searching for files, and checking historical notifications on the computer but also enables quick control of system functions such as network connections, screen brightness, and internet connectivity. The Start menu is the primary launch area for applications, files, and other functionalities. With the release of Windows 11, the system taskbar has undergone significant changes, and we aim to delve into these changes' impact on user experience. Understanding user preferences for the system taskbar in Windows 10 and Windows 11 will provide suggestions for operating interfaces that better meet their needs. This study aims to explore the differences in the system taskbar between Windows 10 and Windows 11 operating systems and evaluate user preferences for both. Through the System Usability Scale (SUS), we will quantitatively assess the usability of the system taskbar to gain deeper insights into user satisfaction and performance perceptions.

Taking into account the aforementioned factors, this study aims to analyze the differences between the system taskbars of Windows 10 and Windows 11 through literature review. Furthermore, it seeks to

DOI: 10.1201/9781003514831-29

understand users' preferences regarding the display state of the taskbar. The scope of the research is outlined as follows:

1. Investigating the functionality of the system taskbar through literature review.
2. Analyzing the evolution of the system taskbar and studying its impact on user usability.
3. Conducting a survey to understand users' preferences regarding the taskbars of Windows 10 and Windows 11 operating systems.

2 EXPERIMENTAL

2.1 *Research on system taskbars*

The system taskbar is a collection of files and applications organized for specific user activities on a computer (Robertson *et al.* 2000). Each active application within this collection has its own icon on the taskbar, which is typically located at the bottom of the screen by default (Sullivan 1996). In multitasking contexts, the system taskbar addresses space constraints by grouping application icons, while also providing basic operations for currently active applications (Smith *et al.* 2003). For example, on the Windows 7 taskbar, hovering the cursor over taskbar items enables basic control and switching operations (Ajinkya 2022a; Ajinkya 2022b). For multimedia playback software, hovering over icons can facilitate basic control and switching (Figure 1). The evolution of each operating system impacts the interface and functionality arrangement of the system taskbar. Next, we will discuss the primary differences in the system taskbars between Windows 10 and Windows 11.

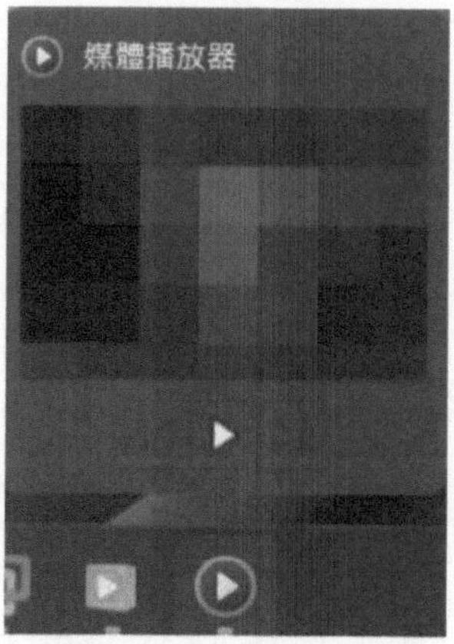

Figure 1. Hovering over the Windows Media Player to play a video.

2.1.1 *Windows 10 system taskbar*

The Windows 10 System Taskbar is located at the bottom of the screen (Figure 2), displaying open applications, system icons (such as Start, Search, and System Notifications ... etc). Users can pin their favorite applications to the system taskbar for quick access and can also use the virtual desktop feature to organize their workspace. Additionally, the system taskbar includes an Action Center, which replaces the previous balloon notifications in the system, allowing users to access system settings and features (such as Bluetooth, Airplane mode, Nearby Sharing, etc.). However, volume control, network control, and battery indicators each have their own separate icons.

Figure 2. Windows 10 system taskbar.

2.1.2 *Windows 11 system taskbar*

Compared to its predecessor, the arrangement of the Windows 11 System Taskbar has changed. It adopts a centered alignment (Figure 3), rather than the left-aligned design of previous versions. Additionally, the Quick Settings panel replaces the Control Center, combining the battery, volume, and network icons, and introduces the Copilot assistive office feature.

Figure 3. Windows 11 system taskbar.

2.2 *Information architecture*

A well-planned information architecture on the system taskbar can reduce cognitive load and minimize operational errors. Information architecture involves organizing and simplifying information to assist people in interpreting and managing it, leading to correct decision-making (Ding and Lin 2010). Turnbull *et al.* (2005) and Li *et al.* (2019) proposed three major elements of information architecture: context, content, and users. Context considers the integration of software and cognition, content is one of the primary elements of the system taskbar, and users need to possess relevant knowledge and experience to understand the conveyance of information architecture. The arrangement of various functional options and icons on the system taskbar is connected in a ribbon-like format, making it easier for users to find the desired functions. The overall design and operation will affect the user experience.

2.2 *Usability*

Usability is crucial for the system taskbar, involving the effectiveness, efficiency, and satisfaction of users in specific contexts (Patayon 2019; Jokela *et al.* 2003). Efficiency in operation includes familiarity with the arrangement and communication form of the system taskbar to ensure accuracy of purpose and minimize errors. User satisfaction relates to the comfort of operation and overall satisfaction with the interface. Effectiveness indicates whether users can achieve expected results and goals during operation, while efficiency refers to the speed at which users can find target functions and the ease of making errors. Satisfaction reflects the overall experience of operating the product.

2.3 *System usability scale*

The System Usability Scale (SUS) is a Likert scale based on forced-choice questions, originally developed by John Brooke in 1986 (Brooke 1996). This scale is used to assess the usability of user interfaces and is widely employed in evaluations of user interface usability (Bangor *et al.* 2009). It serves as a measure of user perception, evaluating user satisfaction with a specific system through questionnaire format. Generally, if the total SUS score exceeds 70, the system's usability is considered acceptable (Bayu 2024). The scale requires respondents to choose from a scale of 5 options and then calculates a total score based on a specific formula to assess the level of system usability (Figure 4).

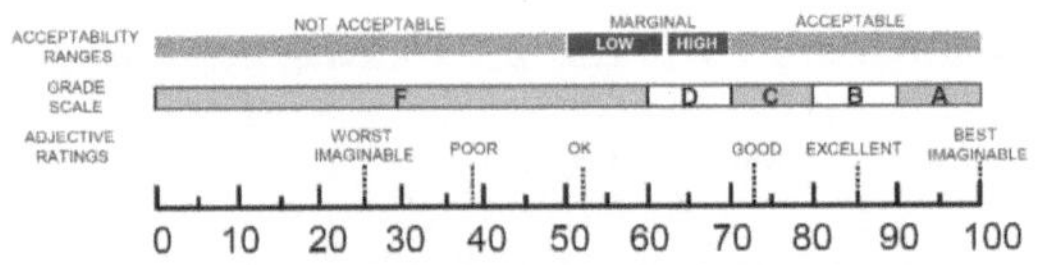

Figure 4. SUS scale score (Bangor & Miller 2009).

2.3 *Questionnaire reliability analysis method*

Cronbach's α is a standard measure of internal consistency reliability, used to estimate the variance caused by multiple variables (Steedle and Shavelson 2010). It is commonly employed to analyze the reliability of Likert scale items in questionnaires, where a Cronbach's α coefficient of 0.7 indicates acceptable consistency reliability (Wu 2018). Therefore, this study implements reliability analysis using this method.

3 RESEARCH METHODOLOGY

3.1 *Research process*

The research conducted a literature review to evaluate users' perceptions of operating Windows 10 and Windows 11 system taskbars. For the evaluation, we utilized a questionnaire that included the System Usability Scale (SUS). The questionnaire separated the two systems into 14 statements, and respondents were required to answer these questions based on their actual experiences. This allowed us to obtain assessments of the usability of the system taskbars and analyze the conclusions and recommendations of the questionnaire (Figure 5).

Figure 5. Research Process.

3.2 *Research limitations*

This study investigates the comparative user experience of the Windows 10 and Windows 11 operating systems, with the limitation that each participant must have used both Windows 10 and Windows 11 for a minimum of two hours each. Additionally, the study focuses on operations closely related to daily office tasks to provide recommendations and references for the layout of the system taskbars.

3.3 *Questionnaire design*

The questionnaire referenced the SUS cognitive variable questionnaire from Umer *et al.* (2017) and Manik *et al.* (2018), which are relevant to system usability and were modified to fit the system taskbars. There are a total of 14 questions, among which 1, 2, 3, 4, 10, 11, 12, 13, 14 are positive questions, and 5, 6, 7, 8, 9 are negative questions. These questions aim to analyze the usability of the system taskbars between Windows 10 and Windows 11 (Table 1).

Table 1. Organizing the questionnaire questions.

Items
Q1. I want to use this system.
Q2. I find the system taskbar easy to operate.
Q3. I am very confident in using the system taskbar.
Q4. I think many people will learn to use the system taskbar quickly.
Q5. I find it very cumbersome to operate the system taskbar. (Reverse question)
Q6. I find the system taskbar doesn't need to be so complicated. (Reverse question)
Q7. I think there are too many differences in the system taskbar. (Reverse question)
Q8. I need to learn a lot before using the system taskbar. (Reverse question)
Q9. I think I need technical support to use this system. (Reverse question)
Q10. I find the arrangement of software blocks on the system taskbar comfortable.
Q11. I think the functions of the system taskbar meet my needs.
Q12. Do you think the arrangement of icons and functions on the system taskbar conforms to your usage habits and expectations?
Q13. Can you easily understand the meanings of various functions on the system taskbar without additional learning or information lookup?
Q14. In your experience using the system taskbar, do you feel that any functions provided by the system taskbar effectively simplify your work?

3.4 *Questionnaire validity analysis discussion*

3.4.1 *Questionnaire collection results*

This study distributed the questionnaire widely online via Google Forms, receiving a total of 319 responses. The age range of participants varied from 11 to 60 years old. Among these, 2 responses were deemed invalid because the answers to the reverse questions contradicted the previous ones, making it impossible to obtain accurate results. Therefore, out of the 317 valid responses, there were 129 female respondents, accounting for 40%; 161 male respondents, accounting for 51%; and 27 respondents who did not disclose their gender, accounting for 8.5%. According to the age distribution, there were 11 responses from the age group of 11 to 15 years old, accounting for 3.4%; 34 responses from the age group of 16 to 20 years old, accounting for 10.7%; 48 responses from the age group of 21 to 25 years old, accounting for 15%; 44 responses from the age group of 26 to 30 years old, accounting for 13.8%; 51 responses from the age group of 31 to 35 years old, accounting for 16%; 41 responses from the age group of 36 to 40 years old, accounting for 12.9%; 36 responses from the age group of 41 to 45 years old, accounting for 11.3%; 26 responses from the age group of 46 to 50 years old, accounting for 8.2%; 22 responses from the age group of 51 to 55 years old, accounting for 6.9%; and 6 responses from the age group of 56 to 60 years old, accounting for 1.9%. Further analysis was conducted on this data.

3.4.2 *Questionnaire reliability analysis*

This study conducted reliability analysis of the SUS scale using Cronbach's Alpha coefficient to ensure the questionnaire possesses stable internal consistency. The analysis results showed that the Cronbach's Alpha coefficient of the SUS scale was 0.781, indicating good reliability of the questionnaire (Table 2).

Table 2. Cronbach's alpha coefficient dimension reliability statistics.

Reliability Statistics	
Cronbach's Alpha	Number of Items
.781	28

3.4.3 *Analysis of system taskbar SUS scores*

Based on the results analysis, the overall questionnaire average for the Windows 11 system taskbar, after weighted calculation, is 90 points, while the Windows 10 system taskbar scored 88 points. The usability of the two system taskbars is very close (Figure 6).

3.4.4 *Results of system taskbar satisfaction*

After collecting and analyzing user questionnaires, we obtained evaluations of the system taskbars of Windows 10 and Windows 11. The usability survey results indicate that users generally find Windows 11 to be more usable and satisfactory compared to Windows 10. Particularly in the positive questions, Windows 11 scored higher overall, suggesting improvements in user experience may have been made in Windows 11. However, in the negative questions, we also observed that the Windows 11 system taskbar is slightly more complex than Windows 10,

Table 3. Analysis of system taskbar usability.

Items	Windows 10	Windows 11
Q1. I want to use this system.	3.66	4.04
Q2. I find the system taskbar easy to operate.	3.63	3.92
Q3. I am very confident in using the system taskbar.	3.60	3.79
Q4. I think many people will learn to use the system taskbar quickly.	3.61	3.76
Q5. I find it very cumbersome to operate the system taskbar. (Reverse question)	2.52(Reverse question 3.48)	2.74(Reverse question 3.26)
Q6. I find the system taskbar doesn't need to be so complicated. (Reverse question)	2.59(Reverse question 3.41)	2.84(Reverse question 3.16)
Q7. I think there are too many differences in the system taskbar. (Reverse question)	2.93(Reverse question 3.07)	3.05(Reverse question 2.95)
Q8. I need to learn a lot before using the system taskbar. (Reverse question)	3.09(Reverse question 2.91)	3.19(Reverse question 2.81)
Q9. I think I need technical support to use this system. (Reverse question)	2.91(Reverse question 3.09)	3.12(Reverse question 2.88)
Q10. I find the arrangement of software blocks on the system taskbar comfortable.	3.75	3.90
Q11. I think the functions of the system taskbar meet my needs.	3.78	3.95
Q12. Do you think the arrangement of icons and functions on the system taskbar conforms to your usage habits and expectations?	3.79	3.90
Q13. Can you easily understand the meanings of various functions on the system taskbar without additional learning or information lookup?	3.76	3.89
Q14. In your experience using the system taskbar, do you feel that any functions provided by the system taskbar effectively simplify your work?	3.73	3.93

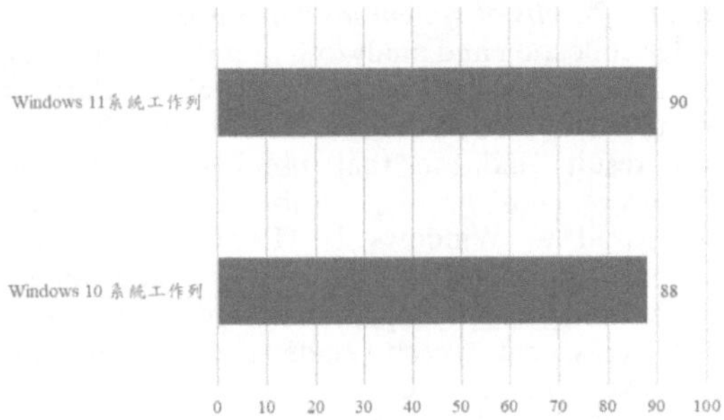

Figure 6. Comparison of the Two System Taskbars.

particularly in question Q6. In question Q7, the perspectives on the taskbars of both systems are more consistent. In question Q8, we observed significantly higher user perceptions of both system taskbars. As for question Q9, the results suggest that the Windows 11 system taskbar may require more technical support to operate.

4 CONCLUSION AND RECOMMENDATIONS

According to the survey results, general users generally prefer Windows 11 over Windows 10 in terms of the taskbar. Windows 11 performs better in usability, operability, confidence and learning curve, and feature satisfaction, compared to Windows 10, and its taskbar design better meets users' needs. Additionally, the cognitive perspectives on operating the taskbar in both systems are more consistent, and users have a higher level of understanding. However, in terms of complexity, the taskbar in Windows 11 is slightly more complex, which may be due to differences in user operating experience, thus requiring some time to get accustomed to. Using the taskbar in Windows 11 may require more technical support. Therefore, we believe that the usage rate of the system may be related to user needs and satisfaction levels. From a functional perspective, Windows 11 has added more features to the taskbar compared to Windows 10, resulting in a more extensive arrangement of interface functions. However, this also leads to a certain degree of functional complexity, which may be confusing for users. Therefore, it is recommended to reduce the complexity of the taskbar, for example: integrating search with Copilot functionality to reduce users' functional choices, while ensuring that users can achieve operational objectives more effectively. Additionally, merging widgets with notifications, allowing users to customize the order of viewing the notification area, further simplifying the information load of the taskbar. Further research on improving the functionality of the taskbar in Windows 11 and its user experience advantages will help to better understand its shortcomings and users' responses to their needs in depth.

REFERENCES

Bangor, A., Kortum, P., and Miller, J. (2009). Determining what individual SUS scores mean: Adding an adjective rating scale. *Journal of Usability Studies*, 4(3), 114–123.

Barve, A. S. (2022). *Application or Tab Switching via Mouse Hover and Scroll.* 1–5. https://www.tdcommons.org/cgi/viewcontent.cgi?article=6503&context=dpubs_series

Barve, A. S. (2022). *Updating the Taskbar with Contextual Actions Relevant to the Application in Focus.* 1–6. https://www.tdcommons.org/cgi/viewcontent.cgi?article=6512&context=dpubs_series

Brooke, J. (1986). *SUS: A Quick and Dirty Usability Scale. Proceedings of the Second Conference of the British Computer Society*, 2. https://www.researchgate.net/publication/228593520_SUS_A_quick_and_dirty_usability_scale

Cullen, A. (2023). *Statcounter. Desktop Operating System Market Share Taiwan Dec 2022 – Dec 2023.* https://gs.statcounter.com/os-market-share/desktop/worldwide

Cullen, A. (2023). *Statcounter. Desktop Windows Version Market Share Taiwan Dec 2022 – Dec 2023.* https://gs.statcounter.com/windows-version-market-share/desktop/taiwan

Ding, W., and Lin, X. (2010). *Information Architecture The Design and Integration of Information Spaces* (No. 169; 2nd ed., Vol. 4). Springer Cham. https://doi.org/https://doi.org/10.1007/978-3-031-02267-8 https://uxpajournal.org/determining-what-individual-sus-scores-mean-adding-an-adjective-rating-scale/

Jokela, T., Iivari, N., Matero, J., & Karukka, M. (2003). The Standard of User-Centered Design and the Standard Definition of Usability: Analyzing ISO 13407 against ISO 9241-11. *Proceedings of the Latin American Conference on Human-Computer Interaction*, 53–60. https://doi.org/10.1145/944519.944525

Kelana, B. (2024). *System Usability Scale Validation from The Expert Perspective. The 3rd International Seminar of Science and Technology*, 483, 1–11. https://doi.org/https://www.e3s-conferences.org/articles/e3sconf/abs/2024/13/e3sconf_isst2024_03008/e3sconf_isst2024_03008.html

Manik, M., Heru, P., Novie, S., & Susilo, D. Y. (n.d.). *Perception Analysis of Microsoft Windows OS Users.* ICENIS, 73. https://doi.org/10.1051/e3sconf/20187312008

Patayon, U. (2019). Operating Systems Usability: A Comparative Study. *JPAIR Multidisciplinary Research*, 36, 92–103. https://doi.org/10.7719/jpair.v36i1.683

Robertson, G., Van Dantzich, M., Robbins, D., Czerwinski, M., Hinckley, K., Risden, K., Thiel, D., & Gorokhovsky, V. (2000). *The Task Gallery: A 3D window manager. Proceedings of the SIGCHI Conference on Human Factors in Computing Systems*, 494–501. https://doi.org/10.1145/332040.332482

Shen, J., Li, L., and Dietterich, T. G. (2007). *Real-Time Detection of Task Switches of Desktop Users.* 07, 2868–2873. https://www.ijcai.org/Proceedings/07/Papers/460.pdf

Smith, G., Baudisch, P., Robertson, G., and Czerwinski, M. (2003). GroupBar: The TaskBar Evolved. 1–10. https://www.researchgate.net/publication/2903374_GroupBar_The_TaskBar_evolved

Steedle, J. T, and Shavelson, R. J. (2010). *Coefficient Alpha* (No. 4; Corsini Encyclopedia of Psychology). https://doi.org/10.1002/9780470479216.corpsy0190

Sullivan, K. (1996). *The Windows 95 User Interface: A Case Study in Usability Engineering*. 96, 473–480. https://dl.acm.org/doi/10.1145/238386.238611

Turnbull, D., and Don, T. (2005). *Information Architecture.* UA Campus Repository, 2006-08-04T00:00:01Z, 1–9. http://hdl.handle.net/10150./105971

Umer, M., Khalid, M., and Saboor, A. (2017). usability and accessibility evaluation of windows 10. *International Journal of Innovative Research in Computer and Communication Engineering*. https://doi.org/2320–9801

System Innovation for an Artificial Intelligence Era – Lam et al. (Eds)
© 2025 the Author(s), ISBN: 978-1-032-84452-7

Feasibility study of integrating immersive gastronomic experience with virtual reality

Liang-Yin Kuo
Department of Education, National Formosa University, Hu-Wei, Yunlin, Taiwan

Jo-Hua Tseng
Department of Education, National Yunlin University of Science and Technology, Yunlin, Taiwan

Chia-fang Hsu
Department of Education, New Era University College

ABSTRACT: When viewed through the lens of Virtual Reality (VR), gastronomy offers a forward-looking culinary experience that transcends conventional sensory perceptions. This immersive journey into virtual gastronomy expands beyond the traditional five senses, incorporating additional sensory dimensions and scientifically engineered stimuli to evoke emotional responses. Such advancements propel the dissemination of VR beyond the realm of gastronomy. In recent times, immersive experiences have garnered considerable attention across diverse industries. By amalgamating real-world settings with elements such as music, olfactory art, installation art, tactile sensations, and dynamic lighting effects, participants are transported into alternate realities, detached from their physical surroundings and fully engaged. This study employs a case study methodology, leveraging panoramic photography to introduce novel experiences and scrutinize the fusion of gastronomy and tourist attractions within VR frameworks. By leveraging panoramic photography to capture diverse perspectives and seamlessly integrating them into VR applications, this approach delivers an immersive and captivating presentation format, offering participants an unparalleled experience that transcends physical constraints.

1 INTRODUCTION

"Immersive experiences" originally emerged as a burgeoning topic in the gaming technology industry. However, with shifts in consumer behavior, "experience" has become an indispensable aspect across various industries in recent years, garnering significant attention and widespread application. Immersive experiences guide participants through sensory engagement, fostering strong connections and resonances with specific contexts. Traditional flat photography methods to showcase food are no longer sufficient to captivate consumers; more dynamic formats such as short videos resonate better. In the post-pandemic era, immersive experiences have further evolved as a burgeoning trend. Through technologies like AR and VR, entirely new interactive modes can be created, transforming brand concepts into more intense sensory stimuli, thereby establishing deeper connections with consumers and reaching a broader potential consumer base.

This research will explore the feasibility of integrating immersive culinary experiences with virtual reality. This may include using virtual reality technology to simulate different dining experiences or incorporating immersive elements into actual dining experiences to enhance consumer perception and satisfaction. Such research holds the promise of bringing new innovations and competitive advantages to the food industry.

2 EXPERIMENTAL

The saying goes, "Food is heaven." The importance of food to people extends beyond mere sustenance; it also involves health, culture, social interactions, and emotional needs, making it an integral part of human life. The seemingly simple question of "What shall we eat today?" often perplexes individuals. Consequently, most people resort to searching and consulting objective reviews to decide where to dine. Additionally, food plays a significant role in social culture. Depending on the purpose of a date, individuals choose dining locations based not only on delicious meals but also on factors like ambiance and service. Therefore, many food businesses frequently employ SEO principles to optimize their marketing strategies. In recent years, most individuals prefer using social media in addition to

DOI: 10.1201/9781003514831-30

server searches to find relevant information. Big data algorithms also periodically insert search results into user pages to reach a broader potential consumer base.

When search results appear alongside content on social platforms, the primary goal is to capture consumers' attention and arouse their curiosity. How can one stand out among numerous competitors? Audiences are typically initially attracted by visuals. However, contemporary individuals seek not only visual but also multi-sensory experiences. With the rise of short videos, traditional text and image-based introductions are no longer sufficient to captivate consumers. Having managed an IG blog for over a year and collaborated with numerous food businesses, I can provide some data to support this claim. According to the data, short videos reach twice as many people as text and image-based introductions, indicating that multi-sensory short videos are more effective in generating traffic. In the future, combining food introductions with virtual reality to enhance the "immersive experience" will undoubtedly create new forms of marketing for food businesses.

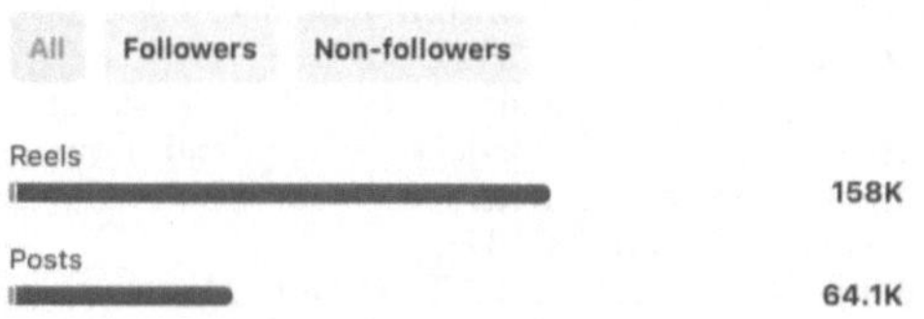

3 RESULTS AND DISCUSSION

Immersion experience was first proposed by Mihaly Csikszentmihaly in 1975. Immersion refers to the state where users are fully absorbed in an activity, automatically filtering out irrelevant perceptions and thoughts, thereby experiencing pleasure. It also pertains to the perception and cognition of virtual environments created and displayed by computer systems, which is one of the core concepts of virtual reality technology. This study aims to explore the possibility of integrating immersive food experiences with virtual reality. By utilizing virtual reality technology, we provide a forward-looking gastronomic experience. This immersive virtual food journey transcends traditional sensory experiences, incorporating additional sensory dimensions and scientific stimuli to evoke emotional responses. We hope to utilize VR to advance the dissemination of gastronomy and provide unique and captivating dining experiences.

Figure 1. The visual landscape of a virtual culinary journey.

Virtual reality gastronomic journeys will offer people a whole new level of experience. Compared to traditional flat photography, virtual reality allows for the presentation of entire scenes in interactive form, enabling viewers to explore from multiple angles and perspectives, thus providing a richer and more immersive viewing experience. For example, by wearing VR headsets, users can virtually immerse themselves in different-dimensional food scenes, witnessing the origin and cooking process of ingredients. By integrating real-world settings with elements such as music, olfactory art, installation art, tactile sensations, and dynamic lighting effects, participants are transported into different realities, fully engaging in the experience.

Figure 2. The visual landscape of a virtual culinary journey.

In the future, virtual food tours can also be used as marketing tools, allowing users to explore specialty cuisines from different regions anytime, anywhere. Through virtual tours, they can learn about the origins, production processes, special features, and even nutritional information of various cuisines, while also enjoying images and videos of the food before deciding whether to dine. Locations and times will no longer be limiting factors, as people can immerse themselves in cuisines from all over the world, from the southern to the northern hemisphere. Similarly, we can also use panoramic photography to showcase various tourist attractions, such as sunrise and sunset at Alishan, beaches in Kenting, and more. Through virtual reality, viewers can experience the magnificence and awe of nature.

In conclusion, virtual reality offers people a more vivid, rich, and diverse immersive experience, thereby fostering empathy connections and enhancing engagement.

Figure 3. The immersive travel scenario visuals.

Figure 4. The immersive travel scenario visuals.

4 CONCLUSION

In summary, the integration of immersive food experiences with virtual reality holds tremendous potential, offering people a whole new culinary journey. By harnessing virtual reality technology, we can transcend traditional sensory boundaries, immersing users in captivating gastronomic adventures. From exploring the origins of ingredients to experiencing the ambiance of dining locations, virtual reality enables unprecedented levels of engagement and immersion. Additionally, virtual food tours serve as powerful marketing tools, allowing easy exploration of culinary gems from around the world. As we continue to innovate in this field, the future of the culinary world promises to be even more immersive, interactive, and enriching, allowing people to delve deeper into the essence of food and culture.

REFERENCES

Csikszentmihalyi, M. (1975). *Beyond Boredom and Anxiety*. San Francisco: Jossey-Bass.

Csikszentmihalyi, M. & Csikszentmihalyi, I. (1988). Itroduction to part IV. In M. Csikszentmihalyi, & I.S. Csikszentmihalyi (Eds.), *Optimal Experience: Psychological Studies of Flow in Consciousness*. New York: Cambridge University Press.

Csikszentmihalyi, M. & LeFevre, J. (1989). Optimal experience in work and Leisure. *Journal of Personality and Social Psychology*, 56(5), 815–822.

Ghani, A. J., & Deshpande, P. S. (1994). Task characteristics and the experience of optimal flow in human-computer interaction. *The Journal of Psychology*, 128(4), 381–391.

Hoffman, D., & Novak, P. T. (1996). Marketing in hypermedia computer-mediated environments conceptual foundations. *Journal of Marketing*, 60(July), 50–68.

Webster, J., Trevino, K. L., & Ryan, L. (1993). The dimensionality and correlates of flow in human-computer interactions. *Computers in Human, Behavior*, 9(4), 411–426.

System Innovation for an Artificial Intelligence Era – Lam et al. (Eds)

Exploring the usability and understanding of flood potential maps: Design recommendations for essential features based on user experience evidence from different levels of disaster prevention literacy

Jia-Xin Guo & Meng-Cong Zheng
Department of Industrial Design, National Taipei University of Technology, Taipei, Taiwan

ABSTRACT: This study analyzes the usability and comprehensibility of flood potential maps. The user operation process was investigated through the SUS, NASA-TLX, UEQ, and semi-structured interviews. The research found that the current flood potential maps are not sufficiently user-friendly. This is due to unclear legends and non-intuitive interfaces, which prevent users from efficiently interpreting risk information about flooding, thereby hindering proper disaster prevention planning. The survey results indicated that the SUS scores failed to meet the usability standard. The UEQ scores are classified as "bad". During the interviews, the map element that users identified 'clearly marked flood-prone areas' as the most critical and essential map element. On this basis, a clearer and more understandable rainfall level classification ranked second, while providing practical prevention recommendations for decision-making in face of risks ranked third. Therefore, the future improvements to potential maps should focus on simplifying legend information, providing prevention guidelines to ensure that users can rapidly and correctly respond in the event of a disaster.

1 INTRODUCTION

Flood disaster is considered one of the natural calamities with the most significant impact on human life and economic losses. The risk of flooding is a hazard with a low probability of occurrence, which tends to happen more easily during specific periods, such as the rainy season. However, for the public, hazards with a low probability are difficult to understand and explain (Bier 2001), leading people to often underestimate these potential risks. Therefore, it is crucial to systematically and comprehensibly communicate flood risk information to the public. Flood risk maps, as tools for spatial planning, local disaster assessment, contingency planning, preparedness measures, and providing public risk awareness(EXCIMAP 2007), have proven effective. A well-designed flood potential map can capture users' attention and motivate them to seek further information and explore more (Hagemeier-Klose & Wagner 2009) Thus, in conveying flood risks to the public, maps serve as an effective medium to fulfill their role (Kellens *et al.* 2009). The significance of potential flood maps includes(Rowan 1991): establishing trust with the information disseminator enhancing risk perception and disaster preparedness awareness, and education. The previous literature emphasizes that communication of flood risks will strengthen people's risk perception, motivate those at risk to take preventive measures, and be prepared for emergencies (Hagemeier-Klose & Wagner 2009). This enables users to make risk assessments based on their circumstances and make wise decisions for their personal safety measures.The information presented on a flood potential map encompasses data related to various types of flooding, such as precipitation levels, flood extents, and inundation depths. These data are crucial for enhancing public awareness of flood risks (Paine J. D. and Watt W. E. 1992). Potential flood maps employ visual imagery to depict foreseeable consequences, thereby strengthening people's understanding of flood-related risks and bolstering their capacity to make informed decisions (Kjellgren S. 2013).

This study centers on flood potential maps as the primary focus of experimental research, exploring disaster prevention information, risk perception, and the development of decision-making behaviors within various scenario-based task designs. Through questionnaires and scales, usability issues and user experience insights are gathered, leading to evidence-based design recommendations for essential features.

2 EXPERIMENTAL

Prior to the study, a preliminary assessment was conducted by disseminating an online questionnaire, which focused on precipitation cognition and disaster prevention literacy. The aim was to gauge the public's comprehension of precipitation data and the concept of rainfall, as well as to collect information on their level of disaster prevention literacy. The online questionnaire amassed 861 responses. After excluding duplicate entries, those completed in

DOI: 10.1201/9781003514831-31

an unusually short time frame, or submissions with identical responses, a total of 775 valid responses were retained. The questionnaire comprised sections on the respondents' basic background information, interactive map usage habits, understanding of rainfall, and metrics of disaster prevention literacy.

In the pre-test phase, participants were enlisted for a usability study. The age range of the participants was 20 to 39 years, and they were all in good physical and mental health. Demographic details of the participants are presented (Table 1). None had prior familiarity with the research topic. Prior to initiating the navigation task, participants provided their basic information, discussed their experience with interactive map usage, reading habits pertaining to interactive maps, and familiarity with disaster prevention tools. Subsequently, they were instructed to operate the flood potential map.

Table 1. Demographic information of participants.

Demographic Characteristics	Classification	Frequency	Percentage
Age	20-29	27	90.0%
	30-29	3	10.0%
Gender	Male	17	56.7%
	Female	13	43.3%
Education-level	High School	1	3.3%
	Undergraduate	8	26.7%
	Graduate	21	70.0%

In the preliminary test, participant selection for a usability study of actual flood potential maps hinged on the average score from the disaster prevention literacy scale. The scale, with a scoring range of 1 to 5, showed no extreme values in participant scores. As a result, an average score of 3.93 (out of 775 respondents) was established as a benchmark to separate high from low literacy levels. Thirty individuals were then chosen for extensive usability tests, among which 15 individuals, scoring above this average, were identified as having high disaster prevention literacy with an average score of 4.35. Conversely, 15 participants scoring below this threshold, averaging at 3.64, were classified as having low disaster prevention literacy.

The experimental interface comprised the 3D disaster potential map developed by the National Science and Technology Center for Disaster Reduction (NCDR) in Taiwan (Figure 1). Participants engaged with the map's interactive features to evaluate flood risks under various scenarios – they interpreted map data, navigated its functionalities, and analyzed risks related to flooding and road safety. They also generated potential analysis reports for aiding disaster prevention decisions. During the assessment, participants utilized the web-based version of the flood potential map on a computer, with their usage patterns and methodologies being closely observed.

The assigned tasks, listed as tasks one through seven (Table 2), were completed by the subjects without the imposition of time constraints.

Table 2. Task operation flow.

Task Number	Task name	Description
Task 1	Use address search	Find a specified location using the address search function in the map. You must find and use the address search button and correctly enter the address.
Task 2	Opening the flood simulation layer	Opening the flood simulation layer
Task 3	Use the measuring tool	Use the measuring tool in the map to draw a range with a radius of 1 kilometer.
Task 4	Activate the key monitoring road section layer	Find and activate the key monitoring road section layer.
Task 5	Open the flood locations of the past five years	Find and activate the point and area layers showing flood locations from the past five years.
Task 6	Use the Taipei City flood simulation function	Find and use the flood simulation layer function for Taipei City.
Task 7	Export decision analysis	Conduct decision analysis after using the Taipei City flood simulation function.

Figure 1. 3D Disaster potential map form NCDR.

To evaluate the usability, efficiency, and user experience of the flood potential map, the experiment adopted the following: (1) System Usability Scale (SUS), (2) The NASA Task Load Index, and (3) User Experience Questionnaire (UEQ). To gain a deeper understanding of the subjects' experiences and perspectives on the use of the flood potential map, semi-structured interviews were conducted after the operational tasks. During the actual interviews, subjects were allowed to freely express their opinions on the map's use and their experiential feelings. The interview process involved follow-up questions based on the subjects' responses to gain deeper insights into their views and feelings.

3 RESULTS AND DISCUSSION

In the survey, we tracked the participants' visual attention patterns regarding the map's 'information' and 'legends'. Attention levels were rated on a Likert scale from 1 (minimal attention) to 5 (maximum attention) (Table 3). We conducted a Pearson correlation analysis to discern the relationship between participants' disaster prevention literacy levels and their visual attention during interaction with the map. The analysis revealed a significant positive correlation between the disaster prevention literacy scores and the average attention level devoted to the map's visual elements (p=0.004). Specifically, there was a moderate relationship between literacy scores and the focus on the 'information' (p=0.002) and 'legends' (p=0.030) within the map.

Table 3. Task operation flow.

N=30	The visual attention habits of interactive maps		
	Information (r)	Legends (r)	Average (r)
Average scores on the prevention literacy scale	**0.002**** (0.54)	**0.039*** (0.39)	**0.004**** (0.50)

*Significance is for the following levels: *p<0.05; **p<0.01

The study carefully parsed the participants' time expenditure across tasks one through seven, categorizing them into seven operational groups. The analysis using an independent sample t-test indicated significant differences in time usage for certain tasks. (Table 4) Participants took significantly different amounts of time to use the address search function (p = 0.042) and access historical flood location map layers (p = 0.039). There were no noted time differences for other specific tasks, such as engaging with flood simulation layers, utilizing the distance measurement tool, accessing critical road section layers, reviewing historical flood layers, or generating decision analysis reports. Observations of user behavior suggested that individuals with higher literacy levels identified functional buttons more swiftly compared to those with lower literacy levels. Thus, the data implies that participants with greater literacy levels exhibit enhanced visual attention towards the interactive map's data, which aids them in rapidly locating necessary functions in an initial task.

Table 4. Task performance in all tasks.

Task performance	Group	Average time (seconds)	SD	p
Task1	High	29.26	9.94	**0.042***
	Low	38.43	13.32	
Task2	High	28.46	23.40	0.298
	Low	38.53	28.30	
Task3	High	50.80	26.15	0.309
	Low	42.66	15.53	
Task4	High	16.20	15.97	0.867
	Low	15.26	14.12	
Task5	High	18.20	18.36	**0.039***
	Low	32.53	17.80	
Task6	High	9.00	7.17	0.943
	Low	8.80	7.96	
Task7	High	15.93	15.00	0.706
	Low	14.20	9.21	

*Significance is for the following levels: *p<0.05; **p<0.01

Upon observing the operation process of participants, it was found that those with higher literacy levels took less time to complete tasks compared to those with lower literacy levels. However, there may be various underlying reasons for this, including unclear methods for activating map layers and the presence of numerous options in dropdown menus that make selection difficult. Since the method for opening map layers varies across different interactive maps, this flood hazard potential map employs a "checkbox" method. Considering not every participant is accustomed to this method of toggling map layers, those unfamiliar with the operation may need more time to consider how to activate it. This study infers that this could be one of the reasons leading to longer operation times. The operation task required participants to select the option "650 mm of rainfall in 24 hours" from a dropdown menu. With many available options, no distinction of time scales, and no scrollbar indicator, participants often spent a significant amount of time meticulously searching for the specified target. In the statistical test, the average operation time of the group with high disaster prevention literacy was much lower than that of the group with low literacy. Upon observing the operation of the participants, it was found that those with high literacy skills became more familiar with the interactive map interface after performing operations one to five, which could be one of the reasons for the shorter operation time. Another possible reason is that the target function of operation five is subtly located, being placed at the very bottom and not easily found. Participants with lower literacy might take more time due to their focus on the information on the interactive map.

During the observation of participants' task performances, this study recorded the performances throughout tasks 1 to 7, including the statistics on the number of operational errors. The study results indicate a significant difference (p = 0.000) in the performance regarding the number of operational errors between groups with high and low prevention literacy, particularly in task 5 (Table 5).

Table 5. Number of failures per participant in tasks.

Task performance	Group	Participant	Mean	SD	p
Task1	High	1	0.06	0.25	0.559
	Low	2	0.13	0.35	
Task2	High	7	0.46	0.51	1.000
	Low	7	0.46	0.51	
Task3	High	10	0.66	0.48	1.000
	Low	10	0.66	0.48	
Task4	High	4	0.26	0.45	1.000
	Low	4	0.26	0.45	
Task5	High	1	0.06	0.25	**0.000*** **
	Low	10	0.66	0.48	
Task6	High	1	0.06	0.25	0.559
	Low	2	0.13	0.35	
Task7	High	1	0.06	0.25	1.000
	Low	1	0.06	0.25	

*Significance is for the following levels: *p<0.05; **p<0.01; ***p<0.001

Performance in other tasks was similar between the two groups. In task five, the study infers that the possible reason for this discrepancy is that the group with low disaster prevention literacy was less familiar with the flood hazard potential map interface, and their understanding of the interactive map information and legend was below that of the high literacy group. This resulted in the low literacy group needing to engage in repeated trial and error to complete the task successfully. The study suggests that the usability of the interactive map interface is poor, indicating significant room for improvement.

According to the SUS scores, the overall average score of participants in the high disaster prevention literacy group was (M = 39.16, SD = 22.09), while the average score of participants in the low literacy group was (M = 43.33, SD = 21.05). The results indicate no significant difference (p = 0.656). This study further combined the SUS scores to explore the usability outcomes of both groups. Both high and low disaster prevention literacy groups fall under the F-grade rating, not meeting the usability pass standard, thereby indicating room for improvement. Additionally, statistics show that the average score given by the high disaster prevention literacy group was lower than that of the low literacy group in terms of usability rating. This study deduces that because participants of the high-literacy group have a higher focus on the information and legends on the interactive map, they have higher standards for usage norms, readability, comprehensibility, and other aspects, hence the lower score ratings.

Based on the weighted scores from the NASA-TLX mental workload scale, there are no significant differences in the individual weighted scores and total weighted scores between the high and low disaster prevention literacy groups. From the statistical bar graph based on the scoring of the high and low disaster prevention literacy groups, it was found that in the weighted scores for "mental, physical, and temporal demands," the group with higher literacy gave themselves higher self-rated scores. This indicates that the high literacy group perceived a higher burden of mental, physical, and time demands when carrying out tasks. It is important to note that in this study's operational tasks, there were no time constraints imposed on the participants. Post-experiment interviews revealed that participants with higher literacy tended to complete tasks as quickly and efficiently as possible, therefore experiencing higher mental and temporal loads. Additionally, in the "performance, effort, and frustration" weights, the low disaster prevention literacy group gave higher burden ratings than the high literacy group, particularly in terms of performance weight where the low literacy group (29.2) rated their effort higher than the high literacy group (23.6), suggesting that the participants with lower literacy generally had a higher self-esteem regarding their task performance. Notably, in terms of the frustration weight, the low literacy group (8.0) felt more frustration compared to the higher literacy group (18.13). This could mean that those with lower literacy might think their performance was better than expected, but still found reading and interpreting the interactive map very challenging and frustrating.

According to the interpretation of 'pragmatic quality' and 'hedonic quality' in the UEQ scale, the values are between -0.8 and 0.8. If the value is greater than 0.8, it indicates a positive evaluation; a value of less than -0.8 indicates a negative evaluation, and the scale ranges from +3 (very good) to -3 (very bad). This study analyzes the self-assessment scores of 30 participants. In the high disaster prevention literacy group, the pragmatic quality received a neutral rating, with a score of 0.633 (between -0.8 and 0.8); hedonic quality received a positive rating, with a score of 1.233, and the overall average of 0.93 is a positive evaluation. In the low disaster prevention literacy group, pragmatic quality (0.006), hedonic quality (0.417), and overall (0.242) scores all received neutral ratings. In addition, the results of the study found that participants with high disaster prevention literacy scored higher than those with low disaster prevention literacy in 'pragmatic quality, hedonic quality'. However, in the independent samples t-test, the pragmatic quality (p = 0.16), hedonic quality (p = 0.07), and overall (p = 0.08) all showed no significance. Looking at the UEQ scores of all participants, the pragmatic quality (0.35) is bad; hedonic quality (0.825) is below average, and the overall score (0.59) is regarded as a bad user experience.

In order to understand user demand and priority for flood maps, this study, following the completion of all operational tasks and scales by the subjects, designed ranking propositions. Participants ranked the functional requirements and importance of flood maps according to their needs. Based on the necessary elements of maps described in Henstra's literature (Henstra *et al.* 2019), the highest and most important rank received 6 points, while the lowest and considered least important received 1 point.

Statistics from the study for participants with high and low disaster prevention literacy showed that the highest functional demand and priority was 'clearly marking potential flood areas,' with the highest scores in both the high literacy (4.93) and low literacy (5.8) groups. This item was also the only one to show significant difference (p = 0.011) in the independent samples T-test as shown in (Table 6). These results suggest that the low disaster prevention literacy group has a higher demand for features that clearly mark flooding areas, as this group already faces a learning curve in interpreting flood potential maps, therefore they desire better visibility and usability of these features on flood potential maps. Furthermore, statistical analysis of the ranking scores from all participants found that the top three priorities for the high disaster prevention literacy group were 'clearly marking potential flood areas (M = 4.93, SD = 1.06), clearer understanding of precipitation level categories (M = 4.46, SD = 1.20), and disaster preparedness suggestions for risk decision-making (M = 3.46, SD = 1.49).' The top three priorities for the low literacy group were 'clearly marking flood potential areas (M = 5.80, SD = 0.56), disaster preparedness suggestions for risk decision-making (M = 4.06, SD = 1.53), and clearer understanding of precipitation level categories (M = 4.00, SD = 1.13).The prioritization of the top three items was identical for both groups of participants, indicating that regardless of their level of disaster prevention literacy, all participants considered these three items to be the most important and necessary features in flood potential maps. This finding will provide an important basis for the future design and development of flood potential maps.

Table 6. Nasa-tlx score for the two groups.

Criterion	Group	Mean	SD	p
Depiction of multiple flood hazrds	High	4.46	1.24	0.29
	Low	4.00	1.13	
Flood zone legibility	High	4.93	1.09	**0.011***
	Low	5.80	0.56	
Historical context	High	2.86	1.45	0.58
	Low	2.60	1.18	
Local context	High	2.80	1.89	0.10
	Low	1.86	1.06	
Explanation of technical terms	High	2.46	1.59	0.70
	Low	2.67	1.23	
Risk reducation advice	High	3.46	1.55	0.29
	Low	4.06	1.53	

*Significance is for the following levels: *p<0.05; **p<0.01

This study compiled key feedback from 30 participants. In the comments on task 2 (opening the flood simulation layer), as many as 86% of participants felt that the visual representation of this feature did not convey the risk posed by flooding. Additionally, nearly 83% of participants were unclear about the severity of precipitation represented by the function menu. Hence, this study suggests that the user experience and usability of the flood simulation feature in this flood potential map is somewhat inadequate. This could be due to the general public's lack of familiarity with amounts of precipitation over particular time periods, leading to confusion when using the flood potential map. Besides, 25 participants (83%) commented that the dropdown menu had too many variables, making selection difficult for users and hindering efficient and effective use of the critical functions of this flood potential map. There were also many legibility issues with the map legends in this flood potential maps, particularly in task 5 (opening the flood locations of the past five years), with 17 participants (56%) feeling that the legend was not prominent, and 25 participants (83%) commenting that they did not understand the legends' meaning. Besides these issues, 10 participants (33%) commented that the help function was not helpful. This study suggests that flood potential maps should prioritize the clarity and understandability of their legends to increase user risk perception and comprehension. In the final task, task 7 (exporting decision analysis), users also encountered difficulties in interpretation. Users indicated that in the presentation of the decision analysis table, 15 participants (50%) believed there was an excessive amount of text, while 10 participants (33%) found it not intuitive to read. Among these participants, 23 participants (76%) suggested mainly using graphical explanations, and 13 participants (43%) recommended reducing the use of specific terminologies related to flooding.

4 CONCLUSION

This research categorized participants into high and low disaster prevention literacy groups based on their scores on a disaster prevention literacy scale to investigate the performance differences in the use of flood potential maps. Through the SUS, NASA-TLX, and UEQ Scale, it was discovered that current flood potential maps still have numerous issues in terms of interface usability and user experience, including hard-to-find functions and poor readability of legends, leading to a non-fluid overall usage process. Suggestions for the future design of flood potential maps are as follows: (1) The design of flood potential maps should consider the needs of users at different levels of disaster prevention literacy. High and low disaster literacy users inherently focus differently on interactive maps. As an interactive map suitable for the general public, it is necessary to meet the usage needs of the high literacy group and care about the usage and comprehension problems of the low literacy group in order to achieve increased disaster prevention efficiency and adaptability to disasters. (2) Use more relatable and optimized, easy-to-understand explanations of flood knowledge and terminology. As not all participants will have professional knowledge about disasters and flooding, especially when using the flood area display layer, nearly 50 comments indicated a lack of clarity about the differences in the textual representation of precipitation. This includes the jargon "hourly rainfall" and the presentation of numbers, making it difficult for users to use the function menu without imaginative space, and even more difficult to select the data that users want to see. Therefore, it is believed that a flood potential maps for public use should have a lower learning cost, providing users with efficient and easy-to-understand map content. (3) Develop map functionalities that meet user needs. The current flood potential maps include various functions and services, such as address search, landmark search, potential layer selection to potential analysis. This study specifically investigates the "Necessary Elements of a Flood Potential Map" proposed by Henstra and others. After compiling the research data, it was found that among the 30 participants, the most important and necessary elements were identified in the following order: "clear identification of potential flooding areas," "preparedness recommendations for risk decision-making," and "a clearer and more understandable classification of precipitation levels." This will provide a constructive design guide for developing directions for flood potential maps.

REFERENCES

Bier V. M. (Feb. 2001). On the state of the art: risk commu-nication to the public, *Reliab. Eng. Syst. Saf.*, vol. 71, no. 2, pp. 139–150, doi: 10.1016/S0951-8320(00)00090-9.

EXCIMAP., P. works Ministry of Transport, and W. M. V&W, *Handbook on Good Practices for Flood Mapping in Europe. Ministry of Transport, Public Works and Water Management*, 2007. [Online]. Available: https://books.google.com.tw/books?id=b3C-NwAACAAJ

Hagemeier-Klose M. and Wagner K. (2009). Evaluation of flood hazard maps in print and web mapping services as information tools in flood risk communication, *Nat. Hazards Earth Syst. Sci.*, vol. 9, no. 2, pp. 563–574.

Henstra D., Minano A. and Thistlethwaite J. (2019). Communicating disaster risk? An evaluation of the availability and quality of flood maps, *Nat. Hazards Earth Syst. Sci.*, vol. 19, no. 1, pp. 313–323, doi: 10.5194/nhess-19-313-2019.

Kellens W., Vanneuville W., Ooms K. and De Maeyer P. (2009). Communicating flood risk to the public by cartography, *presented at the Proceedings of the 24th International Cartographic Conference*, Santiago de Chili.

Kjellgren S. (2013). Exploring local risk managers' use of flood hazard maps for risk communication purposes in Baden-Württemberg, *Nat. Hazards Earth Syst. Sci.*, vol. 13, no. 7, pp. 1857–1872.

Paine J. D. and Watt W. E. (Jan. 1992). Flood risk mapping in canada: 2. Compatible mapping, *Can. Water Resour. J. Rev. Can. Ressour. Hydr.*, vol. 17, no. 2, pp. 139–148, doi: 10.4296/cwrj1702139.

Rowan K. E. (Nov. 1991,). Goals, obstacles, and strategies in risk communication: a problem-solving approach to improving communication about risks., *J. Appl. Commun. Res.*, vol. 19, no. 4, pp. 300–329, doi: 10.1080/00909889109365311.

System Innovation for an Artificial Intelligence Era – Lam et al. (Eds)

Numerical simulation of linear and circular ridge acoustic waveguides

Tai-Ho Yu, Chun-Jen Ku & Tang-Wei Guan
Department of Electronic Engineering, National United University, Miaoli, Taiwan

ABSTRACT: This study applies Hamilton's principle and finite element method to analyze the dispersion characteristics and resonance modes of linear and circular ridge acoustic waveguides using the bi-dimensional finite element method (Bi-d FEM). First, the ANSYS three-dimensional finite element method was used to analyze the circular ridge, calculate the resonance frequency of each mode and the corresponding guided wave phase velocity dispersion curve, and explore the convergence of the elements. The results of Bi-d FEM analysis show that the antisymmetric wave mode of the linear ridge acoustic waveguide will decrease as the vertex angle increases, and the wave speed is smaller than the Rayleigh wave speed, which is consistent with the results of Lagasse's empirical formula. The guided wave energy of linear and circular ridge acoustic waveguides is concentrated at the tip of ridge and is in the form of antisymmetric flexural wave (ASF).

1 INTRODUCTION

Lagasse [1] and Maradudin *et al.* [2] used numerical calculations in 1972 to confirm that the vertex angle of the ridge has a waveguide effect within a certain range of angles. In 1973, Lagasse *et al.* [3] proposed a simple empirical formula to summarize the numerical results. In 1973, Lagasse [4] applied the finite element method to divide acoustic waveguides with different cross-sectional shapes using the triangular elements of the second-order interpolation function to explore the wave propagation behavior of infinitely long linear ridge waveguides and observe the dispersion phenomena of guided waves in different geometric cross-sections.

In 1976, Mckenna *et al.* [5] treated an infinitely long ridge with a small vertex angle as an approximate thin plate. Using the plate and shell theory, the relationship between the phase velocity and wavenumber of the guided wave was derived for ridges with and without truncation. In 1996, Hladky-Hennion [6] used numerical analysis methods, including the Bi-d FEM, to analyze the wave propagation behavior of linear and circular ridge guided waves. The effects of curvature radius, vertex angle, and truncation on the phase velocity of the ridge guided wave were studied.

In 1994, Krylov [7] used geometrical acoustics theory to calculate the wave propagation velocity of guided waves in a ridge in water. In 1999, Krylov [8] derived a simpler and faster analytical solution based on geometric acoustics theory to describe the wave propagation behavior of conical ridges, finding good agreement with the finite element analysis results of Hladky-Hennion [6]. In 2007, Krylov *et al.* [9, 10] experimentally verified the idea of ridge aquatic propulsion for autonomous and crewed vessels that was first proposed about 10 years earlier. The first working prototype of a small catamaran using ridge propulsion via an attached rubber keel was built and tested at Loughborough University.

2 BASIC THEORY

Considering the case in which the surface traction of an infinitely long linear ridge acoustic waveguide is zero, its strain energy U and kinetic energy T are expressed in matrix form as follows:

$$U = \frac{1}{2}\int_{vol} \mathbf{T}^H \mathbf{S} dV = \frac{1}{2}\int_z\int_x\int_y \mathbf{S}^H \mathbf{C}\mathbf{S} dxdydz \quad (1a)$$

$$T = \frac{1}{2}\int_{vol} \dot{\mathbf{u}} H \boldsymbol{\rho}\dot{\mathbf{u}} dV = \frac{1}{2}\int_z\int_x\int_y \dot{\mathbf{u}} H \boldsymbol{\rho}\dot{\mathbf{u}} dxdydz \quad (1b)$$

Following the same approach, considering the case where the surface traction of a circular acoustic waveguide is zero, its strain energy U and kinetic energy T are expressed in matrix form as follows:

$$U = \frac{1}{2}\int_{vol} \mathbf{T}^H \mathbf{S} dV = \frac{1}{2}\int_\theta\int_r\int_x \mathbf{S}^H \mathbf{C}\mathbf{S} r dr dx d\theta \quad (2a)$$

$$T = \frac{1}{2}\int_{vol} \dot{\mathbf{u}} H \boldsymbol{\rho}\dot{\mathbf{u}} dV = \frac{1}{2}\int_\theta\int_r\int_x \dot{\mathbf{u}} H \boldsymbol{\rho}\dot{\mathbf{u}} r dr dx d\theta \quad (2b)$$

where $\mathbf{C}$ is the material stiffness matrix and $\boldsymbol{\rho}$ is a 3×3 diagonal matrix, that represents the density of the elastic acoustic waveguide. The superscript H represents the Hermitian transpose. A dot

DOI: 10.1201/9781003514831-32

represents the partial derivative with respect to time.

If the material stiffness matrix **C** and density matrix **ρ** are symmetric real matrices, then the strain energy and kinetic energy are guaranteed to be positive definite. The Lagrangian functions L of linear and circular acoustic waveguide systems are expressed as follows:

$$L = T - U = \frac{1}{2}\int_z\int_x\int_y (\dot{\mathbf{u}}H\boldsymbol{\rho}\dot{\mathbf{u}} - \mathbf{S}^H\mathbf{C}\mathbf{S})dxdydz \quad (3a)$$

$$L = T - U = \frac{1}{2}\int_\theta\int_r\int_x (\dot{\mathbf{u}}H\boldsymbol{\rho}\dot{\mathbf{u}} - \mathbf{S}^H\mathbf{C}\mathbf{S})rdrdxd\theta \quad (3b)$$

Applying Hamilton's principle, set the first variation of the integral of the Lagrangian with respect to time to zero, as follows:

$$\delta\int_{t_1}^{t_2} Ldt = 0 \quad (4)$$

Then, the equations of motion for linear and circular acoustic ridge waveguides can be derived.

3 RESULTS AND DISCUSSION

First, consider a linear ridge acoustic waveguide with a cross-section aspect ratio $H/W = 3$. The number of elements used is 270 and the number of nodes is 310. The degree of freedom of each node at the bottom is set to zero to calculate its dispersion curve and modes. The dispersion curve in Figure 1 shows that the guided wave has obvious dispersion phenomenon. The ridge section with truncated corners looks like a rectangular section near the top. Therefore, the dispersion curve of the linear ridge acoustic waveguide with truncated corners should also have dispersion.

This study further calculates the resonance modes corresponding to different wave numbers, and the results are shown in Figure 2. The resonance mode characteristics of the rectangular cross-section

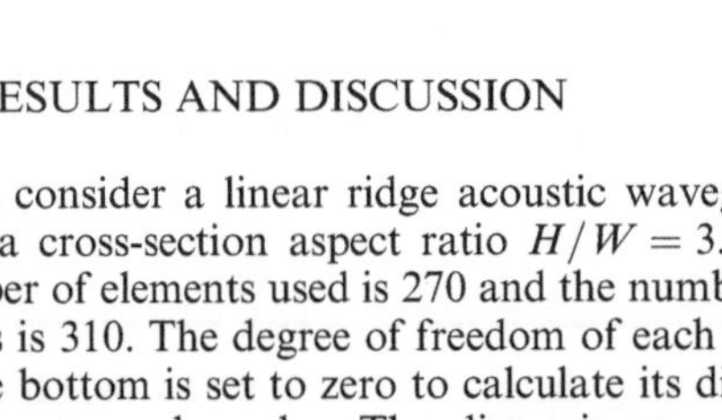

Figure 1. Dispersion curve of a linear ridge ($H/W = 3$).

acoustic waveguide are different from those of the ridge cross-section acoustic waveguide. The rectangular acoustic waveguide does not have sharp corners, and the energy of the wave is distributed throughout the cross section, so the resonance mode of the guided wave is not constrained by the vertex angle. In addition to the form of antisymmetric guided waves, there are also different types of resonance modes such as extensional waves.

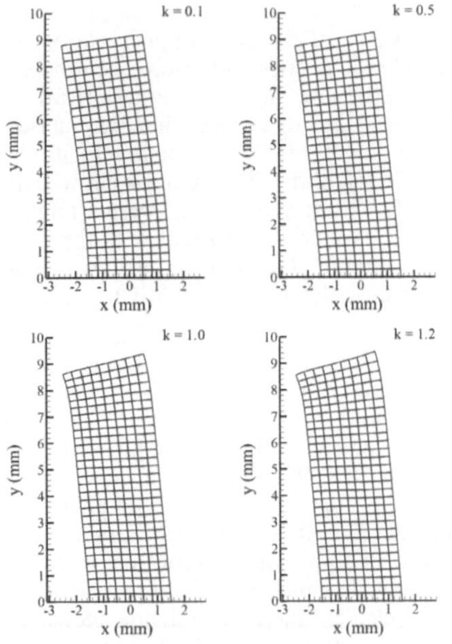

Figure 2. First mode shapes of a linear ridge.

Consider the circular rectangular acoustic waveguide with cross-section aspect ratio $H/W = 3$. The number of cross-section elements is 240, and the number of nodes is 279. The degree of freedom of each node at the bottom is still set to zero, and the dispersion curve and mode of the guided wave are calculated. Figure 3 shows the dispersion curves of the first five modes. The figure shows that the guided waves of this acoustic waveguide are all dispersion waves, and the trend of the dispersion curve is similar to that of the linear ridge acoustic waveguide.

Figure 4 shows the mode shapes of each mode under different wavenumbers. The section of the cylindrical rectangular acoustic waveguide does not have sharp corners, and the energy of the guided wave is distributed throughout the section. Therefore, the guided wave on the section will also have complex vibration modes. Due to the influence of structural curvature, the resonance mode of the circular rectangular acoustic waveguide is slightly different from the resonance mode of the linear rectangular acoustic waveguide.

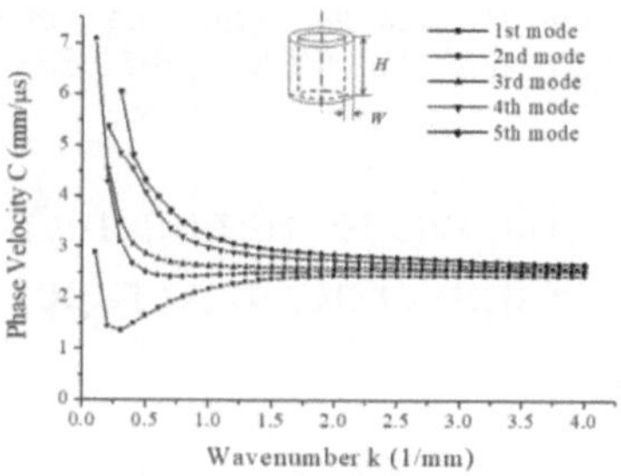

Figure 3. Dispersion curve of a circular ridge ($H/W = 3$).

The curvature of the acoustic waveguide constrains the resonant mode deformation near the inner diameter, and its deformation is smaller than that of the material near the outer diameter.

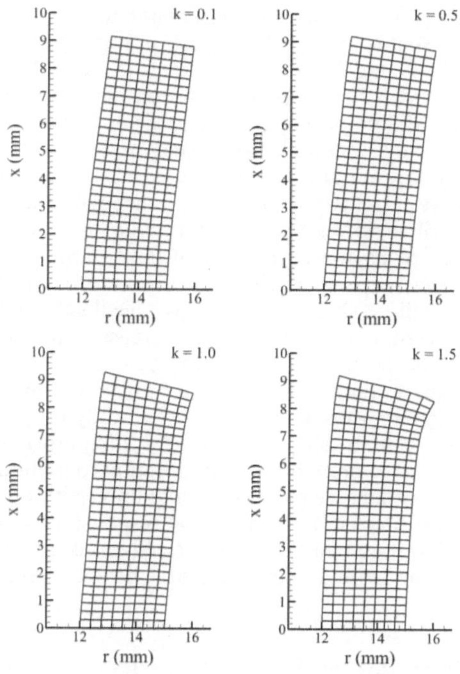

Figure 4. First mode shapes of a circular ridge.

4 CONCLUSION

The obtained resonance mode shapes confirm that the tip of the ridge acoustic waveguide has the largest displacement, that the energy of the wave is concentrated at the tip, and that the guided wave is an ASF wave. The Bi-d FEM provides a powerful analysis method for the study of wave propagation in circular and linear acoustic waveguides.

In the future, the method of this study can be extended to conduct further research on linear and circular ridge acoustic waveguides with different cross-sectional geometries, ridge angles, truncations, and coatings. This could lead to the development of new types of ultrasonic motors and applications in related fields.

ACKNOWLEDGEMENTS

We express our sincere thanks to the National Science and Technology Council in Taiwan (NSTC 112-2221-E-239-002) and the National United University (112-NUUPRJ-05) for their generous support in funding this research.

REFERENCES

Lagasse P. E. (1972). Analysis of a dispersion free for elastic waves, *Electrics Letters*, **8**(4), 372–373.

Maradudin A. A., Walls R. F. and Ballard R. L. (1972). Vibrational edge modes in finite crystals, *Physics Rev*, **B6**, 1106–1111.

Lagasse P. E., Mason I. M. and Ash E.A. (1973). Acoustic surface waveguides analysis and assessment, *IEEE Transactions on Microwave Theory and Techniquesics*, **MTT-21**, 225–236.

Lagasse P. E. (1973). Higher-order finite element analysis of topographic guides supporting elastic waves. *Journal of Acoustical Society of America*, **53**, 1116–1122.

Mckenna J., Boyd G. D. and Thurston R. N. (1974). Plate theory solution for guide flexural acoustic waves along the tip of a wedge, *IEEE Transactions on Sonics and Ultrasonics*, **SU-21**(3), 178–186.

Hladky-Hennion A. C. (1996). Finite analysis of the propagation of acoustic wave in waveguides, *Journal of Sound and Vibration*, **194**(2), 119–136.

Krylov V. V. (1994). Propagation of wedge acoustic waves along wedges imbedded in water, *IEEE Ultrasonics Symposium*, 793–796.

Krylov V. V. (1999), Localized vibration modes propagation along edges of cylindrical and conical wedge-like structures, *Journal of Sound and Vibration*, **227**(1), 215–221.

Krylov V. V. and Pritchard G.V. (2007). Experimental investigation of the aquatic propulsion caused by localised flexural wave propagation in immersed wedges and plates, *Applied Acoustics*, 68, 97–113.

Krylov V. V. and Pritchard G.V. (2007). Experimental confirmation of the propulsion of marine vessels employing guided flexural waves in attached elastic fins, *Journal of Fluids and Structures*, 23, 297–307.

System Innovation for an Artificial Intelligence Era – Lam et al. (Eds)
© 2025 the Author(s), ISBN: 978-1-032-84452-7

Leveraging the Internet of Things for sustainable waste management and improved customer engagement in Taiwanese university beverage culture

Jong-Chao Yeh
Department of Fashion and Performance, Kun Shan University, Tainan, Taiwan

Kherion Jay Miranda & Wei-Cheng Chiu
Department of Mechanical Engineering, Kun Shan University, Tainan, Taiwan

Rong-Bin Lai, Kao Wei & Chen
Green Energy Technology Research Center, Kun Shan University, Tainan, Taiwan

ABSTRACT: This study aims to employ Internet of Things (IoT) technology to formulate sustainable waste management strategies while fostering social responsibility within Taiwanese university (USR) beverage culture and aligning with the United Nations Sustainable Development Goals (SDGs). Using the unique context of Taiwan's beverage culture as a case study, the research focuses on the waste generated from disposable cups in tea and beverage shops. This study not only examines the environmental impact of waste disposal, especially concerning non-biodegradable materials like plastics and polystyrene widely used in Taiwanese communities and university campuses but also integrates a customer relationship management (CRM) perspective to gain comprehensive insights into university students' attitudes toward sustainable dining. The research methodology primarily involves devising a recycling and reusable system to provide university administrators with tangible plans for social responsibility and SDGs implementation. Additionally, it incorporates CRM strategies by personalizing the customers (students) experience at the beverage shops on campus. The machine will be able to test and give feedback on the beverage used in the cup, and its effect before washing cleaning and preparing for reuse. This is to encourage students' active participation in sustainable dining behaviors. Qualitatively, interviews and observations are employed to understand current waste disposal practices and students' attitudes towards sustainability. To better manage student behaviors related to cup usage and disposal, the study proposes the application of IoT devices for real-time monitoring within university campuses. By integrating IoT technology into waste management strategies, this research aims to present a holistic solution to minimize environmental impact and stimulate social responsibility and sustainable practices within universities. The findings have substantive implications for university administrators in terms of social responsibility and achieving SDGs. Simultaneously, the study offers universities more strategic customer relationship management approaches to promote student engagement and identification with sustainable development.

1 INTRODUCTION

As we look at the theme of university Social responsibility and waste management by implementing IoT. The purpose of this research and its potential benefits for universities, in particular, and the community at large must be taken into consideration. Living in Taiwan, you cannot miss the fact that about every 500 meters a tea shop or other beverage establishment can be found. Many people will stop by in the morning and grab a nice warm drink before heading in to work, for lunch they would again shop a cool shake to go along with their food, and then in the evening while heading back they might stop in with a friend or coworker and sip on an evening tea. As we observer this trend, a common question appears, what happen to the thousands of cups that are used on a daily basis after they have supply the many customers' demands? According to Ching-Yan Wu, they recall that Sustainable waste management has been recognized as a core issue in developed countries (Wu, C. Y., Hu, M. C., and Ni, F. C. 2021), and even more so in developing countries that are burdened by waste due to lack of processing capabilities or incentive mechanisms to reduce, reuse, and recycle their waste materials (Awasthi *et al.* 2016). Although Taiwan has been doing well in trying to keep the waste at a minimum level of pollution, there is still so much that can be done to improve the management of reusable and recyclable products.

DOI: 10.1201/9781003514831-33

This can be narrowed down to management of reusable cups that are sold in the many universities around the country. Ching-Yan Wu also mentioned that Taiwan, once nicknamed "Garbage Island," now boasts the world's second-highest effective recycling rate, following Germany (Mei-Chih *et al.* 2021; Yoshida, A. 2022). Thanks to a series of policies and regulation adaptions over the course of Taiwan's industrialization. If this same effectiveness can be trickled down to the schools and universities, where the students would purchase drinks during the day, sometimes disposing the cups as waste because there is not much that can be done with it. Improper waste management, especially of disposable cups, can have detrimental effects on the environment. These cups are typically made of non-biodegradable materials such as plastic or Styrofoam, which can persist in the environment for hundreds of years. When these cups are not properly disposed of, they often end up in landfills or as litter, contributing to pollution and harming wildlife (Häkkinen, T., and Vares, S. 2010). Additionally, the production of these cups requires the extraction of natural resources and the release of greenhouse gases, further exacerbating climate change (Chitra, P., 2022). Therefore, finding effective solutions for waste management is crucial to minimize the environmental impact of our consumption habits. This is where this research seeks to help the situation of waste management through USR promoting a healthier school setting and at the same time contributing to the already improved waste management in Taiwan.

2 IDEA GERENTION

2.1 *USR Cup washer machine*

Implementing the idea of USR means there is a social responsibility that the University has to contribute to in order to fulfill its role. Being socially responsible is very important in today's world and it is the duty of every individual or organization to act for the benefit and development of the society and the environment because the revenue is earned only from the society and the environment so consequently it should be given back (Sharma, and Sharma, 2019). One such way is to work along with their stakeholders, specifically the vendor who sells drinks in their school. Firstly, we will look at the environment of the cafeteria, the role that the universities can play in waste management and sustainable development while implementing the use of technology in the process.

The university in itself is a living organism, with its own unique body of government and people. It's like a small town where teachers, students, staff and other stake holders come to visit for different reasons on a daily. University is the place one build and develop and share ideas and experience with their fellow classmates and friends, and having somewhere that is accessible to all with these or better qualities, we would see an improvement in social and communication skills (Lee, 2007). Besides the exterior appearance and scenery, another very important location is the cafeteria or the place for eating. According to a recent study conducted by Institute of Technology Management, National Tsing Hua University Research Institute, Waste management errors can have a major negative impact on the environment and cause national health crises in both kinds of countries (Chen, 2018). Ninety percent of the plastic products that people use on a daily basis are disposed away after only one use, including straws, coffee cup lids, supermarket bags, plastic wrap, and disposable cutlery. As a result, since the 2000s, ocean plastic pollution has become a significant environmental concern (Obbard *et al.* 2014), leading several nations to implement laws and policies outlawing the use of plastics (Rifat, 2022). Universities in Taiwan collectively generate significant amount of waste each year, with a portion of it being disposable cups from drink purchases. This staggering amount of waste not only contributes to the growing landfill problem but also adds to the carbon footprint of universities. Furthermore, the improper disposal of these cups leads to marine pollution, as many of them end up in rivers and oceans, posing a threat to marine life. By implementing IoT-based waste management systems, universities can significantly reduce their waste generation and mitigate their environmental impact. Everyone has to eat and in keeping healthy and clean, we all want to eat in a place that is fairly appealing and clean (Hsu and Huang 2013). The idea of buying a drink has been around for decades. What if someone could get a drink and sit in a place that is clean, comfortable and appealing to the eyes, drinking from a cup that is steady and strong. I believe that this kind of environment would not only promote more socializing and quality time, but also increase the sales of the number of drinks bought per day.

Now returning to the point of waste management, this is a critical aspect of fostering sustainable practices, and its significance extends far beyond the boundaries of school premises. As we navigate the challenges of the 21st century, the need for responsible waste management in schools becomes increasingly imperative. Beyond its direct impact on the educational environment, proper waste management in schools holds the key to influencing the broader community positively (Sharma and Sharma, 2019). This introduction aims to shed light on the importance of instilling waste management practices within educational institutions and the ripple effect it can have when extended to the community at large. Though the topic of waste management is a broad field, we will focus mostly on using and reusing cups bought for drinks. After a few trail then we can move on to different utensils used in the cafeteria and around the school. The university will be responsible for developing and supplying the reusable cup disinfector and washer. As their part in USR they help reduce and manage the waste of one time used cups and improve the environment if the cafeteria.

2.2 *Creating a reusable cup washer*

There are not much findings on techniques studies to disinfect and wash drinking cups. There are not many washers designed especially for cups. Our goal is to create a reusable cup washer that can efficiently clean, rinse, dry, and store cups for reuse in a matter of minutes. Depending on how many cups are used, the machine can be designed with 20 or 60-cup holders, for example. The cups will be sprayed on the inside and outside to remove any sugar residue or stains from previous drinks. A disinfectant will then be applied to kill any remaining bacteria. From the surface of the cup. Afterwards in order to get rid of any smell from the previous drink or even the disinfectant, it will next be cleaned with a pleasant and new solution. Lastly, the cups will be heated to a high temperature so they may be used again quickly. We will discuss in more detail how the machine will carry out this duty in future study.

2.3 *Implement the use of IoT*

The Internet of Things (IoT) refers to the network of interconnected devices that collect and exchange data through the internet. In the context of waste management, IoT can play a crucial role in optimizing and improving the efficiency of waste collection and disposal processes (Bi and Xu, 2014). One example used is by installing smart sensors in trash bins and recycling containers, this not only reduces the chances of overflowing bins and littering but also allows for more effective resource allocation and cost savings (Korean Society, 2019). We must implement the use of IoT in this operation; this will help us to be more effective and efficient in our research. This will help us to use the right amount of water, disinfectant, cleaner and air. This product should be very beneficial not only to the universities and the cafeterias, but also for external businesses and restaurants, they can adopt this machine to their process. Many people resort to the ancient way of washing cups by hand, but time consuming and tiring. Adopting the cup washer to the daily lives of those that use cups on a regular basis will greatly improve business and our environment in a whole. By incorporating IoT into a cup washing machine, not only can operational efficiency be enhanced, but it also opens up possibilities for data-driven insights and improvements, providing a more intelligent and user-friendly experience. (as shown in Figure 1).

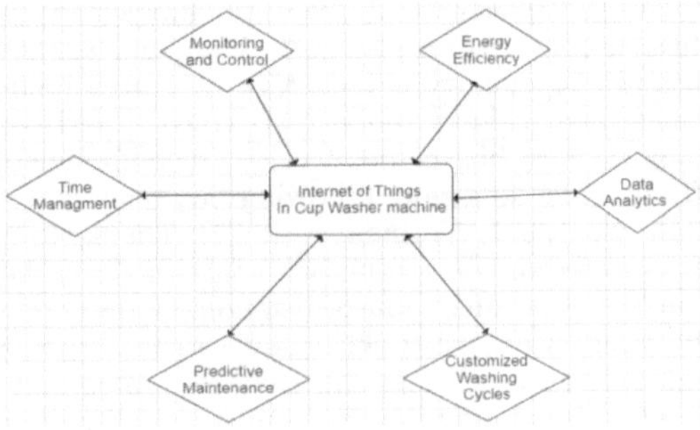

Figure 1. The concept of this study.

2.4 *Customer relationship management*

In the context of sustainable waste management in Taiwanese university beverage culture, the integration of Customer Relationship Management (CRM) strategies plays a pivotal role in this study (Chen and Popovich, 2003). Our research aims to enhance the university's understanding of students' attitudes and preferences regarding sustainable dining practices, while fostering a closer connection between the university and its student body through actively involving them in environmentally conscious initiatives. Through qualitative research methods such as interviews and observations, we delve into the intricacies of student perceptions and behaviors related to sustainable dining. The incorporation of CRM principles allows us to develop targeted communication and feedback mechanisms, encouraging students to actively participate in proposed waste management solutions. This approach strengthens the university's relationship with its student community and creates a shared commitment to sustainable practices.

The implementation of IoT devices for real-time monitoring aligns with CRM principles, providing valuable data insights into student behaviors. This data-driven approach allows for personalized and effective communication strategies, ensuring that the university's initiatives resonate with the student body. In summary, the infusion of Customer Relationship Management into our research framework not only addresses waste management challenges but also establishes a collaborative and communicative relationship between the university and its students, fostering a community-driven commitment to sustainable development (Chen and Popovich, 2003).

3 METHODOLOGY

The conventional methods of washing cups are facing multiple challenges and limitations in today's world. Firstly, they tend to be time-consuming and labor-intensive, as each cup needs to be scrubbed and rinsed individually, which can result in inefficiencies, especially in high-volume settings like cafes or restaurants where many cups need to be cleaned quickly to meet customer demand. This again takes us back to the schools and university system where many students and staff purchase beverages on a daily basis and increase the use of cups. In these high-volume settings large quantities of cups need to be cleaned quickly to meet customer demand.

There is currently a scarcity of cup washing machines that are suitable for use in schools and restaurants (Sawalkar, 2023). Most cup washers available in the market are designed for large factories that produce cups on a massive scale, and as a result, they consume a lot of electricity and water. On the other hand, the small-scale cup washers meant for home use are not ideal for commercial purposes as they can only wash one cup at a time, and often lack the required features to eliminate germs We need a cup washing machine that can provide clean cups in less time, while conserving energy and water. Although there are some options available online, most of them lack modern technology and advanced features like IoT integration, which is essential for efficient operation as the demo we designed.

A second reason to consider alternative cup washing methods is that traditional methods may not always ensure thorough cleanliness and hygiene. Hand washing cups can leave behind residues or bacteria if not done properly, which increases the risk of contamination and foodborne illnesses. Additionally, the use of sponges or brushes that are not adequately sanitized can further contribute to hygiene concerns. Furthermore, traditional cup washing methods may not be environmentally sustainable. They often require large amounts of water for rinsing and cleaning, leading to water wastage and higher utility costs. Additionally, the use of chemical detergents in the washing process can have negative environmental impacts, especially if they are not properly disposed of or contain harmful ingredients.

After careful consideration, it is clear that traditional methods of washing cups have several limitations and challenges in today's world. In order to ensure efficient, hygienic, and sustainable cup-washing practices, individuals, schools and businesses should explore alternative solutions, such as automated dishwashing systems or eco-friendly cleaning products.

3.1 *Demo Model of machine*

This research focuses on the development of an innovative prototype reusable cup washer, designed to enhance efficiency and sustainability in the management of beverage containers. The cup washer incorporates a detailed process, as outlined in the Cup Washer Flowchart (Figure 2). After review of different style of washers, the designs was discussed and chosen to incorporate a unique display, having the door latch opening from the top of the machines for easy on and off load of the cups and also as the water spray upwards into the cups the door latch will hold the cups in place. The following is process of how the machine will work, Cups are loaded upside down into the machine:

1. Pre-Rinse, High-pressure water spray rinses the inside and outside of each cup to remove loose debris and sugar residue.
2. Internal Wash, Rotating cup brushes gently scrub cup surfaces inside and out.

   High-pressure water spray with detergent removes remaining stains and film.
3. Disinfection, Internal spray applies a food-safe disinfectant solution disinfectant circulates and coats all cup surfaces.
4. Deodorization, Interior misting with a pleasant, food-safe deodorizing solution.

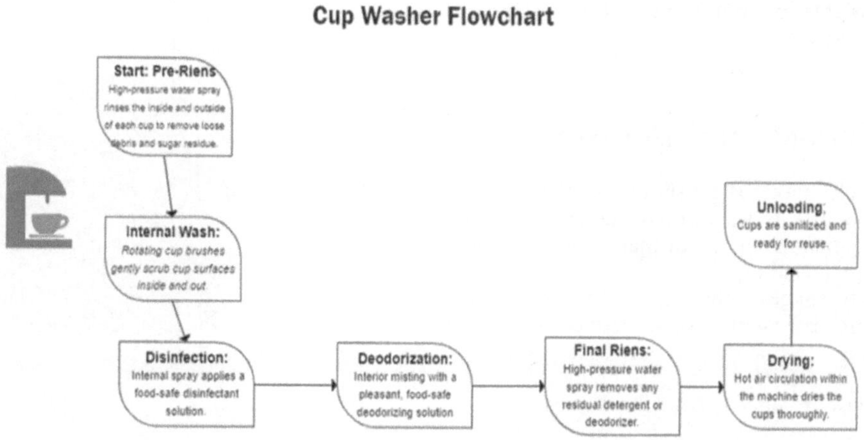

Figure 2. Flow Chart showing cup washer process.

5. Final Rinse, High-pressure water spray removes any residual detergent or deodorizer.
6. Drying, Hot air circulation within the machine dries the cups thoroughly.
7. Unloading, Operator removes clean, dry cups from the machine, cups are sanitized and ready for reuse.

3.3 *Provide an overview of the IoT and its role in connecting devices and collecting data*

The Internet of Things (IoT) plays a pivotal role in connecting devices and collecting data to enhance the functionality and efficiency of the cup washer. In this context, IoT enables seamless communication between the cup washer and various components, facilitating real-time monitoring and control. Through interconnected sensors and data transmission, the IoT empowers the cup washer to optimize its operation and respond to specific conditions, ensuring a more intelligent and adaptive cleaning process.

3.4 *Discuss the data collection and analysis methods that will be used to evaluate the performance and effectiveness of the IoT-enabled washer*

To evaluate the performance and effectiveness of the IoT-enabled washer, data collection will involve monitoring key parameters such as water pressure, temperature, detergent usage, and overall energy consumption during each cycle. This data will be collected through embedded sensors in the cup washer and transmitted to a centralized system for analysis. Statistical methods and analytics tools will be employed to assess the efficiency of the cup washer, identify potential areas for improvement, and measure its environmental impact in terms of resource utilization and waste reduction. The integration of IoT enhances not only the operational aspects but also provides valuable insights for continuous optimization and sustainable practices.

4 PREDICTIONS AND DISCUSSION

The research make suggestions in the implementation of IoT is waste management and discusses the potential of waste management based on the research findings.

The study suggests that using IoT for waste management can significantly reduce environmental impact. As people become more aware of sustainability, this technology is predicted to be used not just in universities but across various industries.

The study also shows a growing concern for sustainability among students, potentially leading to a cultural shift towards environmentally responsible practices.The focus on everyday tasks like cup washing hints at a future where IoT is integrated into daily activities, promoting a more sustainable lifestyle.The positive impact of IoT on waste management suggests increased collaboration between universities, businesses, and communities to implement broader sustainable initiatives.

As sustainability gains importance, governments may introduce regulations and incentives to encourage the use of IoT for waste management. Continuous advancements in IoT technology are expected, leading to even more sophisticated solutions for waste management, potentially involving AI and machine learning for further optimization.Overall, the integration of IoT in waste management has the potential to drive significant changes towards a more sustainable future. Collaboration is key to effectively address environmental challenges. Finally, after relevant discussions, Figure 3 draws the machine model imagined in this study and Figure 4 show the concept of the cup holder on the top. The detailed construction is presented in Figure 5.

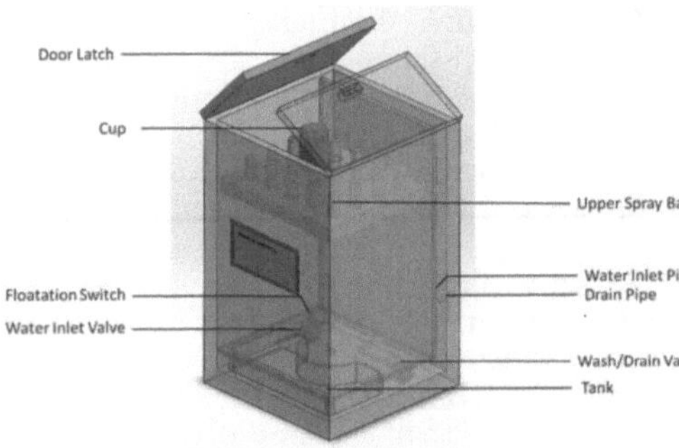

Figure 3. Machine formal styling.

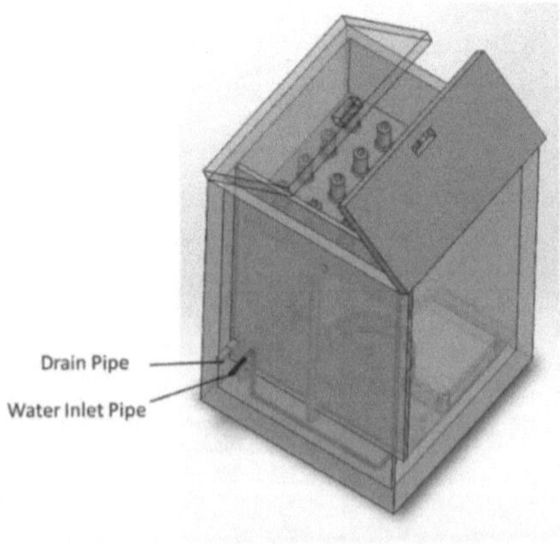

Figure 4. Machine back shape.

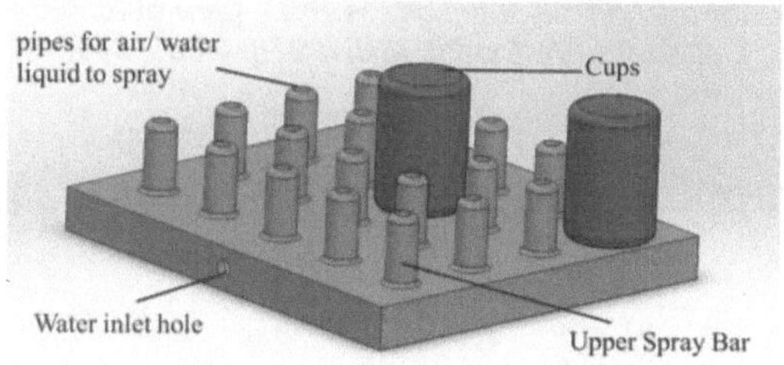

Figure 5. Cup holder placement.

5 CONCLUSION

This study, through an in-depth exploration of Taiwan's beverage culture and the application of Internet of Things (IoT) technology in university social responsibility and waste management, has yielded significant conclusions and insights. Firstly, improper waste disposal, particularly concerning disposable cups, poses a severe environmental impact, presenting a common challenge in both Taiwanese communities and university campuses. The implementation of IoT technology enables real-time monitoring of the usage and disposal of disposable cups, minimizing their environmental impact. This comprehensive solution not only contributes to academic knowledge but also holds practical significance. Our investigation and data analysis reveal a heightened awareness of sustainable development among university students, emphasizing the potential advantages of IoT technology in addressing such challenges.

Crucially, this research underscores the widespread potential adoption of IoT across sustainability-focused industries. The broad application of this technology not only improves university social responsibility and waste management but also fosters sustainable practices in various sectors. Particularly in routine tasks such as cup washing, the application of IoT can simultaneously reduce environmental impact and enhance efficiency.

In summary, this study provides compelling evidence for the importance of IoT in promoting sustainable practices and mitigating the environmental impact of daily tasks. These findings hold practical significance in driving broader IoT applications towards achieving sustainable development goals.

REFERENCES

Alcaraz, C., Lopez, J, (2010). *IEEE Trans.* 40:4, 419–428.
Ali, S., Ahmed, W., Solangi, Y. A., Chaudhry, I. S., Zarei, N. (2022). *Clean Tech.*, 1–7.
Åström, K. J., Hägglund, T. (2006). *Inst. Syst.*
Awasthi, A. K., Zeng, X., Li, J. (2016). *Proce. Envi.*, 35, 676–680.
Barr, S. (2004). *Sust. Deve.*12:1, 32–44.
Bi Z., Xu L. D., Wang C. (2014). *IEEE Tran.* 10:2.1537–154
Chen, I.J., Popovich, K. (2003). *Busi. Proc.* 9:5, 672–688.
Cheng, S. Y., Wang, C. J., Shen, A. C. T., Chang, S. C. (2020). *Annal. Inter.* 173:8, 638–641.
Chitra, P., (2022). *Plastic Waste Management and Behavior Change.*
Gajjar A. G., Patel A. I., Singh R. G., (2015). *Conf. Adv. Ghaziabad, India.*
Häkkinen, T., Vares, S. (2010). *J. Clea.* 18:14, 1458–1463.
Ho, C. C., Chen, M. S. (2018). *Wast.mana.*71, 578–588.
Hsu, C. S., Huang, D. J. (2013). *J.Air & Waste*, 63:7, 796–805.
Internet of Things based Smart Energy Management for Smart Home. 2019
Juvonen, R., Virkajärvi, V., Priha, O., Laitila, A. (2011). *VTT Tied.*, 2599, 107.
Kim, H., Pyun, K. R., Lee, M. T., Lee, H. B., Ko, S. H. (2022). *Adva.Func.* 32(16), 2110535.
Lee, J. C. K. (2007). Routledge.
Li, S., Xu, L.D. Zhao, S. (2015). *Inf Syst Front* 17, 243–259.
Obbard, R. W., Sadri, S., Wong, Y. Q., Khitun, A. A., Baker, I., Thompson, R. C. (2014). *Earth. Futu.* 2:6, 315–320.
Pálsdóttir, A., Jóhannsdóttir, L. (2021). *Sust.*13:16, 8958.
R. I. Rifat, F. B. Haque, T. Jahan, Z. B. Zahir, R. Khan. (2022). *Garb.Mana.* 12 1–10.
Sawalkar, R.S., Undale, S., Muluk, S., Mude, G., Saxena, V.D. Pasumarti, S. (2023).
Sharma, D., Sharma, R. (2019). *J of Emer.* 6:6, 27–35.
Wu, C. Y., Hu, M. C., Ni, F. C. (2021). *Sust. Prod.* 26, 228–238.
Yoshida, A. (2022). *J of Mate.* 24:1, 73–82.

System Innovation for an Artificial Intelligence Era – Lam et al. (Eds)

Design and realization of mosquito trapping device using electronics components

Miguel Amado Gaona Armoa, Danna Gabriela López Almada &
Enrique Sebastián Shrem Smith
Department of Electromechanical Engineering, Universidad Politécnica Taiwan-Paraguay, Asunción, Paraguay

Chen-Yang Lan
Department of Mechanical Engineering, National Taiwan University of Science and Technology, Taipei, Taiwan

ABSTRACT: This technical paper introduces the use of fundamental electronic components, such as resistances, transistors, and diodes; together with solar cell and batteries to develop a prototype mosquito-trapping device with the purpose of further application for sophisticated products in the effort of mosquito-carried disease control by reducing their population nearby. The built mosquito trap uses UV-LEDs to attract the mosquitoes; subsequently the fan traps the mosquitoes into a constant air current and further into a trapping container. In addition, it is designed to possess functions such as adjustable LEDs' brightness and fan speed with common electronic components, providing adaptability for the needs of the user; moreover, its integrated solar cell provides a renewable energy source allowing more independence from electrical outlets and contributing to a low battery replacement rate. In terms of maintenance, its electronic components can be easily detached from their case if occurs any electronical fault. In addition, the trapping container provides easy access for cleaning purposes. In this paper the analytical process for circuit design, the selection of suitable electronic components, the assembly process and testing procedures to achieve the desired features are discussed. Finally, the steps provided will serve as a guide for further applications in the field of electronics.

1 INTRODUCTION

For the concept of this mosquito trapping device, two fundamental stages must be used for the mosquitoes to be trapped, the first stage uses the UV light considering its mosquito attraction effect [3], then the second stage requires the constant air current created by an electric fan for the capture and retention of the mosquitoes within a container. These two stages and other innovative functions such as the usage of solar power for the charge of the batteries; adjustable LED lights; and fan speed control, both for the enhancement of the energy efficiency, along with insights of the device structure will be the aspects discussed throughout this paper.

This paper aims to provide a solution for mosquito disease control and to contribute to further applications for other sophisticated products sharing the efforts or using similar electronic components.

The development of this device was conceived as part of a University Social Responsibility (USR) initiative as a solution for the prevalent high population of mosquitoes around the UPTP facilities in Paraguay.

2 FUNDAMENTALS & DESIGN CONCEPTS

To build an electronic device there are some fundamental concepts that are necessary to know about the components and their usage to consider with respect to their functions and where to apply them during circuit design and analysis.

The electronic components that have a major use in the prototype are resistors, diodes, and transistors. Resistors are devices that function as an impedance for current. Every time current passes through a resistor, a determined voltage will be consumed, this can be defined by Ohms law. In addition, Kirchoff law allowed the calculation of current and voltages in nodes and loops. The diodes are semiconductive devices of P-N junction that prevent the current from flowing a specific direction but still allowing the current to flow the other direction by consuming a determined voltage (activation voltage). Unlike diodes, transistors are semiconducting devices that are constructed with three terminals, base, collector and emissor. For the case of this device, the transistors were implemented as current amplifiers.

DOI: 10.1201/9781003514831-34

Paraguay has an intense sunlight during most part of the year, the daylight last for 7 hours and 15 minutes [1], which makes it an ideal candidate for the implementation of solar cells at small and large scales. As matter of fact, Malaysia that has 6 hours of direct sunlight and much more humidity than Paraguay, it is still suitable for the use of photovoltaic technology [2].

Light traps utilizing ultraviolet (below 380 nm), blue (450–485 nm), and green (500–565 nm) wavelengths consistently demonstrate higher mosquito capture rates compared to those employing longer wavelengths [3].

Studies have been conducted to determine the efficiency of UV light and the effect they have on insects. In Thailand, mosquito traps using UV lights were deployed in residential areas, agricultural sites, and forests. The results suggest that UV light traps are highly effective in capturing mosquitoes, with a significant reduction in mosquito populations observed in the areas where the traps were deployed [4].

3 CIRCUIT ANALYSIS & DESIGN

The design and calculations of this project were carried out based on certain requirements, such as the use of three 2200mA batteries and UV LEDs of 3.4V and 20mA. The selection of the electronic components for the dedicated circuit were based on the ideal model explained below, choosing the most approximated components to adapt them in the design.

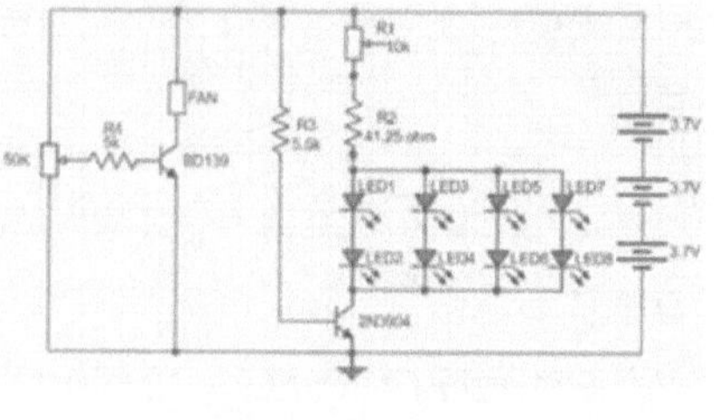

Figure 1. Circuit diagram of mosquito trap.

3.1 *Power supply*

A solar panel was connected to a solar power controller, which has the function to provide voltage to three batteries (the ones used for the circuit) connected in series whenever they discharge.

This configuration will deliver to the circuit a constant voltage to be in continuous operation.

3.2 *LED arrangement*

For the mosquito light attraction, eight ultraviolet LEDs were connected, four branches in parallel, each one with two LEDs in series. A variable resistor enables the control over the current flow through the electric circuit, consequently, the control of brightness in the diodes.

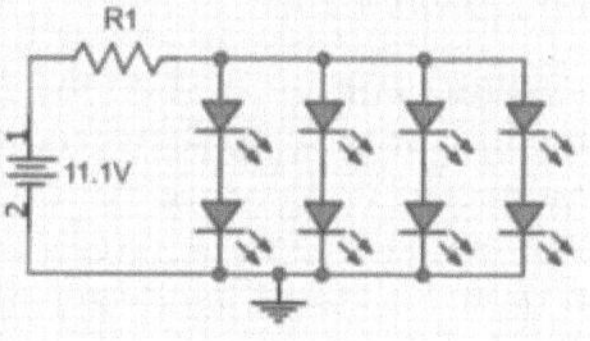

Figure 2. Electrical diagram of LEDs branches.

Each ultraviolet diode needs an activation voltage of 3.4V and a current of 20mA to achieve its brightness limit.

The configuration of a voltage source, resistor and diodes in Figure 2 allows a proper operation; two diodes in series produces a 6.8V drop and the total current is distributed in each branch, causing a 20mA of current flowing in each branch, therefore for an ideal case 80 mA needs to pass through the resistor shown below.

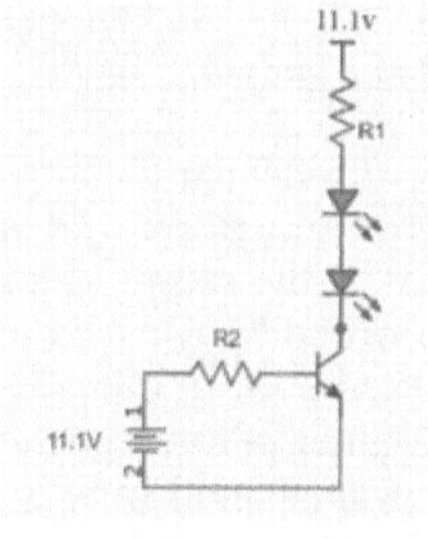

Figure 3. LED-transistor design arrangement.

The BJT device operates as a current amplifier to deliver 80mA. This configuration grants the necessary current and voltage for the diodes to be at their maximum allowed brightness.

The 2N3904 BJT has specific features, including the current gain of $\beta=42$ (obtained by interpolation) and $V_{CE}=1V$, $V_{BE}=0.6V$ [5].

Mathematical equations in (1) and (2) were employed in the electrical circuit analysis, by Kirchoff voltage law:

Loop 1

$$-11.1V + Ic \times R_2 + 6.8V + V_{CE} = 0 \quad (1)$$

Considering that Ic must be 80mA:

$$R_2 = \frac{11.1V - 6.8V - V_{CE}}{80mA} = 41.25\Omega$$

Loop 2

$$-11.1V + (R_2 \times Ib) + V_{BE} = 0 \quad (2)$$

$$Ic = \beta \times Ib$$

$$R_3 = \frac{11.1V - V_{BE}}{Ib} = 5.5k\Omega$$

3.3 *Fan configuration*

For the mosquito capture function, the BD139 BJT operates in common collector configuration, the fan is connected to the positive of the battery and the other end to the collector, the variable resistor is in the base, and the emitter is grounded.

The variable resistor is employed to control the amount of current allowed to flow through the BJT, which subsequently controls the velocity of the fan.

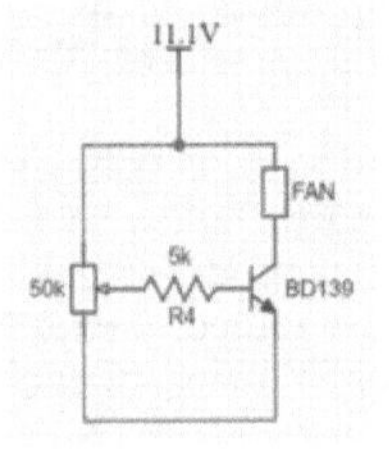

Figure 4. Fan disposition.

4 ASSEMBLY

The structure of the device was designed making allowances for the functions and integrating the components, leaving the required gap for the mosquitoes to be absorbed by the air current provided by the fan with the addition of a sliding module as the trapping container, allowing its cleaning.

The structure was divided in two main modules, the bottom and top, the design of the top part, made of plastic, considered four main functions: to integrate the solar panel within the framework, to contain the battery and PCB, repeal the rainwater from the components fragile to water, and to act as a passage limiting the motion of the mosquitoes for better effectiveness in mosquito capturing task. The bottom part, made of timber, isolates the exterior for not allowing the mosquitoes to escape. Concerning the slipping module for having access to clean the mosquitoes from the bottom by slipping it like a drawer.

The top and bottom parts can be detached and reattached due to the use of screws in charge of holding a flat platform where the PCB and electronics were placed.

Regarding the circuit, due to time and convenience, the use of copper-clad boards has been decided for the PCB (Printed Circuit Board).

The process for the fabrication of the PCB involved transferring the schematic of the circuit to a cooper-clad board, etching the board, drilling holes at designated locations for component placement, assembling the electronic components by soldering, and finally testing and debugging. For the device, two PCB were installed, one for the general circuit components and the other for the LEDs arrangement, this for having more freedom in positioning the LEDs and in distributing the space inside the top of the framework.

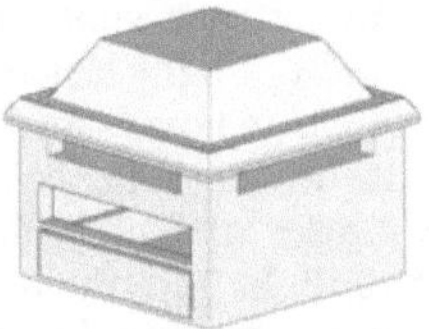

Figure 5. Device framework design.

5 RESULT, DISSCUSSION & CONCLUSION

5.1 *Result discussion*

5.1.1 *Test result*

The number of captured mosquitoes may vary due to several factors, some as location, temperature and the time in which tests are made; Considering this, from 7:00 PM to 9:30 PM at 22°C, five mosquitos were caught in the classroom in a test.

Regarding the batteries, two and a half hours were functioning. Moreover, the capacity of the battery provided by the fabricant may not be precise enough to match the calculated time they should.

5.1.2 *Battery consumption*

For analyzing the battery life, a close approach of the total maximum power consumption (without considering the variable resistances) of the circuit is obtained by calculating the power consumption on each electronic component using the power formulas $P = IV$ or $P = I^2V$:

For R2:

$$P_{R2} = (80.18mA)^2 \times 41.25\Omega = 265.2mW$$

For R3:

$$P_{R3} = (1.91mA)^2 \times 5.5k\Omega = 20.1mW$$

For R4:

$$P_{R4} = (2.12mA)^2 \times 5k\Omega = 22.5mW$$

For the LEDs:

$$P_{LEDs} = \frac{80.18mA}{4} \times 3.4V \times 8 = 545.2mW$$

For the fan:

$$P_{fan} = 84.8mA \times 10.1V = 856.5mW$$

For the 2N3904 transistor:

$$P_{2N3904} = 82.09mA \times 1V = 82.1mW$$

For the BD139 transistor:

$$P_{BD139} = 86.92mA \times 1V = 86.9mW$$

Adding all values obtained above:

$$P_{total} = 1869.2mW$$

Since the total power is of 1869.2mW, a theoretical time for the battery life can be obtained by the product between the specified battery capacity of 2200mA/h and the voltage provided (in series) that was of 11.1V, divided with the calculated total power:

$$Time(h) = \frac{2200\text{mAh} \times 11.1V}{1869.2mW} = 13.06h$$

For this prototype, a single arrangement of three batteries connected in series has been used and therefore the expected or theoretical battery life is of 13.06h or 13h 4min.

5.1.3 *Possible improvements*

After the realization of this prototype, it is necessary to denote the possible improvements that can be applied to the device as well as considerations that were missed during the design process.

Regarding the circuit, the potentiometer was placed in the collector of the BJT transistor, this implies that no matter the intensity of the LEDs, the same power will be consumed. Instead, the potentiometer can be placed in the base of the transistor, keeping its function as current regulator, this decision will allow the adjustability of brightness function to be a battery saving choice. Also, since most of the light emitted is more noticeable in darkness, the implementation of a photocell for the activation of the device during night hours is suggested for the increase of the device efficiency.

Regarding the battery of the device, the battery life can be increased by adding batteries of same voltage in parallel to the actual battery arrangement. Also, the factor of safety can be multiplied to the total power to consider the next mentioned discrepancy that could be with the real battery life.

Regarding the device framework, it is suggested the addition of a preventative mesh, with openings small enough to not allow the entrance of other insects (also attracted to UV light) into the trap without depriving the entrance of mosquitoes.

5 CONCLUSION

This paper demonstrates that by employing UV light attraction and an electric fan powered by batteries and solar power, it is possible to contribute to the effort of reduction of the population of mosquitoes, thus contributing to diseases control efforts. In addition, it was also shown that it is possible to be innovative by adding extra features to the electronic device that will enhance its performance. As a result, through collaborative efforts and the application of fundamental electronic concepts, the development of this prototype exemplifies the potential of technology for addressing critical social challenges while fostering community welfare.

ACKNOWLEDGEMENT

This work was financially supported by the Ministry of Foreign Affairs, Republic of China (Taiwan), under the Universidad Politécnica Taiwan-Paraguay (UPTP) project.

REFERENCES

[1] Rincón A., Jorba O., Frutos M., Alvarez L., Barrios F., and González J., (2018). Bias correction of global irradiance modelled with weather and research forecasting model over Paraguay, *Science Direct*, https://www.sciencedirect.com/science/article/pii/S0038092X18304948ref=pdf_download&fr=RR2&rr=86c76bde7ab84a72

[2] Ghazali M. and Abdul Rahman A. M., (May 2012). The performance of three different solar panels for solar electricity applying solar tracking device under the malaysian climate condition, *Energy and Environment Research*, vol. 2, no. 1, https://doi.org/10.5539/eer.v2n1p235.

[3] Wilson R., Cooper C. E. C., Meah R. J., Wakefield A., Roberts N. W., and Jones G., (Jan. 2023). The spectral composition of a white light influences its attractiveness to Culex pipiens mosquitoes, *Ecology and Evolution*, vol. 13, no. 1, https://doi.org/10.1002/ece3.9714.

[4] Chaiphongpachara T., Laojun S., and Kunphichayadecha C., (Sep. 26, 2018). Effectiveness of ultraviolet (UV) insect light traps for mosquitoes control in coastal areas of Samut Songkhram province, *Thailand*, https://app.periodikos.com.br/article/10.31893/23181265jabb.v7n1p25-30/pdf/jabbnet-7-1-25.pdf

[5] Motorolla Inc, (1996). General purpose transistors NPN sillicon Semiconductor technical data 2N3903.

System Innovation for an Artificial Intelligence Era – Lam et al. (Eds)
 ISBN: 978-1-032-84452-7

Exploring the correspondence of material influence on visual and auditory imagery in over-ear headphones

Tzu-Chia Huang
Graduate student, Department of Industrial Design, National Cheng Kung University, Tainan, Taiwan

Chun-Heng Ho
Associate Professor, Department of Industrial Design, National Cheng Kung University, Tainan, Taiwan

ABSTRACT: The McGurk Effect suggests that visual and auditory information received by the brain interact and influence each other, producing mutual effects. Subsequent studies on audiovisual interaction have emerged, indicating the importance of considering visual imagery when designing products with auditory characteristics. The visual imagery of materials has a significant impact on the appearance of products, and numerous studies have emphasized the sensory characteristics of materials and their influence on product personality. This study primarily focuses on over-ear headphones as a medium, the Semantic Differential (SD) method and subsequent analysis to explore the correlation between visual imagery presented by different materials and auditory imagery presented by different sound performances. The research findings indicate that following factor analysis, the visual imagery presented by the material and the auditory imagery presented by sound performance exhibit differing imagery across the axes, rendering direct correspondence between the two imagery types unfeasible. However, analysis of the data and the distribution of positions in the imagery space still enable the understanding of the affective imagery presented by the visual and auditory performances of various materials.

Keywords: Material Kansei Imagery, Audiovisual Interaction, Over-ear Headphones

1 INTRODUCTION

1.1 *Introduction*

In the product design process, designers convey concepts through functionality and product morphology to define the product's position. Materials, on the other hand, exert a significant influence on the product's appearance, enhancing its visual presentation. As stated in "The Art of Material Selection" by Mike Ashby and Kara Johnson (2003), "In the analysis of product characteristics, the design context defines intention or emotion, shaping, connecting, and completing them; materials and processes create the body and skeleton; ergonomics determine usability; and the aesthetics, connectivity, and perception of the product create its personality." Their repeated emphasis on the sensory characteristics of materials and their impact on product personality underscores the pivotal role of material selection in the design process.

Furthermore, designer Jinsop Lee proposed the Five Senses Theory, which utilizes a "Five Senses Chart" to record experiential ratings of the five senses in daily life. This serves as a crucial indicator during the design process, allowing products to offer sensory experiences beyond vision and touch, thereby making them more ingenious and humane (Jinsop Lee 2013). Additionally, British psychologists Harry McGurk and John MacDonald introduced the McGurk Effect in 1976, suggesting that visual and auditory stimuli received by the brain influence each other, resulting in an interaction. Therefore, it can be inferred that through the interaction of vision and hearing, users should be able to develop a more complete overall perception of the product. Thus, when designing music-related products with auditory characteristics, consideration should be given to the combination of sensory images presented visually.

In the vast market of music-related products, headphones stand out as a representative product. They offer high-quality music experiences without disturbing others, serving as convenient tools for tasks such as voice calls and video conferences. Combining work and entertainment aspects, along with characteristics like portability and comfort, headphones have gradually become an indispensable accessory in modern life. Among the entire product range, over-ear headphones, which provide higher-quality and more stable sound effects, are particularly favored. Moreover, compared to other types of headphones such as in-ear headphones, over-ear headphones have

 DOI: 10.1201/9781003514831-35

a larger size, making their visual impact more significant. Therefore, this study focuses on over-ear headphones to explore the compatibility between visual and auditory imagery.

1.2 *Research purpose*

Related studies have shown a correspondence between visual and auditory presentations, with their interaction enhancing the overall user experience of a product. While visual imagery can stem from characteristics such as form, color, and material, the appearance designs of most brands' over-ear headphones tend to be similar. Therefore, this study focuses on material differences to explore the association between material and the sensory imagery of sound, comparing the sensory imagery evoked by decorative outer shell materials with different sound performances. The research aims to provide insights for designers when considering material choices for headphone products in the future. The specific objectives of this study are as follows:

1. To understand the visual sensory imagery evoked by different decorative outer shell materials.
2. To understand the auditory sensory imagery evoked by different sound performances.
3. To explore the differences and influences of auditory and visual imagery perception, and their corresponding relationships.

1.3 *Research limitations*

The visual imagery of the product can be broadly categorized into three main aspects: form, color, and material. In this study, we have chosen a form commonly produced by most brands, and presented it in grayscale to avoid influencing the perception of material, with the variation in material being the focus of investigation. To prevent confusion among participants when completing the questionnaire due to a large number of experimental samples, this study will select seven commonly used materials for product casing and five sound performance characteristics favored by the market consumers for investigation. The material aspect of the experiment will be presented using visual cards, allowing participants to perceive the sensory imagery conveyed by the material through visual stimuli.

2 EXPERIMENTAL

The research primarily targets individuals with experience using over-ear headphones. The experimental procedure consists of a preliminary preparation phase and a formal experimental investigation phase.

2.1 *Preliminary preparation phase*

As this study primarily investigates the interactive relationship between the visual sensory imagery of materials and the auditory sensory imagery of sound performance, the preliminary preparation phase involves selecting sensory imagery descriptors applicable to both visual and auditory senses, as well as samples of material and sound performance.

The sensory adjectives used in this study were initially gathered in abundance and duplicates were removed. Subsequently, the vocabulary related to the visual sensory imagery of materials and the auditory sensory imagery of sound performance, which shared a common association, was selected through methods such as focus groups and questionnaires. Finally, expert surveys were conducted to select suitable vocabulary (as shown in Table 1) for subsequent experiments.

Table 1. Sensory imagery descriptors screening results.

Ten sets of sensory imagery descriptors				
Warm Cold	Novel Classical	Light-weight Heavy	Rough Delicate	HardSoft
Harmonious Abrupt	Gorgeous Plain	Childish Steady	Simple Rich	Refreshing Chaotic

The material samples were selected from seven common materials commonly used in product appearance in the current market, with emphasis on selecting those that are more representative and exhibit greater visual perceptual differences. These representative materials include fabric, plastic, wood, glass, leather, metal, and bamboo (Xiao 2019). To mitigate the influence of factors such as brand impression and color, the visual characteristics of each brand were aggregated and mixed to generate product sample images, which were then applied with different materials. Finally, the product colors were standardized.

Table 2. Product material application samples.

Product material application samples			
 Plastic	 Metal	 Wood	 Bamboo
 Glass	 Leather	 Fabric (Fur)	

The sound samples were selected from five different mainstream sound performances commonly found in current headphone markets, including

"Treble Boost", "Bass Boost", "Vocal Boost", "Balanced" and "V-shaped". Therefore, it was necessary to select audio sources with full frequency and vocal presence as the basis for the samples. After screening, it was decided to use the chorus section of "Lana Del Rey - Young And Beautiful" as the basic audio source sample. Different frequency intensities were adjusted using audio processing software to create the aforementioned five sound performance samples.

2.2 *Formal experimental investigation phase*

The formal experimental investigation phase aimed to understand the "sensory imagery presented by material visual perception" and the "auditory sensory imagery presented by sound performance." Separate SD semantic differential questionnaires were prepared for the material samples and sound performance samples in the preparatory stage and distributed to the research subjects. Data collected were analyzed using SPSS software to explore the relationship between the visual sensory imagery presented by materials and the auditory sensory imagery presented by sound performance. This analysis provides further discussion and extension regarding the subsequent selection and application of headphone materials.

3 RESULTS AND DISCUSSION

3.1 *Factor analysis of visual sensory imagery for material perception*

The SD semantic differential questionnaire for material visual sensory imagery includes the question: "How do you perceive the degree of OO or OO brought by the OO material for this headphone?" Participants evaluated using a five-point scale. A total of 55 valid questionnaires were collected. To understand the distribution of imagery and factor scores for each material, the average scores for each material imagery were calculated (as shown in Table 3). The numbers highlighted in yellow indicate the maximum and minimum values of the imagery score (which means the two ends of the adjectives). This allows for the comparison of relative imagery among the seven materials. For example, in the case of warm-cold: compared to other materials, metal is relatively cold, while fur is the warmest, and so on.

Overall, most materials exhibit at least one or two prominent visual imagery associations. For example, the visual imagery associated with plastic includes lightweight, harmonious, plain, childish, and simple. However, bamboo material does not exhibit any particularly prominent imagery associations.

Table 3. Average scores of sensory imagery for product materials.

	Plastic	Metal	Wood	Bamboo	Glass	Leather	Fabric (Fur)
Warm Cold	1.02	1.64	-0.26	-0.62	1.13	1.53	1.81
Novel Classical	-0.40	-1.26	0.45	-0.19	-1.57	0.68	1.09
Lightweight Heavy	-1.00	0.06	0.51	-0.43	-0.17	0.43	0.62
Rough Delicate	0.02	0.30	0.57	0.11	1.23	-0.68	-0.25
Hard Soft	-0.36	-1.40	-0.11	0.13	-0.70	1.13	1.38
Harmonious Abrupt	-0.68	-0.47	-0.49	0.32	0.02	-0.19	0.92
Gorgeous Plain	0.98	-0.51	0.62	0.81	-1.02	-0.43	-1.25
Childish Steady	-0.43	0.72	1.30	0.40	-0.04	1.08	-0.36
Simple Rich	-1.38	-0.25	-0.32	0.17	-0.34	0.57	1.28
Refreshing Chaotic	-0.81	-0.79	-0.91	-0.53	-0.98	0.36	1.02

3.2 *Factor analysis of auditory sensory imagery for sound performance perception*

The semantic differential (SD) questionnaire for auditory sensory imagery was designed to avoid bias based on the names of the auditory performances. Therefore, the five types of auditory performances were randomly assigned numbers for identification. Each question was formulated as "How would you rate the degree of OO or OO brought by sound performance O (no.1-5) ?" Responses were evaluated using a five-point scale. A total of 63 valid questionnaires were collected. The average scores for each auditory performance imagery are presented in Table 4, with the maximum and minimum scores for each imagery (representing the two ends of the adjective spectrum) highlighted in yellow for comparative analysis across the five sound samples.

Table 4. Average scores of sensory imagery for sound performance.

	No. 1 (Vocal Boost)	No. 2 (Balanced)	No. 3 (Bass Boost)	No. 4 (Treble Boost)	No. 5 (V-Shape)
Warm Cold	-0.56	-0.10	-0.94	0.77	-0.45
Novel Classical	0.29	0.02	0.45	-0.44	0.15
Lightweight Heavy	0.37	0.08	0.82	-0.66	-0.05
Rough Delicate	0.31	0.27	0.21	-0.02	0.27
Hard Soft	0.47	0.16	0.47	-0.55	0.18
Harmonious Abrupt	-0.79	-0.24	-0.60	0.40	-0.24
Gorgeous Plain	-0.44	0.02	-0.15	-0.21	-0.03
Childish Steady	0.66	0.21	0.71	-0.35	0.24
Simple Rich	0.39	-0.13	0.27	-0.03	0.35
Refreshing Chaotic	-0.34	-0.16	-0.13	0.23	-0.29

As shown in the table, it can be observed that the imagery scores obtained from "Treble Boost" differ significantly from the other four types of sound performances, with more pronounced differences in imagery. In the "hard-soft" adjective vocabulary group, "Vocal Boost" and "Bass Boost" received equal scores, indicating a relatively soft sensory imagery.

3.3 *The correspondence between visual sensory imagery of material and auditory sensory imagery of sound performance perception*

The data obtained from the two questionnaires were subjected to factor analysis using the SPSS statistical analysis software. Principal component analysis (PCA) was employed to extract factors, and analysis was conducted using the maximum variance method with orthogonal rotation. The suitability of the data for factor analysis was evaluated through the Kaiser-Meyer-Olkin (KMO) test and Bartlett's test of sphericity, as shown in Tables 5 and 6. The KMO value for the analysis of visual imagery of materials was 0.689, meeting the adequacy criterion, while for the analysis of auditory imagery of sound performance, the KMO value was 0.715, also meeting the adequacy criterion. Additionally, Bartlett's test of sphericity yielded a significance level of less than 0.001, indicating that factor analysis was appropriate for both datasets.

Table 5. KMO and Bartlett's Test_ visual imagery of product material.

Kaiser-Meyer-Olkin Measure of Sampling Adequacy.		.689
Bartlett's Test of Sphericity	Approx. Chi-Square	934.684
	df	45
	Sig.	.000

Table 6. KMO and Bartlett's Test_ audio imagery of sound performance.

Kaiser-Meyer-Olkin Measure of Sampling Adequacy.		.715
Bartlett's Test of Sphericity	Approx. Chi-Square	960.726
	df	45
	Sig.	.000

Observing the component matrices after rotation for both datasets (Tables 7 and 8), it is evident that the adjectives were divided into two factors in both cases.In the factor analysis of visual imagery of materials presented in Table 7, the first factor comprised five sets of imagery, namely: Refreshing-Chaotic, Warm-Cold, Simple-Rich, Hard-Soft, and Harmonious-Abrupt. The second factor included three sets of imagery: Novel-Classical, Gorgeous-Plain, and Childish-Steady. The other two sets of imagery were excluded due to weaker associations with the two factors.

In the factor analysis of auditory imagery of sound performance presented in Table 8, the first factor comprised four sets of imagery: Warm-Cold, Lightweight-Heavy, Childish-Steady, and Simple-Rich. The second factor included three sets of imagery: Rough-Delicate, Hard-Soft, and Refreshing-Chaotic. Two sets of imagery were excluded due to weaker associations with the two factors, while Harmonious-Abrupt was temporarily excluded from the analysis due to its high correlation with both factors.

Table 7. Rotated component matrix- visual imagery of product material.

	Component	
	1	2
Refreshing Chaotic	.788	
Warm Cold	-.743	
Simple Rich	.728	
Hard Soft	.631	
Harmonious Abrupt	.509	
Rough Delicate		
Lightweight Heavy		
Novel Classical		.761
Gorgeous Plain		.573
Childish Steady		.528

Extraction Method: Principal Component Analysis.
Rotation Method: Varimax with Kaiser Normalization.
a. Rotation converged in 3 iterations.

Table 8. Rotated component matrix- audio imagery of sound performance

	Component	
	1	2
Warm Cold	-.734	
Novel Classical		
Lightweight Heavy	.679	
Rough Delicate		-.740
Hard Soft		-.625
Harmonious Abrupt	-.612	.575
Gorgeous Plain		
Childish Steady	.799	
Simple Rich	.606	
Refreshing Chaotic		.785

Extraction Method: Principal Component Analysis.
Rotation Method: Varimax with Kaiser Normalization.
a. Rotation converged in 3 iterations.

Based on the previous analysis, common factors were identified for both sets of imagery, and variables with similar characteristics were extracted from these factors to serve as axes in the imagery space. The X-axis represents the degree of Hard-Refreshing vs. Soft-Chaotic (higher numbers indicate softer and more chaotic characteristics, while lower numbers indicate harder and more refreshing characteristics), while the Y-axis represents the degree of Childish vs. Steady (higher numbers indicate more childish characteristics, while lower numbers indicate greater stability). By calculating the factor scores for each material and sound performance, the imagery space distribution chart can be obtained (shown in Figure 1).

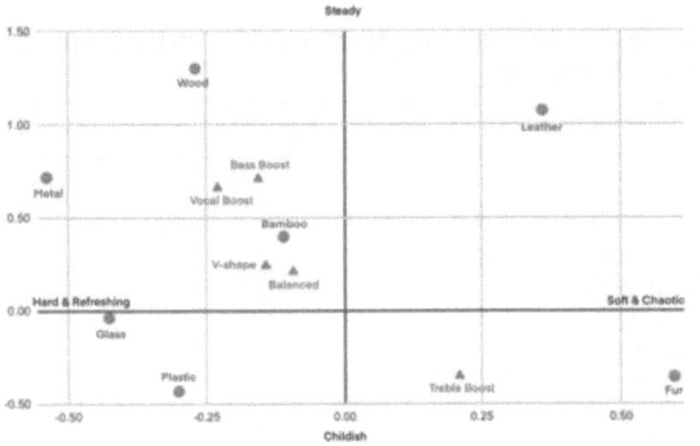

Figure 1. The spatial distribution diagram_imagery of product material and sound performance.

Due to the limited number of imagery axes, there is no direct correspondence between the sound performance and material shown on the graph. However, some level of correspondence between them can still be observed. Taking the second quadrant with the highest concentration of points as an example, if the headphone's sound performance emphasizes vocals or bass, the resulting imagery may lean towards stability, hardness, and freshness. Concurrently, one might consider selecting materials for the headphone's casing that represent similar imagery, such as wood or metal. Conversely, for headphones designed to emphasize mid-high tones as seen in the fourth quadrant, materials like fur could be considered to evoke softness and a more childish imagery.

4 CONCLUSION

The purpose of this study is to investigate whether the selection of materials for the casing of over-ear headphones, in conjunction with sound performance, can enhance users' sensory perception of the product through shared visual and auditory imagery. As discussed in the preceding chapter, the factor analysis of the visual imagery presented by materials and the auditory imagery presented by sound performance revealed that the imagery contained in each axis was not entirely consistent, making it impossible to directly correlate the sensory imagery presented by materials and sound performance. However, insights can still be gleaned from the data and distribution of imagery in the imaginary space, shedding light on the sensory imagery presented by the visual of various materials and auditory performance. It is hoped that these results will serve as a basis for designers to consider enhancing the competitiveness of products by selecting materials for over-ear headphones that evoke similar sensory imagery through visual and auditory performance.

REFERENCES

Chien, Li-Ju. (2003). *A Study on the Sensory Perception of Materials in Products: A Case Study of Desk Lamps.* Industrial Design Institute, Tunghai University, Taichung City.

Hekkert *et al.* (2003). *The Role of Sound in Product Experience.*

Hsiao, Li-Chun. (2019). *A Study on the Correspondence between Product Imagery and Material: A Case Study of the Elderly. Master's Program in Industrial Design,* National Yunlin University of Science and Technology, Yunlin County.

Koichi Ohtomi *et al.* (2009). *Product sound design for diverse customer's kansei*

Mike Ashby and Kara Johnson. (2003). The art of material selection

Nagamachi Mitsuo. (1995). *An Introduction to Kansei Engineering (in Japanese)*, Tokyo, Japan: Japanese Standards Association

Shih, Cheng-Yin. (2011). *The Impact of Material Sensory Imagery on Product Perception.* Industrial Design Institute, National Cheng Kung University, Tainan City.

William Bill Albert, Thomas S. Tom Tullis. *Self-Reported Metrics*

System Innovation for an Artificial Intelligence Era – Lam et al. (Eds)
© 2025 the Author(s), ISBN: 978-1-032-84452-7

Use landmarks as auxiliary positioning information in the presentation of the route navigation interface

Chuan-Hung Yuan & Meng-Cong Zheng
Department of Industrial Design, National Taipei University of Technology, Taipei Tech, Taiwan

ABSTRACT: Current mobile phone navigation systems do not specifically use landmarks as an auxiliary reference in route planning for pedestrians. However, landmarks can be valuable at decision points, assisting users in route interpretation and decision-making. This study analyzes the pedestrian route planning system in mobile phone navigation software and compares the route navigation interfaces of six commonly used mobile phone navigation software in Taiwan. The study summarizes three ways of presenting the route maps and conducts an experimental investigation to understand how they impact the user's wayfinding performance. The results reveal that the location and number of landmarks on the map significantly affect participants' ability to remember landmarks. Participants tend to use corner landmarks as a reference for determining the direction of the route and memorizing them. Participants achieved better results in memorizing landmarks when marked only on the corners of the route on the map, with an average of 4.8 landmarks recalled. In the way of landmarks marked at decision points and in between, the number of remembered landmarks decreased to an average of 3.1. Participants expressed a preference for memorizing essential information while filtering out extraneous details. They also noted that excessive information posed a challenge to recall. These findings offer a critical reference for utilizing landmarks as auxiliary information in route planning systems. The study will provide relevant design recommendations in the future to improve the user's wayfinding performance.

1 INTRODUCTION

Various studies have confirmed that navigation systems can effectively assist users in wayfinding tasks (Fang *et al.*, 2015). Using navigation of smart phone has become a daily routine for people in wayfinding. Although the functions of smart phone navigation tools are improving these days, there is a lack of auxiliary information to assist users in interpreting routes. In Taiwan's Taipei City, streets and alleys intertwine to create a complex environment. This complexity makes it challenging for users to align the navigation system with actual surroundings. Most people need to continuously cross-reference the navigation system with environmental information to ensure the accuracy of their route while on the move. People get lost easily once the information is more complicated than an individual's mind can handle.

Landmarks play a crucial role when aligning maps is challenging because of the environment. Landmarks is the basis for acquiring route knowledge and it can serves as mental anchoring for people of the physical environment (Balaban *et al.*, 2017). Literature states that landmarks are important features in the built environment (Afrooz *et al.*, 2018). Landmarks provide people with directional information to make the environment clear and easy to understand (Winter *et al.*, 2005). It became more useful when they are salient (Raubal and Winter, 2002). It suggests that an object can become a landmark is more salient than other objects around it (Caduff and Timpf, 2008) and it may occupies a salient spatial location geographically (Raubal and Winter, 2002) such as at the intersection of the road. Many reports have noticed that objects placed at decision points are more likely to be remembered longer than non-decision points (Janzen and van Turennout, 2004).

However, existing navigation tools rarely highlight the use of landmarks as auxiliary information in the pedestrian route planning interface. The study investigates existing smartphone navigations and compares performances between different applications of landmarks to learn how to use landmarks as auxiliary information in the pedestrian route planning interface of the smartphone navigation system. Through questionnaires, experimental tasks, and semi-structured interviews, the purpose of this study is to offer design references for future design proposals.

2 EXPERIMENTAL

The study began with a case study of six existing smart phone navigation tools. There are three main ways to display landmarks in the navigation: (1) No landmarks. (2) Presenting landmarks near the route. (3) Displaying landmarks regardless of the range.

DOI: 10.1201/9781003514831-36

Next, the study investigated the performance differences among three landmark applications using an experimental method. The study first selected an existing field in Taipei as the model and created map materials for the experiment, as shown in Figures 1, 2, and 3. Those maps were (A) A map without landmarks. (B) A map with landmarks displayed at corners. (C) A map with landmarks displayed regardless of the range. The study conducted a group experiment to compare the performance differences among three maps. The experiment invited thirty-six participants and divided them into three groups equally. Groups were labeled as Group A, Group B, and Group C according to map types. The experiment was conducted online for the virtual pathfinding task to minimize the impact of real-world experimental variables on task performance.

Each participant completed an SDQ-S questionnaire at the start of the experiment. Participants would then be asked to do a practice scenario to familiarize them with the experimental process before moving on to formal scenarios. Formal scenarios have three levels of difficulty, and each scenario consists of three tasks. Task: 1 Route planning. Task 2: Virtual wayfinding. Task 3: Route recall.

First, participants would be provided with a scenario map of the group and a picture of the starting point of the virtual wayfinding field, as shown in Figure 4. The tester introduced the scenario content simultaneously. Participants had to read two pieces of information and mentally plan a travel route to complete the mission. Once the participants finished reading, they proceeded to Task 1: route planning. In this Task, they could not reread the scenario map and picture. Participants in Task 1 were required to graphically recreate the route planning and information, which included positions of reference landmarks, a travel route, a starting point, and destination.

Participants then started Task 2: virtual wayfinding. Participants instructed the tester to perform a wayfinding task by verbally indicating directions with forward, left, right, and turn around. Instructions have to be conducted at each corner. Participants could also request the tester to observe the environment. Tasks usually require participants to travel from a starting point to one or more destination. The task may accomplished when the destination is reached.

In Task 3, participants retraced and draw the path they made in Task 2 and recalled the locations of landmarks they consider valuable. The participants then progressed to the next scenario, completing the same tasks until all scenarios were finished. Finally, participants would given a semi-structured interview to investigate the participant's knowledge of the navigation system. The tester videotaped the experiment. Participants were required to think aloud during the experiment.

Figure 1. Map of group A: Presenting landmarks near the route.

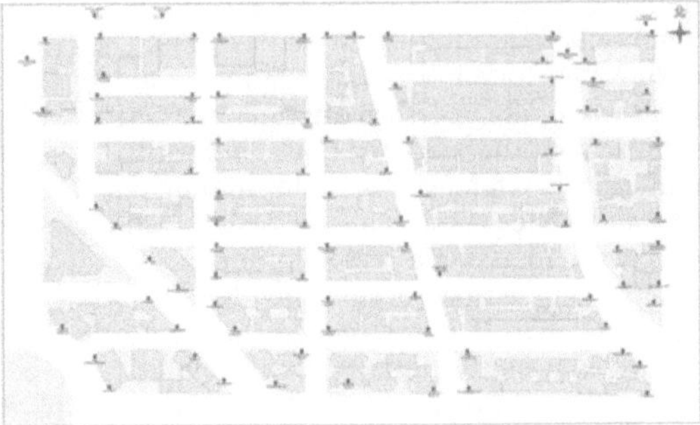

Figure 2. Map of group B: Displaying landmarks at corner only.

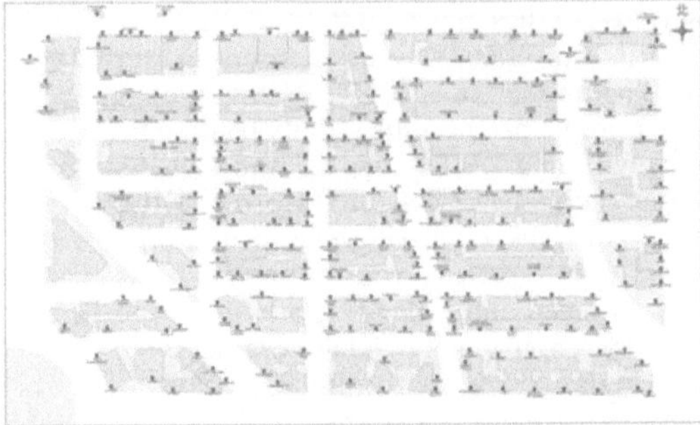

Figure 3. Map of group C: Displaying landmarks regardless of range.

Figure 4. Picture of starting point at virtual environment.

3 RESULTS AND DISCUSSION

3.1 *SDQ-S questionaire*

The SDQ-S questionnaire analyzed two abilities. Orientation perception and memory of daily spatial behaviors. The SDQ-S questionnaire uses a 5-point Likert scale. 1 indicating strong agreement and 5 indicating strong disagreement. Participants were divided into three levels of spatial ability by the score. The mean score of the 36 participants was 67.8. The standard deviation was 11.1. There were six low spatial abilities with scores lower than 56.7 and 7 high spatial abilities with scores higher than 78.9. The remaining 23 participants were medium spatial abilities.

3.2 *Information importance ranking*

A navigation system typically displays various information on the interface, including the distance to the destination, the estimated time of arrival, a route map, landmarks to aid in route interpretation, directions for navigation, and the name of the current or upcoming road. Participants ranked the importance of these information. The results showed that the route map was considered the most valuable information by most participants. The name of the current or upcoming road was the last.

3.3 *Analysis of the correlation between landmarks, route planning, and route memories.*

Table 1 shows the statistics of landmarks memorized in Task 1 and Task 3.

In Task 1, Group A recorded the fewest number of landmarks of 26 landmarks. The average value was 0.7. Group B recorded the highest number of landmarks. 173 landmarks. The average value was 4.8. Task 3 showed a similar result. Group A recorded the fewest number. Group B had the highest number of landmarks.

The average number of landmarks memorized in each scenario was higher in Group B than in Group C than in Group A. Comparing the number of landmarks memorized in the Task 1 and 2. The number of landmarks memorized increased by 0.7 landmarks in Group A and 0.6 landmarks in Group B and Group C.

Group B showed a higher total number and average number of landmarks memorized in both tasks. Group B's map shows 92 landmark at each intersection on all maps. The number is less than Group C and more than Group A. This result may mean that the number of maps in Group B has better memorization and persistence for the participants. The map of Group A does not show any landmark information. Only a few landmarks can be obtained from the phicture of the virtual environment starting point. The results were as expected. Group C's map shows 291 landmark. However, the number of landmarks recorded was less than Group B. It may be too much information on the map, which made it difficult for the participants to read.

Table 1. Number of landmark memories for Task 1 and Task 3.

Task 1					
Group	Scenario 1	Scenario 2	Scenario 3	Total	Average
A	10	5	11	26	0.7
B	46	58	69	173	4.8
C	24	39	46	113	3.1
Task 3					
Group	Scenario 1	Scenario 2	Scenario 3	Total	Average
A	19	13	18	50	1.4
B	53	66	75	194	5.4
C	33	48	51	132	3.7

Table 2. Schematic diagram illustrating the relationship between routes, information reading, and the number of landmarks passed.

Group A	Group B	Group C
	Scenario 1	

0	10	22
	Scenario 2	

0	25	72
	Scenario 3	

0	38	95
	Average	
0	24.3	63

Table 2 captures the shortest route of each scenario as an illustration. It marks the locations of the Table 2. Schematic diagram illustrating the relationship between routes, information reading, and the number of landmarks passed.

Landmarks that the path passes through with green circles. Study how each group is affected by the landmark information when reading the map.

The map of group A does not present any landmark but only provides the location of the starting point and the destination. Although the participants can easily visualize the travel route in their minds, there is a lack of supporting information about the environment. Participants can only memorize the number of intersections and the terrain from the map in order to deduce the route.

The map of Group B passes an average of 24.3 landmarks on the shortest line. Landmarks are all located at intersections in a targeted manner. This location highlights the recognizability of the information. The amount of information presented on the map is smaller but more precise. Even though the number of landmarks passing through the route increases with the complexity of the route, the landmarks are staggered by location. Each landmark gives more spaces in between. It is less likely to be interfered with by the visual reading. Therefore, it makes it easier and less laborious to read.

Group C presents more landmarks than Groups A and B. The route will pass through an average of 63 landmarks. The participants may have enough information to assist them in planning the routes. However, the complexity of reading is higher. The more landmarks the route passes through, the more dense the information is and the more difficult it is to read. Participants need to apply more brain power to observe and filter out useful information. As the result, the performance of landmark memory was less than that of Group B.

Table 3 shows the number of landmarks memorized by groups and scenario difficulty in Task 1 using two-way ANOVA. There was a significant difference in the number of landmarks memorized by groups, $F = 35.293$, $p = 0.00<0.05$. There was a significant difference by scenario difficulty, $F = 3.437$, $p = 0.036<0.05$. Groups and scenario difficulty showed no significant interaction on the number of landmarks remembered, $F = 1.043$, $p = 0.389 > 0.05$. The null hypothesis is not rejected.

Table 4 shows the number of landmarks memorized by groups and scenario difficulty in Task 3 using two-way ANOVA. There was a significant difference in the number of landmarks memorized by groups, $F = 23.011$, $P = 0.00<0.05$. There was no significant difference by scenario difficulty, $F = 0.650$, $P = 0.628>0.05$. Group and scenario difficulty showed no significant interaction on the number of landmarks remembered, $F = 0.650$, $p = 0.628 > 0.05$. The null hypothesis is not rejected.

The two-way ANOVA showed similar results for task 1 and task 3. It suggests that despite the different distances or complexity of the routes, the subjects only memorized the necessary information based on the form of the route. For example, participants likely to memorize landmarks at corners of the route. Other landmarks that were not necessary would be filtered according to personal habits. The number of landmarks and location on the map affected the ability to memorize landmarks. A significant difference was found in the two-way ANOVA($P = 0.00<0.05$). Participants in group B showed a better number of memories. Group C recorded less with more information on the map. It indicates that the amount of information that participants could remember was limited. They had difficulty remembering even when presented with more information.

3.4 *Landmark location*

Table 5 shows the locations of the landmarks recorded by participants. The results indicated that participants preferred to select landmarks at corners in Tasks 1 and 3.

Group A recorded 65% and 74% of the landmarks at corners in both Tasks. Group B's map provided landmarks basically at the corners. Therefore, the memory results were as expected. However, corner landmarks are also significant objects in the increased number of these two tasks. The total number of 192 landmarks recorded in task 3 increased by 19 than Task 1, with 11 landmarks located at corners and 8 landmarks located between road sections. The map in Group C also showed landmarks between road sections and corners. Participants were still likely to memorize

Table 3. Two way ANOVA of number of landmarks memorized for Task 1.

Source of Variation	SS	Degree of Freedom	Mean Square	F-Value	p-Value
Factor A Groups	303.500	2	151.750	35.293	**0.00*** **
Factor B Scenario difficulty	29.556	2	14.778	3.437	**.036***
Groups*Scenario difficulty	17.944	4	4.486	1.043	.389
Within(Error)	425.667	99	4.300		
Total	776.667	107			

*p<.05. **p<.01. ***p<.001.

Table 4. Two way ANOVA of number of landmarks memorized for Task 3.

Source of Variation	SS	Degree of Freedom	Mean Square	F-Value	p-Value
Factor A Groups	289.852	2	144.926	23.011	**.000***
Factor B Scenario difficulty	21.241	2	10.620	1.686	.190
Groups*Scenario difficulty	16.370	4	4.093	.650	.628
Within (Error)	623.500	99	6.298		
Total	950.963	107			

*p<.05. **p<.01. ***p<.001.

Table 5. Landmarks' locations for Task 1 and Task 3.

Task 1						
	Group A		Group B		Group C	
Scenario	Corner	Road Section	Corner	Road Section	Corner	Road Section
1	6	4	46	0	23	1
2	2	3	58	0	38	5
3	9	2	68	1	44	2
Total	17	9	172	1	105	8
Percentage	65%	35%	99%	1%	93%	7%

Task 3						
	Group A		Group B		Group C	
Scenario	Corner	Road Section	Corner	Road Section	Corner	Road Section
1	14	5	51	1	32	1
2	11	2	62	4	42	6
3	12	6	70	4	46	5
Total	37	13	183	9	120	12
Percentage	74%	26%	95%	5%	91%	9%

the locations of the landmarks at the corners to plan their walking strategies.

Overall, most participants showed a preference for memorizing landmarks located at corners. It indicates that the location of the landmarks is more important to participants as a reference value and more useful for route judgment.

3.5 *Task 2 - Virtual pathfinding*

Table 6 shows that Group A and Group C had task failures. 4 in Group A. 1 in Group C. Group A had the highest route errors of 18 times. Group B had the lowest of 8 times.

Table 7 shows the number of route error by groups and scenario difficulty in Task 2 using two-way ANOVA. There was a significant difference in the number of route error by groups, $F = 5.737$, $p = 0.04 < 0.05$. There was no significant difference by scenario difficulty, $F = 0.043$, $p = 0.958 > 0.05$. Groups and scenario difficulty showed no significant interaction on the number of landmarks remembered, $F = 0.690$, $p = 0.600 > 0.05$. The null hypothesis is not rejected.

Group A performed the worst walking performance in scenario 2. Group B performed the best walking distances in all scenario. Group C performed the worst walking performance in scenarios 1 and 3.

Table 8 shows the actual walking routes. Group B had the most uniform routes due to the least detouring behavior. Group A was the least consistent with the most number of detours. Group C's actual walking routes was similar to Group B, but the wayfinding performance was weaker than Group B. Therefore, the route map was a bit messy.

The difference between groups is the amount of landmark information on the map. It also affects wayfinding performance. The map in Group A provides a smaller amount of information. The number of intersections is the priority information to be memorized for route interpretation, making it easy to get lost if the intersections were miscalculated or missed. In Group B, the map displays 92 landmarks at corners. Participants tended to record this information to assist with route planning and interpretation. A mental map of the road was created in the brain by recalling landmarks and cross-referencing their positions with the actual environmental landmarks during wayfinding. This behavior enhances the participants' judgment of direction. Therefore, Group B showed fewer route errors. The map in Group C presents 291 landmarks for the participants to memorize. However, the performance was still slightly lower than that of Group B. The map in Group C hace more landmarks than the map in Group B, leading some participants to misjudge the environment due to the information they retrieved.

Table 9 analyzes the correlation between each participant's performance in task 1, based on landmarks remembered, and task 2, in terms of route errors, using coding and pairing the performance made by each participant. A total of four types of performance were analyzed. Y*L: Participant performed route error(Y) and the number of landmarks recorded was lower than the group mean(L); Y*H: Participant performed route error(Y) but the number of landmarks recorded

Table 6. Task 2: Outcome and number of route error.

	Group A		Group B		Group C	
Scenario	Route Error	Task Failed	Route Error	Task Failed	Route Error	Task Failed
1	5	1	2	0	4	0
2	8	2	2	0	3	1
3	5	1	4	0	2	0
Total	18	4	8	0	9	1

Table 7. Two way ANOVA of number of route error.

Source of Variation	SS	Degree of Freedom	Mean Square	F-Value	p-Value
Factor A Groups	2.463	2	1.231	5.737	**.004***
Factor B Scenario difficulty	.019	2	.009	.043	.958
Groups*Scenario difficulty	.593	4	.148	.690	.600
Within(Error)	21.250	99	.215		
Total	24.324	107			

*p<.05. **p<.01. ***p<.001.

Table 8. Actual route by participants in Task 2.

Group A	Group B	Group C
	Scenario 1	

	Scenario 2	
	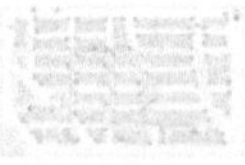	
	Scenario 3	

Table 9. Relationship between landmark memory ability and route error performance.

	Route Error Appeared (Y)		No Route Error (N)	
Group	Low Landmarks Recorded (L)	High Landmarks Recorded (H)	Low Landmarks Recorded (L)	High Landmarks Recorded (H)
A	36.1%	22.2%	30.6%	11.1%
B	13.9%	8.3%	36.1%	41.7%
C	16.7%	8.3%	30.6%	44.4%

was higher than the group mean(H); N*L: Participant did not perform route error(N) but the number of landmarks recorded was lower than the group mean(L); N*H: Participant did not perform route error(N) and the number of landmarks recorded was higher than the group mean(H).

From the explanation, Y*L and Y*H are nagative perfomance. The higher the values, the worse the wayfinding is. N*L and N*H are positive perfomance.The higher the values, the better the wayfinding is.

Group A had the highest Y*L performance, totaling 36.1%. The map used in Group A may have caused the overall performance to be dominated by recording less landmark information and making more route errors. Y*H was also the highest in group A. Although some participants could remember landmark information above the average value, there was a lack of critical landmark information to base their memories on. Hence, there was still a situation of choosing the wrong route.

N*H was the highest performance (41.7%) for group B. The map of group B showed landmarks at corners. The relationship between memorizing a high number of landmarks and making fewer route errors was the most prominent. Group B also showed higher performance in N*L (36.1%). It means some participants can successfully navigate by remembering fewer but crucial landmarks. Group C also showed the highest performance in N*H (44.4%) and second in N*L(30.6%).

To summarize, the success rate of wayfinding is likely to be relatively higher when participants can memorize enough information. On the other hand, the success rate was lower when less information was obtained or memorized. Some participants also demonstrated that they completed the Task 2 with less information. It may indicate that participants do not need excessive information to navigate but only require the essential information to reach their destination.

3.6 *Semi-structured interviews*

The reading style of the navigation varies depending on the mode of transportation. Participants typically rely on the route map as their primary source of information and choose one or more additional pieces of information as supplementary memory aids based on their habits. The following three modes of transportation were used as the primary modes of transportation for participants.

Public transportation: 65% of participants use the navigation system while in motion. 80% of participants only read the route planner mode for navigation. 70% of participants rely more on landmark information.

Riding motor bike: 44.4% of participants read the route map while riding. The rest of the participants used to memorize the route before driving and then rely on their memory to find the way. 38.9% of participants only read the route map because they thought the navigation mode was inaccurate in areas with numerous road forks. 44.4% of participants only memorized a single piece of information to identify the way. 55.6% of participants memorized multiple pieces of information.

Driving car: Participants tend to use the navigation system while moving and prefer the voice navigation mode because it is more convenient for guidance. 87.5% of participants primarily memorize the road names, and 42.9% also memorize the landmarks. They mentioned that when the environment is complex, they will integrate landmark information to assist in determining the route, but road name is considered more crucial than landmarks.

In Group A, 83% of participants believed the map lacked information. In Group B, 66.7% of participants found the presentation of landmarks helpful. 16.7% of participants thought the map had too many landmarks visually, but this did not affect memory. In Group C, 50% of participants indicated that the map marked too many landmarks, making the map visually cluttered and more difficult to filter or memorize information. Furthermore, most of the information was not essential. As a result, Group B's map was recognized by the most participants.

94.4% of participants agreed that it would be more helpful and suitable if landmarks were only presented at the corners of the route map.

4 CONCLUSION

The study concluded with the following results. (1) Presenting the correct number of landmarks along a route is more beneficial for memory retention. (2) Participants are more likely to memorize landmarks at corners, which are crucial references for participants to recognize directions. (3) Different modes of traffic movement will assign varying levels of importance to route information. The information presented by the navigation should be prioritized and organized based on the user's movement. Based on the results of Experiment 1, the study will be followed up by designing and proposing a pedestrian route planning interface and presentation method for smart phone navigation system that incorporates landmarks as auxiliary information in the future.

REFERENCES

Afrooz, A., White, D., & Parolin, B. (2018). Effects of active and passive exploration of the built environment on memory during wayfinding. *Applied Geography*, *101*, 68–74.

Balaban, C. Z., Karimpur, H., Röser, F., & Hamburger, K. (2017). Turn left where you felt unhappy: How affect influences landmark-based wayfinding. *Cogn Process*, *18*(2), 135–144.

Caduff, D., & Timpf, S. (2008). On the assessment of landmark salience for human navigation. *Cognitive Processing*, *9*(4), 249–267.

Fang, Z., Li, Q., & Shaw, S.-L. (2015). What about people in pedestrian navigation? *Geo-spatial Information Science*, *18*(4), 135–150.

Janzen, G., & van Turennout, M. (2004). Selective neural representation of objects relevant for navigation. *Nat Neurosci*, *7*(6), 673–677.

Raubal, M., & Winter, S. (2002, 2002//). Enriching wayfinding instructions with local landmarks. *Geographic Information Science*, Berlin, Heidelberg.

Winter, S., Raubal, M., & Nothegger, C. (2005). Focalizing measures of salience for wayfinding. In L. Meng, T. Reichenbacher, & A. Zipf (Eds.), *Map-based Mobile Services: Theories, Methods and Implementations* (pp. 125–139). Springer Berlin Heidelberg.

System Innovation for an Artificial Intelligence Era – Lam et al. (Eds)
© 2025 the Author(s), ISBN: 978-1-032-84452-7

Research on STEAM toys with attractiveness factors and high customer loyalty

Tz-Ying Lin & Yu-Ling Yang
Department of Industrial Design, National Yunlin University of Science and Technology, Douliu, Yunlin, Taiwan
Department of Industrial Design, College of Design, Yunlin, Taiwan

ABSTRACT: With the declining birthrate and changes of the social structure, toy products in the new era should have a certain degree of sustained appeal to increase customer loyalty and enable products to be continuously or adopted a lot by customers. This study focuses on STEAM toys. It first uses a semi-structured quantitative questionnaire to investigate Taiwanese parents' opinions on STEAM toys and their priorities when purchasing STEAM toys. It attempts to explore differences and commonalities. After that, from the collected 105 responses, select 6 representative families for qualitative research. First, let parents write a "Picture Life Diary of Children and Toys". Next, visit the interviewed families in person and use the "Miryoku Engineering Evaluation Grid Method" to discover the types of STEAM toys that can attract parents to buy again and the attractive factors of sustainable repurchase. Some research and surveys have found that Taiwanese parents fond of buying toys that allow their children to create their own creations, are suitable for all ages, and have self-expanding functions. The top three toys that Taiwanese children love most are LEGO, Qbi toys, and magnetic tiles. This type of STEAM toys has diverse play methods, is easy to operate, and has educational value. In addition, STEAM toys that parents are willing to repurchase must be highly interactive, expandable, attractive to children, playable, and suitable for both dynamic and static play. The results of this study can provide reference for toy designers and the toy industry to develop more attractive STEAM toys.

1 RESEARCH BACKGROUND AND MOTIVATION

The declining birthrate creates business opportunities for darlings. The "2023 Global Toy Market Overview" report predicts that by the end of 2028, the global toy market will increase from US $109.25 billion in 2022 to US$152.41 billion [1]. Even though there are fewer and fewer children, the business opportunities are increasing [2]. This is because modern parents are more willing to spend more money on their children, and modern parents and educators agree that toys can serve as important tools and value for children's cognitive, society and emotional development. Therefore, the benefits of integrating learning and entertainment are Intelligent toys are favored by consumers. But although there are many types of toys, there are only a few toys that children love to play again and again, and only some toys that parents are willing to keep buying again. In view of most toy most toy-related research focuses on the functionality, value, and educational aspects of toys, or on the factors that parents consider when purchasing toys, few directly focus on the emotional aspects of children or parents towards toys. Therefore, this study further uses the "Miryoku Engineering Evaluation Grid Method" to explore the deep psychological desires of parents and children, extracts design elements, and converts them into design features, hoping to give toys and children a higher attractiveness relationship.

2 LITERATURE REVIEW

2.1 *Customer loyalty*

Brown (1952) proposed that customer loyalty is a behavioral pattern that is consistent with customers purchasing a certain brand of products [3]. Day (1977) believed that customer loyalty is the customer's continuous positive purchase behavior of a company's products [4], and the customer is willing to pay a higher price, prefer the original company's product selection, continue to purchase products or services from the company, and increase consumption of the company [5].

2.2 *STEAM toy*

The concept of STEM was first proposed by the National Science Foundation (NFS) in 2001. STEM was originally a combination of the four fields of science, technology, engineering, and

DOI: 10.1201/9781003514831-37

mathematics. Later, it was discovered that in addition to STEM education, children's development also needs to consider humanities and artistic beauty, so Art was added and became STEAM [6] [7]. Taiwan's iconic parent-child magazine "CommonWealth Education Media and Publishing" referred to the certification standards of the American Toy Association, US STEM.org and 7 experts in 2020 to define five major indicators of good STEAM toys: (1) Cross-field, (2) openness/creativity, (3) playability, (4) life application, (5) problem solving [8]. Based on these five indicators, this study carefully selected 100 high-quality toys that conform to the STEAM education concept as the basis for the making of interview charts for this study.

2.3 *Parents' consumption behavior when purchasing toys*

Engel, Blackwell and Miniard (1995) proposed that the consumer behavior model mainly includes five steps: confirming needs, collecting information, program evaluation, purchase decision-making, and post-purchase behavior [9]. W.Z. Xu (2019) found that the factors that affect consumers' purchase of children's toys include: parents' personal background variables, educational functions of toys, toy prices, toy purchase channels, toy usage meanings, toy selection preferences, toy types, packaging styles, product purchase experience, operating experience, toy quality, toy origin, safety, consumption motivation, and purchase decision type [10]. C.S. Fu (1988) found that Taiwanese consumers have nine motivations for buying toys: New Year holidays, rewards, cultivating intelligence, advertising influence, children clamoring for them, fashion trends, in-store display attraction, recommendations from sales staff, teachers, relatives, and friends [11]. Vichasilp (2017) divided the decision-making considerations for purchasing children's toys into seven factors, including gorgeousness, durability, origin, personal sales, advertising, safety, and children's skill development [12]. This study incorporates the above purchase motivations and factors into a semi-structured interview questionnaire to further understand consumers' motivations and considerations for "repurchase".

3 RESEARCH METHOD

This study focuses on STEAM toys. At the beginning of the research, a semi-structured quantitative questionnaire was used to investigate Taiwanese parents' opinions on STEAM toys and their priorities when purchasing STEAM toys. To find representative respondents, 40 parents were invited to help write a "Picture Life Diary of Children and Toys" from the 105 quantitative questionnaires collected (Figure 1).

Figure 1. Example of life diary of children and toys.

This study used the diaries to understand parent-child interactions and parent playtime in representative families within a week. Finally, from the 40 life diaries, we extracted 6 parent-child families that were highly consistent with the purpose of this study, visited the interviewees' homes, and conducted qualitative research using the Evaluation Grid Method (EGM) developed by Sanui to explore the qualitative elements [13] [14]. This study prepared 100 cross-type STEAM toy picture cards as auxiliary tools (Figure 2).

Figure 2. Example of STEAM toy experiment chart.

The subjects make subjective comparisons between the pictures, evaluate their likes and dislikes, and then ask about the abstract reasons and concrete reasons for their preferences, so that they can be classified into upper, mid and lower levels, and the evaluation relationship of each respondent is constructed based on this (Figure 3) [15], from

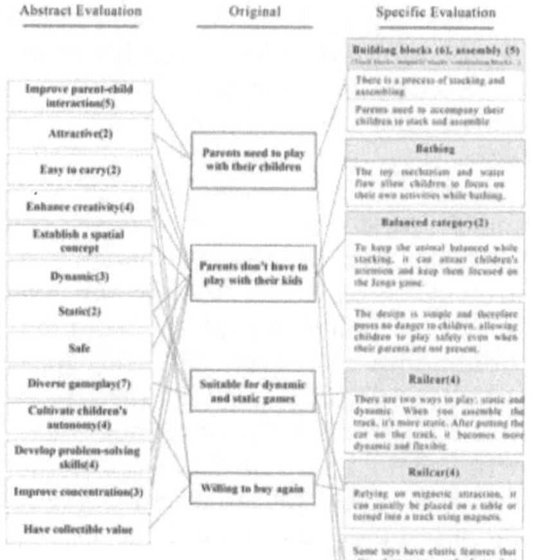

Figure 3. Example of interviewee's STEAM toy evaluation structure chart.

which we can explore the types of STEAM toys that can attract parents to buy again and the attractive factors of sustainable repurchase.

4 RESULTS ANALYSIS AND DISCUSSION

In this study, a total of 105 valid questionnaires were collected in the quantitative survey, and 78 questionnaires from families who would repurchase STEAM toys were selected for analysis. Judging from the gender of the basic information, mainly parents who responded were women (accounting for 85.9%), the age group was 31-40 years old (accounting for 57.7%), and the type of residence was mainly urban (accounting for 78.2%). The educational background is mainly college/university (accounting for 66.7%).

This study further used SPSS analysis to find out that parents believe that the main purpose of STEAM toys is to develop children's multiple abilities in an all-round way, accounting for 21.9%; The aspect that parents are most concerned about when purchasing STEAM toys is that the toys can help children develop "creativity", accounting for 25.1%, followed by "educational", accounting for 21.9%; When it comes to believing that STEAM toys at home are most helpful to their children, 36.5% of parents believe that STEAM toys at home are most helpful to their children by "enhancing children's creativity and imagination." The top three most popular toys for children at home are the LEGO series (including ordinary building blocks, Duplo series, etc.), followed by the Qbi educational magnetic track toy set, and the third is magnetic tile.

In terms of home visits to 6 representative families, this study used the "Miryoku Engineering Evaluation Grid Method" to unearth a total of 37 items of upper abstract feelings, 20 items of median raw evaluations, and 54 items of lower specific evaluations (Figure 4 below). Parents value the impact of toys on their children's growth and development, and the importance of children learning from the process of playing games. "Various ways to play (7)", "Enhancing creativity (5)", "Children preference (4)", "Training children's autonomy (4)" and "Building problem-solving skills (4)" were the top answers among the respondents Items with the number of responses. According to its original reason, among the types of toys, STEAM toys that require children to stack and assemble are more popular among parents. This type of toys can not only enhance parent-child interaction, have diverse and varied play methods, but also allow children to sort and learn to organize toys on their own, training autonomy, the most popular thing among the respondents is especially toys that require assembly of tracks (for example: Qbi toys, Tomica). In addition to the original configuration, these types of toys also have model cars and other related accessories that can be expanded. Even adults can want to collect.

In terms of factors that make them willing to repurchase STEAM toys, most respondents believe that toys that "can be expanded" and "can increase theme variability" will make consumers willing to repurchase them first, and toys that are highly interactive, attractive to children, playability, suitable for both moving and static, easy to store, collectible value, and rail car type are the factors that parents consider when they are willing to buy toys again.

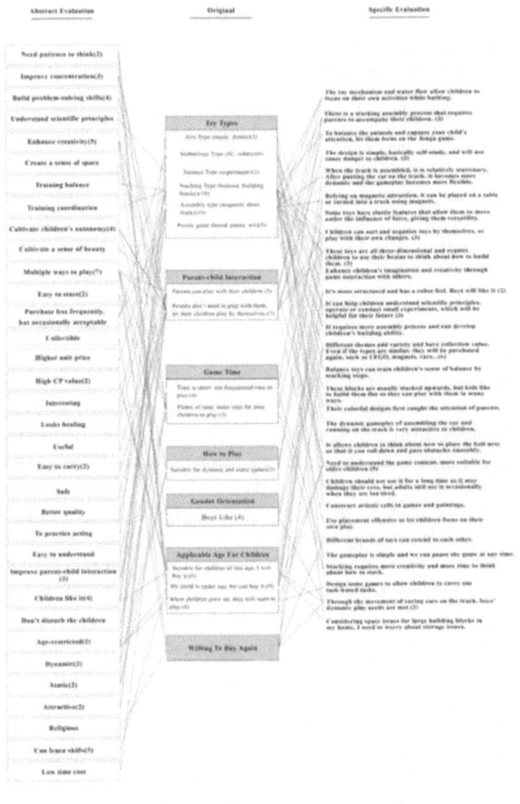

Figure 4. Overall STEAM toy evaluation structure diagram.

5 CONCLUSION

Whether it is a toy brand or a toy manufacturer, when it comes to toy development, they will fall into a deep development bottleneck due to product homogeneity. This study explores the classic STEAM toys that parents and children love, looking for the "charm factor" of these products, and further explores how these STEAM toys can be "continuously repurchased" by consumers.

The "attractiveness factors" of these classic STEAM toy products and the factors that attract consumers to "continue to repurchase" can provide toy designers and related toy development practitioners with a reference to develop more attractive STEAM toys.

REFERENCES

Brown, G. H., (1952). *Brand Loyalty-Fact or Fiction, Advertising Age*, 23. pp.52–55.

Day, R. L. (1977). Extending the concept of consumer satisfaction. *Advances in Consumer Research*, 4(1), 149–154.

Editor. (1992). *Charisma Engineering Research Department. Charisma Engineering*. Kobe: Kaibundo, Japan.

Engel, J. F., Blackwell, R. D., & Miniard, P. W. (1995). Consumer Behavior (8th ed.). The Dryden Press. (pp. 448–460).

Fu C. S. (1988). *Research on Toy Purchasing Behavior in the Domestic Market - Taking Families of Elementary School Children in Taipei City as the Target*, (Master's Thesis, Institute of Business Administration, Chinese Culture University, Taipei City.)

Global Toy Market Factbook (2023 Edition): Analysis By Product Type, By Age, Distribution Channel, By Region, By Country: Market Insights and Forecast (2018-2028). (2023). *Research and Markets.*

Hallinen, J. (2024). STEM. Encyclopedia Britannica. Retrieved from https://www.britannica.com/topic/STEM-education

Huang, Z. E. (2018). "Business Opportunities" the "Pet" economy is coming, and pets and pets will not be lenient in consumption. *Economic and Trade Perspectives Biweekly*, (535).

Kelly, G. A. (1955). *The Psychology of Personal Constructs* (Vols. 1 & 2). Norton, New York.

Qiu, S. W. (2020). 2020 STEAM Good Toys Selection. *Parent-Child World Magazine*, (113).

Sanui, J., & Inui, M. (1986). Repertory grid development techniques. *Collection of Papers and Reports from the Design Department of the Architectural Institute of Japan*, (367).

Stout, P. A., & Rust, R. T. (1993). Emotional feelings and evaluative dimensions of advertising: Are they related. *Journal of Advertising*, 22(1),61–71.

Tang, W. L. (2019). Exploring the development of STEM and STEAM education in the United States. *Curriculum and Instruction Quarterly*, *22*(2), 49–77.

Vichasilp, K. (2017). Factors influencing purchasing decisions of toys for preschool children in the countryside of Thailand. *International Business Management*, 11(12), 2010–2014.

Xu, W. Z. (2019). *Research on the Consumption Behavior, Purchase Intention, Cognitive Value, Product Knowledge and Educational Attitude of Young Children's Parents Towards Toys* (Master's thesis, Child Education and Business Management Industry, Yuda University of Science and Technology. Miaoli.) Retrieved from https://hdl.handle.net/11296/xh2bj2

System Innovation for an Artificial Intelligence Era – Lam et al. (Eds)
© 2025 the Author(s), ISBN: 978-1-032-84452-7

Comparison of input methods in Virtual Reality: Joystick vs. Gesture control modify

Liang-Yin Kuo
Smart Machinery and Intelligent Manufacturing Research Center
Department of Information Management, National Formosa University, Hu-Wei, Yunlin, Taiwan

Yao-Yun Hsiao & Chen-Yu Chuang
Department of Multimedia Design, National Formosa University, Hu-Wei, Yunlin, Taiwan

Nian-Ze Hu
Department of Information Management, National Formosa university, Hu-Wei, Yunlin, Taiwan

Jing-Chung Shen
National Formosa University, Hu-Wei, Yunlin, Taiwan

Chia-fang Hsu
Department of Education, New Era University College

ABSTRACT: This research investigates the operational distinctions between joystick and gesture controls within VR headsets. A virtual reality experiment is planned, utilizing the Meta Quest 3 as the experimental equipment, to compare the effectiveness of these two methods in terms of operability, completion time, and user preference. Participants of different age groups and varying levels of prior VR experience are selected to partake in the experiment, and they are categorized based on these factors. The research process involves consultation using a Likert five-point scale questionnaire and subsequent data analysis. Results reveal that while joystick control allows for faster task completion, gesture operation is perceived as more intuitive and preferred by users. However, prolonged use of gesture controls may induce hand fatigue. These findings contribute to a better understanding of the distinctions in virtual reality operation methods and offer insights for immersive experience design.

1 INTRODUCTION

Common Oculus Quest controllers rely on capacitive sensing to create basic hand gestures for operating buttons, switches, knobs, and sliders in virtual reality (VR). An empirical study was conducted to explore user preferences and text entry performance with four pointing methods (Controller, Head, Hand, Hybrid) combined with two input mechanisms (Swype, Tap). The study revealed that the Controller method outperformed other device-free methods in both text entry efficiency and user experience.

Virtual reality (VR) technology is rapidly advancing, providing immersive experiences with intuitive and efficient interaction methods. Among these methods, joystick control and gesture control represent two primary user interaction approaches in virtual environments. Joystick control typically involves navigation and interaction using physical controllers, whereas gesture control utilizes hand movements and gestures to manipulate virtual objects and navigate within virtual spaces. The Meta Quest 3 VR headset, known for its usability and user-friendly design, was used in this study to compare joystick and gesture controls. Understanding operational differences and user preferences between these input methods is crucial for optimizing user experiences and interface design in VR applications. This study aims to investigate the performance of joystick and gesture controls in terms of operation and user preference. By examining participants of different age groups and VR experience levels, we can assess the effectiveness and user satisfaction of each control method. The study employed Likert scale questionnaires to collect participant feedback and data for quantitative analysis. The results of this research will provide valuable insights into subtle differences in virtual reality interaction methods, offering practical guidance for developers and designers to enhance user engagement and comfort in immersive VR environments. Ultimately, this study aims to provide information for the ongoing

DOI: 10.1201/9781003514831-38

development of VR technology and improve user interface design for optimal user experiences.

2 EXPERIMENTAL

This study compares the performance differences between joystick and gesture input methods using the Meta Quest 3 VR headset in completing virtual tasks. Additionally, it evaluates participants' subjective preferences and satisfaction with these two operation methods. In the experiment, participants are assigned specific virtual tasks, and a third-party notary uses a timer to record and compare the completion times of tasks using joystick and gesture operations.

Figure 1 shows the experimental process flow. During the experiment, participants first wear the Meta Quest 3 VR headset and enter a virtual lobby. Within the lobby, participants access a menu of functionalities, select a designated operational task, and proceed into a pre-set virtual task scene. Within the task scene, participants are required to use either a joystick or gesture controller to grab cubes within the scene and place them in specified locations. Upon task completion, participants open a task window, press a designated button to conclude the task process, and then exit the task scene back to the virtual lobby.

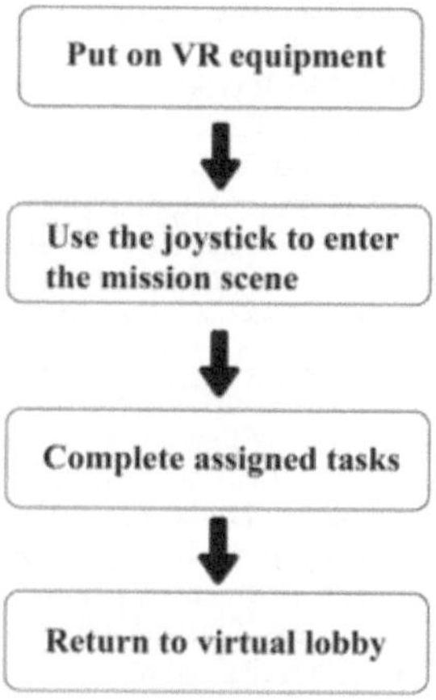

Figure 1. The experimental process flow chart.

The experimental screen shows in Figure 2 where participants complete tasks within the task scene. Upon completion of the experiment, participants will fill out a Likert scale questionnaire to subjectively assess their satisfaction and preferences regarding joystick and gesture input methods. We will analyze the completion times for joystick and gesture operations and compare the performance and user experience of these two methods. These research findings will contribute to a deeper understanding of the differences between various input methods and provide concrete data to guide optimization and improvement of virtual reality interaction design.

This study aims to compare the performance and user experience differences between joystick and gesture input methods using the Meta Quest 3 VR headset in virtual task completion. In the experiment, participants will wear the Meta Quest 3 VR headset and enter a virtual reality environment. Initially, they will navigate to a virtual lobby and select a designated operational task from a menu of functionalities. Subsequently, participants will enter a pre-set virtual task scene and prepare to commence the task. Within the virtual task scene, participants will use either a joystick or gesture controller based on task requirements to grab cube objects within the scene and place them in specified locations. Upon task completion, participants will open a task window and press a designated button to conclude the task process. Throughout the experiment, a third-party notary will time the completion of joystick and gesture operations to objectively assess the performance of both methods. After the experiment, participants will fill out a Likert scale questionnaire to evaluate their satisfaction and preferences regarding joystick and gesture input methods.

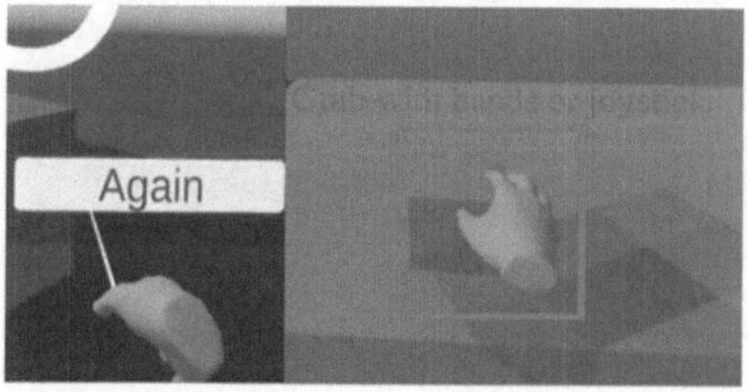

Figure 2. The experimental screen picture.

3 RESULTS AND DISCUSSION

Figure 3 shows the Likert scale questionnaire statistical analysis results for joystick and gesture operations respectively. Based on participant feedback, the majority of participants showed a preference for joystick control, believing it better aligned with their expectations and needs. The precise operation of joystick control makes it a promising and applicable method for future virtual reality interface designs. This study has certain limitations, such as a relatively small and limited sample size. Future research could expand the sample size and delve deeper into exploring the impact of different input methods on various age groups and user types. Furthermore, further research could focus on improving and optimizing gesture control to enhance its applicability and performance in virtual reality. The results contribute to a deeper understanding of differences in virtual reality interaction methods and provide valuable insights for future virtual reality interface design and development efforts.

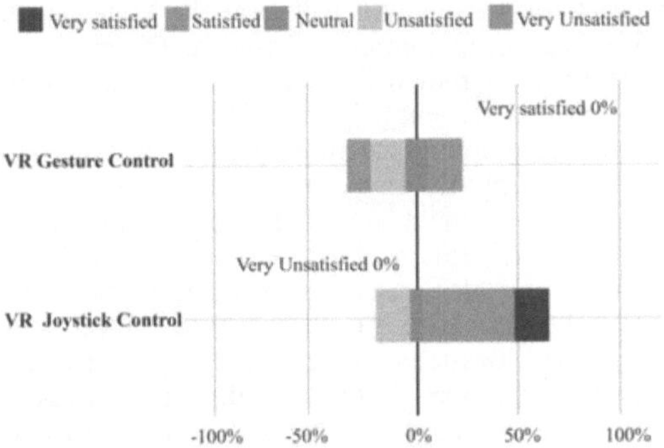

Figure 3. The Likert scale questionnaire.

4 CONCLUSION

This study compares the performance and user preferences between joystick and gesture controls in Meta Quest 3 VR headset. By designing virtual reality tasks and recruiting participants for experiments, we derived the following conclusions: According to our research findings, joystick control exhibits faster and more efficient operation in virtual reality tasks compared to gesture control. This implies that for scenarios requiring rapid task completion, joystick control can be a more effective mode of operation. However, gesture control offers a more intuitive and natural user experience, potentially better suited for users who prioritize operational feel and realism. Additionally, gesture control contributes to a more intuitive user experience. Nonetheless, prolonged use of gesture control may lead to hand fatigue issues for some participants. Finally, our research findings contribute to a deeper understanding of differences in virtual reality interaction methods and provide insights for designing virtual reality interfaces that better meet user needs. Gesture control, as an intuitive and popular mode of operation, should receive more attention and application in future virtual reality applications.

REFERENCES

[1] Tatzgern M. and Birgmann C., (2021). Exploring input approximations for control panels in virtual reality, 2021 *IEEE Virtual Reality and 3D User Interfaces (VR)*, Lisboa, Portugal, pp. 1–9, doi: 10.1109/VR50410.2021.00092.

[2] Xu W., Liang H. -N., He A. and Wang Z., (2019). Pointing and selection methods for text entry in augmented reality head mounted displays, *2019 IEEE International Symposium on Mixed and Augmented Reality (ISMAR)*, Beijing, China, pp. 279–288, doi: 10.1109/ISMAR.2019.00026.

[3] Zhang L., He W., Cao Z., Wang S., Bai H. and Billinghurst M., (2021). Simultaneous real walking and asymmetric input in virtual reality with a smartphone-based hybrid interface, *2021 IEEE International Symposium on Mixed and Augmented Reality Adjunct (ISMAR-Adjunct)*, Bari, Italy, pp. 321–323, doi:10.1109/ISMARAdjunct54149.2021.00072.

[4] Yi D. C., Chang K. -N., Tai Y. -H., Chen I. -C. and Hung Y. -P., (2020). Elastic-move: Passive haptic device with force feedback for virtual reality locomotion, *2020 IEEE Conference on Virtual Reality and 3D User Interfaces Abstracts and Workshops (VRW)*, Atlanta, GA, USA, pp. 40–45, doi: 10.1109/VRW50115.2020.00015.

[5] Van Damme S., Tack J., Van Wallendael G., De Turck F. and Vega M. T., (2023). Are we ready for Haptic Interactivity in VR? *An Experimental Comparison of Different Interaction Methods in Virtual Reality Training, 2023 15th International Conference on Quality of Multimedia Experience (QoMEX)*, Ghent, Belgium, pp. 294–299, doi: 10.1109/QoMEX58391.2023.10178453.

[6] Bustamante M., Livingston M., Hilley J., Tran H., Lovejoy E. and Murray J. T., (2022). ColorBound: Comparing Menu Dynamics in Virtual Reality, *2022 IEEE Conference on Virtual Reality and 3D User Interfaces Abstracts and Workshops (VRW)*, Christchurch, New Zealand, pp. 914–915, doi: 10.1109/VRW55335.2022.00308.

System Innovation for an Artificial Intelligence Era – Lam et al. (Eds)

The study of virtual reality assisted teaching application—basketball assisted teaching: Enhancing flow experience

Kuo Liang Yin
National Formosa University Smart Machinery and Intelligent Manufacturing Research Center

Kitty Leung & Ng Pei Ying
Department of Information Management, Formosa University

Hsu Chia-fang
Department of Education, New Era University College

ABSTRACT: This study employs Virtual Reality (VR) technology to explore and assist users in gaining a comprehensive understanding of various basketball tactics and techniques. The content will provide an immersive experience, allowing users to feel as if they are personally on the basketball court. This holds significant promise for distance education, particularly in the face of potential future pandemics, where such virtual learning environments can offer a safe and effective means of learning.

Expert interviews and experimental methods will be utilized, inviting elementary and middle school students to participate in tests to assess their acceptance of this virtual learning approach and whether any adaptation issues exist. Through these tests, we aim to better understand how to improve the relevant content to make it more suitable for educational purposes.

The findings of this study will contribute to the advancement of virtual reality technology in the field of education, paving the way for new possibilities in future educational models.

Keywords: Virtual Reality, Action Research, Basketball Assisted Teaching, Flow Experience

1 INTRODUCTION

The coronavirus disease (COVID-19) pandemic has precipitated a paradigm shift within the education system, and coerced academic institutions to recalibrate their educating paradigms. The United Nations Education, Scientific, and Cultural Organization (UNESCO) has recognized that the coronavirus pandemic outbreak has impacted the education system in the world (UNESCO 2020b).In facing the COVID-19 pandemic, students could not go out of the house freely and could not walk to the park or playground for exercise, which reduced the time for physical activities but increased the screen time for watching TV or playing with cell phones, which will inevitably bring adverse effects to students' health (Carroll *et al.* 2020; Cuschieri and Grech 2020; Pietrobelli *et al.* 2020). The exigencies imposed by the disease compelled an unexpected departure from the traditional teaching approach of in-person instruction towards non-face-to-face approaches. This metamorphosis induced by the epidemic has notably pervaded the domain of education, compelling students and educators alike to confront its ramifications. As a result, the prevalence of online learning has experienced a significant increase, providing a feasible substitute for conventional face-to-face teaching methods.

Virtual reality (VR) is a simulated experience that employs 3D near-eye displays and pose tracking to give the user an immersive feel of a virtual world, it includes entertainment (particularly video games), education (such as medical, safety, or military training), and business (such as virtual meetings). VR is one of the key technologies in the reality-virtuality continuum (Wikipedia). Currently, augmented and virtual reality (AR/VR) technologies, which perform excellently in providing users with an interactive and immersive environment, have been rapidly recognized and widely used in many fields. The application of VR enables students to carry out immersive simulation learning (Tan *et al.* 2022).

By integrating virtual reality (VR) technology with non-face-to-face instructional methods, the

DOI: 10.1201/9781003514831-39

resultant synthesis promises to engender a markedly compelling and immersive educational experience. Such a convergence not only amplifies pedagogical engagement but also augments the potential for profound cognitive assimilation, thus infusing the learning process with heightened intrigue and efficacy. VR technologies in education are revolutionizing the way students learn and interact with the world around them (Du *et al.* 2020).

This study investigates the efficacy of integrating VR technology with non-face-to-face instructional methods to enhance student learning experiences, focusing on its application within basketball education. This research aims to elucidate how VR-assisted teaching facilitates and augments the learning process, thereby fostering a more immersive and effective educational environment. The integration of VR technology into educational practices presents a powerful means of elevating teaching quality and nurturing student creativity(Hikmawan *et al.* 2019). Furthermore, students who engage with interactive electronic content modules exhibit superior proficiency in cognitive tasks compared to traditional teaching methods(Prabakaran and Saravanakumar 2020).

By leveraging the affordances of VR technology, educational platforms can offer students the flexibility to tailor their learning experiences to individual preferences and engage in experiential learning opportunities (Radha *et al.* 2020). VR technology seamlessly intertwines with basketball instruction, offering an immersive platform that transcends traditional teaching modalities. Moreover, the interactive nature of VR technology fosters active participation and experiential learning, this active engagement not only enhances skill acquisition but also cultivates a deeper understanding of basketball tactics. Thus, the versatility of VR technology allows for personalized learning experiences tailored to individual skill levels and learning preferences, reinforcing fundamental skills while also challenging themselves with more advanced concepts.

As a result, the significance of VR-assisted teaching transcends conventional perceptions, particularly underscored in times of unprecedented challenges such as the COVID-19 pandemic. VR technology offers a resilient solution that enables continuous learning even despite interruptions to traditional educational paradigms. The convergence of VR-assisted teaching and non-face-to-face instructional methods heralds a paradigm shift in educational pedagogy, providing unprecedented opportunities to enhance student learning outcomes. Through an exploration of these emerging methodologies, this study aims to contribute to the advancement of educational practices and the cultivation of optimal learning environments.

Figure 1. "Example A" of user perspective of VR basketball skills application.

2 METHODOLOGY

Research Design: The study will integrate a hybrid methodology, amalgamating action research principles with cutting-edge virtual reality (VR) technology to craft an immersive instructional milieu focused on enhancing comprehension of basketball strategies and proficiencies. During the developmental phase, we encountered a quandary regarding the suitability of augmented reality (AR) or VR for our basketball-assisted teaching application. Upon careful examination, it became evident that VR offers distinct advantages over AR, particularly in its capacity to accurately represent the size and scale of characters engaged in demonstrations. This attribute renders VR a superior choice for our application, ensuring a more authentic and effective learning experience for users.

Figure 2. "Example B" of user perspective of VR basketball skills application.

Literature Review: A comprehensive review of existing virtual reality, and its applications in education will be conducted, with a particular focus on immersive learning environments. Identifying gaps and trends in the research will provide potential prospects for remote education, especially in scenarios where future pandemics may occur. Such virtual learning environments can offer a safe and effective means of learning.

Research Questions and Hypotheses:

- What is the acceptance of immersive teaching among elementary and middle school students?

- What impact does immersive teaching have on students' physical and mental well-being?
- Are students interested in or confident about immersive teaching?
- What is the effectiveness of immersive teaching?

Clear research questions and hypotheses will be formulated based on the identified gaps in the literature review.

Sampling and Participants: Due to the potential for further refinement of the overall application, we anticipate conducting research with our intended target audience, elementary and high school students, in the next phase. Currently, we enlisted college students from National Formosa University of Science and Technology as our subjects. Our selection process primarily focused on students with a limited familiarity with basketball but a keen interest in the sport. Additionally, two VR professionals were invited to assess the feasibility of this technology. They include a master's student from the Multimedia Design Department at National Formosa University of Science and Technology, as well as an associate professor specializing in VR digital content design within the same department. Through their evaluations, insights into the potential development and areas for improvement of this application have been provided, guiding our technical advancements. They will fill out questionnaires after experiencing immersive teaching to assess their acceptance, impact, interest, and confidence in this teaching method. Purposeful sampling techniques will be employed to ensure representation of diverse viewpoints. Currently, the emphasis of our spot checks lies in involving university students without prior basketball knowledge in assessments of their receptivity to this virtual learning method, while also identifying potential compatibility issues. Through these assessments, we aim to gain insights into how to improve the content to successfully integrate this technology into teaching for elementary and secondary students.

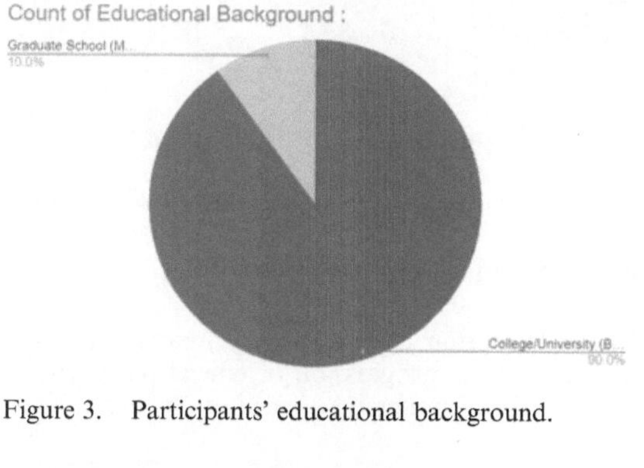

Figure 3. Participants' educational background.

Age	Gender	Educational Background :	Basketball Experience :
21	Male	College/University (Bachelor's Degree)	Beginner
20	Male	College/University (Bachelor's Degree)	Beginner
20	Male	College/University (Bachelor's Degree)	Intermediate
21	Male	College/University (Bachelor's Degree)	Intermediate
21	Male	College/University (Bachelor's Degree)	Beginner
32	Male	Graduate School (Master's Degree)	Beginner
22	Female	College/University (Bachelor's Degree)	Beginner
21	Male	College/University (Bachelor's Degree)	Intermediate
20	Male	College/University (Bachelor's Degree)	Intermediate
22	Male	College/University (Bachelor's Degree)	Beginner

Figure 4. Participants' information.

2.1 *Data collection*

The current phase of the study is focused on assessing the impact of immersive teaching on student learning outcomes. Employing a mixed-methods approach, we combine interviews and questionnaire surveys to analyze two primary categories: basketball skills and virtual reality cognition. These categories are chosen to evaluate whether participants benefit from immersive learning environments.

Interviews will be conducted to delve into participants' understanding of basketball tactics and techniques, as well as their perceptions of immersive teaching. We have designed questions to evaluate students' understanding of overall perceptions of this teaching method before and after experiencing immersive teaching. This portion of the data will be collected through qualitative interviews. (Jin-Ding Lin, Chia-Feng Yen, Mei-Hua Chen. 2005)

Additionally, participants assess their familiarity with virtual reality technology and their perceptions of its potential applications in education. The questionnaire design aims to measure their comfort level with virtual reality, perceived usefulness, and willingness to engage in learning based on virtual reality.

2.2 *Analysis focusing on adolescents*

In future analyses, we will focus on adolescents as the primary research subjects. We plan to recruit adolescent participants and categorize them based on their age and levels of experience with basketball and virtual reality technology. Through such analysis, we will gain deeper insights into the impact of immersive teaching on students of different ages and backgrounds, and explore its potential applications in educational practices.

The data collection items are now as follows:

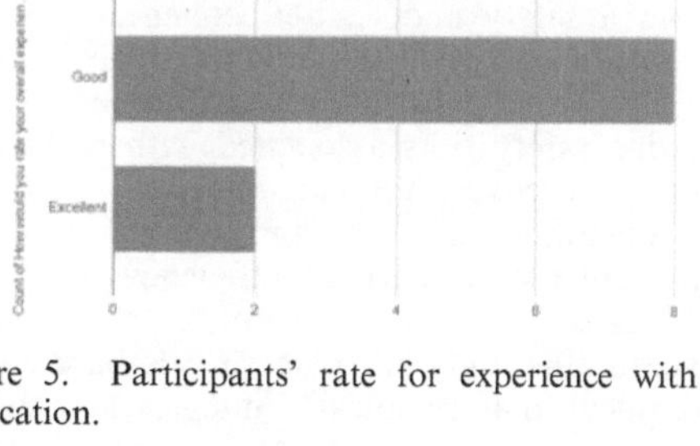

Figure 5. Participants' rate for experience with the application.

Surveys and Questionnaires: Collect quantitative data on participants' fundamental demographic information, and experiences with immersive teaching and basketball learning.

Observations: Conduct observational studies in educational settings implementing virtual reality and immersive learning environments, recording interactions, behaviors, and levels of engagement.

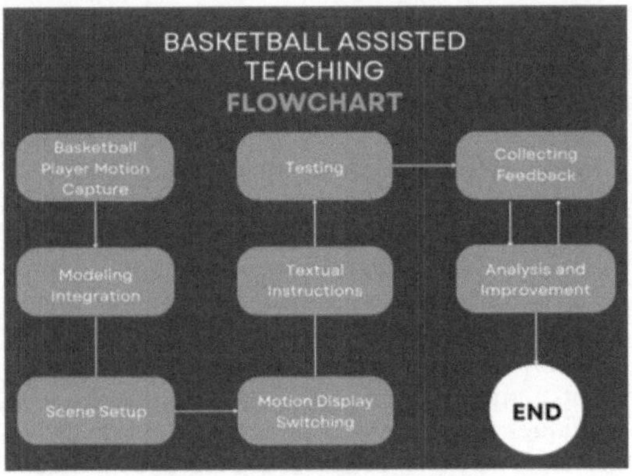

Figure 6. Tutorial software development process.

3 RESULTS

This study presents professional basketball player states captured using motion capture technology. Utilizing virtual reality (VR) technology, users can immerse themselves in a virtual environment and experience these states, providing an experience closely resembling actual gameplay.

Figure 7. Participants' feedback for the experience.

From a safety and cost-effectiveness perspective, this research offers a secure platform for users to closely observe basketball techniques in VR, mitigating safety risks associated with real-world practice. Moreover, VR can alleviate space and time constraints inherent in physical training, thereby enhancing learning efficiency and cost-effectiveness.

Overall, this study introduces a VR-based work environment that seamlessly integrates with the real world. By leveraging VR, flat images are transformed into three-dimensional visions, enabling users to immerse themselves in a vivid and realistic virtual reality environment.

In the first phase, participants showed great enthusiasm and engagement. However, some challenges emerged due to their relative unfamiliarity with virtual reality (VR) experiences or the visual instability caused by the rapid movement of VR equipment. During the instructional process, three participants reported mild dizziness, while others expressed concerns about the clarity of text explanations on the display board. Addressing these issues is crucial for improving visual stability in immersive learning experiences.

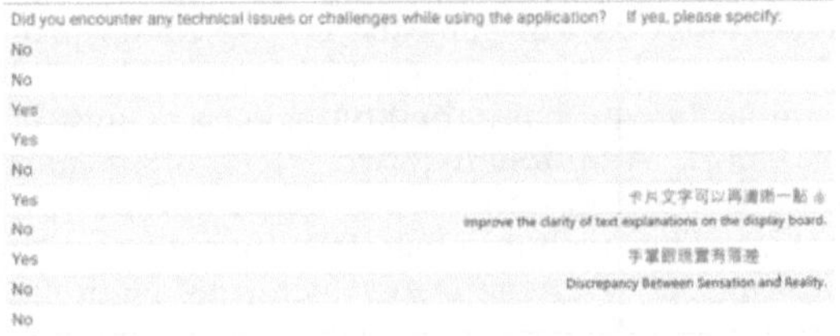

Figure 8. Participants' feedback for the experience.

Despite these challenges, overall interest in immersive learning was high among university students, with over half expressing a desire for increased exposure to such programs in the future.

Concerning teaching effectiveness, although there may be initial challenges in adaptation, there is a noticeable improvement in students' learning attitudes. Participants in immersive learning demonstrate higher precision in executing specific basketball techniques. Future enhancements will focus on refining instructional techniques, such as incorporating narrations with text or refining interface designs to emphasize critical aspects.

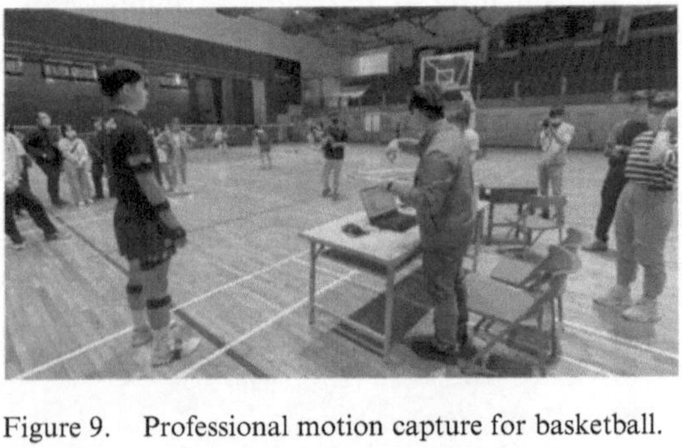

Figure 9. Professional motion capture for basketball.

4 CONCLUSION

The culmination of this inquiry marks a pivotal moment in the progression of education, particularly within immersive learning spheres. This investigation delved into the exploration of virtual learning environments as efficacious educational tools, employing a meticulous blend of methodological rigor and data analysis.

In contrast to AR (Augmented Reality) technology, which may fall short in accurately portraying basketball player's actual height ratios and other crucial aspects, this study focuses on the potential of VR (Virtual Reality) technology to provide a more authentic and effective learning experience for students learning in basketball education. From the findings garnered, profound insights have surfaced regarding the inclination of students towards leveraging metaverse technology for basketball instruction, alongside their receptiveness to this innovative approach. This study presents a substantial stride towards the integration of metaverse technology into basketball pedagogy, elucidating its transformative potential. It underscores the resilience and adaptability inherent in contemporary educational methodologies, especially amid adversities such as the COVID-19 pandemic. By enabling students to partake in self-directed learning, even amidst unprecedented disruptions, this research fosters empowerment.

Nevertheless, it is imperative to recognize that while metaverse technology holds considerable promise for revolutionizing basketball education, there exist notable hurdles to overcome. The technology primarily focuses on representation, accentuating the visual and experiential facets of virtual environments. However, the current infrastructure suffers from limitations necessitating significant refinement before achieving widespread adoption. Thus, the allocation of time and resources becomes indispensable in realizing the complete implementation and efficacy of metaverse technology in basketball education.

This study scrutinizes the inclination of students towards utilizing metaverse technology for basketball tactics, alongside the influencing factors. To achieve the above purpose, the conceptual model in this paper is based on the VR model. By reducing the price variable and expanding the attitude variable, a unique causal path relationship is obtained. Proposing strategies to enhance pedagogical approaches by amalgamating physical education with metaverse technology in forthcoming instructional endeavors. This investigation illuminates the potential of metaverse technology to revolutionize sports education, offering invaluable guidance for future research and educational endeavors.

REFERENCES

Carroll N., Sadowski A., Laila A., Hruska V., Nixon M., Ma D. W. (2020). The impact of COVID-19 on health behavior, stress, financial and food security among middle to high income Canadian families with young children. *Nutrients* 12:2352. 10.3390/nu12082352 [PMC free article] [PubMed] [CrossRef] [Google Scholar]

Du *et al.*, (2020). *DepthLab: Real-time 3D Interaction with Depth Maps for Mobile Augmented Reality*, UIST '20: Proceedings of the 33rd Annual ACM Symposium on User Interface Software and Technology, October 2020, Pages 829–843, https://doi.org/10.1145/3379337.3415881

Jin-Ding Lin, Chia-Feng Yen, Mei-Hua Chen. (2005). Qualitative Research Methods: Analysis of Interview Modes and Implementation Steps. *Journal of Disability Research*, 3(2), 122–136. https://doi.org/10.30072/JDR.200506.0005

Prabakaran B. and Saravanakumar A R. (2020). E-Content module is enhancing the achievement and retention ability in mathematics among high school students. *Wesleyan Journal of Research*, Vol. 13, No. 45 (December 2020).

Radha R., Mahalakshmi K., Dr. Sathish Kumar V., Dr. Saravanakumar A. R. (2020). E-learning during Lockdown of Covid-19 pandemic: A global perspective, *International Journal of Control and Automation*, Vol. 13, No. 4, pp. 1088–1099

Tan Y., Xu W., Li S., and Chen K. (2022). *Augmented and Virtual Reality (AR/VR) for Education and Training in the AEC Industry: A Systematic Review of Research and Applications. Buildings* 2022, 12(10), 1529; https://doi.org/10.3390/buildings12101529.

UNESCO. (2020b). *UNESCO Rallies International Organizations, Civil Society and Private Sector Partners in a Broad Coalition to Ensure #LearningNeverStops.* UNESCO. https://en.unesco.org/news/unesco-rallies-international-organizations-civil-society-and-private-sector-partners-broad

System Innovation for an Artificial Intelligence Era – Lam et al. (Eds)

Sustinnovate or perish: Principles for sustainable innovation

Ingrid Wiederkehr, Denis Tissen, Christian Koldewey & Roman Dumitrescu
Heinz Nixdorf Institute, University of Paderborn, Fürstenallee, Paderborn, Germany

ABSTRACT: Technical systems are becoming increasingly complex. Today, this complexity is increased by the importance of sustainability. Modern companies are struggling to integrate sustainability into their product portfolios. There are two main reasons for this: First, companies lack general design knowledge regarding sustainability and second, they do not know how to integrate design knowledge into the innovation process. Against this background, this paper presents a novel set of sustainability-induced innovation principles designed to support the innovation process and enable companies to integrate sustainability into their product portfolios. The development of these principles followed a cyclical approach, which started with a grey literature review: For this, sustainability reports of sustainability pioneers are scanned. The result of this review was a list of sustainable design options. Next, these design options were abstracted, aggregated, and consolidated into sustainability-induced innovation principles. These innovation principles were then categorized comprehensively documented. A total of 28 sustainability-induced innovation principles were identified. These innovation principles provide a valuable contribution to research and industry: For research, current knowledge is systematically analyzed, and structured in a comprehensible way. In addition, this structured overview of knowledge can be continuously expanded. At the same time, the structured overview supports the innovation process of companies. Companies can be inspired by the sustainability-induced innovation principles and the associated examples and transfer the principles to their own product portfolio.

Keywords: Sustainable Systems, Design Principles, Systems Engineering, Green Engineering, Eco-design

1 INTRODUCTION

Developing technical systems is a complex endeavor (H. Simon 1962). In the past, developing systems was subject to the conflicting priorities of time, cost, and quality (J. Pollack *et al.* 2017). Today, a new factor has emerged: companies are driven by various forces, such as new customer needs or new laws and regulations, to bring innovative and sustainable products to the market (I. Wiederkehr *et al.* 2023). In this context, many companies are integrating the concept of eco-design into their design processes (C. Villamil 2018). Eco-design is the next level of green design, as it goes beyond individual materials and focuses on the entire product's lifecycle (K. Shiels 2022). According to Pigosso *et al.*, eco-design is a proactive approach considering environmental issues in the product development process to minimize environmental impacts throughout the product's lifecycle (D. C. A. Pigosso *et al.* 2015). Charter and Tischner understand the result of eco-design as follows: "*Sustainable solutions are products, [. . .] or system changes that minimize negative and maximize positive sustainability impacts – economic, environmental, social and ethical – throughout and beyond the lifecycle of existing products or solutions while fulfilling acceptable societal demands/needs* "(M. Charter *et al.* 2001). However, simultaneously integrating the environmental, economic, and social dimensions into product development is challenging (F. Paulson *et al.* 2015). Bey *et al.* mention that searching for environmental impact information like materials, alternative components, or manufacturing processes requires too much expertise (N. Bey *et al.* 2013). Paulsen refers to challenges such as the lack of tools to achieve goals at the system and success level, the lack of expertise to implement eco-design, the lack of environmental (or sustainability) oriented requirements in product specifications, and the lack of information on environmental impacts (F. Paulsen 2018).

In summary, there is a lack of design knowledge that shows companies what eco-innovation principles are and how to implement them. Innovation principles are a practical approach to solving these challenges, representing proven design knowledge in solution patterns. Innovation principles generate creative ideas for technical systems, enhance their valuable properties, and eliminate undesirable ones (O. Gassmann *et al.* 2013). Against this background, this article aims to answer the research question: "***How can eco-design specific***

DOI: 10.1201/9781003514831-40

design knowledge be identified, presented in an understandable and structured way, and utilized for the innovation process?**"**. The paper is organized as follows to answer this question: Section 2 outlines the conceptual framework for the design knowledge of eco-design principles. Section 3 presents the methodology used to answer the research question. Section 4 presents the results in the form of cataloged and documented principles for sustainable innovation. Section 5 provides a concluding summary. It addresses contributions to theory and practice and shows limitations and recommendations for future research.

2 CONCEPTUAL BACKGROUND FOR ECO-DESIGN-SPECIFIC DESIGN KNOWLEDGE

As described in section 1, eco-design goes beyond green design: it considers the entire life cycle (see section 1). According to Keivanpour, eco-design practices can be divided into three categories along the life cycle: practices that consider the use phase, those that focus on the end-of-life phase, and life cycle-oriented practices (P. Keivanpour 2022). The author relates these practices to the material, process, and product perspective. For each perspective, the author suggests characteristics that are important in the innovation process, e.g.:

- Material: Clean material, use of material to facilitate recycling, use of recycled material
- Process: Clean production, reduce consumption, improve packaging
- Product: Simplicity, easy access to basics, improved functionality (P. Keivanpour 2022)

The examples show that these categories provide a way to structure design knowledge in an understandable way. Cherifi *et al.* also work with characteristics for eco-design. For this purpose, the authors use *eco-innovation parameters to improve,* such as "material consumption", "energy consumption", "releases", "use parameters" and "ownership of eco-design". ." In addition, they propose 39 other *engineering parameters* such as "weight of moving object" or "energy consumption of moving object". The combination of these parameters provides the foundation for using the TRIZ method, thus ensuring that the eco-design character is considered. (A. Cherifi 2015). TRIZ aims to systematically find creative solutions for technical problems using general principles and patterns derived from a comprehensive analysis of patents (L. Schulyak 2002). Cherifi *et al.* use a matrix to evaluate the relationship between ecological requirements and technical goals. For example, a product's strength can be improved without increasing material consumption. **Accordingly, the methodology for innovative eco-design based on TRIZ provides essential information on identifying design knowledge.**

For concrete application, Pigosso *et al.* have defined ten golden rules for integrating environmental aspects into product development. These ten rules promote sustainable product design by reducing harmful substances, minimizing energy and resource consumption, selecting durable and lightweight materials, and focusing on reparability and durability. They emphasize using simple, recycled materials, an easily upgraded, repaired, and recycled design, and the choice of assembly methods aligned with the product's life cycle to protect the environment. The guidelines are presented below (D.C.A. Pigosso):

- I: Do not use toxic substances and utilize closed loops for necessary but toxic ones.
- II: Minimize energy and resource consumption in the production phase and transport through improved housekeeping.
- III: Use structural features and high-quality materials to minimize weight in products. If such choices do not interfere with necessary flexibility, impact strength, or other functional priorities.
- IV: Minimize energy and resource consumption in the usage phase, especially for products with the most significant aspects in the usage phase.
- V: Promote repair and upgrading, especially for system-dependent products. (e.g., cell phones, computers, and CD players).
- VI: Promote long life, especially for products with significant environmental aspects outside the usage phase.
- VII: Invest in better materials, surface treatments, or structural arrangements to protect products from dirt, corrosion, and wear, ensuring reduced maintenance and longer product life.
- VIII: Prearrange upgrading, repair, and recycling through access ability, labeling, modules, breaking points, and manuals.
- IX: Promote upgrading, repair, and recycling by using few, simple, recycled, not blended materials and no alloys.
- X: Use as few joining elements as possible and use screws, adhesives, welding, snap fits, geometric locking, etc., according to the life cycle scenario (D.C.A. Pigosso).

The rules are generic recommendations for implementing eco-design-specific design know-ledge in innovation.

The presented concepts provide a foundation for eco-design-specific design knowledge and contribute to answering the research question. For example, Keivanpou's categories show how design knowledge can be structured but not how these can be used in innovation. The approach of Cherifi *et al.* provides essential insights into how design knowledge can be identified and abstracted into generic principles. Pigosso *et al.* ten rules are the only approach that shows how design knowledge can be used for the innovation process, but a structured overview is lacking. Therefore, the presented concepts do not fully answer the research question, which means action is needed.

3 METHODOLOGY

The methodology consists of a cyclical iterative procedure. Here, the two generic phases of pattern identification and documentation are addressed according to Kohls and Scheiter (C. Kohls *et al.* 2008). They are detailed in the four steps: *identification of design options, abstraction and aggregation of design options, categorization of innovation principles, and documentation of innovation principles.* Figure 1 shows the procedure.

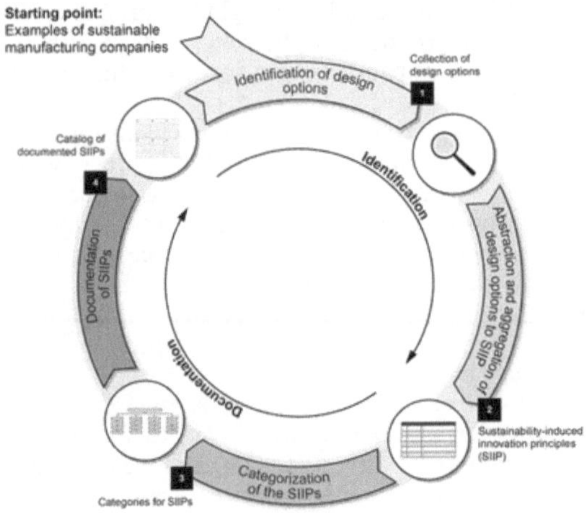

Figure 1. Cyclical procedure for identifying and documenting innovation principles.

The identification of design options is based on Alexander. According to the author, design options are identified by observing and analyzing good examples from the real world (C. Alexander 1979). Good examples from the real world are best analyzed by analyzing companies. Therefore, a grey literature review (GLR) following Garousi *et al.* was conducted. Grey literature can be divided into three categories: the first category has a high outlet control and a high credibility. Examples are books, magazines, government reports, and white papers. Annual reports, news articles, presentations, videos, or wiki articles belong to the second category, while blogs or tweets are literature of the third category. The outlet control and credibility are sinking with subsequent categories (V. Garousi *et al.* 2019).

As mentioned in section 3, design options can be identified best by observing and analyzing good examples from the real world. For this research, annual reports from German companies are analyzed. In particular, sustainability-relevant sections such as products, innovations, and strategic goals were examined. The companies to be analyzed in the GLR are chosen based on a study conducted by stern.de (news portal) and Statista (market research institute). The study evaluated the sustainability of German companies. More than 2,000 companies were analyzed, and 13,000 citizens were surveyed. The assessment was based on Environmental, Social, and Corporate Governance (ESG) criteria. None of the companies could influence their participation in the study or their inclusion on the list, which ensured the neutrality and independence of the data collection (Stern.de 2020). For further analysis, only companies that fit the following selection criteria were chosen: *Is the company a manufacturer of capital goods or complex systems? Is it a manufacturing company, or, for example, does it produce technical systems that have a long and useful life?* This ensured that the company was dealing with technical systems and that the company's analysis would contribute to answering the research question. For example, companies in the automotive or electrical industries were evaluated, while companies such as insurance companies were excluded. In total, 51 relevant companies were identified.

Then, the sustainability reports of these companies were examined for best practices. The result is a collection of design knowledge, e.g., "Using up to 20% recycled pellets in production" or "Reusing stamping waste." These best practices were abstracted and integrated to create sustainability-induced innovation principles: First, this involved reducing design knowledge to essentials and then aggregating similar essentials. This step was repeated several times until understandable, company-independent innovation principles were created. In total, 28 principles emerged. The innovation principles were then grouped into seven thematic categories. These categories are essential in the final step of the cyclical process as they provide a clear and organized structure for cataloging. Within the final step, the innovation principles are also documented. For this, cards are used as they have proven to facilitating creative combinations of information or ideas and provide a common basis for understanding and communication (R. Roy *et al.* 2019). The cards contain the category, the name of the innovation principle, a generic description, the link to related innovation principles, and the best practices. Exemplary best practices for each principle show already established solutions that are understandable and make it easier to transfer innovation principles to other products. Figure 2 shows an example of the cyclic procedure.

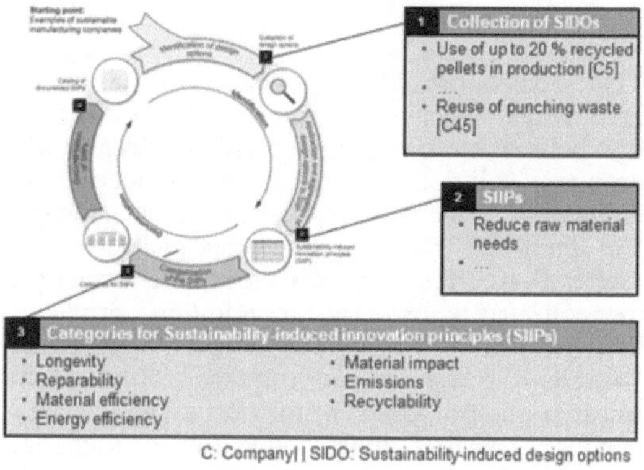

Figure 2. Example of the cyclical procedure.

4 RESULTS

The result from the cyclical procedure is a total of seven categories. These categories are longevity, reparability, material efficiency, energy efficiency, materials impact, emissions, and recyclability. Within these categories, the 28 more detailed innovation principles are subsumed. Table 1 shows the categorized innovation principles.

As stated above, each principle is documented as a card highlighting its most important features. Figure 3 shows the front and back sides of such a card. The front side contains the innovation principle's category and the innovation principle's name. In addition to the description of the principle, there is also a link to related innovation principles. The back side shows (an excerpt of) the best practices from which the innovation principle was derived. For example, the innovation principle "3.2: Use recycled materials" is shown. Related innovation principles are "3.4 Optimize the process", "7.1 Enable reuse of components", "7.3 Avoid mixed materials," and the best practices from which the innovation principles were created. In this case, the card shows Miele and Voith.

The application of the innovation principle is explained below using a fictitious example: Innovation principles represent the link between problems and solutions. Therefore, they represent so-called problems and solutions pairs (F. Buschmann *et al.* 1996). As described in section 1, different forces, such as megatrends, lead companies to consider eco-design aspects.

These forces represent the problem. The following fictitious example is based on the fact that laws have been changed, allowing only repairable household appliances to be put on the market after 2025. A coffee machine manufacturer recognizes the problem that its' coffee machines are currently

Table 1. Catalog of sustainability-induced design options.

Catalog of sustainability-induced design options

- **1 Longevity**
  - 1.1 Emphasize a robust, durable design
  - 1.2 Emphasize a retrofit-compatible design
  - 1.3 Determine the usage period application-oriented
  - 1.4 Increase the usage period
- **2 Reparability**
  - 2.1 Provide care and maintenance instructions
  - 2.2 Ensure availability of spare parts
  - 2.3 Emphasize a repairable design
  - 2.4 Provide a kit for independent repair
- **3 Material efficiency**
  - 3.1 Use renewable raw materials
  - 3.2 Use recycled materials
  - 3.3 Reduce raw material needs
  - 3.4 Optimize the process
  - 3.5 Miniaturize of the product
- **4 Energy efficiency**
  - 4.1 Reduce consumption during use
  - 4.2 Reduce consumption during production
  - 4.3 Use renewable energies
- **5 Material impact**
  - 5.1 Avoid emissions into the environment
  - 5.2 Avoid use of hazardous substances
  - 5.3 Avoid contact with hazardous substances
  - 5.3 Substitute hazardous substances
- **6 Emissions**
  - 6.1 Minimize emission
  - 6.2 Equalize emission
  - 6.3 Avoid emission
- **7 Recyclability**
  - 7.1 Enable reuse of components
  - 7.2 Recycle for a second life
  - 7.3 Avoid mixed materials
  - 7.4 Enable the take-back of old appliances
  - 7.5 Avoid hazardous waste

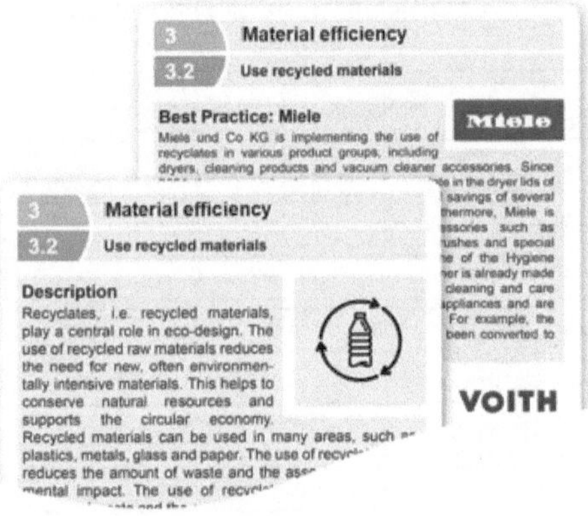

Figure 3. Documented sustainability-induced innovation principles with the best practices from Miele (Miele 2021), (Miele 2023) and Voith (Voith 2023).

not repairable because the components cannot be separated after production, which means that repairability cannot be guaranteed. An appropriate innovation principle to address this challenge is "2.3: Emphasize a repairable design," which belongs to the second category of "Repairability." According to this principle, the design should be adapted to allow for the separation of the individual product components without causing damage. This means, for example, that adhesives should not be used. The card also shows related principles, such as providing a repair kit. This is also a solution to the problem described. Here it is: "2.4: Provide a kit for independent repair" – the description explains what materials are available. Of course, there are other related innovation principles to which the process can be applied. Finally, the innovation principles can be used to identify solution ideas. Figure 4 shows the explained process.

5 CONCLUSION

This article presents 28 principles of sustainable innovation. The innovation principles provide the link between an eco-design-specific problem and a proposed solution. Based on best practices, the innovation principles are clearly understandable and intuitively transferable to other problems. The preparation in the form of a set of cards structures the collected knowledge. In addition, providing links to other principles encourages further ideas. Overall, the innovation principles answer the research question posed at the beginning. Our contribution provides added value for research and practice: For research, current knowledge is systematically analyzed and structured understandably. In addition, this structured overview of knowledge can be continuously expanded. This is necessary, for example, due to the continuous further development of technologies or changes in legislation. The catalog is suitable for this type of expansion and provides an initial classification structure through the categories. At the same time, it forms the basis for future innovations and enables companies to integrate sustainability aspects comprehensively and systematically into their systems. In particular, the best practices provide companies with an understanding of how the principle can be implemented. In addition, effective practices make transferring problem-solving combinations to other technical products easier.

The limitations of this article are related to the starting point of the research design. The list of most sustainable companies on which the research is based stems from 2021 and may contain outdated data. Furthermore, only 51 companies were analyzed. Moreover, these are the only companies from Germany that manufacture complex and durable systems or capital goods. In addition, design knowledge from insurance companies, e.g., can

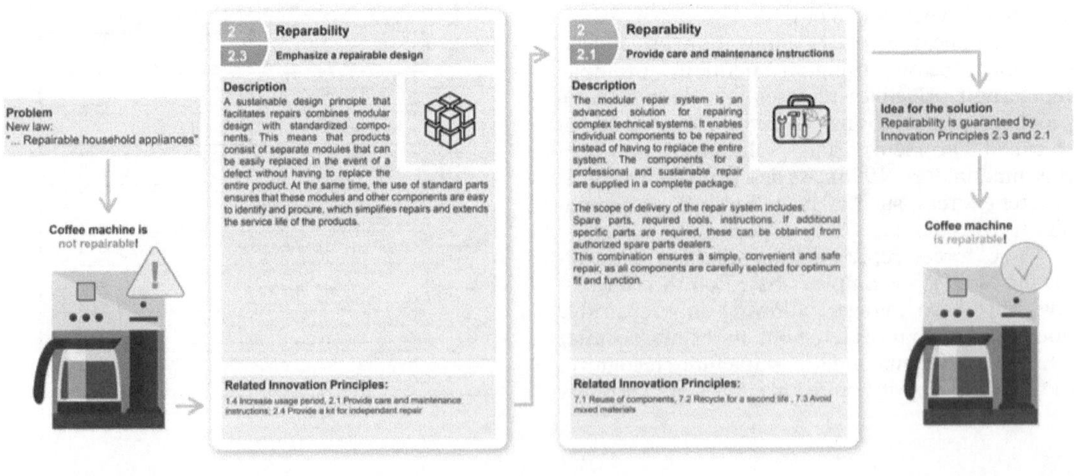

Figure 4. Application of innovation principles.

reveal innovation principles that have not yet been addressed. These considerations also represent the need for future research: First, other sources for identifying design options should be examined, e.g., the search can be extended to Europe. Including other sources for data collection is also possible. Second, a tool for evaluating the applicability of the innovation principles for concrete products is needed. Here, eco-design-specific criteria should be included in the evaluation. Furthermore, companies should consider the implications of implementing the newly identified solution ideas. Exemplary questions could be: *Can the company technically realize the solution idea? Is the implementation of the solution idea economically viable?* And: *How should the implementation be planned? What are the specific requirements resulting from the solution idea?* Third, the innovation principles should be embedded in a structured, methodical approach. This could, for example, include a comprehensive plan for integrating the solution idea.

REFERENCES

Alexander C. (1979). *The Timeless Way of Building.* Oxford University Press, New York.

Bey N., Hauschild M. Z., McAloone T. C. (2013). Drivers and barriers for implementation of environmental strategies in manufacturing companies, *CIRP Annals - Manufacturing Technology*, 62, 43–46.

Charter M., Tischner U. (2001). *Sustainable Solutions, Developing Products and Services for the Future.* Greenleaf Publishing, Sheffield.

Cherifi A., Dubois M., Gardoni M., Tairi A. (2015). Methodology for innovative eco-design based on TRIZ, *Int J Interact Des Manuf* 9:167–175.

Garousi V., Felderer M., Mäntylä M. V. (2019). Guidelines for including grey literature and conducting multivocal literature reviews in software engineering. *Information and Software Technology*, 106, g 101–121.

Gausemeier J., Dumitrescu R., Echterfeld J., Pfänder R., Steffen D., Thielemann F. (2018). *Innovationen für die Märkte vor Morgen - Strategische Planung von Produkten, Dienstleistungen und Geschäftsmodellen.* Carl Hanser Verlag München.

Keivanpour S. (2022). *Approaches, Opportunities, and Challenges for Eco-design 4.0 - A Concise Guide for Practitioners and Students*, Springer Nature Switzerland AG 2022.

Kohls C., Scheiter S. (2008). *The Psychology of Patterns. Proceedings of the 2008 conference on Pattern Languages of Programs (PLoP).* Nashville, Tennessee.

Miele. (2021). *Miele Nachhaltigkeitsbericht 2021.* Accessed 9th April 2024, <https://www.miele.de/media/ex/ce/presseartikel/nachhaltigkeit/miele_nachhaltigkeitsbericht_2021.pdf>.

Miele. (2023). *Miele Nachhaltigkeit Update 2023.* Accessed 9th April 2024, <https://www.miele.com/media/ex/com/Publikationen/miele_Nachhaltigkeits_Update_2023_DEU.pdf>.

Paulson F. (2018). Inclusion of sustainability aspects in product development at manufacturing companies. *Linköping Studies in Science and Technology.* licentiate thesis, no. 1828, LiU-Tryck, Linköping 2018.

Paulson F., Sundin E. (2015). *Challenges and trends within eco-design, Challenges and trends within the Eco-Design International conference on EcoDesign 2015*, Tokyo, December 2–4.

Pigosso D. C. A., McAloone T.C., Rozenfeld H. (2015). Characterization of the state-of-the-art and identification of main trends for ecodesign tools and methods: classifying three decades of research and implementation. *Indian Institute of Science Journal*, 94 (4), 405–427.

Pollack J., Helm J., and Adler D. (2018). *International Journal of Managing Projects in Business* Vol. 11 No. 2, 2018 pp. 527–547.

Roy R., Warren J. P. (2019). *Design Group, School of Engineering and Innovation*, The Open University, Milton Keynes, MK7 6AA, UK.

Schulyak L. (2002). *TRIZ Keys to Technical Innovation by Genrich Altschuller*, Technical Innovation Center, Inc.

Shiels K., LEADING INNOVATION. (2022). *Creative Commons Attribution-NonCommercial-ShareAlike 4.0 International License.*

Simon H. (Dec. 12, 1962). *The Architecture of Complexity, 1962. Proceedings of the American Philosophical Society*, Vol. 106, No. 6. , pp. 467–482.

Stern.de. (2020). *Stern.de – Das sind die 200 nachhaltigsten Unternehmen Deutschlands.* Accessed 9th April 2024, <https://www.stern.de/wirtschaft/news/stern-siegel- die-200-nachhaltigsten-unternehmen-deutschlands-9514568.html>.

Villamil C., Hallstedt S. I. (2018). Sustainability product portfolio: A review carolina villamil1, sophie I. *Hallstedt2, European Journal of Sustainable Development* 7, 4, 146–158.

Voith. (2023). *Voith Nachhaltigkeitsbericht 2023.* Accessed 9th April 2024, <https://voith.com/corp-de/240408_Nachhaltigkeitsbericht_VOITH_2023_de.pdf>.

Wiederkehr I., Schlegel M., Koldewey C., Rapp S., Albers A., Dumitrescu R. (2023). *Interacting Forces for a Resilient, Future-robust Evolution of Product Portfolios.* Conference: 56th CIRP Conference on Manufacturing Systems, CIRP CMS '23.

System Innovation for an Artificial Intelligence Era – Lam et al. (Eds)

Portals across domains: A data analysis task and documentation canvas for data-driven model-based systems engineering

Denis Tissen, Ingrid Wiederkehr, Ruslan Bernijazov, Christian Koldewey & Roman Dumitrescu
Heinz Nixdorf Institute, University of Paderborn, Fürstenallee, Paderborn, Germany

ABSTRACT: Data-driven technologies and methodologies, such as artificial intelligence (AI) and data analytics (DA), are permeating engineering. Faced with a massive increase in data volumes, companies are trying to use these technologies to analyze various data sources in order to optimize their processes, products, and services. A new paradigm in this context is data-driven model-based systems engineering (DDMBSE), which integrates DA with model-based systems engineering (MBSE). DDMBSE improves handling and optimizing systems (products and services) throughout their lifecycle – ranging from the conceptualization, design, production, use, life extension, and disposal – by analyzing and using data and linking data sources directly to the system model. Close collaboration between MBSE and DA experts is critical, but the differences between the domains in terms of terminology and procedures are challenging. Both domains lack adequate tools to communicate efficiently and effectively. To address this gap, this paper presents a data analysis task and documentation canvas for DDMBSE. The canvas is developed by conducting an interview study with experts from the MBSE and DA domains and analyzing the results. The canvas supports MBSE and DA users with a clear description of objectives, problem definition, required data sources, and data analysis requirements. It also helps to summarize and interpret the results of the data analysis. Utilizing the canvas is supported by an application procedure. The results bridge the gap between MBSE and DA, enabling domain approximation and open novel frontiers for DDMBSE and its future research.

Keywords: advanced systems, interdisciplinary collaboration, data-driven system model, cybernetics, anal-ysis assistance

1 INTRODUCTION

Data-driven technologies and methodologies have played a significant role in the global trend of digitization. In recent years, artificial intelligence (AI) and data analytics (DA) have become indispensable fields that are being integrated from domain to domain. As an indicator, the global volume of the big data analytics market is expected to reach approximately $655 billion by 2029 (Statista 2021a). Organizations are investing their resources to gain knowledge about their processes, products, and services. With the parallel growth of data being generated (predicted to reach 181 zettabytes by 2025 (Statista 2021b), organizations are gaining new opportunities and use cases for data and data analytics. By accessing, aggregating, analyzing, and interpreting data, data analytics provides new insights (T. A. Runkler 2016). Promising applications for DA range from finance to biology, manufacturing, and engineering (K. Vassakis *et al.* 2017). With digitalization, more and more companies are turning to digital engineering processes, including new methods and tools. One of these is model-based systems engineering (MBSE). Using modeling languages, methods, and tools, MBSE focuses on a central system model (L. Delligatti 2013). It represents a real-world system in a simplified form that allows different domains and stakeholders to communicate and collaborate in engineering. Therefore, the MBSE practitioner must be an expert in understanding and communicating with different domains (INCOSE 2023). With the rise of needed data analytics experts in engineering, MBSE practitioners have a new domain to communicate with. DA will become an inherent part of MBSE and must be integrated into it (INCOSE 2021). However, support in the form of methods or tools is scarce, creating a research gap. One approach that fills this gap is data-driven model-based systems engineering (DDMBSE) after D. Tissen *et al.* 2023. It combines data-driven aspects with MBSE and represents the interface between the domains. However, as a new field of research, DDMBSE needs further exploration and supporting artifacts for practice.

Especially the communication and collaboration between model-based systems engineers and data analysts need to be supported. Due to the different backgrounds of the domains, the terminologies used vary, and mapping them is not

 DOI: 10.1201/9781003514831-41

trivial. This paper aims to provide a tool to facilitate communication between users while practicing DDMBSE. Therefore, an interview study was conducted to investigate how model-based systems engineers and data analysts communicate tasks. The study consists of 15 interviews with experts from the fields of data analysis and MBSE. The paper unfolds as follows: The scientific background is presented in section 2 and the research design in section 3. The data analytics task and the documentation canvas for DDMBSE and the application procedure are presented in section 4. Section 5 concludes the paper, highlighting limitations and future research needs.

2 SCIENTIFIC BACKGROUND

To better understand the research gap, it is necessary to look at the domain of DA and DDMBSE. Approaches from both domains are briefly discussed below.

2.1 *Data analytics*

DA is the process of examining data sets to identify patterns, draw conclusions, and make decisions based on the data. It encompasses a variety of techniques and processes, from simple statistical analysis to complex machine learning algorithms. DA is used to identify trends, improve performance, understand customer behavior, and predict future events (T. A. Runkler 2016). There are many approaches to proceed and support DA and its execution. A popular and widely used approach is CRISP-DM (see Figure 1.), which includes six phases, from understanding the domain and data, preprocessing the

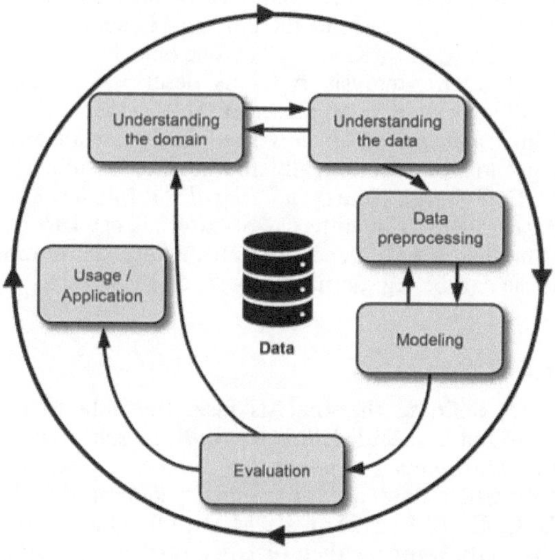

Figure 1. The CRISP-DM Process (adapted from R. Wirth *et al.* 2000).

data, modeling, and evaluation, to the final use of the model (R. Wirth *et al.* 2000). Another approach is knowledge discovery in databases (KDD). KDD consists of several steps, ranging from data selection, preprocessing, and transformation to data mining (application of pattern recognition algorithms) and finally interpretation/evaluation of the results. The focus is on discovering (previously unknown) patterns and relationships in the data (U. Fayyad *et al.* 1996). In the case of supporting tools, A. Kühn *et al.* 2018 propose an analytics canvas for designing data analysis projects. Key requirements and associated infrastructure are documented using a graphical representation. While the approach is interdisciplinary, it does not directly address MBSE as the main application domain, but rather as a general tool for multiple domains. M. Panzner *et al.* 2022 address a related problem with a business-to-analytics canvas. It allows the user to transform business use cases into data analysis tasks for further processing and action. However, there is no support for interpreting the results of the data analysis.

2.2 *Data-driven model-based systems engineering*

As a next step in the evolution of MBSE, data-driven model-based systems engineering (DDMBSE) integrates data analytics approaches into classical MBSE. D. Tissen *et al.* 2023 define DDMBSE as focusing on a data-driven system model that continuously integrates data from multiple data sources. While classical MBSE already uses models to support requirements engineering, domain design, V&V activities, production, system use, and disposal, it now also uses data (INCOSE 2007, INCOSE 2023). D. Tissen *et al.* 2023 describe that standardized interfaces are a key element in evolving a system model into a data-driven system model. With DDMBSE, engineers can gain more insights into the system under development and manage the complexity and processes behind it more effectively and efficiently. As a new approach, DDMBSE faces the challenge of scarce research and a small number of supporting methods and tools.

3 RESEARCH METHODOLOGY

The research methodology in this paper is an interview study. 15 experts and operators in context of MBSE and data analytics were interviewed. The initial data collection was conducted by following the recommendations of M. Saunders *et al.* 2019 and K. M. Eisenhardt 1989 and used to prepare, perform and document the semi-structured interviews. The positive answer rate of interviewees was 15 after contacting 24 persons with an MBSE and/or data analytics background. The anonymized overview of the interviewees is shown in Table 1.

Table 1. Overview of the anonymized interviewees.

Interviewee	Current position	Experience	Interviewee domain	Operational sector
(A)	Head of department	ca. 6 Years	MBSE & Data Analytics	Research
(B)	Senior systems engineer	ca. 10 Years	MBSE	Tool Vendor
(C)	Team leader / consultant	ca. 11 Years	MBSE	Tool Vendor
(D)	Chair member of association	ca. 4 Years	MBSE	Research
(E)	Head of department	ca. 16 Years	Data Analytics	Industry
(F)	Senior systems engineer	ca. 11 Years	MBSE	Industry
(G)	Head of department	ca. 18 Years	MBSE	Tool Vendor
(H)	Senior systems engineer	ca. 11 Years	MBSE & Data Analytics	Industry
(I)	Researcher	ca. 2 Years	MBSE & Data Analytics	Research
(J)	Professor	ca. 16 Years	MBSE	Research
(K)	Senior systems engineer	ca. 7 Years	MBSE	Consulting
(L)	Senior systems engineer	ca. 32 Years	MBSE	Research
(M)	Innovation manager	ca. 10 Years	MBSE	Consulting
(N)	Professor	ca. 13 Years	Data Analytics	Research
(O)	Researcher	ca. 5 Years	Data Analytics	Research

Seperated into six sections, this paper focuses on the sixth section of the interview guideline the "task of a data analysis" with the following input for the interviewees and questions:

Input: *Imagine you were to commission a data analysis. The goal: A system model is to be optimized through a data analysis of system model-internal and system model-external data.*

Questions:

1) *Who would carry out the data analysis(s)? Which departments would be involved?*
2) *How would you set up a data analysis task? What core content would it require?*
3) *Who would evaluate the data analysis results?*
4) *How would a data analysis result be documented? What content should it contain?*

Each interview was held online via Microsoft Teams within a duration of 60 Minutes per person and led by one interviewer. A word-by-word transcription was used to transcribe and later analyze the online recordings. For this a thematic clustering after V. Braun *et al.* 2006, 2012 was conducted. The interview transcripts were coded inductively by two researchers. The code systems of each researcher were compared after the first interview, discussed, and then used to form a general code system for all interviews. After the second interview the procedure was repeated, and differences discussed. The resulted code system was then used for all interviews as a general code system. The code system allows to search and cluster themes within, forming patterns to answer the research questions above. The themes were structured to build the canvas and gave insights for designing an application procedure.

4 RESULTS

Two main results were derived and developed: an **analytics canvas for data-driven model-based systems engineering** and an **application procedure**. The canvas allows to formulate and define a data analysis task (1), necessary data sources (2) and to document the analysis results (3). The analysis canvas contains 13 areas in these three categories with guiding questions in each area for the user (see Figure 2). The guiding questions support the user to formulate the right content in each area. The categories will be described on the following

4.1 *Task definition (1)*

The top part of the canvas defines the analysis task which should be conducted. For this, four areas are essential to know. First the *problem definition and the context* need to be defined (A, B, C, D, E, J, K, O). The user should define "*Which system is in focus of the analysis?*", "*What problem is to be solved?*" and "*What is the environment and context?*" in here. Second the *objective* is defined (A, B, C, D, F, G, I, K, L, M, N). The objective can be of different analysis types as descriptive, diagnostic, predictive or prescriptive (Gartner 2024). Third, *general constraints* need to be considered, e.g. time, or economically and the later results (A, C, D, F, I, K). Fourth, *additional stakeholders and departments* from internal or external need to be listed (A, I, M). Depending on the task definition these can be e.g. domain experts or customers.

4.2 *Data (2)*

After defining the analysis task, the data to by analyzed is specified. It starts with the selection of the *data sources*, whether it is a system, a person, software, a database, or another system model (A, B, C, D, E, F, H, I, J, K, M, N, O). The data of that data source is then *specified* further regarding the content represented, its context, data format and form, and data quality (B, D, E, I, O). Last the

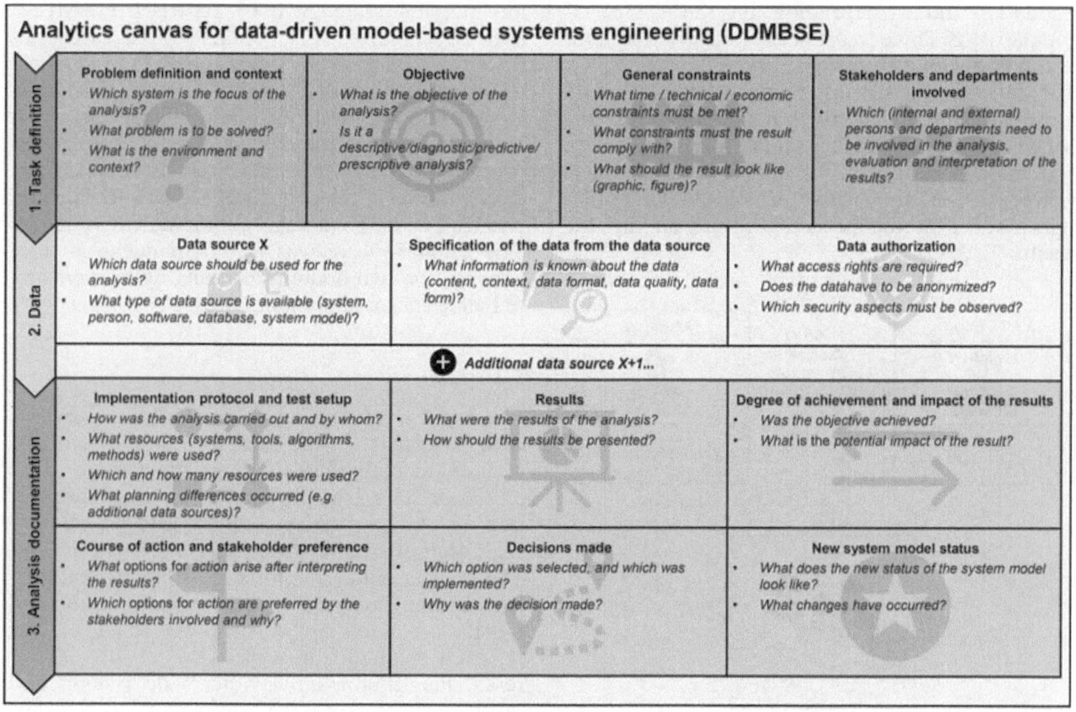

Figure 2. Analytics canvas for DDMBSE.

data *authorization* aspects need to be looked at, whether access is given to the data, if the data needs to be anonymized, or if specific security aspects need to be observed (A, F, M). For each data source, these three areas need to be filled. By this, a clear specification of each data source used is given.

4.3 *Analysis documentation (3)*

The documentation of the results is done in six areas. Starting, the *implementation protocol and test setup* need to be documented (A, E, G, H, I, N, O). Guiding questions here are: *How was the analysis carried out and by whom?, What resources (systems, tools, algorithms, methods) were used?, Which and how many resources were used?, What planning differences occurred (e.g. additional data sources)?* Then the analysis *results* need to be documented in the requested form in combination with the estimated *objective achievement and impact* (A, B, E, F, G, I). Based on this, the *course of action* (*What options for action arise after interpreting the results?*) and *stakeholder preferences* (*Which options for action are preferred by the stakeholders involved and why?*) are listed (E, F, I). When the options for action are reviewed, the *made decisions* need to be documented and the reasoning behind (E, F). The *new system model status* rounds up the third category, by describing the integrated changes (N).

To fill out the analysis canvas, an **application procedure** is shown in Figure 3. The answers of the interviewees were used to understand involved persons/departments and the usual sequences for communication. When asked who was responsible for commissioning data analysis, the picture was mixed: six respondents saw this role as primarily the responsibility of the data analyst (C, E, F, K, N, O). Five respondents emphasized the importance of collaboration between the data analyst and the model-based systems engineer (A, H, I, J, M), while three respondents saw this as the sole responsibility of the model-based systems engineer (B, G, L). There is a clear trend towards the role of the data analyst as the primary responsible party. One respondent felt that, in principle, anyone should be able to perform this task (D). The range of departments involved in this process proved to be very diverse, including the development department (G, E, K), the internal IT department (K, O), and specifically a centralized data analysis department (F). In addition, management levels (M), all departments within the company (D), external consultants, and academic institutions (B) were mentioned. While the execution of the data analysis obtains by the data analyst, seven respondents felt that the evaluation should be performed jointly by the MBSE expert and the data analyst (D, F, G, I, M, N, O). Two respondents felt that either the MBSE expert (D, H) or the data analyst (A, L) should perform the evaluation alone (depending on the degree of complexity). Three respondents favored a joint evaluation with a clear division of labor (B, E, K): The data analyst should create the database and formulate recommendations for action, while the MBSE expert should focus on interpreting the data and making decisions. In addition, three interviewees stressed the importance of involving other stakeholders in the process of interpreting the results (B, J, O).

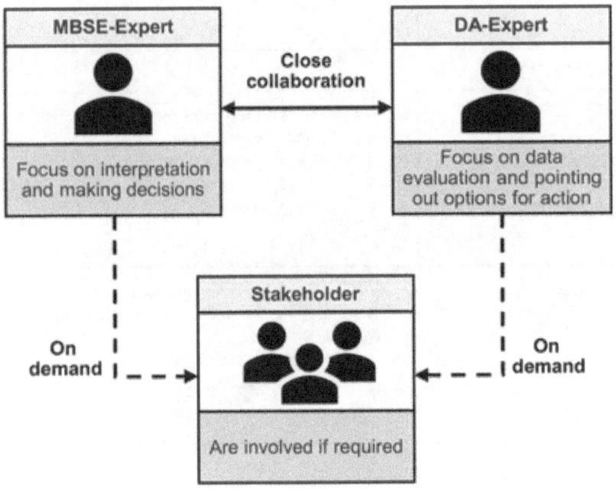

Figure 3. Application procedure for the analytics canvas for DDMBSE.

Thus, three central roles in the evaluation process emerge: the MBSE expert, the data analyst, and other stakeholders. This tripartite division forms the basis for the defined application procedure for interpreting the results. Initially, the MBSE expert and the DA expert work together in close collaboration to fill out the categories of the analysis canvas. If necessary, other stakeholders are involved. When the analysis is done, the results will be evaluated in close collaboration between the MBSE expert and the DA expert. The DA expert hereby has a higher tendency to focus on the data evaluation from the analysis and point out options for action. The MBSE expert has more focus on the interpretation of the results and making decisions based on the options for action. The application procedure is used iteratively starting with the communication between MBSE and DA experts and then involving the stakeholders (A, E, N).

5 DISCUSSION AND CONCLUSION

5.1 *Contribution*

The presented paper describes the derivation of a data analysis task and documentation canvas for data-driven model-based systems engineering and an application procedure. The canvas is a result of an interview study with 15 experts from MBSE and DA. The canvas contains three categories (*task definition, data specification,* and *analysis documentation*) with in total of 13 areas. Each area is supported by guiding questions for the users. With the help of the canvas and the application procedure, MBSE and DA experts can perform data analysis in the context of DDMBSE safe and reliable due to the matching communication terms. The canvas serves as a communication tool between the two domains opening up barriers and reducing misunderstanding.

5.2 *Limitations*

The findings presented are derived from interviews with 15 experts in the areas of model-based systems engineering (MBSE) and data analytics. Due to the domain backgrounds, a different distribution of the participants might lead to another weighting of specific aspects. A more comprehensive understanding might be possible, by conducting further interviews from experts inside and outside of Germany. Also, the bias of the researchers might influence the results. Other researchers might derive other code systems and results. Furthermore, the canvas needs to be tested in practical operation and incorporate possible application improvements. Besides these limitations, the canvas and the application procedure build on a solid database and seem to provide valuable orientation.

5.3 *Future research*

The result of the interview study shows that a close collaboration between MBSE and data analytics experts is possible and needed to perform DDMBSE. Both domains need good communication skills to ease teamwork with further stakeholders and domains. The canvas represents a tool to support this. As a logical construct, the categories and areas of the canvas build a base for a possible software-based solution, allowing to digitalize and automate specific procedures. As an initial aid, it opens the door to further tools for DDMBSE in the future.

REFERENCES

Braun V., Clarke V. (2012). Thematic analysis. In: Cooper, H., Camic, P.M., Long, D.L., Panter, A.T., Rindskopf, D., Sher, K.J. (Eds), *APA Handbook of Research Methods in Psychology, Vol 2: Research Designs: Quantitative, Qualitative, Neuropsychological, and Biological.* American Psychological Association, Washington, pp. 57–71.

Braun V., Clarke V., (2006). Using thematic analysis in psychology. *Qualitative Research in Psychology* 3, pp. 77–101.

Delligatti L. (2013). *SysML Distilled: A Brief Guide to the Systems Modeling Language.* Addison-Wesley Professional, ISBN: 978-0321927866

Eisenhardt, K. M. (1989). Building theories from case study research. *The Academy of Management Review* 14, 532–550.

Fayyad U., Piatetsky-Shapiro G., Smyth P. (1996). *From Data Mining to Knowledge Discovery in Databases. AI Magazine*, 17 (3), 37. DOI: https://doi.org/10.1609/.

Gartner. (2024). *What Is Data and Analytics?* URL: https://www.gartner.com/en/topics/data-and-analytics.

INCOSE, (2007). *Systems engineering vision 2020.* The International Council on Systems Engineering (INCOSE).

INCOSE, (2021). *Systems engineering vision 2035 – Engineering solutions for a better world. The International Council on Systems Engineering (INCOSE).*

INCOSE, (2023). *Systems Engineering Handbook – A guide for system life cycle processes and activities – Fifth edition.* John Wiley & Sons Ltd., Hoboken, New Jersey.

Kühn A., Joppen R., Reinhart F., Röltgen D., von Enzberg S., Dumitrescu R. (2018). *Analytics Canvas – A Framework for the Design and Specification of Data Analytics Projects, 28th CIRP Design Conference,* May 2018, Nantes, France; DOI: https://doi.org/10.1016/j.procir.2018.02.031.

Panzner M., Meyer M., von Enzberg S., Dumitrescu R., (2022). *Business-to-Analytics Canvas – Translation of Product Planning-Related Business Use Cases into Concrete Data Analytics Tasks. 32nd CIRP Design Conference*, Procedia CIRP, Volume 109, 2022, Pages 580–585, DOI: https://doi.org/10.1016/j.procir.2022.05.298.

Runkler T. A., (2016). *Data Analytics – Models and Algorithms for Intelligent Data Analysis, 2nd Edition.* Springer Vieweg Wiesbaden. DOI:https://doi.org/10.1007/978-3-658-14075-5.

Saunders M., Lewis P., Thornhill A. (2019). *Research Methods for Business Students. 8th Edn.* London, New York, Ann Arbor: Pearson; ProQuest Ebook Central, Harlow, England.

Statista. (2021a). *Volume of the global big data analytics market 2021 to 2029 (in billion US dollars).* URL: https://de.statista.com/statistik/daten/studie/1420253/umfrage/big-data-analytics-markt-weltweit/.

Statista. (2021b). *Volume of data/information created, captured, copied, and consumed worldwide from 2010 to 2020*, with forecasts from 2021 to 2025 (in zettabytes). URL: https://www.statista.com/statistics/871513/worldwide-data-created/.

Tissen D., Wiederkehr I., Koldewey C., Dumitrescu R. (2023). Exploring data-driven model-based systems engineering: a systematic literature review, *IEEE International Conference on Technology Management, Operations and Decisions (ICTMOD)*, 22–24 November 2023, Rabat, Marocco, DOI: 10.1109/ICTMOD59086.2023.10438129.

Vassakis K., Petrakis E., Kopanakis I. (2019). *Big Data Analytics: Applications, Prospects and Challenges.* In: G. Skourletopoulos, G. Mastorakis, C. X. Mavromoustakis, C. Dobre, E. Pallis (Eds), *2017. Mobile Big Data – Lecture Notes on Data Engineering and Communications Technologies.* Vol 10. Springer Cham. DOI: https://doi.org/10.1007/978-3-319-67925-9_1

Wirth R., Hipp J. (2000). *CRISP-DM: Towards a Standard Process Model for Data Mining. Proceedings of the 4th International Conference on the Practical Applications of Knowledge Discovery and Data Mining.* 11–13 April 2000, Manchester, UK, ISBN: 1902426088

Research on the application of generative artificial intelligence to the design process of commercial vehicle rear lamp

Tz-Ying Lin & Yu-Ju Lin
Department of Industrial Design, National Yunlin University of Science and Technology, Douliu, Yunlin, Taiwan
Department of Industrial Design, College of Design, Yunlin, Taiwan

ABSTRACT: Taiwan is the world's number one country in terms of productivity of car lights, and rear lamps are the essential items for the car bodies. Their functions are not only warning lighting, but also need to balance safety, visibility, and durability. Many car manufacturers are committed to integrating corporate brand features and the most advanced technology with car lights and conducting series designs to enhance brand competitiveness. However, since the lights of commercial vehicles do not have the restrictions of vehicle body sheet metal, the same type of lights can be installed on different models, and their application is wide. Therefore, this study takes commercial vehicle rear lamps as the research target. First, consult with the lighting experts from Taiwan's largest commercial vehicle lighting company to confirm the appropriate timing for GenAI introduction and construct a GenAI design process based on different GenAI software characteristics. After that, look for 3 car lighting designers and 3 designers from non-car lighting fields to collaboratively use GenAI software to conduct process testing. Finally, look for 10 entry-level designers to verify the GenAI design process to confirm the feasibility and effectiveness of the process. The results of this study confirm that there are three optimal timings to introduce GenAI, and there is a specific GenAI software that is suitable for each timing. In addition, this research process is very suitable for some enterprises to train their new staff and can shorten their development time for the enterprises.

1 RESEARCH BACKGROUND AND MOTIVATION

In the context of rapidly changing market demands, major enterprises are seeking ways to effectively shorten the design and development cycle without compromising product quality. Design creativity is a crucial key to helping products stand out in competitive markets. GenAI serves as a valuable assistant for inspiring, listing, and considering ideas, establishing its extensive knowledge base [1]. Through Generative AI (GenAI), high-quality and commercially valuable images can be quickly obtained by inputting keywords (prompts), significantly enhancing the efficiency of designers in the initial stages of product conceptualization.

This study utilizes front-end tools implemented with algorithms combining natural language processing (NLP) and generative adversarial networks (GAN) based on deep learning. By increasing an "intermediate layer" (Figure 1) [2], the operability, generation speed, and image quality of the results are enhanced. Furthermore, the study integrates these tools into the design process of commercial vehicle headlights, aiming to identify the optimal timing and application of GenAI software.

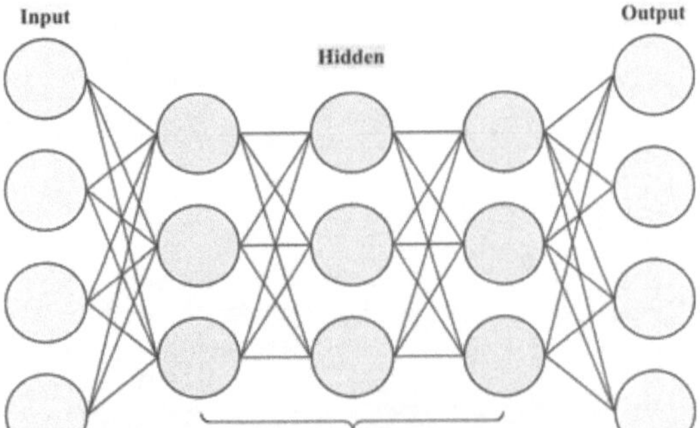

Figure 1. GenAI Deep Learning (DL) process diagram.

2 RESEARCH METHODS AND PROCESSES

This study employed a combination of in-depth interviews and experimental methods, utilizing guided interviews to conduct research, followed by Content analysis for data organization. The research was divided into three stages:

DOI: 10.1201/9781003514831-42

2.1 *Stage one: Consultation with automotive lighting experts*

The traditional design process is labor-intensive, and artificial intelligence can fundamentally change this situation. In the early stages of this study, in-depth interviews were conducted with automotive lighting experts from Taiwan's largest commercial vehicle lighting company, aiming to identify the optimal timing for the introduction of GenAI and selecting the most iconic global automotive lighting brands and products (Table 1), which were then designed as auxiliary experimental cards for automotive lighting (Figure 2).

Table 1. Graphics card generation process chart.

(1) Understand the various functions of the product	(3) Design auxiliary experimental cards
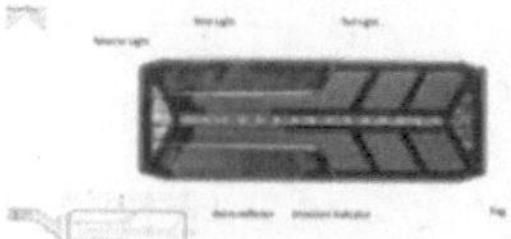	
(2) Re-draw the extracted driving function 	

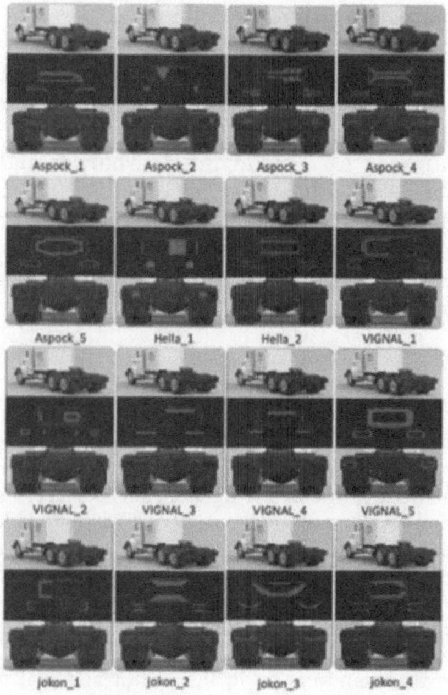

Figure 2. Illustrative product cards of emblematic automotive Combination Rear Lamp brands.

Subsequently, based on the experts' recommendations for the optimal timing to integrate GenAI and the "Double Diamond" theory proposed by the Design Council in 2005, building Gen AI design process.

2.2 *Stage two: Implementation test of GenAI design process*

This study refers to the list of the top 50 GenAI products globally published by the well-known Silicon Valley venture capital company a16z, as shown in Figure 3, to conduct the organization and screening of GenAI software technologies.

Figure 3. Ranking of the top 50 GenAI products by global traffic (Source: a16z forum website of a well-known venture capital company in Silicon Valley, USA).

Subsequently, three automotive lighting designers and three designers from non-automotive lighting fields (with backgrounds in industrial design) were invited to participate in the GenAI generative artificial intelligence algorithm experiment. Through in-depth interviews, insights into the designers' experiences with the use of GenAI were obtained. Additionally, the generated results were further evaluated by consulting experts from Taiwan's largest commercial vehicle lighting company to assess the differences in design outcomes between the two groups of participants and confirm the optimal timing for integrating the outcomes of this stage into the design process.

2.3 *Stage three: GenAI design process validation*

In this stage, ten novice designers were further recruited to validate the GenAI design process, aiming to examine and confirm the feasibility and effectiveness of the GenAI design process. Upon generating the outcomes, further consultations were conducted with automotive lighting experts from Taiwan's largest commercial vehicle lighting company to optimize the design process.

3 RESULTS AND DISCUSSION

The results generated through the GenAI design process in this study (Table 2) indicate that the lighting designs produced by the three automotive lighting designers were superior. They even closely resembled the actual production state, with their appearance and styling being more comprehensive compared to designers from non-automotive lighting backgrounds. This is because automotive lighting designers were able to quickly screen key surfaces that fit the lighting design within limited time frames at different stages. Conversely, designers from non-automotive lighting backgrounds found it more challenging to select key images that fit the lighting design within the same time constraints, hence yielding less satisfactory results.

Table 2. Graphics card generation process chart.

Experimental subjects/GenAI software	Non-automotive designers	Automotive designers
Midjourney Abstract generation	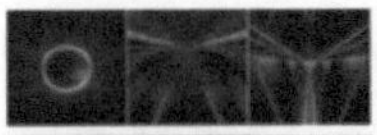	
Midjourney Concrete generation	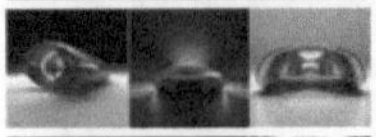	
StableDiffusion Concrete convergence		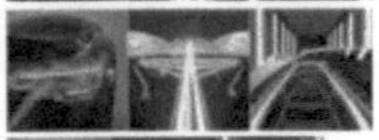
Vizcom Detail adjustment	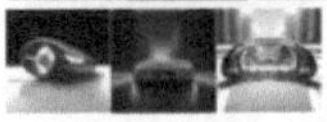	

Table 3. Graphics card generation process chart.

Experimental subjects/GenAI software	Using three students as an example		
	Student 1	Student 2	Student 3
Midjourney Abstract generation		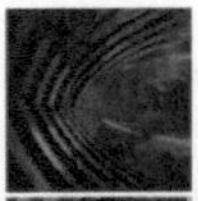	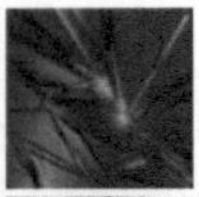
Midjourney Concrete generation			
StableDiffusion Concrete convergence			
Vizcom Detail adjustment			

In the third stage of this study, 10 novice designers were invited to validate the GenAI design process under conditions without time constraints. Summary analysis of the outcomes generated by three participants using GenAI (Table 3) demonstrates excellent

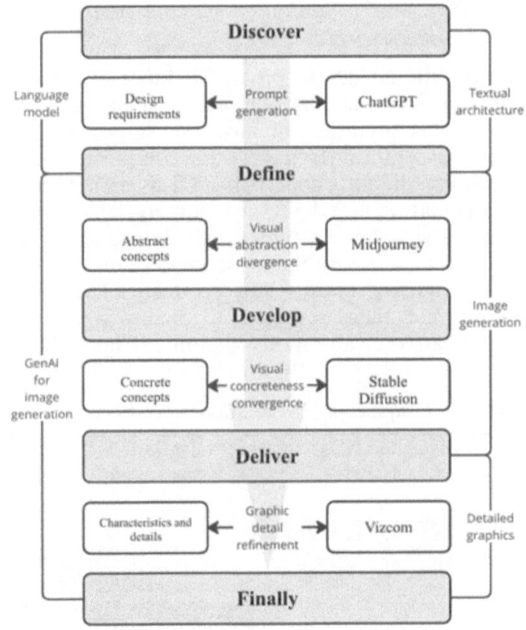

Figure 4. Optimal design process for incorporating GenAI into commercial vehicle Rear Lamp.

results, with designs that closely resemble actual automotive lighting products.

Based on the experimental results mentioned above and insights gathered from in-depth interviews with experts from Taiwan's largest commercial vehicle lighting company, this study proposes the optimal GenAI integration process for commercial vehicle taillight design, as illustrated in Figure 4.

This study considers the characteristics of different GenAI software and optimizes the integration design process. The findings hold significant implications for commercial vehicle lighting designers, providing clear recommendations and auxiliary tools. Moreover, this research process is highly suitable for training new talent within enterprises and can effectively reduce research and development schedule.

REFERENCES

Haase, J., & Hanel, P. H. p. (2023). Artificial muses: Generative artificial intelligence chatbots have risen to human-level creativity. *Journal of Creativity*, *33*(*3*).

Takahashi, Kaito, Tachikawa, Yuuki, Konishi, Koki, Kobayashi, Hiroko, & Ishii, Daisuke. (2023). *Complete Illustrated Guide to Artificial Intelligence*. Tokyo Shoseki.

Verganti, R., Vendraminelli, L., & Iansiti, M. (2020). Innovation and design in the age of artificial intelligence. *Journal of product innovation management*, *37* (*3*), 212–227.

System Innovation for an Artificial Intelligence Era – Lam et al. (Eds)
 ISBN: 978-1-032-84452-7

Greener production of lead sulfide from lead dust through sodium hydrosulfide roasting followed by water leaching

Li-Pang Wang & Yan-Wen Chen
Institute of Environmental Engineering and Management, National Taipei University of Technology, Taipei, Taiwan

ABSTRACT: Lead sulfide (PbS) is an important raw material which can be used for the fabrication of several optical and electrical devices. It has been reported that PbS can be produced using chemical reagent of lead sulfate ($PbSO_4$) through carbothermic reduction method. On the other hand, lead dust is a solid waste collected by the air pollution control equipment in the smelting process of spent lead-acid batteries whose main content is also $PbSO_4$. This study proposed a greener approach to produce PbS from lead dust through sodium hydrosulfide (NaHS) roasting followed by water leaching. The effects of NaHS dosage, roasting temperature, roasting time, liquid-solid ratio, and leaching time, on the conversion rate of $PbSO_4$ to PbS and the PbS grade of the produced PbS were investigated. The results indicated that the optimal conditions of NaHS roasting followed by water leaching were NaHS dosage of 1:1.5, roasting temperature of 400°C, roasting time of 30 mins, liquid-solid ratio of 25 mL/g, and leaching time of 30 mins. The PbS conversion rate of the recovered PbS product reached 99.45% with a PbS grade of 99.93%. The proposed approach is greener than the carbothermic reduction method by which the roasting temperature of higher than 700°C is needed. In addition, the waste reduction and sustainable utilization of lead dust can also be achieved.

1 INTRODUCTION

Lead sulfide (PbS) is an important raw material which can be used for the fabrication of several optical and electrical devices such as photoresistor, laser diode, and optical fiber amplifier (Jahromi 2021). It has been reported that PbS can be produced using chemical reagent of lead sulfate ($PbSO_4$) as the raw material through carbothermic reduction method, in which carbon (Zheng *et al.* 2015) or carbon monoxide was added (Zheng *et al.* 2015) as the reducing agent and heated at a temperature of more than 700°C under oxygen-free circumstances. In order to reduce the energy consumption and greenhouse gas emission, it is needed to search for alternative production method to decrease the heating temperature.

Lead dust is a solid waste collected by the air pollution control equipment in the smelting process of spent lead-acid batteries whose main content is also $PbSO_4$ (Chen *et al.* 2016). Nowadays, most of the lead dust is treated by landfill disposal. In order to reduce the burden of landfill site, it is needed to search for the waste reclamation approach of lead dust.

In this study, a greener approach was proposed to produce PbS from lead dust through sodium hydrosulfide (NaHS) roasting followed by water leaching. The effects of NaHS dosage, roasting temperature, roasting time, liquid-solid ratio, and leaching time, on the conversion rate of $PbSO_4$ to PbS and the PbS grade of the produced PbS were investigated.

2 EXPERIMENTAL

A lead dust sample was obtained from a spent lead-acid battery recycling company in Taiwan. The as-received lead dust sample was pretreated by sulfuric acid (H_2SO_4) washing to remove the impurities. The H_2SO_4 washing conditions were H_2SO_4 of 5 M, washing temperature of 75°C, liquid-solid ration of 5 ml/g, and washing time of 1 hrs, referring to those published by Yang *et al.* (2020). The as-received lead dust sample after H_2SO_4 washing was used in this study, which mainly contained lead (Pb) (63.96%) with other impurities of calcium (Ca) (0.51%), iron (Fe) (0.02%), zinc (Zn) (0.02%), cadmium (Cd) (0.03%), tin (Sn) (0.03%), arsenic (As) (0.01%), barium (Ba) (0.05%), and potassium (K) (0.06%), after digested by aqua regia followed by analyzing using an Inductively Coupled Plasma Optical Emission Spectrometer (ICP-OES) (Optima 8000, PerkinElmer, USA). The main crystal phase of the lead dust before and after H_2SO_4 washing was $PbSO_4$ after confirmed by an X-ray diffractometer (XRD) (BRUKER – D2 Phaser, Germany). The content of $PbSO_4$ in the lead dust after H_2SO_4 washing was 99.91 wt.%, which was determined based on the amount of sulfate ion, i.e., 316.5 mg, according to the method of NIEA W430.51C announced by the Ministry of Environment, Taiwan, after digested by concentrated hydrochloric acid (HCl) followed by analyzing using an Ultraviolet/Visible Spectrophotometer (UV/VIS) (Halo Rb-10, Dynamica Scientific Ltd., UK). NaHS

DOI: 10.1201/9781003514831-43

(purity: 99%) (Sigma-Aldrich, St. Louis, MO, USA) was used as the roasting agent.

1.0 g of the lead dust sample after H_2SO_4 washing was uniformly mixed with NaHS of determined weight and placed in a 100 ml aluminum oxide crucible. The crucible was placed in a muffle furnace (HS-3, Yi-Feng Instrument Co., Ltd., Taipei, Taiwan) and heated with a heating rate of 20°C/min to the determined roasting temperature for the roasting experiment. After the determined roasting time was reached, heating was ceased. After the furnace temperature has cooled to room temperature, the crucible was taken out and 50 ml of deionized water add into the crucible. An ultrasonic disperser was used to completely detach the sample adhering to the crucible. Then, the sample solution was poured into a 250 ml beaker and stirred using a magnetic stirrer at 400 rpm to conduct water leaching experiment. After the designated leaching time was reached, solid-liquid separation was performed using a centrifuge at 3500 rpm. The solid phase obtained was the product of roasting followed by leaching. The product was dried in a 105°C oven for 12 hours until a constant weight is achieved. A portion of the product underwent XRD analysis for crystal phase identification. A portion of that was digested by using hydrochloric acid for the analysis of sulfate ion amount using UV/VIS spectroscopy to calculate the PbS conversion rate. A portion of that was digested by using aqua regia for the analysis of PbS content using ICP-OES to calculate the PbS grade.

3 RESULTS AND DISCUSSION

Figure 1 presents the photographs of the products obtained from NaHS roasting and water leaching under different amounts of NaHS addition. The addition amounts of lead dust sample to NaHS were 1:0.5 (g/g), 1:1 (g/g), 1:1.5 (g/g), and 1:2 (g/g). The lead dust sample after H_2SO_4 leaching was 1.0 g, the roasting temperature was 400°C, the roasting time was 30 minutes, the leaching liquid-solid ratio was 50 ml/g, and the leaching time was 60 minutes. The products with the addition amounts of NaHS from 1:0.5 to 1:2 all exhibited a blue-black color, implying a potential conversion to PbS.

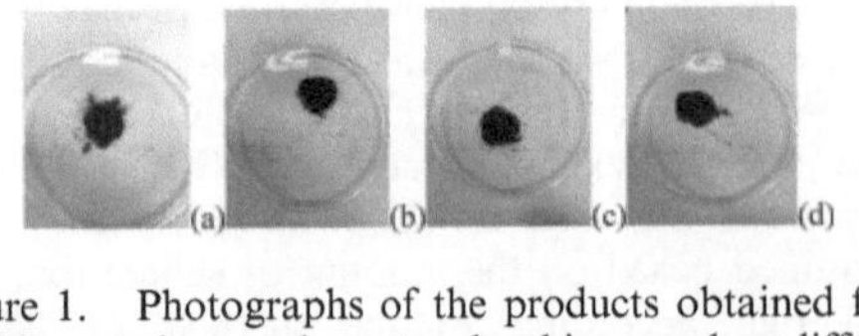

Figure 1. Photographs of the products obtained from NaHS roasting and water leaching under different amounts of NaHS addition (a) 1:0.5 (g/g), (b) 1:1 (g/g), (c) 1:1.5 (g/g), (d) 1:2 (g/g) (Experimental conditions: lead dust of 1.0 g, roasting temperature of 400°C, roasting time of 30 mins, leaching liquid-solid ratio of 50 ml/g, leaching time of 60 mins).

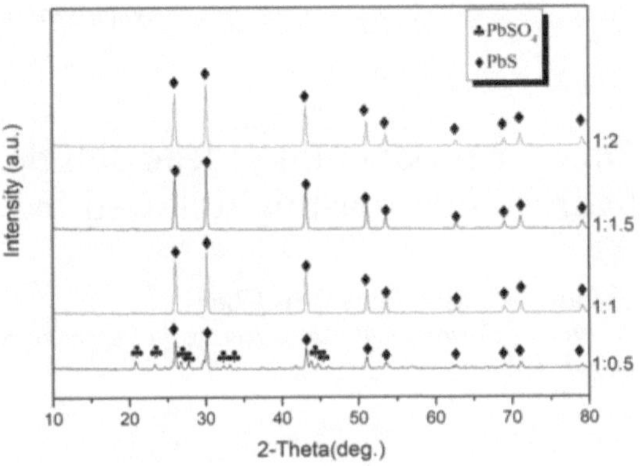

Figure 2. XRD analysis results of the products obtained from NaHS roasting and water leaching under different amounts of NaHS addition (Experimental conditions: lead dust of 1.0 g, roasting temperature of 400°C, roasting time of 30 mins, leaching liquid-solid ratio of 50 ml/g, leaching time of 60 mins).

Figure 2 presents the XRD analysis results of the products obtained from NaHS roasting and water leaching under different amounts of NaHS addition. When the NaHS addition was 1:0.5 (g/g), the characteristic peaks corresponding to $PbSO_4$ and PbS were identified. This indicated that the amount of NaHS added was insufficient to convert $PbSO_4$ to PbS. When the NaHS addition ratio was higher than 1:1 (g/g), only the characteristic peaks corresponding to PbS were observed. This suggested that all $PbSO_4$ has been converted to PbS.

Table 1 presents the sulfate ion amount and PbS conversion rate in the products obtained from NaHS roasting and water leaching under different amounts of NaHS addition. When the NaHS addition was 1:0.5 (g/g), the PbS conversion rate was only 45.81%, indicating insufficient NaHS addition, which cannot completely convert PbSO4 into PbS. When the NaHS addition was 1:1 (g/g) or , the PbS conversion rate increased to 95.98%. When the NaHS addition was 1:1.5 (g/g)-1:2 (g/g), the PbS conversion rate exceeds 99.37% to 99.49%. Therefore, the optimal NaHS addition ratio was determined to be 1:1.5 (g/g).

Table 2 presents the sulfate ion amount and PbS conversion rate in the products obtained from NaHS roasting and water leaching under different roasting temperatures. The roasting temperatures were 100°C, 200°C, 300°C, and 400°C. The lead dust sample after H_2SO_4 leaching was 1.0 g, the NaHS addition amount was 1:1.5 (g/g), the roasting time was 30 minutes, the leaching liquid-solid ratio was 50 ml/g, and the leaching time was 60 minutes. As the roasting temperature increased, the PbS conversion rate also increased. When the roasting temperature was 100°C, the PbS conversion rate was 95.73%. When the roasting temperature increased to 400°C, the PbS conversion rate increased to 99.45%. Therefore, the optimal roasting temperature was determined to be 400°C.

Table 1. Sulfate ion amount and PbS conversion rate in the products obtained from NaHS roasting and water leaching under different amounts of NaHS addition (Experimental conditions: lead dust of 1.0 g, roasting temperature of 400°C, roasting time of 30 mins, leaching liquid-solid ratio of 50 ml/g, leaching time of 60 mins).

NaHS addition ratio (g/g)	(A) Sulfate ion amount in product (mg)	(B) Sulfate ion amount in lead dust (mg)	PbS conversion rate (%) ((B-A)/B)* 100%
1:0.5	171.49	316.5	45.81
1:1	12.71	316.5	95.98
1:1.5	1.59	316.5	99.49
1:2	1.99	316.5	99.37

Table 2. Sulfate ion amount and PbS conversion rate in the products obtained from NaHS roasting and water leaching under different roasting temperatures (Experimental conditions: lead dust of 1.0 g, NaHS addition amount of 1:1.5 (g/g), roasting time of 30 mins, leaching liquid-solid ratio of 50 ml/g, leaching time of 60 mins).

Roasting temperature (°C)	(A) Sulfate ion amount in product (mg)	(B) Sulfate ion amount in lead dust (mg)	PbS conversion rate (%) ((B-A)/B)* 100%
100	13.52	316.5	95.73
200	10.56	316.5	96.66
300	3.77	316.5	98.81
400	1.73	316.5	99.45

Table 3 presents the sulfate ion amount and PbS conversion rate in the products obtained from NaHS roasting and water leaching under different roasting times. The roasting temperatures were 5 minutes, 15 minutes, 30 minutes, 45 minutes, and 60 minutes. The lead dust sample after H_2SO_4 leaching was 1.0 g, the NaHS addition amount was 1:1.5 (g/g), the roasting temperature was 400°C, the leaching liquid-solid ratio was 50 ml/g, and the leaching time was 60 minutes. As the roasting time increased, the PbS conversion rate also increased. When the roasting time was 5 minutes, the PbS conversion rate was 97.82%. When the roasting time was extended to 30 minutes, the PbS conversion rate reached 99.45%. Beyond this point, further increase in roasting time did not significantly enhance the PbS conversion rate. Therefore, the optimal roasting time was determined to be 30 minutes.

Table 4 presents the Pb content and the PbS grade in the products obtained from NaHS roasting and water leaching under different leaching liquid-solid ratios. The leaching liquid-solid ratios were 25 ml/g, 50 ml/g, 75 ml/g, and 100 ml/g. The lead dust sample after H_2SO_4 leaching was 1.0 g, the NaHS addition amount was 1:1.5 (g/g), the roasting temperature was 400°C, the roasting time was 30 minutes, and the leaching time was 60 minutes. When the leaching liquid-solid ratio was 25 ml/g, the Pb content in the product was 84.86 wt.% and the PbS grade was 97.93 wt.%. As the leaching liquid-solid ratio increased, the Pb content and the PbS grade in the products had no significant difference. Therefore, the optimal leaching liquid-solid ratio was determined to be 25 ml/g.

Table 3. Sulfate ion amount and PbS conversion rate in the products obtained from NaHS roasting and water leaching under different roasting times (Experimental conditions: lead dust of 1.0 g, NaHS addition amount of 1:1.5 (g/g), roasting temperature of 400°C, leaching liquid-solid ratio of 50 ml/g, leaching time of 60 mins).

Roasting time (min)	(A) Sulfate ion amount in product (mg)	(B) Sulfate ion amount in lead dust (mg)	PbS conversion rate (%) ((B-A)/B)* 100%
5	6.88	316.5	97.82
15	3.63	316.5	98.85
30	1.73	316.5	99.45
45	2.33	316.5	99.26
60	3.54	316.5	98.88

Table 4. Pb content and PbS grade in the products obtained from NaHS roasting and water leaching under different leaching liquid-solid ratios (Experimental conditions: lead dust of 1.0 g, NaHS addition amount of 1:1.5 (g/g), roasting temperature of 400°C, roasting time of 30 mins, leaching time of 60 mins).

Leaching liquid-solid ratio	Pb content (wt.%)	PbS grade (wt.%)
Lead dust after H_2SO_4 washing	62.38	–
25 ml/g	84.86	97.93
50 ml/g	84.21	97.56
75 ml/g	84.29	97.65
100 ml/g	84.06	97.42

Table 5 presents the Pb content and the PbS grade in the products obtained from NaHS roasting and water leaching under different leaching times. The leaching times were 15 minutes, 60 minutes, 60 minutes, and 60 minutes. The lead dust sample after H_2SO_4 leaching was 1.0 g, the NaHS addition amount was 1:1.5 (g/g), the roasting temperature was 400°C, the roasting time was 30 minutes, and the leaching liquid-solid ratio was 25 ml/g. When the leaching time was 15 minutes, the Pb content in the product was 85.17 wt.%, the PbS grade was 98.23 wt.%, and the Pb recovery was 100%. As the leaching time exceeded

30 minutes, there is no significant difference in Pb content and PbS grade, which was ca. 86.68 wt.% and 99.93 wt.%, respectively. In order to fully remove the byproduct of sodium bisulfate ($NaHSO_4$) after roasting by water leaching, a leaching time of 30 minutes was determined to be the optimal leaching time to ensure high PbS grade and high Pb recovery rate.

Table 5. Pb content and PbS grade in the products obtained from NaHS roasting and water leaching under different leaching times (Experimental conditions: lead dust of 1.0 g, NaHS addition amount of 1:1.5 (g/g), roasting temperature of 400°C, roasting time of 30 mins, leaching liquid-solid ratio of 25 ml/g).

Leaching time	Pb content (wt.%)	PbS grade (wt.%)
Lead dust after H_2SO_4 washing	62.38	–
15 mins	85.17	98.23
30 mins	86.68	99.93
45 mins	86.13	99.88
60 mins	86.07	99.81

4 CONCLUSION

In this study, an alternative approach was proposed to produce PbS from lead dust employing NaHS roasting followed by water leaching method. The optimal conditions were NaHS addition ratio of 1:1.5, roasting temperature of 400°C, roasting time of 30 minutes, leaching liquid-to-solid ratio of 25 ml/g, and leaching time of 30 minutes. The PbS conversion rate of the recovered PbS product reached 99.45%, with a PbS grade of 99.93% and a Pb recovery rate of 100%. The proposed approach is greener than the carbothermic reduction method by which the roasting temperature of higher than 700°C is needed. In addition, the waste reduction and sustainable utilization of lead dust can also be achieved simultaneously.

ACKNOWLEDGEMENTS

The authors appreciate for the National Science and Technology Council, R.O.C. for financially supporting this research under Grant No. NSTC 112-2221-E-027-023-MY2.

REFERENCES

Chen C. S, Shih Y.J., Huang Y.H. (2016). *Waste Manag.*, 52, 212–220.

Jahromi H.D. (2021). *Optik* 231, 166503.

Yang T., Xie B., Liu W., Zhang D., Chen L., (2020). *Sep. Purif., Technol.* 233, 116035.

Zheng Y., Liu W., Qin W., Jiao F., Han J., Yang K., Luo H., (2015). *J. Cent. South Univ.*, 22(8), 2929–2935.

Zheng Y., Liu W., Qin W., Han J., Yang K., Luo H., (2015). *Physicochem. Probl. Miner. Process* 51(2) 535–546.

System Innovation for an Artificial Intelligence Era – Lam et al. (Eds)
 ISBN: 978-1-032-84452-7

An intelligent swimming recording and replay system based on computer vision and accelerometer

Chih-Hsien Huang, Fu-Sung Lin, Ying-Hsien Chen, Pu-Chun Liu, Hsiao-Ching Lin, Chun-Yuan Chen & Han-Wei Lian
Department of Electronic Engineering, National Cheng Kung University, Tainan, Taiwan

ABSTRACT: In recent years, applying computer vision and machine learning to improve the effectiveness and safety of athletic training has been a rising topic. However, not many studies discussed using these technologies in training swimmers. Hence, this research reports an intelligent swim training system that could automatically provide qualitative and quantitative information after each practice. This study evenly positioned eight cameras in a standard 25-meter swimming pool and an accelerometer (Movella DOT, Xsens) in the swim cap to record visual and inertial data, respectively. The authors employ the YOLOv4-tiny as the pre-trained model to train a swimmer identification model using 150,000 frames from multiple swimmers with different swim styles. The model achieved a final Intersection over Union (IoU) of 78.36% with 99.61% mean average precision (mAP). After that, the variation of the bounding box coordinate and acceleration data will be synchronized based on the timing of the first stroke after the underwater dolphin kick. With the synchronized position data, the proposed system can detect the number of strokes and the duration of the underwater dolphin kick with an accuracy of 97.92% and 96.23%, respectively. Finally, the synchronized inertial data and the statical results would be replayed with the video simultaneously to provide feedback to the swimmers and coaches immediately.

1 INTRODUCTION

Athletic training and assessment were critical to improving athletic performance and reaching peak performance. Whether for professional athletes or amateur enthusiasts, effective training methods and accurate assessment tools are key to achieving outstanding performance. Traditional methods of athletic assessment typically relied on subjective observation or manual measurements, which could lead to subjective and inconsistent evaluation results. At the same time, traditional methods often provided limited data and failed to provide in-depth motion analysis, highlighting the need for innovative methods and technologies.

In recent years, target tracking based on computer vision has emerged as a key application in the field of computer vision [1]. This technology has attracted much attention in sports competitions and training [2,3], demonstrating its versatility across various sports disciplines. By analyzing various data collected from athletes, coaches could use this information to customize training programs, adjust intensity levels, and use algorithms to predict potential sports-related injuries [4].

Recent studies have pointed out that swimming is not only an excellent aerobic exercise [5] , but also helps reduce body fat and increase muscle mass [6,7]. In addition, swimming can improve health, such as helping asthma patients relieve symptoms and manage chronic diseases [8]. Many machine learning studies have shown that analyzing swimming-related data can provide a clearer understanding of various aspects of swimming performance [9]. These studies highlight the potential of data analytics to improve understanding of swimming performance. Through extensive data analysis, coaches and swimmers can gain valuable insights to improve techniques and training plans, enabling more evidence-based, personalized swimming training.

This study aimed to use deep learning technology to develop a multi-object tracking system to automatically record swimmers' swimming time in the swimming lane. This automation was designed to improve swimmers' training efficiency. Display and segment timing technology had fueled the development of machine vision systems that could not only continuously track swimmers but also collect data for a variety of analyses. Coaches and swimmers could use these systems to check performance on a computer and develop more effective training methods.

2 METHOD

As Figure 1 illustrates, this study evenly positioned eight cameras in a standard 25-meter swimming pool, and the cameras' frame rates were 30 Hz.

DOI: 10.1201/9781003514831-44

Four cameras are located above the water's surface, and the others are in the water. An accelerometer (Movella DOT, Xsens) is placed in the swim cap to record the 3-axis acceleration data with a frame rate of 30 Hz. This system employs the YOLOv4-tiny model as the swimmer identification model. To eliminate the effect of the splash, this study filmed multiple swimmers completing a swimming lap in different swimming postures. After that, 150,000 frames were extracted from the videos to train the YOLOv4-tiny model. The coordinate information from the bounding box of the YOLOv4-tiny model will be synchronized with the acceleration data to identify the movements or swimming stages of the swimmers. Besides, the swimmer's time when pass-ing specific distances such as 0m, 15m, 25m, and 50m are also identified using the bounding box. Other than that, the bounding box is also used to re-alize the panning cam of the swimmer to allow the users to observe the movements more closely. Fi-nally, the information about split time and the pan-ning cam would be processed and integrated into the replay videos to provide visual feedback to the users.

The proposed system is capable of detecting the number of strokes. Initially, the accelerometer data along the Z-axis undergoes Fast Fourier Transform (FFT) to convert the signal from the time domain to the frequency domain. Subsequently, after passing through a low-pass filter to eliminate noise, the sys-tem identifies the number of peaks, thereby deter-mining the number of strokes. Besides, the analyti-cal data extracted from the coordinates of the bounding box and acceleration measurements of the inertial sensor would provide long-term statistical results to the users as the reference for modifying training and competition approaches.

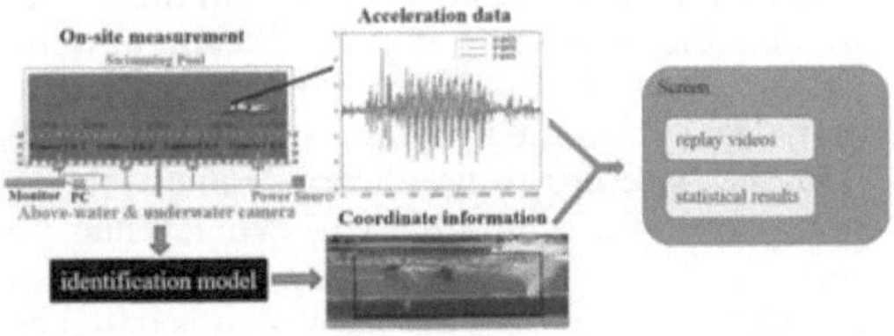

Figure 1. Hardware architecture and process flow of the proposed sensor fusion swimming recording system.

3 RESULTS AND DISCUSSION

More than 150,000 frames of image data were extracted from the recorded videos and split into training, validation, and test sets in an 8:1:1 ratio. After 20,000 training iterations, the model achieved a final Intersection over Union (IoU) of 78.36%. The mean average precision (mAP) at an IoU threshold 0.5 reached 99.61%. The results demonstrate that the model accurately recognizes swimmers even with splashes.

As shown in Figure 2, the proposed system accurately detected posture variations of the swimmer and marked them up with the bounding box. Also, the numbers displayed above the bounding box were the confidence scores, higher scores represented better accuracy for the posture recognition.

As Figure 3 illustrates, the playback system not only replayed the processed video, but also displayed numerical information including split time, total time, acceleration data, and panning.

Figure 4 presents the processed accelerometer's z-axis signal. After the local maximum method was applied, the number of stroking could easily be found by counting the number of red spots in the middle section.

Figure 5 presents the synchronized 3-axis acceleration data and y-coordinates of the bounding box's center (ymid). The patterns of each swimming stage can be identified using both data, including pushing off the wall, diving, and stroke. It reveals the potential of using statistical or machine learning methods to quantify swimming performance, such as the number of strokes, the traveling distance of each stroke, and the diving length. In the future, a 5G Wi-Fi environment will be built in the swimming pool of this study. The processed videos and motion sensing analysis will be displayed on a giant screen beside the swimming pool and uploaded to a cloud data center for further research.

Figure 2. The result of a swimmer's posture detection, appended with a bounding box and confidence score.

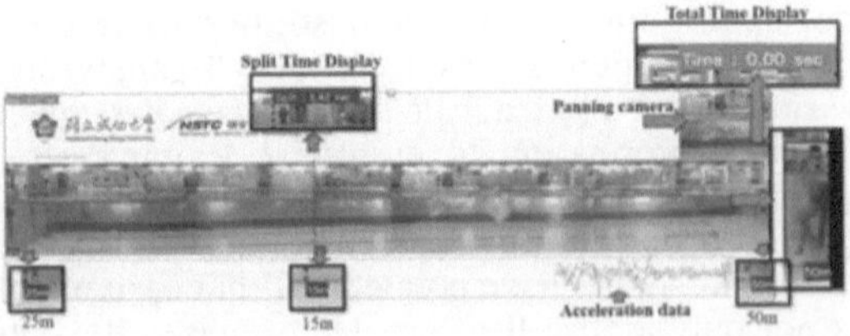

Figure 3. The playback of the processed video includes recordings of the swimmer's split time, total time, accelerometer data, and panning camera.

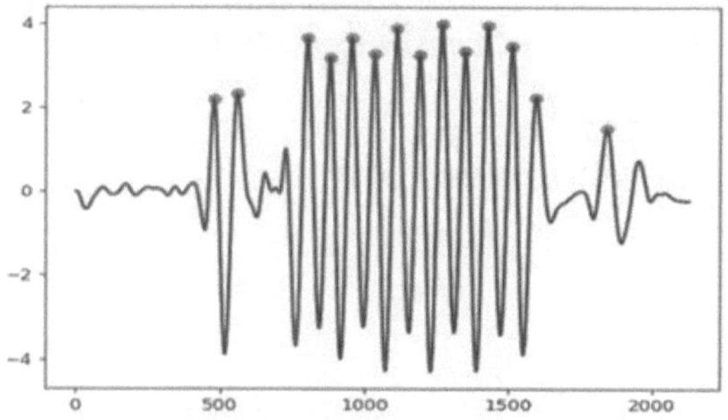

Figure 4. The filtered z-axis acceleration signals.

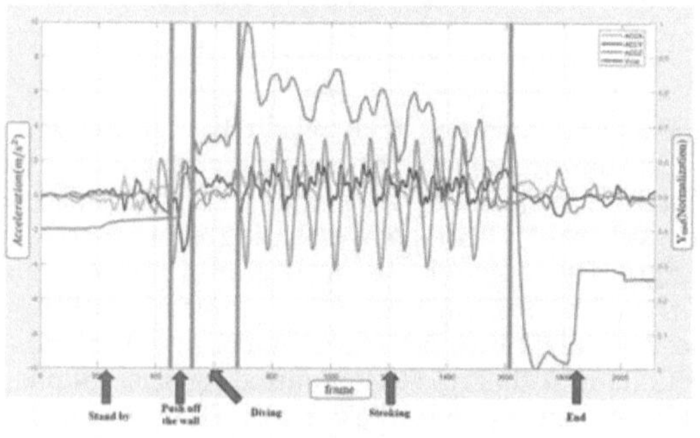

Figure 5. The synchronization of the recordings from the accelerometer and coordinates of the bounding box.

4 CONCLUSION

This paper presents a comprehensive swim training assistance system that integrates computer vision and accelerometer technology to analyze and enhance swimming performance. Not only could the system automatically provide swimmer tracking information during swimming, but also segment different stages during swimming in real-time. By strategically positioning 8 cameras around the swimming pool and utilizing a swimmer tracking model based on the YOLOv4-tiny algorithm, the system effectively identifies swimmers despite the interference of splashes. Besides, using the accelerometer measurement and the bounding box further enhances the accuracy of identifying swimming posture and stages.

REFERENCES

[1] Cao Z., Simon T., Wei S.-E., and Sheikh Y. (2017). Realtime multi-person 2d pose estimation using part affinity fields, in *Proceedings of the IEEE Conference on Computer Vision and Pattern Recognition*, pp. 7291–7299.

[2] Richter C., O'Reilly M. and Delahunt E. (2021). Machine learning in sports science: challenges and opportunities, *Sports Biomechanics*, pp. 1–7.

[3] Bunker R. and Susnjak T. (2022). The application of machine learning techniques for predicting match results in team sport: A review, *Journal of Artificial Intelligence Research*, vol. 73, pp. 1285–1322.

[4] Rossi A., Pappalardo L. and Cintia P. (2021). A narrative review for a machine learning application in sports: an example based on injury forecasting in soccer, *Sports*, vol. 10, no. 1, p. 5.

[5] Van Eetvelde H., Mendonça L. D., Ley C., Seil R. and Tischer T. (2021). Machine learning methods in sport injury prediction and prevention: a systematic review, *Journal of Experimental Orthopaedics*, vol. 8, pp. 1–15.

[6] Sigurdson H. and Chan J. H. (2021). Machine learning applications to sports injury: A review, *icSPORTS*, pp. 157–168.

[7] Lahart I. M. and Metsios G. S. (2018). Chronic physiological effects of swim training interventions in non-elite swimmers: A systematic review and meta-analysis, *Sports Medicine*, vol. 48, pp. 337–359.

[8] Celik O., Salci Y., Ak E., Kalaci A. and Korkusuz F. (2013). Serum cartilage oligomeric matrix protein accumulation decreases significantly after 12 weeks of running but not swimming and cycling training–A randomised controlled trial, *The Knee*, vol. 20, no. 1, pp. 19–25.

[9] LieberD. C., Lieber R. L. and Adams W. C. (1989). Effects of run-training and swim-training at similar absolute intensities on treadmill VO2max, *Medicine and Science in Sports and Exercise*, vol. 21, no. 6, pp. 655–661.

[10] Bar-Or O. and Inbar O. (1992). Swimming and asthma: benefits and deleterious effects, *Sports Medicine*, vol. 14, pp. 397–405.

[11] Nualnim N., Parkhurst K., Dhindsa M., Tarumi T., Vavrek J., and Tanaka H. (2012). Effects of swimming training on blood pressure and vascular function in adults> 50 years of age, *The American Journal of Cardiology*, vol. 109, no. 7, pp. 1005–1010, 2012.

[12] Tanaka H., Bassett Jr D. R., Howley E. T., Thompson D. L., Ashraf M., and Rawson F. L. (1997). Swimming training lowers the resting blood pressure in individuals with hypertension, *Journal of Hypertension*, vol. 15, no. 6, pp. 651–657.

[13] Worsey M. T., Pahl R., Espinosa H. G., Shepherd J. B., and Thiel D. V., (2021). Is machine learning and automatic classification of swimming data what unlocks the power of inertial measurement units in swimming?, *Journal of Sports Sciences*, vol. 39, no. 18, pp. 2095–2114.

[14] Xie J., Xu J., Nie C., and Nie Q. (2016). *Machine Learning of Swimming Data Via Wisdom of Crowd and Regression Analysis, Mathematical Biosciences & Engineering*, vol. 14, no. 2, pp. 511–527.

[15] Woinoski T., Harell A. and Bajic I. V. (2020). Towards automated swimming analytics using deep neural networks, *arXiv preprint arXiv*:2001.04433.

[16] Powell L., Polsley S., Casey D. and Hammond T., (2023). The real-time classification of competency swimming activity through machine learning, *International Journal of Aquatic Research and Education*, vol. 14, no. 1, p. 6.

System Innovation for an Artificial Intelligence Era – Lam et al. (Eds)

Applying brainwave information to dynamic image form for constructing a novel interactive art

Hung-Ying Lee
Graduate School of Design, Yunlin University of Science and Technology, Yunlin, Taiwan
Department of Multimedia and Game Development, Mingsin University of Science and Technology, Hsinchu, Taiwan

Chao-Ming Wang
Department of Digital Media Design, Yunlin University of Science and Technology, Yunlin, Taiwan

ABSTRACT: In the digital generation, the development of art creation is increasingly merging with high technology. How to present new technological artwork has become one of the most crucial issues in contemporary art. This study proposes the use of EEG (electroencephalogram) emotion signals in conjunction with interactive technology to create artworks. When EEG brainwave sensing is activated, it transforms into a creatively immersive interactive design. The research develops a prototype of an interactive art system to design the artwork "*Sunset in Tamsui*" by Taiwan's renowned painter, Tan Teng-pho. It analyzes operational usability, effectiveness, and enjoyment through expert interviews and audience questionnaires following experimentation. The EEG data in this research identifies four emotions: Attention, Meditation, Anxiety, and Calmness. As audiences engage with "*Sunset in Tamsui*" on the interactive wall, these brainwave signals generate corresponding dynamic image. The design concept aims to leverage real-time computing to produce consequent interactive art effects. Through this experience, this research aims to bridge the gap between art and the general public, deepen audience involvement, and enhance aesthetic appreciation.

Keywords: Brainwave Sensing, Motion Image, Interactive Art, New Media Art, Human-Machine Interface

1 INTRODUCTION

The "Digital Native" Z generation is refining interactive art creation to be engaging and popular in our current lives. We can observe an increasing number of technologies being applied in various fields, with art being no exception. Contemporary artists are exploring high technology to create immersive environments for audiences. Merely standing in front of a painting or installation in the traditional manner no longer satisfies audiences; they crave more perceptual and interactive experiences. In this research, we aim to interact with the audience, which is why we utilize EEG (electroencephalogram) technology to create an immersive world. The new trend in the art world involves visitors not only observing artwork in front of artwork but also actively participating in and immersing themselves in the experience.

In the past, fine art was often classified into incomprehensible categories. Combined with the current cultural connotation transmission and intelligent development trend, we need to study and exert its value and improve the development speed of digital media art (Ciguli Wu 2021). However, there has been a shift towards making art more accessible to the public. Fortunately, artists have discovered new media art as a solution to this problem. Consequently, this study aims to combine electroencephalography and interactive technology to create a new form of interactive artistic expression through motion image.

This approach allows us to connect more closely with the audience and demonstrate that art is not as difficult to understand as previously thought.

The purposes of this study are as follows.

(1) How can we incorporate new technology into the creation of interactive motion image for art?
(2) How can we effectively evaluate the impact of brainwave signal information?
(3) How can we explore the structure and presentation of motion image?

1.1 *Motion image*

"Motion Image" originally refers to continuously changing pictures or photos in movies, animations,

DOI: 10.1201/9781003514831-45

videos, media, etc. This study used "*Sunset in Tamsui*" by Tan Teng-pho, an important early painter in Taiwan, to create the prototype of the motion image. Compared with pure visual or auditory stimulation, watching a video film can simultaneously stimulate people's visual and auditory senses and, thus, can give them a stronger sense of substitution and better emotional evocation. (Zhuang *et al.* 2019). This study hopes that this prototype will help to observe and detect different emotions of different audiences, and be used to record how visual art affects people's emotional perceptions.

Figure 1. Tan Teng-pho 's "*Sunset in Tamsui*".

1.2 *Case analysis*

1.2.1 *"E.E.G. KISS"*

Figure 2. The actor and actress are in front of audience on performs.

This is a public installation and performance ongoing from 2014 to 2022. The artist poses special questions to the audience: "*Can we measure a kiss and what do kissers feel together? Do we want to save our private kisses in a transparent database to be used by others*?" People are accustomed to thinking of kissing as a shy, private, and personal activity, rarely engaging in it in front of others except in special cases such as newlyweds at weddings. The installation materializes actors' and actresses' physiological reactions to the audience, providing clues to everyone's physiological state. 'Spontaneous' EEG interaction between participants is part of the design, making brain signals during intimate interactions explicit to facilitate a shared experience of social touch through reflection and dialogue. Brain signals during intimate interaction of kissing are made explicit to facilitate a shared experience of social touch through reflection and dialogue (Lancel *et al.* 2019). The artist suggests that emotions can be quantified and should not be a source of embarrassment, as they are all natural physiological reactions. External views do not affect the true influence of emotions within us.

1.2.2 *British Museum and galleries museum-goers' brainwaves into a simplified real-time visualization*

Figure 3. The audience is wearing EEG to collect brainwave when they are watching artwork.

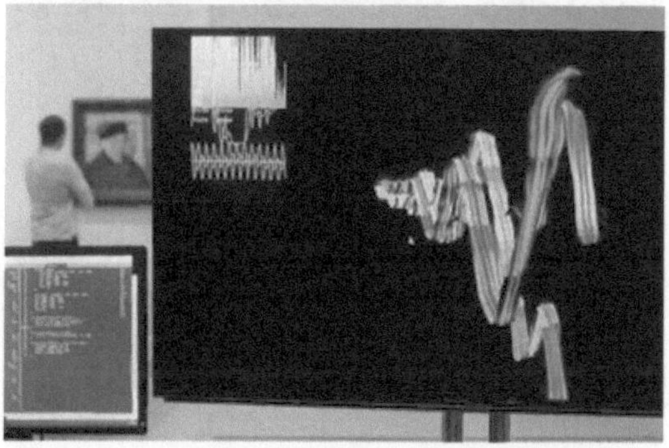

Figure 4. The signal of EEG to show the brainwave.

After COVID-19, The Mill, funded by Art Fund director Jenny Waldman, launches a scheme to encourage people back to museums and galleries. They collaborate with interactive artist Seph Li to create 3D graphics showing real-time brainwave computing on the screen when audience wear an

EEG to view abstract, traditional fine art oil paintings, or the other artwork. Similarly, it resembles an EEG or other equipment to immediately demonstrate our real physiology. The graphic physiological activity is a brand-new experience for the audience. They never imagined brainwaves would be recorded and displayed as 3D motion images of their physiology by EEG on the screen. As far as results are concerned, it's successful. Their scheme fascinates some people who want to experience EEG for personal recording of 3D motion images. People not only stand in front of the artwork but also receive their special 3D colorful motion image. That would make visiting galleries and museums more interesting.

1.2.3 *Paint with Your Mind – Designing EEG – based Interactive Installation for Traditional Chinese Artworks*

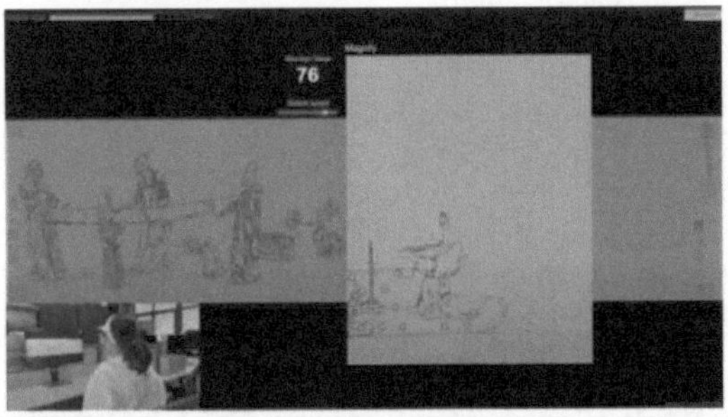

Figure 5. The participant is using EEG to paint the artwork.

Figure 6. The participant is using EEG to paint the artwork

"*Paint with Your Mind*" is an art installation exhibition with EEG to paint the traditional Chinese artwork – *"Court Ladies Preparing Newly Woven Silk"* by Xuan Zhang. They use Attend mode of EEG. In the begging when audience wears EEG in front of the screen in museum. The EEG would collect the signal of brainwave from visitors. If the signals reflect the people's Attend mode concentration level. The three steps of the installation are designed: sketch, coloring and movement-control. Designing EEG-based interactive exhibition could be a promising solution for engaging non-professional and non-Chinese visitors in Chinese painting appreciation (Chen Z, *et al.* 2021). Because of people not really understand these traditional artworks, sometimes audience would skip treasure in the museum, the sturdy makes people enjoy the traditional art that is their purpose. They want to through the installation to let audience participate the ancients' painter step by step. It's not hard but the audience willing to stay in the traditional fine art painting in the museum more time. Audience might want to learn deeper of this artwork after experiment. As a result, the target is successful. People spend to comprehend more ancient artwork in the museum. People spend to comprehend more ancient artwork in the museum. The study achieved the goal the goal.

2 RESEARCH METHODS

2.1 *Interactive painting system development*

The research employs EEG to create a prototype motion image system. Throughout the experiment, it highlights the Human-Computer Interaction between the audience and the artwork. By utilizing real-time EEG computing, the equipment not only analyzes the changes in brainwave signal strength but also enables the audience to perceive their own virtual status and emotional strength through brainwave imagery following the motion image display. This imagery accurately reflects the physiological reactions of people and provides additional information about emotions, such as introducing "Sunset in Tamsui" by Tan Teng-pho. The research aims for the audience to engage with the artwork, fostering an understanding of its inherent qualities and facilitating immersive experiences that exchange emotion through the prototype system.

2.2 *Questionnaire*

Figure 7. Audience wears EEG to watch the painting.

One of the research methods employed is a questionnaire survey. For credibility and validity, the questionnaire must be evaluated by two empirical experts after designing it, and then at least 30 participants should be invited to take part. The questionnaire should cover aspects such as usability, entertainment, and emotional pleasure. After collecting the questionnaires, they will be analyzed to explain the relationship between emotion and EEG, thus demonstrating the completeness and credibility of the research.

2.3 *Expert interview*

The research would invite 2–3 different types of experts to conduct in-depth interviews. These interviews would cover various aspects to discuss the reactions of different people to the motion image. Experts in related fields would offer professional insights to delve deeper into the research and experiments, potentially mentioning future directions for the research. These experts might also discuss new technologies and different methods, contributing to enhancing the quality and effectiveness of the research.

3 RESULT AND ANALYSIS

Through the prototype developed in this research, the interactive experience of the audience is provided. The audience can intuitively know the brainwave activities and emotional feelings. Interactive installation art can evoke a range of positive and negative emotions, and designers and curators must understand the role of emotion in the aesthetic experience of interactive installation art to create impactful experiences for viewers (Chen and Ibrahim 2023). It applies to tech art artwork to explore more creative possibilities. When different motion images are shown on the screen, people begin to focus on their psychology and become curious about the interaction between EEG and the motion image. This allows the creative and prominent aspects of EEG to have more potential. The research highlights the important and valid value of EEG.

In the developmental prototype of the research, "Sunset in Tamsui" would be in standby mode for the audience on the screen initially. The standby mode displays scenes such as sea waves, people moving, and different angles of the sunset playing in a loop. When the audience wears the EEG in front of the screen, it starts to read the brainwave signals of the audience. There are four corresponding emotional types of motion images: Attention, Meditation, Anxiety, and Calmness. Attention involves noise and more crows showing on the screen; Meditation gradually transitions to a darker night view; Anxiety displays a storm; Calmness shows a sun appearing at sea level with changing colors. Finally, each motion image returns to the standby mode.

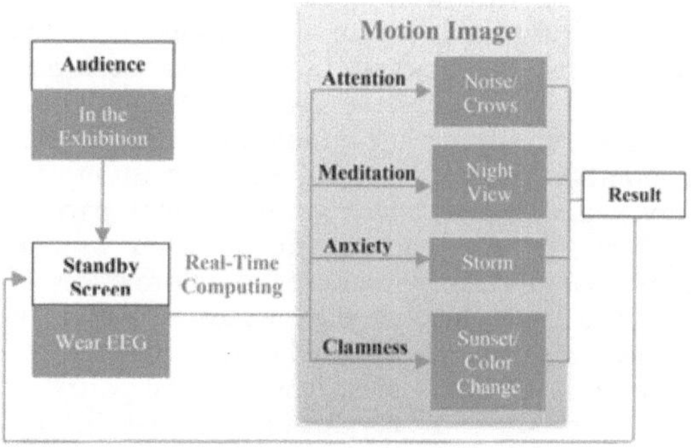

Figure 8. The system flow of our research.

One of the evaluation methods for EEG-based emotion recognition in this research is a questionnaire survey. After collecting basic information such as gender, age, or prior EEG experience, the survey includes questions about mental state, followed by an analysis of EEG signals to assess relevance or external performance of truthfulness. The second part assesses the participants' experiences using a 5-point Likert scale ranging from Strongly Agree to Strongly Disagree to express their perceptual experience. Traditional behavioral and affective analysis methods often rely solely on self-reporting, which may limit the expression of the complete emotional panorama (Jimenez, *et al.* 2022). The experiment aims to involve at least 30 participants to ensure the credibility of the research.

The four emotions identified through EEG apply to the appreciation of fine art. By classifying these emotions, EEG allows the audience to understand their own emotions while viewing artwork and evaluating the collected brainwave data. This has implications for both perceptual and motor processing, potentially influencing how we perceive art pieces or prepare for goal-oriented movements (Jesus G. Cruz-Garza *et al.* 2017). When art creation interacts with the audience, it expands the new realm of tech art, inspiring viewers to connect their senses, sight, and emotional expression for a deeper understanding.

4 CONCULTION AND DISCUSSION

In the art of the new media environment, it does not rely on the new media platform to disseminate or merely display cutting- edge art but implies a new art mainly involving interaction, which seems incomprehensible at the vanguard and general social level (Guo and Zhang 2022). This research explores the potential of technological art through prototype development. It records brainwave activity, perception, and emotions during the experience. The use of summary statistics in scene perception not only gives observers an immediate sense of the visual world, but also provides a foundation for further exploration. (Michael A. Cohen *et al.* 2016). Summary statistics aid in scene perception, providing observers with immediate visual insights and laying the groundwork for further exploration (Wang, X. *et al.* 2023). Questionnaire survey and expert interview are helpful part of analysis. In the part of questionnaire, might emphasis to classify background information of audience (participants), e.g., education level or family background to conduct further analysis.

In terms of limitations, it will be the part that adjusts the variables: (1) the audience has feelings after watching the painting, or (2) the emotions they have already carried into the space before they watch the interactive art? If we want to eliminate the interference of adjustment variables, we may be able to extend standby mode time of the prototype to make audience have deeply experience in the art by themselves before using the EEG to detect and analyze emotions.

This research offers a fresh perspective, intertwining emotion with fine art, where EEG detects emotions, enhancing the interactive art experience. Physiological-based emotion recognition, difficult to manipulate deliberately, lends credibility Physiological-based emotion recognition is considered more reliable as it is difficult for individuals to deliberately control their physiological signals (Li, *et al.*, 2023). This convergence fosters synergy between technological and traditional art forms, with EEG prototypes enriching public appreciation and understanding of creative artwork backgrounds.

REFERENCES

Castiblanco Jimenez, I. A., Nonis, F., Olivetti, E. C., Ulrich, L., Moos, S., Monaci, M. G., ... & Vezzetti, E. (2023). Exploring User Engagement in Museum Scenario with EEG—A Case Study in MAV Craftsmanship Museum in Valle d'Aosta Region, Italy. *Electronics*, *12*(18), 3810.

Chen, X., & Ibrahim, Z. (2023). A Comprehensive Study of Emotional Responses in AI-Enhanced Interactive Installation Art. *Sustainability*, *15*(22), 15830.

Chen, Z., Liao, J., Chen, J., Zhou, C., Chai, F., Wu, Y., & Hansen, P. (2021, February). Paint with Your Mind: Designing EEG-based Interactive Installation for Traditional Chinese Artworks. In Proceedings of the Fifteenth International Conference on Tangible, Embedded, and Embodied Interaction. pp. 1–6.

Cohen, M. A., Dennett, D. C., & Kanwisher, N. (2016). What is the bandwidth of perceptual experience?, *Trends in cognitive sciences*, *20*(5), 324–335.

Cruz-Garza, J. G., Brantley, J. A., Nakagome, S., Kontson, K., Megjhani, M., Robleto, D., & Contreras-Vidal, J. L. (2017). Deployment of mobile EEG technology in an art museum setting: Evaluation of signal quality and usability. *Frontiers in human neuroscience*, *11*, 527.

Guo, L., & Zhang, L. (2022). Exploration on the application of new media interactive art to the protection of traditional culture. *Scientific Programming*.

Lancel, K., Maat, H., & Brazier, F. (2019). EEG KISS: shared multi-modal, multi brain computer interface experience, in public space. *Brain Art: Brain-Computer Interfaces for Artistic Expression*, 207–228.

Li, J., Pan, W., Huang, H., Pan, J., & Wang, F. (2023). STGATE: Spatial-temporal graph attention network with a transformer encoder for EEG-based emotion recognition. *Frontiers in Human Neuroscience*, *17*, 1169949

Teresa Nowakowski (2023, November 15). See What Your Brain Does When You Look at Art. https://www.smithsonianmag.com/smart-news/this-headest-shows-you-what-your-brainwaves-do-when-you-look-at-art-180983261/

Wang, X., Ren, Y., Luo, Z., He, W., Hong, J., & Huang, Y. (2023). Deep learning-based EEG emotion recognition: Current trends and future perspectives. *Frontiers in psychology*, *14*, 1126994.

Wu, C. (2021). Application of digital image based on machine learning in media art design. *Computational Intelligence and Neuroscience*.

Zhuang, J. R., Guan, Y. J., Nagayoshi, H., Muramatsu, K., Watanuki, K., & Tanaka, E. (2019). Real-time emotion recognition system with multiple physiological signals. *Journal of Advanced Mechanical Design, Systems, and Manufacturing*, *13*(4), JAMDSM0075–JAMDSM0075.

System Innovation for an Artificial Intelligence Era – Lam et al. (Eds)
© 2025 the Author(s), ISBN: 978-1-032-84452-7

Image cropping based on photographic composition

Tsu-Kang Fang & Tsorng-Lin Chia
Department of Applied Artificial Intelligence, Ming Chuan University, Taiwan

ABSTRACT: In photography, the precision in image composition and cropping processes can effectively identify optimal strategies for subject geometric adjustment, thereby enhancing images' overall compositional integrity and visual appeal. In light of this, the present study introduces a novel deep learning system for image cropping, predicated on the principles of photographic composition, termed the Composition-based Image Cropping Network (CICNet). This system initially employs Convolutional Neural Networks (CNN) to identify key visual elements within images, subsequently utilizing deep learning techniques to classify photographic composition styles and generate heatmaps depicting the distribution of principal objects within the images. Following this, based on the composition style most congruent with the image, the system applies a deep cropping network to trim the original image, ensuring that principal objects are positioned in locations that align with visual aesthetic principles. Through a series of empirical validations, the system developed in this research has demonstrated its efficacy in enhancing the compositional beauty of images via automated cropping processes.

Keywords: Photographic composition, Image cropping, Composition Recognition, Deep learning

1 INTRODUCTION

The rapid development and ubiquity of smartphones and digital cameras have significantly lowered the barrier to photography. However, amateur photography enthusiasts often struggle to achieve high-quality photos due to a lack of technical skills and aesthetic sense. Since aesthetic judgment is subjective, involving emotions and personal taste (B. Martinez *et al.* 1998), mastering aesthetics poses a significant challenge.

Photographic composition can endow a photo with beauty and follow specific rules, making it a crucial factor in the aesthetics of an image. Effective photographic composition can clearly convey an image's intended content and atmosphere to viewers, while different compositions elicit varied feelings and effects. This study employs compositions identified from the literature listed in Table 1, including the Rule of Thirds, vertical, horizontal, diagonal, curves, triangles, center, symmetry, pattern, frame, and vanishing point. The Golden Ratio and radiating lines are excluded from discussion due to their similarity to the Rule of Thirds and vanishing points, respectively, and the scarcity of literature discussing them. This research defines the 11 types of composition as follows:

(1) *Rule of Thirds Composition* (*RoT*): Utilizes four imaginary lines to divide the image horizontally and vertically into three equal parts, placing the subject or theme on these lines, around them, or at their intersections. This type creates a strong, balanced photo that guides the viewer's attention to the subject, as shown in Figure 1(a).
(2) *Center Composition* (*CC*): Positions the subject in the center of the image to emphasize it. Often, the background is adjusted to fall out of focus, highlighting the subject's details and concentrating the viewer's attention on the subject, as shown in Figure 1(b).
(3) *Symmetric Composition* (*SC*): Uses reflections from water, mirrors, metals, etc., to create a symmetrically arranged image. Typically, a vertical or horizontal dividing line appears in the image, which is a basis for displaying symmetry above and below or to the left and right, as shown in Figure 1(c).

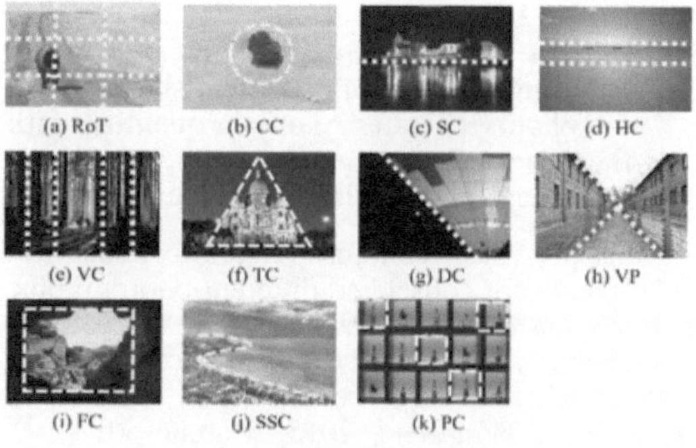

Figure 1. Types of photographic composition.

DOI: 10.1201/9781003514831-46

(4) *Horizontal Composition* (*HC*): Displays a visual effect parallel to the ground or horizon, conveying a sense of calm, stability, and serenity. It is often used in landscape photography, as shown in Figure 1(d).
(5) *Vertical Composition* (*VC*): Vertical elements can make a photo more appealing, using vertical objects in the photo to create depth and a sense of space, emphasizing foreground, middle ground, and background, as shown in Figure 1(e).
(6) *Triangle Composition* (*TC*): The triangular shapes of the main objects can show direction and orientation, creating effects of perspective and depth. Incorporating different triangular compositions in the image can serve various purposes, such as guiding the viewer's gaze forward or backward depending on the triangle's orientation, as shown in Figure 1(f).
(7) *Diagonal Composition* (*DC*): Arranges objects along the diagonal direction of the image or divides the frame diagonally, separating different areas and color blocks. This type creates a sense of extension, imbalance, and dynamic, uneasy, or majestic perceptions, as shown in Figure 1(g).
(8) *Vanishing Point Composition* (*VP*): Lines (parallel lines in physical space) or linear objects in the image extend towards the background, converging at a point or area to form a vanishing point. These lines guide the viewer's gaze to the vanishing point, as shown in Figure 1(h).
(9) *Frame Composition* (*FC*): Uses elements in the foreground as a frame surrounding the subject, similar to a picture frame, drawing the viewer's attention to the subject within. The frame's shape is not limited to rectangles; common frames include windows, caves, or doorways that are monochromatic foreground objects, as shown in Figure 1(i).
(10) *S-Curved Composition* (*SCC*): The image uses smooth curves to guide the viewer's gaze, enhancing the liveliness of the original image and maintaining the viewer's focus. Often used in landscape images, the S-shaped composition gives viewers the feeling of following the curve through the scenery, as shown in Figure 1(j).
(11) *Pattern Composition* (*PC*): By repeatedly featuring elements of consistent shape within the photo, it creates beauty in quantity, with the repeated elements becoming the most prominent part, as shown in Figure 1(k).

Most previous techniques related to visual aesthetics have not considered different photographic composition rules. For instance, they subjectively or objectively assess the beauty of photos without a thorough analysis of composition rules (Luo *et al.* 2011; Tian *et al.* 2015; Marchesotti *et al.* 2011). Therefore, if it were possible to automatically determine the composition rules of photos, the efficiency of these image applications could be further enhanced. Table 1 presents recent literature on image cropping and enhancement based on photographic composition.

Table 1. Photographic composition types considered in the literature on image applications based on photographic composition

Ref.	RoT	CC	SC	HC	VC	TC	DC	VP	FC	SCC	PC	Number of types
L.Jiu et al. 2010	✓			✓			✓					3
P.Obrador et al. 2010	✓											2
R. Abdullah et al. 2011	✓	✓		✓			✓		✓			5
Y. W. Guo et al. 2012	✓											1
J. Park et al. 2012	✓											1
C. L. Wen et al. 2012	✓	✓	✓			✓	✓	✓	✓			7
R. Gallea et al. 2013	✓						✓					2
Mavridaki et al. 2015	✓			✓					✓			3
M. T. Wu et al. 2017	✓								✓		✓	3
Y. F. Li et al. 2020	✓	✓		✓		✓		✓	✓			6
C. Hong et al. 2021	✓	✓	✓	✓	✓	✓	✓			✓	✓	9
X. Li et al. 2022	✓											1
S. Ni et al. 2022		✓	✓	✓	✓		✓					6
D. Sakaguchi et al. 2022	✓	✓	✓				✓					4
J. Hou et al. 2023	✓	✓	✓	✓	✓		✓		✓			7
Y. Wang et al. 2023	✓	✓	✓	✓	✓	✓	✓			✓	✓	9

This study introduces a deep network model for image cropping based on photographic composition. This model learns various photographic composition rules and identifies the optimal arrangement of subjects within photos. Finally, it crops the photos based on this information to produce images that conform to the aesthetics of photographic composition.

2 METHODS

2.1 *System structure*

This study focuses on developing a deep learning model capable of recognizing photographic composition and performing optimal aesthetic cropping of photos, thereby ensuring the outputted images exhibit aesthetics in line with photographic composition. This study proposes a deep learning system comprised of three main modules: *Key Visual*

Elements Extraction (*KVEE*), *Photo Composition Recognition* (*PCR*), and *Composition-Based Cropping* (*CBC*). Based on the design principles outlined above, the operational framework of the proposed *Composition-based Image Cropping Network* (*CICNet*) is illustrated in Figure 2.

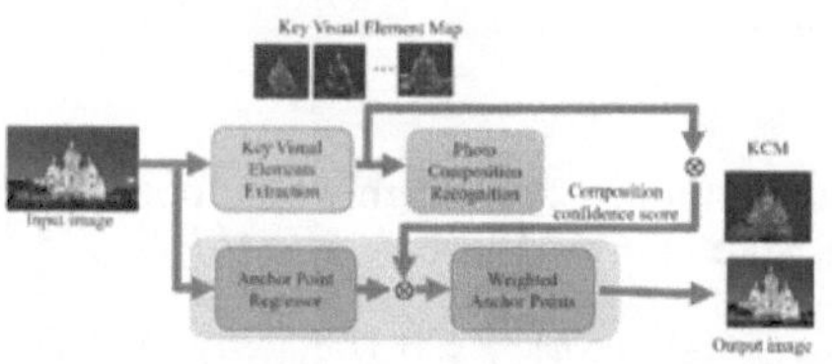

Figure 2. Operational framework of the Composition-based Image Cropping Network (*CICNet*).

2.2 *Key visual elements extraction*

Objects in photographs can be regarded as elements within the image. These elements' placement, shape, size, and color directly or indirectly affect the content the photo aims to convey, making them crucial to the image. Among these elements, a few are particularly important for the impact on the visual content of the image and can be referred to as key visual elements. To discover these key visual elements in images, this study is inspired by *Class Activation Maps* (*CAM*) (Zhou *et al.* 2016) and will utilize traditional backbone networks for effective feature extraction and fusion, forming a *Key Visual Element Map* (*KVEM*). This study employs the concise VGG16 as the backbone network to extract key visual elements of different compositions and incorporates the U-Net architecture to make the generated feature maps more closely resemble the original image. The *KVEE* network architecture is illustrated in Figure 3.

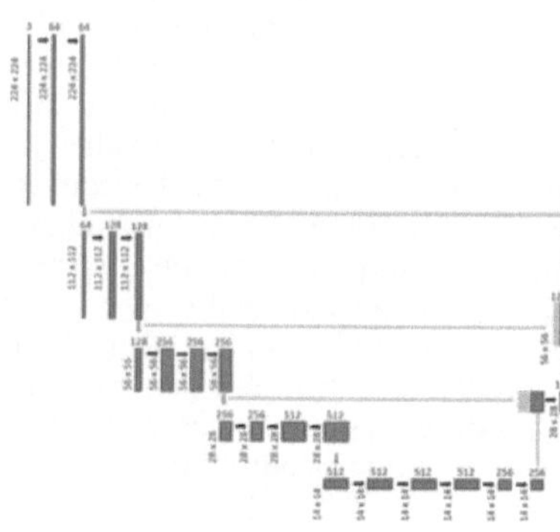

Figure 3. *KVEE* network architecture.

2.3 *Photo composition recognition*

After extracting feature maps of each image through *KVEE*, *PCR* processes these feature maps through a Global Average Pooling (*GAP*) layer and fully connected layers to determine which compositions are present in the input image and to score the image's relevance to the 11 types of composition. The *PCR* outputs two results: confidence scores for the 11 types of composition in the image and a *Key Composition Map* (*KCM*) indicating the distribution of the main compositions in the input image. Since an image may contain more than one type of composition, to preserve key compositions, the final convolutional layer's feature maps are used to generate *CAMs* for the 11 types of composition via *GAP*, which are then globally weighted and averaged based on their scores to produce the *KCM*. The *PCR* architecture and the schematic representation of *KCM* synthesis are illustrated in Figure 4. The *KCM* is generated through the calculation:

$$KCM = \frac{1}{K}\sum_{k=1}^{K}\sum_{c=1}^{11} s_{kc} \cdot U(FM_k), \qquad (1)$$

where K is the number of photographic composition types considered, which is 11 in this study, FM_k represents the k-th input feature map, S_{kc} is the score for the c-th composition characteristic from the k-th input feature map, c refers to the names (numbers) of the 11 types of composition, and $U(\cdot)$ is a normalization function. The content displayed by the *KCM* is considered the primary area influencing the image's composition. For images with different photographic compositions, the corresponding KCMs are shown in Figure 5.

2.4 *Composition-based cropping*

This study approaches image cropping by combining expert skills with the principles of compositional

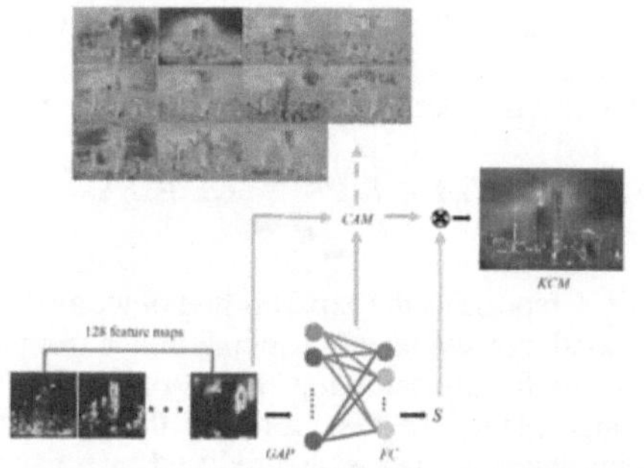

Figure 4. *PCR* architecture and *KCM* generation flow diagram. The input consists of 128 feature maps, and *S* represents the composition characteristic score vector.

aesthetics. A deep learning model is utilized to learn the cropping methods of experts, with multiple *cropping image frame* candidates defined by their corners, length, and width parameters. Through referring to the contents of the *KCM*, multiple cropping image

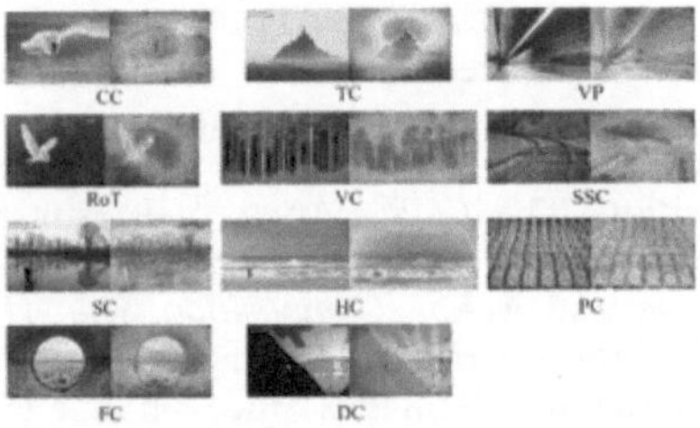

Figure 5. *KCM*s obtained from images of different composition types.

frames are filtered and selected for aesthetic optimization; the final output is a cropped image with a high degree of compositional aesthetics.

For cropping a rectangular area from an image with dimensions $W \times H$, existing image cropping databases label the crop through manual expert intervention in a four-dimensional definition (i, j, w, h), where (i, j) represents the top-left corner of the crop rectangle, commonly referred to as the anchor-point, and (w, h) represents the width and height of the cropping rectangle. This labeling method could produce $O(W^2H^2)$ possibilities, which, when applied to a deep learning model, is significantly demanding in terms of hardware and training time. To overcome this challenge, this study down-samples the four-dimensional labels into spatial $p = (i, j)$ and size $b = (w, h)$, initially learning the expert cropping skills at a lower resolution. The cropping labels are adjusted as follows:

$$CL : (p, b) \rightarrow \left(\lfloor \frac{p}{S} \rfloor, \lfloor \frac{b}{P} \rfloor \right), \tag{2}$$

where S represents the spatial sampling interval on the image plane, and P denotes the sampling interval for the width and height of the cropping rectangle. Therefore, the deep network used for learning image cropping skills can learn with fewer weight parameters and a smaller output layer, effectively acting as a transformation function between the input image I and the rectangular cropping image frame:

$$CM : I \rightarrow CL, I \in R^{W \times H \times 3}, CL \in R^{\frac{W}{S} \times \frac{H}{S} \times \frac{W}{P} \times \frac{H}{P}}, \tag{3}$$

where CL represents the expert rating outcome for the corresponding rectangular cropping image frame.

After uniformly sampling the cropping labels on the image plane, the resolution of the rectangular cropping image frames is also reduced by a factor of S in both the horizontal and vertical directions. This process means that further fine-tuning of the rectangular cropping image frame's position within an $S \times S$ range will be necessary. Initially, this study up-samples the CL containing expert ratings in both the spatial and size domains to the same resolution as the input image plane, $W \times H$, and then incorporates compositional information carried by the *KCM*. It performs point-by-point weighted processing on rectangular cropping image frames with the same anchor points but different sizes within *CL*:

$$CW(i,j,w,h) = \frac{KCM(S \cdot (i + 0.5), S \cdot (j + 0.5))}{\sum_{a=0}^{S-1} \sum_{b=0}^{S-1} KCM(S \cdot (i - 1) + a, S \cdot (j - 1) + b)},$$
$$i, w \in [0, W], \; j, h \in [0, H], \tag{4}$$

thereby obtaining the Weighted Cropping Anchor-Point Map (WCAM):

$$WCAM(i,j,w,h) = CW(i,j,w,h) \cdot KCM(i,j) \tag{5}$$

The anchor point of the rectangular cropping image frame will be estimated within an $S \times S$ range through a local displacement vector $d = (\Delta i, \Delta j)$, while the size of the rectangular cropping image frame will be estimated within a $P \times P$ range through an extension vector $e = (\Delta h, \Delta w)$ based on length and width. The selection criteria are based on the degree of overlap between the rectangular cropping image frame and the content of the *KCM*. When the current rectangular cropping image frame can encompass the maximum content of the *KCM*, it indicates that the cropping area contains the main objects in the image without losing the original intention of photography in subject or composition. Therefore, after being weighted by the *KCM*, the anchor point of each rectangular cropping image frame undergoes an optimization search for all rectangular anchor points and sizes:

$$CB = \arg \max_{(i,j,w,h)} WCAM(i,j,w,h) \cdot Overlap((i,j) + d, (w,h) + e), \tag{6}$$

where *Overlap*() calculates the average value within the *WCAM* for the rectangular cropping image frame defined by a specific anchor point and size.

When the *Overlap*() of the rectangular cropping image frame *CB* exceeds a threshold value, it indicates that this cropping frame meets the requirements of photographic composition rules and aligns with expert cropping techniques. Consequently, an image output is generated, including the type of photographic composition and cropping labels. When an image has multiple photographic compositions, the system will produce an optimally cropped image for each type of photographic composition.

3 RESULTS AND DISCUSSION

3.1 *Datasets used in the experiment*

The study utilizes the KU-PCP (Lee *et al.* 2018) dataset as the training set for composition discrimination, with its composition types reclassified into 11 categories. The reclassification comparison is illustrated in Figure 6.

The FCDB image cropping dataset containing 1,706 images was used for the cropping experiments, of which 1,395 were used for training and 311 for testing. Each image includes information on 10 types of cropping frames and the number of subjective recognitions for each frame, with the frame having the highest number of subjective recognitions labeled as the cropping frame label.

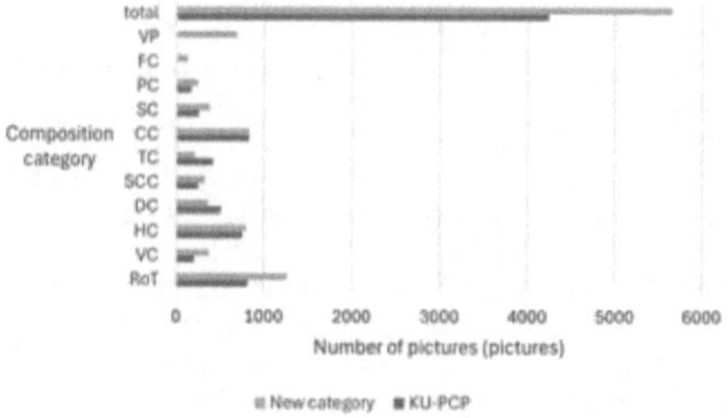

Figure 6. Statistics of composition types for training composition recognition.

3.2 *Image composition recognition experiment*

In this study, the *KCM* serves to guide the cropping work and verify whether the system has learned the rules of photographic composition. The *KCMs* generated by the system for various compositions refer to Figure 5, indicating that the system has indeed learned different rules of photographic composition. The deeper red areas within the *KCM* signify sections of greater importance to the model, representing the key areas for composition judgment. The composition discrimination accuracy for the *KUPCP* dataset is 84.07%, with the accuracy for different compositions shown in Figure 7.

3.3 *Image cropping experiment*

In this study, the performance of the cropping experiment is evaluated using the Intersection over Union (IoU) method, which calculates the overlap

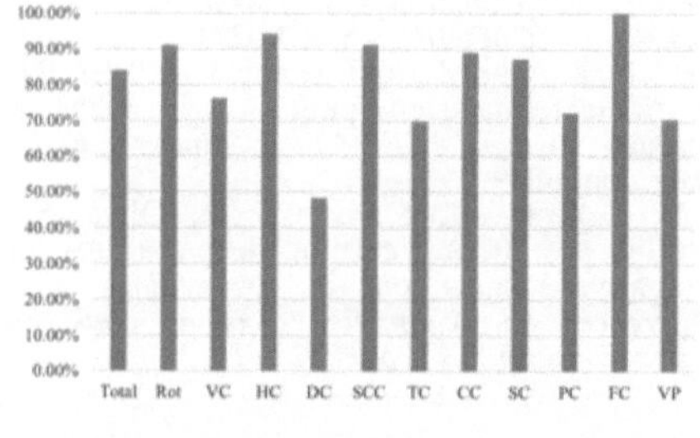

Figure 7. Overall recognition accuracy and accuracy for various compositions.

Table 2. Performance comparison of *CICNet* and existing methods.

Methods	IoU
VEN (Z. Wei *et al.* 2018)	0.73
VFN (Y. L. Chen *et al.* 2017)	0.68
VPN (Z. Wei *et al.* 2018)	0.71
GAIC (H. Zeng *et al.* 2019)	0.67
CICNet (our)	0.70

between the predicted cropping frame and the actual cropping frame to assess the efficiency of the cropping model. Compared with existing methods, as shown in Table 2, the *CICNet* may not perform the best, but it delivers quality results.

To explore whether the cropping system endows photos with compositional aesthetics, this study assesses the alignment with theoretical expectations by comparing the confidence scores of photo compositions before and after cropping. Figure 8 shows that the confidence scores for photo compositions increase after cropping, proving that the system enhances the compositional aesthetics of the photos.

To explore whether the cropping system enhances the aesthetic composition of photos, this study compares the confidence scores of photo

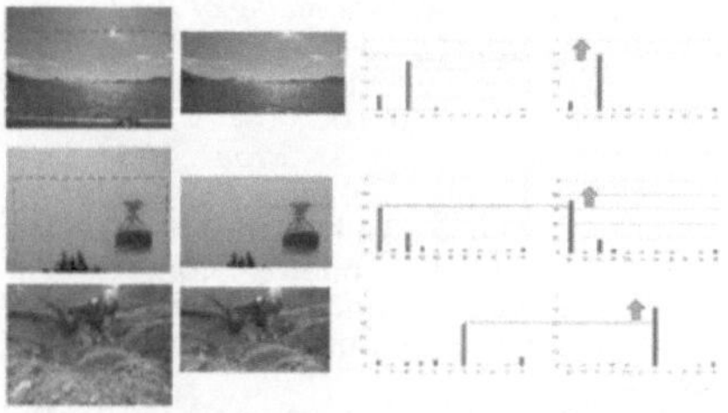

Figure 8. From left to right, the images are the original image, the result after one crop, the composition score of the original image, and the composition score of the cropped image. The red dashed line on the original image indicates the cropping output area.

compositions before and after cropping to assess their alignment with theoretical expectations. Figure 8 shows that the confidence scores of photo compositions have increased after cropping, demonstrating that the system enhances the aesthetic composition of the photos.

4 CONCLUSION

This paper proposes an image cropping method based on photographic composition, comprising two modules: Photo Composition Recognition and Image Cropping. The Photo Composition Recognition module can identify 11 common photographic composition styles and generate scores indicating the degree of association of the input image with these 11 compositions, as well as produce a *KCM* to mark the distribution of subjects within the image, serving as a basis for subsequent image cropping. Experimental results demonstrate that the proposed system architecture can effectively learn different rules of photographic composition, achieving an accuracy of 84.07% in tests on the KUPCP dataset.

The proposed method can determine the appropriate image cropping options in the image cropping module based on photographic composition scores and the *KCM*. Validation experiments using the FCDB image cropping dataset not only realize a cropping method based on composition but also prove that the system can automatically enhance the aesthetic composition of photos.

In the future, this study will extend the recognition and cropping results based on photographic composition into automated image enhancement processing, allowing images to receive correct adjustments in both composition and color simultaneously.

REFERENCES

Abdullah R., Christie M., Schofield G., Lino C., and Olivier P., (July 2011). Advanced composition in virtual camera control. In *Proceeding of the International Symposium on Smart Graphics*, pp. 13–24.

Chen Y. L., Klopp J., Sun M., Chien S. Y., and Ma K. L. (2017). Learning to compose with professional photographs on the web. In *Proc. ACM Int. Conf. Multimedia*, 2017, pp. 37–45.

Gallea R., Ardizzone E., and Pirrone R. (Sept. 2013). Automatic aesthetic photo composition. In *Proceeding of the International Conference on Image Analysis and Processing*, Springer, Berlin, Heidelberg, pp. 21–30.

Guo Y. W., Liu M., Gu T. T., and Wang W. P. (Sept. 2012). Improving photo composition elegantly: Considering image similarity during composition optimization. *Computer Graphics Forum*, vol. 31, no. 7, pp. 2193–2202.

Hong C., Du S., Xian K., Lu H., Cao Z., and Zhong W. (2021). Composing photos like a photographer. In *Proceedings of the IEEE/CVF Conference on Computer Vision and Pattern Recognition*, pp. 7057–7066.

Hou J., Lin W., Fang Y., Wu H., Chen C., Liao L. and Liu W. (2023). Towards transparent deep image aesthetics assessment with tag-based content descriptors. *IEEE Transactions on Image Processing*, doi: 10.1109/TIP.2023.3308852.

Lee J. T., Kim H. U., Lee C., and Kim C. S. (2018). Photographic composition classification and dominant geometric element detection for outdoor scenes. *Journal of Visual Communication and Image Representation*, vol. 55, pp. 91–105.

Li X., Zhang G., Wu Y., Li X., and Zhang Y. (2022). Rule of thirds-aware reinforcement learning for image aesthetic cropping. *The Visual Computer*, pp. 1–17.

Li Y. F., Yang C. K., and Chang Y. Z. (2020). Photo composition with real-time rating. *Sensors*, vol. 20, no. 3, article 582, pp. 1–29.

Liu L., Chen R., Wolf L., and Cohen-Or D. (2010). Optimizing photo composition, *Computer Graphics Forum*, vol. 29, no. 2, pp. 469–478.

Luo W., Wang X., and Tang X., Content-based photo quality assessment. In *Proceeding of the Conference on Computer Vision and Pattern Recognition (CVPR)*, 2011, pp. 2206–2213.

Marchesotti L., Perronnin F., Larlus D., and Csurka G. (2011). Assessing the aesthetic quality of photographs using generic image descriptors. In *Proceeding of the IEEE International Conference on Computer Vision*, pp. 1784–1791.

Martinez B. and Block J. (1998). *Visual Forces, an Introduction to Design*. Prentice-Hall, New York.

Mavridaki E. and Mezaris V. (Sept. 2015). A comprehensive aesthetic quality assessment method for natural images using basic rules of photography. In *Proceeding of the 2015 IEEE International Conference on Image Processing (ICIP)*, pp. 887–891.

Ni S., Shao F., Chai X., Chen H., and Ho Y. S. (2023). Composition-guided neural network for image cropping aesthetic assessment. *IEEE Transactions on Multimedia*, vol. 25, pp. 6836–6851.

Obrador P., Schmidt-Hackenberg L., and Oliver N. (Sept. 2010). The role of image composition in image aesthetics. In *Proceeding of the 2010 IEEE International Conference on Image Processing*, pp. 3185–3188.

Park J., Lee J. Y., Tai Y. W., and I. S. Kweon. (Sept. 2012). Modeling photo composition and its application to photo re-arrangement. In *Proceeding of the 2012 19th IEEE International Conference on Image Processing*, pp. 2741–2744.

Sakaguchi D., Takimoto H., and Kanagawa A. (2022). Study on relationship between composition and prediction of photo aesthetics using CNN. *Cogent Engineering*, vol. 9, no. 1, article 2107472, pp. 1–15.

Tian X., Dong Z., Yang K., and Mei T. (2015). Query-dependent aesthetic model with deep learning for photo quality assessment. *IEEE Transaction on Multimedia*, vol. 17, no. 11, pp. 2035–2048.

Wang Y., Ke Y., Wang K., Guo J., Yang S. (Feb 2023). Spatial-invariant convolutional neural network for photographic composition prediction and automatic correction. *Journal of Visual Communication and Image Representation*, vol. 90, article 103751.

Wei Z., Zhang J., Shen X., Lin Z., Mech R., Hoai M., and Wei Z., Zhang J., Shen X., Lin Z., Mech R., Hoai M., and Samaras D. (2018). Good view hunting: Learning photo composition from dense view pairs. In *Proc. IEEE Conf. Comput. Vis. Pattern Recog.*, pp. 5437–5446.

Wen C. L. and Chia T. L., The fuzzy approach for classification of the photo composition. In *Proceedings of the 2012 International Conference on Machine Learning and Cybernetics*, vol. 4, July 2012, pp. 1447–1453.

Wu M. T., Pan T. Y., Tsai W. L., Kuo H. C., and Hu M. C. July (2017). High-level semantic photographic composition analysis and understanding with deep neural networks. In *Proceeding of the 2017 IEEE International Conference on Multimedia & Expo Workshops (ICMEW)*, pp. 279–284.

Zeng H., Li L., Cao Z., and Zhang L. (2019). Reliable and efficient image cropping: A grid anchor based approach. In *Proc. IEEE Conf. Comput. Vis. Pattern Recog.*, pp. 5949–5957.

Zhou B., Khosla A., Lapedriza A., Oliva A., and Torralba A. (2016). Learning deep features for discriminative localization. in *Proceedings of IEEE Conference on Computer Vision and Pattern Recognition (CVPR)*, pp. 2921–2929.

System Innovation for an Artificial Intelligence Era – Lam et al. (Eds)
© 2025 the Author(s), ISBN: 978-1-032-84452-7

Deep learning-based yoga pose assessment system

Yu-Ting Tsai & Tsorng-Lin Chia
Department of Applied Artificial Intelligence, Ming Chuan University, Taoyuan, Taiwan

ABSTRACT: This study aims to develop a home-applicable yoga posture assessment system. Built on the MMPose deep learning framework, the system utilizes a general camera to instantaneously estimate the practitioner's bodily joint coordinates in three-dimensional space. By conducting relative corrections of these three-dimensional coordinates, the system swiftly calculates changes in the bending angles of various joints. Moreover, this research employs Dynamic Time Warping (DTW) technology to compare the spatiotemporal discrepancies between user and instructor movements, thereby generating movement scores and corrective suggestions. Verified through empirical experimentation, the system developed in this study accurately provides movement scoring and offers specific recommendations for improvement.

1 INTRODUCTION

Yoga, originating from India, is a health-oriented physical practice that emphasizes harmony between the body and mind. Numerous studies have confirmed the efficacy of yoga in maintaining physical and psychological health and facilitating spiritual evolution (B. K. S. Iyengar. 1965). In recent years, with the rise in health consciousness and technological advancements, home fitness has emerged as a popular trend, especially during the pandemic, as individuals seek safe and convenient exercise. This exercise has led to the critical need to develop a simple equipment-based home yoga training system.

Many individuals opt for self-learning yoga online, yet the absence of professional guidance can lead to injuries during practice. Research indicates incorrect yoga postures may result in musculoskeletal system injuries, including sprains and muscle strains (H. Cramer *et al.* 2019). Hence, a system that provides accurate posture and technique guidance is crucial for injury prevention. Furthermore, such a system should be capable of offering personalized recommendations and a scoring mechanism based on the practitioner's abilities to encourage safe and sustained practice.

Existing human pose estimation technologies, such as 2D human pose estimation tools, including OpenPose (S.-W. Shen *et al.* 2022), MediaPipe (A. Lobo *et al.* 2022), and AlphaPose (M. Chasmai *et al.* 2022), offer detection of 2D keypoints. However, due to the lack of depth information in 2D data, accurately detecting some complex human postures or movements can be challenging. For instance, in 2D data, detecting keypoints may be prone to errors when body parts overlap, posing difficulties for further human motion analysis and its applications. Moreover, 2D information cannot precisely describe the position and rotation of the human body in three-dimensional space, impacting applications that require accurate positional and rotational information. In yoga applications, accurately measuring the body's angles and positions is crucial to ensure the safety and effectiveness of the practice.

In recent years, integrating artificial intelligence (AI) with yoga training has seen a gradual increase in research to enhance practitioners' skill levels and prevent injuries. In prior studies, Wu *et al.* utilized wearable sensors to implement a yoga posture identification and quantitative assessment method, offering improved feedback and guidance for yoga learners (Z. Wu *et al.* 2019). Sharma *et al.* introduced a system named 'Surya Namaskar,' focusing on a posture estimation study of a yoga sequence encompassing eight yoga poses (A. Sharma *et al.* 2022). This system provides real-time feedback by extracting key points using MediaPipe and employing deep learning techniques (CNNs) to detect correct yoga postures. It assists yoga practitioners in executing movements accurately, thereby improving practice effectiveness and safety. These studies have primarily focused on identifying individual movement types without considering the overall dynamic poses. Yoga is a training method that requires coordinated movement of the entire body, where each action not only affects individual parts but also impacts the overall posture and balance of the body. Therefore, in researching yoga training, the identification of single movements can only offer limited information and often fails to reflect the accuracy and efficiency of the overall motion. Compared to the identification of single movements, the recognition of overall dynamic poses can capture the motion information of yoga practitioners more comprehensively, thus enhancing the accuracy and efficiency of training.

In conclusion, developing a yoga movement assessment system suitable for home environments

DOI: 10.1201/9781003514831-47

can aid practitioners in evaluating whether their movements are standard, thus preventing exercise-related injuries. It also provides a safe and efficient learning environment for practice.

2 THE PROPOSED METHOD

2.1 *Human joint coordinate extraction*

In this study, the MMPose toolkit was utilized to extract human posture information from yoga movement videos to conduct quantitative analysis and evaluate practitioner poses. MMPose is an open-source 3D human pose estimation toolkit based on the PyTorch deep learning framework, developed by OpenMMLab, capable of achieving high-precision 2D/3D human pose estimation. MMPose identifies human regions in the input video using object detection algorithms and extracts each human region. It estimates 2D keypoints for each region using 2D human pose estimation algorithms. These 2D key point sequences from each region are then used as input to estimate 3D keypoints in the human body's 3D space through 3D human pose estimation algorithms. The coordinate values of these keypoints are standardized based on the overall width and height of the video, with the value range set between -1 and 1. This standardization ensures that the position of the keypoints is not affected by the specific size of the video, thereby accommodating different video sizes. As shown in Figure 1, the results include a horizontal axis representing time and a vertical axis displaying the standardized coordinate values (X, Y, Z) of keypoints in three-dimensional space. MMPose also transforms the 3D key point time series output into a video of the 3D skeleton. MMPose can identify 17 key points in yoga movements, as shown in Figure 2, providing a basis for precise analysis of yoga poses and capturing the continuity of each movement, with detailed records of joint positions at different time points.

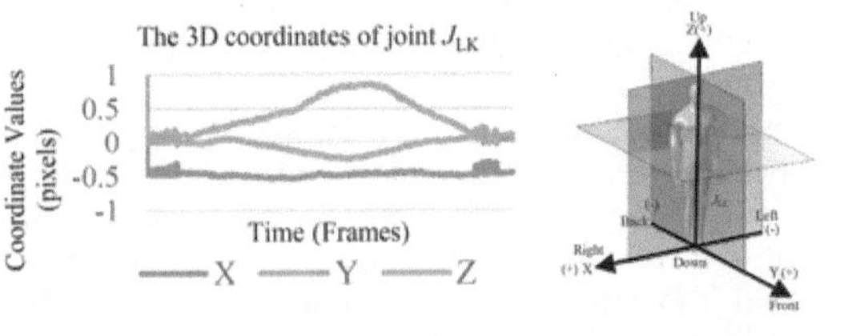

Figure 1. 3D coordinate curve and motion trajectory of joint J_{LK}.

Joint Number	Joint Name	Joint Number	Joint Name
1	Hip(HP)	10	Neck/Nose(NE)
2	R_hip(RP)	11	Head(HE)
3	R_knee(RK)	12	L_shoulder(LS)
4	R_foot(RF)	13	L_elbow(LE)
5	L_hip(LP)	14	L_wrist(LW)
6	L_knee(LK)	15	R_shoulder(RS)
7	L_foot(LF)	16	R_elbow(RE)
8	Spine(SP)	17	R_wrist(RW)
9	Thorax(TH)		

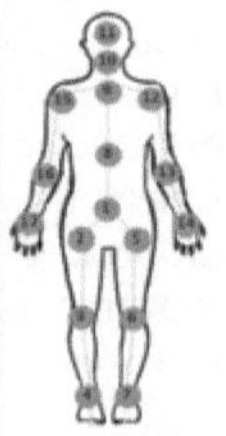

Figure 2. Numbering and naming of joint positions in MMPose.

2.2 *Joint angle calculation*

By extracting human posture information from image data using the MMPose toolkit, the extracted joint coordinate data will be utilized to calculate the angles between joints. This angle calculation is crucial for understanding the accuracy of yoga movements. The angle data can be compared with standard movements, identifying deviations in the practitioner's movements thereby providing targeted improvement suggestions. Therefore, this study opts to calculate angles based on the three-dimensional coordinates of the joint points using the cosine theorem and obtains full-range angles through manual adjustment strategies. This approach aids in capturing the completeness and dynamic changes of movements.

We analyze yoga movements using nine joint points shown in Figure 3, with angles calculated based on the following definitions: The bending angle θa is defined by the joint point J_{ao} and its two adjacent joint points (J_{a1} and J_{a2}). Considering the joint point coordinates obtained over time from the video, the 3D coordinates of the joint points are represented as $J^{(i)} = [X^{(i)}, Y^{(i)}, Z^{(i)}]$, where $i = 0, 1, ..., N$ is the number of data points. Thus,

$$V_{a1}^{(i)} = J_{a1}^{(i)} - J_{a0}^{(i)} = \left((X_{a1}^{(i)} - X_{a0}^{(i)}), (Y_{a1}^{(i)} - Y_{a0}^{(i)}), (Z_{a1}^{(i)} - Z_{a0}^{(i)})\right) \quad (1)$$

$$V_{a2}^{(i)} = J_{a2}^{(i)} - J_{a0}^{(i)} = \left((X_{a2}^{(i)} - X_{a0}^{(i)}), (Y_{a2}^{(i)} - Y_{a0}^{(i)}), (Z_{a2}^{(i)} - Z_{a0}^{(i)})\right) \quad (2)$$

and

$$\theta_a^{(i)} = COS^{-1} \frac{V_{a1}^{(\mathrm{i})} \cdot V_{a2}^{(\mathrm{i})}}{\left|V_{a1}^{(\mathrm{i})}\right| \cdot \left|V_{a2}^{(\mathrm{i})}\right|} \quad (3)$$

If the X-axis is considered the primary axis, with the angle range limited to 0 to 360 degrees, then the adjustment of joint angles is as follows:

$$\theta_a^{(i)} = \begin{cases} \theta_a^{(i)}, & if X_{a0}^{(i)} < X_{a1}^{(i)} \\ 360^\circ - \theta_a^{(i)}, & otherwise \end{cases} \quad (4)$$

This method initially determines the angle between joints based on the three-dimensional spatial positions of three joint points, applying the cosine theorem. The theorem is utilized by calculating the 3D coordinates of the joint itself, J_{ao}, and its two adjacent joints (J_{a1} and J_{a2}). When the position of J_{ao} on its primary axis of motion is lower than the adjacent joint J_{a1}, the angle calculation remains as the original value obtained from the cosine theorem. Conversely, if the position of J_{ao} is higher than J_{a1}, the obtained angle is subtracted from 360 degrees to yield a complete angle value within the range of 0 to 360 degrees. Such adjustment ensures that the angle values accurately reflect the actual direction of joint movement in three-dimensional space.

Angle Name θ_a	Angle Notation J_{a0}: J_{a1}, J_{a2}	Angle Position
Right Knee Jointθ_{RK}	J_{RK}:J_{RF},J_{RP}	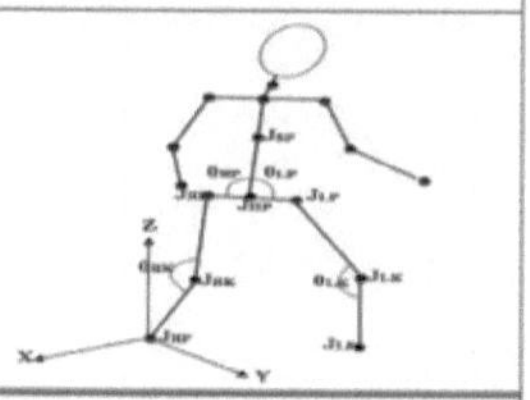
Left Knee Jointθ_{LK}	J_{LK}:J_{LF},J_{LP}	
Right Hip Jointθ_{RP}	J_{HP}:J_{SP},J_{RP}	
Left Hip Jointθ_{LP}	J_{HP}:J_{SP},J_{LP}	
Right Wrist Jointθ_{RE}	J_{RE}:J_{RW},J_{RS}	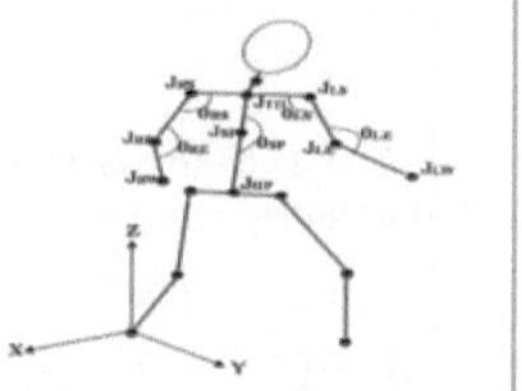
Left Wrist Jointθ_{LE}	J_{LE}:J_{LW},J_{LS}	
Spinal Joinθ_{SP}	J_{SP}:J_{TH},J_{HP}	
Right Shoulder Jointθ_{RS}	J_{RS}:J_{RE},J_{TH}	
Left Shoulder Jointθ_{LS}	J_{LS}:J_{LE},J_{TH}	

Figure 3. Numbering and naming of joint positions used in this study.

The original calculation results from the cosine theorem can only provide a range of 0 to 180 degrees. This study introduces a manual adjustment mechanism to encompass the full range of human movements. Based on the primary axis of motion, angles are adjusted to a range of 0 to 360 degrees to obtain complete information on movement angles. This adjustment ensures the accuracy of the data, allowing the analysis to more closely match actual variations in yoga poses and providing key features for subsequent motion comparison.

2.3 *Smoothing process*

When processing yoga movement videos using the 3D skeletons and joint coordinates output by MMPose, occlusions between limbs can produce discontinuous and distorted coordinate values. This study employs filtering to smooth the angle data shown in Figure 4. Initially, angles for each joint are calculated based on *XYZ* coordinates, and then a median filter is used to remove noise caused by

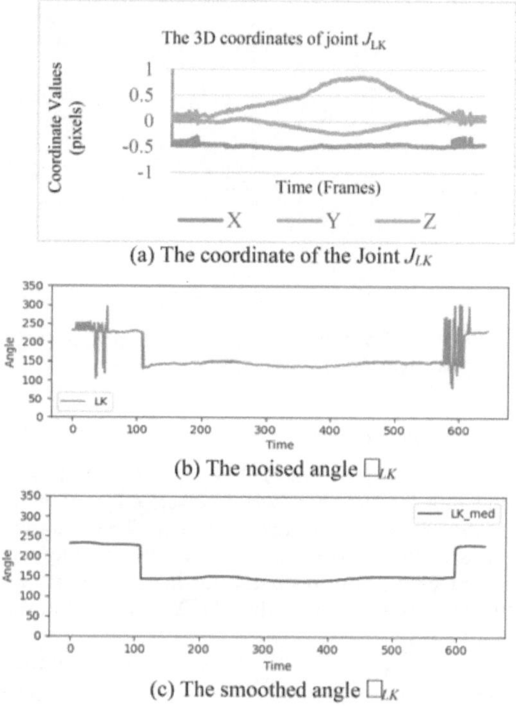

Figure 4. Smoothing effect for the Plank pose θ_{LK}

limb occlusion, ensuring data continuity and accuracy. The median filter is a nonlinear filter that effectively removes extreme value noise from the signal, preserving the details of the time series signal and offering superior performance in handling discontinuous spike noise. Therefore, a median filter is selected to smooth the joint coordinate data, reducing noise interference during motion.

2.4 *Posture assessment*

In assessing the similarity among yoga movements, accurate posture assessment is crucial. The DTW method, widely used for comparing the similarity of time series (P. Senin 2008), is adept at handling various deformations and scale transformations. Moreover, it accurately measures their similarity by calculating the distance between two time series. As shown in Figure 5, the minimum distance warping path in the cost matrix presents the optimal match between time series TS_1 and TS_2. This matching allows for stretching and compressing between sequences, thus measuring their similarity even in asynchronous situations. This study employs the DTW technique to assess the similarity of yoga movements. It matches sequences and calculates their distance by extracting the angles of nine joint points in yoga movements to form time series as input for the DTW algorithm. This method quantifies the similarity between yoga movements.

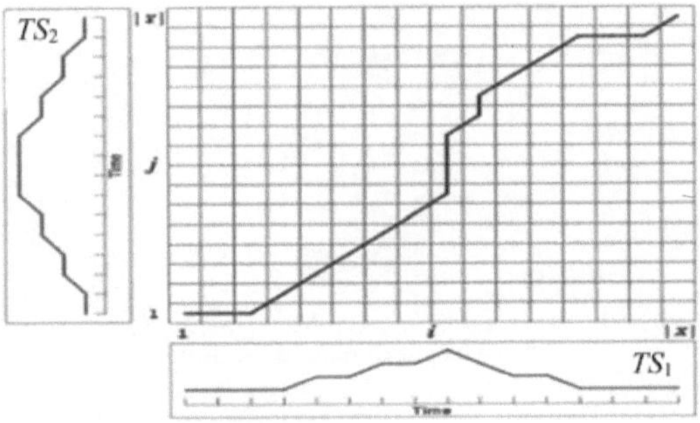

Figure 5. DTW cost matrix and optimal path for time series matching (P. Senin 2008).

2.5 *Posture scoring*

This study employs the DTW algorithm to assess the similarity between two time series by calculating the DTW distance between them. This distance indicates the degree of similarity between the two sequences, and on this basis, further scoring of the output results from the DTW algorithm is conducted.

A scoring metric S measures the similarity between two time series. When $S = 0\%$, it indicates that the two sequences are entirely different; when $S = 100\%$, it indicates that they are identical. The calculation method for the metric is as follows: Assume there are two time series, TS_1 and TS_2, with lengths L_1 and L_2, respectively.

$$S = \frac{x - l_0}{l_1 - l_0} \cdot 100\% \qquad (5)$$

where

$$l_0 = L_1 + L_2, l_1 = \begin{cases} L_1 + (\sqrt{2} - 1)L_2, & L_1 > L_2 \\ (\sqrt{2} - 1)L_1 + L_2, & L_1 \leq L_2 \end{cases},$$

$$x = \begin{cases} \sqrt{2}L_1, & L_1 > L_2 \\ \sqrt{2}L_2, & L_1 \leq L_2 \end{cases}$$

The DTW scoring analysis uses a right-angled triangular ratio of $1 : 1 : \sqrt{2}$ to simulate the similarity calculation between time series shown in Figure 6.

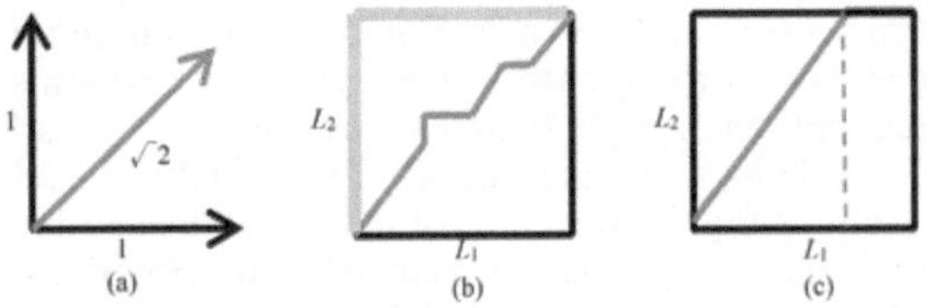

Figure 6. Three illustrations of DTW scoring.

2.6 *Movement recommendations*

DTW path diagrams are employed to provide more in-depth recommendations. By analyzing the variations in the DTW path diagrams, five types of changes are summarized to offer more specific guidance, including correct movement, incorrect movement, too fast, and too slow.

Through the DTW path diagrams, the practitioner's movements can be compared with the standard instructor's movements, providing feedback and suggestions for improvement. Figure 7 displays the five main types of analysis using DTW path diagrams for movement accuracy, guiding practitioners toward more precise movement corrections.

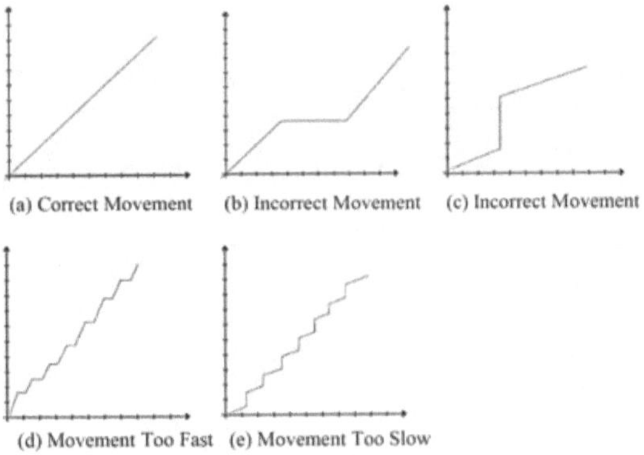

Figure 7. Types of variations in movement assessment path diagrams.

3 RESULTS AND DISCUSSION

3.1 *Acquiring posture information with MMPose*

Initially, the Faster R-CNN object detection model is used to detect the location of individuals in the video, facilitating further posture estimation. Subsequently, the HRNet posture estimation model detects key points of human posture in the video. Lastly, the VideoPose3D algorithm generates a 3D skeleton from the results of 2D posture estimation.

After extracting the body skeleton using MMPose, this study delves deeper into analyzing the relative angles between key points. Calculating the angles between key points makes quantifying each movement within yoga poses possible, mapping these dynamic and static posture transitions to a set of angle values.

This study employs smoothing techniques after extracting angle information to ensure the accuracy and consistency of angle data. Random noise and outliers in the data are filtered using a median filter, optimizing the curve of angle changes to enhance its smoothness and continuity.

Figure 8 illustrates the overall process of yoga movement analysis, using poses such as Mountain, Camel, and Cobra as examples. The first column shows the 2D skeleton output from the original

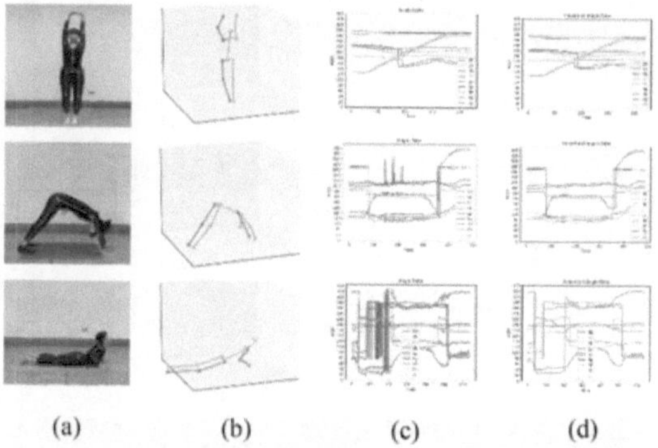

Figure 8. Yoga posture analysis process: from video to smoothed angle data. (a) Yoga posture (Mountain, Camel, and Cobra), (b) 3D skeleton output, (c) Calculation results of the angles for nine utilized joints and (d) Smoothed angle curve.

video, the second column displays the 3D skeleton output, and the third column presents the original calculation results of the angles for nine utilized joints, where some outliers and noise are evident. After smoothing, the correct angle variations are revealed, as shown in the fourth column.

3.2 *Posture assessment*

Six yoga poses are tested to simulate actual practice scenarios, examining the scoring system's ability to identify movement correctness and speed variations. These tests include:

- Test 1: patterns of perfect alignment,
- Test 2: fast-stop-fast,
- Test 3: slow-stop-slow,
- Test 4: overall faster movements,
- Test 5: overall slower movements, and
- Test 6: incorrect movements.

The assessment results for each pattern are displayed through the DTW technology match path diagrams, which include statistical graphs of horizontal and vertical line segments. These statistical charts effectively reflect the time differences between practitioners and standard movements and the frequency of movement delays or accelerations. The full results are shown in Figure 9.

In a series of experiments utilizing Dynamic Time Warping (DTW) analysis to explore the alignment of movement sequences, various patterns emerged, elucidating the relationship between the movement tempos of practitioners and instructors. These findings were visually represented in DTW path diagrams and accompanying statistical charts, which provided a detailed comparison of movement dynamics across different tests.

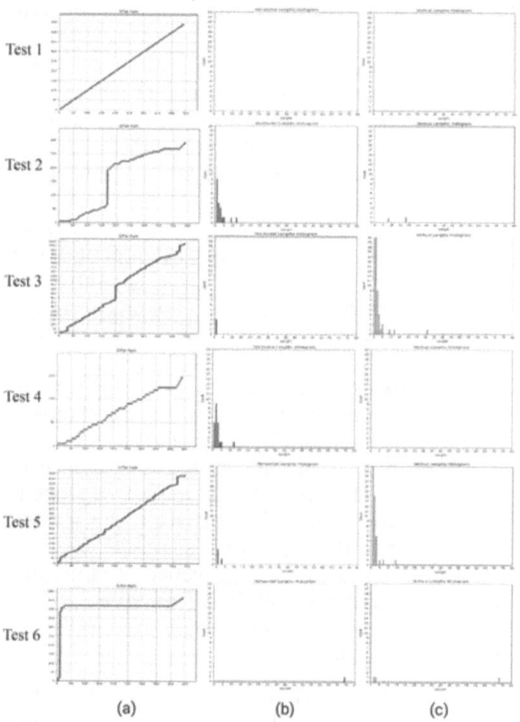

Figure 9. Following the order of rows, experimental results for Tests 1 to 6. (a) DTW cost matrix, (b) and (c) the histograms of horizontal and vertical segments in the DTW path.

- Test 1 illustrated a quintessential alignment of sequences, as evidenced by the diagonal trajectories within the DTW path diagram. This lack of horizontal or vertical deviations indicated a seamless correspondence between the practitioner's and the instructor's movements, leaving no distinct segments to be noted in the statistical chart.
- Test 2 uncovered distinct movement disparities within the frame ranges of 0 to 50, where short horizontal lines signified the practitioner's accelerated motion relative to the instructor. The period between frames 50 and 200 was marked by extended vertical lines, denoting prolonged pauses in the practitioner's actions in contrast to the instructor's continuous movement. A subsequent acceleration by the practitioner was suggested by the reappearance of short horizontal lines between frames 200 and 300, with the statistical chart revealing an increased prevalence of short horizontal segments alongside a singular elongated vertical line.
- Test 3 was characterized by numerous short vertical segments from frames 0 to 400 and 550 to 1050, indicating a deceleration in the practitioner's movement pace compared to the instructor's. A significant hiatus in the practitioner's motion, relative to the instructor's, was depicted through longer vertical segments between frames 400 and 550. The statistical

representation consisted of many short and longer vertical lines.

- Test 4 demonstrated the practitioner's rapid movement through a sequence of short horizontal lines, with the statistical chart densely populated by similar horizontal indicators, suggesting a consistent speed surpassing that of the instructor.
- Test 5 presented a path diagram filled with short vertical lines, indicative of the practitioner's slower pace. Correspondingly, the statistical chart had short vertical lines, underscoring the reduced velocity of the practitioner's movements.
- Test 6 uniquely commenced with the practitioner's movements aligning with the standard motions, as indicated by a vertical ascent in the DTW path diagram, suggesting an absence of subsequent matches. The attempt to reconcile the temporal disparities between the practitioner's and the instructor's sequences was visually marked by the emergence of elongated horizontal and vertical lines, mirrored in the statistical chart by a single long horizontal and a long vertical line.

Through these analytical endeavors, the DTW analysis provided insightful distinctions and alignments within the kinetics of motion, contributing significantly to the academic discourse on movement analysis and optimization.

3.3 *Scoring and recommendations*

When scoring test data, the study first utilizes DTW technology to analyze the motion trajectory of practitioners. The distance calculated by DTW is mapped to a scoring range from 0 to 100, providing a quantified indicator to evaluate the practitioners' movement performance shown in Figure 10.

Ultimately, time series plots indicate correct and incorrect movements in different colors, as shown in Figure 11. Green segments represent correctly executed movements, orange segments indicate too-slow movements and blue segments denote too-fast movements. Red segments reveal moments of incorrect (omitted) movements.

Test Data	Score
Test 1	100
Test 2	47.13
Test 3	71.73
Test 4	70.79
Test 5	71.96
Test 6	16.53

Figure 10. Scoring Results for Tests 1 to 6.

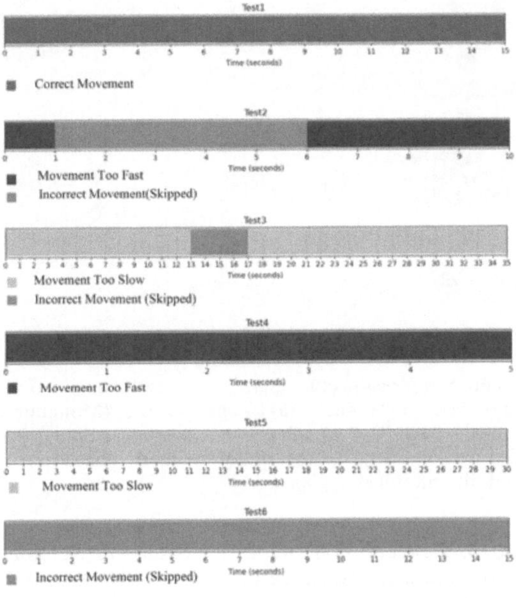

Figure 11. Representation of Suggested Feedback for Tests 1 to 6.

4 CONCLUSION

This study has developed a yoga posture assessment system based on deep learning technologies, which precisely extracts human posture information from yoga movement videos, enabling quantitative analysis and evaluation of practitioner postures. Extracting 3-D skeletons and joint coordinates from 2D images and then calculating the angles between joint points provides a solid foundation for an in-depth assessment of the accuracy of yoga movements.

The smoothing of angle data reduces discontinuous distortion values caused by limb occlusion, further enhancing the data's continuity and accuracy. Additionally, introducing a median filter removes noise while effectively retaining the details of temporal signals, enhancing the precision of movement assessment.

The application of DTW technology allows for an in-depth similarity analysis between the practitioner's movements and standard movements, accurately measuring their similarity by calculating the distance between time series. Experimental results demonstrate that the system can effectively identify movement correctness and speed variations, conduct comprehensive movement assessments and scoring, and provide targeted improvement suggestions.

REFERENCES

Chasmai M., Das N., Bhardwaj A., and Garg R. (Sep 2022). A view independent classification framework for yoga postures, *Social Netw. Comput. Sci.*, vol. 3, no. 6, pp. 1–15.

Cramer H., Quinker D., Schumann D., Wardle J., Dobos G., and Lauche R. (2019). *Adverse effects of yoga: A national cross-sectional survey, BMC Complementary and Alternative Medicine*, vol. 19, no. 1, article 190, pp. 1–10.

Iyengar B. K. S. (1965). *Light on yoga: The Definitive Guide to Yoga Practice*, Schocken Books, New York.

Lobo A., Lobo J., Joy A., and Mirza I. A. (2022). Yoga correction using machine learning, in *Proceeding of 2022 2nd Asian Conference on Innovation in Technology (ASIANCON)*, pp. 1–6.

Senin P. (2008). *Dynamic time warping algorithm review, Inf.* Comput. Sci. Dept., Univ. Hawaii Manoa Honolulu USA, Technical Report 855, pp. 1–23.

Sharma A., Sharma P., Pincha D., and Jain P. (Sept. 2022). Surya Namaskar: Real-time advanced yoga pose recognition and correction for smart healthcare, *arXiv preprint arXiv:2209.02492*, pp. 1–11.

Shen S.-W., Huang W.-C., Anggraini I. T., Funabiki N., and Fan C.-P (2022). Exercise and performance learning assistant system for self-practice dynamic yoga by openpose and fuzzy based design, in *Proceeding of 2022 10th International Conference on Information and Education Technology (ICIET)*, pp. 16–21.

Wu Z., Zhang J., Chen K., and Fu C. (2019). Yoga posture recognition and quantitative evaluation with wearable sensors based on two-stage classifier and prior Bayesian network, *Sensors*, vol. 19, no. 23, article 5129, pp. 1–19.

System Innovation for an Artificial Intelligence Era – Lam et al. (Eds)
 ISBN: 978-1-032-84452-7

The study on how AI is used in early childhood education in Taiwan

Chia-fang Hsu
Associate Professor, New Era University College, Malaysia

Liang-Yin Kuo
Associate Professor, National Formosa University, Hu-Wei, Yunlin, Taiwan

ABSTRACT: Artificial Intelligence (AI) technologies have been applied in various domains in Taiwan, including education field. Integration of AI technology is a recent significant trend since Taiwan's Ministry of Education is promoting AI learning by reshaping the school environment to make AI an integral part of students' lives; apart from that, from 2019, it is compulsory to teach AI-related topics in public schools from elementary to high school. There are more and more studies and relevant literatures of AI in education realm, however, there is a lack of studies that discuss how the AI is used in early childhood education (ECE). In this paper, the researchers aim to provide an up-to-date and in-depth overview of how the AI technologies have been implemented in ECE in Taiwan through detailed literature review and analysis, in which the current representative works will be summarized, the trends and challenges will be outlined, the ethical issues will be discussed, and the insightful recommendations for future research will be provided. This study significantly contributes to provide an up-to-date and in-depth literature review that is suitable as introductory material for future researchers interested in AI application in ECE in Taiwan.

Keywords: AI Early Childhood Education (ECE)

1 INTRODUCTION

Today is the Era of Artificial Intelligence (AI). In the past decades, it has revolutionized human being's daily routines by waving itself as part of every domain of human life, from voice assistants to smart home devices, AI technologies have provided better comfort and efficiency to perform tasks in less time.

AI systems are not only fundamentally changing our lifestyle but also affecting present and future generations of children. Nowadays, children are already interacting with AI technologies both directly and indirectly in different ways; they are embedded in toys, virtual assistants, video games, social media, game machine, and are used to drive chatbots and adaptive learning software. In addition to these direct interactions between children and AI, children's lives and well-being are also indirectly impacted by automated decision-making systems such as health care, and families' housing applications, and education access and management system (Druga *et al.* 2018; Su and Yang, 2022). The life of children from this decade onwards are immersed in the impact of AI in every occasion, including all levels of educational contexts.

Due the reason above, currently, there are more and more studies showing the growing interest of research in the topic of AI in early childhood education (ECE) in Taiwan. However, to date, there is a lack of comprehensive review of the condition and impact of AI in ECE concerning in which dimension of ECE that AI has been mostly employed so far in Taiwan. Therefore, in this paper, the researcher aimed to conduct a comprehensive literature review elaborating the above issue and providing recommendations for future research. This paper significantly contributes to provide an up-to-date review that is suitable as introductory material for beginners to AI in ECE in Taiwan, as well as supplementary material for future research.

2 AI IN ECE IN TAIWAN

2.1 *AI policy in Taiwan*

The application of AI technology in all levels of education realms is a significant trend worldwide (Chen *et al.*, 2020, 2022). In the past few decades, the role of technology and AI to young children have been widely discussed and researched in western world, and relevant educational policy as well as guidelines have been enacted in many countries, such as U.S, U.K. and New Zealand,

 DOI: 10.1201/9781003514831-48

etc., to set the roadmap for how AI and information technology should be all time properly integrated into teaching and learning for K-12, including early learners (Wu, 2023). In Taiwan, the concern of the need to accelerate the official policy making for AI in ECE is also raised and discussed (Su, 2015; Wu, 2023). So far, the guidelines for ECE are not yet enacted in Taiwan; however, researchers have tried to construct pedagogical instructions for preschoolers (Lin, 2016).

Despite of the lack of national level policy, Taiwan government has set a goal to develop smart city policy initiatives. Some local authorities have already started setting up the AI application in educational sector from K-12. For example, in New Taipei City, the Chatbot has been introduced to auxiliary students language learning; while in HsinChu county, the "All-Kids City project" has been launched and implemented since 2021 by integrating various AI devices into K-12 kids learning as well as kids in special needs.

2.2 *AI in ECE in Taiwan*

Studies have pointed out that there are quite some AI applications incorporated for education purposes, such as intelligent tutoring systems for special education, natural language processing for language education, educational robots for AI education, educational data mining for performance prediction, discourse analysis in computer-supported collaborative learning, neural networks for teaching evaluation, affective computing for learner emotion detection, and recommender systems for personalization, have been incorporated for improving interaction with students and satisfying various educational needs (Herodotou, 2018; Liu and Hwang, 2020; Mantilla and Edwards, 2019; Roblyer and Doering 2007).

As for ECE in Taiwan, previous studies have mostly investigated the aspects of AI in cutting-edge research, including how AI devices are implemented as teaching and learning auxiliaries in kindergarten; how AI can assist the learning and development of children in special needs; teachers and parents' attitudes and perceptions; AI in management. Issues regarding advantages and risks of AI accessibility; and the ethical concerns, have also been raised and addressed.

2.3 *AI in teaching and learning*

It has been a worldwide trend to have information technology integrated in ECE for a decade, especially during the disease outbreak and the pandemic of covid-19, the global schooling was interrupted, and the remote learning emerged as a necessary option. There is thus an growing amount of research attention appeared on how information or AI technologies can benefit young children's development and learning achievement. For preschool educators, it is ubiquitous to integrate information technology, including digital media devices multimedia resources, communication devices and so on, into daily pedagogy in Taiwan (Chuang, 2023; Huang, 2023; Sun, 2022; Wang, 2017).

Some studies point out that compared to traditional education, AI technologies provide students benefits not only by inspiring students' motives for creative activities, but also avoiding shame and embarrassment in front of their peers and assisting students with personalized self-adaptive learning (Prentzas, 2013; Xu *et al.*, 2021). And therefore, AI technology, such as robots, AR, VR, electronic books, ChatGPT, chatbot and so on, have been used in children's learning process in both formal and informal educational settings.

In recent years, the researchers in Taiwan tend to investigate the involvement of assistive and educational robots with various topics of learning (Chang, 2022; Lin, 2011; Wang, 2021). Chen (2019) probes into the influence of AI assistive robots for teaching emotion picture books on children's emotion competence, and the result points out that the combination with picture books and AI robots not only encourage the learning interests, but also enhances the emotion competence of awareness, recognition, understanding, expression, and regulation of young children. Huang (2017) investigated how educational robots influence on young children's English learning, and the result shows that the involvement of interactive AI robots enhances young children's English skill development significantly. Sheng (2021) conducted the study that proves the significant influence of social robots on both children's social emotional competence development as well as language learning. Some studies have investigated on the impact of AI devices on children's reading ability. It is shown that the adoption of assistive story-telling robots or virtual reality robots during reading session is also beneficial to enhance children's reading ability by attracting interests and promoting focus while reading (Guo, 2009; Lee, 2022; Liu, 2023).

There has also been increasing research attempts on the application of AI to improve the development of children with sensory integration issues, developmental delay, and special needs. Some researchers adopt multimedia techniques as well as robots with functions of social interaction and storytelling, to assist children with autism to improve the social interaction, communication skills, expression ability, comprehension, and subject learning. The results show that using robots not only raise children's social interaction capabilities, but also improve the effect of learning (Hsieh, 2014; Huang, 2019; Wu, 2020; Yin, 2014).

2.4 *AI in school management*

The digital transformation and AI technology is indispensable not just to teaching and learning, but also the whole ECE industry, as they have changed the education ecology, and school operation and management. With the auxiliary of AI technology, the on-site job as well as administrative tasks would be able to be more organized and convenient (Tang, 2022). When referring to kindergarten management, 2 dimensions are involved, the classroom management and the whole school management.

On classroom management basis, the quality of communication between teachers and parents is a crucial element. Some researchers have conducted in-depth studies investigating the application of information technology as a bridge of parent-teacher communication (Jiang, 2022; Liu, 2020; Su, 2019). The results show that apart from traditional contact book, the AI technology, including electronic contact book, as well as online communication apps, have been widely used in order to provide more effective, reliable, and prompt communications (Jiang, 2022; Liu, 2020).

On the whole kindergarten management basis, the application of AI into management is beneficial by automating the basic administrative tasks and thus educators, administrative personnels would be able to spare more time and energy on caring job. Tang (2022) has researched and introduced digital transformation, digitalization and AI technology to kindergartens and found that, the incorporation of AI system can help to reduce administrative errors and increase efficiency, allowing schoolers and teachers to better manage resources and focus on providing quality education for students. Overall, it ensures that administrative work is done accurately and efficiently and allows teachers to reduce manual effort in repetitious admin tasks, such as basic grading, assessment, recording student attendance, and sending notice, etc. And therefore, teachers can create more time with pupils and to produce quality teaching and learning experiences as well as communicating with parents.

3 CONCERN AND OPPORTUNITY

Despite of the strengths, convenience and effectiveness that AI technology has proven to bring to teaching, learning and administrative areas in previous studies, some factors such as how important principals and teachers perceive towards digital transformation of schools; the high cost of building AI system, and the attitudes of parents, might hinder the possibility of implementing digital transformation of kindergartens (Sun, 2022).

Moreover, some studies have pointed out that parents and teachers are concerned about the negative influences, the ethical issues that AI might cause to young children (Wang, 2018); furthermore, the regional inequality that might occur. However, providing to the fact that AI is a global trend and will cause drastic long-term impact on human beings, old educational paradigms must be revisited with a futuristic perspective. When referring to AI or information technology, children should be correctly guided with well-prepared guidelines and national strategies, rather than forbidden; so that young generation will be able to stay up with times and master the abilities and literacy required by rapidly changing world (Sun, 2022; Wang, 2018).

4 CONCLUSION

As mentioned previously, in the recent years, the application of AI technology has been gaining popularity in the fields of education; however, the employment of AI in ECE seems on a rather initial stage and thus it is found that the amount of relevant research is far fewer comparing to those in elementary and secondary sectors. As a result, in the present study, as the introductory material for future research, previous studies are reviewed providing a comprehensive understanding of the current status of AI in ECE in Taiwan. There are other crucial topics, such as ethical issues of adopting AI in ECE, teachers and parents' perceptions and attitudes, teachers' technology literacy, and national guidelines and strategies etc., need more future investigation and discussion.

REFERENCES

Chang, Y. (2022). *Advancing Young Children's Computational Thinking with KIBO Robotics (unpublished master's thesis)*. National Taitung University. Taitung.

Chen, S. (2019). *The Influence of Using Ai Assistive Robots for Teaching Emotion Picture Books on Children's Emotional Competence (Unpublished master's thesis)*. National Pingtung University. Pingtung.

Chen, X., Xie, H., Zou, D., & Hwang, G. J. (2020). Application and theory gaps during the rise of artificial intelligence in education. *Computers in Education: Artificial Intelligence*, 1, 100002. https://doi.org/10.1016/j.caeai.2020.100002

Chen, X., Zou, D., Xie, H., Cheng, G., & Liu, C. (2022). Two decades of artificial intelligence in education: Contributors, collaborations, research topics, challenges, and future directions. *Educational Technology & Society*, 25(1), 28–47.

Chen, Y.H. (2021). *Integrating Information Technology into the Early Childhood Curriculum: The Implementation and Teachers' Perspectives (unpublished master's thesis)*. Chaoyang University of Technology. Taichung.

Guo, Y.T. (2009). *A Storytelling Robot as a Teaching Assistant (unpublished master's thesis)*. ChungHua University. Hsinchu.

Hsieh, C.T. (2014). *An Interactive System Based on a Robotic Companion for the Children With Autism (unpublished master's thesis)*. National Central University. Taoyuan.

Herodotou, C. (2018). Young children and tablets: A systematic review of effects on learning and development. *Journal of Computer Assisted Learning*, 34(1), 1–9. https://doi.org/10.1111/jcal.12220

Huang, T. (2017). *A Study on Robot-Based English Learning for Infancy*. National Kaohsiung Normal University (unpublished master's thesis). Kaohsiung.

Huang, Y.C. (2019). *Action Research on Improving Children's Digital Learning Performance with Robot Assisted Instruction (unpublished master's thesis)*. Nanhua University. Chiayi.

Huang, Y.F. (2023). *Research on Integrating Information Technology Into the Curriculum and the Design of Distance Teaching Activities in the Preschool (unpublished master's thesis)*, Chaoyang University of Technology. Taichung.

June Druga, S., Williams, R., Park, H. W., & Breazeal, C. (2018). How smart are the smart toys? Children and parents' agent interaction and intelligence attribution. In *Proceedings of the 17th ACM Conference on Interaction Design and Children* (pp. 231–240). Trondheim, Norway: IDC. https://doi.org/10.1145/3202185.3202741.

Jiang, F.Z. (2022). *A Study of Correlation between Parent's Technology Acceptance Patterns and Parent-Teacher Communication Efficiency in Childcare Institutions unpublished master's thesis)*. Cheng Shiu University of Technology. Kaohsiung.

Lin, C.H. (2011) *The Research on the Application of Virtual Reality Robots on English Learning for Young Children (unpublished master's thesis)*. National Defense University. Taipei.

Lin, M. (2016). *The Construction of Technology Education Guidelines for Kindergarten Teachers (unpublished PhD thesis)*. National Tainan University. Tainan.

Liu, C., & Hwang, G. J. (2020). Roles and research trends of touchscreen mobile devices in early childhood education: *Review of Journal Publications from 2010 to 2019 Based on the Technology-Enhanced Learning Model. Interactive Learning Environments*, 1–20. https://doi.org/10.1080/10494820.2020.1855210

Li, T.Y. (2022).*The Research of Applying Robot to Assist Preschoolers in Activities of Reading Picture Book (unpublished master's thesis)*. National Yunlin University of Science and Technology. Yunlin.

Liu, Y.C. (2020). *A Study of Effectiveness on Preschool Parent and Teacher Communication by LINE (unpublished master's thesis)*. Nanhua University. Chiayi

Liu, P.Y. (2023). *Exploring the Use of Picture Book Reading in Parent-Child Reading (unpublished master's thesis)*. National Pingtung University. Pingtung.

Prentzas, J. (2013). Artificial intelligence methods in early childhood education. In X.- S. Yang (Ed.), *Artificial Intelligence, Evolutionary Computing and Metaheuristics* (pp. 169–199). Springer.

Roblyer, M.D., & Doering, A.H. (2013). *Integrating Educational Technology into Teaching*. Boston, MA:.

Su, L.W. (2019). *A study on Kindergarten Parents of Tainan Using Technology Acceptance Model to Conduct Parent-Teacher Communication (unpublished master's thesis)*. National Tainan University. Tainan.

Sheng, S.Y. (2021) *Using Social Robots to Teach Children with Social Emotional Learning for Exploring the Influence of Emotional intelligence and Language Learning Achievementm (Unpublished master's thesis)*. National Taiwan Normal University. Taipei.

Su, J., & Yang, W. (2022). Artificial intelligence in early childhood education: A scoping review. *Computers and Education*. Artificial Intelligence.

Sun, C.M. (2022). *The Study of Preschool Technology Literacy And Information Technology Integrated into Instruction in Tainan City (unpublished master's thesis)*. National Tainan University. Tainan.

Tsai, (2019). *Action Research on Word Recognition of Elementary School Students with Learning Disabilities to Combine an Intelligent Robot with Radical-Group-Text Teaching Method (unpublished master's thesis)*, Nanhua University. Chiayi.

Tang, F.M. (2022). An innovative digital transformation in preschools during the era of digital intelligence. *Early childhood Education, 334*, p 54–65

Wang, S.C. (2017). A Survey of Information Technology Integrating Into Kindergarten Teaching In Taipei city (unpublished master's thesis) Taipei City University. Taipei.

Wang, W.Y. (2021). The Study of Ai-Robot Assisted Health Education for Preschoolers to Develop Health Habits In Pingtung Area Kindergarten (unpublished master's thesis). National Pingtung University of Science and Technology. Pingtung.

Wu, Y. (2023). The prospect of technology education: from the viewpoint of ECE. *Taiwan Education, 202304* (740).

Xu, Y., Liu, X., Cao, X., Huang, C., Liu, E., Qian, S., ...Zhang, J. (2021). Artificial Intelligence: A Powerful Paradigm For Scientific Research. *The Innovation*, 2(4), Article 100179.

Yin, T.C. (2014). *Design and Evaluation of Applying Robots to Assisting and Inducing Children with Autism in Social Interaction (unpublished master's thesis)*. National Taiwan University of Science and Technology. Taipei.

Yu, L.W. (2021). *Public Preschool Teachers' Beliefs of the Application of Technology-Instruction Integration in Taiwan (unpublished master's thesis)*. National Taichung University of Education.

System Innovation for an Artificial Intelligence Era – Lam et al. (Eds)
© 2025 the Author(s), ISBN: 978-1-032-84452-7

Research on carbon dioxide adsorption and desorption experiments of 0.5mm membrane containing 50:50 PBAT/PBS

Yang Yi-Ling & Tsai Jung-Chin
Department of Chemical and Biochemical Engineering, Ming Chi University of Technology, New Taipei City, Taiwan

ABSTRACT: Polybutylene adipate terephthalate (PBAT) is an advanced environmentally friendly polymer with excellent biodegradability and excellent mechanical properties. It combines the properties of PBA and PBT, including good ductility and elongation at break, as well as good heat resistance and impact properties. Polybutylene succinate (PBS) is produced by the condensation polymerization of succinic acid and butylene glycol. The resin is milky white, odorless and tasteless. It is easily catabolized by various microorganisms in nature or enzymes in animals and plants, and finally decomposed into carbon dioxide and water. It is also a typical fully biodegradable polymer material. It has good heat resistance, and the heat distortion temperature and product use temperature can exceed 100°C. This experiment combines the advantages of two biodegradable materials and processes them in a PBAT/PBS (0.5mm membrane) ratio of 50:50. In this experiment, carbon dioxide was used as a adsorbent to conduct adsorption and desorption experiments. Under the studied pressure and temperature conditions, it can be found that the adsorption amount first increases with the increase of time and then reaches an equilibrium state. The results show that the adsorption capacity of carbon dioxide under high temperature conditions is significantly lower than that under low temperature conditions, and the adsorption capacity under high pressure is also significantly higher than that under low pressure conditions.

1 INTRODUCTION

In the contemporary landscape of material science, the imperative for environmentally sustainable alternatives has become increasingly urgent. With the detrimental impacts of conventional plastics on ecosystems and human health coming into sharp focus, the quest for biodegradable polymers has gained significant traction. Among these, Polybutylene adipate terephthalate (PBAT) and Polybutylene succinate (PBS) stand out as promising contenders, embodying a balance of biodegradability and mechanical robustness.

PBAT, a polymer amalgamation of PBA and PBT, exhibits remarkable biodegradability alongside admirable mechanical properties, including ductility, elongation at break, heat resistance, and impact resistance. PBS, on the other hand, synthesized from succinic acid and butylene glycol, epitomizes biodegradability through microbial or enzymatic decomposition pathways, presenting a compelling ecological alternative to traditional polymers.

In the pursuit of sustainable material solutions, this study endeavors to harness the synergistic potential of PBAT and PBS by formulating a novel PBAT/PBS membrane in a 50:50 ratio. This composite material is envisioned to inherit the desirable traits of its constituent polymers while offering enhanced biodegradability and mechanical performance. Moreover, the incorporation of carbon dioxide as an adsorbent introduces an innovative dimension to the investigation, exploring its adsorption-desorption kinetics on the PBAT/PBS membrane under varied temperature and pressure conditions.

The primary objective of this study is to unravel the intricate interplay between temperature, pressure, and carbon dioxide adsorption capacity on the PBAT/PBS membrane. Through systematic experimentation and analysis, we aim to delineate the temporal evolution of adsorption and desorption processes, shedding light on the underlying mechanisms governing carbon dioxide capture. Furthermore, we seek to elucidate how temperature and pressure variations influence the adsorption behavior, thereby informing the design and optimization of future carbon capture technologies.

By elucidating the adsorption kinetics of carbon dioxide on the PBAT/PBS membrane, this research not only contributes to the fundamental understanding of gas-solid interactions but also holds implications for practical applications. The insights gleaned from this study are poised to inform the development of sustainable polymer materials tailored for diverse environmental contexts, ranging from carbon capture and sequestration to biodegradable packaging and beyond.

In essence, this interdisciplinary endeavor endeavors to bridge the chasm between polymer science

 DOI: 10.1201/9781003514831-49

and environmental sustainability, offering a holistic approach to material innovation. By marrying the principles of biodegradability, mechanical robustness, and carbon capture, we aspire to pave the way for a greener, more sustainable future. Through collaborative efforts and relentless pursuit of knowledge, we endeavor to catalyze positive change in the realm of materials science, driving towards a circular economy that harmonizes with the natural world. mechanism of enhancement conductivity for the PEDOT: PSS films by adding different molar dilute concentrations of H_2SO_4 were further studied. These experimental results provided further evidences for our proposed mechanism.

2 EXPERIMENTAL

2.1 *Materials*

PBAT, or Polybutylene Adipate Terephthalate, is a biodegradable polymer commonly used in the production of biodegradable plastics. It is formed by polymerizing monomers such as butylene glycol, adipic acid, and terephthalic acid. When placed in a composting environment, microbes break down PBAT into simpler compounds, ultimately disappearing from the environment with minimal residue. This makes PBAT a popular alternative plastic, especially in applications requiring biodegradability, such as single-use packaging materials. PBAT exhibits good flexibility, tensile strength, and impact resistance, making it suitable for various applications like biodegradable films, bags, packaging materials, and disposable products.

PBS, or Polybutylene Succinate, is a biodegradable polymer. It is synthesized by polymerizing monomers such as butylene glycol and succinic acid. When placed in a composting environment, PBS can be broken down by microorganisms, eventually transforming into simpler compounds and ultimately disappearing from the environment with minimal residue. This makes PBS a popular choice for biodegradable plastics, particularly in products requiring biodegradability, such as single-use packaging materials. PBS exhibits good flexibility and heat resistance, making it suitable for various applications, including biodegradable films, bags, packaging materials, and disposable products.

The carbon dioxide (CO_2) with a purity of 99.9% refers to carbon dioxide gas that is almost entirely free of impurities, with a purity level of 99.9%. This means that 99.9% of the gas consists of carbon dioxide molecules, while the remaining 0.1% may consist of other gases or trace impurities. CO_2 with such high purity is commonly used in various industrial applications, such as food and beverage manufacturing, medical applications, and environmental control systems. Its high purity ensures consistency and reliability in these processes where precise gas composition is crucial.

2.2 *Sample preparation*

Ensure that both PBAT and PBS materials are in the same form, such as pellets or powder. Additionally, make sure they have similar particle sizes for better mixing. Add the pre-measured PBAT and PBS materials in a 50:50 ratio into the mixing equipment. This can be a rotary mixer, twin-screw extruder, or any other type of mixing device. Ensure the mixing process is uniform and continuous to thoroughly blend the two materials. Heat the blended materials to the appropriate temperature to melt them. This is typically done above the materials' glass transition temperature to ensure proper blending. After reaching the desired temperature, pass the molten materials through a compounding or extrusion machine to further mix and homogenize them. This helps ensure that PBAT and PBS are blended together in a 50:50 ratio. Cool the compounded materials to solidify them into the desired shape, such as pellets, films, or other forms. The cooling process can be done in a cooling bath or other cooling equipment.

2.3 *Sorption and desorption tests*

At specific temperature and pressure conditions, PBAT/PBS 50:50 samples were exposed to CO_2 using a high-pressure vessel, and the weight loss over time was measured. The adsorption of CO_2 was determined by subtracting the initial weight of the PBAT/PBS 50:50 samples from the total mass of PBAT/PBS 50:50 and CO_2 measured. To minimize gas diffusion losses, the time interval between removing the samples from the high-pressure vessel and conducting weight measurements was set to 30 seconds. However, due to potential diffusion losses of CO_2 during sample transfer, the measured adsorption quantities may not be entirely accurate. Therefore, linear fitting and extrapolation of desorption curves were performed to obtain more accurate adsorption data.

3 RESULTS AND DISCUSSION

Through adsorption and desorption experiments, we determined the appropriate saturation time and evaluated the solubility of CO_2 in the polymer under different pressures and temperatures. Solubility of CO_2 refers to the amount of CO_2 dissolved in the polymer at specific temperature and pressure conditions. In the adsorption experiments, there may be a slight loss of gas diffusion due to the short time interval between gas release into the high-pressure vessel and weight measurement. Consequently, the measured adsorption quantity may be lower than the actual adsorption. To obtain accurate adsorption data, desorption experiments were conducted to calculate the actual adsorption under certain conditions. In desorption experiments, we observed a linear relationship between residual adsorption and the square root of desorption time, following Fick's diffusion law. Hence, the actual adsorption at td = 0 can be determined by extrapolating the desorption curve.

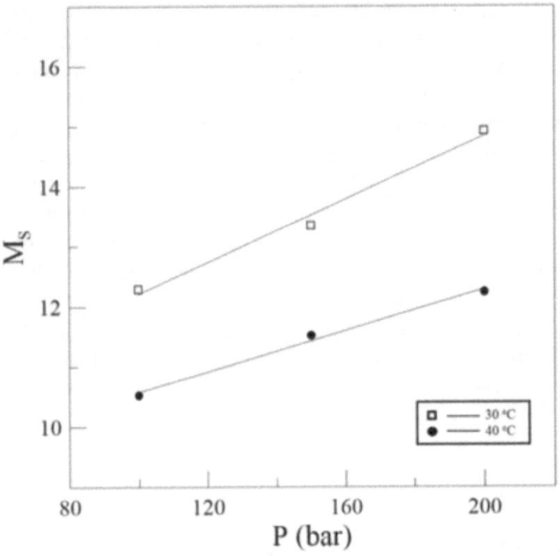

Figure 1. The relationship between adsorption capacity and pressure at different operating temperatures.

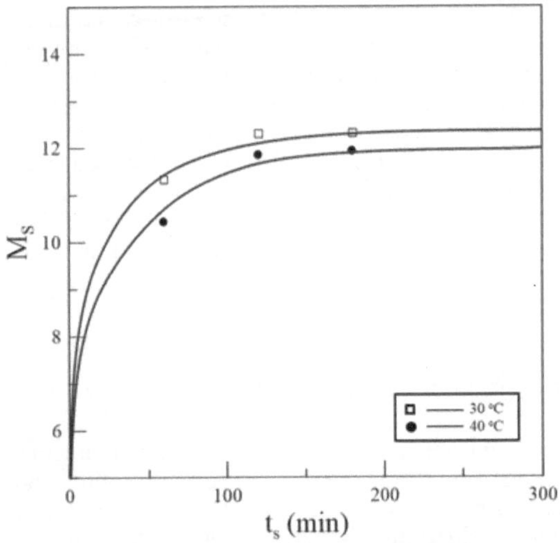

Figure 2. The relationship between adsorption capacity and adsorption duration at different temperatures under 10 MPa pressure.

Figure 1. We can see that as saturation pressure increases, the solubility of CO_2 exhibits an approximate linear growth. This implies that under different pressure conditions, the adsorption of CO_2 varies accordingly. As saturation pressure increases, the solubility of CO_2 exhibits an approximate linear growth. This implies that under different pressure conditions, the adsorption of CO_2varies accordingly.

In Figure 2. we notice that as saturation pressure increases, the solubility of CO_2 exhibits an approximate linear growth. This implies that under different pressure conditions, the adsorption of CO_2 varies accordingly.

4 CONCLUSION

Our study demonstrates that the adsorption process of carbon dioxide on PBAT/PBS(0.5mm membrane) follows a typical pattern where the adsorption amount initially rises with time and eventually reaches equilibrium. Furthermore, we observed that the adsorption capacity of carbon dioxide is inversely related to temperature, with lower temperatures favoring higher adsorption capacities. Conversely, higher pressures lead to increased adsorption capacities. These findings highlight the importance of temperature and pressure control in optimizing the adsorption process for carbon dioxide capture using PBAT/PBS membranes. Future research could explore further optimization strategies to enhance the efficiency of carbon dioxide adsorption under varying environmental conditions.

REFERENCES

Pei-Hua Chen, Chin-Wen Chen,* Hsu-I Mao, Chi-An Dai, Chie-Shaan Su, Jung-Chin Tsai & Feng-Huei Lin,* (2023), Bio-Based PLA/PBS/PBAT Ternary Blends with Added Nan ohydroxyapatite: A Thermal, Physical, and Mechanical Study

Yue Wang, Luyao Huan, Haiyan Liang, Xuejia Ding & Jianguo Mi. (2022). Foaming biocompatible and biodegradable PBAT/PLGA as fallopian tube stent using supercritical carbon dioxide, *Chinese Journal of Chemical Engineering Volume* 47, Pages 245–253

Zhaozhi Wang, Guilong Wang, Zhaorui Xu, Chengming Yang & Guoqun Zhao. (2023). Structure-tunable poly (butylene adipate-co-terephthalate) foams with enhanced mechanical performance derived by microcellular foaming with carbon dioxide as blowing agents, *Journal of CO2 Utilization Volume* 72, 102495

System Innovation for an Artificial Intelligence Era – Lam et al. (Eds)
© 2025 the Author(s), ISBN: 978-1-032-84452-7

Precise advertising based on people flow analysis using low-power mobile devices

Chun-Liang Lee
Department of Computer Science and Information Engineering, School of Electrical and Computer Engineering, College of Engineering, Chang Gung University, Taoyuan, Taiwan

ABSTRACT: Precise advertising allows advertisers to deliver more personalized experiences to consumers, resulting in increased engagement rates, enhanced conversion rates, and ultimately, improved returns on investment. With the rapid development of artificial intelligence, tools for people flow analysis can provide fast and accurate results. In this paper, we propose a framework for precise advertising based on people flow analysis. In this framework, a mobile advertising unit (e.g., a billboard truck or motorcycle) is outfitted with a video camera, a GPS receiver, and sensors to collect useful data for providing tailored advertisements. Images captured by the camera are analyzed either online or offline to extract information about the people viewing the advertisements, such as the number of people, their age and gender. Advertisements displayed on the LED display panel of the unit can be dynamically changed based on the results of people flow analysis, current location, and sensor data. By utilizing off-the-shelf low-power mobile devices like Raspberry Pi or Arduino, the hardware on the mobile advertising unit is both power-efficient and cost-effective, making it suitable for large-scale deployment.

1 INTRODUCTION

Traditional advertising focuses on reaching large audiences through mass media channels such as television commercials, radio spots, billboards and direct mail, which tend to be one-size-fits-all and less interactive. Precise advertising, also known as precision advertising or targeted advertising, refers to a data-driven approach to marketing that focuses on reaching specific individuals or segments with highly personalized and relevant messages. By leveraging advanced data analytics, artificial intelligence algorithms, and consumer behavior data, it delivers tailored advertisements and promotions to the right audience at the right time (X. Yang *et al.* 2021; S. Gao *et al.* 2017; C. Zhao *et al.* 2023).

One application that has received a lot of attention is people flow analysis—a sophisticated method of harnessing the power of artificial intelligence to derive insights from the movement patterns of individuals in public places. By analyzing data collected from sensors, cameras, and other devices, people flow analysis provides advertisers with valuable information about potential target audiences, including the gender and age distribution of crowds and real-time preferences.

Against this backdrop, we propose our precise advertising framework, firmly grounded in the principles of people flow analysis. At its core, our framework leverages the latest advances in artificial intelligence and sensor technology to create a dynamic, responsive advertising ecosystem capable of delivering tailored messages to consumers with unparalleled accuracy and efficiency.

Our framework is crafted for maximum accessibility and scalability, achieved through the utilization of readily available, low-power mobile devices like Raspberry Pi or Arduino. By leveraging these affordable and widely available components, we ensure that the hardware powering our mobile advertising units is both power-efficient and cost-effective—making it suitable for large-scale deployment in urban environments around the world.

The remainder of this paper is structured as follows: In Section 2, we delve into related work concerning this paper. Section 3 introduces our proposed framework. Following that, Section 4 addresses implementation issues and outlines future research directions. Finally, Section 5 provides the conclusion to this paper.

2 RELATED WORK

2.1 *People flow analysis*

People flow analysis provides valuable insights into the movement patterns and behavior of individuals within physical spaces (Y. C. Kuo *et al.* 2018). By leveraging data from sensors, Wi-Fi-tracking, and other technologies, advertisers can gain insights into people flow patterns, dwell

DOI: 10.1201/9781003514831-50

times, and demographic profiles in specific locations. This information allows them to tailor their advertising messages more accurately to target audiences and deliver them at the right time and place.

2.2 *Face detection*

Face detection plays a crucial role in various fields, including entertainment, marketing, security, and authentication (M.K. Hasan *et al.* 2021). Traditional face detection methods often relied on handcrafted features such as Haar cascades (P. Viola *et al.* 2001) or Histogram of Oriented Gradients (HOG) (N. Dalal *et al.* 2005). While these methods were effective to some extent, they had limitations in handling variations in lighting, pose, and occlusions. The advent of deep learning, particularly Convolutional Neural Networks (CNNs), has significantly improved face detection accuracy and robustness. Techniques like the Viola-Jones algorithm have been replaced by CNN-based architectures such as Single Shot Multibox Detector (SSD) (W. Liu *et al.* 2016), Faster R-CNN (H. Wang *et al.* 2017), and You Only Look Once (YOLO) (J. Redmon *et al.* 2016). A variety of face detection tools are available that can be used on edge devices like Raspberry Pi. Popular tools include Open Source Computer Vision Library (OpenCV), Dlib, Multi-task Cascaded Convolutional Networks (MTCNN), and Pi-FaceDet.

2.3 *Age and gender classification*

Gender and age detection using deep learning have become highly accurate and reliable, enabling a wide range of applications across different domains (M. Duan *et al.* 2018; G. Levi *et al.* 2015). These technologies enable personalized user experiences and demographic analysis, helping businesses tailor their products and services more effectively. The tools used for face detection can also be used for age and gender classification with their pre-trained models.

2.4 *Object tracking*

Object tracking in computer vision refers to the process of locating and following a specific object or multiple objects in a sequence of frames in a video or series of images. The goal of object tracking is to continuously monitor the movement and position of objects over time as they move within a scene. When tracking human faces, it can be used to count the number of people in a video. Simple Online Real-Time Tracking (SORT) (A. Bewley *et al.* 2016) tracks objects by predicting their trajectories using Kalman filters and Euclidean distances. Deep SORT (N. Wojke *et al.* 2017) proposed further improvements, incorporating a depth metric for visual recognition in addition to the Euclidean distance and Kalman filter. Therefore, for an object to be considered the same in consecutive frames, it must not only be in the same neighborhood and move roughly in accordance with the projected trajectory but also be consistent with that trajectory.

3 PROPOSED FRAMEWORK

In this section, we illustrate the proposed framework as depicted in Figure 1. The main components are described below:

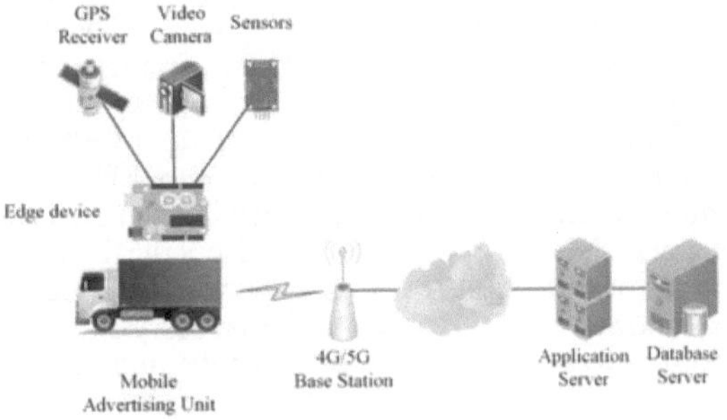

Figure 1. Proposed framework.

- Mobile advertising unit: This is the key component of our framework. It can be a billboard truck or motorcycle equipped with an LED display panel, serving as mobile canvases for advertisers to showcase their messages to a diverse range of audiences. It is also equipped with an edge-computing device, high-resolution camera, GPS receiver, and sensors that will be explained below.
- Edge-computing device: This is a low-power mobile device like Raspberry Pi or Arduino, providing the capability to process images captured by the camera and collect data from sensors. To support advertisers in dynamically adjusting their messaging and uploading results in real-time, this device is equipped with 4G/5G mobile network capabilities.
- Video camera: This component is utilized to capture images continuously for people flow analysis, enabling various analyses such as headcount, age and gender statistics.
- GPS receiver: This component obtains the location information of the mobile advertising unit.
- Sensors: This component includes a variety of sensors designed to assist in delivering target advertisements. These sensors may include temperature, humidity and raindrop detection capabilities.
- Application server: This component serves as the middleware between mobile advertising

units and the database server. It also provides an interface for advertisers to obtain the status of a mobile advertising unit on demand or to control the playback of advertisements.

- Database server: This component stores the results obtained by all mobile advertising units. Results can be analyzed to improve the effectiveness of advertisements. For example, based on the gender results and the number of people that were viewing an advertisement video, the advertiser can identify their target customers. With this result, the advertisement playing on the mobile advertising unit can be changed dynamically according to the gender results analyzed.

4 DISCUSSIONS

The framework illustrated in Section 3 can be used to support other types of services. For example, for mobile advertising units that are not equipped with enough computing power, captured videos can be saved for further offline analysis if high-resolution videos are required. While advertisers may not be able to dynamically adjust the advertisement displayed in the mobile advertising unit based on the results of people flow analysis, they can use the results obtained by GPS receivers and sensors.

One of the results derived from people flow analysis is the number of people who viewed the advertisement. To avoid double counting, the number of people cannot be counted on a frame-by-frame basis. Instead, multiple people should be detected and tracked throughout the frame stream using object tracking algorithms, as discussed in Section 2.

To provide highly accurate counting results, the camera needs to capture high-resolution images at a high frame rate. Since low-end mobile devices have limited processing power, there is a trade-off between frame resolution and frame rate. People flow analysis can only be performed using either high-resolution video at a low frame rate or low-resolution video at a high frame rate. High-resolution video provides highly accurate age and gender classification, while high frame rates provide highly accurate head counts. If the accuracy of age and gender classification and the accuracy of headcount are to be met, then the mobile advertising unit can save high-resolution videos at high frame rates. The saved video can be processed offline on return.

To further improve the accuracy of counting people viewing an advertisement, the detected face landmarks can be used to exclude some people who are not facing the display panel on the mobile advertising unit. Most face detection tools return not only all face in an image, but also facial landmarks in each detected face. Common facial landmarks include the left eye, the right eye, the nose, and both corners of the month. These landmarks can be used to calculate the approximate angle of the face towards the camera. Of course, this feature consumes additional computing power and requires more powerful edge-computing devices.

5 CONCLUSION

Precise advertising is a transformative approach that empowers advertisers with enhanced strategies for targeted advertisement placement. In this paper, we introduced a comprehensive framework designed specifically for precise advertising, leveraging advanced techniques in people flow analysis. Our proposed framework not only promises a more refined and effective way to reach target audiences but also ensures cost-effectiveness and scalability through the utilization of off-the-shelf low-power mobile devices. By harnessing the capabilities of these devices, advertisers can achieve widespread deployment and maximize their outreach potential. Additionally, we addressed potential issues that may arise when utilizing this framework.

ACKNOWLEDGMENT

This work was supported by the National Science and Technology Council, Taiwan (NSTC 112-2221-E-182-033) and Chang Gung Memorial Hospital (BMRP 942).

REFERENCES

Bewley A., Ge Z., Ott L., Ramos F., Upcroft B. (2016). Simple online and realtime tracking, *in Proc. IEEE International Conference on Image Processing (ICIP)*, 3464–3468.

Dalal N., Triggs B. (2005). Histograms of oriented gradients for human detection, in *Proc. IEEE Conference on Computer Vision and Pattern Recognition (CVPR)*.

Duan M., Li K., Yang C., Li K. (2018). A hybrid deep learning CNN–ELM for age and gender classification, *Neurocomputing*, 275, 448–461.

Gao S., Yang C. (2017). Urban human flow data analytics for advertising applications: A case study in New York City. in *Proc. IEEE International Conference on Big Data*, 1444–1453.

Hasan M.K., Ahsan M.S., Abdullah-Al-Mamun; Newaz S.H.S., Lee G. M. (2021). Human face detection techniques: A comprehensive review and future research directions, *Electronics*, 10(19), 2354.

Kuo Y. C., Chen Y. H., Cheng C. Y. (2018). Improving precision in targeted advertising using location-based services data, *Journal of Business Research*, 89, 1–8.

Levi G., Hassncer T. (2015). Age and gender classification using convolutional neural networks, in *Proc. IEEE Conference on Computer Vision and Pattern Recognition Workshops (CVPRW)*, 34–42.
Liu W., Anguelov D., Erhan D., Szegedy C., Reed S., Fu C.-Y., Berg A. C., Liu W. (2016). *SSD: Single shot multibox detector*, in *Proc. European Conference on Computer Vision*, 21–37.
Redmon J., Divvala S., Girshick R., Farhadi A. (2016). You only look once: Unified, real-time object detection, in *Proc. IEEE Conference on Computer Vision and Pattern Recognition (CVPR)*, 779–788.
Viola P., Jones M. (2001). Rapid object detection using a boosted cascade of simple features, in *Proc. IEEE Conference on Computer Vision and Pattern Recognition (CVPR)*.
Wang H., Li Z., Ji X., Wang Y. (2017). Face R-CNN, [Online]. Available: arXiv:1706.01061.
Wojke N., Bewley A., Paulus D. (2017). Simple online and realtime tracking with a deep association metric, in *Proc. IEEE International Conference on Image Processing (ICIP)*, 3645–3649.
Yang X., Li H., Ni L., Li T. (2021). Application of artificial intelligence in precision marketing, *Journal of Organizational and End User Computing*, 33(4), 209–219.
Zhao C., Ogawa Y., Chen S., Oki T., Sekimoto Y. (2023). People flow trend estimation approach and quantitative explanation based on the scene level deep learning of street view images, *Remote Sensing*, 15(5), 1362.

System Innovation for an Artificial Intelligence Era – Lam et al. (Eds)
© 2025 the Author(s), ISBN: 978-1-032-84452-7

Adsorption and desorption experiments using carbon dioxide as the adsorbent for membranes containing 30:70 pbat/pbs

Hsieh Yueh-Ling & Tsai Jung-Chin
Department of Chemical & Biochemical Engineering, Ming Chi University of Technology, New Taipei City, Taiwan

ABSTRACT: This study focuses on the experimental investigation of the carbon dioxide (CO_2) adsorption-desorption behavior and mechanical properties of a blend consisting of Polybutylene Adipate Terephthalate (PBAT) and Polybutylene Succinate (PBS). The experiment explores various temperature, pressure, and time conditions to understand the intricate relationship between the CO_2 adsorption process and the mechanical properties of the PBAT-PBS blend. PBAT and PBS are biodegradable plastics. PBAT is polymerized from monomers such as adipic acid, 1,4-butanediol and terephthalic acid. The molecular structure of PBAT gives it soft and malleable properties, making it suitable for applications requiring greater flexibility. Due to its relatively low melting point and high melt flowability, it is easy to process and is suitable for methods such as blow molding and injection molding. PBS is composed of monomers such as adipic acid, 1,4-butanediol, and succinic acid. It is relatively hard and possesses a certain level of strength, making it suitable for applications requiring higher mechanical performance. It also has a higher melting point and heat resistance, performing well in applications with elevated temperatures. The biodegradability properties of the two plastics complement each other and can improve the overall biodegradation properties when blended. The experiment concludes that with a composition of 30% PBAT and 70% PBS, higher pressure results in increased adsorption-desorption quantities, and lower temperatures lead to higher adsorption-desorption quantities.

1 INTRODUCTION

In recent years, the development of advanced membrane technologies has gained significant traction due to their vital role in addressing environmental challenges and enhancing industrial processes. Among these advancements, the utilization of carbon dioxide (CO_2) as an adsorbent for membranes containing a blend of 30% poly(butylene adipate-co-terephthalate) (PBAT) and 70% polybutylene succinate (PBS) presents a promising avenue for innovation. This paper delves into the exploration of adsorption and desorption phenomena within such membranes, shedding light on their potential applications and contributions to sustainable separation processes.

The blend of PBAT and PBS in a 30:70 ratio represents an intriguing material composition for membrane fabrication. PBAT, a biodegradable polyester, and PBS, a bio-based and biodegradable aliphatic polyester, offer a synergistic combination of mechanical strength, biocompatibility, and environmental sustainability. By leveraging these properties, membranes incorporating this blend exhibit favorable characteristics for various separation applications, including gas separation, wastewater treatment, and environmental remediation.

Carbon dioxide emerges as a versatile adsorbent for membrane-based separation processes. Its abundance, low cost, and environmentally benign nature render it an attractive option for industrial applications. Moreover, CO_2 demonstrates tunable adsorption properties, enabling selective capture and release of target molecules within membranes. Leveraging CO_2 adsorption and desorption processes within PBAT/PBS membranes holds promise for enhancing the efficiency and selectivity of gas separation processes while minimizing environmental impact.

This study aims to investigate the adsorption and desorption behaviors of CO_2 within PBAT/PBS membranes through comprehensive experimental analyses and theoretical modeling. By elucidating the underlying mechanisms governing these processes, valuable insights can be gained into the design and optimization of membrane-based separation systems. Furthermore, understanding the performance of these membranes under various operating conditions will facilitate their practical implementation in diverse industrial settings, ranging from carbon capture to gas purification.

The significance of this research lies in its potential to advance both membrane technology and sustainable separation processes. By elucidating the adsorption and desorption mechanisms of CO_2 within PBAT/PBS membranes, this study aims to contribute to the development of more efficient, cost-effective, and environmentally

DOI: 10.1201/9781003514831-51

friendly separation technologies. Ultimately, the insights gained from this research endeavor hold promise for addressing pressing environmental challenges and promoting the transition towards a more sustainable future.

Expanding upon this introduction, it is essential to delve deeper into the specific aspects of membrane technology, carbon dioxide adsorption, and the blend of PBAT and PBS. Membrane technology has emerged as a versatile tool for various separation processes due to its ability to selectively allow certain substances to pass through while retaining others. This property is particularly useful in industries such as wastewater treatment, gas purification, and chemical separations. Traditional separation methods often involve energy-intensive processes such as distillation or chemical treatments, which can be costly and environmentally harmful. Membrane-based processes offer a more sustainable alternative by requiring less energy and producing fewer harmful by-products.

In recent years, there has been a growing interest in utilizing renewable resources and environmentally friendly materials in membrane fabrication. PBAT and PBS are two such materials that have gained attention due to their biodegradability and sustainability. PBAT is a copolyester derived from adipic acid, terephthalic acid, and 1,4-butanediol, while PBS is a polybutylene succinate derived from succinic acid and 1,4-butanediol. Both polymers are biodegradable under appropriate conditions, making them attractive options for eco-friendly applications. By blending PBAT and PBS in a specific ratio, membranes can be tailored to exhibit desired properties such as mechanical strength, flexibility, and biocompatibility.

The incorporation of carbon dioxide as an adsorbent in PBAT/PBS membranes adds another dimension to their functionality. CO_2 is abundant in the atmosphere and is a by-product of many industrial processes, making it readily available and inexpensive. Additionally, CO_2 can be selectively adsorbed onto certain surfaces, allowing for targeted separation of gas mixtures. This property is particularly useful in carbon capture and storage applications, where CO_2 needs to be separated from other gases before sequestration or utilization. By leveraging the adsorption and desorption properties of CO_2 within PBAT/PBS membranes, it is possible to design efficient and selective separation processes with minimal environmental impact.

In conclusion, the exploration of adsorption and desorption phenomena using carbon dioxide as the adsorbent for membranes containing a blend of PBAT and PBS represents a significant advancement in membrane technology. By combining renewable materials with environmentally friendly processes, it is possible to develop sustainable separation technologies that address pressing environmental challenges while enhancing industrial efficiency. Through comprehensive experimental analyses and theoretical modeling, this study aims to contribute to the ongoing efforts towards a more sustainable future.

2 EXPERIMENTAL

2.1 *PBAT/PBS*

PBAT is a copolyester synthesized from butylene adipate and terephthalate monomers. It possesses excellent mechanical properties and can degrade under appropriate conditions into non-toxic by-products, making it an environmentally friendly alternative to conventional plastics. Additionally, PBAT is biocompatible, meaning it can be used in contact with food and medical products without adverse effects..

On the other hand, PBS is a bio-based and biodegradable aliphatic polyester derived from succinic acid and 1,4-butanediol. Like PBAT, PBS offers good mechanical properties and biodegradability, making it suitable for a wide range of applications. PBS is particularly attractive for applications requiring flexibility and elasticity, such as packaging films and compostable products.

On the other hand, PBS is a bio-based and biodegradable aliphatic polyester derived from succinic acid and 1,4-butanediol. Like PBAT, PBS offers good mechanical properties and biodegradability, making it suitable for a wide range of applications. PBS is particularly attractive for applications requiring flexibility and elasticity, such as packaging films and compostable products.

When blended together in a 30:70 ratio, PBAT and PBS create a material matrix that combines the desirable properties of both polymers. This blend exhibits enhanced flexibility, mechanical strength, and biodegradability compared to either polymer alone. Moreover, the blend's composition can be tailored to meet specific application requirements, making it a versatile material for various industries, including packaging, agriculture, and biomedical applications.

PBAT and PBS represent promising materials in the realm of biodegradable polymers, offering a sustainable alternative to traditional plastics. Their blend combines the advantages of both polymers, making it a versatile option for environmentally conscious applications.

2.2 *Sorption and desorption*

Adsorption and desorption experiments were conducted using carbon dioxide (CO_2) under various pressures and temperatures to measure the adsorption capacity at different time intervals and observe its variations. Additionally, studying the impact of different pressure and temperature conditions helps us understand how environmental factors influence the adsorption process, allowing for more effective utilization of CO_2.

3 RESULTS AND DISCUSSION

Figure 1 shows that the adsorption quantities of CO_2 increases with increasing pressure, and the adsorption capacity at 40°C is better than at 30°C.

Figure 2 shows that the adsorption quantities of CO_2 increases over time until reaching equilibrium.

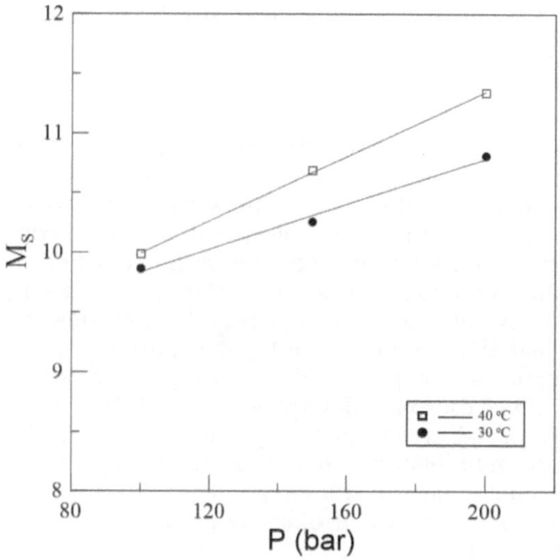

Figure 1. Adsorption quantities in CO_2 different pressures and temperatures.

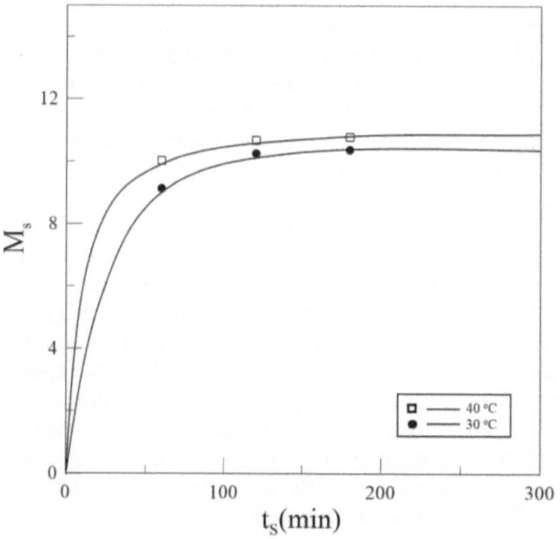

Figure 2. adsorption quantities in CO_2 changes with time.

4 CONCLUSION

In this paper, for a mixture containing 30% PBAT and 70% PBS, the adsorption of CO_2 increases under high-pressure conditions. This may be because higher pressure enhances the interaction between the gas and the adsorbent material. Additionally, lower temperatures also result in higher adsorption-desorption quantities, possibly due to the increased availability of active sites on the adsorbent material surface, thus enhancing CO_2 adsorption capacity. Therefore, this conclusion supports the idea that, under specific compositions, adjusting pressure and temperature can modulate CO_2 adsorption properties. This has potential significance for addressing climate change and environmental protection.

REFERENCES

Bingling Yuan, Guoxiong Zhan, Lei Xing, Yuchen Li, Zhoulan Huang, Zhen Chen, Lidong Wang, Junhua Li. (2024). Boosting CO_2 absorption and desorption of biphasic solvent by nanoparticles for efficient carbon dioxide capture, *Separation and Purification Technology, Volume 329*, 125108

Dongdong Hu, Kun Xue, Zhen Liu, Zhimei Xu & Ling Zhao. (2022). The essential role of PBS on PBAT foaming under supercritical CO_2 toward green engineering, *Journal of CO_2Utilization, Volume 60*, 101965

Zhaozhi Wang, Guilong Wang, Zhaorui Xu, Chengming Yang, Guoqun Zhao. (2023). Structure-tunable poly (butylene adipate-co-terephthalate) foams with enhanced mechanical performancederived by microcellular foaming with carbon dioxide as blow-ing agents, *Journal of CO_2 Utilization, Volume 72*, 102495

System Innovation for an Artificial Intelligence Era – Lam et al. (Eds)
© 2025 the Author(s), ISBN: 978-1-032-84452-7

Enhanced biodegradable polyester adsorption-desorption materials: A study of 1.0 mm membrane containing 50:50 PBAT/PBS

Tsai Yu-Chia & Tsai Jung-Chin
Department of Chemical & Biochemical Engineering, Ming Chi University of Technology, New Taipei City, Taiwan

ABSTRACT: This study revolves around polybutylene succinate-co-butylene succinate (PBS), employing cutting-edge polymer micro-porous adsorption-desorption technology with carbon dioxide as an elegant desorption agent. We successfully crafted a series of PBS adsorption-desorption materials, exhibiting exceptional mechanical properties and adjustable structures. Carefully adjusting the adsorption static time during experiments vividly showcased the subtle controllability of adsorption and desorption effects of PBS materials under specific experimental conditions. Introducing polyhexanedioic acid into poly(butylene adipate-co-terephthalate) (PBAT), a pivotal component significantly impacting the microstructure and mechanical properties of PBS materials. Simultaneously, incorporating polyhexanedioic acid into PBAT not only expanded the application scope but also markedly enhanced the adsorption efficiency and mechanical performance of PBS adsorption-desorption materials.It is noteworthy that a 50:50 additive ratio (PBAT:PBS = 50:50) was adhered to during experiments, and results indicated a positive correlation between increased pressure and enhanced material adsorption capacity. In summary, this abstract's thorough exploration provides an elegant and efficient.approach for preparing biodegradable polyester adsorption-desorption materials. Additionally, it offers valuable insights into studying environmentally friendly biodegradation characteristics.

1 INTRODUCTION

The growing concerns regarding environmental degradation and the urgent need for sustainable alternatives have propelled extensive research in the realm of materials science. Among the myriad of materials under investigation, biodegradable polymers have garnered significant attention due to their potential to address pressing environmental issues while catering to various industrial needs. In this context, the development of advanced biodegradable polyester adsorption-desorption materials emerges as a promising avenue for both environmental remediation and industrial applications.

Polymer micro-porous adsorption-desorption technology represents a frontier in materials science, offering unprecedented control over material properties and functionalities. At the heart of this technology lies the intricate interplay between polymer matrices and adsorbates, wherein tailored microstructures enable efficient adsorption and desorption processes. By leveraging the unique properties of polymers, particularly those of polybutylene succinate-co-butylene succinate (PBS) [1], researchers have endeavored to fabricate adsorption-desorption materials with enhanced performance and versatility.

A notable advancement in the field involves the utilization of carbon dioxide (CO2) as an elegant desorption agent. CO2, owing to its environmentally benign nature and readily adjustable properties, presents a compelling alternative to traditional desorption agents. Through meticulous experimentation and optimization, researchers have harnessed the desorption capabilities of CO2 to achieve efficient and reversible adsorption-desorption processes, thereby enhancing the sustainability and practicality of the developed materials.

The incorporation of polyhexanedioic acid into poly(butylene adipate-co-terephthalate) (PBAT) [3] represents a significant innovation in enhancing the microstructure and mechanical properties of PBS materials. By strategically modifying the polymer matrix, researchers have not only expanded the application scope but also augmented the adsorption efficiency and mechanical performance of the resulting materials. This synergistic approach underscores the importance of molecular engineering in tailoring material properties to meet specific requirements.

The experimental methodology adopted in this study is characterized by its systematic approach and meticulous attention to detail. Through a series of controlled experiments, researchers have elucidated the intricate interdependencies between various parameters, such as adsorption static time, additive ratios, and pressure levels, on the adsorption-desorption behavior of PBS [2]

DOI: 10.1201/9781003514831-52

materials. The results obtained underscore the remarkable controllability and versatility of the developed materials, reaffirming their potential for diverse applications.

The findings presented in this study hold significant implications for both academic research and industrial applications. The development of advanced biodegradable polyester adsorption-desorption materials not only contributes to the ongoing discourse on sustainable materials but also offers practical solutions to environmental challenges. Moving forward, future research endeavors may focus on further optimizing material properties, exploring novel synthesis techniques, and elucidating the underlying mechanisms governing adsorption-desorption processes. Additionally, efforts towards scaling up production and integrating these materials into real-world applications are warranted to realize their full potential.

2 EXPERIMENTAL

2.1 *Materials*

The materials used in the experiment include polybutylene succinate-co-butylene succinate (PBS), polybutylene adipate-co-terephthalate (PBAT), carbon dioxide, etc. These materials are essential raw materials required for the experiment, and through precise blending and processing, the desired polymer micro-porous adsorption-desorption materials weresynthesized.

2.2 *Experimental instrumentation*

The primary instrument utilized during the experiment is the adsorption-desorption apparatus manufactured by ISCO (International Scientific Communications, Inc.). This apparatus features a sophisticated control system and highly sensitive sensors, enabling precise control and real-time monitoring of the adsorption and desorption processes. The ISCO adsorption-desorption apparatus provides a stable experimental platform, facilitating researchers in accurately measuring the adsorption and desorption properties of the samples.

2.3 *Determination of diffusion coefficients [4]*

To delve deeper into the adsorption-desorption properties of the materials, it is imperative to determine their diffusion coefficients. In the experiment, a series of standardized measurement methods were employed, coupled with the principles of mass balance and diffusion theory, to accurately determine the diffusion coefficients of the materials. This endeavor serves as a crucial foundation for further understanding the adsorption-desorption mechanisms of the materials.

3 RESULTS AND DISCUSSION

The solubility of carbon dioxide at different pressures and temperatures was determined using adsorption and desorption experiments. In the adsorption experiment, because the gas is released very quickly and the gas diffusion loss is very small, the measured adsorption amount needs to be adjusted to the actual adsorption amount. To this end, we performed desorption experiments to calculate the actual adsorption amount Figure 1.

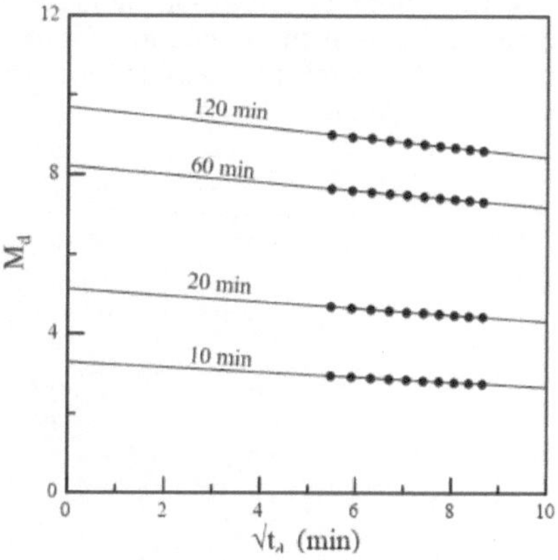

Figure 1. Residual adsorption capacity (M_d) and the square of desorption time ($\sqrt{t_d}$).

Based on the experimental results, we drew Figure 2 to show the effect of pressure on carbon dioxide solubility at different temperatures. It was observed that as time increased, the adsorption amount gradually increased and reached equilibrium. As the temperature increases, the time required to reach equilibrium decreases, suggesting that carbon dioxide diffuses more easily at high temperatures.

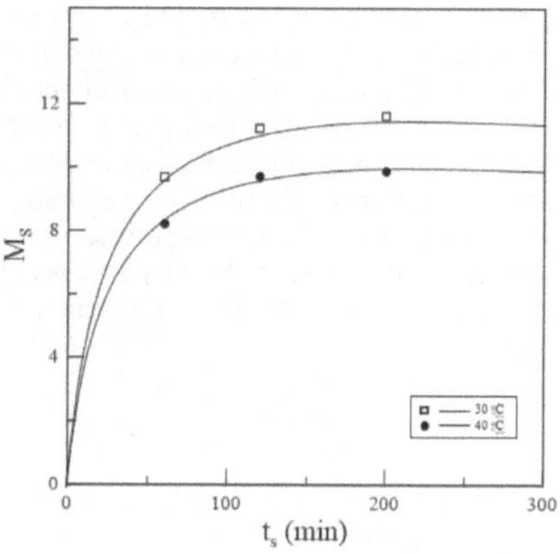

Figure 2. Effect of carbon dioxide solubility at 30°C and 40°C.

It can also be seen from Figure 2 that as the temperature increases, the solubility of carbon dioxide decreases significantly. This is because rising temperatures cause bonds to break, making it easier for carbon dioxide to be released.

To gain a deeper understanding of the relationship between solubility and pressure, we plotted Figure 3, which shows changes in carbon dioxide solubility as a function of temperature and pressure. A linear increase in carbon dioxide solubility was observed with increasing pressure, consistent with Henry's law. As the temperature increases, the solubility of carbon dioxide gradually decreases because the bonds are more easily broken at high temperatures, causing more carbon dioxide to be released.

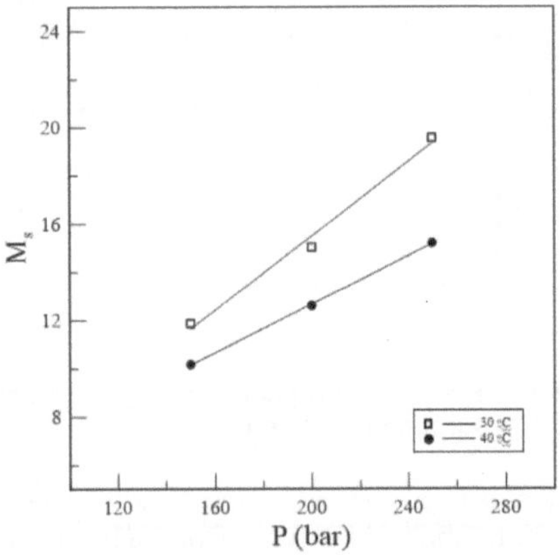

Figure 3. Effect of pressure on carbon dioxide solubility.

4 CONCLUSION

This study revolves around polybutylene succinate-co-butylene succinate (PBS), employing cutting-edge polymer micro-porous adsorption-desorption technology with carbon dioxide as an elegant desorption agent. We successfully crafted a series of PBS adsorption-desorption materials, exhibiting exceptional mechanical properties and adjustable structures. Carefully adjusting the adsorption static time during experiments vividly showcased the subtle controllability of adsorption and desorption effects of PBS materials under specific experimental conditions. Introducing polyhexanedioic acid into poly(butylene adipate-co-terephthalate) (PBAT), a pivotal component significantly impacting the microstructure and mechanical properties of PBS materials. Simultaneously, incorporating polyhexanedioic acid into PBAT not only expanded the application scope but also markedly enhanced the adsorption efficiency and mechanical performance of PBS adsorption-desorption materials. It is noteworthy that a 50:50 additive ratio (PBAT:PBS = 50:50) was adhered to during experiments, and results indicated a positive correlation between increased pressure and enhanced material adsorption capacity. In summary, this abstract's thorough exploration provides an elegant and efficient approach for preparing biodegradable polyester adsorption-desorption materials. Additionally, it offers valuable insights into studying environmentally friendly biodegradation characteristics. This research holds promise in offering new perspectives and solutions for better addressing current environmental challenges and sustainable development goals.

REFERENCES

[1] Massimiliano Barletta, Clizia Aversa, Muhammad Ayyoob, Annamaria Gisario, Kotiba Hamad, Mehrshad Mehrpouya & Henri Vahabi. (2022). Poly(butylene succinate) (PBS): Materials, processing, and industrial applications. *Progress in Polymer Science Volume 132*, 101579

[2] Dongdong Hu, Kun Xue, Zhen Liu, Zhimei Xu & Ling Zhao. (2022). The essential role of PBS on PBAT foaming under supercritical CO2 toward green engineering. *Journal of CO2 Utilization Volume 60*, 101965

[3] Zhaozhi Wang, Guilong Wang, Zhaorui Xu, Chengming Yang & Guoqun Zhao. (2023). Structure-tunable poly(butylene adipate-co-terephthalate) foams with enhanced mechanical performance derived by microcellular foaming with carbon dioxide as blowing agents. *Journal of CO2 UtilizationVolume 72*, 102495

[4] Pengke Huang, Yaozhuo Su, Haibin Luo, Xiaoqin Lan, Yunkai Chong, Fei Wu & Wenge Zheng. (2022). Facile one-step method to manufacture polypropylene bead foams with outstanding thermal insulation and mechanical properties via supercritical CO2 extrusion foaming. *Journal of CO2 Utilization Volume 64*, 102167

System Innovation for an Artificial Intelligence Era – Lam et al. (Eds)

Developing micropillar array of microfluid smart sensor for Carbon Footprint on-line monitoring

Yun-Peng Yeh & Martin Kwek
Department of Mechanical Engineering, National Taiwan University of Science and Technology, Taipei, Taiwan

Alan Tobias Torres Velazquez, Tais Nathaly Machado, Ana Cristina Ojeda Caceres & Tamara Raquel Von Glasenapp
Department of Electromechanical Engineering, Universidad Politécnica Taiwan-Paraguay, Asunción, Paraguay

ABSTRACT: Greenhouse gas emissions caused globe warming and the frequent occurrence of extreme weather in recent years and hence monitoring carbon emissions has become a global imperative. Microfluidics is an interdisciplinary field that involves the manipulation of fluids in a small system that operates at the microscale level; channels inside the microfluidic are usually in the size of micrometers and can only be passed by tiny volumes of fluids. In this study, we have demonstrated the high aspect ratio micropillars in microfluid system are designed in polydimethylsiloxane (PDMS) flexible form, which partakes in effecting the characteristic sensing features. The prelimary result shows that the spacing between micropillars depends on the diameter of itself and fluid flows with different speed, if the spacing is too narrow, there will be insufficient space for the micropillar to manipulate. Therefore, finding best ratio of diameter to spacing, which can not only avoid the aggregation between the micropillars but also increase the sensitivity of carbon adsorption for monitoring.

1 INTRODUCTION

The utilization of microfluidics chips has become increasingly prevalent across various domains, offering capabilities for real-time and precise measurement within sensing systems. These chips find application in pH control, cost-effective drug delivery, and the generation of chemical gradients through diffusive fluid mixing. Facilitated by the integration of microstructures such as microchannels, micropumps, valves, and microelectronic components, these lab-on-a-chip devices have revolutionized biomedical applications. Signi-ficantly, they are instrumental in swiftly and precisely detecting greenhouse gases, effectively mitigating environmental challenges related to climate change and resource management [1–4].

Microfluidic-based greenhouse gas, for example, carbon dioxide (CO_2) capture represents an innovative technology aimed at regulating and directing the absorption process of CO_2 gas by a chemical solvent within miniaturized absorbers or reactors characterized by channel dimensions on the micro or millimeter scale. Extensive research has demonstrated that the gas-liquid two-phase flow within microchannel reactors, when appropriately configured and managed, can manifest several advantageous attributes. These include augmented mass transfer facilitated by a heightened volumetric interfacial surface area, enhanced repeatability and stability, reduced risk of leakage enabling safe handling, and simplified scalability through the implementation of a numbering-up approach.

The design and fabrication of microfluidic channels have proven to be both practical and cost-effective, particularly in high-demand biomedical contexts. Soft lithography techniques are commonly employed for microchannel fabrication, often supplemented by printed circuit board (PCB) tech-nology to create master molds for smart sensing systems. Additionally, efforts are directed towards minimizing the dimensions of microfluidic devices to reduce sample consumption and waste generation, thereby enhancing portability and disposal ease while maintaining cost efficiency.

Contemporary advancements in microscale fabrication have empowered the development of diverse microstructured devices tailored for numerous biochemical applications, seamlessly integrating microelectronic components. Among these innovations, micropillar arrays have garnered significant attention since the 2000s, primarily due to their capability to effectively capture and identify various types of cells, including normal and cancer cells, across a spectrum of sizes [5–7].

The theoretical and practical significance of micropillars in biomedical engineering and energy-related fields is widely recognized. Despite

DOI: 10.1201/9781003514831-53

persistent challenges associated with their topographical features, the imperative to enhance micropillar quality remains paramount. Researchers endeavor continuously to refine parameters such as stability, bending stiffness, and material selection to surmount these obstacles. This study proposes the integration of High-Aspect-Ratio (HAR) micro-pillars into microfluidics chips. Leveraging COMSOL Multiphysics Simulation software, we conduct a comprehensive analysis of the micro-structural characteristics of these micropillars. Our investigation seeks to dynamically explore the variation in HAR micropillar designs within microfluidic systems, with the overarching goal of advancing the production of high-quality micro-pillars tailored for carbon online monitoring applications.

2 EXPERIMENTAL

This study commenced by assimilating insights from extant theoretical literature, followed by the conceptualization and design of a micropillar array using COMSOL Multiphysics simulation software. The primary aim was to acquire proficiency and expertise in utilizing the simulation platform. Subsequently, a comprehensive examination of individual micropillars was conducted, encompassing analyses of variations in shape, aspect ratio, material composition, and fabrication parameters. The design of microfluid channel provide minimal simulation error results and similar knowledge but is adequate. One of the reasons is also because of the massive quantity of domain, boundary, and edge elements. Unless a supercomputer is exclusively used for the simulation, it is important to reduce the massive elements that will lag or hang when simulating on a normal computer.

A fundamental attribute of micropillars is their high aspect ratio (HAR), defined as the ratio between the pillar's height (h) and its diameter (d). The utilization of HAR micropillars is pivotal for optimizing the efficiency of individual pillars. However, challenges arise due to constraints in fabrication techniques and the impact of external forces. Consequently, it is imperative to ascertain the most suitable parameters for implementation. Simulation of micropillar growth facilitates the examination of pillar deflection. Moreover, micropillar designs extend beyond cylindrical structures, encompassing rectangular and conical shapes, each characterized by distinct aspect ratios that influence pillar stability. Figure 1 depicts the design of a single cylindrical micropillar, featuring a substrate measuring 0.2x0.2 mm and a pillar with dimensions of 0.1 mm in diameter and 0.15 mm in height, resulting in an aspect ratio of 1.5.

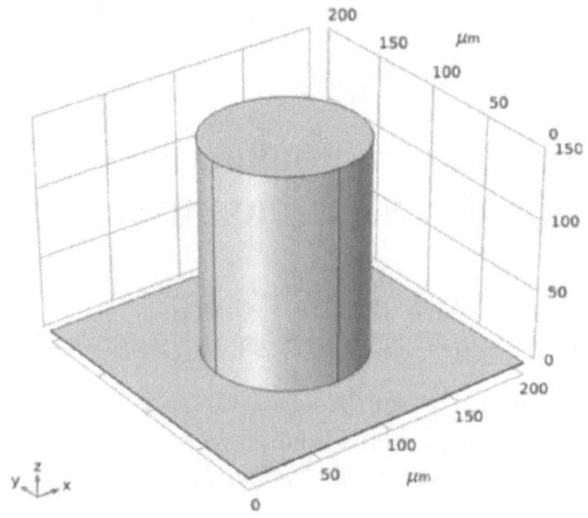

Figure 1. A micropillar designed for growth simulation.

3 RESULTS AND DISCUSSION

Polydimethylsiloxane (PDMS) was used in this study for its biocompatible properties and tunable composition for the carbon trapping applications. Establishing a predictive framework to anticipate experimental outcomes using COMSOL Multiphysics is essential for simulating real-world experiments. Preliminary investigation of material properties is crucial, as these properties significantly influence simulation results. The setup involves defining boundary conditions for fluid-structure interaction, integrating laminar flow and solid mechanics analyses. Within the laminar flow framework, all surrounding walls are designated as part of the flow domain, with certain walls designated as inlets and outlets. Velocity inlet conditions are specified, and outlet conditions are set to zero pressure. In the solid mechanics analysis, micropillars serve as the domains, with fixed constraints applied to ensure designated parts remain stationary and unaltered during simulation. Key parameters configured for simulation include wall slip, deforming domains, fixed constraints, and preset velocities.

This study investigates the impact of fluid flow on micropillars through simulation employing fundamental boundary conditions. The physical and chemical properties of water and polydimethylsiloxane (PDMS) at standard room temperature are detailed in Table 1. Within the simulation framework, solid deformation evolves over time due to the dynamic nature of the solid field.

Table 1. Physical-chemical properties of PDMS.

Properties	PDMS Sylgard 184 at 24°C	Unit
Density	982	kg/m^3
Dynamic Viscosity	–	Pa*s
Young's Modulus	$1.2x10^6$	Pa
Poisson's Ratio	0.46	1

The fluid velocity, characterized by various magnitudes of oscillation, exerts an influence on the aspect ratio of columnar micropillars, as shown in Figures 2 and 3, where that by bending deformation are able to find the content for carbon trapping.

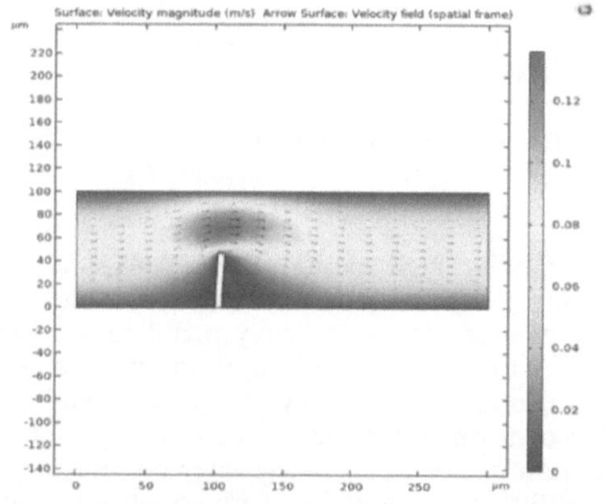

Figure 2. Fluid flow effect for PDMS side bending.

Consequently, the design of micropillars in polydimethylsiloxane (PDMS) with an aspect ratio of 1:7 is deemed optimal for controlling fluid velocity across different ranges. Analysis of the results reveals that the spacing between micropillars is contingent upon the diameter of the microstructures; excessively narrow spacing may impede manipulation of the microstructure due to insufficient clearance. Hence, a diameter-to-spacing ratio of 1 is recommended to mitigate interaction and aggregation among micropillars.

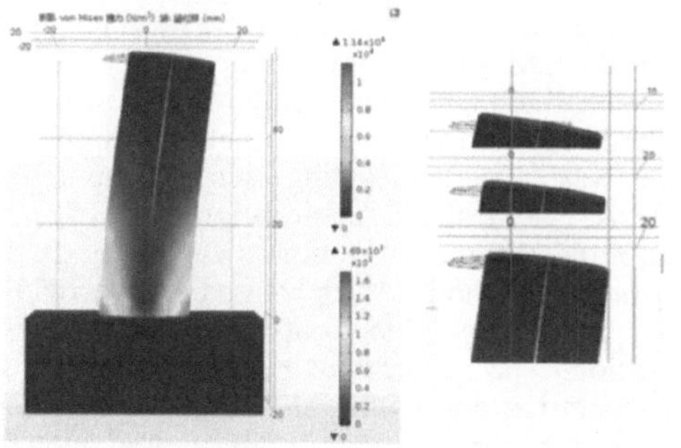

Figure 3. Side deformation of PDMS micropillar from various fluid speed.

4 CONCLUSION

In summary, our study has demonstrated the design and implementation of high-aspect-ratio (HAR) micropillars within microfluidic systems, with a focus on their flexibility, as for carbon reactive absorption, entails the absorption of a carbon dioxide into a liquid phase through a combination of chemical reaction and absorptive mass transport.. We have also conducted simulations and designs of micropillar array chips with this flexibility. The choice of material composition, particularly polydimethylsiloxane (PDMS), significantly influences the characteristic features of these micropillars due to its tunable properties. Looking ahead, advancements in modern fabrication techniques hold promise for enhancing micropillars, allowing for higher aspect ratios, reduced clustering, and the potential to maintain a balance between cost-effectiveness and quality. Leveraging simulation software to explore various combinations of microstructural parameters, material compositions, and multiphysics simulations offers significant potential for the development of high-quality, high-aspect-ratio micropillars.

ACKNOWLEDGEMENT

This work was financially supported by the Ministry of Foreign Affairs, Republic of China (Taiwan), under the Universidad Politécnica Taiwan-Paraguay (UPTP) project.

REFERENCES

[1] Koronaki I.P. *et al.* (2015). *Renew Sustain. Energy Rev*, Vol. 50, p 547–566.

[2] Abdeen F.R.H. *et al.* (2016), *Chin. J. Chem. Eng.*, Vol. 24, Iss. 6, p 693–702.

[3] Yin Y. *et al.* (2019), *Chem. Eng. Sci*, Vol. 201, p 264–273.

[4] Ma D. *et al.* (2020), *Sep. Purif. Technol.*, Vol. 242, p 116795.

[5] Lai T., *et al.* (2014), *Adv. Mater. Res.*, Vol. 941–944, p 2149–2153.

[6] Sattari-Najafabadi M. *et al.* (2018), *Chem. Eng. Process. Intensif.*, Vol. 127, p 213–237.

[7] Aghel B. *et al.* (2022), *Int. J Greenh. Gas. Control*, Vol. 119, p 103715.

System Innovation for an Artificial Intelligence Era – Lam et al. (Eds)

Study on the graphene oxide absorbent to treat heavy metal nickel wastewater

Shan-Yi Shen & Yuan-Kai Chang
Department of Environmental Engineering, Da-Yeh University, Dacun, Changhua, Taiwan

ABSTRACT: Due to its excellent chemical and thermal stability, high specific surface area, and high electron mobility, graphene oxide is widely used in different types of wastewater treatment. This study prepared graphene oxide adsorbent and explored the removal efficiency of heavy metal nickel from wastewater. Moreover, we test different operation parameters to obtain the most appropriate operating conditions. From microscopic observation, it shows that the surface morphology of graphene oxide is a granular structure with many pores. Elemental analysis has confirmed that the material mainly contains carbon and oxygen components. According to the adsorption test of nickel wastewater (10 mg/L), the removal efficiency of nickel can be increased from 0.5 g/L of 30% to 5 g/L of 84% with an adsorbent in a reaction time of 10 mins, and the adsorption capacity of nickel is increased approximately three times. This indicates that the amount of adsorbent will significantly affect the adsorption of nickel in wastewater. The initial pH value of wastewater also affects the adsorption behavior of nickel, increasing from 64% at pH 3 to 87% at pH 7, indicating that neutral conditions are conducive to the adsorption of nickel. The wastewater temperature has no significant effect on the removal efficiency of nickel. Based on the above, the graphene oxide in this study can effectively remove heavy metal nickel and has a high potential for treating heavy metal wastewater in the future.

1 INTRODUCTION

Clean water is one of the most precious and indispensable natural resources for life on earth. The pollution of water by heavy metal ions from different sources has become an environmental issue of global concern. Heavy metal contaminants in industrial wastewater or natural water sources will cause serious environmental pollution and health risks due to their high toxicity and non-biodegradable properties (Tian *et al.* 2022). Moreover, heavy metals in industrial wastewater have been listed as priority pollutants by the United States Environmental Protection Agency and can bioaccumulate through the food chain (Kamali *et al.* 2019, Xu *et al.* 2018, Zhou *et al.* 2016). Once the discharged wastewater from factories is not effectively treated, it will be directly injected into urban surface runoff, rivers, and other receiving water bodies. Food and drinking water will be polluted, and human health will face major risks (Wiśniewska and Nowicki, 2019). With the simultaneous rise of urbanization and industrialization, the use of heavy metals in industry has increased significantly over the past few decades. Metals are used in production processes across a variety of industries, including electroplating, pigments, mine drainage, and so on. These companies discharge large amounts of wastewater containing various metals into the environment, causing significant pollution to our water bodies. Therefore, quickly and effectively removing heavy metals from wastewater is an environmental issue of great concern.

So far, there have been many reports on various methods for removing heavy metal contaminants from wastewater, including adsorption, reverse osmosis, chemical coagulation, ion exchange, and membrane filtration. Among these methods, the adsorption method is widely used in the field of water purification due to its low cost and easy operation. The adsorption method is suitable for the separation and collection of metal ions. It has been proven to be an economical and effective treatment method and has the potential to be used on an industrial scale. Among these adsorption materials, graphene oxide (GO) has a high surface area, high current density, and light transmittance, making it a suitable material for adsorbing toxic compounds and has been widely used in wastewater treatment. Graphene oxide is the oxidized form of graphene obtained from graphite through chemical exfoliation. It has a single monolayer containing oxygen functional groups such as carboxyl, carbonyl, epoxy, or hydroxyl groups (Jahan *et al.* 2022). These features widen the separation between layers

DOI: 10.1201/9781003514831-54

and make the material hydrophilic. It has been reported that GO is an effective adsorbent for heavy metal ions due to its surface area, high water solubility, and oxygen-containing surface functional groups. Because graphene oxide has the above properties, it has the potential to become a candidate material for removing various pollutants from wastewater. This study provides a safe, environmentally friendly, and effective separation method for heavy metal-contaminated wastewater, and demonstrates new applications in the field of heavy metal nickel ion-contaminated wastewater.

2 EXPERIMENTAL

The study is to prepare graphene oxide adsorbent, and the prepared graphene oxide for treating nickel-contaminated wastewater has been evaluated. The production procedure for graphene oxide is as follows: First, we slowly add 46 mL of sulfuric acid solution (purity of 98%) and 0.6 g of graphite powder in an ice bath at 4°C and stir for 30 minutes; then, add 1 g of sodium nitrate and stir for 15 minutes, and add 6 g of potassium permanganate and stir for 30 minutes. Next, take it into an oil bath at 35°C, and slowly add 96 mL of deionized water and raise the temperature to 98°C for 2 hours; add 280 mL of deionized water and 30 mL of hydrogen peroxide (purity of 30%) in sequence and stir for 30 minutes, respectively. Turn off the heating device after completion. Add 70 mL of hydrochloric acid (purity of 36%) stir for 30 minutes, and stand for 1 day. Remove the supernatant and rinse the GO sol with deionized water until neutral quality. Finally, drain it in an oven at 70°C for 6 hours.

The experimental steps for adsorption of nickel contaminated wastewater are as follows: This study uses nickel nitrate to prepare nickel-containing metal wastewater. During the process, prepare a nickel-containing metal solution with a concentration of 100 mg/L as a stock solution. Then, the concentration of 100 mg/L was diluted to 10 mg/L to carry out the subsequent experiments. The wastewater with different initial pH values was adjusted by 1.0 M sodium hydroxide and 1.0 M nitric acid, respectively. First, put the 20 mL of wastewater with a nickel concentration of 10 mg/L into a 250 mL Erlenmeyer flask, and then add a fixed dose of graphene oxide (0.5~5 g/L) to carry out an adsorption reaction for 3 hours. Wastewater was sampled at fixed time intervals, and the sample was filtered with a 0.45 μm filter. Finally, the nickel metal concentration in the wastewater is detected using an atomic absorption spectrometer. In addition, the variation in water pH and conductivity during the adsorption reaction are also monitored simultaneously. The operating parameters of this experiment are mainly include the solid-liquid ratio, initial pH value, and operation temperature.

3 RESULTS AND DISCUSSION

Figure 1 shows the surface morphology and elemental analysis of graphene oxide prepared in this study. It is known from FESEM microscopic observation that the surface of graphene oxide has a particle structure with many pores, which is conducive to the adsorption reaction of pollutants. The elemental analysis of the EDS spectrum shows that the material mainly contains carbon and oxygen elements. From microscopic observation, it shows that the surface morphology of graphene oxide is a granular structure with many pores. Elemental analysis has confirmed that the material mainly contains carbon and oxygen components.

Figure 2 shows the removal efficiency of nickel concentrations with different contents of graphene oxide adsorbent. The results show that the removal efficiency of nickel concentration has significantly increased with the content of graphene oxide, from 30% at 0.5 g/L to 84% at 5 g/L. This result indicates that the adsorbent content affects the removal efficiency of nickel contaminants. It can provide more adsorption sites when adding a high proportion of graphene oxide, leading to an increase in the removal efficiency of nickel concentration.

(a)
(b)

Figure 1. The FESEM (a) and EDS pattern (b) of graphene oxide.

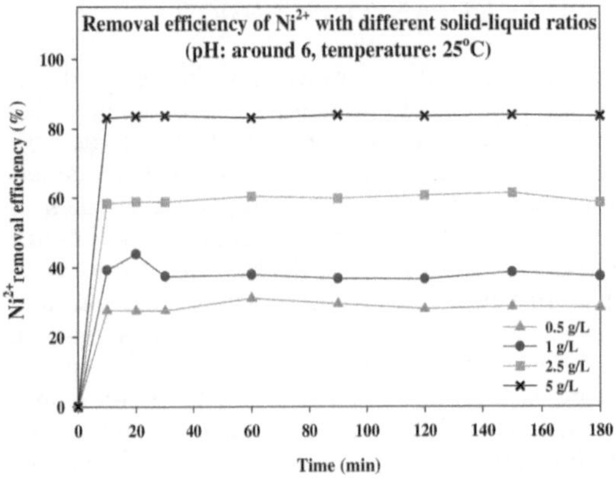

Figure 2. The removal efficiency of nickel concentrations with different contents of graphene oxide adsorbent.

Figure 3 shows the removal efficiency of nickel concentrations under different initial pH conditions. The results show that the removal efficiency is lower at pH 3 (64%). When the pH value is raised to 4, the removal efficiency has a significantly improved trend. However, when the pH value is continuously raised to 7, the removal efficiency is not significantly improved. The initial pH value of wastewater affects the adsorption behavior of nickel, increasing from 64% at pH 3 to 87% at pH 7, indicating that neutral conditions are conducive to the adsorption of nickel.

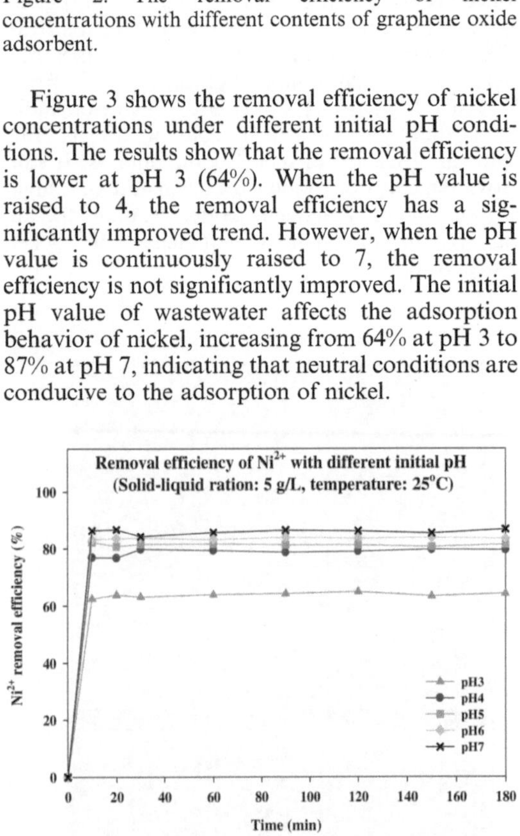

Figure 3. The removal efficiency of nickel concentrations with different initial pH values.

Figure 4 shows the removal efficiency of nickel concentration under different wastewater temperature conditions. The results show that the removal efficiencies are very similar under different wastewater conditions, and all can reach removal efficiencies of more than 80%. This result indicated that graphene oxide can remove nickel pollutants at any wastewater temperature.

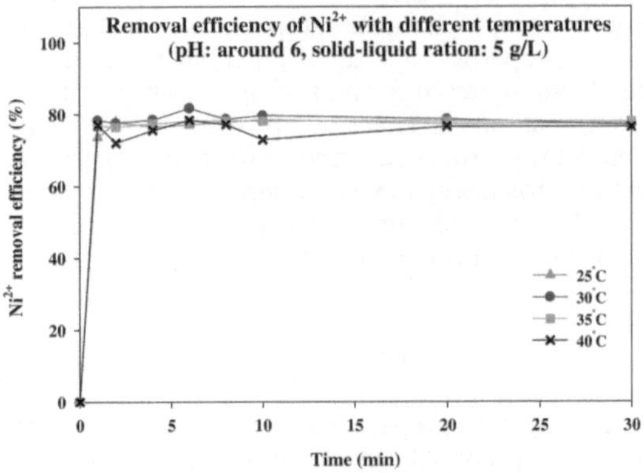

Figure 4. The removal efficiency of nickel concentrations with different wastewater temperatures.

4 CONCLUSION

According to the experimental results, the following conclusions can be drawn:

1. This study has successfully prepared graphene oxide adsorbent. The granular structure with many pores is beneficial for nickel adsorption removal.
2. Adjusting the solid-to-liquid ratio of graphene oxide significantly affects the removal of nickel concentration, which can increase from 30% at 0.5 g/L to 84% at 5 g/L.
3. Under different initial pH values of wastewater and wastewater temperature conditions, this study has obtained relatively appropriate operating parameters.
4. Graphene oxide can effectively and quickly remove nickel heavy metal from wastewater and is a highly potential adsorbent material.

REFERENCES

Kamali, M., Suhas, D.P., Costa, M.E., Capela, I., Aminabhavi, T.M. (2019). *Chem. Eng. J.* 368, 474–494.

Wei Tian, Yaqin Rong, Dali Li, Jin Tian, Naiming Lin, Zhongde Wang. (2022). *Journal of Physics and Chemistry of Solids* 162 110516

Wiśniewska, M., Nowicki, P. (2019). *J. Mol. Liq.* 296, 111806.

Xu, L., Yang, W., Jiang, F., Qiao, Y., Yan, Y., An, S., Leng, X. (2018). *J. Coast. Conserv.* 22, 209–215.

Zhou, Y., Tang, L., Zeng, G., Zhang, C., Zhang, Y., Xie, X. (2016). *Sens. Actuators B Chem.* 223, 280–294.

System Innovation for an Artificial Intelligence Era – Lam et al. (Eds)

New style of interactive navigation system of Taipingma Museum of Fuxing Templein Xiluo

Shih-Mo Tseng
Graduate School of Design, Nation Yunlin University of Science and Technology, Yunlin, Taiwan

Chao-Ming Wang
Department of Digital Media Design, Nation Yunlin University of Science and Technology, Yunlin, Taiwan

ABSTRACT: With the advancement of technology, interactive guided tours have gradually replaced personal guided tours, and new technologies have been applied to present museum information and improve communication efficiency. This study takes the Taiping Mazu Cultural Relics Museum of Fuxing Palace in Xiluo as the site, and creates an interactive navigation system for the important cultural relics in the cultural relics collection. Through public display, the general public and experts can actually participate in the interface operation experience and watch the audio-visual content of cultural relics. By inviting palace and temple experts to actually participate in the device operation experience, expert experience will be used as a reference for future development of new interactive navigation systems.

Keywords: digital media, interactive technology, interactive navigation system, local cultural relics museum

1 INTRODUCTION

With the growth of social economy and improvement of education, cultural tourism has become one of people's holiday leisure options. According to UNWTO (United Nations World Tourism Organization) estimates in 2017, cultural tourism accounts for 39% of all international tourist arrivals (G. Richards, 2018). Use holidays to participate in some cultural visits and experience activities, such as visiting palaces, temples, ancient houses, ruins and Museums, or experiencing unique local folk activities and markets are all types of cultural tourism. Considering the frequency of major folk religious festivals in Taiwan, there are opportunities to participate in local cultural tourism almost throughout the year. Among them, the incense-offering processions and circumambulation activities organized by Mazu temples across Taiwan attract tens of thousands of people to participate, demonstrating the influence of Mazu belief on Taiwanese people. Fuxing Temple in Xiluo is a relay station for many Mazu activities around the country. It is also one of the important Mazu belief centers in the Yunlin area. Fuxing Temple also demonstrates its lofty status by preserving many ancient cultural relics of temples. The Taipingma Museum was established in 2022 to display important ancient cultural relics of temples, and to establish the cultural asset value of the cultural relics through expert research. This study takes the Fuxing Palace Taiping Ma Art Museum as the research object, and uses digital media to tailor an interactive navigation system for the museum to provide believers and tourists who visit Fuxing Palace with an in-depth understanding of the museum's collections. The interactive installation of the cultural relics uses a touch screen to provide visitors with on-demand audio and video guided tours, thereby promoting Taiping Mazu's spirit of protecting all living beings.

2 LITERATURE

2.1 *Cultural tourism*

Regarding the definition of cultural tourism, the definition in Western society is not clear. Cultural tourism is not just visiting places of interest in the traditional sense, it also includes consuming the lifestyle of different destinations(S.S. Mousavi, *et al.* 2016). Generally speaking, the purpose of cultural tourism is to understand and become familiar with the history and lifestyle of a specific place. Tourists are tourism activities whose fundamental motivation is to learn, discover, experience and consume the tangible and intangible culture of the tourist destination. According to research by Western scholars, purposeful attempts to use cultural activities to improve local image have proven to be effective (G. Richards, *et al.* 2004). Urry (1990) believes that

DOI: 10.1201/9781003514831-55

"tourism is culture". These assumptions have led some scholars to believe that cultural tourism is a new thing or a postmodern phenomenon. In Europe, travel helps increase cultural understanding and rebuild broken economies (G. Richards, 2018). By using culture as a medium and tourism as a method, we can enhance the cultural literacy and self-cultural identity of Chinese people.

2.2 *Local cultural center*

In recent years, the government has promoted the "Local Cultural Center Project", which uses the revitalization and reuse of existing buildings or spaces, combined with community creation through cultural life circles, and the integration of local residents and social groups to allow cultural construction to take root in the local area. Cultural Centers It provides opportunities for the public to participate in artistic and cultural activities, and can become a base for cultural tourists to visit and learn about local culture. The Musee d'Orsay in Paris, France, was formerly a train station. It was successfully transformed into a world-renowned art museum by reusing the old space. Many communities in the United States have reused vacant space buildings, including schools, factories, warehouses, and churches, to build arts and culture-related facilities (J. Delconte, *et al.* 2016). Many local cultural centers in Taiwan have also repurposed old spaces, thus presenting various appearances, emphasizing the preservation of local culture and the promotion of local spirit. The Taipingma Museum, the subject of this study, is affiliated to Xiluo Fuxing Temple. The historical relics preserved by the palace and temple are combined with government guidance and partial resource injection. The original wing room is used to create the highest standard cultural relics preservation space with constant temperature and dehumidification, avoiding the environment of a museum. Improper control has resulted in many ancient cultural relics being damaged or even destroyed (X. Luo, *et al.* 2015).

2.3 *Interactive guided tours*

An increasing number of museums are working to enhance visitors' museum experiences in-house in personal, intensive and engaging ways(I. Roes, *et al.* 2009). The purpose of museums has changed from merely providing information about static collections to providing personalized services to different visitors around the world, consistent with the visitor's personal characteristics, goals, tasks, and behaviors(Y. Wang, *et al.* 2009). Therefore, the plan to build a museum based on the spirit of interactive exploration, friendly design, open data downloads and personalized tours, and integration of digital planning resources was born. Interactive tours depend on the media used. Commonly used are multimedia audio-visual tours and AR/VR tours. The user interface is usually nothing more than a mobile device, a touch screen or a head-mounted display. It allows users to instantly watch digital video content with a touch screen, or immerse themselves in a virtual space environment that integrates reality and virtuality through a head-mounted display. These two methods are quite common in museum space navigation applications. These two methods are quite common in museum space navigation applications. They can transmit image scenes of the museum and exhibits to remote visitors through the network. It can independently choose the exhibits to be visited and the viewing conditions, and provide comprehensive guidance assistance for museum visitors groups or individuals (P. Trahanias, *et al.* 2000).

3 METHODS

3.1 *Text analysis*

Text analysis can be divided into two major types: linguistics treats text as an object of analysis; sociology treats text as a window into human experience(H.R. Bernard, *et al.* 1998). This study will focus on the methods of collecting and analyzing written texts, including the four major cultural relic research books provided by Fuxing Temple, information and academic research papers provided on the official website of Fuxing Temple. In order to convey the profound cultural value of ancient cultural relics, Fuxing Palace Planning has published a series of research books on the local history of Xiluo and temple cultural relics, giving researchers the opportunity to have a glimpse of the historical records related to Fuxing Temple. Among them, the research on the four major cultural relics of "Having Good Faith and Congfeng Plaque", "No Disrespect to Relatives Plaque", "Eagerly Offering Sacrifices to the Table", and "Taiping Ma Nantou Pottery Incense Burner" (Figures 1–4) are introduced in detail in special books. This study uses four major cultural relics books as the basis, plus the official website of Fuxing Temple, academic research papers and related reports, to plan a set of professional guide content, removing difficult academic terms and cumbersome historical eras, and using the most vernacular way Present the most important information of each cultural relic and convey the spirit and meaning behind the important cultural heritage of the Xiluo region.

Figure 1. "Having Good Faith and Congfeng" plaque.

Figure 2. "No disrespect to relatives" plaque.

Figure 3. Eagerly offering sacrifices to the table.

Figure 4. Taiping Ma Nantou Pottery Incense Burner.

3.2 *Prototyping*

Prototypes are often seen as new attempts to bring future reality into the current situation. A prototype is a representative example or unfinished product, usually an incomplete physical installation, made for a purpose to achieve the conceptual structure of a final product. A prototype is a preconceived model of some aspect of a product concept or final design. Prototyping is intertwined with nearly all product, service, and system development efforts(B. Viswanathan, *et al.* 2017). Therefore, prototype design will guide the final appearance of the focus and future product design. This research is oriented towards user experience and describes "experience prototyping" as a form of prototyping that enables design team members, users and customers to gain experience in existing or future uses by actively participating in prototyping. First-hand information from the author's experience(M. Buchenau, *et al.* 2000). It was concluded from the experience in the case study: understanding existing experience, exploring design ideas and communicating design concepts as three important principles for prototyping.

3.3 *Expert interviews*

Expert interviews are a method often used in policy analysis, qualitative semi-structured or open interviews with "expert knowledge" (L. Van Audenhove, *et al.* 2019). The advantage of face-to-face interviews is that the interviewer and the interviewee can grasp more social clues(I. Mergel, *et al.* 2019) and collect more useful information to help the researcher obtain clearer information. This study selected temple officials and opinion leaders who have active concerns about religious cultural relics in the Xiluo area as interview subjects (Table 1). Among them, P1 is the chairman of the Xiluo Local Foundation and has been actively involved in local cultural preservation and local creation for more than 30 years. P2 is a director of a well-known local temple. He has served as a consultant to the local county government and a municipal government for many years. He is very familiar with local affairs and the needs of temples. P3 is the CEO of a well-known local temple and is in charge of the internal affairs of the temple and the management of the cultural relics museum. This study first invites experts to experience the navigation system, and then asks the experts to make a comprehensive evaluation of the interactive navigation content and express their opinions, which will serve as a reference for this study to evaluate the quality of the system design. The highlights of this expert interview include (Table 1):

1. Does the digital tour content produced in this study meet the actual needs of the exhibition site?
2. Does the guided tour of cultural relics clearly describe the characteristics and value of the cultural relics?
3. Is your organization willing to adopt the same interactive tour system?

Table 1. List of qualifications of interviewed experts.

No.	Professional title	Relevant qualifications	PS
P1	Chairman of the Foundation	engaged in local creation for 30 years	
P2	Director of a well-known temple	consultant to the county government, consultant to the municipal government	
P3	CEO of a well-known temple	CEO of Cultural and Educational Foundation and temple for many years	

4 RESULTS

4.1 *Digital media applications meet the needs of the exhibition site*

In the past, in addition to the physical display of cultural relics, the display of local cultural relics relied only on static pictures, text descriptions and staff guides. Under the constraints of limited space and time, only a few people could actually participate in high-quality and content-rich cultural learning. With the development of electronic technology, traditional forms of museum tours are no longer able to attract tourists. Some scholars believe that it is very important for visitors to participate in the process of cultural learning and experience rich experiences(Y. Tian, 2023). The integration of digital audio-visual and interactive technologies provides for the exchange of knowledge between visitors and historical relics. Taking advantage of digital audio-visual media, visitors learn knowledge through dynamic audio-visual introductions and an interactive mechanism where they can choose the content of the tour independently. Affected by the COVID-19 epidemic, the educational functions of museums have been severely restricted due to travel bans, and the demand for online exhibitions and non-personnel guided tours has continued to grow. This study uses digital audio-visual media for the production of ancient cultural relics and a vertical touch screen as a solution for human-machine interface design (Figure 5). Interviewed expert P1 said that although the foundation she serves does not have a collection of cultural relics, it does collect a lot of old photos and local cultural and historical materials. However, there has been a lack of professional assistance to produce digital audio-visual files for use. Although the temple to which expert P2 belongs collects many cultural relics, it has not established a cultural relics museum for the study and collection of cultural relics. It currently lacks specific cultural relic research text content, let alone the production of digital audio and video content. Expert P3 temple has already set up a cultural relics museum, and has completed the research and research on the four major cultural relics and published a set of special books. However, it still has not actively produced digital audio and video guide content for the cultural relics. Instead, it wants to develop AR (augmented reality) experience system in the future. At this stage, the digital audio-visual content produced in this research is still affirmed.

Figure 5. Vertical touch screen.

4.2 *Interactive tours will become a new form of cultural relics tours*

In the past, cultural museums, large and small, introduced the cultural relics in their collections through guided tours or audio tours. The unknown stories behind each cultural relic could only be explained verbally. Only with the help of detailed introductions and explanations by the tour guides can the hidden stories be understood. After the interviewed experts P1 and P2 experienced the touch screen developed by our institute, they both praised the user-friendly design of the interactive interface. They definitely introduced the content of cultural relics through digital audio-visual methods in a clearer and more detailed way than the human guide. Moreover, visitors can click on the content they like to read through the touch screen. In the future, they are willing to give priority to the interactive navigation system produced by our institute as the navigation mechanism when building a cultural navigation system. The interviewed expert P3 believes that the interactive design function of the touch screen and the introduction of the four major cultural relics are very complete. Even without a tour guide, visitors can first understand the information about the cultural relics in the collection. When entering the cultural relics museum, visitors can pay more attention observe the details of the cultural relics. The position of the temple is to further develop AR tour guide method, so that visitors can feel the high-tech atmosphere while entering the traditional temple.

4.3 *Digital content production provides complete information on cultural relics*

Fuxing Temple began to conduct textual research on the four major cultural relics in 2016. It took five years to publish five special books to record the four major cultural relics and the history of Fuxing Temple (Figure 6). However, only sporadic media reporting videos were retained, and there wasn't focus on each individual cultural relic to create any audio-visual for introduction. Based on the contents of the four major cultural relic research books and other relevant research reports, this research produces an exclusive audio-visual guide video for each cultural relic. Introduce the origin, current situation, characteristics and importance of each cultural relic in 3 to 5 minutes. The content includes pictures of the cultural relics, audio and video recordings, narration subtitles and audio introductions. It will be provided to Fuxing Temple and the navigation installation developed by this research. . At the same time, we also hope to establish a new navigation model to provide reference for relevant museums and temples.

Figure 6. Fuxing Temple publishes four major cultural relic research books

4.4 *Interactive tours can become a new option in the future*

Small local museums often do not have dedicated staff to handle the daily visitors, and they often need to make an appointment in advance before they are open for viewing. Since the collections in the museum are not open to visitors all year round and are usually free to visit, there are not enough professional tour guides stationed to provide tour services at any time. Therefore, the interactive navigation device developed by this research institute can make up for the manpower needs of local cultural relics museums in a timely manner. Before entering the museum, visitors can take a guided tour to understand and learn about the cultural relics they are interested in. After entering the museum, they can further observe the important details of the physical cultural relics. The navigation device interface is mounted on a touch screen (Figure 7). Visitors can click on the required content to watch according to their personal needs, fully enjoying the freedom of content selection. Since the 55-inch screen is large enough for multiple people to read at the same time, it can become a reliable assistant for tour guides. The tour content can also be updated at any time, making it a good medium for small local museums.

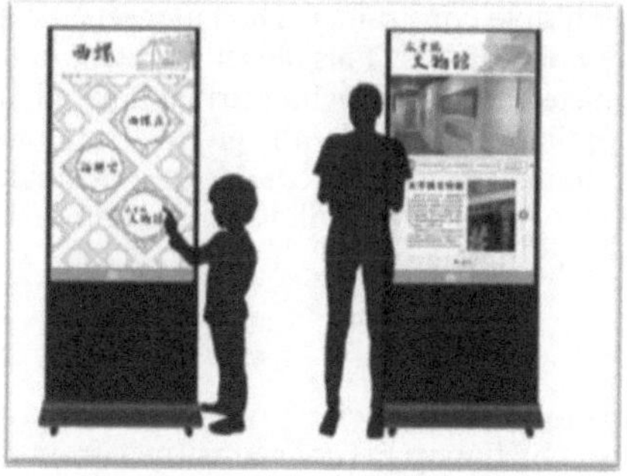

Figure 7. Interactive navigation interface operation diagram.

5 CONCLUSION

Based on the discussion on the three aspects of cultural relics guidance needs, cultural relic content communication and visitor learning, and integrating the above expert opinions and analysis results, the conclusions of this study are as follows:

1. Historical relics are the cultural assets left by the ancestors. Through display, they can create spiritual communication with modern people. Passive information transmission through human guides is no longer appropriate. Using digital media for active communication can better convey the wisdom of the ancestors.
2. Physical display and graphic explanations are not attractive to modern people. Audio-visual media can activate the display content to attract the audience's attention and present more aspects of cultural relics.
3. The combination of digital media production and interactive technology can explore the rich connotation of local heritage through human-computer interface operation, which can show a new style of cultural learning.

REFERENCES

Bernard H.R., Ryan G., (1998). Text analysis. *Handbook of Methods in Cultural Anthropology*, 613.

Buchenau M., Suri J.F. (2000). *Experience Prototyping. In PROCEEDINGS of the 3rd Conference on Designing Interactive Systems: Processes, Practices, Methods, and Techniques* 424–433.

Delconte J., Kline C.S., Scavo C. (2016). The impacts of local arts agencies on community placemaking and heritage tourism. *Journal of Heritage Tourism*, 11(4), 324–335.

Luo X., Gu Z., Li T., Meng X., Ma T., Yu C. (2015). Environmental control strategies for the in situ preservation of unearthed relics in archaeology museums. *Journal of Cultural Heritage*, 16(6), 790–797.

Mergel I., Edelmann N., Haug N. (2019). Defining digital transformation: Results from expert interviews. *Government Information Quarterly*, 36(4), 101385.

Mousavi S.S., Doratli N., Mousavi S.N., Moradiahari F. (2016). *Defining cultural tourism. In International Conference on Civil, Architecture and Sustainable Development* Vol. 1, No. 2, 70–75.

Richards G. (2018). Cultural tourism: A review of recent research and trends. *Journal of Hospitality and Tourism Management*, 36, 12–21.

Richards G., (2018). Cultural tourism: A review of recent research and trends. *Journal of Hospitality and Tourism Management*, 36 12–21.

Richards G., Wilson J. (2004). The impact of cultural events on city image: Rotterdam, cultural capital of Europe 2001. *Urban Studies*, 41(10), 1931–1951.

Roes I., Stash N., Wang Y., Aroyo L. (2009). A personalized walk through the museum: *The CHIP interactive tour guide. In CHI'09 Extended Abstracts on Human Factors in Computing Systems* 3317–3322.

Tian Y. (2023). Feasibility study on AR guided tours in the National Museum of China. *Probe-Media and Communication Studies*, 5(3).

Trahanias P., Argyros A., Tsakiris D., Cremers A., Schulz D., Burgard W., Sophias V., (2000). May. *Tourbot-interactive museum tele-presence through robotic avatars. In Proc. of the 9th International World Wide Web Conference.*

Van Audenhove L., Donders K. (2019). Talking to people III: Expert interviews and elite interviews. *The Palgrave Handbook of Methods for Media Policy Research*, 179–197.

Viswanathan B., Linsey J., Anderson D., Jensen, Crawford D., Wood K. (2017). Design prototyping methods: state of the art in strategies, techniques, and guidelines. *Design Science*, 3, e13.

Wang Y., Stash N., Sambeek R., Schuurmans Y., Aroyo L., Schreiber G., Gorgels P. (2009). Cultivating personalized museum tours online and on-site. *Interdisciplinary Science Reviews*, 34(2–3), 139–153.

System Innovation for an Artificial Intelligence Era – Lam et al. (Eds)
© 2025 the Author(s), ISBN: 978-1-032-84452-7

IQLS: Framework for leveraging metadata to enable large language model based queries to complex, versatile data

Sami Azirar
RWTH Aachen University, Aachen, Germany

Hossam A. Gabbar
Faculty of Engineering and Applied Science, Ontario Tech University, Oshawa, Canada

Chaouki Regoui
Digital Technologies, National Research Council Canada, Ottawa, Canada

ABSTRACT: As the amount and complexity of data grows, retrieving it has become a more difficult task that requires greater knowledge and resources. This is especially true for the logistics industry, where new technologies for data collection provide tremendous amounts of interconnected real-time data. The Intelligent Query and Learning System (IQLS) simplifies the process by allowing natural language use to simplify data retrieval . It maps structured data into a framework based on the available metadata and available data models. This framework creates an environment for an agent powered by a Large Language Model. The agent utilizes the hierarchical nature of the data to filter iteratively by making multiple small context-aware decisions instead of one-shot data retrieval. After the Data filtering, the IQLS enables the agent to fulfill tasks given by the user query through interfaces. These interfaces range from multimodal transportation information retrieval to route planning under multiple constraints. The latter lets the agent define a dynamic object, which is determined based on the query parameters. This object represents a driver capable of navigating a road network. The road network is depicted as a graph with attributes based on the data. Using a modified version of the Dijkstra algorithm, the optimal route under the given constraints can be determined. Throughout the entire process, the user maintains the ability to interact and guide the system. The IQLS is showcased in a case study on the Canadian logistics sector, allowing geospatial, visual, tabular and text data to be easily queried semantically in natural language.

1 INTRODUCTION

Due to its increasingly complex domain (Sandberg *et al.* 2022), logistics is a knowledge-intensive field that requires complex and diverse data, which has to be consolidated depending on the use case/application under consideration, such as route planning. The current research mainly focuses on optimizing models to fulfill this task (Borgi, Zoghlami, and Abed 2017), given the required data in a structured format. Retrieving, consolidating and applying the right data is, therefore, a huge task. Relational Databases (RDB), using SQL (Codd 1970), are the industry standard in most domains. Advancements in natural language processing (NLP), as well as the emerging field of process mining (Van Der Aalst 2012), have created the need to use the data more efficiently. Yet, with the increasing complexity of the data, retrieving it with traditional Query Languages requires an increasing level of human expertise, both regarding the database query language and the domain, as well as great effort to create complex and long queries.

Therefore, approaches of natural language queries, which map predefined entities from a pre-built model in the query to entities in the database, have emerged, e.g. in managing instruction texts (Gabbar *et al.* 2021a, 2021b). Still, one of RDBs limits is that they require complex queries when faced with highly interconnected data. Graph Databases tackle this issue, e.g. SPARQL for knowledge management (Barbieri *et al.* n.d.). NoSQL databases, such as Neo4j, are therefore used for handling large and diverse datasets, with applications including route planning (Raj *et al.* 2022). However, they require high levels of effort to set up and maintain since they rely mainly on rule-based reasoning and structured data. Moreover, they need more capability when dealing with a wide variety of data and datatypes (Palovská 2015). Also, RDB's have little flexibility for adapting to new, differently structured data. With the emergence of Large Language Models (LLM), (Vaswani *et al.* 2017) new possibilities in data and knowledge utilization have emerged, most commonly by querying data with vector

DOI: 10.1201/9781003514831-56

databases (Anon [2022] 2023; Guo *et al.* 2022). However, these approaches deliver one-shot solutions, i.e. they make one thinking step to retrieve the data. This requires increasingly big models while having limited capabilities regarding context awareness when it comes to domain specifics since they are mainly trained on general data. In this paper, the IQLS leverages data schema to tackle this issue. The paper is structured in the following manner: Firstly, it provides a brief overview of recent approaches in complex data querying. Next, it explains the IQLS on a high level and proceeds to describe our case study. After that, the IQLS is explained step by step, followed by a brief analysis of its capabilities and limits. It concludes with directions for future work.

2 RELATED WORK

To make use of existing databases without altering them, text-to-SQL-Query approaches have emerged, using semantic trees (Giordani and Moschitti 2010) or knowledge graphs for the mappings (Ahmad, Khan, and Ali 2009). In (Das and Balabantaray 2019), the authors also utilized the database schema to parse correct SQL queries from natural text. Still, one ongoing challenge is the fuzziness of queries, which makes static mapping difficult (Mama and Machkour 2021). Therefore, the emergence of vector embeddings, i.e. embedding the semantic meaning of a text into a latent space, has made more dynamic and versatile text-to-SQL-Query approaches possible by encoding the data into vectors and using the vector similarity to enable semantic parsing of natural texts to SQL Queries (Wang *et al.* 2021).This is also done by mapping the query vectors directly to the encoded data (Anon n.d.-g). The vector search got further extended with the capabilities of multimodal encoding of complex data into fully vector-based databases (Anon [2022] 2023; Guo *et al.* 2022). In combination with LLM-based chatbots (Anon n.d.-d; Ansari 2023), and improvements in focused retrievals like Retrieval Augmented Generation (RAG) (Lewis *et al.* 2021) data querying has become simpler and more accurate.

3 IQLS

While current methods effectively process unstructured data, they fall short of fully exploiting the knowledge embedded in the metadata of complex and differently standardized but structured data (Becker and Intoyoad 2017). Additionally, there is an absence of a comprehensive framework to iteratively guide Large Language Models (LLMs) through data, incorporating the methodologies of Knowledge Graphs for more efficient data management. Moreover, while frameworks like Langchain (Anon n.d.-f) make it possible to extend LLMs to new data sources easily without alteration, data source selection has to be made by the initial setup, i.e. it requires human effort to include different databases. Common approaches retrieve data in a one-shot process, meaning they directly fetch data in response to a user query. This process requires LLMs to process all data simultaneously, which, despite data schema assistance, can be computationally intensive with numerous databases and is constrained by token length limitations. This enables very little user control over the query process. This is particularly challenging in domain-specific applications like logistics, where such one-shot retrieval may miss crucial domain-specific contexts or in use cases where private hosting is required.

Additionally, the output of these implementations is typically limited to one datatype, most commonly text, which may only partially exploit the data's potential. In this paper, we present IQLS, a framework that leverages the ability of LLMs to be versatile and dynamic, with the structure of other NLP-based approaches leveraging metadata. Instead of one-shot data retrieval, the IQLS guides the LLM stepwise from query understanding through data filtering to the final data retrieval and application. Through this stepwise approach, the IQLS is easily adaptable to new data and is capable of processing complex queries on complex data from various data sources while autonomously selecting the needed ones, thus increasing the retrieved data relevance to the query and minimizing the required computational and human resources required. The IQLS not only retrieves the data needed but also applies the data to the required use case.

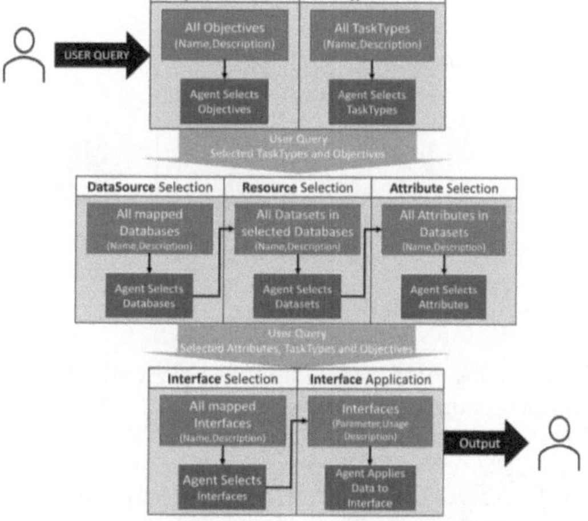

Figure 1. IQLS Framework, agent wraps the LLM.

3.1 *IQLS process*

The IQLS uses an LLM-powered agent. This agent makes decisions based on available choices within an environment connected to a graph database that stores data abstractions. The environment serves as a bridge between the user, the agent, and the actual data. Instead of directly accessing the data, an abstraction is stored in a graph database with different levels of data, descriptions, and metadata. The process of the IQLS is divided into three main steps, as shown in Figure 1. The first step is to classify the user query based on the *TaskType* and the Objective. Limiting the agent to a predetermined scope of tasks and objectives will set the context to guide the agent toward applicable results. The second step is the data filtering. The agent selects the data sources based on the user query and the abstraction of the data. A *DataSource* represents the highest level of the data hierarchy, such as a database. The environment then searches the Graph Database for the corresponding Resources. A Resource refers to midlevel elements, i.e. the datasets, such as an SQL database table. The agent selects the appropriate resource, after which this process is repeated for the Attributes which correspond to the lower-level elements, like Columns.

In the final step, the interface gets selected and applied. Interfaces are the tools used by the agent to actually retrieve and apply data as the user requires. The interfaces include information retrieval, law retrieval, internal document querying, route monitoring, and route calculation under constraints, as depicted in

3.2 *System design*

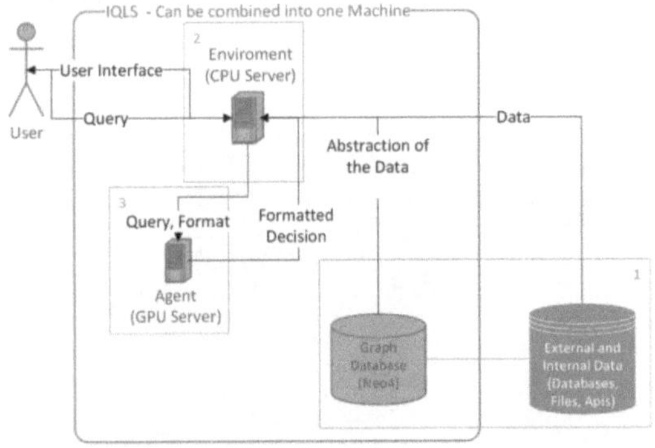

Figure 2. IQLS system design.

The Intelligent Query Language System (IQLS) comprises three key components as illustrated in Figure 2: (1) a graph database for data abstraction, (2) a CPU server for environmental control, and (3) a GPU server for agent operation. Component, the graph database, employs Neo4J (Anon n.d.-h) and stores essential data elements such as labels, descriptions, paths, and links. It is the primary component that requires updates or additions for new data integration. To utilize structured data, users must import it into a Neo4J database conforming to the defined schema. Component 2, which covers all interfaces, manages the user interfaces, and provides methods to interact with various components. It operates using a Python script, a configuration file, and a mapping file. Modifications to this module are necessary only for system improvements or additions, such as updating Neo4J paths or refining agent prompts. It initializes an agent by selecting an LLM and specifying its parameters, then communicates with the agent by delivering a query and the desired output in a Pydantic (Anon n.d.-m) class format. This approach allows for versatility, as any function expressible via a Pydantic class can be implemented. The environment enhances the agent's understanding of queries through context learning, incorporating examples and additional contexts like *TaskTypes* and guidance. Component 3 houses a GPU that drives a large language model. In this paper, the Mixtral 8x7B-Instruct-v0.1 (Jiang *et al.* 2023) is used. This compatible with all models on Hugging Face (Anon 2024), as long as chat templates are available, and the model supports altering the output logits. Outputs are formatted using the 'lm-format-enforcer' and the Langchain framework (Anon n.d.-i). Fundamental changes to the IQLS System require updates to the agent.

This architecture allows for easy adaptation to new data while minimizing the need for extensive modifications, ensuring stability in system deployment phase. In particular, the agent only accesses data abstractions, making the IQLS suitable for privacy-sensitive applications, with the option to host the agent on the cloud while keeping the data local. For faster inference, the IQLS supports all APIs that follow the OpenAI schema, though this limit altering logits to function calls only, potentially reducing the precision of the filtered data and increasing the error rate. Not all APIs support complex refinements and validators in their function calls. Therefore, the IQLS uses an iterative refinement process for its output. It starts with an initial try, and if the parsing fails, it identifies the attributes that violate constraints or validators. A new Pydantic class is then constructed, consisting solely of the attributes with errors. These are reprocessed by the agent alongside the initial queries and corresponding error messages. If this approach fails, the Langchain *RetryParser* is utilized.

4 LOGISTICS CASE STUDY

This paper presents a case study in the Canadian logistics sector, selected for its demand for a straightforward system and the varying complexities in queries and data.

4.1 *Query complexity*

The IQLS system is designed to handle various complex and diverse queries with ease. Its ability to handle language ambiguity is one of its key features, which enhances user experience by accurately.

Table 1. Objective definitions.

Objective	Description
Time	Queries about time aspects, like the fastest route.
Resources	Queries focusing on resource efficiency, such as the cheapest route.
Regulations	Ensuring adherence to local, provincial, and national regulations.
Safety	Queries concerning the safety of operations, personnel, and cargo.
Distance	Calculating and minimizing travel distances.
Certainty	Queries about data reliability, like routes with minimal rain risk.

Interpreting natural language queries. Even when dealing with vague expressions or imprecise adjectives like" near," IQLS maintains effective communication with users. Moreover, the system caters to the diverse objectives inherent in logistics-related queries while maintaining the consistency of the underlying data, a key strength of IQLS. This feature enables the system to adapt autonomously to various query goals, eliminating the need for users to navigate between different systems to find answers. The IQLS seamlessly integrates these objectives, ensuring a smooth and efficient user experience. The specific objectives addressed by the IQLS are described in Table 1. The final complexity involves task type diversity. Regardless of the objective, queries may target different needs. To address these needs, we defined six task types for the IQLS, as shown in Table 2. These task types and objectives determine the scope of the queries the IQLS is developed for.

Table 2. *TaskType* definitions.

TaskType	Description
Information Retrieval	Tasks for finding or gathering data related to a specific objective.
Route Planning and Design	Tasks for planning routes based on specific objectives or constraints.
Estimations and Forecasting	Tasks for estimating probabilities and forecasting impacts.
Supervision & Monitoring	Tasks combining compliance assurance with monitoring updates.
Objective Calculation & Evaluation	Tasks for calculating or evaluating metrics for a given objective.
Requirement Determination	Tasks for determining requirements to fulfill a specific objective.

4.2 *Data*

The IQLS can deal with not only complex queries but also complex data. The data used in the case study was unaltered. However, a prerequisite is that the data must be structured, as this is the main leverage point of the IQLS. This leaves the data complexity in the diversity of data sources, and data formats while being able to be queried semantically across the data types. For this, the case study focuses on four main Data Sources, the first one being the Canadian National Road Network (NRN) Data, which is distributed in thirteen provincial or territorial datasets and includes two linear entities (Road Segment and Ferry Connection Segment) and three punctual entities (Junction, Blocked Passage, Toll Point) [NRN 2023]. These entities are characterized by a range of descriptive attributes, e.g., House Number, Street, Functional Road Class, and Pavement Status. While it is documented, it needs to be clearly interlinked, e.g. the road segments are not natively linked (Anon n. d.-a). However, the validation of the data shows that the geometry borders serve as a sufficient border. The next data source is the 511 API. It includes real-time and static information on Events, Construction, Road Conditions, Transit Hubs, Carpool Lots, Ferries, Service Centres, Travel Info Centres, Truck Rest Areas, Inspection Stations, Roundabouts, Seasonal Loads, Alerts. It can be accessed as an API and returns the data in JSON format. It is available for Ontario as well as Alberta, British Columbia, Yukon, and Prince Edward Island provinces (Anon n.d.-j, Anon n.d.-b, Anon n.d.-c, Anon n.d.-l), but for practical reasons, the case study is limited to the Ontario 511 data source. For regulation data, the CanLII API is utilized, which includes different databases for up-to-date Canadian provincial and federal statutes, regulations, and laws (JSON format). Lastly, this case study included various documents to account for potential internal documents in the application (Anon [2019] 2023).

5 IQLS-PROCESSES

5.1 *Query classification and data filtering*

First, the user inputs a query to the system. Upon receiving the query, the system retrieves all nodes labelled as *TaskTypes*. Each node contains a name and description of *TaskTypes*, according to the details provided in Table 2. Based on the *TaskTypes* present, the system assigns agents to classify the query. While most queries require only one task type, the agent can also utilize multiple task types. Next, the system identifies the objectives based on Table 1. The data filtering process begins by retrieving the *DataSources* connected to the selected *TaskTypes*. For each *DataSource*, the agent receives the relevant name and description. Each description contains a summary of the data content, as well as the included Resources and potential applications. If the *TaskType* indicates that route planning is necessary, the NRN Database is also retrieved. The process for the Resources and Attributes is equal. At each step, the lower layer's information is contained in the description. Additionally, at each layer, the user can modify the choices if necessary. This approach allows for filtering large datasets without relying on encoding them into latent spaces, enhancing explainability and user control.

5.2 *Interfaces*

The environment retrieves the possible interfaces and their descriptions based on the chosen *DataSource*, as listed in Table 3. The interfaces are not semantically connected to the *DataSources* but based on the data type. For example, only spatial data can be used to plan routes. When the agent needs to fulfill queries that require multiple interfaces, they choose which interfaces to use and in what order. For instance, in route planning, the agent is advised to use all necessary interfaces.

5.2.1 *Information retrieval*

The Information Retrieval Interface is designed to retrieve various types of data. First, the environment retrieves the path associated with the *DataSources*. The path could be stored directly, with variables for the resources if they are relevant to the path, e.g. if the path is an API endpoint or as a direct link to the *DataSource*. For each chosen resource of the database, the full path will be built if applicable. For all other paths leading to a database or geopackage requiring specific resources, the layer or table will be specified. Finally, for each path, the data will be retrieved and filtered by the chosen Attributes. The Attribute selection is made before data retrieval, if possible, as with gkpg, CSV, or JSON files, otherwise, it's done after loading the data. The data will be converted either to pandas DataFrame or GeoDataFrame (Anon n.d.-k; Jordahl *et al.* 2019). If the files contain geospatial data, the agent will be asked to decide whether the spatial element is relevant. If it is, the agent also has to specify a location, e.g. Toronto, and a radius of interest. The location is converted to coordinates with the GeoPy module (Anon n.d.-n).

The geospatial data is loaded as a geoDataFrame. If the geometry isn't automatically retrieved, it's converted from the corresponding column, such as from EncodedPolyLine format in the 511 API. Now that the spatial filtering is done, all files will be converted to pandas DataFrames. Some data contains also visual data, e.g. camera recordings. For this, the environment tasks the agent to formulate a question based on the visual input, e.g. asking whether a traffic jam is visible. This question is then processed by the ViLT model (Kim, Son, and Kim 2021), which analyzes the visual input to provide an answer, subsequently stored as a new column mapped to the corresponding row, e.g. camera data. The ViLT model can be hosted on the GPU or on the CPU server, however, with a longer inference time. This will be done for all entries and added as a new column to the table. All tables are stored in a newly initialized SQLite database by the environment, enabling compact, easy, and quick querying. Finally, the SQLite Database will be queried by the agent with respect to the metadata and user query, using the Langchain SQL module.

5.2.2 *Law interface*

The Law Interface is a platform that showcases the use of data sources that have too many attributes or attributes that can't be easily transformed into tabular data. It directly utilizes the CanLII API and goes through the databases and resources. When selecting attributes, the normal approach of choosing from a list is not feasible because there are thousands of potential attributes, such as documents. Instead, a different approach is used, combining retrieval augmented generation with BGE embeddings (Zhang *et al.* 2023) and (George and Rajan 2022) vector search. This approach can also be CPU-hosted, allowing for a quick selection of attributes while maintaining control over the results and data. The selected regulations are retrieved from the CanLII website, not provided natively by the API. These regulations are then converted into texts and queried by the agent using RAG.

5.2.3 *Internal document retrieval*

The internal document retrieval interface should only be used if the user specifically requests it or if the other interfaces fail to provide the requested information. This system uses RAG in combination with BGFE embeddings and FAISS vector search to query the relevant documents efficiently.

5.2.4 *Route monitoring interface*

The Route Monitoring Interface is a tool that can answer questions about specific routes, like finding alerts between points A and B. To do this, the agent first figures out the starting and ending locations that the user is interested in by using the Nominatim module. Next, it imports NRN Road Data as a GeodataFrame and transforms it into a graph using the momepy library (Fleischmann 2019). This graph shows the road segments as directed edges, which reflect the traffic flow direction, and the boundaries of these segments (represented by LineStrings) serve as nodes. The length of each edge is calculated based on the coordinates within the LineString. After dividing the graph into connected components, the largest component is selected for further processing as long as it encompasses over 99.99% of the edges. If not, the system will trigger an error. Once the graph is refined, the agent identifies the nodes closest to the start and end locations. The environment then computes up to ten of the fastest routes, and the chosen road segments are recorded in a database. Finally, the system retrieves additional data in a similar way to the Information Retrieval Interface but without using spatial filtering. Instead, it leverages the Road NID where possible, or it uses the geometry of the road segments to filter information.

Table 3. Interfaces and descriptions.

Interface	Description
Route Planning Interface	The interface for planning routes under constraints and various objectives. It can be utilized with preselected data to perform various types of route planning. This interface can be used after another interface, if necessary, to perform route planning with the knowledge of the other interface.
Law Interface	Includes data about various laws and statutes from different provinces of Canada. It can be used before another interface if needed. This interface does not contain information about specific places or road segments, but only laws and statutes, and should be used solely for this purpose.
Specific Information Retrieval Interface	This interface can be used if a specific piece of knowledge is needed. It can be used to retrieve specific data from the previously filtered data, such as the average speed of a road segment, the weather in a certain area, any restrictions on a road segment, something captured on a camera, etc.
Road Monitoring	This can be used to check if there are any problems on a route, such as accidents or construction. It can be used after another interface if the user wants to know about potential problems on the route. It can also be used standalone if there is no complex route planning needed and the user wants to know if there could be problems on a route or in a certain area. Additionally, it can be used to monitor a specific area, like a city or a province.
Internal Document Retrieval	This can be used if the user specifically needs some internal documents. It can also be used if the required knowledge cannot be found elsewhere. It should be used after the other interfaces have been tried or if the user specifically requests internal documents.

This approach checks for intersections with the route segments, guaranteeing an accuracy of over 10 meters, which is consistent with the NRN Data Specifications.

5.2.5 *Conditional route planning*

This interface showcases the applicability of the IQLS and its advantages over standard approaches due to the complexity of the task. This interface, however, is only suited for spatial data; for non-spatial data, the Information Retrieval Interface has to be applied first, e.g. to get a regulation, which then gets applied using the Route Planning Interface. The Data is retrieved following the same approach as the Information Retrieval Interface, only without spatial filtering. When loaded into GeoDataFrames, they are merged into one large GeoDataFrame based on spatial intersections with the road segments. This will be done regardless of the geometry. With this GeoDataFrame, a graph is built with momepy, where all the attributes are edge properties, and the nodes are propertyless. Elements spanning multiple segments are duplicated. To prevent errors, all Nans are filled either with 0 or "None". This graph serves as the model for the road network. To really perform multi-constrained multi-objective path calculation, the agent has to model the pathfinding. For this, the IQLS utilized Pydantic models. The IQLS uses the Driver and the *ConstraintActions* classes. The Driver is modelled through the route network with attributes, while the *ConstraintActions* determine how the Driver interacts with its environment. A modified version of the Dijkstra algorithm is used, as shown Algorithm 1 To determine the best route based on the given constraints and objectives, for each edge for each edge the attribute values of the driver are calculated, skips edges that violate constraints and selects the one with the smallest objective weight. Initially, the Driver has to be determined, so the environment tasks the agent with generating a list of Driver attributes.

All the Driver's attributes should be float values, initialized with 0, i.e. it always should sound up or down, e.g. the time since rest should be counted, not the remaining time until rest. This is important, to simplify the problem towards a minimum problem. The Driver's attributes should be determined based on the user query and a description of each should be generated. The description helps the agent to reason correctly and for later steps to not misunderstand. Now, the start and endpoint get determined, as well as the objective attribute, which must be one of the created Driver attributes. The next step involves defining actions to be executed at each step. For this, the agent gets information about which edge attributes are available and their description. Attributes that do not change during each step should have a non-effective action, yet the iterating ensures that for each defined attribute, the agent thinks about its use accordingly. To improve reasoning and enable better later use, each action should be named and described briefly. Each action also includes an operation, e.g. multiply and a Value. The Value could be an Operand object or a string or number. Operand Object defines the attribute of a source, i.e. either the edge or the Driver. The first validator ensures that the source is either the edge or the Driver, and the second validator confirms that the attribute belongs to the source. The action validators ensure that the attribute is indeed the right Driver's attribute. The second action validator ensures that the operation is one of the predefined ones Each action is performed by applying the operator on the Driver's attribute and the Value, e.g. adding the travel time of an edge to the total time of the Driver. Based on the user query, edge and driver attributes, the constraints are defined by the agent. This is a three-stage process. First, a list of constraint names and descriptions is defined by the agent, i.e. the concept of the constraints. This enforces the reasoning of the agent. In the second stage, the constraints are defined by iterating over each name-description pair in the list. For each constraint, the operator, e.g. greater than Operands 1 and 2, gets defined. Operand1 is either a driver or an edge attribute, while Operand2 can also be a text or number. This enables the agent to determine what should be the trigger, e.g. if the travel time without a rest area is greater than 5h or if the road is wet. The action is not defined at this stage, but the name and description, again to reason accordingly and think about the concept first, then the implementation. Finally, the third step determines the action. For the constraints, the action skip edge is always predefined. This skips edges that violate the constraint, e.g. if the travel distance without a fuel station would be large. Again, the operator gets validated. Finally, these get aggregated by the environment into the *ConstraintActions* and the initial Driver object. The *ConstraintActions* objects have the main methods do step, i.e. calculating the new Driver object based on the edge attribute values, and the evaluate condition method, which evaluates all conditions and executes the corresponding actions if the constraints are violated. As shown in 1, this is performed at each edge, always for the next connected edges.

Algorithm 1 Optimized Pathfinding Considering Driver Constraints

Let $G = (V, E, w)$ be a weighted directed graph, where $w : E \to \mathbb{R}$.
Define driver attributes D and associated state D_s.
Define constraints set $\mathcal{C}$ and actions set $\mathcal{A}$.
Let $s, t \in V$ be the start and target nodes, respectively.
Initialize a min-priority queue Q as $\{(0, s, D_s)\}$.
Initialize a map Visited : $V \to \mathbb{R} \times V$ with Visited$(s) = (0, \text{null})$.
while Q is not empty **do**
    $(\_, v, D_v) \leftarrow$ dequeue the element from Q with the smallest key.
    if $v = t$ **then**
        break
    end if
    for each $u \in V$ such that $(v, u) \in E$ **do**
        Update D_v according to $\mathcal{A}$ and edge attributes of (v, u).
        if D_v satisfies $\mathcal{C}$ **then**
            Let $\delta = w(v, u) + \text{weight}(D_v)$.
            if $u \notin$ Visited or $\delta <$ Visited$(u).weight$ **then**
                Visited$(u) \leftarrow (\delta, v)$.
                enqueue (δ, u, D_v) to Q.
            end if
        end if
    end for
end while

5.3 *User integration*

The user integration in this system is adeptly managed through a Gradio interface (Abid *et al.* 2019), which is designed to offer two operational modes: automatic and control. In automatic mode, the user inserts a query and the IQLS processes without further interaction. Control mode returns the output of each step to the user, who then corrects and validates the selections. Depending on the query, the output provided to the user varies: for spatial and path data queries, a Folium visualization (Anon n.d.-e) accompanied by textual information is presented, whereas for other types of queries, the output is text-based. An example of Route Monitoring is shown in Figure 4.

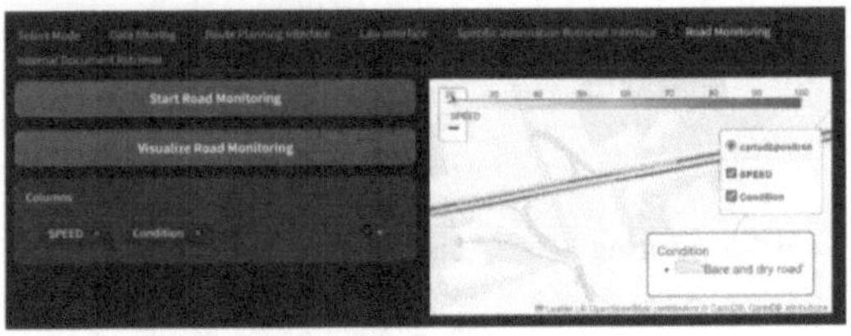

Figure 3. Road monitoring UI.

6 EVALUATION

When compared with standard querying methods, the IQLS transforms the task of writing long, complex queries towards making natural language queries and dropdown menu selections. This enables access to complex data for users not familiar with databases. It enables the user to query across databases, independent of the type and origin of data and delivers aggregated results. For example, the query" I am transporting livestock with a truck from Toronto to Ottawa. What do I have to check. I also want to avoid ice on the roads," would require the user to check the internal documents for the checklist. Then, the user would have to get the road weather data and manually construct a route without ice. With the

IQLS, the user has only to check the agent's selections and correct them via the Dropdown menu if needed to receive a visualization of the route, as well as the required Information. The IQLS does not require the user to know what is in the available data beforehand, i.e. the user can make queries and then receive the Information and where to find it. In comparison to other LLM-based approaches, the IQLS enables the user to take advantage of the versatility of LLMs while still enabling the user to take control over the querying process in a user-friendly way.

Also, the IQLS can be utilized for any domain and is bounded only by the data structure requirements. Depending on the hardware and query, the local inference time ranges between 5 and 20 minutes with a 40GB Nvidia A100, while API deployment inference takes less than 3 minutes. However, the local deployment enables more control over the output since it enables more precise query results by modifying the length penalty and better output parsing through the output logits. Figure 4. Road Monitoring UI The IQLS extends the capabilities of LLMs to more complex tasks like route planning while still making use of the available data structures. Combined with the output parsing via Pydantic classes, as well as the improved refinement process makes the IQLS versatile in its capabilities. Still, the adaptability of the models strongly depends on the data. Structured data, along with the corresponding metadata, can be inserted naively by importing the data schema in the Neo4J database. More complex APIs, or new database types can be included, yet require programming capabilities. For more complex tasks, like conditional route planning, user involvement increases, for very different use cases. The IQLS enables the user to make queries without using domain-specific language. However, since the IQLS maps the user query to the data based on the schema, the wording of the query also influences the results. For example, the IQLS often included the" Route Number" data when queried for routes since the name is in it.

7 FUTURE WORK

To further explore the capabilities and limitations of the IQLS, practical testing in real use environment is necessary. Additionally, the models presented in this paper have yet to be finetuned, leaving room for further improvements. Future research could identify more error-proof ways to parse complex structured outputs from LLMs to enable more versatile applications. Future research can leverage the IQLS Framework in other domains to enable the use of LLM-powered data retrieval on structured data. Furthermore, the integration of ontologies can expand the scope of the IQLS to include unstructured domain-specific data. Finally, we could leverage the IQLS to enable complex use cases on smaller LLMs.

ACKNOWLEDGEMENT

This project was supported in part by collaborative research and funding from the National Research Council of Canada under the Artificial Intelligence for Logistics Program as well as the NRC-MITACS-RWTH Globalink Research Award Program and by The Smart Energy Systems Lab (SESL) at Ontario Tech University

REFERENCES

Abid, Abubakar, Ali Abdalla, Ali Abid, Dawood Khan, Abdulrahman Alfozan, and James Zou. (2019). *Gradio: Hassle-Free Sharing and Testing of ML Models in the Wild.*

Ahmad, Rashid, Mohammad Abid Khan, and Rahman Ali. (2009). Efficient Transformation of a Natural Language Query to SQL for Urdu.

Anon. [2019]. (2023). Canlii/API_documentation.

Anon. [2022]. (2023). *Chroma-Core/Chroma.*

Anon. (2024). *Hugging Face – The AI Community Building the Future.* Retrieved 15 January 2024 (https://huggingface.co/).

Anon. (n.d.-a). *6. Data Exchange Model — NRN Home / Accueil RRN Documentation.* Retrieved 23 June 2023 (https://nrn-rrn.readthedocs.io/en/latest/source/en/product_documentation/data_exchange_model.html).

Anon. (n.d.-b). *511 Alberta.* Retrieved 6 November 2023 (https://511.alberta.ca/developers/doc).

Anon. (n.d.-c). *511 Yukon.* Retrieved 6 November 2023 (https://511yukon.ca/developers/511yukon.ca).

Anon. (n.d.-d). *ChatGPT.* Retrieved 10 August 2023 (https://chat.openai.com).

Anon. (n.d.-e). *Folium — Folium 0.1.Dev1 + g4e6ebfe Documentation.* Retrieved 15 January 2024 (https://python-visualization.github.io/folium/latest/#).

Anon. (n.d.-f). *LangChain.* Retrieved 15 January 2024 (https://www.langchain.com/).

Anon. (n.d.-g). *MongoDB Atlas Vector Search. MongoDB.* Retrieved 15 January 2024 (https://www.mongodb.com/products/platform/atlas-vector-search).

Anon. (n.d.-h). *Neo4j Graph Database & Analytics – The Leader in Graph Databases. Graph Database & Analytics.* Retrieved 15 January 2024 (https://neo4j.com/).

Anon.(n.d.-i). Noamgat/Lm-Format-Enforcer: Enforce the Output Format (JSON Schema, Regex Etc) of a Language Model. Retrieved 15 January 2024 (https://github.com/noamgat/lm-format-enforcer).

Anon. (n.d.-j). *Ontario 511.* Retrieved 9 August 2023 (https://511on.ca/developers/511on.ca).

Anon. (n.d.-k). *Pandas – Python Data Analysis Library.* Retrieved 15 January 2024 (https://pandas.pydata.org/).

Anon. (n.d.-l). *Prince Edward Island 511.* Retrieved 6 November 2023 (https://511.gov.pe.ca/developers/511.gov.pe.ca).

Anon. (n.d.-m). *Pydantic*. Retrieved 15 January 2024 (https://docs.pydantic.dev/1.10/).

Anon. (n.d.-n). *Welcome to GeoPys Documentation! — GeoPy 2.4.1 Documentation*. Retrieved 15 January 2024 (https://geopy.readthedocs.io/en/stable/).

Ansari, Tasmia. (2023). UC Berkeley Releases Open LLaMA, an Open-Source Alternative to Meta's LLaMA. *Analytics India Magazine*. Retrieved 30 June 2023 (https://analyticsindiamag.com/uc-berkeley-release-an-open-source-alternative-to-metas-llama/).

Barbieri, Luciana, Edmundo Madeira, Kleber Stroeh, and Wil van der Aalst. (n.d.). *A Natural Language Querying Interface for Process Mining*.

Becker, Till, and Wacharawan Intoyoad. (2017). Context Aware Process Mining in Logistics. *Procedia CIRP* 63:557–62. doi: https://doi.org/10.1016/j.procir.2017.03.149.

Borgi, Tawfik, Nesrine Zoghlami, and Mourad Abed. (2017). Big Data for Transport and Logistics: A Review. Pp. 44–49 in *2017 International Conference on Advanced Systems and Electric Technologies (IC_ASET)*.

Codd, E. F. (1970). *A Relational Model of Data for Large Shared Data Banks*. 13(6).

Das, Alaka, and Rakesh Chandra Balabantaray. (2019). MyNLIDB: A Natural Language Interface to Database. Pp. 234–38 in *2019 International Conference on Information Technology (ICIT)*.

Fleischmann, Martin. (2019). Momepy: Urban Morphology Measuring Toolkit. *Journal of Open Source Software* 4(43):1807. doi: 10.21105/joss.01807.

Gabbar, Hossam A., Sk Sami Al Jabar, Hassan A. Hassan, and Jing Ren. (2021a). An Intelligent Experience Retention System: Challenges and Limitations for Operation and Maintenance in Nuclear Power Plants. *IEEE Systems, Man, and Cybernetics Magazine* 7(4):31–34. doi: 10.1109/MSMC.2021.3098981.

Gabbar, Hossam A., Sk Sami Al Jabar, Hassan A. Hassan, and Jing Ren. (2021b). Development of Knowledge Base Using Human Experience Semantic Network for Instructive Texts. *Applied Sciences* 11 (17):8072. doi: 10.3390/app11178072.

George, Godwin, and Rajeev Rajan. (2022). A FAISS-Based Search for Story Generation. Pp. 1–6 in *2022 IEEE 19th India Council International Conference (INDICON)*.

Giordani, Alessandra, and Alessandro Moschitti. (2010). Semantic Mapping between Natural Language Questions and SQL Queries via Syntactic Pairing. Pp. 207–21 in *Natural Language Processing and Information Systems, Lecture Notes in Computer Science*, edited byH. Horacek, E. Métais, R. Muñoz, and M. Wolska. Berlin, Heidelberg: Springer.

Guo, Rentong, Xiaofan Luan, Long Xiang, Xiao Yan, Xiaomeng Yi, Jigao Luo, Qianya Cheng, Weizhi Xu, Jiarui Luo, Frank Liu, Zhenshan Cao, Yanliang Qiao, Ting Wang, Bo Tang, and Charles Xie. (2022). Manu: A Cloud Native Vector Database Management System. *Proceedings of the VLDB Endowment* 15(12):3548–61. doi: 10.14778/3554821.3554843.

Jiang, Albert Q., Alexandre Sablayrolles, Arthur Mensch, Chris Bamford, Devendra Singh Chaplot, Diego de las Casas, Florian Bressand, Gianna Lengyel, Guillaume Lample, Lucile Saulnier, Lélio Renard Lavaud, Marie-Anne Lachaux, Pierre Stock, Teven Le Scao, Thibaut Lavril, Thomas Wang, Timothée Lacroix, and William El Sayed. (2023). *Mistral 7B*.

Jordahl, Kelsey, Joris Van den Bossche, Jacob Wasserman, James McBride, Jeffrey Gerard, Jeff Tratner, Matthew Perry, Carson Farmer, Micah Cochran, Sean Gillies, Matt Bartos, Lucas Culbertson, Nick Eubank, maxalbert, Martin Fleischmann, Geir Arne Hjelle, Dani Arribas-Bel, Christopher Ren, Sergio Rey, Martin Journois, Levi John Wolf, Aleksey Bilogur, Nick Grue, Joshua Wilson, YuichiNotoya, Leah Wasser, Filipe, Chris Holdgraf, Adam Greenhall, and John Trengrove. (2019). *Geopandas/Geopandas: V0.4.1*.

Kim, Wonjae, Bokyung Son, and Ildoo Kim. (2021). ViLT: Vision-and-Language Transformer Without Convolution or Region Supervision.

Lewis, Patrick, Ethan Perez, Aleksandra Piktus, Fabio Petroni, Vladimir Karpukhin, Naman Goyal, Heinrich Küttler, Mike Lewis, Wen-tau Yih, Tim Rocktäschel, Sebastian Riedel, and Douwe Kiela. (2021). *Retrieval-Augmented Generation for Knowledge-Intensive NLP Tasks*.

Mama, Rachid, and Mustapha Machkour. (2021). Fuzzy Querying with SQL: Fuzzy View-Based Approach. *Journal of Intelligent & Fuzzy Systems* 40(5):9937–48. doi:10.3233/JIFS-202551.

NRN. (2023). *NRN*.

Palovská, Helena. (2015). What can NoSQL serve an enterprise. *Journal of Systems Integration* 6:44–49. doi: 10.20470/jsi.v6i3.240.

Raj, Pethuru, Abhishek Kumar, Vicente García Díaz, and Nachamai Muthuraman, eds. (2022). *Demystifying Graph Data Science: Graph Algorithms, Analytics Methods, Platforms, Databases, and Use Cases*. Institution of Engineering and Technology.

Sandberg, Erik, Pejvak Oghazi, Koteshwar Chirumalla, and Pankaj C. Patel. (2022). Interactive research framework in logistics and supply chain management: Bridging the academic research and practitioner gap. *Technological Forecasting and Social Change* 178:121563. doi: 10.1016/j.techfore.2022.121563.

Van Der Aalst, Wil. (2012). Process Mining: Overview and Opportunities. *ACM Transactions on Management Information Systems* 3(2):1–17. doi: 10.1145/2229156.2229157.

Vaswani, Ashish, Noam Shazeer, Niki Parmar, Jakob Uszkoreit, Llion Jones, Aidan N. Gomez, Lukasz Kaiser, and Illia Polosukhin. (2017). *Attention Is All You Need*. doi: 10.48550/ARXIV.1706.03762.

Wang, Bailin, Richard Shin, Xiaodong Liu, Oleksandr Polozov, and Matthew Richardson. (2021). *RAT-SQL: Relation-Aware Schema Encoding and Linking for Text-to-SQL Parsers*.

Zhang, Peitian, Shitao Xiao, Zheng Liu, Zhicheng Dou, and Jian-Yun Nie. (2023). *Retrieve Anything To Augment Large Language Models*.

System Innovation for an Artificial Intelligence Era – Lam et al. (Eds)
© 2025 the Author(s), ISBN: 978-1-032-84452-7

Research on the packaging design of local cultural specialties in Taiwan—taking Chiayi Xingang Candies as an example

Huang Hsiao Chien & Hwang Shyh Huei
National Yunlin University of Science and Technology, Yunlin, Taiwan

ABSTRACT: I In recent years, the trend of "cultural industrialization and industrial culturalization" has been continuously discussed, which has promoted the industrialization of local cultural characteristics and has become the main driving force for local economic development. Various regions in Taiwan have transformed "local specialties" into mainstream industries to bring promotion and business opportunities to local industries. The century-old "Xingang Candy" has always been a local specialty of Xingang Township, Chiayi County. However, the cultural characteristics of Xingang Township are not visible on the packaging of "Xingang Candy". Through this research, we will explore "Xingang Candy" from a cultural perspective. Packaging design reshapes the packaging image design to convey the local characteristic industries of Xingang Township.The main research method of this study is to use qualitative research through semi-structured interviews to collect the rich history and stories of Xingang Township, and then use quantitative research to understand the local characteristics and development of Xingang through local residents and tourists, and summarize the unique characteristics and development of Xingang Township through quantitative research. The four major aspects of authenticity, locality, storytelling and commemoration are used in the packaging design text, color, graphics and shape to produce the construction of a series of 16 packaging themes with four major themes of "Xingang Yi"."Key words: Local culture, local specialties, packaging design, Xingang Township.

1 INTRODUCTION

1.1 *Research background motivation*

In recent years, Taiwan's government and private sector have actively promoted activities with local characteristics in cities, towns, and towns. Since 1978, the Small and Medium Enterprises Division of the Ministry of Economic Affairs has already implemented the "Local Characteristics and Community Small Business Mentoring Program" to promote one township, one specialty, referred to as " OTOP" means "One Town One Product" (OTOP Taiwanese local characteristic website, 2012).

Taking townships, towns, and cities as local areas, combined with the local development of historical, cultural, and unique characteristic industries, the government commissioned a private creative design team to comprehensively plan and re-evaluate, and then adapt the products to the current lifestyle. It can be packaged and put on the market, which can promote local characteristic culture and create a new economic market with high added value. It can also strengthen the local image and form an alternative local brand marketing.

Xingang Township has first-, second-, and third-grade historic sites. The number of historic sites in the township ranks first in Chiayi County. Fengtian Temple, a famous religious resort in the country, attracts about three million pilgrims every year, making Xingang Township develop a good cultural tourism industry. Conditions, however, Xingang Township's famous product "Xingang Candy" has a long history and tradition like Xingang Fengtian Palace. Xingang Candy, which is sweet and soft, has a good taste and aroma, has been a local specialty that Xingang people are proud of since the Qing Dynasty. It has played an important role in the economy This product won the 2008 Taiwan Souvenir Award—Taiwan Food Souvenir Representative and Taiwan's Top 15 Souvenir Products. It was also the best gift among Chiayi County's Top 10 Souvenirs in 2012. Such a high-quality specialty has not yet established a packaging design that represents the characteristics of Xingang Township. This study will focus on the local specialty "Xingang Candy" in Xingang Township, Chiayi County, and connect the local characteristics of Xingang Township with the image of the township to complete the local specialty "Xingang Candy" Packaging construction design thinking development.

1.2 *Research purpose*

This study applies Taiwanese local cultural characteristics to specialty product packaging design. Through the research method, it is expected to

DOI: 10.1201/9781003514831-57

achieve the following three points as the main purpose: 1. Discuss the existing Xingang candy packaging imagery and local industry characteristics and culture, and formulate the positioning and design direction of product packaging. 2. In-depth understanding of various local characteristics such as products, nature, and humanities owned by local industries, and establish the structure of "Xingang Candy" product packaging design. 3. Convert local tangible and intangible cultural assets into cultural codes that can be used in design, so that products can promote local industrial characteristics.

1.3 *Research scope and objects*

Taking Xingang Township's local industry "Xingang Candy" products as the research object for packaging design, we analyzed the differences in cultural impressions and packaging pros and cons between the past and the present, developed new creative ideas, and implemented product packaging design.

2 LITERATURE DISCUSSION

2.1 *Documents related to local cultural characteristic industries*

2.1.1 *Local culture*

Local culture must be historical, local, and have a cultural background and living space where residents live together. Therefore, local culture is a collection of characteristics such as lifestyles, values, beliefs, and customs formed by local residents in the context of history. These characteristics include Its uniqueness cannot be copied elsewhere (Zhang Qiaofang, 2001). The development of local culture has five meanings and values: historical: historical and cultural inheritance, memory and trajectory; life: the provision of leisure activity space for ordinary people and the provision of multicultural space; social: the cohesion of community awareness, enriching the place Socially accumulated resources for the development of cultural industries; uniqueness: highlighting the particularity of regional space, so as to gather consensus and identity on local development; economy: providing economic power for the sustainable development of local culture through the development of special cultural industries (Gu Yiling, 2007).

2.1.2 *Cultural assets*

According to the concept of local resource exploration proposed by Miyazaki Kiyoshi, local cultural resources are divided into five categories: land, scenery, people, culture, and products to provide locally available resources, as shown in Table 1:

Table 1. Local asset discovery table (Source: Kiyoshi Miyazaki, 1996).

Land-Natural Resources	The natural resources contained in a place, such as minerals, climate, etc.
Scenery-Landscape Resources	The special landscapes possessed by places, such as canyons, plains, mountain views, etc.
People – human resources	The humanities, history and ethnic groups possessed by the place, such as elders, celebrities, etc.
Text – cultural resources	The crafts, technologies, beliefs and activities inherited by the place, such as handicrafts, festivals, etc.
Production – production resources	Items and crops produced locally, such as crops, food, etc.

2.1.3 *Definition of local characteristic industries*

In the "Green Paper on Local Characteristic Industries", scholars from many domestic research institutes, as well as relevant local government personnel and industry representatives, also have many different interpretations of "local characteristic industries". Zhu Zihao, Li Genyong, Lin Zetian, Chen Chiji Scholar Xie Haoming compiled the literature on the definition of local cultural characteristic products and divided the characteristics of local cultural characteristic industries into the following four major characteristics:1. Uniqueness: Characteristics that are different from others and have irreplaceable qualities.2. Locality: It is connected to a certain local culture and conveys that the local characteristic culture comes from various things related to people and the environment in daily life.3. Storytelling: A series of events or story culture composed of related people and things are presented in the product.4. Commemorative: Through a visit here, people can remember, miss and share the feeling.

2.2 *Packaging design related literature*

2.2.1 *Packaging visual design elements*

According to psychological theory, more than 70% of the information that humans receive from external stimuli on a daily basis is obtained through visual organs. Information collected through visual organs has a higher recall in the human memory bank. value. Therefore, in addition to focusing on the protective function, packaging must also take into account the overall visual function, and focus on visual requirements to shape the brand image.

In commercial activities, products are promoted through differentiation of packaging to generate profits, so packaging has become an important factor. According to various scholars (Huang Ronghui, 1990; Li Tianlai, 1992; Zhu Chenchuntian, 1993;

Kaneko Shuya, 1996) When organizing the visual elements of packaging, their definition and interpretation can achieve the explanatory power of the product itself through text, color, graphics, shape and other visual communication elements, as shown in Table 2:

Table 2. The aspects of packaging visual design elements (Source: Organized by this study).

The aspects of packaging visual design elements	
Text	The product names on the packaging include Chinese standard characters, English standard characters, slogans, slogans, standard characters, etc. and combined characters.
Color	The main standard color and auxiliary standard color used on the packaging convey the image of the product.
Graphics	The decorative effect of visual elements such as main patterns, symbolic shapes, and auxiliary patterns on the packaging.
Shape	The packaging expresses the visual effect of the contents through shape and material.
Overall	The overall visual experience and performance style of the packaging are presented.

2.2.2 *Case analysis of packaging of local specialties in Taiwan*

Through Huang Yaqing and Yao Cunxiong, 2000; Zhu Ruxing, 2003; Huang Lingyi, 2005; Liu Junnan, 2007; Huang Yaqing, 2009; Li Fangyi, 2011, the design performance of local specialties in Taiwan is summarized: 1. Text use: traditional calligraphy is the most popular Calligraphy fonts are the best way to shape the local image and text. 2. Use of colors: Red and purple can enhance product image, while brown, yellow and white can create local cultural image. Traditional packaging is often dominated by festive red. 3. Graphic application: Integrating traditional materials into current design techniques, the selection of images can mainly use traditional imagery, local characteristics, and product tips. 4. Modeling application: Paper is the most widely used material and is more environmentally friendly. It can be used in different forms, materials, and capacities to form unique products. 5. Overall image: Beautiful, high-end and exquisite can improve the brand image of local specialties. Series, family or individual multi-original products can be used to present brand advantages.

Based on the above-mentioned research on local specialties in Taiwan, it is concluded that most local specialties can be presented using old traditional packaging materials and combined with local elements through packaging improvement and design, which can improve the brand image and popularity of local specialties.

3 RESEARCH METHODS

This research method starts from qualitative research – in-depth interviews with local residents to collect more representative characteristics of Xingang Township, and then from quantitative research – questionnaire survey to understand how local people and tourists think about Xingang candy packaging and local characteristics, as a packaging Reference basis for design construction.

3.1 *Depth interview*

The interviews were conducted with the Xingang Township local associations and Xingang restaurant operators. There were 8 interviewees in total. The interviewees included the Xingang Charming Business District Association of Chiayi County, the board of directors of Xingang Fengtian Temple, a legal person consortium, the Xingang Township Culture and Leisure Industry Development Association, and the consortium Legal person Xingang Cultural and Educational Foundation and Xingang candy store: Jin Changli, Tianguanzhen, Yishengtang, Zhengmeixiang, etc. The verbatim transcript of the interview is included in the text. Each person was formally interviewed once, each time lasting approximately one hour. The interview content is summarized below as shown in Table 3:

Table 3. Summary of interviews in Xingang Township (Source: Organized by this study).

Summary of interview content	
Xingang Candy Specialties	Xingang candy is the most representative souvenir of Xingang Township. Xingang candy products have a century-old history. Many pilgrims and tourists will come to buy Xingang candy as tribute to Mazu for blessing.
Unique	Xingang Township is home to the "Fengtian Palace" temple, a national third-level historic site, which brings three to four million pilgrims to Xingang every year. "Fengtian Palace-General Hu Ye" can only be qualified to sit on the table if he is recruited by the emperor. The worship is also a unique deity in the province.
Locality	The Centennial Renaissance Iron Bridge was once the longest existing iron bridge for transporting sugarcane in Taiwan. However, the bridge was broken during a storm. Gradually, Taiwan's sugar was also lost. However, the people of Bantou Village near the Iron Bridge hoped that one day the Centenary Renaissance Iron Bridge would be restored. Will be back.
Storytelling	"Kaitai Mazu" is the belief center of local residents. It is popular with incense and has spirits all over the world. Every year, millions of pilgrims come to pay homage and pray to Mazu to bless the country and the people, and everything goes well.
Commemorative	When tourists visit Xingang Township, there are many places worth recalling, and

(*continued*)

Table 3. Continued

	they are all experiences worth remembering and sharing.
Packaging concept	If you can find elements that represent Xingang Township or Xingang Candy, it will be the best design that can maintain the history and culture of Xingang Candy and design creative and textured packaging through packaging.
Packaging expectations	I hope that Xingang Candy will be positioned as a high-quality product so that more different ethnic groups, not only Taiwanese but also foreign tourists, can see that Xingang Township has the local specialty of Xingang Candy.

3.2 *Questionnaire*

A quantitative study was conducted on local people and tourists to understand their intentions towards Xingang Township. The questionnaire was divided into three major items: 1. Basic information; 2. Travel experience; 3. Survey on local specialty Xingang glutinous rice. A total of 250 copies of the paper part were issued, with an actual recovery rate of 243 copies, and a total of 300 valid questionnaires, as shown in Table 4:

Table 4. Overall recovery status of questionnaires (Source: Organized by this study).

A total of 250 questionnaires were distributed			
Questionnaire	Local people	Tourist	Total
Return	116	127	250
Lnvalid	16	27	43
Efficient	100	100	200

Summary of the statistical analysis of the survey respondents. The data of the subjects were consolidated. In terms of questionnaire statistics, the key points are summarized as follows, as shown in Table 5:

Table 5. Summary table of survey elements (Source: Organized by this study).

Collection of questionnaires on Xingang Township's image and packaging design	
Basic information	1. Mostly "female". 2. Most people know Xingang Township through "friend recommendation". 3. The average number of people who choose to travel by "self-guided tour", "group tour" or "walk and see" are the majority.
Travel experience	1. Among tourist visits, "Fengtian Palace" accounts for the most historical monuments. 2. The largest proportion of local attractions have been to "Xingang Railway Park", "Xingang Fragrance Art" and "Ban Tao Kiln". 3. Most of them participate in the

(*continued*)

Table 5. Continued

Collection of questionnaires on Xingang Township's image and packaging design	
	special festivals of "Dajia Mazu's pilgrimage" and "Mazu's birth". 3. Regarding craftsmanship, most people think that the handicraft of "cut and stick Cochin pottery" is a local art.
Xingang Candy Packaging	1. 93.5% of Xingang locals and tourists have purchased Xingang candies. 2. The main purposes of purchasing Xingang candy are "gift giving" and "personal use". 3. The main purchasing principles are "taste", "packaging" and "brand". 4. In the packaging part, the testers' views on packaging are more "ordinary". 5. If the packaging design can be re-designed, the majority of the respondents prefer "traditional style" and "nostalgic style". It can be seen that Xingang candy is a local specialty with a history of more than 100 years in the minds of the respondents.

3.3 *Packaging design construction and creation*

Through the aforementioned in-depth interviews and questionnaire surveys in Xingang Township, this study understands local characteristics and experiences, and cooperates with the discussion of relevant literature on local cultural characteristic industries and packaging design-related literature to summarize the elements and local characteristics of the packaging design of the local specialty "Xingang Candy" The distinctive design themes that can be used include four major themes and 16 series of packaging themes, as shown in Figure 1:

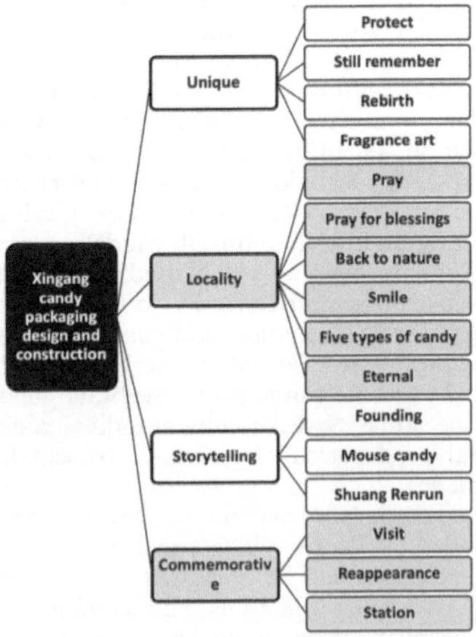

Figure 1. Local cultural industry design and construction (Source: Organized by this study).

This research explores and creates the packaging design of "Xingang Candy" local specialty products from four aspects: uniqueness, locality, storytelling, and commemoration. The overall structure is shown in Table 6 below:

Table 6. Xingang candy packaging design (Source: Organized by this study).

Uniqueness – Packaging Design		
Name	Forming	Design Notes
Protect		**Color:**red,yellow **Image:**Tiger Lord, Tiger Teeth, Golden Flower **Shape:**fold into table
Still remember		**Color:**green red blue orange **Image:**Iron bridge, train, countryside **Shape:**cross vision
Rebirth	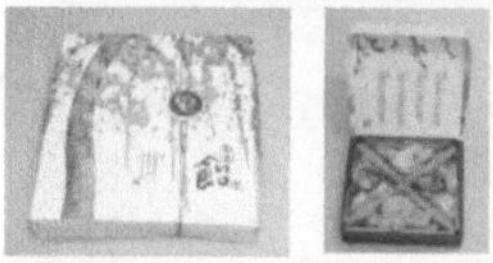	**Color:**coffee, white **Image:** Four Seasons, Cochin Pottery **Shape:**Gift box style
Fragrance art		**Color:**red orange coffee green **Image:**fragrant lamp **Shape:**Fragrance art styling

Locality – Packaging Design		
Name	Forming	Design Notes
Pray		**Color:**yellow red blue black **Image:**Fengtian Palace **Shape:**safety lamp
Pray for blessings		**Color:**red orange green blue **Image:** Opening the Mazu stage and embroidering the shrine **Shape:**firecracker shape
Back to nature		**Color:**green blue orange pink **Image:** plains, rivers, **Shape:**Interlaced graphics and text
Smile		**Color:**pink blue green orange **Image:** Smiling faces, people, careers, **Shape:** round sphere

(*continued*)

Table 6. Continued

Locality – Packaging Design		
Name	**Forming**	**Design Notes**
Five types of candy		**Color:** Kraft paper **Image:** Five steps for Xingang Yi **Shape:** hexagon
Eternity		**Color:** red yellow pink **Image:** Lisianthus and Mazu hat **Shape:** Mazu head

Storytelling – Packaging Design		
Name	**Forming**	**Design Notes**
Founding		**Color:**black and white **Image:**Mr.Lu,rain pole **Shape:** Interlacing effect
Mouse candy		**Color:**Bright contrasting colors **Image:** mouse, festive **Shape:** rat head
Shuang Renrun		**Color:** vivid colors **Image:** Wedding candies, gems, **Shape:** candy packaging

Commemorative – Packaging Design		
Name	**Forming**	**Design Notes**
Visit		**Color:** pastel colors **Image:** Fengtian Palace, Ban Tao Kiln, Railway **Shape:** rectangle
Reappearance		**Color:** Kraft paper **Image:** Ticket booth, local museum **Shape:** Sending form
Station		**Color:** orange red yellow red **Image:** Railway five-minute train, Fengtian Palace **Shape:** temple shape

4 FINDINGS AND DISCUSSION

In Taiwan, which pays attention to etiquette, "reciprocity" has always been a sincere exchange between Taiwanese people. Through the beauty of the product itself or the packaging, local characteristics and humanistic life are closely integrated. In order to enhance the design aesthetics of local characteristic products, this study The packaging design of "Xingang Candy" in Xingang Township is used to convey the life culture and local spirit of Xingang Township. The following conclusions can be drawn:

4.1 *Xingang candy packaging design positioning can be divided into four major themes and 16 series of packaging themes*

This study sorted out the four major aspects of uniqueness, locality, storytelling, and commemoration through literature review, and used "in-depth interviews" with local organizations and Xingang candy operators to sort out the characteristics that represent Xingang Township, so that Packaging design can enrich the content through creativity and ingenuity. Then, a "questionnaire survey" was used to understand the opinions of tourists and locals on the cultural characteristics of Xingang Township and the packaging of Xingang Candy. The quantitative and qualitative results were incorporated into the packaging design, and four major themes and 16 series of packaging themes were sorted out to summarize Xingang Candy. Packaging imagery and local industry characteristics and culture are used to formulate the positioning of packaging design.

4.2 *Through the four major aspects of uniqueness, locality, storytelling, and commemoration, the text, color, graphics, and shape of the packaging design are used to create the structure of the "Xingang Candy" product packaging design*

The scope of local cultural industry characteristics is vast, and all processes and contexts must be recorded. Therefore, the research of relevant literature at four levels, the development of creative concepts through packaging design aspects, and the execution of packaging design creation are helpful in designing. Zhongneng can organize it more systematically.

4.3 *Follow-up research directions*

4.3.1 The designed results can be reviewed and enhanced through in-depth interviews and questionnaires. It is hoped that the works can be cooperated with manufacturers and circulated on the market in the future, and convey the local cultural characteristics of Xingang Township. This makes the paper not only an academic research, but also practical for mass production.

4.3.2 The presentation of product packaging must be observed and communicated through the triangle relationship between designers, businesses, and consumers. If we can have a deeper understanding of the segmentation of the consumer market and understand the potential problems caused by the industry, we can further improve the works created through packaging. Through communication, it will be more able to make product packaging more in line with the needs of consumers and businesses.

REFERENCES

Albrecht Rothacher. (2006). *The Story Behind the Brand: Brand Management Strategy And Corporate Culture*, translated by Li Xiaoxu, Jiushi Cultural Industry Co., Ltd.

Chen Junhong. (2001). *Introduction to Visual Communication Design, Yang Dongmin,* Quanhua Science and Technology Book Co., Ltd.

Chiayi County Singang Township Office Global Information Network. (2012). Retrieved from: http://singang.gov.tw

Daryl Travis. (2001). Experience Brand: How to Build Custo mer Loyalty Advantages, translated by Zhu Puxuan and Zifeng, Chengbang Cultural Industry Co., Ltd.

Duane Knapp. (2001). *Brand Thinking-Five Strategies for Buil ding Advantageous Brands, translated by Yuan Shipei, American McGraw*. Hill International Co., Ltd. Taiwan Branch.

Geoffrey Randall. (2000). The power of brand building, translated by Cai Peijuan, Xiaozhitang Cultural Industry Co., Ltd.

Huang Lingyi. (2005). *Research on the Application of Lukang Characteristic Culture in Specialty Product Packaging*, Master Class of Visual Communication Design Department, National Yunlin University of Science and Technology.

Huang Ronghui. (1990). *Research on Graphic Design of Commercial Packaging*, Big Society and Cultural Affairs Press.

Huang Yaqing. (2009). *Research on the Packaging Design Strategy of Rice Gift Boxes in Taiwan*, master's thesis, Institute of Business Design, National Taichung Institute of Technology.

Huang Yiting. (2012). *Research on Bisexual Poster Creation*, Master's Class, Department of Visual Communication Design, National Yunlin University of Science and Technology.

Huang, F. S. (2000). *The Humanistic Meaning of Local Comm*

Li Fangyi. (2011). *Research on the Packaging Design and Image of Local Specialties – Taking the Top Ten Souvenirs in Fucheng as an Example*, Master's Class of the Department of Visual Communication Design, National Yunlin University of Science and Technology.

Li Tianlai. (1992). *Packaging Points, Lines and Surfaces*, New Image Publishing Co., Ltd.

Lin Qin. (2012). *Local Cultural Product Design Strategy, Master's Class.* Department of Industrial Design, National Yunlin University of Science and Technology.

Liu Junnan. (2007). *Research on the Packaging Design of Mid-Autumn Cake Gift Boxes in Taiwan's Pastry Industry*, Master's Thesis, Institute of Business Design, National Taichung Institute of Technology.

Masaaki Tanaka. (1999). Visual Communication Design, translated by Su Shouzheng, Liuhe Publishing House.

Shuya Kaneko. (1996). *Packaging Design, translated by Liao Zhizhong*, Fu Yuan Publishing Co., Ltd.

Thomas Gad. (2003). *Brand Code-Interpreting the Gene Map of Successful Brands*, Chen Zhiming, Pearson Education Publishing Co., Ltd., Taiwan.

Zhu Chen Chuntian. (1993). *Packaging Design*, Jinguan Publishing House.

Zhu Ruxing. (2003). *The Impact of Local Specialty Product Packaging Design on Brand Image*, Master's Thesis, Institute of Design Management, Mingchuan University.

System Innovation for an Artificial Intelligence Era – Lam et al. (Eds)

Control resource set and slot-based resource allocation in 5G new radio networks

Ya-Ju Yu & Hao-Chun Kuan
Department of Computer Science and Information Engineering, National University of Kaohsiung, Kaohsiung, Taiwan

ABSTRACT: New radio (NR) is an important technology for fifth-generation and following sixth-generation networks to support enhanced mobile broadband (eMBB) and massive Internet of Things (IoT) devices. Before receiving data in a physical downlink shared channel (PDSCH), a device should monitor candidates in physical downlink control channels (PDCCHs) of control resource sets (CORESETs) to decode downlink control information blindly. When the CORESET resource is exhausted, the PDSCH resource is not usable, and vice versa. However, massive IoT devices consume more CORESET resources, while eMBB devices need more PDSCH resources. There is a resource trade-off between massive IoT and eMBB devices. To solve the above issue, this paper targets the CORESET and slot-based resource allocation problem in NR networks. The objective is to minimize slot consumption, subject to the data requirements. We first propose an algorithm to determine two CORESETs, one for IoT devices and the other for eMBB devices. Then, based on the two CORESETs, this paper presents an algorithm to utilize the CORESETs and PDSCH in each slot efficiently. Simulation results show that the proposed algorithms can reduce slot consumption by 21% compared to the two baselines.

1 INTRODUCTION

New radio (NR) is an essential technology for fifth-generation (5G) and sixth-generation (6G) networks to support diverse applications such as massive machine-type communication, also referred to as Internet of Things (IoT), ultra-reliable low-latency communication, and enhanced mobile broadband (eMBB) (H. Ji *et al.* 2018). IoT devices generally require small data sizes, while eMBB devices need high data rates. NR provides flexible radio resource configurations to satisfy various requirements of applications.

In cellular networks, before a device can receive data in a physical downlink shared channel (PDSCH), it must monitor candidates in a physical downlink control channel (PDCCH) to blindly decode downlink control information (DCI). Massive IoT devices with small data require more resource element groups (REGs) in PDCCH than eMBB devices. On the other hand, eMBB devices need more REGs in PDSCH to accommodate their high data requirements. The number of devices in a network affects the size of PDCCH, while the data requirement influences the size of PDSCH. Due to the limited radio resources in a slot, there is a trade-off between the PDCCH and PDSCH requirements.

Long-term evaluation (LTE) has a fixed PDCCH position across the entire bandwidth, whereas NR can flexibly determine PDCCH positions in a slot through CORESETs (H. Ji *et al.* 2018). A base station also can adjust the bandwidth of a CORESET. Once CORESETs are decided, PDSCH uses the remaining resources of a slot. Balancing CORESETs and PDSCH in a slot is critical for improving resource utilization. If the bandwidths of CORESETs are not appropriately configured, it can lead to resource wastage. Therefore, this paper examines how to determine the bandwidths of CORESETs and allocate the resources of CORESETs and PDSCH.

Recently, some related works have studied resource allocation considering numerologies of 5G NR. Akhtar *et al.* (A. Akhtar *et al.* 2018) proposed an adaptive numerology resource allocation algorithm to schedule different non-orthogonal numerologies for improving throughput. Bag *et al.* (T. Bag *et al.* 2019) proposed a dynamic resource allocation scheme with mini-slots to reduce the average scheduling latencies. Elias *et al.* studied the admission control and resource allocation problem in multi-connectivity scenarios, considering different requirements and NR features (J. Elias *et al.* 2021). Some recent works considered the aspect of PDCCH and CORESETs in 5G NR. Miao *et al.* (H. Miao *et al.*) studied the performance of different constructions for REG bundles in channels with diverse frequencies. Braun *et al.* (V. Braun *et al.* 2019) gave a design overview of PDCCH and CORESETs and proposed an enhancement to the

DOI: 10.1201/9781003514831-58

NR PDCCH hash function. Mozaffari *et al.* (M. Mozaffari *et al.* 2021) analyzed the NR PDCCH blocking probability and showed that CORESET sizes affect the blocking probability. However, related works have not addressed our target problem and the resource trade-off between PDSCH and CORESETs for the different resource demands of IoT and eMBB devices.

This paper explores the joint control resource set and slot-based resource allocation problem in NR networks. The objective is to minimize the slots consumed to satisfy the downlink data requirements of IoT and eMBB devices. The contributions of this paper are summarized as follows:

- This paper points out the trade-off of control and data resources between massive IoT and eMBB devices.
- To solve the target problem, this paper first proposes an algorithm for determining control resource sets according to the control resource requirements of IoT and eMBB devices. The algorithm decides two CORESETs, one for IoT devices and the other for eMBB devices. Except for the REGs of the CORESETs, the other REGs can be used for PDSCH. Then, based on the two CORESETs, we propose an algorithm to efficiently utilize the two CORESETs and PDSCH of each slot.
- Compared with the two baselines, the simulation results show that the proposed algorithms can reduce slot consumption by 21% to satisfy the data requirements of devices. The performance improvement of the proposed algorithms is more evident when more eMBB devices reside in a cell.

The remainder of the paper is organized as follows. In Section 2, we describe the system model and problem formulation. In Section 3, we explain the proposed algorithms. Section 4 presents the simulation results, and we conclude this paper in Section 5.

2 SYSTEM MODEL AND PROBLEM FORMULATION

2.1 *System model*

This paper focuses on a base station to serve IoT and eMBB devices' downlink data requirements using a bandwidth part. In each time slot, 14 symbols are present, while the number of resource blocks (RBs) in the frequency domain depends on the numerology and bandwidth used. The paper considers that the numerology and bandwidth are fixed. An RB comprises 12 subcarriers, while one REG consists of one symbol and one RB.

A base station can configure the bandwidth of each CORESET. In the time domain, the duration of a CORESET can be one, two, or three symbols. For each CORESET, the paper considers a fixed duration of three symbols starting from the first symbol. A control channel element (CCE) comprising six REGs is a basic resource unit that transmits DCIs for devices in a CORESET. Depending on the channel's quality, the device needs a minimum number of CCEs used for a PDCCH candidate to receive a DCI, referred to as an aggregation level. Before receiving data in PDSCH, a device should monitor PDCCH candidates in its CORESETs. The paper ignores the candidate number's limitation and considers the CCE amount as a resource limitation. Specifically, each CCE can be the start of a PDCCH candidate.

Apart from the REGs used for CORESETs, the other REGs can operate as PDSCH for downlink data transmission. The paper considers slot-based resource allocation. A base station should determine a start symbol and symbol length (type A) in the time domain for a device to receive data. In the frequency domain, the base station decides a start RB and RB length (type 1) for a device. The paper investigates how to determine suitable bandwidths of CORESETs and allocate the CORESET and PDSCH resources.

2.2 *Problem formulation*

This paper studies the control resource set and slot-based resource problem in NR. The objective is to minimize the consumed slots to satisfy devices' data requirements and increase resource utilization.

We consider an NR base station to serve$D = |D_E| + |D_I|$ devices, comprised of eMBB device set D_E and IoT device set D_I. The data requirement of device d is $⊠_d$. The SNR between the base station and device d is $10\log_{10}(\rho_d/\sigma)$, where ρ_d is the received power at device d and affected by the path loss, shadowing, fast fading, transmit power, and bandwidth. σ is the noise spectral density. Based on the SNR, device d suffers a bit error rate and requires a minimum aggregation level η_d to achieve reliability for receiving a DCI. The aggregation level is the necessary number of CCEs. According to the Shannon capacity, one REG can provide W_d data bits for device d. Because this paper considers slot-based mapping type A and type 1 for PDSCH, we should continuously allocate T^{data} symbols at one RB in a slot for a device. The number of REGs assigned for device d to receive data is denoted as R_d^{data}.

For a CORESET, we should determine its resource in the time and frequency domains. A base station can configure multiple CORESETs in a bandwidth part. T^{COR} is the symbol number (i.e., duration) in the time domain and F_i^{COR} is the RB number in the frequency domain for the i-th CORESET. We consider a fixed duration for each

CORESET in this paper. The number of CCEs in the i-th CORESET is $\frac{F_i^{COR} \times T^{COR}}{6}$.

We define an indicator $I_{d,s,x,y}$, which is 1 if the REG at the x-th symbol and y-th RB in the s-th slot is allocated to device d, and 0 otherwise. The number of symbols is N^{Symb} and RBs is N^{RB} in a slot. The consumed slots to satisfy the data requirements of all devices are S.

Table 1. Summary of notations.

Symbol	Description
D	Number of devices
ψ_d	Data requirement of device d
S	Number of consumed slots
η_d	Number of required CCEs of device d
W_d	Data bits provided by one REG for device d
N^{Symb}	Number of symbols in a slot
N^{RB}	Number of RBs in a slot
R_d^{data}	Number of REGs allocated for device d to receive data
T^{data}	Allocated symbols for data access
T^{COR}	Duration (symbols) of a CORESET
F_i^{COR}	RB number of the i-th CORESET
$I_{d,s,x,y}$	Indicator function which is 1 if the REG at the x-th symbol and y-th RB in the s-th slot is allocated to device d, and 0 otherwise.

Requirement Constraint: The base station should allocate sufficient REGs to satisfy the data demand of each device d.

$$R_d^{\text{data}} \times W_d \geq \psi_d, \forall d \tag{1}$$

Resource Constraints: Each REG can be allocated to only one device. The allocated symbols cannot exceed N^{Symb} at each RB, and the allocated RBs cannot surpass N^{RB} at each symbol in each slot.

$$\sum_{d=1}^{D}\sum_{s=1}^{S}\sum_{x=1}^{N^{\text{Symb}}}\sum_{y=1}^{N^{\text{RB}}} I_{d,s,x,y} \leq 1 \tag{2}$$

$$\sum_{d=1}^{D}\sum_{x=1}^{N^{\text{Symb}}} I_{d,s,x,y} \leq N^{\text{Symb}}, \quad \forall y, s \tag{3}$$

$$\sum_{d=1}^{D}\sum_{y=1}^{N^{\text{RB}}} I_{d,s,x,y} \leq N^{\text{RB}}, \quad \forall x, s \tag{4}$$

The objective of this paper is to minimize the number of consumed slots to serve the devices and the objective function is formulated as follows.

$$\min_{F_i^{\text{COR}},\, I_{d,s,x,y}} S, \tag{5}$$

subject to the constraints of Equation (1)-(4). The notations are summarized in Table 1.

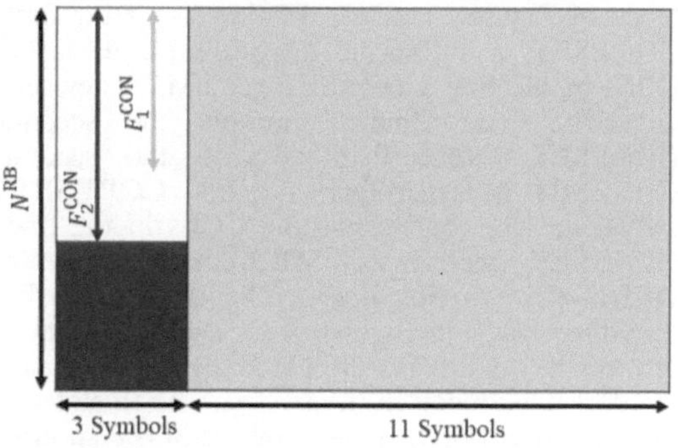

Figure 1. Design concept of the proposed algorithms.

3 CONTROL RESOURCE SET AND RESOURCE ALLOCATION

Section 3.1 describes the design concept of the proposed algorithms. In Section 3.2, we propose an algorithm to flexibly determine two CORESETs, one for IoT and the other for eMBB devices. In Section 3.3, we propose a CCE and REG allocation algorithm based on the two CORESETs to use the radio resources of each slot efficiently.

3.1 *Design concept of the proposed algorithms*

This section introduces the design concept of the proposed two algorithms, as shown in Figure. 1. Algorithm 1 evaluates the required CCEs at each slot by pseudo-scheduling the devices and flexibly determines two CORESETs with the same duration of three symbols. The bandwidth of the first CORESET is set referring to the required CCE number, which is the most used during the scheduling. The bandwidth of the second CORESET is used according to the longest required CCE number. We design the second CORESET longer than the first CORESET because we can assign the CCEs in the second CORESET to use the remaining RBs with the three symbols marked as the dark-blue block shown in Figure 1, even if the first CORESET is wholly utilized. The first CORESET is designed for IoT devices with small data requirements, while eMBB devices with big data requirements monitor the second CORESET.

The two CORESETs returned by Algorithm 1 are used for Algorithm 2. Algorithm 2 allocates the CCEs in the CORESETs and the REGs in the PDSCH for devices. We first assign the CCEs in the first CORESET and the RBs with the 11 symbols in the PDSCH, marked as the light-blue block in Figure 1, for IoT devices. If the IoT devices are satisfied, the REGs can be used for eMBB devices. Then, Algorithm 2 allocates the remaining RBs and the three symbols, marked as the dark-blue block in Figure 1, for eMBB devices. If the first CORESET is sufficient, Algorithm 2 uses the CCEs in the first CORESET for eMBB devices. Otherwise, Algorithm 2 adopts the second CORESET. Note that because the second CORESET is larger than the first CORESET, eMBB devices can receive the CCEs in the first CORESET. Because an eMBB device receives a DCI in the s-th slot, it generally cannot use the dark-blue block in the s-th slot. The base station can use the DCI to inform it to access the REGs of the dark-blue block in the next $(s+1)$-th slot via the slot offset policy defined by the 3GPP specifications (3GPP TS 38.331 2022). The resources of the dark-blue block are generally wasted in traditional 4G networks or if we do not efficiently use them. This paper designs the two algorithms to increase resource utilization in 5G NR networks.

3.2 *Control resource sets*

The pseudo-code of the proposed control resource set determination is shown in Algorithm 1. This algorithm flexibly decides two CORESETs by evaluating the CCE and data requirements. The first CORESET is used for IoT devices, and the second CORESET is assigned for eMBB devices. The second CORESET is generally larger than the first one.

In Lines 1-4, we initialize some parameters. In Line 1, S (the total consumed slots) is set as zero. In Line 2, the allocated symbols T^{data} to each device for data access are set as 11. In Line 3, the duration T^{COR} for the CORESET is set as 3 (symbols), and the frequency of the CORESET is the whole bandwidth (i.e., N^{RB}). In Line 4, the number of used CCEs N_s^{CCE} at the s-th slot is initialized as zero, $\forall s$. In Lines 5-20, we evaluate the used CCEs N_s^{CCE} of each slot by pseudo-scheduling the devices until the data requirements of all devices are satisfied. In Line 6, we start to allocate the radio resources of a slot, and the number of consumed slots is increased by one (i.e., $S = S + 1$). The variable y, initialized as zero, is the number of allocated RBs in the frequency domain (Line 7).

While the number of allocated RBs is less than the total RBs (i.e., $y < N^{\text{RB}}$), we first schedule IoT devices (Lines 9-13). If the used CCEs add the required CCEs of device d is less than or equal to the total available CCEs and its data requirement is not satisfied, $\exists d \in D_I$, the used CCEs are increased by the required CCEs of device d (i.e., $N_S^{\text{CCE}} = N_S^{\text{CCE}} + \eta_d$) in Lines 9-10. Then, we can allocate the REGs in the PDSCH for the IoT device to receive data. When there are remaining RBs and the data requirement of device d has not been satisfied, we allocate T^{data} symbols and one RB. The number of allocated REGs R_d^{data} for device d adds T^{data}, and y (the number of allocated RBs) is increased by one (Lines 11-13).

Algorithm 1: Control Resource Set Determination

Input: D, D_I, D_E ψ_d, η_d, N^{RB}

1 $S = 0$
2 $T^{\text{data}} = 11$
3 $T^{\text{COR}} = 3$
4 $N_s^{\text{CCE}} = 0, \forall s$
5 **while** $\psi_d > R_d^{\text{data}} \times W_d, \exists d, 1 \le d \le D$ **do**
6 $\quad S = S + 1$
7 $\quad y = 0$
8 $\quad$ **while** $y < N^{RB}$ **do**
9 $\quad\quad$ **if** $N_S^{CCE} + \eta_d \le \lfloor \frac{T^{COR} N^{RB}}{6} \rfloor$, *and* $\psi_d > R_d^{\text{data}} \times W_d$, $\exists d \in D_I$ **then**
10 $\quad\quad\quad N_S^{\text{CCE}} = N_S^{\text{CCE}} + \eta_d$
11 $\quad\quad\quad$ **while** $y < N^{RB}$ *and* $\psi_d > R_d^{\text{data}} \times W_d$ **do**
12 $\quad\quad\quad\quad R_d^{\text{data}} = R_d^{\text{data}} + T^{\text{data}}$
13 $\quad\quad\quad\quad y = y + 1$
14 $\quad\quad$ **else if** $N_S^{CCE} + \eta_d \le \lfloor \frac{T^{COR} N^{RB}}{6} \rfloor$, *and* $\psi_d > R_d^{\text{data}} \times W_d$, $\exists d \in D_E$ **then**
15 $\quad\quad\quad N_S^{\text{CCE}} = N_S^{\text{CCE}} + \eta_d$
16 $\quad\quad\quad$ **while** $y < N^{RB}$ *and* $\psi_d > R_d^{\text{data}} \times W_d$ **do**
17 $\quad\quad\quad\quad R_d^{\text{data}} = R_d^{\text{data}} + T^{\text{data}}$
18 $\quad\quad\quad\quad y = y + 1$
19 $\quad\quad$ **else**
20 $\quad\quad\quad$ break
21 Find the CCE number N^{CCE} that is used the most times in $N_s^{\text{CCE}}, 1 \le s \le S$
22 $F_1^{\text{COR}} = \lceil N^{\text{CCE}} / T^{\text{COR}} \rceil + 1$
23 $F_2^{\text{COR}} = \left\lceil \max_{1 \le s \le S} N_s^{\text{CCE}} / T^{\text{COR}} \right\rceil$
24 **return** F_i^{COR}

If the IoT devices are fulfilled, we then serve the eMBB devices in the same way (Lines 14-18). If the CCEs in the CORESET or the REGs in the PDSCH are exhausted at the s-th slot, we break the while loop and go to the next slot (Lines 19-20).

When the data requirements of all devices are satisfied, this algorithm determines two CORESETs. From the S slots, we find the CCE number N^{CCE} used the most times in the S slots (Line 21). The number of RBs for the first CORESET is set as $F_1^{\text{COR}} = N^{\text{CCE}} / T^{\text{COR}} + 1$ (Line 22). The number of CCEs for the second CORESET is set as the maximum CCEs used during the S slots. Thus, the number of RBs for the second CORESET is F_2^{COR} (Line 23). This algorithm outputs the two CORESETs for the proposed Algorithm 2 (Line 24).

3.3 *CCE and REG allocation*

The pseudo-code of the proposed CCE and REG allocation is shown in Algorithm 2. At each slot, this algorithm first assigns the CCEs in the first CORESET and uses the RBs with the 11 symbols in the PDSCH for IoT and eMBB devices as the light-blue block of Figure 1 (Lines 7-22). Then, if the first CORESET is not sufficient to utilize the remaining resources, we use the second CORESET and the remaining RBs with the three symbols in the PDSCHfor eMBB devices to access the dark-blue block of Figure 1 (Lines 23-35).

Algorithm 2: CCE and REG Allocation

Input: F_i^{COR}, D, D_I, D_E ψ_d, η_d, N^{RB}
1 $S = 0$
2 $T^{\mathrm{COR}} = 3$
3 **while** $\psi_d > R_d^{\mathrm{data}} \times W_d, \exists d, 1 \le d \le D$ **do**
4 $\quad S = S + 1$
5 $\quad y = 0$
6 $\quad N_S^{\mathrm{CCE}} = 0$
7 $\quad T^{\mathrm{data}} = 11$
8 $\quad F^{\mathrm{COR}} = F_1^{\mathrm{COR}}$
9 $\quad$ **foreach** $d \in D_I$ **do**
10 $\quad\quad$ **if** $N_S^{\mathrm{CCE}} + \eta_d \le \lfloor \frac{T^{\mathrm{COR}} F^{\mathrm{COR}}}{6} \rfloor$ *and* $y < N^{RB}$ *and* $\psi_d > R_d^{\mathrm{data}} \times W_d$ **then**
11 $\quad\quad\quad N_S^{\mathrm{CCE}} = N_S^{\mathrm{CCE}} + \eta_d$
12 $\quad\quad\quad$ **while** $y \le N^{RB}$ *and* $\psi_d > R_d^{\mathrm{data}} \times W_d$ **do**
13 $\quad\quad\quad\quad R_d^{\mathrm{data}} = R_d^{\mathrm{data}} + T^{\mathrm{data}}$
14 $\quad\quad\quad\quad I_{d,S,x,y} = 1, \forall T^{\mathrm{COR}} < x \le T^{\mathrm{data}}$
15 $\quad\quad\quad\quad y = y + 1$
16 $\quad$ **foreach** $d \in D_E$ **do**
17 $\quad\quad$ **if** $N_S^{\mathrm{CCE}} + \eta_d \le \lfloor \frac{T^{\mathrm{COR}} F^{\mathrm{COR}}}{6} \rfloor$ *and* $y < N^{RB}$ *and* $\psi_d > R_d^{\mathrm{data}} \times W_d$ **then**
18 $\quad\quad\quad N_S^{\mathrm{CCE}} = N_S^{\mathrm{CCE}} + \eta_d$
19 $\quad\quad\quad$ **while** $y \le N^{RB}$ *and* $\psi_d > R_d^{\mathrm{data}} \times W_d$ **do**
20 $\quad\quad\quad\quad R_d^{\mathrm{data}} = R_d^{\mathrm{data}} + T^{\mathrm{data}}$
21 $\quad\quad\quad\quad I_{d,S,x,y} = 1, \forall T^{\mathrm{COR}} < x \le T^{\mathrm{data}}$
22 $\quad\quad\quad\quad y = y + 1$
23 $\quad$ **if** *the first CORESET is exhausted* **then**
24 $\quad\quad F^{\mathrm{COR}} = F_2^{\mathrm{COR}}$
25 $\quad\quad y = F_2^{\mathrm{COR}} + 1$
26 $\quad$ **else**
27 $\quad\quad y = F_1^{\mathrm{COR}} + 1$
28 $\quad T^{\mathrm{data}} = 3$
29 $\quad$ **foreach** $d \in D_E$ **do**
30 $\quad\quad$ **if** $N_S^{\mathrm{CCE}} + \eta_d \le \lfloor \frac{T^{\mathrm{COR}} F^{\mathrm{COR}}}{6} \rfloor$ *and* $y < N^{RB}$ *and* $\psi_d > R_d^{\mathrm{data}} \times W_d$ *and* $\sum_{x=1}^{N^{Symb}} \sum_{y=1}^{N^{RB}} I_{d,S,x,y} = 0$ **then**
31 $\quad\quad\quad N_S^{\mathrm{CCE}} = N_S^{\mathrm{CCE}} + \eta_d$
32 $\quad\quad\quad$ **while** $y \le N^{RB}$ *and* $\psi_d > R_d^{\mathrm{data}} \times W_d$ **do**
33 $\quad\quad\quad\quad R_d^{\mathrm{data}} = R_d^{\mathrm{data}} + T^{\mathrm{data}}$
34 $\quad\quad\quad\quad I_{d,S,x,y} = 1, \forall 1 \le x \le T^{\mathrm{data}}$
35 $\quad\quad\quad\quad y = y + 1$
36 **return** $I_{d,s,x,y}$

In Line 1, the number of consumed slots is initialized as zero (i.e., $S = 0$). The duration for each CORESET is set as three (i.e., $T^{\mathrm{COR}} = 3$) in Line 2. In Lines 3-35, we allocate the CCEs in the CORESETs and the REGs in the PDSCH until the data requirements of all devices are satisfied. Whenever the consumed slots are increased by one, we initialize the number of allocated RBs ($y = 0$) and the number of assigned CCEs ($N_S^{\mathrm{CCE}} = 0$) at the current new slot (Lines 4-6). The algorithm first allocates the RBs with the 11 symbols in the PDSCH and uses the first CORESET (Lines 7-8). The CCEs in the CORESET are allocated for IoT devices and then for eMBB devices.

We allocate the CCEs and REGs for the IoT devices (Lines 9-15). If there are sufficient CCEs and REGs and the data requirement of device $d \in D_I$ is still required, we allocate η_d CCEs for the device (Lines 10-11) and set the corresponding $I_{d,S,x,y} = 1$ for the used REGs of the CCE positions. Then, the algorithm assigns an RB one by one for the device until the RBs are exhausted, or the data requirement of the device is fulfilled (Line 12). Because we consider the slot-based resource allocation, we should continuously allocate T^{data} symbols when assigning one RB. The data requirement of the device can be decreased by $T^{\mathrm{data}} \times W_d$ (Line 13). The corresponding REG positions at the indicator function $I_{d,S,x,y}$ are set as one, and the number of allocated RBs is increased by one (Lines 14-15). If the IoT devices are satisfied, the CCEs can be used for the eMBB devices (Lines 16-22), and the resource steps for the IoT devices are the same as Lines 9-15. Then, we start to utilize the remaining RBs with the three symbols (Lines 23-35). If the CCEs in the first CORESET are not sufficient, we use the second CORESET ($F^{\mathrm{COR}} = F_2^{\mathrm{COR}}$), and the allocated RB index y for data access is set as $F_2^{\mathrm{COR}} + 1$ (Lines 23-25). Otherwise, this algorithm can allocate the CCEs in the first CORESET, and the assigned RB index y starts from $F_1^{\mathrm{COR}} + 1$ (Lines 26-27). The allocated symbols are $T^{\mathrm{data}} = 3$ (Line 28). For each eMBB device $d \in D_E$, if device d is not satisfied and scheduled and the resources are sufficient, the number of used CCEs is increased by η_d (Lines 29-31). We allocate $1 \times T^{\mathrm{data}}$ REGs for the device until its data requirement is satisfied or the RBs are all used (Lines 32-35). Finally, the algorithm returns the allocation result $I_{\mathrm{d,s,x,y}}$ (Line 36).

4 PERFORMANCE EVALUATION

4.1 *Simulation setups*

The proposed algorithms, namely the *Control Resource Set and Resource Allocation (CRSRA)* algorithm, are compared with two baselines. The first baseline, namely the *Legacy* scheme, employs all RBs of the bandwidth as the CORESET with the duration of three symbols and the other REGs as the PDSCH. The second baseline, denoted as the *Single Control Resource Set (SCRS)* method, adopts our Algorithms 1 and 2, but only the second CORESET is used for Algorithm 2.

Our simulation is based on the specifications (3GPP TS 36.213 2020, 3GPP TS 36.213 2020). We consider a base station that serves devices with downlink data requirements. The subcarrier spacing is set to 15 kHz, and the bandwidth is 50 MHz. A slot has 270 RBs in the frequency domain and 14 symbols in the time domain (3GPP TS 38.101-1 2023).

Our simulation involves a varying number of eMBB devices, ranging from 10 to 40. The eMBB applications include audio and video, with each eMBB device randomly selecting one or the other. Audio data rates range from 100 to 300 kbps, while video data rates fall between 2000 to 6000 kbps as references (C. Chen *et al.* 2015, YouTube 2023).

In addition to eMBB devices, we also consider massive IoT devices with a Poisson arrival rate of λ at each slot. These devices have small data requirements of 200, 400, or 1500 bytes, with corresponding probabilities of 65%, 25%, and 10% (A. Roy *et al.* 2015).

The base station's transmission power is 46 dBm. The pass loss model is $128.1 + 37.6\log_{10}(\alpha)$ dB, where α is in kilometers. Thermal noise density is -170 dBm/Hz. The shadowing fading uses a log-normal distribution with the mean of 0 and standard deviation of 8 dB. Then, we can calculate the SNR and compute the bit error rate of each device. In order to achieve a DCI transmission reliability of 99%, we should select a minimum aggregation level of 1, 2, 4, or 8 for each device.

4.2 *Simulation results*

In Figure 2, we examine the impact of eMBB device numbers on slot consumption under IoT devices with a Poisson arrival rate of $\lambda = 10$. As the number of eMBB devices increases, so does the number of consumed slots. This is because more devices require more data, which in turn requires the base station to consume more slots to serve those devices. Our proposed SCRS and CRSRA algorithms can effectively determine the CORESETs and use the REGs of each slot, which helps to reduce the number of slots required to meet device requirements. In contrast, the Legacy method fails to balance control and data resources, resulting in wasted slots. By comparison, our proposed algorithms can reduce slot consumption by 21% for data transmission, with more significant improvements observed when more eMBB devices attend.

In Figure 3, we can observe how the arrival rates of IoT devices impact the network throughput when there are ten eMBB devices. As the number of small data devices increases, the throughput decreases. This is because each small data device requires CCEs, and as more IoT devices are present, they consume more CCEs than eMBB devices. However, IoT devices require fewer REGs than eMBB devices for data requirements. It's important to note that CCEs are control resources and do not contribute to the network throughput. The proposed algorithms yield a 19% improvement in the network throughput compared to the Legacy method. The improvement is more evident when there are fewer small data devices.

Figure 4 reveals the impact of the eMBB device numbers on the average waiting time for devices to be scheduled in IoT devices with a Poisson arrival rate of $\lambda = 10$. The average waiting time gradually increases with the number of devices because the base station should organize more devices and consume more time. The proposed algorithm is

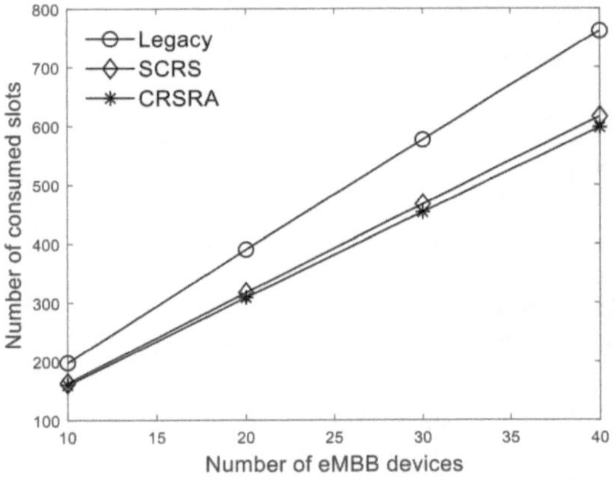

Figure 2. Effect of the eMBB device numbers on the number of consumed slots.

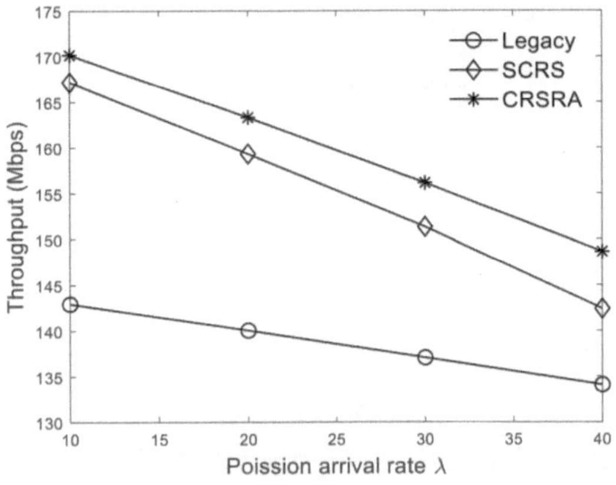

Figure 3. CCE and REG allocation.

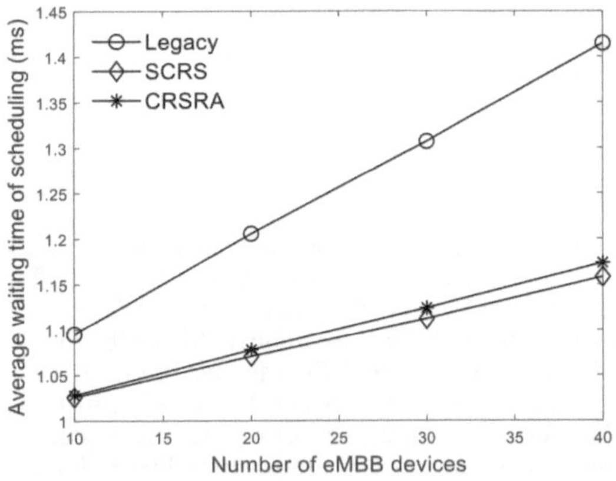

Figure 4. Effect of the eMBB device numbers on the average waiting time of devices to be scheduled.

more efficient than the Legacy method and reduces the average waiting time to be scheduled to below 1.2 ms.

5 CONCLUSION

This study focuses on the control resource set and slot-based resource allocation in NR cellular networks. The objective is to minimize the number of slots used to satisfy the downlink data requirements. We indicate the trade-off of resource requirements between massive IoT and eMBB devices. Unsuitable CORESETs result in radio resource wastage, so this paper proposes two algorithms to solve the CORESET and slot-based resource allocation problem. The first algorithm identifies two suitable CORESETs, and the second algorithm allocates each slot's CORESET and NPDSCH resources. By using these algorithms, the simulation results show an efficient reduction in slot consumption to complete data transmission compared with the Legacy method.

ACKNOWLEDGEMENT

This work was supported by the National Science and Technology Council (NSTC), Taiwan, under Grant 110-2221-E-390-010-MY3.

REFERENCES

Akhtar A. and Arslan H., (2018). Downlink resource allocation and packet scheduling in multi-numerology wireless systems, in *IEEE Wireless Communications and Networking Conference Workshops (WCNCW)*, pp. 362–367.

Bag T., Garg S., Shaik Z., and Mitschele-Thiel A., (2019). *Multi-Numerology Based Resource Allocation for Reducing Average Scheduling Latencies for 5G NR Wireless Networks, in European Conference on Networks and Communications (EuCNC)*, pp. 597–602.

Braun V., Schober K., and Tiirola E. (2019). 5G NR physical downlink control channel: Design, performance and enhancements, in *IEEE Wireless Communications and Networking Conference (WCNC)*, pp. 1–6.

Chen C., Zhu X., de Veciana G., Bovik A. C., and Heath R. W. (2015). Rate adaptation and admission control for video transmission with subjective quality constraints, *IEEE Journal of Selected Topics in Signal Processing*, vol. 9, no. 1, pp. 22–36, 2015.

Elias J., Martignon F., and Paris S., (2021). *Optimal Split Bearer Control and Resource Allocation for Multi-Connectivity in 5G New Radio, in Joint European Conference on Networks and Communications & 6G Summit (EuCNC/6G Summit)*, pp. 187–192.

Ji H., Park S., Yeo J., Kim Y., Lee J., and Shim B. (2018). Ultra-reliable and low-latency communications in 5G Downlink: Physical Layer Aspects, *IEEE Wireless Communications*, vol. 25, no. 3, pp. 124–130,

Miao H. and Faerber M. (2018). Configurable distributed physical downlink control channel for 5G new radio: resource bundling and diversity trade-off, in *IEEE Wireless Communications and Networking Conference Workshops (WCNCW)*, pp. 368–372.

Mozaffari M., Wang Y.-P. E., and Kittichokechai K. (2021). Blocking Probability Analysis for 5G New Radio (NR) Physical Downlink Control Channel, IEEE International Conference on Communications (ICC), pp. 1–6.

Roy A., Zeng H., Bagga J., Porter G., and Snoeren A. C. (2015). Inside the Social Network's (Datacenter) Network, *SIGCOMM Computer Communication Review*, vol. 45, no. 4, pp. 123–137.

YouTube, Live Encoder Settings, Bitrates, and Resolutions. (2023). [Online]. Available: https://support.google.com/youtube/answer/2853702?hl=en

3GPP TS 38.101-1 Technical Specification Group Radio Access Network; NR; User Equipment (UE) radio transmission and reception; Part 1: Range 1 Standalone (Release 18). (March 2023).

3GPP TS 38.331 V17.3.0 Technical Specification Group Radio Access Network; NR; Radio Resource Control (RRC) protocol specification (Release 17). (Dec. 2022).

3GPP TS 36.213 V16.0.0 Technical Specification Group Radio Access Network; Evolved Universal Terrestrial Radio Access (E-UTRA); Physical Layer Procedures. (Jan. 2020).

3GPP TS 36.331 V16.0.0 Technical Specification Group Radio Access Network; Evolved Universal Terrestrial Radio Access (E-UTRA); Radio Resource Control (RRC); Protocol Specification. (Apr. 2020).

System Innovation for an Artificial Intelligence Era – Lam et al. (Eds)

A virtual MIMO transmission scheme using network coding for B5G C-V2X communication systems

Cheolwoo You
Department of Information and Communication Engineering, Myongji University, Yongin, South Korea

ABSTRACT: In this paper, a cooperative transmission scheme for B5G moving vehicle networks is proposed to acquire high transmission efficiency and coverage extension under multipath fading environment. The proposed scheme uses a concept of virtual multiple-input multiple-output (MIMO) transmission using a Network coding scheme and the space-time block code (STBC) for performance improvement. Simulation results have shown that the proposed scheme can provide signal-to-noise ratio (SNR) improvement and reduce transmission time compared to the conventional scheme. Also, the proposed scheme can alleviate inter symbol interference (ISI) and inter channel interference (ICI). This is because the proposed scheme can simultaneously get spatial diversity and coding gain by using the virtual MIMO configuration. The proposed scheme is expected to be efficiently utilized in B5G cellular vehicle-to-everything (C-V2X) communication systems.

1 INTRODUCTION

Recently, relays have been used in various fields such as the communications between the source and destination via some cooperation protocol as well as the extension of the range (A. Reichman 2008). In multi-user systems, different users can also act as cooperative partners or relays to share resources and assist each other in information transmission, thereby creating a cooperative network. The way the nodes share information is by tuning into each other's transmitted signals and by processing information that they overhear. For example, the two-hop cooperative system has half rate transmission speed. But in a cooperative network, two or more nodes can share their information and transmit jointly as a virtual antenna array. This enables them to obtain higher data rates and diversity than those they could achieve individually (A. Sendonaris *et al.* 2003, J. N. Laneman *et al.* 2004). In result, it significantly improves data throughput, spectral efficiency, and link range without additional bandwidth or transmission power. In this paper, we propose a virtual MIMO transmis-sion scheme for moving networks which has high transmission efficiency and accuracy through cooper-ation. The proposed scheme partially adopts the STBC scheme of (M.E Celebi *et al.* 2006).

2 COOPERATIVE COMMUNICATION SYSTEMS

With the advent of research on moving networks in which the whole network is mobile, it is expected that public transportation will communicate each other while in movement. In this way, a moving network consists of one or more mobile routers with a number of devices connected to them (C. Politis *et al.* 2004).

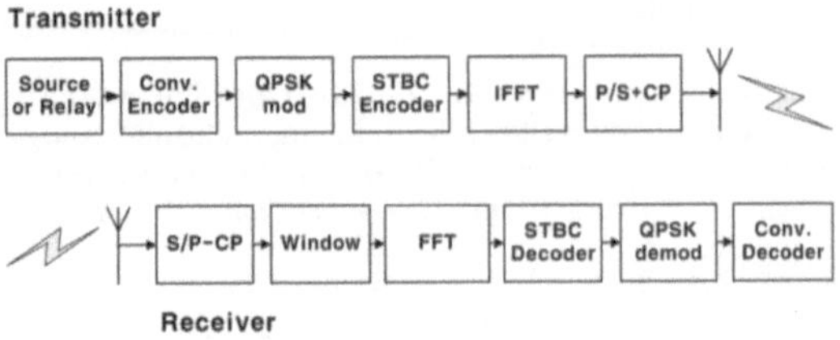

Figure 1. System model.

In this paper, we consider a moving network that operates in a cooperative wireless inter vehicle communication (IVC) environment using the orthogonal frequency division multiplexing (OFDM) transmission as illustrated in Figure 1. That is, the moving network is the set of nodes that are sources or destinations with mobility. We also assume that they can communicate with each other for sharing information without inter-user interference. This model can also be used for future Intelligent Transportation System (ITS) scenario.

Assuming that there are n source and/or relay nodes which transmit modulated symbols to the destination node j, the transmitted data can be represented $\mathbf{s}_j = (s_1, s_2, \ldots, s_n)$ during one symbol time interval. At this moment, the fading channel coefficients vector between the n transmit nodes and

 DOI: 10.1201/9781003514831-59

the destination node j is defined as $\mathbf{h} = (h_{j1}, h_{j2}, \ldots, h_{jn})^{\mathrm{T}}$, and thus the received signal r_j at the destination node j is given as $r_j = \mathbf{s}_j\mathbf{h}/\sqrt{n} + n_j$, where n_j is the additive white Gaussian noise (AWGN) that is a complex zero-mean sample with variance σ^2 per dimension at the destination node j.

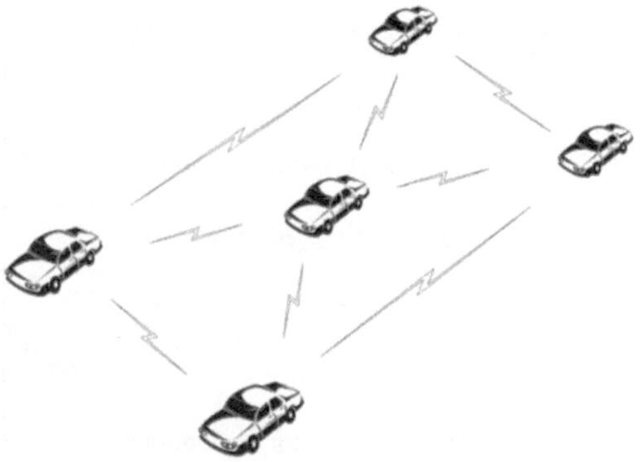

Figure 2. An example of moving networks.

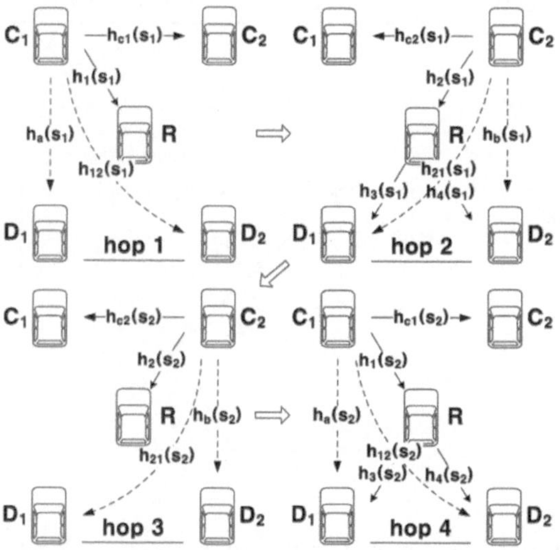

Figure 3. Data transmission of the conventional scheme.

Table 1. Transmission procedure of the conventional scheme.

Hop 1	$C_1 \xrightarrow[h_1]{s_1} R$
Hop 2	$R \xrightarrow[h_3,h_4]{s_1} D_1, D_2$
Hop 3	$C_2 \xrightarrow[h_2]{s_2} R$
Hop 4	$R \xrightarrow[h_3,h_4]{s_2} D_1, D_2$

The moving network considered in this paper is illustrated in Figure 2. This cooperative communication system consists of two sources, two destinations and one relay. Each mobile user employs one antenna at the transmitter and receiver, respectively. The relay uses the decoded-and-forward (DF) protocol. Data bits are mapped by the quadrature phase shift keying (QPSK).

When the nodes of moving networks try to communicate each other, the (conventional) transmission scheme shown in Figure 3 can be used to send and receive information without interference. In this figure, the solid lines represent direct-path and the dotted lines represent overhearing by other nodes. This scheme is an expanded version of classic cooperative communication scheme. At the first hop, C_1 can send the information s_1 to relay R where the hop duration is one symbol time. Then the relay sends its received data s_1 to the two destinations (e.g., D_1 and D_2) at the same time during the second hop. The procedure is completed over two hops as shown in Table 1. It is necessary to use four hops if D_1 and D_2 want to receive s_1 and s_2, respectively. At the destinations D_1 and D_2, the received signal y in four hop periods can be written as

$$D_1 \ : \ \begin{cases} y_1 = h_a s_1 + n_1, \\ y_2 = h_{21} s_1 + h_3 s_1 + n_2, \\ y_3 = h_{21} s_2 + n_3, \\ y_4 = h_3 s_2 + h_a s_2 + n_4, \end{cases} \tag{1}$$

$$D_2 \ : \ \begin{cases} y_1 = h_{12} s_1 + \widehat{n}_1, \\ y_2 = h_b s_1 + h_4 s_1 + \widehat{n}_2, \\ y_3 = h_b s_2 + \widehat{n}_3, \\ y_4 = h_{12} s_2 + h_4 s_2 + \widehat{n}_4, \end{cases} \tag{2}$$

where h_a, h_b, h_{12}, and h_{21} represent the direct path channel gains between the source nodes and the destination nodes, as shown in Figure 3. Also, the channels h_1, h_2, h_3 and h_4 are related to the source to relay and the relay to destination. In (2), y_t represents the received signal during the t-th hop period. n and $\widehat{n}$ are also AWGN at each time.

3 THE PROPOSED SCHEME

When the number of users is increasing, the conventional scheme needs additional hops in order to receive data without interference. This decreases data rate. Also, if we use a broadcasting system, it may raise inter-user interference that gives rise to performance degradation. In this section, we propose the enhanced scheme that can minimize data transmission time, eliminate the inter user interference, and get additional diversity through cooperation.

Figure 4 shows the proposed transmission scheme for moving networks. In this scheme, during the first hop period, C_1 sends data s_1 to R.

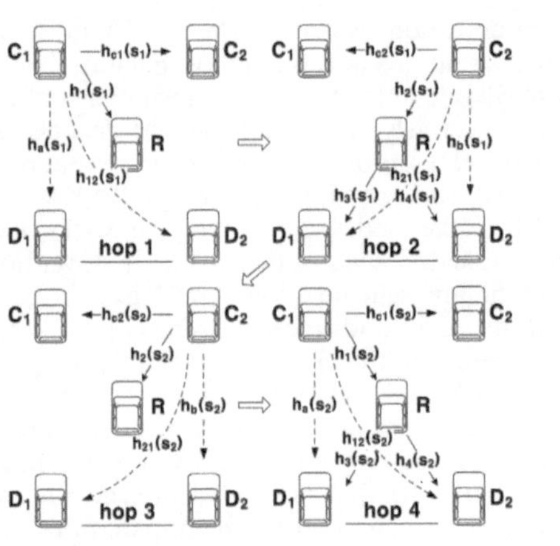

Figure 4. Data transmission of the proposed scheme.

Table 2. Transmission procedure of the proposed scheme.

hop 1	$C_1 \xrightarrow[h_1]{s_1} R$
hop 2	$C_2 \xrightarrow[h_2]{s_2} R,$
	$R \xrightarrow[h_3,h_4]{F\times s_1} D_1, D_2$
hop 3	$C_1 \xrightarrow[h_1]{s_2^*} R,$
	$C_2 \xrightarrow[h_1]{-s_1^*} R,$
	$R \xrightarrow[h_3,h_4]{F\times s_2^*} D_1, D_2$

Then C_2 sends data s_2 to relay and the relay sends C_1's data $F \times s_1$ with channel feedback bit F to the destinations D_1 and D_2 at the same time during the second hop period. Finally, C_1, C_2 and R transmit their encoded data s_2^*, $-s_1^*$ and $F \times s_2^*$ at the same time, as shown in Table 2. Let us consider the STBC pair of C, defined as

$$C = \begin{pmatrix} s_1 & s_2 & F\times s_1 \\ s_2^* & -s_1^* & F\times s_2^* \end{pmatrix}. \tag{3}$$

At D_1, the received signal y in three hop periods can be written as

$$\begin{aligned} y_{1+2} &= h_a s_1 + h_{21} s_2 + h_3(F\times s_1) + n_{1+2}, \\ y_3 &= h_a s_2^* - h_{21} s_1^* + h_3(F\times s_2^*) + n_3. \end{aligned} \tag{4}$$

Stacking y_{1+2} and y_3 in one vector, we obtain

$$\begin{bmatrix} y_{1+2} \\ y_3 \end{bmatrix} = \begin{bmatrix} s_1 & s_2 & F\times s_1 \\ s_2^* & -s_1^* & F\times s_2^* \end{bmatrix} \begin{bmatrix} h_a \\ h_{21} \\ h_3 \end{bmatrix} + \begin{bmatrix} n_{1+2} \\ n_3 \end{bmatrix}. \tag{5}$$

The estimates of s_1 and s_2 can be obtained by linear processing,

$$\begin{aligned} \widehat{s}_1 &= (h_a^* + F\times h_3^*)y_{1+2} - h_{21}y_3^*, \\ \widehat{s}_2 &= h_{21}^* y_{1+2} + (h_a + F\times h_3)y_3^*. \end{aligned} \tag{6}$$

Finally, we can obtain

$$\begin{aligned} \widehat{s}_1 &= \frac{s_1}{\sqrt{3}}\left[\left(|h_a|^2 + |h_{21}|^2 + |h_3|^2\right) + \right. \\ &\quad \left. + F\times 2\mathrm{Re}\{h_a h_3^*\}\right] + \widehat{n}_1, \\ \widehat{s}_2 &= \frac{s_2}{\sqrt{3}}\left[\left(|h_a|^2 + |h_{21}|^2 + |h_3|^2\right) + \right. \\ &\quad \left. + F\times 2\mathrm{Re}\{h_a h_3^*\}\right] + +\widehat{n}_2. \end{aligned} \tag{7}$$

where

Table 3. System parameters for simulations.

Carrier Frequency	5GHz
Sampling Time	1μs (BW = 1MHz)
Mobile Speed	30km/h, 300km/h
Doppler Frequency	139Hz, 1.39KHz

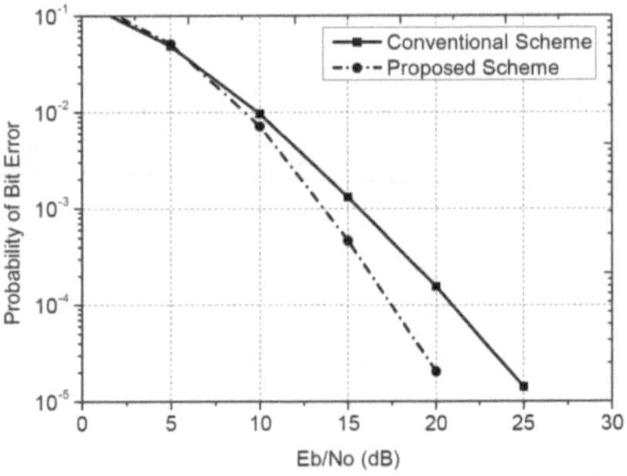

Figure 5. Uncoded performance under multipath fading.

$$\begin{aligned} \widehat{n}_1 &= (h_a^* + F\times h_3^*)n_{1+2} - h_{21}n_3^*, \\ \widehat{n}_2 &= h_{21}^* n_{1+2} + (h_a + F\times h_3)n_3. \end{aligned} \tag{8}$$

In (3), if we choose $F = 1$ or $F = -1$ when $2|\mathrm{Re}\{h_a h_3^*\}|$ is positive or negative, respectively (i.e., $F = sgn|\mathrm{Re}\{h_a h_3^*\}|$), the cross-product term in (7) will always positive and it will be greater than sum of magnitude square of all path gains. Finally, we can get the term $(|h_a|^2 + |h_{21}|^2 + |h_3|^2)$ related to the diversity gain and the term $2|\mathrm{Re}\{h_a h_3^*\}|$ related to the coding gain. It has been well known that path diversity and coding gain improve the effective SNR of fading channels.

4 SIMULATION RESULTS

The bit error probabilities of the proposed scheme were evaluated by computer simulation. The 64-point FFT and the 16-point cyclic prefix (CP) were used for the QPSK-OFDM modulation. Other system parameters are described in Table 3. We considered a two-ray multipath channel model, where path gains are statistically independent. It is assumed that the maximum delay spread is smaller than the CP length and all nodes have perfect channel state information. Various scenarios were examined according to the quality of channel (e.g., the average received SNR) between cooperating users.

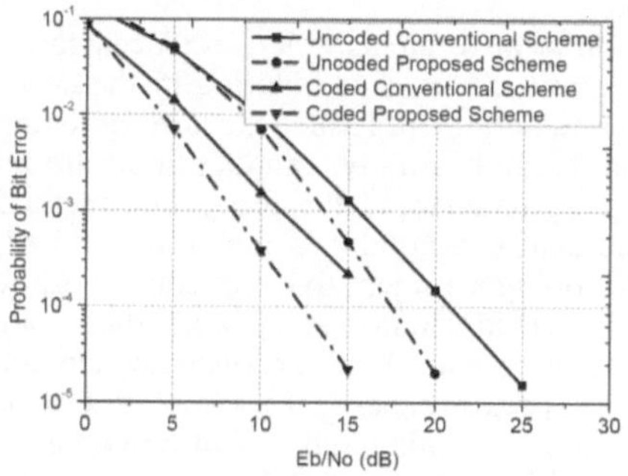

Figure 6. Performance of the coded and uncoded systems.

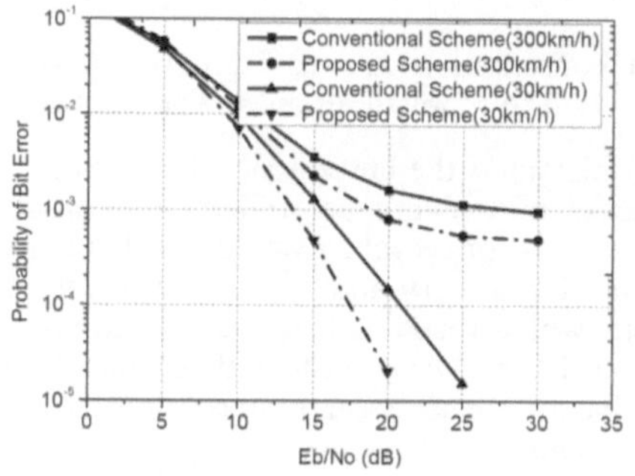

Figure 7. Performance at various velocities.

Figure 5 shows the uncoded bit error probabilities of the proposed and conventional schemes. It is shown that the uncoded proposed scheme clearly outperforms the conventional scheme. When compared with the conventional scheme, the proposed scheme provides an E_b/N_0 gain of 5dB for a bit error rate (BER) of $P_b = 10^{-4}$. Also, the proposed scheme has one hop faster transmission time than the conventional scheme.

Figure 6 shows the BER performance when $r = 1/3$ convolution coding is used for both systems. Both schemes get performance improvement because of coding gain. Still the proposed scheme provides an E_b/N_0 gain of 5dB at $P_b = 10^{-4}$. Figure 7 shows the BER comparison according to various mobile speeds. When the relative mobile speed is 30km/h, the performance trend of two schemes is similar to the previous cases. But when the relative mobile speed is 300km/h, the signal is distorted by Doppler spread and thus the performance is saturated at about 20dB.

5 CONCLUSIONS

In this paper, new cooperative transmission scheme has been proposed for wireless moving networks. To improve the performance, we configured the virtual MIMO transmission scheme using the cooperation and the STBC technique. Especially, the proposed scheme aims to maximize diversity gain. Moreover, it also acquires the coding gain. Thus, the proposed scheme has strong tolerance against the Doppler effect. When compared with the conventional scheme, the proposed scheme reduced transmission time from four hops to three hops by using the virtual MIMO transmission scheme. So the transmission efficiency has been improved without additional bandwidth. Finally, simulation results have shown that the proposed scheme has better performance (one hop faster and 3.5dB gain at 10-4) than the conventional scheme.

ACKNOWLEDGMENTS

This work was supported by Institute for Information & Communications Technology Planning & Evaluation (IITP) grant funded by the Korean Government (MSIT) (No. 2021-0-00368, Development of the 6G service targeted AI/ML-based autonomous-regulating medium access control (6G STAR-MAC)).

REFERENCES

Celebi M.E, Sahin S., and Aygolu U., (2006). Increasing diversity with feedback: balanced space-time block coding, *Proceeding of ICC'06*.

Laneman J. N., Tse D. N. C., and Wornell G.W., Cooperative diversity in wireless networks: Efficient protocols and outage behavior, *IEEE Trans. on Info. Theory*, 50–12 (2004) 3062–3080.

Politis C., Oda T., Dixit S., Schieder A., Lach H. Y., Smirnov M. I., Uskela S., and Tafazolli R., (2004). Cooperative networks for the future wireless world, *IEEE Comm. Mag.* 42-9 (2004) 70–79.

Reichman A. (2008). Cooperative communication overview, *Proceeding of the IEEE International Conference on Microwaves, Communications, Antennas and Electronic Systems*.

Sendonaris A., Erkip E., and Aazhang B., (2003). User cooperation diversity – Part I. System description, *IEEE Trans. on Comm.*, 51–11, 1927–1938.

System Innovation for an Artificial Intelligence Era – Lam et al. (Eds)

Research on recycle and innovation of discarded desks and chairs

Li-Mei Chen
Department of Visual Communication Design, Tainan University of Technology, YongKang, Tainan, Taiwan
Department of Digital Media Design, Asia University, Taichung, Taiwan

Cheng-Jia Chuang
Department of Industrial Design, National Yunlin University of Science and Technology, Yunlin, Taiwan

ABSTRACT: Taiwan is currently facing a severe declination of birthrate as well as replacement of old desks and chairs, resulting in many discarded wooden desks and chairs accumulating in the corner of elementary and junior high school campuses, to reuse wooden material and reduce the need of logging to produce products, this research utilize the method of Action research. Firstly, conduct literature review and field investigation to carry out current situation survey, to understand the processing method for recycling of discarded wooden material as well as desks and chairs. Secondly, conduct test and experiments using desks and chair to redesign and repurpose, developing a project that integrates both forest conservation education and the recycling of reusable materials called "Action story box", the remaining wood shaving are then use to produce "Forest paper" and "Action story box" products, which contain action story box, innovative notebook and pencil , further incorporating them into school and involving students in the creative redesign of discarded desks and chairs, the learning and design outcomes of the students not only enable the reuse of discarded desks and chairs but also inspire children towards creative reuse of waste and raise awareness about forest conservation. The established "Recycling for creative design of discarded desks and chairs" can be applied to related issues.

1 INTRODUCTION

In the early days of Taiwan, trees were cut down to make wooden desks and chairs (referred to as desks and chairs). Over time and with declining birth rates, the old and worn-out desks and chairs are gradually being phased out and transferred to corners of school campuses, awaiting disposal and recycling by cleaning squads, leading to a waste of wooden material. Additionally, in recent years, with the accelerated urbanization, various climate disasters such as earthquakes, typhoons, floods, droughts, etc., have threatened the environment, impacting the natural and economic development of cities. Therefore, scholars, scientists, designers, and the public have become more aware of environmental conservation. Currently, many teams in Taiwan are recycling discarded desks and chairs to give them a second uses, using them as teaching materials for many different courses. In summary, after deepening our understanding of the problem of discarded desks and chairs in schools, our research aims to effectively utilize the waste wooden material by using discarded desks and chairs to create story box related to forest conservation issues. Introducing these products into schools and engage students in redesigning discarded desks and chairs, ultimately inspiring public awareness of forest conservation and the importance of recycling wood to innovate upon it. The purpose is to:

(1) Understand the current situation of discarded desks and chairs as well as the processing methods for recycling waste wooden material.
(2) Develop a story and interactive reading strategy with a focus on forest conservation.
(3) Develop an "Action Story Box" that combines education with waste material innovative and recycling.
(4) Establish "Recycling for creative design of discarded desks and chairs".

2 LITERATURE REVIEW

From the website of the Forest of Central Cross-Island Highway and Development Administration (https://www.fcea.gov.tw/), it can be understood that in 1968, in order to implement 9 years of compulsory education, veterans and their families worked day and night to provide wooden desks and chairs for students nationwide. According to the Veterans Forest Conservation Administration, 90% of the desks and chairs used for "developing and improving national education" were provided by them.

 DOI: 10.1201/9781003514831-60

In recent years, to improve the quality of education, schools have been replacing old desks and chairs that are often unbalance or have uneven tabletops with new ones. The old desks and chairs are often left in dark corners of classrooms, waiting for the school to arrange for the township cleaning squad to dispose and recycle them. For example, in 2020, Taitung County replaced desks and chairs for all first graders, and starting in 2021, they will implement a plan to replace desks and chairs for all elementary and junior high schools in the county.

Meng-Chian Jwo (2017) categorized and analyzed global warming, valuable forests, landscape changes, renewable energy, air pollution, and garbage recycling, created environmental conservation-themed cards, reminding readers that small actions in their surroundings can lead to habitat destruction and urging them to consider whether item is reusable or not.

Currently, most domestic environmental conservation education products primarily consist of picture books or paper-based products, with little utilization of wooden material with volumetric forms in storytelling and presentation. In order to utilize the "Action Story Box" designed from recycled desks and chairs for a secondary purpose, the story content should align with forest conservation issues. The following analysis focuses on relevant forests and wood utilization as well as animal ecology. Utilization of waste wood to reproduce Russet Passer cinnamomeus:

Zhi-Wei Qian (2014) pointed out that while Passer cinnamomeus are commonly seen in urban or rural areas, the similar-looking "Russet Passer cinnamomeus" are nowhere to be found. In 2008, they were listed as a protected species by the Forestry Bureau, with the Taiwan Wild Bird Federation estimating that there are fewer than 500 left nationwide. Russet Passer cinnamomeus in Guanziling, Tainan, almost always use power poles as nesting sites. To conserve Russet Passer cinnamomeuss, the Wild Bird Society cooperated with contractors of the Tsengwen Reservoir to add power poles in the reservoir area and attempt to attract Russet Passer cinnamomeuss to nest and reproduce using "artificial nest boxes," with initial results already achieved. Jing Hui Liao (2019) investigated the reasons behind this, attributing them to the fact that Russet Passer cinnamomeus grow in mountainous areas at altitude of 600 to 2,000 meters, where they frequent villages, surrounding farmland and grasslands. Their main diet consists of plant seeds, grains, and insects. Originally, they bred in tree cavities, but due to their small size and limited number of tree cavities, their competitiveness was poor. Therefore, setting up artificial nest boxes for breeding Russet Passer cinnamomeuss highly dependent on human activities has enabled successful propagation, with community recognition and support being key.

3 RESEARCH METHODS AND PROCEDURES

Interview on Discarded Desk and Chair Related Knowledge: Interviews with elementary school teachers regarding the disposal of discarded desks and chairs in elementary schools on December 30, 2021.

(1) The basement of the elementary school is filled with discarded desks and chairs.
(2) Desks and chairs are prone to dampness and require maintenance.
(3) The disposal method for these desks and chairs is to destroy and get rid of a large number of them.
(4) Secondary damage caused by burning and destruction results in atmospheric pollution.
(5) Our team can access these discarded desks and chairs.

Additionally, on December 6, 2021, the study visited a woodworking master's studio seeking professional guidance on the reuse of wood.

(1) If untreated, raw wood is susceptible to warping (cupping) due to moisture absorption, needing the use of waterproof paint and other tools to compensate for the shortcomings of raw wood.
(2) When constructing box structures, attention must be paid to using 3D software to model internal structures and narratives, repeatedly testing the practicality of appearance and functionality.
(3) Post-processing of discarded wood involves using laser cutting to produce some parts.
(4) Illustration designs can be applied to the surface of the box, utilizing CNC machining for relief.
(5) The box structure is linked to the narrative chapters of the internal storybook.

The method proceeds through four stages of progress:

Stage One: Problem and Situation Investigation

First explore relevant issues and knowledge regarding the reuse of discarded desks and chairs, wood recycling, and forest conservation through literature review and field investigation.

Stage Two: Plan and Strategy Formulation

Based on the identified problems and situations, develop plans and strategies for product development (action story box and innovative notebook). The educational purpose of the story is to use discarded desks and chairs to create "artificial nest boxes" to conserve the habitat for "Russet Passer cinnamomeuss". The target audience includes elementary school children and parent-child education.

Stage Three: Product Development and Testing

In the production process, first test the use of raw wood and study woodworking principles by reading related books. Then attempt to manufacture action story box and utilize the remaining wood shaving to create the " innovative notebook" using papermaking techniques. The experiments and testing of raw wood and residual wood pulp are detailed in in Tables 1 and 2, respectively. The two products of this study are the "Action story box" and the "Innovative notebook".

Table 1. Experimental and testing steps for wood recycling in "action story box" and paper making from waste materials after wood recycling.

"Action Story Box" Testing		Paper Making Testing	
	1.Collect discarded desks and chairs.		1.collects wood shavings from workshop that cut raw wood
	2.Dismantle the desks and chairs with a hammer, and remove nails.	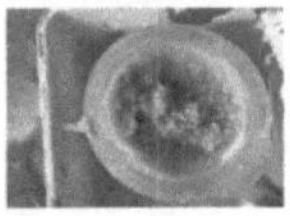	2.Sieve the larger particles of wood shavings to facilitate better paper manufacturing during the paper making process.
	3.Adjust the values for the log and the Z-axis.		
	4.Apply moisture-resistant wood paint.	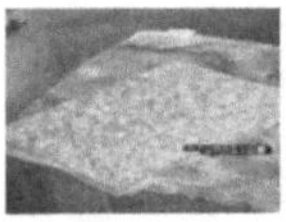	3.Test the pulp composition and disperse the pulp for a certain period of time.

Stage Four: Implementation in Campus and Innovative Design

Introduce the "Action story box" into elementary schools to convey the conservation story of "Russet Passer cinnamomeuss" with education. Through the creativity of students in designing the "innovative notebook" with recycled materials, our team produces the students' designed items, enabling the creative reuse of discarded desks and chairs to continuously cycle and innovate through the program.

Table 2. Production process and finished product of the "action story box".

Shapr3D、Rhinoceros 3D Software Application Modeling		
	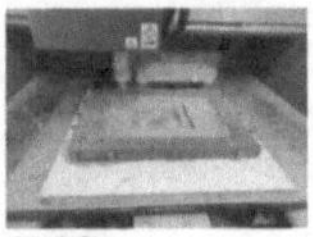	
Using a saw table with homemade tools to vertically cut the tabletop	CNC machining	Hand planed the edges

Using a scroll saw machine to cut the remaining waste wood	Smoothing the edges	Completing the semi-three-dimensional story cover

4 RESULTS

Production and Gameplay Settings of the "Action Story Box": The "Action Story Box" aims to educate everyone about ecological conservation and the importance of waste utilization through storytelling and interactive experiences. Children can read the chapters of the story and then manually assemble key parts (Table corners) to create desks and chairs, adding an element of fun. This nurture children's problem-solving skill, logical thinking, and spatial awareness (Figure 1). The target audience is elementary school children and parent-child education.

Each chapter of the story utilizes wood carving to create a semi-three-dimensional cover. Upon opening the cover, the story unfolds in the form of

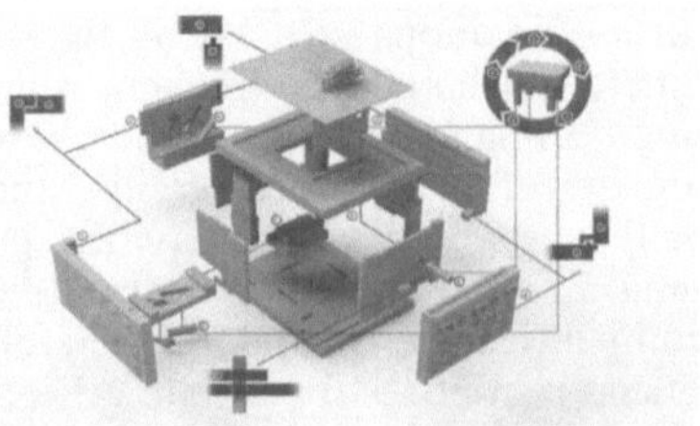

Figure 1. Overall story and desk assembly association diagram for each chap.

fold-out pages. The story depicts Taiwan's forests being heavily deforested in the past, leaving many forest-dwelling animals homeless, such as Russet Passer cinnamomeuss. Additionally, due to declining birth rates, a large number of desks and chairs face disposal.

In the story, trees in the forest are cut down by workers, leading to landslides. However, the wood carried down by the mud was picked up by A bird enthusiast, with some crafted into a nesting place for Russet Passer cinnamomeuss, and others into desks and chairs for the school, However, after several years, even these desks and chairs faced the fate of being discarded.

The Storybook Fold and Illustration Design picture: The story foldout is designed to complement the Outline of the "Action Story Box" with different folding and pull-out mechanisms, along with assembly parts for the tabletop. Here are the details:

Table 3. Story and Interactive Strategies for the "Action Story Box".

Story Outline Content and Interaction Linking Strategy	Completion of Semi-Three-Dimensional Wood Carving
Chapter One: Silent Lament (Reasons for Deforestation) Synopsis: Trees in the forest are being cut down, causing Russet Passer cinnamomeuss to leave their habitat. The wood is then transported to be made into desks and chairs. **Interaction:** After reading Chapter Five, students can assemble the components from Chapter One, including table legs, Passer cinnamomeuss, and the "Voice of the Forest" team logo.	
Chapter Two: That Night, the Mud Flowed (Consequences of Deforestation) Synopsis: Deforestation causes soil erosion and mudslides, carrying away remnants of trees. **Interaction:** After reading Chapter Two, students can flip open the assembly kit containing table legs and Passer cinnamomeus nest box components for assembly.	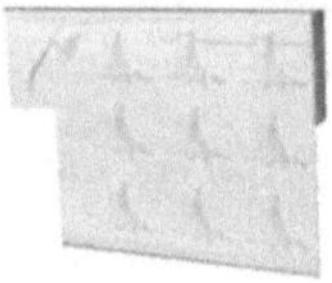
Chapter Three: Melodic Harmony (Human Utilization of Wood for Forest Eco-Restoration) Synopsis: A bird enthusiast picks up a piece of driftwood and crafts it into an artificial nest for Russet Passer cinnamomeuss, which is then tied to a utility pole. **Interaction:** Students can insert the Passer cinnamomeus nest box from Chapter Two onto a corner utility pole and wait for the return of the Passer cinnamomeuss.	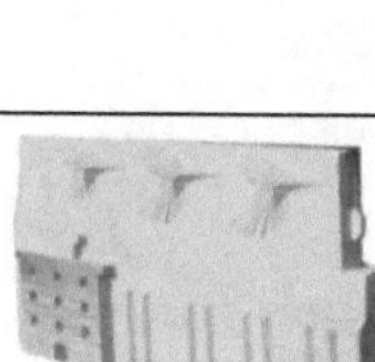

(*continued*)

Table 3. Continued

Story Outline Content and Interaction Linking Strategy	Completion of Semi-Three-Dimensional Wood Carving
Chapter Four: Crackling Flames (Humans Burning Wood) Synopsis: The neighboring elementary school replaced its desks and chairs, and the discarded ones were gathered for burning. **Interaction:** After reading, the students pulled out the table legs that were not yet burnt in the flames.	
Chapter Five: Memories of the Bell Synopsis: The desks and chairs in the bonfire are burned, and the days of joyfully accompanying students in their studies are now shattered in the flames. Memories of the cheerful atmosphere in the classroom come flooding back, inspiring a desire to continue contributing to education (thus, utilizing the desks and chairs again, allowing the discarded items to return to education once more...). **Interaction:** Pulling out the top pangolin assembly board, the back features illustrations of children studying in class. Next, pull out the assembly board for the first chapter and assemble it, forming the mission of reintegrating the wood back into education.	
Assembly of Completed Action Story Product	

4.1 *Chapter One: The silent lament*

Designed as a 7-page, 6-fold horizontal right-pull foldout with vertical text layout, the content depicts a tranquil forest suddenly disrupted by the buzzing sound of chainsaws. Passer cinnamomeuss flee from the felled trees as workers in work uniforms indiscriminately cut down large quantities of trees in the forest. They then select useful logs to transport downhill, where they are transformed into raw wooden desks and chairs, leaving behind broken and damaged wood remnants (Figures 2 and 3).

Figure 2. Illustration for Chapter One "Silent Lament" - 6-fold horizontal right-pull page.

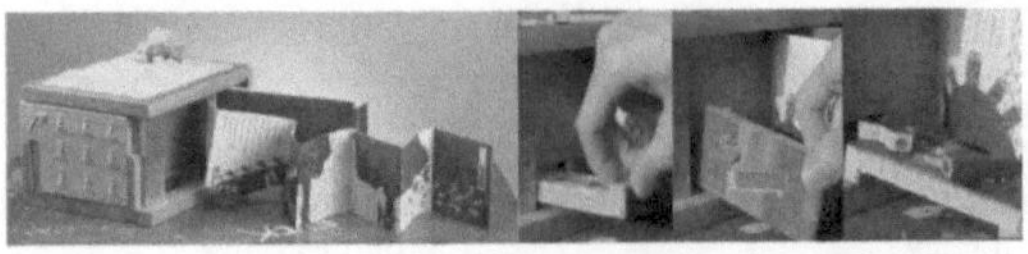

Figure 3. Real scene and assembly parts of chapter one.

4.2 *Chapter two: That night, The mud flowed*

Designed as a 5-page, 4-fold vertical downward-pull foldout with vertical text layout, this chapter describes a night when heavy rain began to fall, rumbling loudly as it loosened the soil and rocks, causing a mudslide. This resulted in a large amount of abandoned wood and driftwood being washed downstream. The disaster prompted local residents to consider the importance of forest conservation and soil conservation (Figures 4 and 5).

Figure 4. Illustration for Chapter Two "That Night, The Mud flowed" - 5-page 4-fold downward pull page.

Figure 5. Real scene and assembly parts of chapter two.

4.3 *Chapter three: Melodic harmony*

Designed as a 4-page, 3-fold horizontal right-pull foldout, this chapter depicts how the trees from the mountains turned into driftwood due to mudslides. A bird enthusiast, picks up one of these driftwood pieces and skillfully crafts it into an artificial nest for Passer cinnamomeuss. Using wires, the worker then securely attaches the nest to an electrical pole (Figures 6 and 7).

Figure 6. Illustration for chapter three "Melodic Harmony" - 4-page 3-fold horizontal right-pull page.

Figure 7. Real scene and Assembly Parts of Chapter Three.

4.4 *Chapter four: Crackling flames*

Designed as a single page with a flip-down section for text, this chapter integrates with the "Action Story Box" wooden carving mechanism. The neighboring elementary school decided to replace its old desks and chairs, discarding them in a corner of the classroom. However, as renovations were needed, the discarded desks and chairs were sent to a recycling plant and burned, crackling loudly. It was then that the observant children rescued the burnt legs of the desks (Figures 8 and 9).

Figure 8. Illustration for chapter four "Crackling Flames" - single-page with flip-down text.

Figure 9. Real scene and Assembly Parts of Chapter Four.

4.5 *Chapter five: Memories of the bell*

Designed as a single page, this chapter complements the wooden carving on the previous page of the "Action Story Box" and integrates with the assembly of various chapter parts to form the tabletop, creating a cohesive visual narrative. The desks and chairs in the bonfire are burnt to a charred black, symbolizing the days they spent accompanying students in their studies, now shattered by the flames (Figures 10 to 12.). Memories of joyful classroom atmospheres flood back. Inspired, he decides to contribute further to education. (realizing that even discarded desks and chairs can be repurposed to serve education once again...).

Figure 10. Illustration for Chapter Five "Memories of the Bell" - Single-page illustration.

Figure 11. Real scene and Assembly Parts of Chapter Five.

The Production of Promotional Items and Tools from Remaining Wood Shavings After Wood

Figure 12. Complete assembly of the "Action Story Box".

Carving: This study collects the leftover wood shavings from the wood carving of the "Action Story Box" and transforms them into paper materials named "Forest Paper." These materials are used to create the necessary promotional items for the overall project execution, including the story foldout book, "Innovative Notebook," project introduction, and pencils. The production process involves collecting the residual wood chips and using a shredder to mix them with pulp to create handmade paper (Table 4 and Figure 13).

Table 4. Production process and finished products of story folding book paper material, "Innovative Notebook," introduction, and pencils made from wood chips after wood carving.

Collect wood shavings	Using a shredder to make pulp	Paper making	Drying paper	Completion of "Forest paper"
		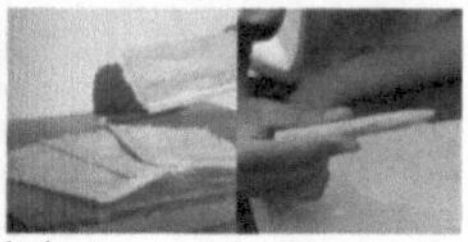		
Pressing the "Innovative Notebook" and pencils using a mold	Gluing the cover of the "Innovative Notebook" with "Forest Paper"	Completion of the "Innovative Notebook" and pencils		

Figure 13. "Innovative Notebook", Introduction, and Pencils.

Circular Design for Repurposing Discarded Desks and Chairs: To achieve a circular system that combines forest conservation education with waste material recycling, an on-campus educational approach has been implemented using recycled wooden desks and chairs to create the "Action Story Box" as an example, evoking emotions in students and encouraging them to unleash their creativity by designing recycled wooden products. Practical designs are selected from the students, and products are produced using recycled desks and chairs, thus closing the recycling loop and fostering an understanding of waste reuse among students. Ultimately, this initiative aims to raise awareness of forest conservation education and the circular significance of waste recycling among the public. The "Circular Design Mechanism" (Figure 14) is established with the following steps:

(1) Creating the "Action Story Box" from Discarded Desks and Chairs: Designing discarded desks and chairs, plan for disposal, into educational "Action Story Boxes" to be used in classrooms.
(2) Introducing the "Action Story Box" into School Education: Using the "Action Story Box" to interact with students and tell stories that help them understand the reasons behind discarded desks and chairs, evoking memories and emotions related to these items.
(3) Distribute the "Innovative Notebook" for Student to Design: Teaching students through collaborative drawing sessions to creatively sketch designs for products using recycled wood. Our team then selects practical designs for production.
(4) Recycled Desks and Chairs Reutilization: Implementing this circular design allows us to transform discarded desks and chairs into a variety of products, enhancing students' knowledge and application of waste utilization while promoting environmental awareness and protection through ecological recycling.

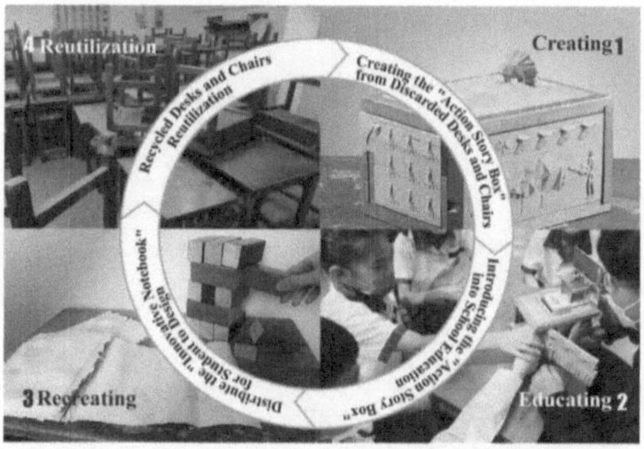

Figure 14. The "Circular Design Mechanism".

5 CONCLUSION

Early in Taiwan's history, forests were cut down to create wooden desks and chairs, which over time became worn and destined for disposal. This research aims to effectively utilize these discarded desks and chairs by developing the "Action Story Box," which integrates environmental education and the recycling of waste materials. The goal is to see a mountain of discarded desks and chairs from elementary schools reused through the "Action Story Box" and to gradually instill in students an awareness of forest conservation as they engage in manual reading with the box. Students are encouraged to unleash their creativity by designing the discarded desks and chairs, and the team implements the production of student-designed products.

While there are limitations to mass-producing the "Action Story Box," the team can replicate several units and introduce them to different schools at different times. By combining storytelling and symbolism in the "Action Story Box," children can learn about the current environmental challenges and the importance of conserving ecosystems while having fun. This initiative aims to build awareness that simple actions and thinking can contribute to forest conservation, empowering individuals to take part in environmental protection.

REFERENCES

Executive Team: Cheng-Jia Chuang, Tzu-Li Yen, Jian-Yu Chen, Ya-Fen Guo, Chen-Yu Wang.

Jing Hui Liao (Russet Sparrow Conservation) *Bait Stations, Rice Bran Spreading, Friendly Farming Methods, and Nest Boxes Progress Together*. https://e-info.org.tw/node/224797

Meng-Chian Jwo. (2017). *Teaching Cards on the Theme of Environmental Conservation*. Master's thesis, National Taichung University of Education, Taiwan. https://hdl.handle.net/11296/z4ytng

Taitung County Government Education Bureau. *Promoting Friendly Campuses and Improving Classroom Desk and Chair Equipment for Elementary and Junior High Schools*. https://www.taitung.gov.tw/News_Content.aspx?n=13370&s=59299

Zhi-Wei Qian, (2014). *Our Island Animal Chapter 754*. https://ourisland.pts.org.tw/content/1578

System Innovation for an Artificial Intelligence Era – Lam et al. (Eds)
© 2025 the Author(s), ISBN: 978-1-032-84452-7

Manufacturing supply chains for electric vehicles solutions: Identifying themes and opportunities using a systematic literature review

Adrian E. Coronado Mondragon
School of Business and Management, Royal Holloway University of London, Egham, UK

ABSTRACT: The push for electric vehicles (EVs) around the world represents an important change of paradigm for the automotive industry. The manufacturing of EVs require access to materials and energy to manufacture complex components. This work investigates current developments in EVs manufacturing supply chains using a systematic literature review to identify patterns, research trends and foundational themes. A search of the SCOPUS database identified 188 articles published between 2003 and early 2024. Bibliometric tools including VOSviewer and SciMat were used in this work. The analysis showed that almost two thirds of the articles have been published in the last 5 years. The analysis revealed clusters with core items comprising supply chains, automotive industry, lithium-ion batteries, sustainability, electric vehicles, and permanent magnets. Among the core items identified, supply chains had the strongest and most numerous connections that included links to optimization, hybrid vehicles, environmental management, battery manufacturing, batteries supplies, and rare earth among others. The findings can be used to define future research paths and to assist policy makers in the formulation of initiatives for the consolidation of EVs and electromobility manufacturing supply chains.

1 INTRODUCTION

Electric vehicles (EVs) in general play a key role in the sustainable and green agendas adopted in several countries around the world. For example, Canada has developed policies to support end-to-end supply chains for EVs (Invest in Canada, 2024). In countries like China, automobile electrification is seen as a viable strategy for achieving carbon neutrality goals (Qiao *et al.* 2024).

The turn to electromobility is based on advanced technologies (e.g., lithium-ions batteries) that tied its development to many functional materials with problematic supply (Schuster *et al.* 2023). Electromobility and the evolution of autoparts and components for EVs represent a significant industry that is poised to continue its steady growth and adoption in various countries.

In their work, Torres-Rivera *et al.* (2021) expressed that electromobility is emerging as a viable solution to tackle demand for environmentally sustainable transportation by integrating renewable energies into the supply chain. The authors claimed that by lowering greenhouse gas emissions and minimizing the environmental effect of the sector, incorporating renewable energy into the supply chain may support sustainability at every stage.

The manufacturing supply chain of EVs require a comprehensive approach that embraces sustainability principles. Torres-Rivera (2021) indicated that in the case of producing EVs and charging infrastructure, it implies the acquisition of raw materials, logistics, and waste management. Gulley *et al.* (2019) highlighted that the Chinese government has listed EVs manufacturing as a strategic emerging industry for addressing the ambitious targets of reducing CO_2 emissions and fossil fuel imports. In China, companies like Biyadi (BYD) sells battery-electric vehicles for the equivalent of US$26,000 (Flanakin, 2024). BYD is an example of vertically integrated manufacturing supply chain, as the organization makes its own batteries, semiconductors, and seal upholstery, and its nearly 30,000 patents owned or filed puts the company ahead of many competitors (Flanakin, 2024).

In India, government initiatives are directing expenditure of significant resources and guiding efforts to achieve EVs technology growth in the country in pursuit of reaching emission net zero by the year 2070 (Prasad *et al.* 2023).

Although the automotive sector has received most of the focus because of EVs and electromobility solutions, electrification is expected to impact all transportation modes (Schuster *et al.* 2023). Nonetheless, the size and the economic importance of the automotive sector makes it among the prime industries where the development of new battery and powertrain technology can be implemented. However, to make this happen it is important to develop a supply chain network that will be able to support the transition from internal combustion engine (ICE) vehicles to EVs and other electromobility solutions. The electric

DOI: 10.1201/9781003514831-61

vehicle market is very competitive which includes first movers like Tesla, BYD and legacy automakers VW and GM among others. Nonetheless, promising companies like Arrival entered administration as it failed to launch electric van (de Prez, 2024). Companies like Lordstown Motors and Electric Last Mile Solutions have already declared bankruptcy.

Given the important role supply chains play in the future of EVs and electromobility solutions, two research questions have been formulated based on the trends and patterns observed in the literature, these are as follows:

RQ #1: What publication patterns appear in extant EVs manufacturing supply chains?

RQ #2: What are the foundational themes, recent research trends, and potential future frontiers for EVs manufacturing supply chains research?

To present this systematic literature review involving the manufacturing supply chains of EVs, first we present an overview of important developments in the area with an especial emphasis on the supply networks responsible for sourcing the parts and components needed by the industry. The next section outlines the methodology used which is based on bibliometric tools. The analysis answers the research questions by presenting the performance analysis of the field, followed by the analyses of major themes, their development over time, and future research directions.

2 AN OVERVIEW OF THE SUPPLY CHAIN FOR ELECTRIC VEHICLES AND ELECTROMOBILITY SOLUTIONS

Around the world it is possible to find examples about projects addressing the manufacturing supply chains of EVs and electromobility solutions. India is a country that is investing heavily in growing its EV technology. In their work, Prasad *et al.* (2023) identified the risks and barriers linked to supply chain, technology, finance, and policy for the promotion of growth of EVs in India. The researchers found that the barriers comprising availability of battery packs, raw materials, charging network, and interoperability of batteries are the most critical supply chain barriers to implementing EV technology in India.

Batteries represent one of the main components, as lithium-ion batteries are extensively utilized in consumer electronics (Qiao *et al.* 2024). He and Chen (2024) recognized that Chinese electric vehicle battery manufacturers are experiencing difficulties evaluating suppliers' green collaboration. The authors developed a system that included: *a)* the ability to meet buyer needs; *b)* the relationship between manufacturers and suppliers; *c)* responsibility and commitment, and *d)* the implementation of environmental management. According to the authors, the first one, the ability to meet buyer needs, is the most important criterion.

In the UK, Bridge and Faigen (2023) stated that efforts have been focusing on seeking to onshore lithium-ion battery production and build a national battery supply chain. The authors emphasized the importance of engaging governmental, industrial and research actors in securing mineral materials for battery manufacturing as well as developing battery manufacturing capacity, as the country is not part of the EU anymore and there is the perceived 'global battery race' where new and old partners are determined by geopolitical interests.

In North America, significant developments are taking place in setting manufacturing centers in various locations in Canada, the US and in particular the state of Texas and Northeast Mexico. The state of Texas is fast become a major hub for the manufacture of EVs. According to the Office of Governor, in the year 2019 the electric-transportation sector contributed close to $700 million to the Texas economy (Texas Economic Development & Tourism – Office of Governor, 2021). The same office highlighted that in the state of Texas more than 1,200 electric-transportation-related businesses employed more than 7,100 people and by the end of the year 2024, the electric transportation workforce in the state is expected to expand to 13,500. According to the same source, with the aid of EV policies, the state of Texas is achieving substantial progress in this fast-growing sector. The state of Texas is already host to major OEMs, first-tier manufacturers and many other suppliers in the automotive sector representing a robust supply chain.

It can be appreciated that the move towards electrification in the automotive sector is happening in many regions around the world. The next section presents the methodology used in this work.

3 METHODOLOGY

To carry out this research, we performed a search in early March 2024 on the SCOPUS Database, using the following keyword strings: "Electric Vehicle" OR "EV" AND "supply chain" AND "manufacturing". This initial search resulted in a data set of 188 publications. The number of articles identified in the search somehow demonstrated that this area is relatively young with most of the articles published in the past five years. Two different tools bibliometrics were used including VOSviewer and SciMat. VosViewer is a software tool used for making and visualizing bibliometric networks (VOSviewer, 2023). SciMat SciMAT (Science Mapping Analysis software Tool) is an open-source science mapping software tool which includes among other things methods, algorithms, and measures for mapping workflows, from preprocessing to the visualization of the results (SciMat, 2024).

4 ANALYSIS

The analysis revealed that out of the 188 articles published, almost two-thirds of them have been published in the last five years. The histogram showing the distribution of publications by year is presented in Figure 1.

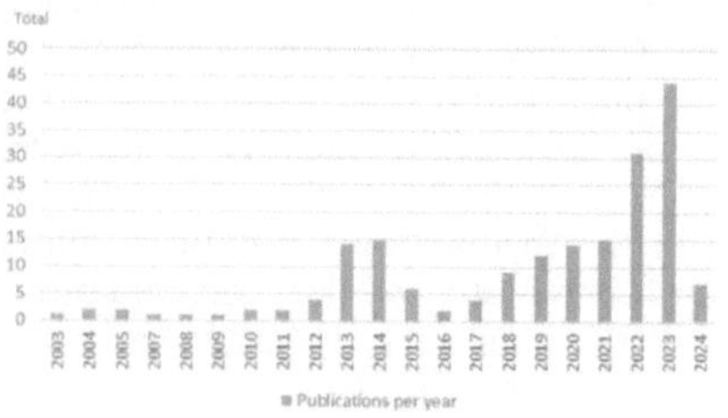

Figure 1. Histogram showing the number of publications between the years 2003 – 2024.

In Figure 1 it is possible to appreciate that the year 2023 was the most prolific year in terms of articles published on EVs and the related manufacturing supply chains. The analysis gave the opportunity to identify 1682 items, then the minimum number of occurrences of an item was set at five, hence using the VOSviewer tool allowed the identification of 61 items that met the threshold. To perform the analysis based on bibliographic data or text data, very often data cleansing is required. VOSviewer's thesaurus file was used to merge different variants of keywords like lithium-ion battery to lithium-ion batteries or electric automobiles to electric vehicles.

The results of the cluster analysis identified six major groups, with cluster one grouping 14 items, cluster two 14 items, cluster three 13 items, cluster four 11 items, cluster five 7 items and cluster six 2 items. Along with the cluster analysis, the use of density visualization gives the opportunity to identify the items that had more occurrences during the search as shown in Figure 2.

With the use of density visualization, it is possible to appreciate items with the highest occurrence levels like supply chains, electric vehicles, lithium-ion batteries, automotive industry, lithium-ion batteries, and life cycle. Then, using cluster analysis it was possible to identify items that emerged as the most important based on their scored weight. In cluster 1, supply chains arose as the most important one among the other 14 items grouped in that cluster. In cluster 2, automotive industry was the item that showed the highest score based on weight recorded among 14 items. In cluster 3, lithium-ion batteries and recycling were the items that showed the highest scores. In cluster 4, among 11 items, the item that showed the highest score based on weights recorded was sustainability. In cluster 5, electric vehicles arose as the item that showed the highest score among seven items. In cluster 6, permanent magnets were the item that showed the highest score among two items.

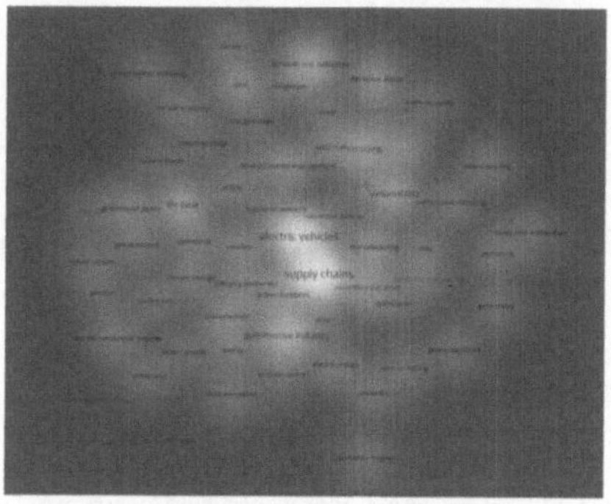

Figure 2. Representation of the density visualization showing items with more occurrences in the search.

Clusters are non-overlapping and cluster numbers are an example of an attribute (VOSviewer, 2023). The use of cluster analysis in conjunction with network visualization gives the opportunity to clearly distinguish the items with the highest scores based on weights while showing the and highest number of links in each cluster. Figure 3 shows the network visualization diagram for the items analyzed.

In the network visualization map depicted in Figure 3 it is evident that the weight of an item is directly related to its importance; hence, items with a higher weight are regarded as more important than those items with lower weights. Following this principle, it is possible to identify a handful of items with the highest weights in the analysis. Here, supply chains emerged as the item with the highest weight and links of all the items analyzed. The item supply chains showed strong connections to a significant number of items within its own cluster, number one, but also with other items present in the other five clusters. Figure 4 shows the network visualization for the item supply chains.

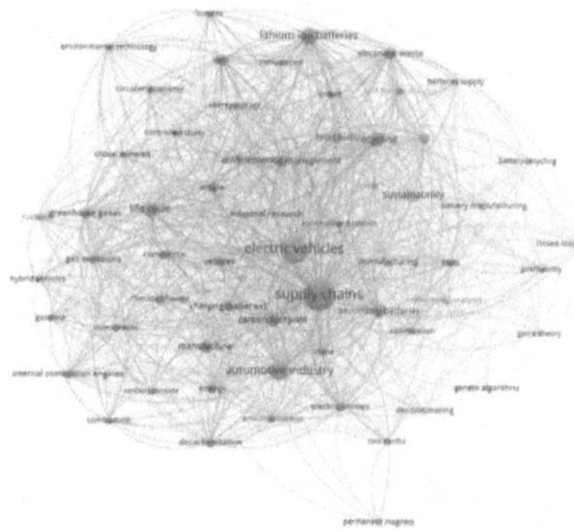

Figure 3. Network visualization diagram based on the number of item occurrences identified in the analysis.

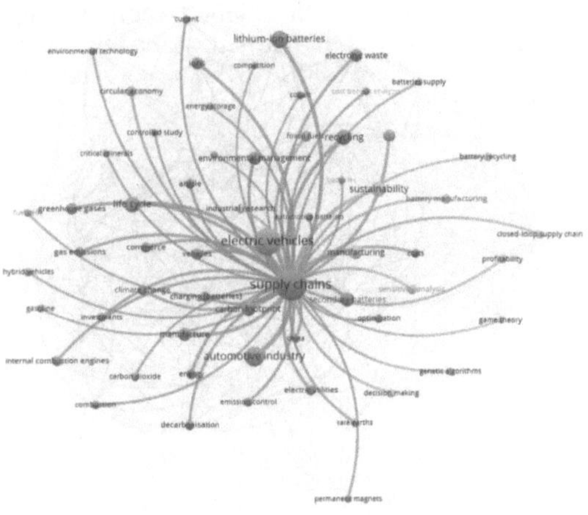

Figure 4. Supply chain was identified as the item with the highest weight and number of connections.

The network diagram clearly shows the item supply chains is strongly linked to items within its own cluster like optimization, electric utilities, decarbonization, emission control, and energy among others. Additionally, the supply chains item connections with cluster two showed links to hybrid vehicles, gasoline, internal combustion engines, and environmental technology among others. The connections of supply chains with cluster number three showed links to automotive batteries, environmental management, energy storage, and electronic waste among others. Furthermore, connections with cluster number four showed links to battery manufacturing, battery recycling, industrial research, and decision making among others. Connections of supply chains with cluster number five showed links to circular economy, batteries supply, and competition. Finally, supply chains connections with cluster number six showed links to rare earths and permanent magnets.

As manufacturing clusters/hubs of EVs/electromobility manufacturing continue to grow and consolidate in different parts of the world, the results of the systematic literature review analysis show important items linked to supply chains supporting the manufacturing of EVs/electromobility solutions. It is worth pointing out that during the analysis, China was the only country related to an item identified using the condition of five occurrences. This somehow supports the idea that China is a major player in the development and manufacturing of EVs and electromobility solutions.

5 CONCLUSION

This work attempts to review and map the growing body of literature comprising the manufacturing supply chains of EVs, by relying on published work during a span of 21 years (2003 – 2024) and containing 188 articles from the SCOPUS database. The use of systematic literature review analysis gives the opportunity to summarize the body of articles found in the search by integrating various elements and therefore giving the opportunity to identify foundational themes, recent research trends, and potential future frontiers shaping the EVs manufacturing supply chains.

The findings of this systematic literature review analysis can be used to identify future research paths and give the possibility to assist policy makers in the formulation of industry initiatives that can contribute to the consolidation and growth of EVs manufacturing and electromobility solutions and some of its key components like lithium-ion batteries and all the components and materials behind it. Additionally, it may help to identify trends in the development of efficient and cost-effective recycling methods that will reflect on sustainable management of EVs, electromobility solutions and the batteries powering them.

Electric vehicles represent a growing industry that will continue to draw the attention from business, practitioners, and academia. Hence, the supply chain plays a fundamental role in the support and future consolidation of EVs and electromobility solutions. This is highly relevant as the globalized and extended model of supply chains is replaced by regional supply networks that will support the EVs manufacturing hubs in China, North America, and Europe.

REFERENCES

Bridge G., Faigen E. (2023). Lithium, Brexit and Global Britain: Onshoring battery production networks in the UK, *Extrac Ind Soc*, 16, 101328.

De Prez, M. (2024). Arrival enters administration as it fails to launch electric van. *Fleet News*, 5 February 2024. https://www.fleetnews.co.uk/news/arrival-enters-administration-as-it-fails-to-launch-electric-van

Flanakin D. (2024). *As China Builds Yugos, EVs May Be the New Edsels*. February 27. https://realclearwire.com/articles/2024/02/27/as_china_builds_yugos_evs_may_be_the_new_edsels_1014698.html

Gulley A.L., McCullough E.A., Shedd K.B. (2019). China's domestic and foreign influence in the global cobalt supply chain. *Resour. Pol.* 62, 317–323. https://doi.org/10.1016/j.resourpol.2019.03.015

He Q.-R., Chen P.-K., (2024). Developing a green supplier evaluation system for the chinese electric vehicle battery manufacturing industry based on supplier willingness to participate in green collaboration. *IEEE Trans on Eng Man*, 71, 3098–3116.

Invest in Canada. 2024, Accessed: 01/03/2024. https://www.investcanada.ca/industries/ev-supply-chain?

Mane P., Digalwar A., Adhithyan C.S. (2023). Modeling the supply chain risk and barriers to electric vehicle technology adoption in India. *IFIP Adv in Inf and Comm Tech*, 699 AICT, 202 – 214.

Qiao D., Ma Y., Bao Y., Hong Y., Batunacun, Narenmandula, Dai T. (10 March 2024). Exploring the potential impact of electric passenger vehicle battery recycling on China's cobalt supply and demand under the goals of carbon peaking and carbon neutrality during 2010–2060. *Journ. Clean Prod.* Volume 444, 141139. https://doi.org/10.1016/j.jclepro.2024.141139

Schuster V., Ciacci L., Passarini F. (2023). Mining the in-use stock of energy-transition materials for closed-loop e-mobility, *Res. Policy.*, Volume 86, Part B, October 2023, 104155.

SciMat. (2023). *Science Mapping Analysis Tool.* https://sci2s.ugr.es/scimat/description.html

Texas Economic Development & Tourism – Office of Governor. (2021). June. https://gov.texas.gov/uploads/files/business/Automotive_2021.pdf

Torres-Rivera A.D. McNamara Valdes A.d.J., Florencio Da Silva R. (2023). The *Resilience of the Renewable Energy Electromobility Supply Chain: Review and Trends. Sustainability*, 15, 10838. https://doi.org/10.3390/su151410838

VOSviewer, Visualizing scientific landscapes. (2024). https://www.vosviewer.com/

System Innovation for an Artificial Intelligence Era – Lam et al. (Eds)
© 2025 the Author(s), ISBN: 978-1-032-84452-7

AI-empowered IoV architecture for real-time application in heterogeneous 6G

Linyu Zhu
College of Software Engineering, Sichuan University, Chengdu, China

Song Bao & Shoupeng Lu
College of Computer Science, Sichuan University, Chengdu, China

ABSTRACT: With the advent of the Internet of Everything (IoE) era, the Internet of Vehicles (IoV) is playing a pivotal role in modern society. IoV facilitates communication and data exchange between vehicles, equipping both vehicles and drivers with enhanced information and intelligent functionalities. This contributes to improved traffic safety, efficiency, and convenience, as well as positively impacting urban traffic management and sustainable development. However, connected car technologies are currently confronted with challenges such as big data management, high latency, and mobility issues. One of the most promising solutions to these challenges is the utilization of Cloud-Edge-Terminal collaborative architectures and next-generation wireless network technologies. In this paper, we first introduce the concept of the Cloud-Edge-Terminal collaboration architecture, provide a detailed overview of its components, including terminals, edge devices, and cloud platforms, and analyze their collaboration methods and techniques based on their characteristics. Furthermore, we explore the application scenarios of artificial vehicle networking and illustrate its practical applications in real-world contexts. Finally, we discuss the key challenges and opportunities that Cloud-Edge-Terminal collaborative computing may face, along with potential directions for future research.

1 INTRODUCTION

The evolution of artificial intelligence (AI) and IoV, along with advancements in next-generation wireless technology like 6G (Yrjölä, *et al.* 2023) and sensor devices, are driving the development of traditional connected car services towards intelligent connected car services. The AI-based Internet of Vehicles (AIoV) connects vehicles to the internet, enabling data exchange and communication between vehicles, infrastructure, and service providers. It leverages wireless communication technology and intelligent sensing devices to provide innovative tools and solutions for stakeholders in road traffic safety.

In the context of IoV, AI plays a crucial role. AI analyzed large amounts of IoV data, through techniques such as machine learning and deep learning (Liang, *et al.* 2022), to extract useful information and knowledge.

However, the transmission latency of Level 3 and above autonomous driving, which ranges from 10 to 20 milliseconds (Litton, *et al.* 2024), and the high-speed processing of real-time data at speeds of up to 2GB/s pose challenges in terms of data transmission, fast processing, and data storage while ensuring privacy protection (Khan, *et al.* 2024). Traditional cloud computing paradigms are no longer suitable for AIoV. Centralized approaches cannot provide the required scalability and responsiveness, and they would impose significant traffic loads on communication networks.

Therefore, a new Cloud-Edge-Terminal AI inference architecture has been proposed. This architecture distributes data processing across the cloud, edge devices, and vehicles themselves to achieve more efficient real-time applications. By offloading some tasks from the cloud to edge devices or vehicles themselves, data transmission and processing latency can be reduced. Additionally, combining edge computing with AI algorithms can improve the response speed and decision accuracy.

The rest of this article is organized as follows. In section 2, we propose the Cloud-Edge- Terminal collaborative architecture and collaboration approaches, and discuss the key technologies involved. In section 3, we deeply introduce the application scenarios of AIoV based on Cloud-Edge-Terminal collaborative architecture, and demonstrate its application possibilities in the future. In section 5, we discuss the future directions of this research theme. Finally, we conclude the article.

 DOI: 10.1201/9781003514831-62

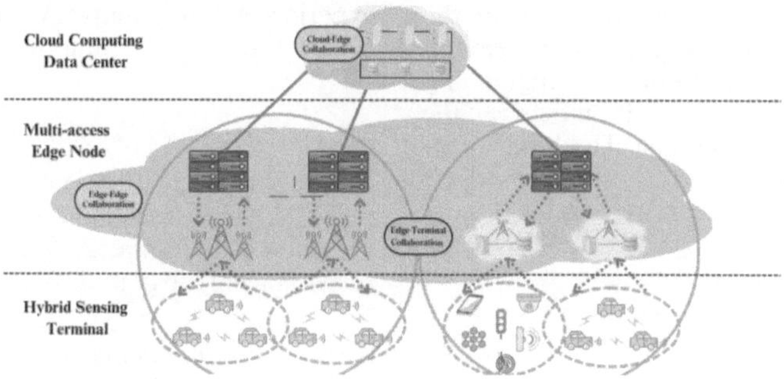

Figure 1. Cloud-edge-terminal collaborative architecture.

2 CLOUD-EDGE-TERMINAL COLLABORATION ARCHITECTURE AND TECHNOLOGY

In this section, we will give a brief introduction to the proposed Cloud-Edge-Terminal collaboration architecture and discuss the specific key technologies involved.

2.1 *Cloud-edge-terminal collaboration architecture*

The proposed architecture shown in Figure 1 extends from the data sources located in intelligent vehicles or their surroundings to the service providers. It consists of the following major components:

- *Hybrid Sensing Terminal:* Hybrid sensing terminals include the vehicle end and the road end. The vehicle end refers to the sensing devices distributed inside or outside of the connected and autonomous vehicles (CAVs), which are used to obtain information like the distance and driving speed. On the other hand, the road end refers to the Roadside Units (RSUs), which are used to obtain information about road congestion, vehicle density, traffic signal status, and other road conditions.
- *Multi-access Edge Node (MEN):* There are two types of edge nodes (Toczé and Nadjm-Tehrani 2018). The first type is the Edge Server, which connects to other parts of the network, including the cloud. Edge servers are typically located in fixed positions and have relatively high computing capacity (Ahn, *et al.* 2023). The second type is the Coordinator Edge, which acts as coordinators among other edges, collecting and managing information from other edges. Coordinator edges can be mobile and typically have lower computing capacity and bandwidth compared to edge servers.
- *Cloud Computing Data Center:* Cloud computing data centers comprise a multitude of high-performance servers, storage devices, network equipment, and power supplies, enabling efficient data storage and complex data analysis. Notable examples of cloud computing data centers include Amazon data centers, and Apple data centers.

In IoV, considering various Quality of Service (QoS) jobs and heterogeneity of devices, types of collaborations among clouds, edges and terminals are proposed to suit different situations, which is also shown in Figure 1.

2.1.1 *Cloud-oriented cloud-edge collaboration*

The substantial data generation at terminals necessitates intensive computational resources for model training, resulting in large-scale models. However, edge devices are unable to independently complete complex model training tasks because of limited resources. To address this issue, a cloud-based cloud-edge collaboration architecture is proposed. In this framework, edges are responsible for real-time analysis. The cloud is responsible for training more complex machine learning models, and regularly updating the models of the edges. Offloading the training tasks to the cloud reduces the computational workload on the edge nodes.

2.1.2 *Edge-oriented cloud-edge collaboration*

Uploading a large amount of data to the cloud cannot meet the user's need for privacy protection (Cao, *et al.* 2021). As a result, an edge-oriented cloud-edge collaboration architecture is proposed. The cloud is responsible for training generic models, and transmitting the models to edge devices. Edge devices fine-tune the models based on local data. The main difference between edge-oriented cloud-edge collaboration and cloud-oriented cloud-edge collaboration is whether the edge devices optimize the models. This requires edge to have high computing and storage capabilities. It is especially useful for offline scenarios as it reduces the continuous need for internet connectivity and real-time performance.

2.1.3 *Dual-oriented cloud-edge collaboration*

Mixing edge-oriented and cloud-edge collaboration, the edge can send parameter updates based on local data back to the cloud to adjust global patterns or update generic models. Additionally, the cloud can aggregate and consolidate data from the edge and end devices to periodically or on-demand train and update generic models (Wang, *et al.* 2022). This bidirectional collaboration between the cloud and edge aims to combine the cloud's capabilities in generalized learning with the edge's capabilities in personalized learning, enabling more efficient collaborative learning and inference (Cao, *et al.* 2022). This bidirectional collaboration mode allows the entire system to adapt to evolving application demands more flexibly and efficiently.

2.1.4 *Vertical edge-edge collaboration*

Traditional centralized resource management strategies are facing significant challenges (Benzaoui 2020). To address this issue, researchers propose an edge-to-edge collaborative framework. In this framework, edges are equal to execute subsets of a computational task and can communicate with each other through wireless networks. This strategy aims to achieve a balanced distribution of load and optimal utilization of computing resources through intelligent task assignment and resource scheduling mechanisms. The key advantage of this architecture lies in its efficient adaptability to heterogeneous edge devices and its ability to significantly enhance task processing speed through collaboration between edge devices without relying on centralized cloud services.

2.1.5 *Horizontal edge-edge collaboration*

Resource configuration is of paramount importance in edge computing. Therefore, a horizontal edge-to-edge collaboration mechanism is proposed, where edge devices include a coordinator edge and an edge server. The coordinator edge intelligently assigns tasks to the most suitable edge server based on QoS and real-time resource availability. Moreover, the coordinator edge can serve as a parameter server, responsible for aggregating parameters from edge servers and performing updates to the global model. The core advantage of this mechanism lies in its exceptional scalability, enabling it to adapt to the growth and variability of device numbers and to achieve flexibility in resource management (Chen, *et al.* 2022).

2.1.6 *Hybrid edge-edge collaboration*

In this framework, it is the mix of Vertical Edge-Edge Collaboration and Horizontal Edge-Edge Collaboration.

2.1.7 *Edge-terminal collaboration*

In this collaborative approach, terminal devices gather data and transmit it to the edge (Shao, *et al.* 2021), where centralized computing is performed on the received data. The outcomes and instructions are then sent back to the terminal devices, which carry out the specific operations directed by the edge. Edge-Cloud collaboration is well-suited for tasks that require quick response times but have low computational demands. Because edge experiences lower network latency and relies less on the Internet, resulting in faster response times.

2.2 *Technologies in collaborative architecture*

In this part, we will primarily focus on key technologies deployed in collaborative architectures. They are shown in Figure 2. We did not consider the relevant technology in the field of communications.

2.2.1 *Task offloading*

Offloading refers to the process of transferring tasks from resource-poor locations to resource-rich ones. The aim is to maximize resource utilization and latency. Tasks can be offloaded from the edge to the cloud, from the terminal to the edge, or between edges. Thus, which task to offload and where to offload them, i.e., the offloading strategy, is a hot topic of current research. When considering task offloading, the strategy needs to balance request scheduling latency and service migration costs, ensuring an effective balance between the real-time response capabilities of edge devices and the powerful computing power of cloud servers. So, most of them are mathematical optimization (Hsieh, *et al.* 2023, Wang, *et al.* 2019) and deep learning methods (Fan, *et al.* 2023).

2.2.2 *Model compression*

Model compression can be a part of the preparation of the offloading process. By reducing the model size, it aims to improve communication efficiency while maintaining the quality of model as much as possible. There are two main methods of model compression (Ahmmed, *et al.* 2021): frontend compression and backend compression. Frontend compression techniques (such as knowledge distillation, lightweight network design, and filter-level pruning) allow compression without altering the original network structure. Backend compression techniques, including low-rank approximation, unstructured pruning, parameter quantization, and binary networks, focus on minimizing the model size but may involve modifications to the network structure.

2.2.3 *Federated learning*

Federated Learning (FL) enables AI model training across distributed clients in edge computing, with each client updating the model using local data and uploading parameters to a central server for aggregation into a global model. Two primary aggregation strategies are employed (Ahmed, *et al.* 2021): synchronous, which waits for all clients, and asynchronous (Chen, *et al.* 2020), which processes updates as they arrive. The choice depends on network health and client capabilities, with researchers now exploring hybrid strategies for efficiency and model performance (Feng, *et al.* 2023).

2.2.4 *Transfer learning*

Transfer learning addresses virtual concept drift (VCD) (Arif and Perera 2023) in IoV to solve the problem of AI model performance degradation due to time-varying input characteristics. It is basically divided into three methods (Zhang, *et al.* 2022): model based, data based and query, which focus on model structure adjustment, source domain data utilization and online information retrieval respectively. The development direction aims to improve cross-domain adaptability and generalization ability achieve fast and effective knowledge transfer (Tao, *et al.* 2022), and pay attention to deep knowledge transfer and privacy security (Lin, *et al.* 2023).

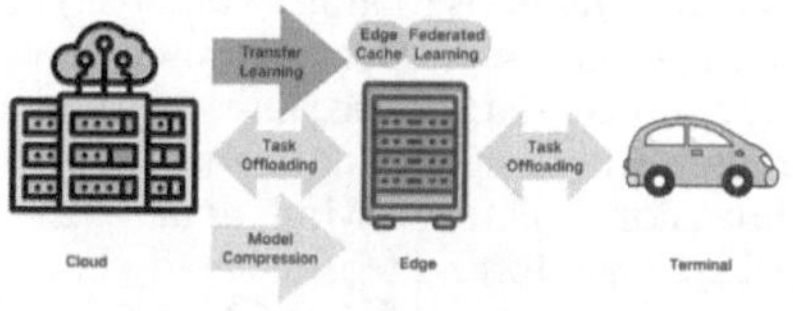

Figure 2. Key technologies in collaborative architecture. in Cloud-Edge-Terminal Collaboration Architecture.

2.2.5 *Edge caching*

Edge caching refers to the practice where edge servers deployed at or near the user's location cache certain content to reduce retrieval latency and alleviate network congestion (Zeng, *et al.* 2023). Cache usually cache supporting data, which is road condition data from other CAVs or sensors. Because of limited cache capacity of edge servers, cache replacement schemes are very important to improve cache hit rates and reduce service latency. Currently, cache replacement schemes can be classified based on different criteria, such as the type of cached content, the caching policy, and the caching location. In particular, reinforcement learning-based cache replacement strategies are representative, as they have the learning ability to perceive and capture large-scale user requests in dynamic networks. (Sun, *et al.* 2021)

3 AIOV USE CASES AND APPLICATION SCENARIOS

In this section, we will present several specific application scenarios of Cloud-Edge-Terminal collaboration in AIoV, as shown in Figure 3.

3.1 *Autonomous driving*

In a Cloud-Edge-Terminal collaborative autonomous driving system, we highlight two of the most widely researched functions:

- *Intelligent Driving Decision-Making:* The cloud-edge-terminal architecture optimizes intelligent driving decisions through distributed computing resources. For instance, VECFrame utilizes vehicular edge computing to process data from on-vehicle sensors, enabling indirect communication and edge-enabled cooperative object detection (Tang, *et al.* 2021). It allows data fusion between vehicles, enhancing the perception of the surrounding environment, leading to more accurate driving decisions. Moreover, real-time processing at the edge reduces reliance on cloud resources, lowers latency, and improves the timeliness of decision-making.
- *Route Planning*: In terms of route planning, the cloud-edge-device architecture facilitates collaborative planning of autonomous vehicle routes by deploying sensors and computing units along roadsides. As described in (Huang, *et al.* 2021), edge computing technology can analyze and calculate road traffic conditions, large-scale vehicle guidance strategies, and intelligent traffic scheduling, thereby optimizing the vehicle's route. This collaborative planning not only reduces the need for individual vehicle sensors but also lowers the computational costs per vehicle, while also providing the possibility for remote manual takeover in complex traffic scenarios.

3.2 *Intelligent transportation*

Cloud-Edge-Terminal based collaborative transportation system offers intelligent, efficient, and safe traffic management solutions. The applications of cloud-edge collaboration in Intelligent Transportation are as followed:

- *Signal Light Control:* In urban road networks, signal lights are critical for traffic flow and safety. A new modeling approach using Timed Petri nets ensures efficient signal control (Huang, *et al.* 2021), prioritizing emergency vehicles and reducing delays. This system analyzes traffic data to optimize signal timing, enhancing traffic efficiency and safety at busy intersections.
- *High-definition Travel Video Surveillance:* Road infrastructure and vehicle-mounted cameras capture real-time video data (Liu, *et al.* 2022). Edge computing and cloud storage analyze the data for vehicle behavior analysis (Gao and Zhang 2023), incident detection (Wan, *et al.* 2021), and identification of abnormal behavior. This enhances traffic safety, accident prevention, and provides effective monitoring for traffic management.
- *Drones Utilization:* Thanks to the high mobility, easy deployment, and robust interconnectivity of drones, they can gather data from wireless sensors within RSUs. In Intelligent Transportation Systems (ITS), sensors generate a vast amount of data daily.

A flying strategy using drones as data collection and computing nodes for RSUs has been proposed (Arthurs, *et al.* 2022). Additionally, drones have importance in ground traffic monitoring, especially to prevent emergencies. Applications such as accident reporting, aerial police oversight, and mobile RSUs can be facilitated by drones equipped with Edge Intelligence (EI) (Gong, *et al.* 2023).

3.3 *Intelligent ambulance*

Intelligent Ambulances use AIoV technology to monitor traffic conditions, plan optimal routes, and reduce delays (Zhai, *et al.* 2021). IoV shares ambulance location with nearby vehicles via wireless transmission (Ahmed, *et al.* 2023), ensuring a clear path. Integration with traffic signals provides advance notifications, minimizing waiting time at intersections. Medical devices in the ambulance capture vital signs and patient information, transmitted to the cloud platform for doctors and hospitals via IoV. This enables doctors to prepare for treatment and gives medical staff access to historical records. Overall, Cloud-Edge-Terminal collaboration and AIoV enhance emergency medical services and patient outcomes.

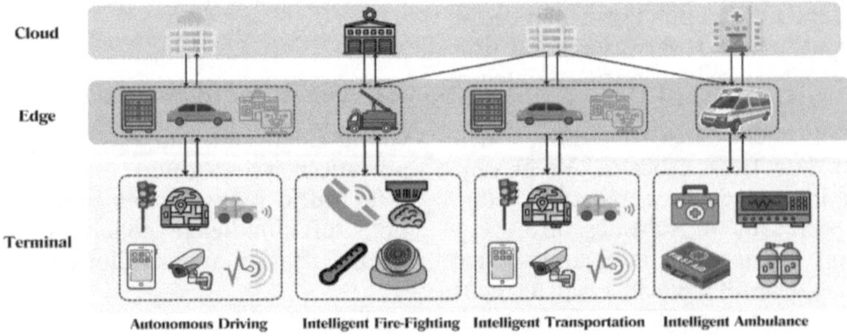

Figure 3. Four potential AIoV scenarios supported by cloud-edge-terminal collaborative architecture.

3.4 *Intelligent fire-fighting*

During emergency response, the vehicle starting system on the edge, mounted on the fire truck, automatically plans the arrival time and required route based on user requirements. This planning incorporates geographic information, route planning, driving navigation, and time estimation to efficiently reach the fire scene. The cloud center, connected to the fire truck through cloud servers and platforms, enables real-time monitoring(Dai, *et al.* 2023), remote command, data analysis, and forecasting. It provides fire risk assessment and decision support. The edge side of the system collects sensor data and engages in real-time communication through devices mounted on fire trucks, such as smoke sensors and OBUs. It transmits data and interacts with the cloud for data transmission and instruction interaction. Terminals, including intelligent networked walkie-talkies, water pump systems, tank systems, and others, facilitate data transmission, reception, display, and operation.

4 FUTURE DIRECTIONS AND WORKS

4.1 *Tensor-based multimodal data compression*

The exponential growth of IoV devices is a cause for concern. The multimodal data generated by in-vehicle cameras, sensors, and other devices, such as sensor readings, and location information (Yang, *et al.* 2024), are not only voluminous but also diverse in types. So efficient and sophisticated multimodal analysis (MA) methods (Yang, *et al.* 2024) are needed. Tensors can be used to represent and store multidimensional data, especially those involving temporal sequences, spatial data, or spatiotemporal multidimensional features (Man, *et al.* 2023, Yang, *et al.* 2022). Therefore, a tensor-based general data processing framework for MA is promising, enabling deep compression and acceleration of model networks on edge devices (Liu, *et al.* 2020). Since compression is based on approximations that may reduce accuracy, balancing precision and compression is an optimization problem and a direction for future research.(Dai, *et al.* 2023)

4.2 *Large and small models collaboration*

Driven by computational resources, large-scale models have evolved into a new round of "arms race" in the field of AI. The energy consumption and edge deployment issues that arise from it have also limited the practical application of large-scale models. Therefore, the current demand lies in developing a computational collaborative model that can fully leverage the advantages of cloud computing while also mobilizing the agility of edge-side computing. The ultimate goal is to advance AI towards cognitive capabilities and intelligence close to human levels. In the future, as we commit to the industrial application of large-scale models, high-precision compression of large-scale models and the realization of cloud-edge-device collaborative evolution will become new research issues (Ding, *et al.* 2022).

4.3 *Hardware-Software integrated edge intelligence acceleration*

Traditional hardware design (Aman and Sikdar 2021) and algorithm development are often conducted independently, leading to performance bottlenecks and resource waste. Therefore, future AIoV systems need to consider the algorithm requirements (Lata, *et al.* 2022) during the hardware design phase and enable algorithms to fully leverage hardware features for efficient computation and data processing. First, hardware design should take into account the computationally intensive demands of algorithms, such as integrating dedicated AI accelerators in vehicle sensors and actuators to support real-time data processing and decision-making (Rodríguez, *et al.* 2022). Second, the development of intelligent algorithms should be closely coordinated with the hardware architecture to achieve algorithmic hardware-friendliness (Oltra-Oltra, *et al.* 2021). This includes algorithm lightweighting, parallelization, and integration of hardware acceleration libraries, enabling algorithms to run on limited hardware resources while maintaining low latency and high throughput.

5 CONCLUSION

This paper envisions AIoV enhancing system performance through a Cloud-Edge-Terminal architecture, addressing data processing and energy constraints. It introduces key collaboration technologies for advanced inference tasks and sustainable services, presents practical applications for daily convenience, and identifies challenges and opportunities for further research in AIoV collaborative systems.

REFERENCES

Ahmed K. M., Imteaj A. and Amini M.H. (2021). *IEEE International Conference on Machine Learning and Applications.* 1146–1152

Ahmed S. T., Basha S. M., Ramachandran M., et al., (2023). *IEEE Internet of Things Journal.* 10, 11497–11506

Ahmmed A., Paul M. and Pickering M. (2021). *Data Compression Conference.* 335–335

Ahn Y. J., Kim M., Lee J., et al. (2023). *IEEE Network.*1–8

Aman M. N. and Sikdar B., (2021). *IEEE Globecom Workshops.* 1–6

Arif M. and Perera D. G. (2023). *IEEE International Sympo-sium on Circuits and Systems.* 1–5

Arthurs P., Gillam L., Krause P., et al., (2022). *IEEE Transactions on Intelligent Transportation Systems*, 23, 6206–6221

Benzaoui N. (2020). *Optical Fiber Communications Conferen-ce and Exhibition.*1–3

Cao K., Hu S., Shi Y., et al., (2021). *IEEE Transactions on Industrial Informatics.* 17, 7806–7819

Cao Y., Lv P., Zhou X. et al. (2022). *IEEE Conference on Telecommunications, Optics and Computer Science.* 24–28

Chen L., Wu J., Zhang J., et al. (2022). *IEEE Transactions on Cloud Computing.* 10, 2451–2468

Chen Y., Sun X. and Jin Y. (2020). *IEEE Transactions on Neural Networks and Learning Systems.* 31, 4229–4238

Dai C., Huang Y. and Chien W.-C., (2023). *Future Generation Computer Systems*, 143, 51–60

Dai C., Zhang Y. and Zheng Z., (2023). *IEEE Transactions on Intelligent Transportation Systems.* 1–12

Ding C., Zhou A., Liu Y., et al., (2022). *IEEE Transactions on Cloud Computing.* 10, 1489–1499

Fan W., Yang F., Wang P., et al. (2023). *IEEE Transactions on Network and Service Management.* 20, 4478–4493

Feng C., Yang H. H., Wang S. et al. (2023). *IEEE Transac-tions on Communications.* 71, 7008–7022

Gao Y. and Zhang B., (2023). *International Conference on Image Processing, Computer Vision and Machine Learning.* 1100–1103

Gong T., Zhu L., Yu F. R. and Tang T., (2023). *IEEE Transactions on Intelligent Transportation Systems.* 24, 8919–8944

Hsieh C. Y., Ren Y. and Chen J. C. (2023). *IEEE Transac- tions on Wireless Communications.* 22, 7158–7171

Huang J., Xiao S., Xiao L., et al., (2021). *2021 IEEE 5th Information Technology,Networking,Electronic and Automation Control Conference.* 137–142

Huang L., Liu Y. and Huang D., (2021). *CAA Symposium on Fault Detection, Supervision, and Safety for Technical Processes.* 1–5

Khan L. U., Elhagry A., Guizani M. and Saddik A. E. (2024). *IEEE Internet of Things Magazine.* 7, 120–126

Lata K., Chhabra S. and Saini S., (2022). *IEEE Consumer Electronics Magazine.*11 92–97

Liang S., Wu H., Zhen L., et al., (2022). *IEEE Transactions on Intelligent Transportation Systems.* 23, 25345–25360

Lin J. W., Kim T., Tu H. J. and Kuo P. H. (2023). *Interna-tional Conference on Data Science and Information Technology.* 56–60

Litton M. L., Drusinsky D. and Michael J. B. (2024). *IEEE Reliability Magazine.* 1, 37–46

Liu G., Shi H., Kiani A., et al., (2022). *IEEE Transactions on Intelligent Transportation Systems.* 23, 12027–12038

Liu H., Yang L. T., Ding J., et al., (2020). *IEEE Transactions on Computational Social Systems.* 7, 873–885

Man C., Chang C., Ding C., et al., (2023). *Design,Automation & Test in Europe Conference & Exhibition.*1–2

Oltra-Oltra J. A., Madrenas J., Zapata M., et al., (2021). *IEEE International Symposium on Circuits and Systems.*1–5

Rodríguez A., Otero A., Platzner M. and Torre E. d. l., (2022). *IEEE Transactions on Computers*, 71, 2903–2914

Shao J., Zhang H., Mao Y. and Zhang J. (2021). *IEEE International Conference on Acoustics, Speech and Signal Processing.* 8488–8492

Sun Y., Chen W. and Yuan Y. (2021). *2021 3rd International Academic Exchange Conference on Science and Technology Innovation.* 521–524

Tang S., Chen B., Iwen H., et al. (2021). *IEEE International Conference on Edge Computing.*68–77

Tao Y., Qiu J. and Lai S. (2022). *IEEE Transactions on Cloud Computing*, 10, 56–71

Toczé K. and Nadjm-Tehrani S. (2018). *Wireless Communications and Mobile Computing.* 1–23

Wan S., Xu X., Wang T. and Gu Z., (2021). *IEEE Transac- tions on Intelligent Transportation Systems.* 22, 4487–4495

Wang G., Luo T., Gao C. et al. (2022). *3rd International Conference on Computer Information and Big Data Applications.* 1–9

Wang Q., Guo S., Wang Y. and Yang Y., (2019). *IEEE In- ternational Conference on Communications.* 1–6

Yang W., Wang Z., Ni J., et al., (2022). *IEEE International Conference on Image Processing.* 3738–3742

Yang Z., Xiong B., Chen K., et al. (2024). *IEEE Internet of Things Journal.* 11, 256–267

Yrjölä S., Matinmikko-Blue M. and Ahokangas P. (2023). *Joint European Conference on Networks and Communications & 6G Summit.* 705–710

Zeng F., Zhang K., Wu L. and Wu J., (2023). *IEEE Transactions on Vehicular Technology.* 72, 2468–2481

Zhai Y., Xu X., Chen B., et al., (2021). *IEEE Network.* 35, 190–196

Zhang Y., Zhang K. and Wu F. (2022). *IEEE 22nd International Conference on Communication Technology.* 373–378

System Innovation for an Artificial Intelligence Era – Lam et al. (Eds)
© 2025 the Author(s), ISBN: 978-1-032-84452-7

Machine learning-based prediction model of thermal-induced error in mechanical press

Tsung-Liang Wu, Yu-Tang Cao & Jing-Yao Huang
Department of Mechatronics Engineering, National Kaohsiung University of Science and Technology, Kaohsiung, Taiwan

ABSTRACT: High-speed press machines are widely used in product manufacturing, including the creation of lead frames, electromagnetic shielding, armatures, and relay holders. While there is much research being conducted on mold design, the deterioration of molds or components, and quality control of workpieces, there are fewer studies focused on the deviation of the mold clamp issue caused by the thermal effect in the press machine. This deviation can be an inevitable contributor to the aforementioned research topics. In this particular study, we received a solid model of a crank-type mechanical press from a press manufacturer. We defined thermal sources in five critical locations and used the thermal-deformation analysis results of the finite element model of the press to train a prediction model for revealing the deviation of mold clamping position. To conduct a more comprehensive study, we computed five clamping positions to obtain the three-dimensional tilting posture of the clamping molds. We used the Levenberg-Marquardt optimizer to evaluate our prediction model, aiming for training and validating accuracy between 71% to 80% to prevent overfitting. The final testing accuracy was obtained with an accuracy of 87%.

1 INTRODUCTION

Since lots of industry have embraced the manufacturing processes, such as pressing, stamping, piercing, and drawing, because of the benefits of high efficiency and accuracy, research topics related to mechanical press attract many scholars. These topics include deterioration of molds, prediction model for forming, punch forces and strain, and so on. The topic of degradation of quality of mold is a well popular topic (Fang, Nagato *et al.* 2020, Gitschel, Hering *et al.* 2023). Gitschel *et al.* (Gitschel, Hering *et al.* 2023) proposed a counter-pressure concept to alleviate the stress triaxiality, which is proofed that is an indicator for the evolution of ductile damage in mold. For a precision product, such as non-grain-oriented electrical steel lamination, the clearance between production molds does matter. Fang *et al.* (Fang, Nagato *et al.* 2020) conducted experiments with extreme small clearance of 1 and 5 μm while controlled parameters of punching speed, counterforce, and others, which can help motor designers to balance the punch-induced changes of magnetic properties and dimensional accuracy.

The prediction model for forming can benefit for mold designer (Li, Huang *et al.* 2019, Kubik, Hohmann *et al.* 2021, Gan, Li *et al.* 2022). Gan *et al.* (Gan, Li *et al.* 2022) proposed the Digital Twin model, which can acquire the real-time punch force to update the friction coefficients through the Particle Swarm Optimization with Differential Evolution algorithm. Their goal is to predict the mold wear by continuous updating the forming parameters with real status. Kubik *et al.* (Kubik, Hohmann *et al.* 2021) focused on the feature extraction from the force-displacement curves for establishing correlations with produced components. Li *et al.* (Li, Huang *et al.* 2019) estimated the energy demand for drawing process by identifying the major contributor with parameters of punch velocity, blank holder force, and drawn depth. By adopting the results, a grey relational analysis was used to build the correlation between the forming quality and the process energy.

The punch forces and strain are the cause-effect factors for the pressing motion. Gan *et al.* (Gan, Li *et al.* 2022) acquire continuously punch forces inline because that stress varies in accordance with the wear condition of mold, i.e. altering of the contact area. Fang *et al.* (Fang, Nagato *et al.* 2020) concluded that higher punch speed will result in the higher strain-hardening level, which implies the higher shear strain rate, i.e. a cleaner cut but the clearance ratio should set to 2% or below.

Although all the scholars pointed out the crucial parameters which should be addressed. However, the effects of thermal-deformation are rare considered in the research work. Although the thermal-deformation error are quite attractive topic in the field of turning machine in the last two

 DOI: 10.1201/9781003514831-63

decades, the capability of precision manufacturing of the mechanical press is revealed within a decade. In this study, we will study the thermal-induced error in mechanical press. Then, we will propose an efficient approach to train the prediction model for the position error and deviations of error on the platform will also discussed.

2 EXPERIMENTS AND SIMULATIONS

This segement interprets the prelimiary thermal experiments conclusion, simulations results, and predicition model training process. For the experiment conclusion, we can comment that the anomaly themral conditions are difficut to replicate in the commercial press machine. Figure 1 denotes the set up locations of thermcouples and IR sensors and the psoition deviations are barely observed from optical ruler. Thus, we need a more comprehensive approach to study the themral-induced error in mechanical press.

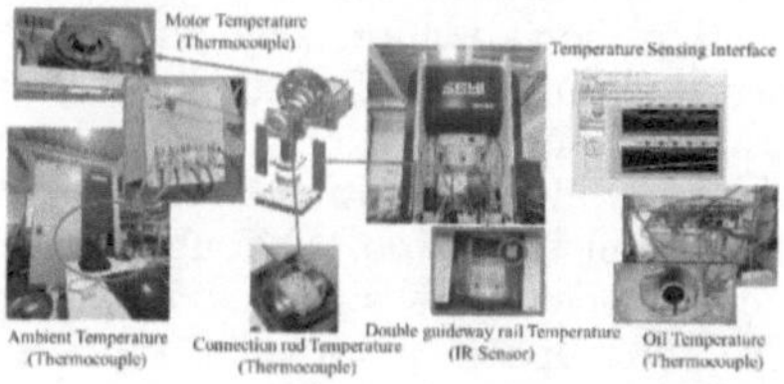

Figure 1. Experimental temperature measuring locations in the mechanical press.

The most effective way to fulfill the above need is the finite element analysis (FEA). We have received the solid model of one type of mechancial press from Taiwan-based manufacturer, Shieh yih Machinery Industry Co., LTD. Figure 2 presents the proper boundary conditions and five thermal locations, including Left and right guideways rails (T_{11} and T_{12}), connection rod (T_{21}), ball socket (T_{22}), and Motor (T_{31}), respectively. There are also five moniting locaitons for position errors on the upper platform and marks as N_i, i=1,2,...,5. The diagram of the platform and initial coordinates are also shown.

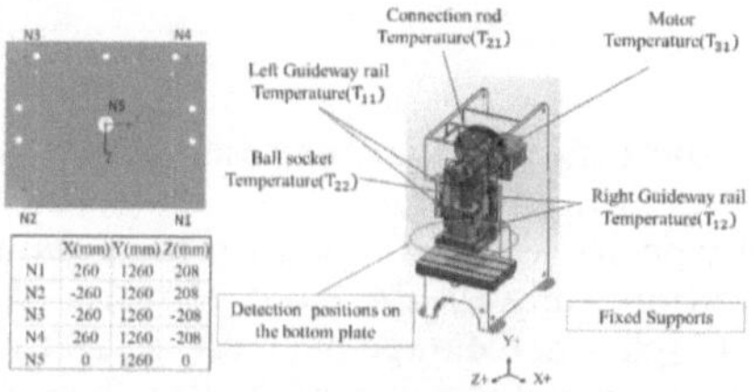

Figure 2. Simulated temperature heating and position measuring locations.

For more systematic demonstrating the posture of the platform, a 3D-axis plot is adopted to explain the simulation results, shown in Figure 3 and Figure 4. Since we have defined five thermal spots, the contributions to the platform are depicted with 3D planes. Once connection rod is heated and other components maintaining the room temperature, the side formed by N1 and N2 points descends, referring to Figure 3 left side. The very similar phenomena is also shown in thermal ball socket, shown in Figure 3 right side. The postures of the platform descending present slightly different because of the boundary conditions of rod and ball socket. The variation of position is ranged from 0.02 to 0.12 mm.

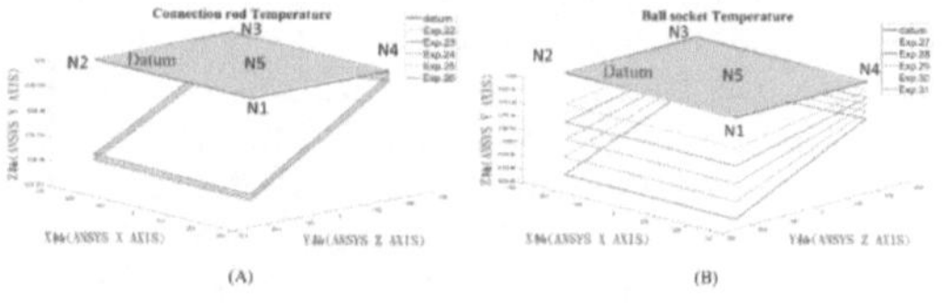

Figure 3. Position deviations with thermal connection rod and ball socket respectively.

By exploring more results of FEA, the thermal guideway rails expectantly cause more significantly deviation in positions, shown in Figure 4. The variation of position is about 0.2 mm. Meanwhile, the rotating axis shifts from edge to unexpected position on the platform. We can further confirm the complexity of the thermal-induced position error by observing the Figure 4 right. When all thermal spots emerge, the deviations of position alter by the summation of every surrounding thermal expansions.

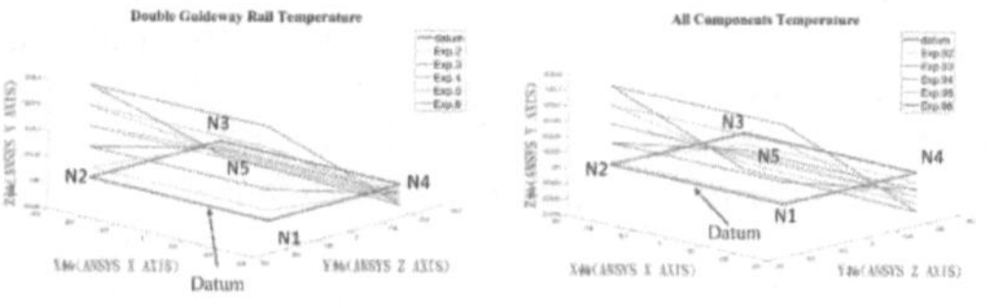

Figure 4. Position deviations with thermal guideway rails and all components.

3 THERMAL-INDUCED ERROR PREDCITION

Although the FEA can analyze different kind of complicate models and thermal conditions, need for high computation and well trained computer-aid engineering engineer work in the field is impractical. In this study, we try to fill the gap between the engineering and manufacturing. The machine learning approach is one of the feasible approach for this need. As we mentioned in the previous section that a comprehensive simulation can perform in the software instead of experiment in the field.

We have designed 111 combinations of thermal conditions for each components and all sets of thermal setup are conducted the steady-state thermal deformation analysis in ANSYS, presented in Appendix Table 1. For the plan of training, validation, and testing, the 96 sets of results are used to train the model and preserve 15 sets for model testing. For the training process, the 96 sets data is randomly divided into 82 sets (85%) and 14 sets (15%) for each epoch and the maximum epoch is set to 100.

Two issues arise when we start to train our prediction model in the beginning stage. First, the thermal-induced position error is between 0.01 to 0.2 mm and a proper normalization is needed. In our case, the min-max normalization is used for the issue of model convergence. The normalized input does easy the training effort in computation when it is applied to the task of classification. However, in this study, the numerical deviations of position need to convert from the normalized results, which the accuracy of prediction is not guaranteed. That is the second issue. Thus, to evaluate the prediction results, we have set up a threshold for position error prediction, which is the true value instead of the normalized one.

Following the above description, the flow chart of the model training is elaborated in Figure 5. Nevertheless, for assuring the complexity of the prediction model, five thermal value of all five locations, which serve as the input to the neural network, are applied to predict sole one position error. Thus, there are five models corresponding to five normalized position error, which are also shown in Figure 5 left side. For more detail of our Backward Propagation neuron network, Figure 6 depicts the scheme of the design, which are consist of hidden layers with [11-13-2] structure and the Levenberg-Marquardt optimizer is adopted to update the weighting during the training process. The Backward Propagation structure is

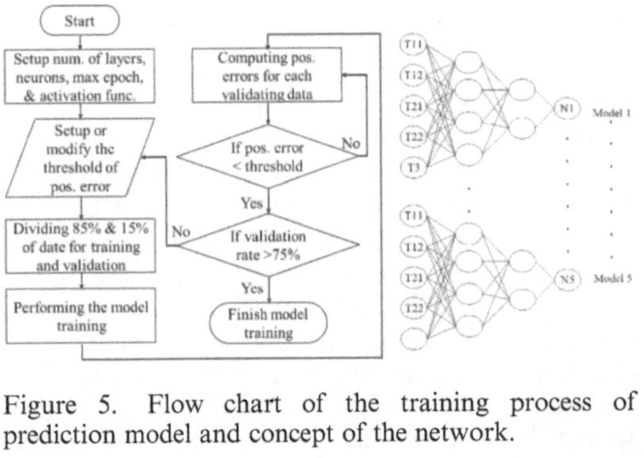

Figure 5. Flow chart of the training process of prediction model and concept of the network.

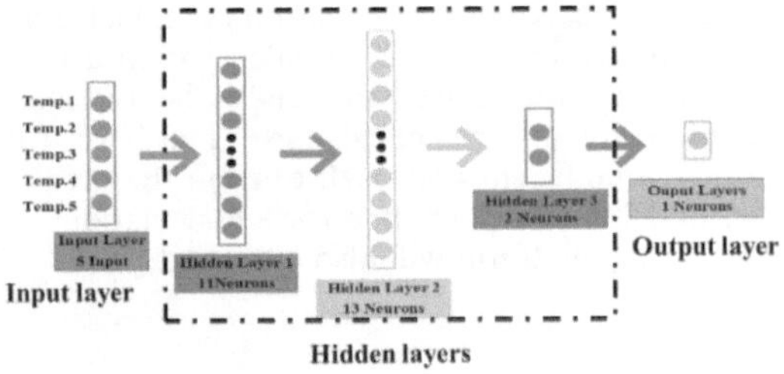

Figure 6. Scheme of the neuron network.

4 RESULTS AND DISCUSSION

There are five machine learning-based models for prediction thermal-induced positon error and, in this study, the threshold is set to 10%. It defines that the prediction of the normalized value need to convert to the true value of poisiton error in unit of millimeter. The difference between the results of the thermal-deformation from FEA shall smaller than 10%. Our code can compute the converting process within the validation stage and the results are interpreted in Figure 7. All five validation processes are presented and only the further evaluated with true value of position error is computed as one successful validation. Although all the five models can reach 100% accuracy of validation within the 100th epoch, referring to our previous work, we retrieve the model with accuracy between 71 to 80%. In this study, we have adopted five prediction models with the same stucture and weightings.

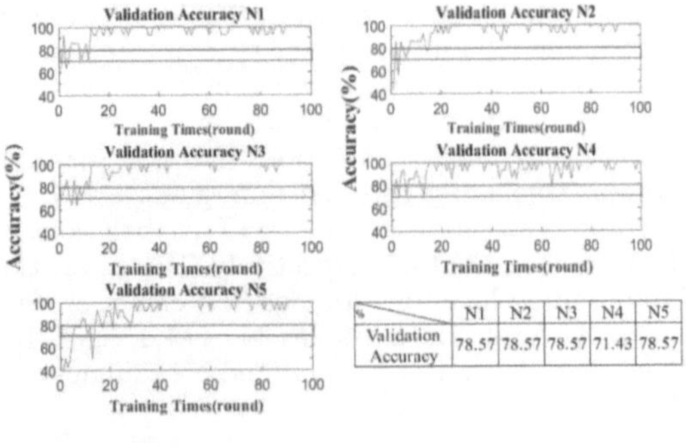

Figure 7. Validation results in the five locations.

For the testing stage, the reserved 15 sets of data are utilized. Figure 8 presents the error percentage for each of the 15 sets of data. By observation, only the 14th and 15th sets of data show a higher percentage of error, i.e. 20 to 29%. Hence, we conclude our prediction model can predict up to 87% in accuracy, i.e. 13/15 = 0.87%.

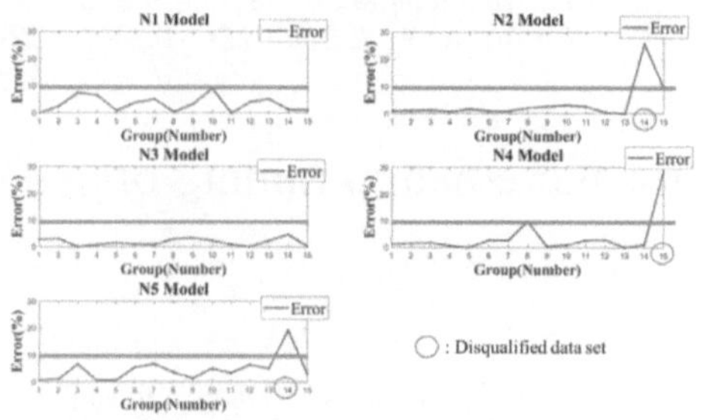

Figure 8. Testing results and disqualified data labeled.

5 CONCLUSION

In this study, we have accomplished the follows:

1. The thermal-deformation analysis can provide a more comprehensive observation for position error in mechanical press.
2. A compact deep learning model with 10% error allowance can predict up to 87% in accuracy.
3. A proper accuracy of validation models, i.e. 71 to 80%, may help to predict the unknown thermal-deformation results with higher accuracy.

6 APPENDIX

Table 1. Selected combination of temperature sets.

	T_{11}	T_{12}	T_{21}	T_{22}	T_{31}
Exp.2	44	44	45	45	45
Exp.3	63	63	45	45	45
Exp.4	82	82	45	45	45
Exp.5	101	101	45	45	45
Exp.6	120	120	45	45	45
Exp.22	25	25	60	45	45
Exp.23	25	25	75	45	45
Exp.24	25	25	90	45	45
Exp.25	25	25	105	45	45
Exp.26	25	25	120	45	45
Exp.27	25	25	45	60	45
Exp.28	25	25	45	75	45
Exp.29	25	25	45	90	45
Exp.30	25	25	45	105	45
Exp.31	25	25	45	120	45
Exp.92	44	44	60	60	60
Exp.93	63	63	75	75	75
Exp.94	82	82	90	90	90
Exp.95	101	101	105	105	105
Exp.96	120	120	120	120	120

REFERENCES

Fang, Z., et al. (2020). Micro-level clearance punching on NGO electrical steel. *Journal of Manufacturing Science and Engineering*, **142**(8).

Gan, L., et al. (2022). Digital twin-driven sheet metal forming: modeling and application for stamping considering mold wear. *Journal of Manufacturing Science and Engineering*, **144**(12).

Gitschel, R., et al. (2023). Controlling damage evolution in geometrically identical cold forged parts by counterpressure. *Journal of Manufacturing Science and Engineering*, **145**(1).

Kubik, C., et al. (2021). Exploitation of force displacement curves in blanking—feature engineering beyond defect detection. *The International Journal of Advanced Manufacturing Technology*, **113**(1–2): 261–278.

Li L., et al. (2019). Variations of energy demand with process parameters in cylindrical drawing of stainless steel. *Journal of Manufacturing Science and Engineering*, **141**(9).

System Innovation for an Artificial Intelligence Era – Lam et al. (Eds)
© 2025 the Author(s), ISBN: 978-1-032-84452-7

An efficient clique-based tree approach for incremental mining of co-locations from spatial database

Ye-In Chang
Department of Computer Science and Engineering, National Sun Yat-Sen University, Kaohsiung, Taiwan

Chen-Chang Wu
Department of Biotechnology, Fooyin University, Kaohsiung, Taiwan

Jia-Zhen Qiu
Department of Computer Science and Engineering, National Sun Yat-Sen University, Kaohsiung, Taiwan

ABSTRACT: While several algorithms exist for mining spatial co-location patterns, the clique-based approach has drawbacks, such as not creating all possible candidates or the maximum candidate first, leading to significant space and time requirements. To address this, Bao *et al.* proposed a clique-based algorithm for mining spatial co-location patterns in static databases. However, their method requires intersection and union operations for calculating the participation index (*PI*) value, with additional sorting processes for non-prevalent patterns. To overcome these challenges, a new algorithm has been proposed using the clique-based tree approach. Additionally, a *CDH* table method has been introduced for efficient mining of prevalent co-location patterns, along with a shuffling strategy to mitigate time-consuming processes. Furthermore, the new method has been extended to handle incremental databases, enabling efficient detection of new prevalent patterns and changes in pattern prevalence. Experimental results demonstrate the superior performance of the new algorithm compared to Bao *et al.*'s algorithm.

1 INTRODUCTION

Spatial data mining, particularly co-location pattern mining, plays a crucial role in uncovering frequently occurring patterns within large spatial datasets. Existing approaches are categorized into three main types. The first type, typified by algorithms from Shekhar *et al.* [2004] and Yoo *et al.* [2009], adopt an Apriori-like methodology, systematically handling patterns of varying sizes. Meanwhile, the second type incorporates clique-based techniques, starting with larger patterns and moving towards smaller ones. Examples include algorithms proposed by Kim *et al.* [2011] and Tran *et al.* [2019].

However, algorithms falling under the first category often face inefficiencies when dealing with dense datasets due to the abundance of small-sized patterns, leading to prolonged processing times. While algorithms in the second category are more efficient, they encounter challenges in identifying maximal cliques. To overcome these limitations, Bao *et al.* [2019] introduced a novel approach—a clique-based algorithm (the third type)—for mining spatial co-location patterns, which focuses on identifying cliques of any size, thereby addressing the drawbacks of the second type.

In our paper, we propose a novel clique-tree algorithm aimed at enhancing Bao *et al.*'s method. Our algorithm generates clique trees with fewer nodes, prioritizing instance counts in ascending order, thereby reducing both space and time requirements. Additionally, we introduce the *CDH* table method with a shuffling strategy to efficiently mine prevalence patterns. Our algorithm eliminates the need for reconstructing the tree structure upon inserting new data, resulting in shorter processing times and reduced storage space compared to Bao *et al.*'s method, across dense and sparse spatial datasets.

The paper is structured as follows: Section 2 surveys existing co-location pattern mining algorithms. Section 3 introduces our proposed algorithm. Section 4 presents a performance evaluation, including a comparison with Bao *et al.*'s method. Finally, Section 5 offers concluding remarks.

2 RELATED WORKS

2.1 *Co-location patterns mining*

The co-location pattern mining problem was introduced by Huang *et al.* [2004] in 2004, integrating association rule mining with spatial relationships via a join-based approach akin to Apriori. Despite its completeness and accuracy,

 DOI: 10.1201/9781003514831-64

this method suffers from inefficiency. In response, Yoo *et al.* proposed more efficient alternatives: the partial-join [J. S. Yoo *et al.* 2004] and join-less [J. S. Yoo *et al.* 2009] approaches. The former supplements the *Apriori* technique with instance join for unidentifiable instances, while the latter reduces reliance on joins by utilizing star neighborhood structures for spatial dataset storage. However, with increasing spatial data volumes, these approaches face efficiency challenges due to their dependency on *Apriori*-like frameworks.

2.2 *Incremental co-location patterns mining*

To tackle incremental challenges in co-location pattern mining, He *et al.* [2008] introduced the *IMCP* algorithm, utilizing 'cross' to transform updated datasets into three neighborhood sets. Subsequently, Lu *et al.* [2015] proposed the *IMPCA* algorithm, which pre-cuts unaltered co-locations for accurate solutions. Additionally, Wang *et al.* [2017] presented the *UUOC* algorithm, integrating high utility and incremental techniques, achieving effective results by categorizing changed co-location patterns into four cases and employing utility ratios to assess unchanged patterns.

Moreover, Andrzejewski *et al.* [2019] developed the *INC-EUCOLOC* algorithm, leveraging distributed techniques and parallel computing. Employing the *iCPI*-tree and an efficient GPU-based model, their approach demonstrated remarkable performance. However, these methods rely on *Apriori-like* frameworks, which diminish efficiency with large or dense datasets.

In contrast, the clique-based approach offers a top-down technique in co-location pattern mining, enabling rapid extraction of prevalence co-location patterns using clique patterns. Kim *et al.* [2011] pioneered this technique with the *AGSMC* algorithm, enhancing star neighborhood materialization and tree-type data structures for maximum clique expression. Subsequent refinements by Cheng *et al.* [2013] and Tran *et al.* [2019] introduced the maximal-clique technique, manifested respectively in the *MCEBOSON* and *OMCP* algorithms. However, generating maximal cliques remains time-consuming.

To address these challenges, Bao *et al.* [2019] proposed a clique-based approach surpassing the efficiency of previous algorithms like *MCEBOSON* and *OMCP*. They introduced two mining strategies (*IDS* and *NDS* approaches) for clique pattern generation, augmented by a *C-hash* structure for efficient clique information storage.

3 PROPOSED ALGORITHMS

This section presents our novel algorithm, which utilizes the count-clique tree (*CCL* tree) and clique-double-hash table (*CDH* table) to efficiently mine candidate patterns. Our approach also boasts versatility, as it can seamlessly handle incremental databases.

3.1 *The problem statement*

Co-location patterns denote patterns frequently occurring in close geographic proximity. Input data for co-location pattern mining typically includes spatial feature types, instance *IDs*, and their respective locations. Each feature type possesses its set of instances, where an instance is denoted by the feature type followed by an instance *ID* (e.g., *E.7* represents the seventh instance of feature *E*). In co-location pattern mining, two data objects are considered to have a distance relationship if they belong to different feature types and their Euclidean distance is less than or equal to a predefined distance threshold.

Consider Figure 1 as an illustrative example. It showcases four feature types: *A*, *B*, *C*, and *D*, with respective instance counts of 5, 4, 4, and 4. The solid black lines connecting objects represent the presence of a distance relationship between them. The primary objective of co-location pattern mining is to identify all prevalent patterns. Prevalence patterns are determined based on the participation ratio (*PR*) value and the participation index, as described by Huang *et al.* [2004].

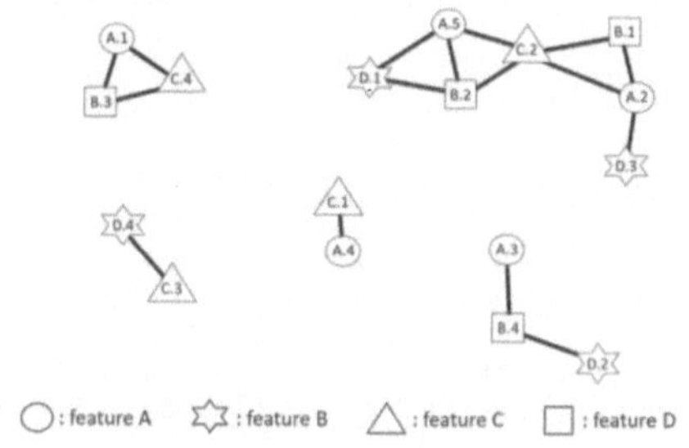

Figure 1. An example for the definition of Participation Index (PI).

For a size-*k* co-location pattern $CLP = \{f_1, ..., f_k\}$ and a data object f_i belonging to *CLP*, the participation ratio $PR(CLP, f_i)$ represents the fraction of instances of feature f_i within *CLP*. Additionally, the participation index of *CLP* is defined as the minimum *PR* value among its constituent features.

When determining if a co-location pattern qualifies as a prevalence pattern, the participation index (*PI*) value serves as a crucial metric. For example, consider the pattern [*A*, *C*]. The instance pairs associated with this pattern are [*A*.1, *C*.4], [*A*.2, *C*.2], [*A*.4, *C*.1], and [*A*.5, *C*.2]. To calculate the participation ratio (*PR*) value for feature *A* within pattern [*A*, *C*], we compute *PR*(*A*, {*A*, *C*})

= 4/5 = 0.8. Similarly, we obtain $PR(C, \{A, C\})$ = 3/4 = 0.75. Consequently, the *PI* value of pattern [*A*, *C*] is determined as 0.75, calculated as the minimum *PR* value among its constituent features, *i.e.*, $PI(\{A, C\}) = min(0.8, 0.75) = 0.75$. Suppose the prevalence threshold is set at 0.6. In this scenario, since the *PI* value of pattern [*A*, *C*] (0.75) exceeds the prevalence threshold, the pattern is classified as a prevalence pattern.

3.2 *The grid partition table*

The input data for co-location pattern mining consists of a triple set containing <*feature type*, *instance ID*, *location*>, along with the user-defined distance threshold. To efficiently compute all distance relationships, we utilize a Grid Partition Table (*GP*), where the grid side length equals the distance threshold. This allows us to calculate distances only for the surrounding eight grid cells, rather than all data points.

Consider Figure 2(a) as an example of input data, where each grid cell stores multiple data points or remains empty. By scanning the database once, we organize the data into the form shown in Figure 2(b). For instance, *grid* [3, 3] contains two data points (*B*.4 and *E*.3), while *grid* [3, 4] is empty.

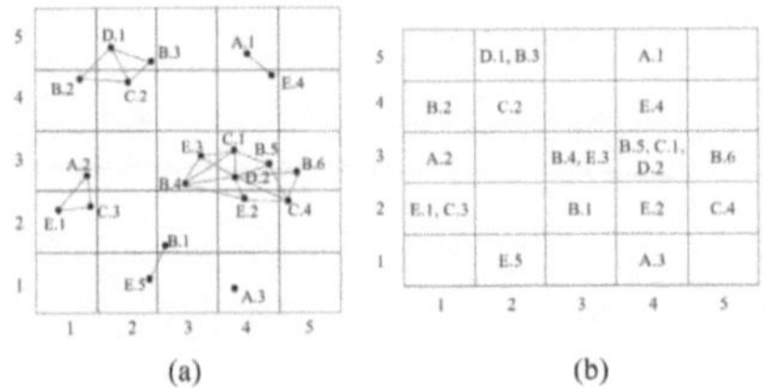

Figure 2. The generation of the *GP* table: (a) An example of the grid partition with the length of the side equal to the value of the distance threshold (b) The *GP* table after organizing the data.

3.3 *The simplified star neighborhood table*

To effectively record distance relationships, we adopt an improved version of the star neighborhood proposed by Bao *et al.* [2019]. We organize the star neighborhood structure in increasing order by count and employ the Simplified Star Neighborhood table (*SSNI* table) to store the data. The *SSNI* table comprises three columns: *InstID*, BNs, and *CountN*. *InstID*: Records the instance *ID* in the star neighborhood relationship. *BNs*: Represents the "bigger neighborhoods with large order," containing all data points attached to the *InstID* with a higher increasing order by count. *CountN*: Denotes the number of data points in the *BNs* column. Figure 3 depicts the *SSNI* table corresponding to Figure 3(a).

InstID	BNs	CountN
D.1	[C.2, B.2, B.3]	3
D.2	[C.1, C.4, E.2, E.3, B.4, B.5, B.6]	7
A.1	[E.4]	1
A.2	[C.3, E.1]	2
A.3	-	0
C.1	[E.3, B.4, B.5]	3
C.2	[B.2, B.3]	2
C.3	[E.1]	1
C.4	[E.2, B.5, B.6]	3
E.1	-	0
E.2	[B.4]	1

InstID	BNs	CountN
E.3	[B.4]	1
E.4	-	0
E.5	[B.1]	1
B.1	-	0
B.2	-	0
B.3	-	0
B.4	-	0
B.5	-	0
B.6	-	0

Figure 3. The *SSNI* table of Figure 3(a).

3.4 *The count-clique tree*

In the *CCL* tree structure, each row from the *SSNI* table corresponds to a node in the tree. The *CCL* tree is rooted at a node, where each subsequent node represents an instance *ID* from the *SSNI* table. For instance, Figure 4 depicts complete *CCL* trees derived from the *SSNI* table shown in Figure 3.

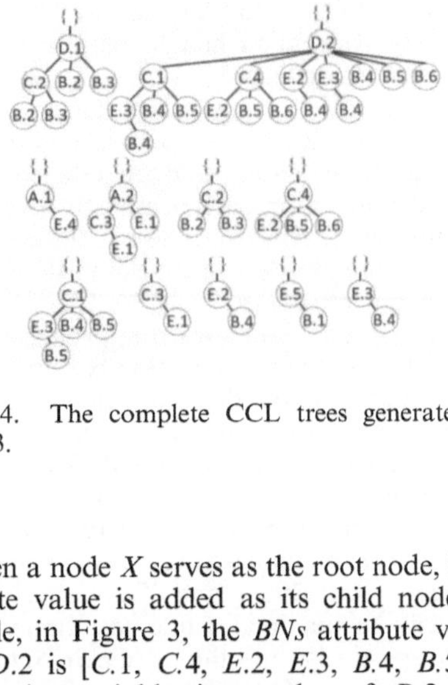

Figure 4. The complete CCL trees generated from Figure 3.

When a node *X* serves as the root node, its *BNs* attribute value is added as its child nodes. For example, in Figure 3, the *BNs* attribute value of node *D*.2 is [*C*.1, *C*.4, *E*.2, *E*.3, *B*.4, *B*.5, *B*.6], representing neighboring nodes of *D*.2. These nodes are directly added as child nodes of *D*.2 in the *CCL* tree. However, if node *X* is not the root node, its child node is determined by the intersection of its *BNs* attribute value and *RS* attribute value. The *RS* attribute value represents relative nodes sharing the same parent node as node *X* and possessing a higher increasing order by count. For instance, consider node *C*.1, with *RS* attribute value [*E*.2, *E*.3, *B*.4, *B*.5, *B*.6], as these nodes are neighboring nodes of *C*.1 with a higher increasing order by count. The child nodes of *C*.1 are *E*.3, *B*.4, and *B*.5, determined by the intersection of the *BNs* attribute value and *RS* attribute value.

The *CCL* tree exhibits a notable property: every traversal from the root node to a leaf node forms a clique set. This property emerges from the construction process, where child nodes are determined by the intersection of *BNs* attribute values and *RS* attribute values. The *BNs* attribute ensures that the child node maintains a relationship with its parent node, while the *RS* attribute ensures a relationship with ancestor nodes. As a result, nodes added as child nodes maintain connections with nodes higher up in the tree.

For example, consider the clique set [*D*.2, *C*.1, *E*.3, *B*.4] of *size*-4. In Figure 3(a), these four points are indeed connected. Similarly, [*D*.2, *C*.4, *E*.2] and [*D*.2, *B*.4] represent clique sets of *size*-3 and *size*-2, respectively, where each node within the set is connected to all other nodes within the set.

3.5 *The clique-double-hash table*

The *Clique-Double-Hash* table (*CDH* table) builds upon the C-hash structure introduced by Bao *et al.* [2019], enhancing the efficient calculation of the participation index (*PI*) value for each pattern and facilitating the rapid identification of prevalence patterns. The *CDH* table utilizes clique sets generated by the *CCL* tree and employs a double hash data structure.

In the first layer of the *CDH* table, the hash table's *HKey* directly uses pattern *X* as the key, where pattern *X* represents the candidate pattern. The *CValue* contains each feature in the key, along with a list value that stores the clique relationship. In the second layer, the hash function's *HKey* corresponds to the feature type, and the associated *CValue* represents the total number of instances of that feature type appearing in the pattern. Figure 5 illustrates a *CDH* table derived from the data presented in Figure 4.

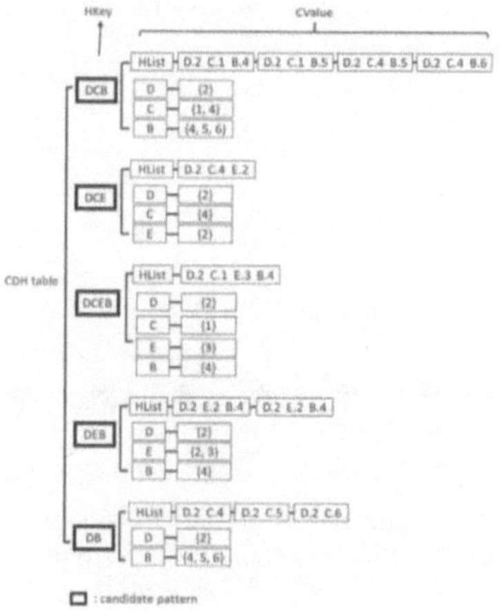

Figure 5. The CDH table of Figure 4.

3.6 *The mining process*

Our mining process involves four key steps: "*Construction of Neighbor Relationships,*" "*Cliques Generation,*" "*Candidate Co-location Patterns Generation,*" and "*Prevalent Co-location Patterns Filtering.*"

3.6.1 *Construction of neighbor relationships*

- Generate the *GP* table with *d_thres* as the grid side length.
- Compare each point *X* in the grid with points *Y* in the eight surrounding grids and the grid itself.
- Insert points into the *SSNI* table if the distance between *X* and *Y* is smaller than or equal to *d_thres.*
- Determine the increasing order by count and insert larger order points (*X*) into the *BNs* column of smaller order points (*Y*).

3.6.2 *Cliques generation*

- Utilize the *CCL* tree structure to store data efficiently.
- Initialize the *CCL* tree structure with each *InstID* field in the *SSNI* table as a root node.
- Construct each *subtree* recursively, ensuring that nodes from root to leaf form a clique set.

3.6.3 *Candidate co-location patterns generation*

- Use clique sets generated in the previous step to construct the *CDH* table.
- Merge and sort clique sets based on their features to obtain candidate patterns.
- Traverse the *CDH* table to calculate *PR* values quickly and efficiently.

3.6.4 *Prevalent co-location patterns filtering*

- Filter candidate patterns from *size-*(*k*) to *size-*(*k-1*) to efficiently identify prevalent patterns.
- Calculate the *PI* value for each candidate pattern and compare it with the *min_prev* threshold.
- Remove non-prevalent patterns and their subpatterns from the candidate patterns.
- Iterate until no candidate patterns remain, obtaining the prevalent co-location patterns.

For example, in the filtering step, patterns are filtered from the largest size bucket to the smallest size. Each pattern's *PI* value is calculated, and if it exceeds the *min_prev* threshold, it is considered a prevalent pattern. Subpatterns of prevalent patterns are removed from the candidate patterns, and the process continues until no candidate patterns remain. Finally, prevalent co-location patterns are identified, such as "*DCB,*" "*DCE,*" "*EB,*" "*CB,*" "*CE,*" "*DB,*" "*DC,*" "*AE,*" and "*DE.*"

3.7 *The processing step for mining incremental database*

The processing step for mining incremental databases is crucial for efficiently identifying new prevalence co-location patterns. When dealing with incremental data, three key cases must be considered:

Case 1: New Edge Formation with Existing Items: When an inserted item "*X*" forms a new edge with an existing item.
Case 2: New Edge Formation between Newly Added Items: When an inserted item "*X*" forms an edge with another newly added item "*Y*".
Case 3: No New Edge Formation: When the newly inserted item "*X*" does not form any new edge with any other item.

To illustrate, consider Figure 6, an incremental database derived from Figure 3(a). Here, five inserted data items (*A*.4, *D*.3, *E*.6, *E*.7, and *E*.8) introduce various new relationships.

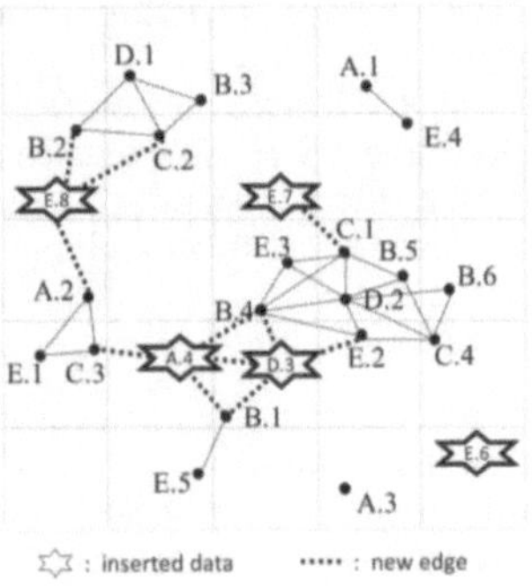

Figure 6. The incremental dataset of Figure 1, where the star points are the inserted data and the dash lines are the new distanced relationship.

Our method uses previously established data structures to design an efficient incremental data mining algorithm. The process involves:

(1) Insertion into the *GP* Table: Newly inserted data is added to the *GP* table to identify new neighborhood relationships quickly.
(2) Updating the *SSNI* Table: The *SSNI* table is updated without altering the existing order. New relationships are recorded based on the order of inserted data items.
(3) Generation of the Inserted *CCL* Tree: Using the updated *SSNI* table, an inserted *CCL* tree is efficiently created. Only the new rows are used to reconstruct the tree, minimizing computation.
(4) Insertion into the *CDH* Table: New patterns identified from the inserted *CCL* tree are inserted into the *CDH* table. Utilizing the same function used for initial *CDH* table creation ensures quick updates.
(5) Mining New Prevalence Co-location Patterns: Patterns potentially affected by the inserted data items are identified by examining the *CDH* table. Follow the filtering process, with adjustments to accommodate the limited range.

By following these steps, our method efficiently handles incremental data mining, enabling the identification of new prevalence co-location patterns while leveraging existing data structures for swift computation.

4 EXPERIMENTAL RESULT

In this section, we compare the performance of our approach with the clique-based approach [X. Bao *et al.* 2019] using synthetic datasets. The parameters considered in the experiments are:

- *d_thres*: Threshold of the neighbor distance.
- *min_prev*: Threshold of minimum prevalence, ranging from 0 to 1. A pattern with a *PI* value greater than *min_prev* is considered a prevalence co-location pattern.
- |*En*|: Number of features.
- |*NIn*|: Number of incremental instances.

We conduct experiments on sparse synthetic datasets to evaluate processing time under different *min_prev* values. The synthetic dataset used contains 24 features and 600 instances, covering a spatial area of 600x600 units.

Figure 7 displays the experiment results, showing the processing time comparison for a sparse dataset with varying *min_prev* values. We set *min_prev* from 0.1 to 0.8 and *d_thres* to 45. Our algorithm outperforms Bao *et al.*'s algorithm in terms of processing time. This superiority is attributed to the reduced computational overhead of our approach, as Bao *et al.*'s algorithm requires a larger number of intersection and union operations during *PI* value calculation.

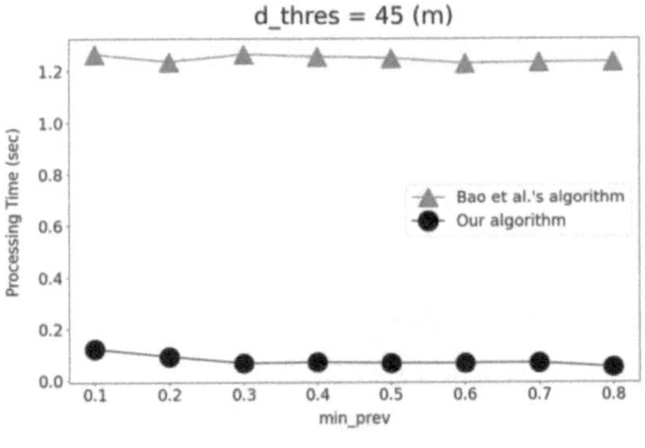

Figure 7. A comparison of the processing time of a sparse dataset under the change of the *min prev* value (*Den*=0, *d thres* = 45, *En* = 24, *In* = 600).

In addition to the sparse synthetic dataset, we also evaluate the performance of our algorithm and Bao *et al.*'s algorithm on a dense synthetic dataset. This dense dataset comprises 24 features and 1200 instances, covering a spatial area of 600x600 units, similar to the sparse dataset.

Figure 8 illustrates the performance results under varying *min_prev* values. Even with the denser dataset, our algorithm consistently outperforms Bao *et al.*'s algorithm across different *min_prev* values. As the dataset becomes denser, both algorithms generate more clique sets for the same *d_thres*, leading to an increased number of candidate patterns. Despite this increase in computational complexity, our algorithm maintains superior performance compared to Bao *et al.*'s algorithm.

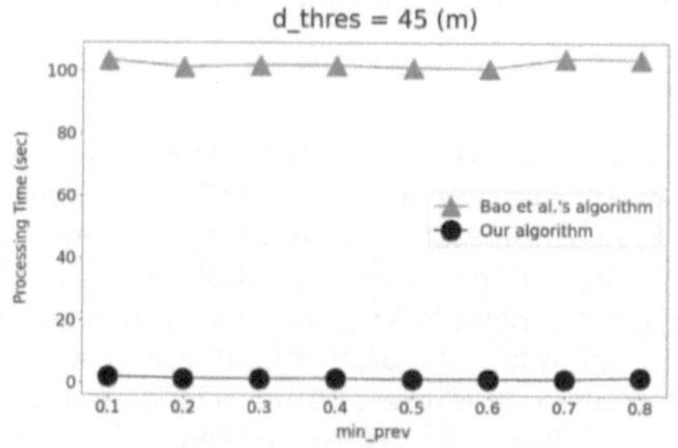

Figure 8. A comparison of the processing time of a dense dataset under the change of the *min prev* value (*Den* = 1, *d thres* = 45, *En* = 24, *In* = 1200).

In the upcoming experiments, we focus on evaluating the performance under incremental datasets. We simulate incremental scenarios by randomly adding 100 data points at a time. We then compare the processing time difference between our algorithm and Bao *et al.*'s algorithm. Notably, incremental scenarios are not explicitly handled in Bao *et al.*'s algorithm. Therefore, the processing time for Bao *et al.*'s algorithm is determined by re-executing the process each time new data is added.

Figure 9 presents the results of this experiment. It is evident from the figure that our algorithm outperforms Bao *et al.*'s algorithm consistently across incremental scenarios. This superiority underscores the efficiency and effectiveness of our approach in handling incremental datasets.

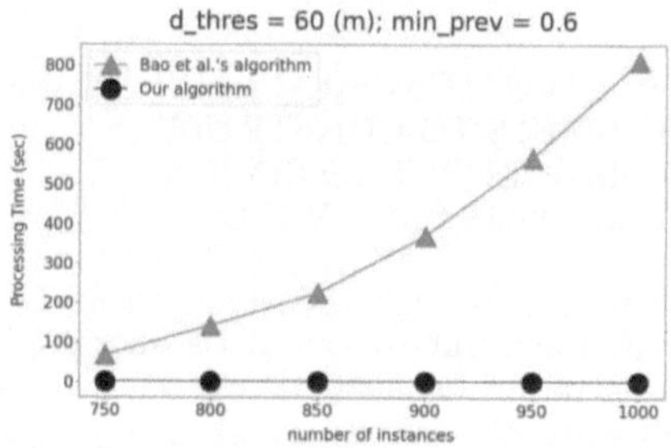

Figure 9. A comparison of the processing time of the synthetic dataset under the change of the number of instances (*min prev* = 0.6, *d thres* = 60, *En* = 24).

5 CONCLUSION

In summary, this paper introduces a novel approach for mining spatial co-location patterns in incremental spatial databases, leveraging cliques and trees for efficiency. A comparative performance study with Bao *et al.*'s algorithm was conducted across various dataset types and settings, including synthetic and real datasets, static and incremental scenarios, and sparse and dense datasets. The results consistently demonstrated the superior efficiency of our algorithm over Bao *et al.*'s algorithm.

ACKNOWLEDGEMENTS

This research was supported by grants MOST 109-2221-E-110-060 from the National Science and Technology Council, Taiwan.

REFERENCES

Andrzejewski W. and Boinski P. (2019). *Information Sciences*, 496, 485–505.

Bao X. and Wang L. (2019). *Information Sciences*, 490, 244–264.

Cheng Y., Tianjun Z., and Junli L. (2013). *Conf. Intelligent Human-Machine Systems and Cybernetics*, 1, 164–167.

He J., He Q., Qian F., and Chen Q. (2008). *IEEE Int. Conf. Data Mining Workshops*, 399–407.

Huang Y., Shekhar S., and Xiong H. (2004). *IEEE Transactions on Knowledge and Data Engineering*, 16 (12), 1472–1485.

Kim S. K., Kim Y., and Kim U. (2011). *FTRA Conf. Secure and Trust Computing, Data Management, and Application* 241–250.

Lu J., Wang L., Xiao Q., and Shang Y. (2015). *Conf. Fuzzy Systems and Knowledge Discovery* 612–617.

Tran V., Wang L., and Zhou L. (2019). *IEEE Conf. Mobile Data Management*, 467–472.

Wang X. and Wang L. (2017). *IEEE Conf. Big Data and Smart Computing* 215–222.

Yoo J. S. and Shekhar S. (2009). *IEEE Transactions on Knowledge and Data Engineering*, 18, 1323–1337.

Yoo J. S., Shekhar S., Smith J., and Kumquat J. P. (2004). *ACM Int. workshop Geographic information systems*, 241–249.

System Innovation for an Artificial Intelligence Era – Lam et al. (Eds)

The quantum brain hypothesis that self-entraps in the two-layer body-mind memory model dilemma—the existence of free will in conscious and unconscious psychological activities

Jung-Jung Wen
Ph. D. Program, Department of Mechanical Engineering, Kun Shan University, Tainan, Taiwan

ABSTRACT: Because determinists who are physicists and neuroscientists have not accurately analyzed mental activities within psychological structures, this has led to varying opinions on whether consciousness possesses free will. Neuroscientists underscore the premise of brain automatization, asserting that memory is stored in the brain while overlooking the existence and functioning of the overall psychological functional structure. Quantum biology has recently emerged as an interdisciplinary field, linking quantum physics with the life sciences. Therefore, there is growing concern about whether the quantum brain hypothesis can progress toward quantum neurobiology and effectively tackle the two challenges mentioned above. The components of the quantum brain hypothesis include the material neural system and the conscious psychological domain. This paper argues that this proposition lacks logical coherence in addressing the different aspects of memory, and aims to provide evidence against the notion that the two-layer memory model is the storage location for memory. This paper also explores the concurrent existence and operation of free will in consciousness and the unconscious within the psychological functional structure.

Keywords: Consciousness, Unconscious, Free will, Memory storage, Quantum brain, Ontology, Primordial being.

1 EXPLANATION

Western idealist philosophers typically use mental statements to describe the realm of subjective consciousness, considering it an internal world; although lacking spatial characteristics, it possesses subjective evidence and certainty, leading them to believe that the ontology of value propositions is the ideal pursuit in life (B. R. Du *et al.* 2020). However, they often need help to provide clear explanations of how consciousness is born, exists, and operates. Similarly, materialists are also unable to clearly explain the encoding, storage, and retrieval of memory. Physicists who follow materialist views assert that the universe comprises particles and fields. Initially, they propose that the principles of quantum mechanics apply to theories of all scales, attempting to establish Quantum Gravity theory as a grand unified theory to explain all physical phenomena. Additionally, in 1989, physicist Edward Witten introduced the topological theory to describe how two-dimensional gravity allows for changes in spatial topology. By integrating these perspectives, it is suggested that the entire universe is described by a "quantum wave function (S. Carlip, 2017)". However, a wave function is only one possible interpretation of the material nature of the universe; it cannot serve as the storage location for life memory, and it also fails to explain all the phenomena of life. Therefore, applying the quantum brain hypothesis of quantum mechanics—even by using the characteristics of the neural system, which simultaneously possesses a dissipative structure with a constantly increasing entropy and a topological structure with homogeneity—it is insufficient to fully construct an ontology of existence that encompasses the complexity, variability, and computational capacity stemming from quantum events in neuronal dynamics. In essence, the nature of the quantum brain hypothesis still embodies the connotations of cosmology and consciousness value ontology. It is neither a solution to personal identity theory nor a plausible explanation for the origin of universal life. Similarly, it doesn't explain the storage location of memory.

2 THE QUANTUM BRAIN HYPOTHESIS ATTEMPTS TO CONSTRUCT A PROBABILITY THEORY AND CONSCIOUS FREE WILL

The quantum brain explains the fundamental reality of the physical world from a materialistic perspective and expresses the fundamental true subjective mental experience of idealism through spoken language and textual description. Therefore, the quantum brain hypothesis encompasses the probability theory of the material body and the introduction of conscious free will. It adheres to a compromise between

DOI: 10.1201/9781003514831-65

the dualism of mind and body, neither denying nor arguing the precedence of the other. However, the mediating interface between these two narratives considers the nervous system as a nested hierarchical structure of molecules, cells, microcircuits, and the complex nonlinear network of the brain (P. Jedlicka, 2017). Therefore, this conceptual thinking aligns with the topological structure system, providing an alternative description that doesn't adhere strictly to geometric transformations of homomorphic properties. Because the initiation of neural action potentials follows an "all or none" rule, memory events are considered neurophysiological operations in the neural network, resulting in the manifestation of Probability Theory. Furthermore, since consciousness entails subjective mental experiences that can be articulated, there is free will rather than being solely governed by probability. Therefore, the quantum brain hypothesis aligns with a balanced dualism of mind and body, suggesting that each mental event is associated with a physical state in the nervous system, and their existence and operation are unified as an interactive whole. Currently, it is possible to observe a rough correspondence between neuronal activity in the cortical neural network of the brain using functional magnetic resonance imaging, indicating an interdependent relationship between the two.

2.1 *The neurobiological foundation of the quantum brain, based on the probability theory, violates the law of physiological constancy*

In 1932, von Neumann proposed using the rules of quantum mechanics to explain the functioning mechanism of the brain, specifically the behavior of ions and conductive substances. Since photons, electrons, and ions exhibit wave-particle duality, applying quantum laws to electromagnetic theory is appropriate. Therefore, the entire brain can be described using quantum mechanics, including the probability of all activities of the brain's constituent elements (J. M. Schwartz *et al.* 2005). Based on the generation of action potentials in brain neurons and the subsequent initiation of electrical impulses, the process operates according to the "all or none" law. If psychological events or the observer's brain are included in the quantum system, describing the wave function of its material brain state will collapse in response to new knowledge, transitioning from a superposition state to a single state; this then becomes the only reality. Therefore, if the quantum brain hypothesis is understood through the Probability Theory of the quantum process using the wave collapse of the Schrödinger equation, this means that the superposition state of various brain activities may collapse due to two reasons: The possibility or impossibility of calcium ions reaching the position to trigger synaptic vesicles, or the possibility or impossibility of neurotransmitters being released (J. M. Schwartz *et al.* 2005). Simultaneously, due to the extremely microscopic and narrow channels and spaces where the "all or none law" occurs—with the ion channels of neurons being less than 0.001mm and the distance between dendrites and the cell body being around 10^{-5}cm—Heisenberg's uncertainty principle and the probability theory of the wave collapse in Schrödinger's wave equation apply. However, if discussed based on the dissipative structure system in physics that biological individuals possess, and considering observations in line with the second law of thermodynamics, all physical phenomena occurring within the system operate in an ordered stable state under deviation from equilibrium. Therefore, the above-mentioned probability theory, observed under the departure from the system's ordered state, actually violates the discourse based on the law of physiological homeostasis in biology.

2.2 *The quantum brain hypothesis introduces conscious free will on a time scale that cannot manifest a continuous temporal axis*

According to the Copenhagen interpretation proposed by physicists such as Heisenberg, when consciousness is introduced into quantum theory, it replaces objective reality. It is also stated that without the introduction of the concept of consciousness, the quantum mechanical rules of the brain would have no basis (J. M. Schwartz *et al.* 2005). Psychologists particularly emphasize the effect of conscious Focus, defining it as a series of rapid consents. William James (1842-1910) explicitly stated that focus can be external stimuli or internal thoughts. He argued that focus can prolong the time a fleeting idea stays in consciousness, thus suggesting the genuine possibility of willpower in science (J. M. Schwartz *et al.* 2005). Therefore, maintaining sustained focus is an expression of conscious free will. Only through the learning of conscious focus can cognitive encoding function, leading to the acquisition of different information contents in both short-term and long-term memory. Therefore, the probability theory of the quantum brain must incorporate the free will of conscious focus learning; otherwise, it cannot convincingly explain the factual existence of memory. However, the intermittent and continuous nature of consciousness exists and operates only in moments of time; it has no continuity on the temporal axis, and thus it cannot be a storage location for memories. Therefore, the quantum brain hypothesis must address the interdependence between discontinuity and continuity, proposing how to resolve the necessity of this logical principle.

2.3 *Free will must exist in both conscious and unconscious psychological activities*

The existence and operation of psychological functions are more visibly observed in the structural aspects, roughly encompassing conscious and unconscious psychological activities. From the perspective of developmental biology, scholar Tsai Li-Cheng distinguishes the occurrence of consciousness and unconsciousness on this basis, demonstrating the sequence of the two and further elucidating that the implication of this unconsciousness aligns more closely with what Searle refers to as "deep unconsciousness." He also argues that the "unconscious" mentioned by Libet in 1985, regarding the "unconscious initiative of the brain," does not deviate from the framework of the body-mind two-layer structure (L. C. Tsai, 2024). Upheld under the premise of complete brain automatization, proponents of physical determinism argue that the theory of free will is necessary. However, proponents of neuroscientific determinism deny the existence of free will. In 1978, through the study and understanding of Social Neuroscience, David Premack suggested that humans inherently possess the ability of the Theory of Mind, enabling them to comprehend and interpret the desires, intentions, and beliefs of others (M. S. Gazzaniga, 2013). This paper observes patients with hemiplegia caused by a cerebral vascular accident (CVA) during their three-stage treatment with traditional Chinese acupuncture, and analyzes how under different conditions, patients can make different choices within the framework of free will. Acupuncture treatment emphasizes De Qi (obtaining Qi) and maintaining the De Qi process' operation. Sensations commonly associated with De Qi include soreness, swelling, tingling, numbness, and pain. Sustaining the De Qi effect can lead to appropriate neurophysiological responses, resulting in positive treatment feedback. The three stages of the manipulating needles to achieve the De Qi acupuncture sensation are as follows:

(1) Doctor inserts needle: At this moment, there is no pain, but the patient anticipates the imagined symptoms of De Qi - soreness, swelling, tingling, numbness, and pain. This state is comparable to the "readiness potential" observed in Libet's EEG experiment results. The presence of this anticipated state involves three associations: Memory encoding, channel extraction, and memory storage. Therefore, conscious intentional mental activities such as thinking, analyzing, and summarizing coexist with unconscious mental activities like information extraction and supply. (2) Doctor twists needle to achieve De Qi: At this moment, the appearance of De Qi symptoms occurs through the transmission of sensory nerve pathways to the brain's somatosensory cortex. Hence, the presence of the body's De Qi location and cortical neural activity is evident. The operational and imaginative aspects of "needle insertion to achieve De Qi" are integrated into one. At this point, the perceptual nature of conscious intention and the continuity of unconscious mental activity coexist simultaneously. (3) Patient verbalizing De Qi: At this moment, due to the discomfort of experiencing De Qi, the patient expresses various sensations such as soreness, pain, or numbness. This condition indicates that the second action potential of Libet's brainwave chart (EEG) and verbalized De Qi symptoms have equivalent significance. During this time, the conscious intent of distinguishing cognitive and unconscious psychological activities is actively and simultaneously present.

Typically, when patients with stroke-related hemiplegic symptoms seek medical assistance, common treatment approaches include Western rehabilitation and acupuncture therapy. The treatment regimen may involve scalp acupuncture combined with body acupuncture, or a plan integrating both methods. Since free will is a selective behavior, during communication between the physician and the patient, the patient inherently possesses the free will to choose either approach or a combination of both treatments. Next, in the three-stage acupuncture treatment process mentioned above, sustained concentration is required to maintain the De Qi phenomenon. Between stages (1) and (2), patients can freely decide whether to continue each needle treatment based on past experiences with discomfort or not. Additionally, during a 60-minute session, patients are instructed to maintain the intensity of De Qi by tapping their teeth, moving their fingers, and stimulating the neck. They are also advised to seek the doctor's assistance if the De Qi symptoms gradually diminish. However, both of these doctor-prescribed conditions can be determined by the patient's free will to follow or not. In observing the De Qi acupuncture process mentioned above, because consciousness has intermittent characteristics such as absent-mindedness, forgetfulness, interruptions, and changes of mind, patients are required to enhance their sustained concentration during the treatment to maintain the therapeutic effects of De Qi symptoms. From this perspective, patients have many opportunities during this period to express their choice of whether to continue or discontinue maintaining the De Qi state. Therefore, from the perspective of intention, consciousness exhibits free will on this timeline, but in terms of initiative and persistence, it falls under the domain of unconscious mental functions. Thus, whether from the perspective of Libet's experiments or the observation of the acupuncture De Qi process, the manifestation of free will must occur in synchronous coordination between conscious and unconscious mental activities.

3 THE TWO-LAYER BODY-MIND MODEL IS NOT A STORAGE LOCATION FOR MEMORY

The nature of memory must include accurate, complete, and detailed records. Therefore, there must be no forgetting or alteration during the process of memory events. Through observing this fact, a comparison can be made to demonstrate the extraction in short-term and long-term memory. Therefore, the occurrence of memory events must involve the processes of memory encoding, memory storage, and memory retrieval. The interaction and operation of these three can be observed and verified. Similarly, the birth, existence, and operation of consciousness do not happen without cause. Therefore, neither the material body (which undergoes metabolism and variation) nor consciousness (which cannot sustain continuous mental functioning) can be where memories are storied. Hence, the arguments of those who advocate for full brain automation are debatable.

3.1 *The material body is not a storage location for memory*

Today, neurophysiologists believe that the only place where neural activity can induce changes is at the synapse. When the action potential's electrical impulse is generated, it transmits neural messages through the axon. These messages then form effective synapses through connections between the axon and countless dendrites, gradually establishing neural circuits with the transmission of information between numerous neuron synapses. Ultimately, the connections between countless neural circuits reveal that meaningful psychological phenomena can be described. Therefore, strengthening synapses can enhance changes in short and long-term memory. Hence, it is said that synapses are the storage location for memory. Similarly, the larger physiological changes in the posterior hippocampus of London taxi drivers are attributed to the connection of place cells and grid cells, presenting cognitive environmental mental maps and to the belief that neurons with memory encoding exist in the CA1 region of the hippocampus. Therefore, the hippocampus is considered the storage location for both short-term and long-term memory. Network neuroscience believes that mental maps can be analyzed and structured using mathematical graph theory to examine interaction patterns between the neurons of the brain. Neurophysiologists, neurophysicists, and neuro-linguists also use graph theory to construct functional networks to map the neurophysiological pathways of cognitive functions (M. Bertolero *et al.* 2019; O. Ore, 2007). Graph theory utilizes nodes and edges to simulate the data imaging of functional magnetic resonance imaging (fMRI), where nodes – specifically neurons in the brain – represent units in the network. Edges refer to the connections between nodes, representing the connections between neurons. Constructing network connections for connector hubs and local hubs provides the integrated capability for complex computations across modules in the neural networks of the brain cortex.

This paper suggests that the significance of graph theory lies in visualizing numerous neural circuits as two-dimensional geometric shapes, allowing for analysis in three-dimensional space and time. In this context, regional hubs on the neural network cortex can integrate local procedural memories and form chunk-style habitual neural circuits. Connector hubs, on the other hand, develop synchronously to integrate the functions of different local hubs, manifesting as coordination of body movements and clarity of sensory perception. However, the differences in the formation, connections, and elimination of synapses during infancy, adolescence, and adulthood are very apparent, especially in newborns; and all physiologically functional neural pathways are already complete and no longer changing. Therefore, the brain's plasticity can only occur at the narrow neural fiber terminals to establish connections with synapses anew, and it is impossible to alter the physiological structures that already exist and no longer change. Therefore, combining neurophilosophers' emphasis on synapses, neural circuits, the hippocampus, and neural networks as memory storage locations, along with the theory of mental maps advocated in network neuroscience, fundamentally falls under the materialistic discourse. It fully encompasses the characteristics of impermanence, variability, and the limitations of time and space. Therefore, according to the principle of "impermanence must rely on the permanence," a material body that possesses perishable characteristics is definitely impossible to be the storage location for memories (J. J. Wen *et al.* 2020).

3.2 *The mind's consciousness is not a storage location for memories*

Dumchev argues that memory is a psychological function that involves both the processes of remembering and forgetting. If we consider information storage to be an independent process, further analysis becomes meaningless, as memory inherently implies storage (A. A. Dumchev, 2020). However, it is incorrect to initially attribute memory to the function of consciousness and then assign the storage of memory to consciousness. All of these are erroneous thoughts that have not been verified in the overall structure of psychological functions. This paper presents several viewpoints: (1) From embryological studies, we know that the smooth cortical layer of the brain begins to change and develop wrinkles around the 25th week of embryonic development. Scientists consider this to be objective evidence for the birth of consciousness. Based on this relative evidence, it can be proven that consciousness, at birth, existence, and operation, possesses temporality and spatiality. Therefore, the nature of consciousness does not have continuity but is a fact that can be born as "originally nonexistent, now existing". (2) Consciousness, since it arises during the embryonic period and ceases during five stages (dreamless deep sleep; syncope; thorough death; the absorption of non-perception; and the absorption of total cessation), is thus discontinuous and impermanent. According to the logical principles of "impermanence must rely on permanence" and "the false must base on reality," when considering that the material body is constantly in a state of trauma, variation, and decay, and that there is no existence of the material body after a certain period of life and death, this facts indicate that consciousness is a temporarily existent phenomenon and cannot be the storage location for memory. (3) Materialist scientists believe that mental phenomena are all activities of the brain. Looking at the quantum processes of the quantum brain described by the Schrödinger wave equation, when the wave function collapses from a superposition state to a single state, conscious mental activity occurs. Therefore, what this answer tells us is that mathematical probability determines the occurrence of conscious mental events (J. M. Schwartz *et al.* 2005). However, since the activities of consciousness fluctuate, it inevitably includes spatial and temporal events that can be described and expressed. In addition, consciousness itself has intentional and coherent volition, which is clearly not determined by probability. (4) Neurophysiologists and psychologists consider memory to be a function of consciousness and establish the location of memory within the neural networks of synapses, the hippocampus, basal ganglia, and the cerebral cortex. However, this is only an intuitive interpretation of the physical foundation recognized through conscious perception from statistical data in magnetic resonance imaging; it does not explain internal first-hand experiences. If memory is a function of consciousness, then consciousness cannot exhibit the "seven sins" described by Daniel L. Schacter: Absent-mindedness, transience, blocking, misattribution, suggestibility, bias, and persistence (D. L. Schacter, 2002).

Contemporary scientists and philosophers generally categorize memory into short-term and long-term, and approach the study of past-life memories with an open-minded perspective without outright denial. Therefore, when examining the birth of consciousness, the locus of memory, and the continuity of memory, hypotheses such as "idealism," "physicalism," or "body-mind dualism" must all be categorized as theories of cessation. Therefore, based on analytical reasoning, there must be an imperishable essence for it to be a possible storage location for memory. Further, memory retrieval also requires an intact and continuous pathway as a route. Neither the impermanent nature of material bodies nor the discontinuity of mental consciousness meet these conditions. Therefore, the processes of memory encoding, storage locations, and retrieval pathways operate separately and must not be confused. At the same time, these three elements are intricately interconnected, forming an inseparable whole. The above examples all illustrate that consciousness lacks the constant, imperishable, and unchanging stability necessary to be a stable repository for memories. Similarly, memories cannot be stored in neurophysiological structures with metabolic processes that lack stability.

4 THE QUANTUM BRAIN HYPOTHESIS IS ENTANGLED WITH THE PERPLEXITY OF MEMORY ASSOCIATIONS

In 2010, Dante Chialvo pointed out that, based on Darwinian evolution, the brain operates neural network activities at critical points. He initially explained that in a subcritical state, everything is consistently ordered, and there might not be a critical and plastic brain, as memory could also be unchanging (P. Jedlicka, 2017, D. R. Chialvo, 2010). If this were the case, working memory and procedural memory would not need to be exercised, and all efforts to concentrate on learning would be in vain, unable to be retained. This perspective contradicts educational learning and deviates from reality, making it logically unreasonable. Chialvo further explained that in the supercritical world, everything is constantly changing without any regularity to follow, so long-term plasticity and memory are of no assistance (P. Jedlicka, 2017, D. R. Chialvo, 2010). If that were the case, memory of experiential events and the knowledge acquired through educational learning would also be impossible. However, clinical experiments conducted by Edward Taub, who advocated the theory of learned nonuse in 1968, have shown that forcing the Silver Spring monkeys with useless injured limbs to use them and rehabilitating stroke patients to use their affected limbs demonstrates the plasticity of neurons and the reorganizational capacity of the cerebral cortex. Finally, it is believed that even in a critically complex world where surprising events occur, there is also regularity. Therefore, the brain needs to record and update stored memories (D. R. Chialvo, 2010; P. Jedlicka, 2017). If this is the case, the aforementioned short-term and long-term memories would be constantly updated and no longer exist. However, this explanation is not plausible. After all, we can often recall countless memories and simultaneously experience the reality of diverse emotional fluctuations associated with those memories, which are not fictitious events. In 2012, Beggs and Timme, based on their observations, proposed the existence of three types of dynamic states in the brain: (1) ordered subcritical dynamics characterized by high coordination and low variability; (2) stochastic supercritical dynamics, characterized by low coherence and high variability; (3) complex critical dynamics in the brain, characterized by high coherence and high variability (P. Jedlicka, 2017, J. M. Beggs *et al.* 2012). However, since the quantum brain hypothesis introduces the concept of consciousness, the correlations displayed by these three states are similar to the process of consciousness thinking, analyzing, discerning, and knowing; for memory events, it only extends to the encoding process and does not reach the storage location of memory. Strictly speaking, incorporating the variability of high and low coordination already falls within the selective "sins" mentioned by Schacter, namely, bias, suggestion, misattribution, and persistence (what Schachter calls the sins of commission). Furthermore, if observed through the lens of the second law of thermodynamics, dissipative structures rarely exist in a true equilibrium state, and consumption and decline are inevitable. Although the transition from order to disorder demonstrates instabilities during the occurrence process. However, within other localized regions of the system, order can still be maintained based on certain specified processes; for example, fluid dynamics, chemical reactions, molecular folding and unfolding, and ion channels in cell membranes are all processes that occur significantly away from equilibrium. Therefore, based on the principles of biological physiological homeostasis, this paper argues that observing the three types of quantum dynamics in the brain is merely a manifestation of physiological phenomena in the material body. This must not be confused with causality by resorting to all-or-none probability or a compromising approach to imply a connection between quantum dynamics and memory. Imposing classification or linkage with memory may lead to deviation from fact and falls into the realm of pure imagination. Moreover, physicists consider continuity and connectivity of space to arise from the entanglement of quantum mechanics, so time and space emerge from fundamental quantum phenomena. However, quantum states are transient and perishable, and once an entangled state is disrupted, information stored in quantum bits will be completely erased and cease to exist. Of course, there will be no quantum rebirth in the future to revealing any continuity (M. A. Nielsen, 2011). Therefore, the Quantum Memory Hypothesis cannot explain the coherence of human memory, and Temporal Emergence cannot address the conditions of the universe's creation. Hence, physicists like Stephen Hawking believe that relativity will be replaced by a more profound theory that doesn't exist within the framework of spacetime (S. Carlip, 2017). In 2007, scholar Tsai Li-Cheng proposed the existence of a non-dimensional "original being," which is both knowable and verifiable and serves as the storage location for memory. Tsai further extrapolates three judgment principles with operational significance—permanence, non-substance, non-separation from substance"—based on the nature of the "original being" to distinguish between properties of impermanence and permanence (L. C. Tsai, 2007). However, the proposition of this "original being" is not discussed in this paper.

5 CONCLUSION

First, based on the observation and analysis of the psychological functional structure, this paper asserts that free will must exist in both conscious and unconscious mental activities. Similarly, the interaction of memory encoding, memory retrieval, and memory storage can also be discussed and confirmed through conscious and unconscious mental activities. Next, material structures (including synapses, the cerebral cortex neural network, and the hippocampus) fundamentally possess characteristics of impermanence, variability, and spatiotemporal limitations. Therefore, they cannot have the function of storing memories. The proponents of the quantum brain hypothesis integrate subjective conscious experiences with the objective physical world of the material body. But in fact, this argument does not clearly differentiate between consciousness and the unconscious, highlighting the differences in the psychological functions they exhibit within the operation of the material body in a complete psychological framework. After all, quantum theory itself is based on the principle of probability, and Quantum Spacetime is manifested by the "emergence" of quantum entanglement, thus demonstrating this theory's impermanence, variability, and instability. Therefore, the quantum brain hypothesis can neither refute the destructibility of the material body, nor prove that mental consciousness is permanent and indestructible, nor analyze from the perspective of free will in the psychological functional structure how conscious and unconscious mental activities synchronously influence the operation of the material body. In summary, due to the vast amount of audiovisual information in daily life related to vision, hearing, touch, and more, the current two-layer memory model framework of material body and mental consciousness is inadequate to appropriately explain the phenomena of human memory operations. Therefore, how to validate the existence of an ultimate entity beyond the framework of time and space must be a continued endeavor for humanity in the future.

REFERENCES

Beggs J. M., Timme N. (2012). Being critical of criticality in the Brain. *Front. Physiol.* 3: 163. doi: 10.3389/fphys.2012.00163

Bertolero M., Bassett D. S. (2019). *How the Mind Emerges from the Brain's Complex Networks.* Scientific American Chinese Edition, 214, 42–49.

Carlip S. (2017). Quantum gravity in 2+1 dimensions. In Y. Q. Gao (Trans.), *Scientific American Bo Xue Zhi (Chinese Version)*, 16–23.

Chialvo D. R. (2010). Emergent complex neural dynamics. *Nat. Phys.* 6, 744–750. doi: 10.1038/nphys1803

Du B. R., Chen R. H. (2020). *Introduction to Philosophy.* Taipei: Wu-Nan Book Inc.

Dumchev A. A. (2020). *Remember Everything: A Practical Guide to Improving Your Memory.* In W. Q. Lei (Trans.), *Cheng Du: Si Chuan Wen Yi.*

Gazzaniga M. S. (2013). *Who's in Charge? Free Will and The Science of The Brain.* In P.G. Zhong. (Trans.), Taipei: Owl Publishing House.

Jedlicka P. (2017). Revisiting the Quantum Brain Hypothesis: Toward Quantum (Neuro)biology? *Front. Mol. Neurosci.* 10: 366. doi: 10.3389/fnmol.2017.00366

Nielsen M. A. (2011). Quantum computation and quantum information. In Y. Q. Gao (Trans.), *Scientific American Bo Xue Zhi (Chinese Version)*, 126–137.

Ore O. (2007). The application with dot and line graphs. *Hsinchu*, Taiwan: Fanyi Publishing.

Schacter D. L. (2002). *The Seven Sins of Memory. In M. Lee (Trans.)*, Taipei: Locus Publishing.

Schwartz J. M., Begley S. (2005). In M. H. Zhang (Trans.), *The Mind and The Brain- Neuroplasticity and The Power of Mental Force*, Taipei: China Times Publishing Co.

Tsai L. C. (2007). The Definition of Being in The Agama Sutras. *Journal of True Enlightenment*, (1), 23–48.

Tsai L. C., (2024). Decoding the Unconscious for Insight into Free Will and Memory Storage - A Comparative Analysis of the Three-Layer Psychological Structure and the Two-Layer Body-Mind Memory Model, Japan, Kyoto.

Wen J. J., Chou H. M. (2020). An Exploration of the personal identity dilemma and its solutions. In *Innovation in Design, Communication and Engineering*, 128–132.

System Innovation for an Artificial Intelligence Era – Lam et al. (Eds)

Examining the meaning of "Truly Attaining the Path of Vision" in Mahāyāna based on Master Xuanzang's comment about "The Inability of Confused Persons to Understand Complementarity": The Buddha Dharma of Mahāyāna was indeed expounded by the Buddha himself

Ping-Shi Xiao
Buddhist True Enlightenment Practitioners Association, Taipei, Taiwan

Ling-Tze Chao
Ph. D. Program, Department of Mechanical Engineering, Kun Shan University, Tainan, Taiwan

ABSTRACT: Before Dharma Master Xuanzang's journey to the West, he had already enjoyed a good reputation for his profound knowledge of Buddhism during the Tang Dynasty. Once, during his stay in the Nalanda monastery, he said the following to a renowned Madhyamaka master who raised an objection to the Consciousness-Only doctrines (Vijñaptimātratā): "The inability of confused persons to understand complementarity would lead them to assert the existence of a contradiction between the Dharmas; thus, the fault lies with the transmitters, not with the Dharma." This key point and substantive connotation of Master Xuanzang's statement of "the inability of confused persons to understand complementarity" deserves in-depth exploration.

According to research on ancient scriptures and treatises, relying on the penetrative complementarity virtues stemmed from the personal realization of prajñā-related wisdom, the essence and crux of Master Xuanzang's aforementioned statement about the notion of "complementarity" is whether an individual has truly attained the "Path of Vision" in Mahāyāna and thereby relies on the stage of "vision of characteristics" and the "proficiency stage" that post-awakening Buddhists pursue to advance their cultivation of particular aspects of wisdom pertaining to prajñā and the knowledge-of-the-aspects-of-paths pertaining to Consciousness-Only. As such, this also explains why Master Xuanzang was able to fully comprehend the Dharma and subdue all the dharma masters from five parts of India, set forth the principle of the True Pramāṇa of Consciousness-Only, and authored Cheng Weishi Lun to benchmark the meta-value resulting from the implementation of the Yogācāra teachings—by subsuming the empirical Buddha Dharma under the eighth consciousness (vijñāna), the reality-suchness. This paper also presents scriptural proofs to justify that the Buddha Dharma of Mahāyāna was indeed expounded by the Buddha Himself and not constructed, created, and evolved by Buddhists after the period of Sectarian Buddhism.

Keywords: Master Xuanzang, truly attaining the path of vision, Enlightenment, Cheng Weishi Lun.

1 FOREWORD

This paper investigates the complementarity and interconnectedness of the Three-Vehicle Bodhi's contents in various Buddhist canonical sources, drawing from the Āgama Sūtras, Prajñā Sūtras, and Consciousness-Only Vaipulya Sūtras, all of which can be traced back to the core of tathāgatagarbha as a single lineage of these expositions. During the Tang Dynasty, Master Xuanzang restored his valid knowledge (pramāṇa) abilities related to deliverance by wisdom, the Mahāyāna awakening to the True Mind (tathāgatagarbha), and was able to see the Buddha-nature with his physical eyes. The mainstream Dharma teachers of the Great Tang Dynasty had not personally witnessed the true meaning of the Buddha Dharma themselves, and the existing Buddhist sūtras were also incomplete, leading to the propagation of disputes and contradictions in the Buddhist teachings. Master Xuanzang was dead set on traveling to India to get the complete roadmap for post-awakening cultivation regarding the sequential cultivation stages and contents of the Path to Buddhahood. His main goal was to obtain the entire Yogācārabhūmi-śāstra text as it was the key to settling once and for all the long-running debates back home over the true definitions of the Dharma (Da Tang Gu San Zang Xuan Zang Fashi Xing Zhuang, CBETA 2023, P.S. Xiao, 2023). During his studies at the renowned Nalanda Monastery, Master Xuanzang engaged in a crucial dialectic with Madhyamaka master Sijha-raśmi, who lodged a critique against Xuanzang's promulgation of the Consciousness-Only doctrine, viewing it as antithetical to the tenets of the Madhyamaka tradition. In response to this challenge, Xuanzang said the following: "If certain erudite

DOI: 10.1201/9781003514831-66

people think that there are apparent contradictions between the early and later discourses of the Buddhist Dharma, this dissonance must be attributed to the inadequacies of those entrusted with its exegesis and transmission, as well as the limitations of those engaged in its study, whose inability to synthesize and reconcile these teachings has engendered such discrepancies. How can this be the problem of the Dharma itself? (A Biography of the Tripitaka Master of the Great Ci'en Monastery of the Great Tang Dynasty, CBETA 2023)" During his sojourn in India, Master Xuanzang convened an "unhindered assembly of doctrinal debate" in the city of Kanyakubja, where he promulgated the Mahāyāna tenet that all Dharmas were Consciousness-Only. This exegetical endeavor was undertaken with the overarching imperative of establishing the definitive meaning of the Buddha Dharma. Master Xuanzang's ultimate Mahāyāna teachings earned him veneration from both Buddhists and non-Buddhists in the five parts of India, leading him to obtain the honorifics of "Heaven of Liberation" and "Heaven of Mahāyāna," respectively. His monumental act served to affirm and settle the practical value of the "Consciousness-Only school" in Buddhism. The practical value of the Consciousness-Only school is the practitioner's reliance upon the empirical realization of the ālayavijñāna, the eighth consciousness, as a threshold, thereby facilitating a personal experience of the incipient unfolding of the prajñā wisdom. Following this actual realization, the practitioner will be capable of penetrating the profound teachings of the Buddhist sūtras accordingly and will have a coherent understanding of the Three-Vehicle Bodhi. Therefore, while the Consciousness-Only school has a theoretical value, its meta-value lies in the actual realization of the ālayavijñāna (J.W. Yu *et al.* 2017).

A German Buddhist scholar, Mr. Lambert Schmithausen, published a paper in 2015 titled "On the Problem of the External World in the Ch'eng wei shih lun." Schmithausen argues that the Yogācāra school's view of Consciousness-Only theory cannot be found at all, or at best can be found only sporadically, in some early Yogācāra works (especially the Yogācārabhūmi-śāstra, but also others, such as the Abhidharmasamuccaya) (Lambert Schmithausen, 2015). In addition, Schmithausen points out that Master Xuanzang also states in Ch'eng wei shih lun that the independent existence of matter (rūpa) in the so-called external (i.e., physical) world must not be denied, and Schmithausen agrees with another Buddhist scholar, Lusthaus, that the Consciousness-Only teachings are not basically theoretical in nature but are a kind of "therapeutical device," a metaphysical strategy directed at attachment and appropriation in Buddhism. Schmithausen concludes that Xuanzang's Ch'eng wei shih lun can also be understood as an attempt to reestablish the conservative Abhidhamma theory (Lambert Schmithausen, 2015). Moreover, the 14th Dalai Lama evinces in his book of interviews, Gentle Bridges, that there exists an apparent dichotomy between the early and late discourses of the Dharma by the Buddha Himself; there are palpable contradictions shown across the Three Turnings of the Dharma Wheel (Ph.D. Jeremy *et al.* 2001).

However, the author relies on the prajñā wisdom of "truly attaining the Path of Vision" to delve deeply into the sūtras. In many of my Buddhist publications, I reiterate that truly attaining the Path of Vision (enlightenment) in Buddhism is actually the most important task for practitioners. Enlightenment serves as the apodictic key, enabling practitioners to penetrate the profound depths and understand the recondite meanings ensconced within scriptural teachings. However, in the Dharma-ending age, ordinary scholars and academics bereft of the virtues stemming from seeing the Path of Vision have formulated an imaginary Madhyamaka based solely on the six-sense consciousness model. This mistaken view has gained wide currency in the world, leading to a misinterpretation of the profound prajñā teachings from the Second Turning of the Dharma Wheel. Mere linguistic expression analysis indicating that all Dharmas are empty has resulted in the complete loss of transmission of the True Dharma—the secret meaning of the eighth consciousness, the diamond-like mind, as expounded in sūtras such as the Diamond and Heart Sūtras (P.S. Xiao, 2013). It is for this very reason that in the book series The Correct Meanings of the Āgama Sūtras: Exploring the Origin of the Doctrine of Consciousness-Only (P.S. Xiao, 2007), I deliberately drew solely from the Four Āgama Sūtras, eschewing citations from the Mahāyāna sūtras. This is to conclusively demonstrate that the profound teachings present in the Mahāyāna sūtras and Vaipulya Consciousness-Only texts do not, in fact, contradict the Path to Liberation expounded in the Four Āgamas. This book series also substantiates, through extensive citations from sūtras and śāstras, that the core Mahāyāna tenets have not evolved over time. In addition, it indicates that, when elucidating the notion of nirvāṇa within the original Buddha Dharma, the Buddha, to preclude falling into the nihilistic views of non-Buddhist proponents, invariably taught the secret meaning using implicit terms about the existence of the eighth consciousness (Originally Existent Dharma). Therefore, the remainderless nirvāṇa of the Two-Vehicle Dharma is, in fact, an expedient teaching ultimately subsumed under the Mahāyāna Dharma as its foundation. Without the foundational principle of the Mahāyāna school—the eighth consciousness, tathāgatagarbha—the Two-Vehicle nirvāṇa will not be able to escape the ridicule of having the view of annihilation, and its essence will become nihilistic emptiness.

This paper delves into the vast chasm that separates the sublime heights of realization attained by Buddhist practitioners who have "truly attained the Path of Vision" through their personal realization of nirvāṇa, including the essential connotations of the liberating wisdom of the mundane and supramundane worlds pertaining to prajñā, from the miscomprehensions of the sūtras by academic researchers lacking any virtuous wisdom born of "truly attaining the Path of Vision."

2 THE WISDOM OF THE PATH TO BUDDHAHOOD ENCOMPASSES THE PATH TO LIBERATION

Buddhism involves empirical realization, whose connotation transcends the state of contemplative thought of mental consciousness. Buddhist teachings can be verified by practitioners themselves through direct observation of their actual progress in the cultivation of different stages, the depth and breadth of their wisdom, and the differences in their accumulation of meritorious virtues. It cannot be grasped merely by the academic study of texts and intellectual postulation based on appearance and logic.

2.1 *Liberation wisdom in Buddhism*

According to the Yogācārabhūmi-śāstra, Buddhist wisdom is roughly divided into two categories—worldly wisdom and supramundane wisdom—or can be explained as the five branches of insight: inner knowledge, the insight of logic, the insight about language, the insight about medicine prescription, and the insight about manual arts (Yogācārabhūmi, CBETA 2023). In Buddhism, worldly wisdom refers to the process of open learning by "students" acquiring the knowledge to navigate and accept this phase of life from birth. Starting from a state of complete ignorance, one gradually enriches one's worldly wisdom from the ground up through experience and new scientific knowledge in the three realms. Worldly wisdom is in constant flux, changing and evolving with each passing day. Yet, it serves as the foundation on which one can embark on learning Buddhist teachings in the future. Supramundane wisdom is the exclusive, transcendent wisdom unique to Buddhism. It refers to the wisdom that frees one from being bound to the three realms of existence. It is based on the Buddha's ultimately perfect mundane and supramundane enlightened wisdom as the central axis along which the stages of cultivation are established for the Buddha's disciples in Figure 1.

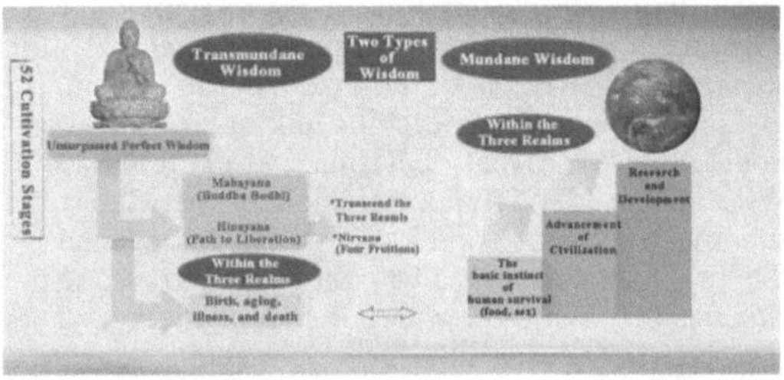

Figure 1. Two types of wisdom. A chart designed by L. Z. Chao (On the right, worldly wisdom develops upwards. On the left, Buddhist transmundane Wisdom passes downwards).

In other words, the wisdom leading to liberation from the three realms of existence and culminating in Buddhahood follows 52 well-defined successive closed stages established by the Buddha Himself. Providing answers to the reality of life, the Path to Buddhahood allows practitioners to attain realization progressively and in the proper sequence. The fundamental and indispensable requirement for embarking on the Buddha Bodhi Path is the attainment of the pure Dharma-eye of the Two Vehicles. Only then can practitioners achieve "enlightenment and awakening to the True Mind," thereby igniting the prajña wisdom. The Buddha taught the Three Turnings of the Dharma Wheel to accommodate the different capacities of sentient beings since the Four Āgama Sūtras of early Buddhism. It is clear that the Dharma of no-self was taught, indicating that the five-aggregate body and mind are the result of causes and conditions coming together. However, the existence of a real self was also implicitly taught in all the sūtras, and this "self" was described as nirvāṇa or the "vijñāna" that conditions name and form. During the Second Turning of the Dharma Wheel, the principle of Buddhism—the limit of intrinsicality nirvāṇa—is also said that such a mind is the mind of the sign of reality. In the Third Turning of the Dharma Wheel, the Lankāvatāra Sūtra and the Ārya-saṃdhi-nirmocana-sūtra, and so on, it is clearly stated that the eighth consciousness is termed as tathāgatagarbha, ālaya-vijñāna, vipāka-vijñāna, reality-suchness, etc.

2.2 *The "observation of renunciation" and the "observation of tranquility" have already been mentioned in the Āgama period during the First Turning of the Dharma Wheel*

In the Long Āgama Sūtra, during the First Turning of the Dharma Wheel, had already mentioned that all Buddhas practiced the methods of "observation of renunciation" and "observation of tranquility" to perfectly attain the Path to Buddhahood (Long Āgama Sūtra, CBETA 2023). Meanwhile, in the Āgama Sūtras, the Buddha had already taught the general principle of the practicing method of "observation of tranquility" to establish nirvāṇa based on it, so that the Two-Vehicle saints would not be afraid of destroying the five skandhas and would seek liberation from the transmigration of birth and death in the three realms to attain nirvāṇa. The followings are expounded in the sūtras by the Buddha:

(1) "I will speak of corruptible and indestructible dharmas to you. Listen attentively, and think about what I have said. Form is a corruptible dharma; after its destruction, nirvāṇa remains an indestructible dharma. Sensation, perception, formation, and consciousness are also corruptible dharma; after consciousness is destroyed, nirvāṇa remains an indestructible dharma (Saṃyukta Āgama, CBETA 2023)".

In this passage from the Saṃyukta Āgama, the Buddha explains the gist of the Dharma and clearly defines the corruptible and indestructible Dharmas. The five skandhas belong to the corruptible Dharma. After the five skandhas and six-sense consciousnesses are completely extinguished, there will be nirvāṇa, which is an indestructible Dharma.

(2) "One must understand the origin of the Dharma to be able to comprehend nirvāṇa (Saṃyukta Āgama, CBETA 2023)". Based on the Twelve Links of Dependent Origination, the World-Honored One explained each link in reverse order to guide the intellectual Susima, who tried to plagiarize His teachings (Dharma). The Buddha then asked Susima, "Will you be able to stay away from the evil Dharma of the desire realm, continue to progress, and achieve nirvāṇa—liberation from the transmigration of birth and death in the three realms—when you observe the Twelve Links of Dependent Origination in this way?" Susima replied, "No, World-Honored One!" This reply indicates that Susima knew that a Dharma exists permanently so that the contemplative observation of the Twelve Links can be completed; otherwise, doubts will inevitably arise, and one will wonder whether there will be nihilistic emptiness when there is no longer any future existence. One will become an ordinary person who experiences "fear due to external phenomena and fear due to internal phenomena (Madhyama Āgama, CBETA 2023)". Therefore, one who practices the Twelve Links of Dependent Origination without understanding the Ten Links of Dependent Origination will not know or recognize that there is an everlasting, abiding Dharma. One who does not know that a Dharma abides eternally will not be able to fully accept the teachings of the Twelve Links of Dependent Origination: that the five-aggregate self must be completely annihilated, that only remainderless nirvāṇa remains. Thus, the ignorance that ought to be annihilated through practicing "observing causes and conditions" cannot be duly eliminated. Therefore, what will remain after the renunciation of the five aggregates of this life, when there is "no longer any future existence," is not nihilistic emptiness but the original abiding Dharma—the eighth consciousness, tathāgatagarbha. That is why the Buddha said to Susima, "This is what I mean when I say, 'One must understand the origin of the Dharma to understand nirvāṇa.'" According to this sūtra's teachings, if Buddhist disciples could annihilate the five aggregates of the self and mind to attain nirvāṇa, it is evident that during the First Turning of the Dharma Wheel, they would have already mastered the Dharma paths for overcoming greed, anger/hatred, and delusion. Through these paths, they would have been capable of liberating themselves from the three realms of existence. Could it be, as Schmithausen suggests, that the Buddhist Consciousness-Only teaching is a metaphysical strategy intended to counteract the practice of attachment described in the Buddha's teachings? This would falsely characterize the Consciousness-Only school as a "therapeutic devise" rather than a set of theoretical doctrines.

(3) "Ananda! What is meant by 'Upon the condition of vijñāna (consciousness), name and form (nāmarūpa) become'? If vijñāna does not enter the mother's womb, would there be name and form?" Ananda replied, "No (Long Āgama Sūtra, CBETA 2023)". This passage from the Long Āgama Sūtra shows a dialogue between the Buddha and Ananda indicating that depending on the "vijñāna," name and form are generated. What does this mean? If "vijñāna" does not enter a mother's womb, will there be any name and form? Ananda answered, "No." Next, the Buddha confirmed this and said, "If vijñāna enters the womb without being born, would there be name and form?" Ananda answered, "No!" This sūtra already indicates that there is another consciousness outside name and form, and the seven consciousnesses. To differentiate it from the seven-consciousness self of sentient beings, the World-Honored One also said that "another consciousness" must enter and remain in the womb for name and form to be born, and that "this consciousness" must reside in the womb for a sufficient period of time to be perfectly ready for name and form to be born.

As evidenced by the aforementioned sūtras, even during the period of the First Turning of the Dharma Wheel, the Buddha clearly and explicitly expounded the originally abiding Dharma in many instances. This original abiding Dharma, as the fundamental cause of name and form, serves as the basis for the establishment of nirvāṇa. While emphasized in the Four Āgama Sūtras, it was elaborated only in terms of the "observation of renunciation" for attaining the Path to Liberation within the Two-Vehicle Bodhi. However, the connotation of the sign of reality present in the "observation of tranquility" of Mahāyāna was not expounded in detail. Only the terms were mentioned, there were no explicit explanations of the entity nature of the sign of reality for all Dharmas within the Dharma realm. The connotation of the intrinsic limit on which remainderless nirvāṇa depends was also never mentioned. If one departs from the wondrous principle of the eighth vijñāna, tathāgatagarbha, taught in the Mahāyāna sūtras, then the state of remainderless nirvāṇa attained in the Path to Liberation taught in the Four Āgama Sūtras of the Two Vehicles will surely fall into the view of annihilation and become the nihilistic view of non-Buddhists. Consequently, the compilers of the Four Āgamas could summarize only the Mahāyāna tenets they had heard into sūtras pertaining to the Path to Liberation. However, this had no relation to the Buddha Bodhi Path of Mahāyāna. They covered only the dharmic teachings of the Two Vehicles, which enabled transcending the delimited existence within the three realms. The detailed practices of the tranquility path for attaining Buddhahood were never mentioned. No one could achieve the Path to Buddhahood based on those teachings alone, and the "observation of tranquility" in the Mahāyāna dharma could not be included. Detailed explanations would not be provided until a later period, when the Mahāyāna bodhisattvas from the four assemblies had finished compiling the Vaipulya and Consciousness-Only sūtras. It was only after the compilation of the Mahāyāna sūtras that fully expounded the Path to Buddhahood that the "observation of tranquility" previously taught by the Buddha in the Four Āgamas could be completed. Only then could the propagation of the World-Honored One's Path to Buddhahood be perfected.

Moreover, the academic claims of the Buddhist scriptures' fifth or even sixth compilations are incorrect. Only the First Council of the Five Hundred (Arhats) and the later legendary and superficially documented grand Buddhist council of a thousand people outside Saptaparna Cave (S. Saptaparṇaguhā) are mentioned in the Āgama Sūtras. The so-called later compilations referred to by academics have no relation to the compilation of scriptures and should not be called such.

3 THE MAHĀYĀNA BUDDHIST TEACHINGS WERE STRAIGHT FROM THE SOURCE, SPOKEN BY THE BUDDHA HIMSELF

A comprehensive examination of the contents of the aforementioned scriptural teachings during the compilation of the Four Āgamas at the First Buddhist Council revealed that although many of the sūtras originally contained the tenets of Mahāyāna teachings, because the śrāvakas who heard the Buddha's words could not gain supreme understanding, they were unable to accomplish the mental concomitant of memory regarding the meanings of the Mahāyāna Dharma. Consequently, from the perspective of the Path to Liberation, they compiled the Āgamas into the sūtras of the Two-Vehicle Bodhi. Therefore, the Mahāyāna sūtras obtained from the initial compilation by the śrāvakas inevitably became the contents of the sūtras on the Path to Liberation of the Two-Vehicle Bodhi. The bodhisattvas were certainly not satisfied with the compilation results and would naturally have expressed at that time the need for another compilation. From the cited scriptures and their true meanings discussed above, it can be deduced that the Mahāyāna sūtras were indeed spoken by the Buddha Himself and not fabricated by later people. Claiming that they were fabrications by later people would be a grave mistake. The reasons are as following:

(1) It was the Buddha's teachings that gave rise to the saints of the Path to Liberation and the Two-Vehicle Bodhi. As the Dharma meanings in the extant Mahāyāna sūtras far surpass those of the Four Āgama Sūtras, if someone claims that the Mahāyāna sūtras were written by later people, it would imply that the wisdom of later people exceeds that of the Buddha—which is a grave mistake.
(2) The contents of the Four Āgama Sūtras do not expound the Path to Buddhahood. It was only in the Vaipulya and Consciousness-Only Mahāyāna sūtras that the Path to Buddhahood began to be fully taught. This would reveal that the Four Āgama Sūtras did not fully expound the Buddha's teachings, leaving an extremely large portion of the Buddha Dharma yet to be taught in the latter Mahāyāna sūtras.

Moreover, there is no evidence whatsoever that Buddhist scholars and researchers both in this country and abroad can prove that the Mahāyāna sūtras were created and compiled hundreds of years after the Buddha's parinirvāṇa by later Sectarian Buddhists from the sound-hearer Dharma. For this reason, Pingshi authored The Correct Meanings of the Āgama Sūtras: Exploring the Origin of the Doctrine of Consciousness-Only in seven volumes using solely the words of the Buddha's teachings from the Four Āgama Sūtras to reveal the true meaning and principle of the Three-Vehicle Bodhi. Pinghsi has cited the golden words of the World-Honored One from the Long Āgama Sūtra about there being three Turnings of the Dharma Wheel as proof of the authenticity of the Buddha's teachings to the three assemblies. This demonstrates that the sūtras of the Three Turnings of the Dharma Wheel were all personally spoken by the Buddha, not gradually created and compiled by later people over a long period. In this way, it has been proven that the Mahāyāna sūtras were originally spoken by the World-Honored One Himself and not fabricated by later people.

4 CONCLUSION

The great Master Xuanzang, during the ancient era of India and the Tang dynasty, conclusively affirmed and settled the true principle of the Buddha's teachings—that all Dharmas are mere Consciousness-Only. This means that all Dharmas arise and are accomplished by the Mind—the eight consciousnesses—and that the first seven derivative consciousnesses arise from the eighth consciousness; apart from the eighth consciousness, there is no single Dharma or state that can be attained. Upon his return to the Tang Kingdom, Master Xuanzang took up the monumental task of translating a vast corpus of sūtras and treatises, imbuing them with his profound insight into the Buddha Dharma, as directly expounded by the Tathāgata himself. Later, upon the request of the Buddhist community, he also authored Cheng Weishi Lun and the Verses Delineating the Eight Consciousnesses, elucidating therein, with breathtaking depth and clarity, the intrinsic virtuous qualities of the "eight consciousnesses." His exegeses plumbed the unfathomable depths of these teachings with unparalleled mastery, revealing vast vistas of supreme profundity hitherto unsurpassed. These peerless, magnificent expositions have been reverentially studied and circulated by generations of disciples until this very day across lands and ages. The Buddha's teachings in the First Turning of the Dharma Wheel focused on enabling Two-Vehicle practitioners to personally attain liberation from transmigration of birth and death. The Second Turning of the Dharma Wheel, represented by the Prajñā Sūtras, aimed to help Bodhisattva practitioners gain a thorough understanding of the sign of reality. Lastly, the Third Turning of the Dharma Wheel, known as Vaipula and Consciousness-Only, emphasized the knowledge-of-all-aspects for the perfect accomplishment of the Buddha Bodhi. All the teachings in the three Turnings of the Dharma Wheel consistently revolve around the core Dharmic gist of the eighth consciousness, the tathāgatagarbha, revealing the nuanced differences in depth and breadth in the Three-Vehicle Bodhi without any contradictions. This is why Master Xuanzang commented on scholar-monk Sijha-raśmi as follows: "The inability of confused persons to understand complementarity would lead them to assert the existence of a contradiction between the Dharmas; thus, the fault lies with the transmitters, not with the Dharma (A Biography of the Tripitaka Master of the Great Ci'en Monastery of the Great Tang Dynasty, CBETA 2023)". For this reason, we know that if Buddhist disciples want to have the virtues stemming from understanding complementarity in the Three-Vehicle Bodhi, they must truly attain the Path of Vision—awakening to the True Mind. The watershed is the personal realization of the Mind, which relies on the prajñā wisdom stemmed from truly attaining the Path of Vision, cultivating the post-awakening stage of the "Characteristic Path of Vision," and advancing to the Proficiency stage to accomplish the knowledge-of-specific-aspects pertaining to prajñā and knowledge-of-the-aspects-of-paths pertaining to Consciousness-Only. This is the basis of the practical value of personally realizing the eighth consciousness to experience the Consciousness-Only school beyond its theoretical value. Its meta-value lies in personal realization. However, all practitioners of the first fruition (srotaāpanna) who have eliminated the three fetters, including arhats in the Two-Vehicle Bodhi, can directly witness that the teachings in the Mahāyāna sūtras are far more wondrous and profound than those in the Four Āgama Sūtras. The Mahāyāna sūtras represent the complete teachings of the Buddha spanning the Three Turnings of the Dharma Wheel across the early and later periods. They encapsulate, in their entirety, the consummate Path to Buddhahood—comprising all the causes and conditions for Buddha Śākyamuni's manifestation in the Saha world to liberate sentient beings. As such, Mahāyāna is a teaching directly delivered by the Buddha Himself. When certain scholars or practitioners posit contradictions or conflicts between Mahāyāna scriptures across different eras or allege that these sublime sūtras were mere posthumous compilations, they betray a fundamental lack of personal realization of the Dharma's truth. They remain unawakened, akin to the ordinary persons the Diamond Sūtra speaks of—fixated on textual study with their "naked eyes," while bereft of the penetrating "wisdom-eye" and "Dharma-eye" that can bring forth true Buddhist wisdom. Individuals in such a state cannot possibly fathom the essence of Buddhist sūtras and commentaries, nor can they gain the profound wisdom therein. However, this is not on account of Buddhism itself but of their own inability to recognize the coherence permeating the Great Vehicle in Buddhism.

REFERENCES

A Biography of the Tripitaka Master of the Great Ci'en Monastery of the Great Tang Dynasty, CBETA 2023. Q3, T50, no. 2053, pp. 244c2–3.

Da Tang Gu San Zang Xuan Zang Fashi Xing Zhuang, CBETA (2023). Q4, T50, no. 2052, pp. 214c6–10.

Jeremy Ph.D., Hayward W.W., *et al.* (2001). *Gentle Bridges: Conversations with the Dalai Lama on the Science of Minds, U.S.A.*, pp. 31, EBAY.

Lambert Schmithausen. (2015). *On the Problem of the External World in the Ch'eng wei shih lun, Electronic Edition with Corrections and Additions*, Tokyo, The International Institute for Buddhist Studies

Long Āgama Sūtra, CBETA 2023. Q4, T01, no. 1, pp. 8b9–10

Long Āgama Sūtra, CBETA 2023.Q4, T01, no. 1, pp. 61b7–10.

Madhyama Āgama, CBETA, T01, no. 26, pp. 765, a26.

Saṃyukta Āgama, CBETA 2023. Q4, T02, no. 99, pp. 12b26–29.

Saṃyukta Āgama, CBETA 2023. Q4, T02, no. 99, pp. 97c1–2.

Xiao P. S. (2013). *Mastering the Essence of the Diamond Sūtra*, Taiwan, True Wisdom Publishing Center, pp. 2.

Xiao P.S. (2007). *The Correct Meanings of the Āgamas: Exploring the Origin of the Doctrine of Consciousness-Only*, pp. 10–30.

Xiao P.S., (2023). Commentary on Cheng Weishi Lun, Taiwan, True Wisdom Publishing Center, pp. 6–8.

Yogācārabhūmi, CBETA 2023. Q4, T30, no. 1579, pp. 345a22–23

Yu J. W., Chou H. M. (2017). Notice of retraction: The meta-value of Yogācāra Buddhism, *2017 International Conference on Applied System Innovation (ICASI), Sapporo*, Japan, pp. 1048–1051.

System Innovation for an Artificial Intelligence Era – Lam et al. (Eds)
 ISBN: 978-1-032-84452-7

A study on psychological defense mechanisms and memory storage in multiple personality disorder: An interpretation using the three-layer psychological structure memory model

Wei-Hsuan Fan
Ph. D. Program, Department of Mechanical Engineering, Kun Shan University, Tainan, Taiwan

ABSTRACT: Multiple Personality Disorder (MPD) is a most extreme pathological dissociative disorder, usually characterized by activation of patient unconscious psychological defense mechanisms in response to severe physical and mental trauma. This causes the patient to enter various mental states, resulting in multiple alternating personalities with varying memory information regarding age, gender, thoughts, behaviors, languages, and expertise.

However, how do the unconscious psychological defense mechanisms of patients with MPD activate and function? Furthermore, where do patients with MPD store the large amounts of memory information that they have never experienced or acquired? Beyond the physiological basis, consciousness is only the most superficial and explicit component of the three-layer psychological structure memory model constructed by the consciousness and two types of unconsciousness. Apart from consciousness, we must differentiate two types of unconscious physiological functions: Type one is the memory channels and spontaneity functions, and type two has the memory storage functions. This paper argues that the psychological defense mechanism of MPD falls within type one's unconscious memory channel and spontaneity functions, and the different personality memories that emerge in MPD are caused by type two's unconscious memory storage function.

Keywords: Multiple Personality Disorder (MPD), Unconscious, Psychological defense mechanisms, Memory storage, Three-Layer Psychological Structure Memory Model

1 INTRODUCTION

In psychology, scholars have different definitions and perspectives on personality. Allport defines personality as the dynamic organization of an individual's inner psychological and physiological systems, determining their unique thoughts and behaviors (G. W. Allport, 1961). Phares and Chaplin define personality as a unique pattern of thoughts, emotions, and behaviors specific to an individual, differing from the patterns of another person, and maintaining consistency across different times and situations (E. Jerry Phares, *et al.* 1997). While scholars differ in their definitions, characteristics common to the concept of the personality can be observed from the above definitions. Personality is seen as a unique pattern of thoughts, emotions, memories, and behaviors (including habits and attitudes) generated within the operation of an individual's internal physiological and psychological systems. This pattern maintains consistency and stability across different times and contexts. Through these commonalities, distinctions can be made between the personality patterns of one individual and another.

However, if the same individual exhibits two or more personalities or identity states, alternating control over thoughts, emotions, memories, and behaviors; two or more independent and unique patterns of personality or identity exist; different behavioral patterns at various times and contexts exist; and the individual is unable to maintain consistency and stability states, then this condition becomes a pathological psychological disorder known as Multiple Personality Disorder (MPD), later renamed Dissociative Identity Disorder (DID) (American Psychiatric Association, 2007). However, for reading convenience and understanding, this paper will continue to use the term "Multiple Personality Disorder (MPD)".

Based on their different definitions of personality, many psychologists have developed different personality theories, aiming to provide systematic statements on the concept of the personality, and understand its operational aspects and interactive influences, each with their own emphases.

Sigmund Freud, a key figure in psychoanalytic theory, argued that the overall structure of personality consists of three parts: The id, ego, and superego. Conflicts and coordination occur between these three parts, influencing an individual's psychology and behavior. Freud proposed that the ego and superego are partly within the individual's consciousness and partially in the

 DOI: 10.1201/9781003514831-67

unconscious (also referred to as the subconscious), the id belongs entirely to the unconscious (subconscious). However, Freud's unconscious (subconscious) still belongs to the overall structure of consciousness. Therefore, Freud's psychoanalytic theory still confines itself to one of the traditional two-layer body-mind structure, with a physiological foundation and consciousness as the core. This paper argues that the two-layer body-mind structure memory model lacks power to interpret the phenomenon of multiple personalities.

Going beyond the traditional two-tier body and mind structure model with a physiological foundation and consciousness as the core, Lichen Tsai proposes, from the perspective of developmental biology, that beyond consciousness there are two unconscious psychological functions, forming a three-layer psychological structure memory model. The unconscious can be divided into two psychological functions: The first is the expandable psychological function, that is, the portion where psychological functions and effects can increase as the fetus grows and develops. The second is the fixed and non-expandable psychological function, that is, the portion where psychological functions and effects do not increase as the fetus grows and develops (L.C. Tsai, 2024). This paper attempts to cite the three-layer psychological structure memory model to interpret the phenomenon of MPD.

It is generally believed that the development of MPD is extremely likely to be triggered by the activation of personal psychological defense mechanisms. Why does the personal psychological defense mechanism lead to the manifestation of personalities distinct from the original one? MPD can be explained as an individual's psychological defense mechanism being activated due to severe trauma, and perceptions and emotions related to traumatic events being removed from consciousness, giving rise to another personality with distinct memories, emotions, and behaviors. These phenomena seem to indicate that beyond consciousness, humans possess unconscious psychological functions with active mental agency to control the activities of consciousness, giving rise to alternate personalities. At times, these may not be under the conscious control or choice of individuals with MPD. Therefore, the functionality of personal psychological defense mechanisms and the memory storage of alter egos' personalities cannot be attributed to consciousness.

The question of how to interpret and understand the phenomenon of MPD touches on in-depth understanding what structures are necessary for human psychological functions. Only with an interpretative psychological structure for the phenomenon of MPD can we approach to understanding the true structure of human psychology. Therefore, exploring psychological defense mechanisms and memory storage associated with the phenomenon of MPD must hold significant meaning for understanding the unique structures that may be present in human psychological functions.

2 THE FIRST TYPE OF UNCONSCIOUS PSYCHOLOGICAL DEFENSE MECHANISM

In psychoanalytic theory, the concept of psychological defense mechanisms was first introduced by Freud in his 1894 paper "The Neuro-Psychoses of Defense" (Sigmund Freud, 2014 July). Freud argued that psychological defense mechanisms are unconscious (or subconscious) psychological processes that individuals use to cope with external pressures, and to resolve emotional anxieties and trauma. When an internal conflict arises in an individual, the unconscious employs different behavioral patterns to defend the individual self. These behavioral patterns are what we refer to as "psychological defense mechanisms (Sigmund. Freud, 2014 October)".

While Sigmund Freud and others proposed "unconscious" psychological defense mechanisms, Freud conceptualized consciousness and the unconscious as something like an iceberg, where consciousness is only the visible part above the water, and the vast unconscious is hidden beneath the surface. Therefore, consciousness and the unconscious are fundamentally part of the overall consciousness. The 'water surface' of the Freudian iceberg metaphor currently lacks specific content and indicators to delineate the boundary between consciousness and the unconscious. However, a three-layer psychological structure memory model from the perspective of developmental biology allows for a clear distinction between consciousness and the unconscious, with further analysis of the functions of the unconscious, while providing a rational explanation for the phenomenon of disorder multiple personalities.

In a diagram of the three-tier psychological structure memory model in Figure 1 (L.C. Tsai, 2024), in addition to the physiological foundation

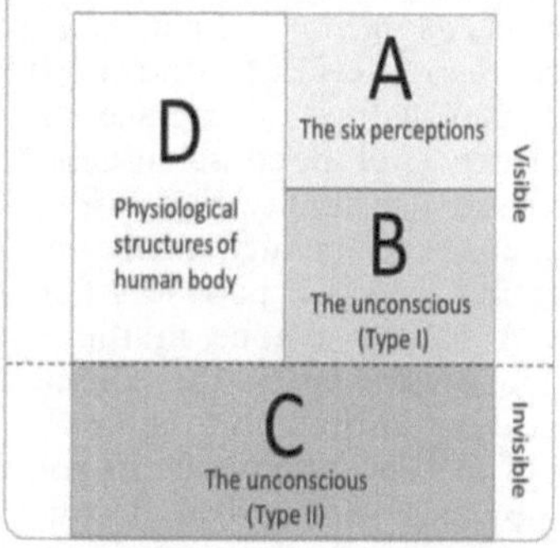

Figure 1. The diagram of the three-layer psychological structure.

D, humans have three layers of psychological structures: A represents the six perceptions (consciousness, sight, hearing, smell, taste, and touch); B represents the first type of unconscious function, proactive psychological functions like survival and reproductive instincts, and the psychological defense mechanisms discussed in this paper; and C represents the second type of unconscious function, namely the memory storage function.

The consciousness A (including consciousness, and the vision, hearing, smell, taste, and touch sensory perceptions) and the unconscious B are entirely different psychological domains. The most significant difference is that while consciousness A is the most superficial, outward, and easily known to others, it is also a psychological function that only appears in the later stages of fetal development as the brain is gradually maturing. It is interrupted during deep sleep and reemerges during dreaming. On the other hand, the unconscious B already exists during fetal development, and regardless of whether the consciousness A is interrupted during deep sleep or emerges during dreaming, the unconscious B continuously operates in conjunction with the physiological foundation D, persisting throughout life without interruption. Therefore, the unconscious B is the active force behind controlling the appearance or disappearance of consciousness A in mental activities. Consciousness A has no control over when it should appear or disappear. Therefore, after humans fall into deep sleep, consciousness A is already interrupted without any awareness. However, the unconscious B and the physiological foundation D continue to operate uninterrupted. The unconscious B instinctively continues to exert control behind the scenes, determining when to let the consciousness A reappear, which is how waking up occurs.

MPD is a most extreme phenomenon in pathological dissociation, which separates the consciousness A from the real world. When the consciousness A faces stress or trauma that is too painful to endure, the unconscious B, which serves as the channel for memory storage and retrieval, activates a defense mechanism to protect itself. It actively shuts down the retrieval of memories related to the original personality, interrupts the consciousness A's memory of painful experiences, and triggers the retrieval of memories from other alter egos' personalities to replace the relevant painful experiences of the consciousness A with the newly retrieved personality memories. Since the newly retrieved personality memories are not encoded from the real experiences that the consciousness A has undergone in this world, the phenomenon of separating the consciousness A from the real world occurs in the newly retrieved memories of the unconscious B. In any case, the defense mechanism's survival instinct of the unconscious B controls the activities of the alter egos' consciousness A. Whether the alter egos are aware or unaware of each other's existence, and whether or not they interact, it is all a manifestation of the unconscious B serving as the memory channel linking the unconscious C and the consciousness A.

3 THE SECOND TYPE OF UNCONSCIOUS IS METAPHORICALLY A MEMORY STORAGE LOCATION WITHOUT LIMITATIONS ON CAPACITY OR TIME

For individuals with MPD who possess numerous personality memories, questions like "Where do these vast personality memories come from?" and "Where are they stored?" are worthy of further exploration. Currently, research on the storage of personality memories in individuals with MPD still follows the general hypotheses about memory storage. Scholars have varying perspectives on the storage location of memories. For instance, neuroscientists in the 1950s mostly argued that short-term memory is stored in the hippocampus, while long-term memory is stored in the cerebral cortex. 21st-century research suggests that when an event occurs, both short-term memory in the hippocampus and long-term memory in the cerebral cortex are generated simultaneously. However, many recent studies indicate that memory storage is not confined to specific regions, but is distributed throughout the brain, particularly within specific neural cell groups and their synapses in the brain's neural networks.

This paper posits that regardless whether memory data is stored in the hippocampus, cerebral cortex, or neural synapses, the hypothesis that memory data is stored in the physiological foundation D cannot reasonably explain the phenomenon of MPD patients alter egos exhibiting distinct sets of memories, thoughts, emotions, and behaviors. The formation of a human personality involves not only the physiological foundation D, but also requires a conscious learning process from childhood to adulthood. Through long-term learning, individuals gradually become aware of and understand external social culture and engage in cognitive learning. This enables the encoding, storage, and retrieval of memories, leading to the development of a stable personality. After experiencing trauma, individuals with MPD suddenly acquire many different personality memories. If memory data were stored in the physiological foundation D of the brain, the personality memory contents encoded from experiences of different personalities in this lifetime, which the conscious A has never undergone, would not be possible to be stored in the physiological basis D of the brain. The sudden appearance of a unique set of memories, thoughts, emotions, and behaviors associated with other alter egos in the hippocampus, cerebral cortex, or neuron synapses of the brain's physiological foundation D for individuals with MPD is naturally an impossible phenomenon.

This is because the physiological foundation D undergoes gradual development from the embryo, encoding, storing, and retrieving personality memories of this life in collaboration with conscious A. Therefore, we can reasonably infer that the alter ego memory data observed in individuals with MPD is not extracted from the brain's physiological foundation D. Instead, it is very likely stored in the psychological function of the unconscious memory storage location C. This storage contains memories of different personalities across multiple lifetimes. Thus, the unconscious memory storage location C is likely to possess eternal psychological functions that transcend multiple life periods, and its memory capacity may have no limitations.

Therefore, this paper argues that the commonly accepted notion of memory storage in the brain's physiological foundation D cannot reasonably explain the issue of where the memory data of different personalities in MPD is stored. Apart from the physiological foundation D, through the awareness and understanding of the consciousness A, humans cognitively learn, encode memories, or retrieve memories. Additionally, we need the unconscious B as an intermediary memory channel function to store memory data in the unconscious C. The memory data stored in the unconscious C is then provided as the memory content that the consciousness A wants to retrieve. The unconscious B plays a crucial role as the intermediary memory channel. Therefore, the three-layer psychological structure memory model is reasonable and possesses explanatory power.

4 CONCLUSION

In the research and understanding of MPD, there are still some controversies and areas that require further investigation. Current research on the phenomenon of MPD attributes psychological defense mechanisms to the unconscious, but the division between consciousness and the unconscious lacks clear indicators or standards. From the perspective of developmental biology, the psychological functions of the unconscious are complex and require further differentiation. Therefore, attributing psychological defense mechanisms solely to the unconscious cannot enhance our understanding of the phenomenon of MPD.

In addition, current research on the phenomenon of MPD has found that, while the general population typically possesses a single, stable set of personality memories, individuals with MPD have multiple personality memories of unknown origin that recur, resulting in a phenomenon where a single physiological foundation hosts distinctly different sets of personality memories. The phenomenon of different personality memories coexisting with the original personality appears to be closely related to the concepts of memory channels and memory storage. In particular, the mechanisms of switching between different personalities, and the origin and storage location of alter egos outside the original personality, remain subjects of investigation. Current research on MPD still follows the general hypothesis of memory storage, but lacks explanatory power with regard to these questions.

This paper utilizes the three-layer psychological structure memory model to explore how, with individuals with MPD, the switching of different personality memories is attributed to the activation of psychological defense mechanisms. Psychological defense mechanisms are considered to belong to the first type of unconsciousness, which is characterized by its proactive and memory channel functions. This attribution is essential for the phenomenon of personality memory switching to occur. As for the personality memories of the individual, they must be stored in the second type of unconsciousness, which possesses memory storage psychological functions. In individuals with MPD, due to the activation of the psychological defense mechanism, the first type of unconsciousness extracts other pre-existing alter ego memories from the second type of unconsciousness to the consciousness A, resulting in the phenomenon of multiple personalities. This paper argues that in MPD case studies, in addition to having the physiological foundation D and consciousness A, humans must also possess active mental activities initiated by the first type of unconsciousness B's defense mechanisms, and the actual storage location C of memories in the second type of unconsciousness. These elements provide a reasonable interpretation of the phenomenon of multiple personalities. Therefore, the three-layer psychological structure memory model is a model worthy of further consideration and exploration in the fields of psychology and philosophy.

REFERENCES

Allport G. W. (1961). *Patterns of Growth in Personality*. New York: Holt, Rinehart and Winston.

American Psychiatric Association. (2007). DSM-IV-TR diagnostic and statistical manual of mental disorders, 4th ed. *Translated by Kung Fan-chung*. Hochi Publishing.

Jerry Phares E., William F. Chaplin. (1997). *Introduction to personality*. New York: Longman.

Sigmund Freud. (2014 July). *The Neuro-Psychoses of Defence*, Paperback, White Press.

Sigmund. Freud. (2014 October). *Five Lectures on Psycho-Analysis*. Translated by Li Yun, Shanghai Academy of Social Sciences Publishing House.

Tsai L. C. (2024). *Decoding the Unconscious for Insight into Free Will and Memory Storage - A Comparative Analysis of the Three-Layer Psychological Structure and the Two-Layer Body-Mind Memory Model*, ICASI, Kyoto.

System Innovation for an Artificial Intelligence Era – Lam et al. (Eds)

Decoding the unconscious for insight into free will and memory storage – A comparative analysis of the three-layer psychological structure and the two-layer body-mind memory model

Li-Chen Tsai
Ph. D. Program, Department of Mechanical Engineering, Kun Shan University, Tainan, Taiwan

ABSTRACT: Current memory models are categorized into a traditional two-layer body-mind structure where is the foundation and consciousness is the core. This structure does not reasonably explain common psychological functions such as free will and memory storage, nor phenomena like forgetting and recall. This is because, beyond consciousness, there exists a structural unconscious function. However, modern science currently lacks specific analytical capabilities to understand unconscious functions, resulting in a lack of explanation for psychological functions such as free will and memory storage. This paper argues that based on the facts of developmental biology, consciousness is a psychological function that gradually forms in the later stages of fetal development. The existence and operation of unconscious psychological functions precede conscious functions and persist throughout life. Therefore, beyond the physiological foundation there should be a three-layer psychological structure, where consciousness is merely the outermost visible part of the structure. This paper suggests that unconscious psychological functions should be divided into those that can change with bodily development (possessing memory channels and initiative) and those related to memory storage (which do not change with bodily development). The three-layer psychological structure memory model, composed of consciousness and two types of unconsciousness, explains human psychological functions such as free will and memory storage. This model facilitates a clearer analysis and presentation of unconscious functions, contributing to the advancement of modern psychological research.

Keywords: Unconscious, Free will, Memory storage, Developmental biology, Two-layer body-mind memory model, Three-layer psychological structure memory model

1 INTRODUCTION

With the progress of science, humans can use technological devices and techniques to explore our psychological functions. The topics of free will and memory storage remain two unresolved and important subjects of debate. Physics determinists view life as mechanistic, where psychological functions are simply manifestations of the body and do not exist independently of the physiological foundation. Therefore, free will is an illusion. In 1983, Benjamin Libet (1916-2007) and other scholars jointly published a famous experiment, the readiness potential (RP), which indicates that brain activity preparing for motor action precedes the reported time of conscious intention to act, even when performed voluntarily by the subjects (Benjamin Libet *et al.* 1983). The result of this experiment further strengthens determinists' denial of free will.

For indeterminists, the quantum brain hypothesis suggests that brain function also adheres to the rules of quantum mechanics (J. M. Schwartz *et al.* 2005). For example, the behavior of brain ions and neurotransmitters, whether calcium ions can reach position to stimulate synaptic vesicles, and the release/non-release of neurotransmitters, etc., generates a superposition of brain activities that can be expressed through Schrödinger's wave function. Therefore, probability theory becomes a new basis for making free will a reality.

Libet was a supporter of free will and found it difficult to accept the interpretation of his previous experimental results as suggesting the non-existence of free will. Therefore, in 1985, he published a scientific paper titled "Unconscious cerebral initiative and the role of conscious will in voluntary action", proposing the concept of "free won't (Benjamin Libet, 1985)". In the experiment, subjects exhibited electroencephalographic readiness potential 0.55 seconds before movement, while their conscious awareness of the intention to move occurred 0.2 seconds before the action. Libet

DOI: 10.1201/9781003514831-68

stated that in conclusion, the initiation of spontaneous voluntary actions in the brain begins unconsciously. However, the research found that in the remaining c. 150 milliseconds after the appearance of a specific conscious intention, the final decision to act could still be consciously controlled. In fact, subjects could "veto" their planned actions within 100-200 milliseconds before the act (Benjamin Libet, 1985). As a result, there has been an ongoing controversy as to whether Libet's free will experiments support the existence of indeterminism, or the nonexistence of determinism.

Both the mechanical explanation of physical determinism or the probabilistic explanation of indeterminism lack a comprehensive discourse and understanding when it comes to denying or affirming the existence of free will. This is because all mental activities involve the storage and retrieval of memory, enabling the continuous operation of psychological functions. The operation of free will, whether illusory or not, invariably requires the continuous encoding, storage, and retrieval of a substantial amount of memory information at every moment of coherent psychological functioning. Therefore, both the mechanical and probabilistic explanations fall significantly short of understanding psychological functions. This paper argues that this limitation is due to modern memory models being confined to the two-layer body-mind structure of physiology and consciousness. Although quantum mechanics operates on probability and introduces the concept of observer consciousness, leading to the collapse of superposition and attempting to break free from the two-layer body-mind structure, the hypothesis of the quantum brain still can't escape the constraints of the physiological consciousness-based two-layer body-mind structure.

In his 1985 paper, Libet mentioned "unconscious cerebral initiative". This paper argues that the semantics of "unconscious" are vague and prone to ambiguities. Libet's usage of the term "unconscious" does not break away from the two-layer body-mind structure. In this paper, the "unconscious" denotes a structural aspect of psychological functions, an "unconscious" anatomically separable from consciousness – one that closely aligns with what Searle describes as the "deep unconscious", encompassing states and contents that are essentially inaccessible to consciousness (J. M. Schwartz *et al.* 2005). Furthermore, a deeper analysis of the "unconscious" surpasses the body-mind two-layer memory model constructed of a physiological foundation plus consciousness. Instead, it gives rise to a three-layer psychological structure memory model, offering a reasonable resolution and interpretation for the existence of free will and memory storage.

2 DILEMMAS OF THE TWO-LAYER BODY-MIND MEMORY MODEL

The body-mind two-layer structure, with physiological foundations and consciousness at its core, has a long history. In the 19th century, discussions on the philosophy of the unconscious emerged, but the content still revolved around physiology foundation and consciousness as the core. For example, Eduard Von Hartmann (1842-1906) in Philosophy of the Unconscious stated that consciousness only acquires value through memory; that is, leaving a lasting impression through brain oscillations, or molecular changes that are more easily triggered under the same oscillations as before (Eduard Von Hartmann, 2000). Hartmann believed that memory exists in the brain's physiological substance, and stated that for the "unconscious", we can say there is no memory. Understanding recent events is possible because impressions reside in the brain. With brain damage, memory may be entirely or partially lost, temporarily or permanently (Eduard Von Hartmann, 2000). Hartmann's concept of the "unconscious" is the essence of the universe, a primitive state preceding will and reason. He considered the "unconscious" to be synonymous with having no memory, indicating a complete absence of psychological functions beyond consciousness in humans. Since memory is stored in the brain's molecules, brain damage may result in total or partial, temporary or permanent memory loss. Therefore, Hartmann is a typical representative of this two-layer body-mind structure thought.

However, after Hartmann, in The Interpretation of Dreams, Sigmund Freud (1856-1939) proposed that the unconscious is the repository of forgotten dreams (Freud, Sigmund, 1996). When analyzing dreams, Freud discovered that a significant portion of dreams is forgotten, but through psychoanalytic methods, sometimes forgotten dreams can be retrieved. If memory is stored in the brain's molecules, forgetting would mean the brain molecules lack records and permanently disappear, making it seemingly impossible to recover or reconstruct memories through psychoanalysis. Therefore, Freud believed that the unconscious is the repository of forgotten dreams; the content of dreams did not disappear, but rather clues were lacking for retrieval. However, in Freud's semantics, consciousness and the unconscious are like an iceberg, with consciousness being the visible part and the vast unconscious that stores all memories hidden beneath the surface. Only the whole composed of both consciousness and unconsciousness is the true, complete consciousness. Hence, Freud's constructed iceberg of consciousness still falls within the framework of the two-layer body-mind structure.

The two-layer body-mind structure is not logical because, based on the facts from developmental biology, consciousness is a psychological function that gradually forms in the later stages of fetal development. Before the physiological foundation supporting conscious operations matures, all fall within unconscious psychological functions, and these unconscious psychological functions persist throughout one's life. Therefore, the existence of unconsciousness in psychological structures beyond consciousness aligns with the facts of developmental biology. Since Freud and other psychoanalytic schools fundamentally adhere to the two-layer body-mind structure, which has given rise to a myriad of interpretations, it becomes evident that the two-layer body-mind structure lacks explanatory power for psychological functions. This has inevitably led to differences of opinion and debates between various schools of thought.

In Libet's experiment, he referred to the activation of readiness potential in the brain 0.55 seconds before the action, hence the term "unconscious cerebral initiative". This paper argues that the activation of "initiative" by the physiological material of brain neurons is, in fact, a structurally unconscious psychological function. This function operates independently of consciousness; or put another way, this unconscious function is the foundation on which consciousness exists and functions. Therefore, from fetal development when consciousness emerges, unconsciousness continuously and closely coordinates with consciousness throughout life. The psychological functions of the unconscious begin developing in the fetal stage, preceding the existence of consciousness. Therefore, the phenomenon of unconscious psychological functioning's proactive readiness potential manifesting earlier than the conscious intentional awareness is a reasonable occurrence. Consciousness must ratify the actions of the unconscious, fully aligning with the fact that unconscious functions exist before conscious functions.

Regarding the quantum brain hypothesis attempting to utilize probability theory as the practice of free will, Wen Jung-jung states that the probability theory of the quantum brain contradicts the free will of conscious focus learning (J. J. Wen, 2024). This is because the encoding and focus of consciousness rely on the continuous initiative of the unconscious. Probability can only apply to a single event, so probabilistically speaking, maintaining continuous encoding and focus for a significant duration should be impossible. Additionally, the encoding and focus of consciousness must also have the memory storage function of the unconscious to achieve the free will of applying focus to learning. The continuous initiative of the brain and memory storage must be functions of the unconscious, not functions achievable by consciousness alone. Therefore, the two-layer body-mind structure lacks a reasonable explanation power for the free will of conscious focus learning. This paper argues that for a more reasonable explanation of free will and memory storage issues, we must abandon the narrow two-layer body-mind structure and analyze the implied psychological functions of the unconscious to move toward a more reasonable explanation.

3 THE THREE-LAYER PSYCHOLOGICAL STRUCTURE MEMORY MODEL CAN EXPLAIN PHENOMENA REASONABLY

From the perspective of developmental biology, since the psychological functions of the unconscious precede those of consciousness, how should we reasonably categorize the psychological functions and effects of the unconscious? Reflecting on developmental biology, the functions and effects must be classified into two types. The first type is expandable functions and effects – those psychological functions and effects that increase as the fetus grows and develops. The second type is fixed and non-expandable functions and effects – those psychological functions and effect that do not increase with the growth and development of the fetus (Lichen Tsai, 2011).

First, the second type of fixed and non-expandable unconscious psychological function must be the memory storage function. As we undergo development, growth, and aging, our bodies go through stages of growth, weaken, and atrophy. However, our memories, operating on the basic physiological foundation, continue to accumulate without loss. Some memories that we believe to have been completely forgotten can suddenly be recalled given certain clues or conditions. Therefore, the memory storage of the unconscious must be a fixed and unchangeable psychological function. Thinking about the psychological functions of life from the perspective of modern computer science, protecting data is the most essential task of all computer systems. Similarly, in terms of the functions and operations of life, the continuous accumulation and non-loss of memory are also the most precious and irreplaceable aspects of life. As Hartmann stated, "consciousness only gains its value through memory", and life is no exception.

In other words, memory content can continually increase, but the psychological function for storing memories remains fixed and unchanged. When asked "How much information can long-term memory store? How long can this information be maintained?" and other questions related to memory storage capacity, cognitive psychologist Robert J. Sternberg candidly admitted: "We don't know, and we don't know how to find the answers (Robert J. Sternberg *et al.* 2013)". We can design experiments to measure the limits of short-term memory, but when it comes to long-term memory, which may span an entire lifetime, we are unsure how to measure its limits and determine its capacity. However, in terms of the rationality of dividing unconscious psychological functions, we should consider the memory storage function independently.

The first type of expandable unconscious psychological function, from the perspective of developmental biology: the gradual division of the fertilized egg, ultimately leading to the birth of the baby after basic human tissues and organs are formed. Therefore, the consciousness supported by the body's tissues, organs, and brain neurons must rely on the unique and proactive psychological functions of the unconscious. Examples include instincts like survival and reproduction, which don't require learning, and the proactive psychological function of exploring the world and focusing on learning to adapt to the environment. Although these psychological functions are closely tied to consciousness, from the standpoint of developmental biology, much of the instincts and proactivity possessed by consciousness originate from unconscious psychological functions. Especially considering that the manifestation of these conscious psychological functions requires encoding, storing, and retrieving memories, the first type of expandable unconscious psychological function must include the function of linking consciousness and the second type of memory storage channel. Therefore, proactivity and functioning as a memory storage channel can be summarized as characteristics of the first type of expandable psychological function.

Figure 1 is a diagram of the three-layer psychological structure, where A represents the six perceptions: consciousness, vision, hearing, smell, taste, and touch. B represents the first type of unconscious function, C represents the second type of unconscious function, and D represents the overall physiological organization and organs of humans. A, B, and C, from top to bottom, illustrate the three-layer psychological structure, but this doesn't imply D's absence or non-function. In the two-layer body-mind structure, only A and D are present, which is a partial structure of the three-layer psychological structure rather than a complete psychological structure. In the overall three-tier psychological structure, A, B, and D are observable components. It is challenging to discover and observe C unless one possesses the learned and trained psychological anatomical skills that give consciousness the observational ability needed for the level of consciousness (L. C. Tsai *et al.* 2019).

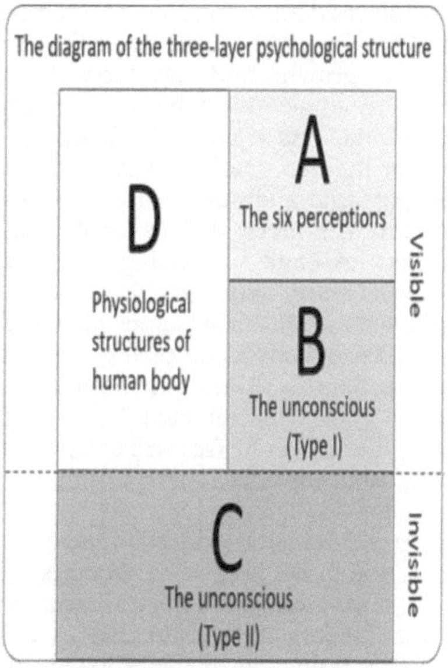

Figure 1. The diagram of the three-layer psychological structure.

B below A represents the first type of unconscious function as the foundation for the operation of the six perceptions. C below BD indicates that the functions of ABD all require the support of C. The D beside AB signifies the material physiological foundation, indicating that the functional operations of consciousness A and unconsciousness B both require the cooperative operation of the material D physiological foundation. Therefore, if the physiological functions of D are impaired, this may lead to a decline and impairment in the operation of the six perceptions or the functions of unconsciousness B.

For example, when brain D is affected by disease, causing a memory decline, it is not the disappearance of the memory data itself but a decrease and impairment in the functioning of the unconscious B as the memory channel. This results in the inability to access the memory content of C below. What people commonly refer to as "memory decline" is a phenomenon caused by poor coordination and cooperation between the functions of A and B. Because the function of B as a memory channel is not smooth, it inevitably leads to difficulties of consciousness A in memory encoding or retrieving, manifesting as "memory decline". If the brain D function is restored through proper rest or treatment, the memory capabilities of both A and B working together can be restored. The ability of humans to recall forgotten memories indicates that the unconscious B indeed possesses the function of a memory channel.

The survival instinct of the unconscious B function exists and operates actively from the initial state of the fertilized egg, just like the bottom-level resident programs (daemons) in a computer, continuously cycling the will to survive, enabling the growth and development of life. Therefore, spontaneous voluntary actions are initiated by the unconscious B function, which exists earlier than consciousness A, activating the readiness potential and notifying consciousness A to coordinate the perceptual psychological functions. This facilitates encoding a "perceived action" into memory data, with the unconscious B storing the data in the unconscious C. The ABCD whole collaboratively completes spontaneous voluntary actions and the memories requiring recording. Thus, self-directed and sustained focused actions driven by free will exist in humans, all accomplished by life as a whole through the three-layer psychological structure.

In summary, the unconscious B function holds a crucial position in life because it occupies a central role in the three-layer psychological structure. The unconscious function B is connected to C and D from the very beginning of life, allowing the psychological functions of consciousness A to gradually grow and develop into the top layer of overt psychological functions. The attention-grabbing conscious psychological functions overshadow the most central – yet difficult-to-discover and to observe – unconscious psychological functions in life.

4 CONCLUSION

The two-layer body-mind structure, with the physiological foundation and consciousness as the core, can be traced back to the exploration of the soul in the era of Plato and Aristotle. However, the body-mind two-layer structure represents a modern examination of life, initiating research into the "unconscious" beyond consciousness. Unfortunately, various schools of psychoanalysis within the body-mind two-lay structure, despite delving into the content of the "unconscious", still revolve around the concept of consciousness itself. For instance, Jung proposed the "spiritual entity beyond consciousness (Jung, Carl Gustav, 1989)", but Jung's "spiritual entity" remains a hypothesis or an unknowable force outside individual humans, lacking scientific verification. Therefore, the exploration of the soul or spiritual entity lacks a perspective based on factual modern scientific research.

This article argues that while modern science attempts to explore the deeper unconscious from the perspective of consciousness, adopting research methods of modern scientific experiments, the interpretations of scientific research are still confined to the viewpoint of consciousness. This limitation arises because familiar psychological functions are usually based on conscious experiences, often obstructing the view and making it challenging to identify unconscious psychological functions. The three-layer psychological structure discussed in this article draws on the concept of "three kinds of transformations" from Yogacara Buddhism (Venerable Pingshi, 2024). This is because the three-layer psychological structure aligns well with the developmental biology description of life progressing from an unconscious state to gradually developing conscious psychological functions. Utilizing the three-layer psychological structure provides a reasonable interpretative framework for realizing free will and memory storage propositions. This article contends that the memory model within the three-layer psychological structure offers a clearer delineation of unconscious psychological functions, contributing to further exploration and research into human psychological functions.

REFERENCES

Benjamin Libet, (1985). Unconscious cerebral initiative and the role of conscious will in voluntary action, *The Behavioral and Brain Sciences* (1985) 8, Printed in the United States of America, pp. 529–566.

Benjamin Libet, Curtis A. Gleason, Elwood W. Wright and Dennis K. Pearl, (1983). Time of conscious intention to act in relation to onset of cerebral activity (Readiness-Potential), *Brain* 103, pp.623–642.

Eduard Von Hartmann. (2000*). Philosophy Of The Unconscious, Translated by William Chatterton Coupland,* London:Routledge. (original work 1869, original translated work published 1931), pp. 54.

Freud, Sigmund, (1996). *The Interpretation of Dreams, Freud's Collected Works, Translated by Lai Qiwan and Fu Chuanxiao*, edited by Xia Guangming and Wang Lixin; Anhui Literature and Art Publishing House.

Jung, Carl Gustav. (1989). *Psychology of the Unconscious.* Translated by Liang Luqi, China International Radio Press (Beijing), pp. 29.

Lichen Tsai. (2011). A brief discussion on the defect in the logic of history and its correction—A clue from the logic vision of life-and-death studies and psychology, *Journal of Kum Shan University* 8, pp. 117–139.

Robert J. Sternberg, *Renzhi X. X. (Cognitive Psychology), Trans. Lee Y. S., Chiang W. C.* (2013). 5th ed. Taipei: Yeh Yeh Book Gallery.

Schwartz J. M., Begley S. (2005). In M. H. Zhang (Trans.), *The Mind and The Brain- Neuroplasticity and The Power of Mental Force*, Taipei: China Times Publishing Co.

Tsai L. C., Fan W. H., Pai C. W. (2019). A preliminary exploration of psychological classification and a standard ruler of the conscious mind, *Engineering Innovation and Design* - Lam *et al.* (Eds), pp 354–358.

Venerable Pingshi. (2024). *Commentary on Cheng Weishi Lun*, Vol. 2, The Wisdom Publishing Center (Taipei), pp. 113–134.

Wen J. J. (2024). *The Quantum Brain Hypothesis That Self-Entraps in the Two-Layer Body-Mind Memory Model Dilemma—The Existence of Free Will in Conscious and Unconscious Psychological Activities*, ICASI 2024, Kyoto, Japan.

System Innovation for an Artificial Intelligence Era – Lam et al. (Eds)
© 2025 the Author(s), ISBN: 978-1-032-84452-7

A study about generative design of product form based on the PUR model

Ruoyu Wang
School of Intelligent Media and Design Art, Tianjin Ren'ai College, Tianjin, China

Yue Zong
College of Design and Art, Hebei Academy of Fine Arts, Shijiazhuang City , Hebei Province, China

Yi Xu
Zhejiang Normal University Technological Innovation Research Institute Jinhua Co., Ltd., Jinhua, Zhejiang Province, China

Foundation items: *Tianjin Research Innovation Project for Postgraduate Students (No.2021KJ087)*

ABSTRACT: The application of generative design based on parameterization has been extended from digital art, graphic design and architectural design to product design, and some cases of commercialization have appeared. With the development of parametric design tools and additive manufacturing, generative design has covered almost all fields of product research and development. This article explores the process and method of product form generative design through the parameterized model technology combined with the quantitative demands based on ergonomics, kansei engineering and PUR (Product-Use-Requirement) model, in order to help designers further improve the quality of product form and user experience.

Keywords: Product form, PUR model, Quantitative demands, Generative design

1 INTRODUCTION

With the continuous development of the relationship between humans and computers, design has also become a teamwork of human and program collaboration. Computers are gradually transformed from auxiliary tools to artificial intelligence instead of human calculations, and even act as decision-makers to a certain extent.

Generative design is a design method in which images, sounds, models, animations, etc. are generated by predetermined rules or algorithms. Designers and engineers construct the design problems they want to solve, enter goals, constraints and priorities in the software. The system is based on parametric modeling, through artificial intelligence and cloud computing, to quickly explore possible solutions in order to pursue the best good result. Generative design is bottom-up, based on relevant data and physical laws, and naturally generates models. The initial candidate plan is determined by the computer. In summary, the steps are as follows:

1. Design space: input requirements based on constraints such as volume, weight, material, strength, technology, cost, and time;
2. Design evaluation: The computer uses algorithms and AI to generate a large number of programs, and conducts performance analysis and evaluation of each program;
3. Design optimization: The designer studies options, changes the design goal when necessary, and enters the design iteration loop optimization.

Generative design can handle complex problems faster than humans and provide more potential solutions. Used in various design fields such as art, architecture, clothing and products.

2 OVERVIEW OF THE DEVELOPMENT OF GENERATIVE DESIGN

The application of generative design in the field of architecture is earlier than product design, mainly because it is not limited to the technical limitations of open-mold mass production. The repeated deliberation process of complex building structure conforms to the definition of generative design. Fashion design industries such as jewellery and clothing have a large number of style requirements and apply generative design to create unique and artistic works. At the same time, even if there is an industrialized part, there are still manual or small-scale production ingredients as Figure 1.

In the fields of digital art, vision, plane, color, etc., generative design is also being widely used.

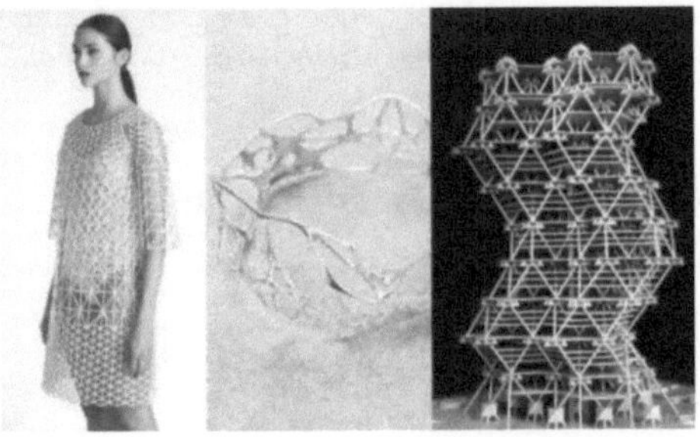

Figure 1. Generative design in architecture and fashion design.

For example, mathematician Benoit Mandelbrot creates complex computer fractal geometric images from mathematical formulas. Variable fonts cooperated with brand consulting company DN&CO to generate a new and dynamically changing brand image of Broadgate in a unique way. The designer Kenya Hara realized the logo design concept of the Xiaomi brand "Alive" through the mathematical equation $|x|^n + |y|^n = 1$ (selecting parameter $n = 3$). Designer Katharina Brunner's graphic graphics and print design and Jon McCormac's color ecosystem "Colour field" based on evolution software, as Figure 2.

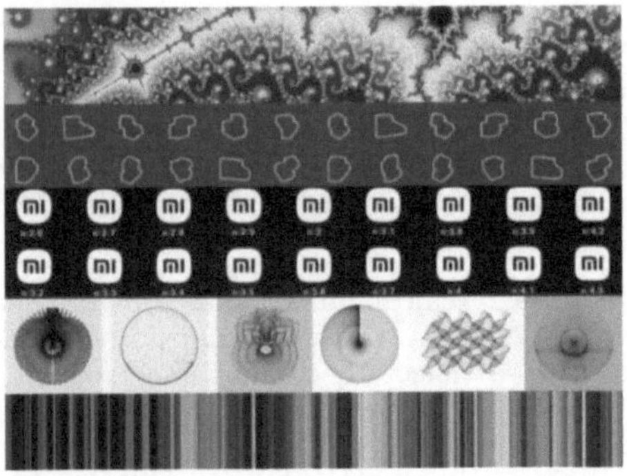

Figure 2. Generative design in digital art, vision, plane, color, etc.

Commonly used parameters for generative design include plane image tools DBN and Processing, 3D tools Grasshopper, Kangaroo, Autodesk Fusion 360, Dynamo, and Rhino's 3D modeling environment support. Generative design is the process of programming modeling. For example, scripting and programming-related knowledge (basic C language, Python or VB) helps to establish logic when using Grasshopper.

In the field of product design, the application of generative design often needs to start from a plane, such as generating a plane figure of a wheel hub, and then rotating and copying it to obtain a complete wheel hub scheme. But if you want to generate a 3D model, you need to use Rhino or Solidworks to model again. For bottom-up planar results, it is difficult to deal with this complicated three-dimensional process. At the same time, it is also affected by the development of digital production technologies such as 3D printing. Therefore, although generative design is gradually entering the field of product design, it is not as much applied in vision and architecture.

3 THE STATUS QUO OF GENERATIVE PRODUCT DESIGN

Most products optimized using generative design have two requirements: lightweight and high strength. And this is where the advantage of generative design lies. Under the condition of ensuring the strength, the structure is optimized through the program, so that the weight of the product is reduced, and the material cost and transportation cost are greatly reduced.

General Motors uses Autodesk Fusion 360 generation technology to design a functionally optimized seat bracket. The traditional seat bracket requires 8 components and several suppliers, while the generative design proposes more than 150 alternatives. The new bracket requires only one component, and the weight is reduced by 40%, and the strength is increased by 20%.

PIX created the world's first car AI generative design algorithm, which is 10 times faster than traditional vehicle design methods. After confirming the parameters, combined with the additive manufacturing equipment, the product automatically enters the manufacturing stage. Maintaining an integrated body, integrating multiple components into one component, reducing hundreds of parts, thereby reducing assembly procedures, and greatly shortening the product development process, as Figure 3.

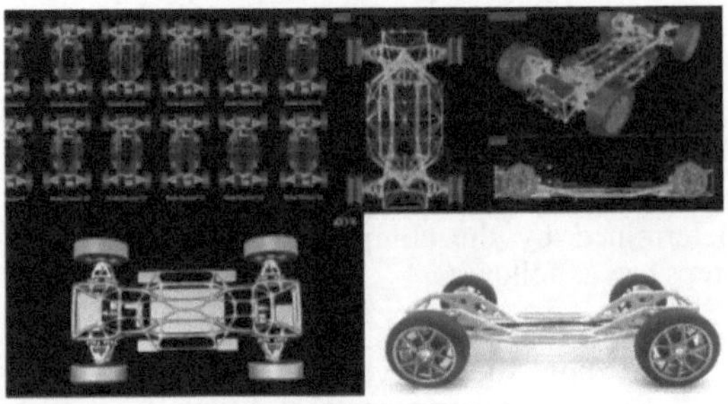

Figure 3. Generative design in car design.

Customization is also part of the requirements for generative product design. For personal customization, even simple products will have many differences, and it is impossible to rely on designers to tailor them for everyone. With the help of generative formula, the problem can be simplified, and the product form can be directly changed by only changing individual parameters based on the user's size. Several possibilities for customization requirements:

1. The existing products are not suitable or very uncomfortable to use.
2. I want to be customized due to emotional factors or personal style, similar to fashion design.
3. Commercialization factors, such as the company's need to have customized product designs.

Generative design is also applied to product surface textures. The common formulas as Table 1:

Table 1. Product surface generation formula.

Type	Name
Hole array	Cross hole array
	Braided triangle hole array
	Spiral function hole array
	Diamond grid hole array
	Diamond grid array
	Circular gradient hole array
	Laser pattern
	Function interference hole array
	Coordinate product hole array
	Tyson polygonal organic hole array
Texture	Random tyson segmentation
	Simulate bionic texture
Weave	Free form woven
	Wave flexible texture
Epidermal interference	Surface UV dynamic skin interference
	Magnetic dynamic skin interference
Parameter mapping	Grid parameter mapping
	Surface parameter mapping
Surface	Surface range conversion
	Segment reconstruction
	Flow principle analysis
Data control	Interference control
	Random filter delete
Mesh	Mesh inlay molding
	Grid parameter mapping
	Mesh subdivision

It can be seen that generative design has been used in product design to a certain extent, but most of the requirements are centered around a part or a certain design element, and a comprehensive method and process have not yet been formed.

4 ANALYSIS OF FORM DESIGN ELEMENTS BASED ON PUR MODEL

Form is the bridge and medium between function and use, and the relationship between form and function is based on use. All parts that are in contact with people's perceptions, thoughts, actions, and emotions are called user interfaces. It includes the visual interface of graphic text, symbols and various defined thinking interfaces and behavior-related action interfaces.

Color and texture often affect human emotions (visual, thinking, and emotional interfaces are called Soft Interface); the use interface of form is mainly perceived by human touch (mainly acting on the surface of form and the contact surface of behavior, which can be called For Hard Interface); so we can focus on form design on Form surface design.

Use is a continuous dynamic process, in this process, vision, movement, touch and emotion cooperate with each other to form a complex organic whole. Generative product form design must consider this dynamic experience process.

According to the use process and use environment of the form, understand the relationship between the user's behavior, lifestyle, thinking mode and form, so as to determine the elements of generative form design.

Among the existing research results, the PUR model has been constructed grounded on Suh's AD (the Axiomatic design) model and Gero's FBS (Function-behavior-Structure) model to make designers understand the process of user experience again when designers are studying the user's needs. AD model points out that the beginning of designing product is to get user's needs,but it is only to study user's functional requirements. FBS model points out that the beginning of the design should build prototype, and it emphasizes to reuse designers' experience and knowledge, not to acquire new knowledge.

User's functional requirements and the process of user experience can affecting the user's needs. The user's functional requirements can only meet the user's goals, but can't meet user's expectations, we must consider various factors affecting user's experiences (function, structure, human-machine size, emotional appeal, product DNA, etc).

Experience is a dynamic, complex and subjective process, and it depends on the multi-level perception of design to screen and understand the context under the content of the environment.

A major feature of digital design methods is that in addition to designers, consumers and makers can participate in the design process and decision-making.

It will be more effective to create a good experience by goal-oriented design based on the PUR model analysis, such as Figure 4.

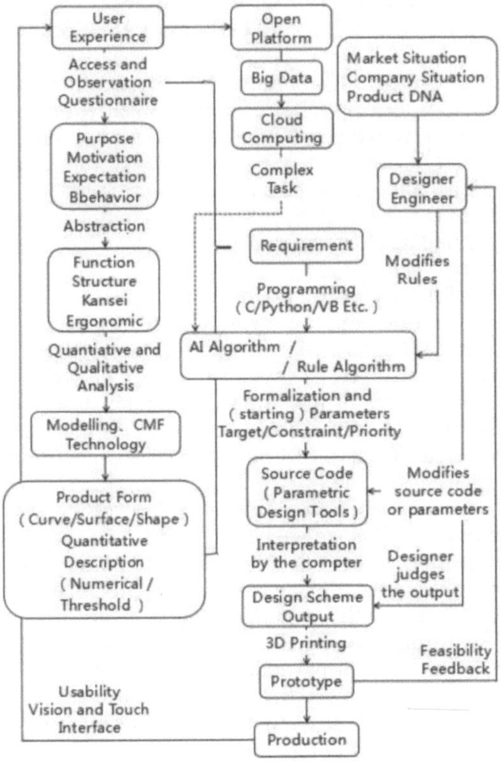

Figure 4. Process method diagram of generative product form design.

5 DEMAND QUANTIFICATION BASED ON ERGONOMICS

Based on the ergonomics of physiology, psychology, anthropometry, biomechanics, etc. around operating behaviors and action methods, relatively mature quantitative research methods have been formed, such as: human body volume experience formula V = 1.015W - 4.937; and The dynamic measurement parameters of the body related to the work space; consider the size of the muscle strength of different parts, the flexibility and accuracy of the action, the shape of the operating device and the compatible movement characteristics and the design of the operating device, etc. Possess various parameter information needed to interface with generative design.

For example, in the design of a steering wheel, the product needs to withstand external forces as well as internal stresses, and some basic data such as the position of the input force, load-bearing force, and supporting force are required. Generate an idealized structure through a parametric tool (Millipede component of Grasshopper). The structure of the honeycomb is based on predetermined rules. Where the stress is stronger, the cell size and shape will also change with the data.

The parameterized sports shoes are automatically generated according to the user's own data, the weight data changes the structure of the sole (thickness, density, etc.), and the shoe size data changes the product size and shape.

6 DEMANDS QUANTIFICATION BASED ON DESIGN DNA AND KANSEI ENGINEERING

6.1 *The main elements of design genetic analysis*

The design gene refers to the characteristic unit of the modelling form or the general functional unit and the relationship between the units that inherit the modelling attributes of the product. These units should be verified by practice and have the information of the product design. Realizing the innovation of the overall function and principle by reorganizing the unit is a process of designing genetic manipulation, duplication, and reorganization, and its essence embodies the ideas and laws of heredity and optimization.

There are many factors that affect the shape of a product. Characterization mainly includes structure, component, size, material, surface treatments and so on, and recessive genes mainly include product function, performance, and correlation.

In order to achieve the establishment of a quantitative system and subsequent design, according to the extracted design gene composition morphological design plan, the combination principle in the design method is used. The basic concept has three aspects:

6.1.1 *Basic information characteristics*

Refer to the basic components that reflect the overall information of the product from different angles, such as functional characteristics, principle characteristics, morphological characteristics, layout characteristics, structural characteristics, etc. These characteristics constitute the design genes of the product.

6.1.2 *Cutting*

The operation of separating M genes from a product containing N design genes of a certain type is called a cutting of such design genes. If M = 1, it is called simple cutting, and M $\geq$ 2 is called For compound cutting.

6.1.3 *Combination*

The process of re-merging the design genes from n products to generate a new product is called combination. Among them, if n = 1, the combination is called self-combination, otherwise it is called compound combination.

6.2 *Quantitative analysis of Kansei demands*

"Kansei" perceptual appeal can be interpreted as: the feeling or image that people hold toward the product is a kind of psychological expectation and feeling about the product, which mainly exists in the feeling, perception, and recognition generated after the external stimulus is transmitted to the sensory organs. In the process of a series of interactions between people and products, such as knowledge, emotion, movement, performance, etc.

Kansei can be divided into vision, hearing, smell, touch, movement, body balance, etc. What Kansei seeks to express is something similar to human experience, such as light, sound, temperature and humidity, vision, touch, smell, taste and beauty, Wabi-sabi and other abstract sensory experiences, as well as changes and mixing between them. The ambiguous state even has an element of imagination.

People will inevitably have feelings about the product when using the product. The use process of the product can not only fulfill its functional needs, but also needs the spiritual level to satisfy its perceptual appeal. The product becomes a carrier for users to realize their self-worth.

6.2.1 *The measurement of Kansei demands information*

In order to calculate the degree to which product modeling expresses a specific appeal or sensibility, Veryzer Jr. proposes a calculation equation:

$$S = [(Ip/C) + (Ip/It)]/2$$

Among them, S is the perceptual quantity of a single styling factor or a styling combination (product), Ip is the number of times the testee mentions a specific perceptual appeal vocabulary for this styling, C is the number of testees, and It is the testee's suggestion for this styling The total number of Kansei demands vocabulary.

When performing perceptual measurement, it is necessary to rely on the oral evaluation of the product by the user of the product, and other similar products should be added for comparison during the evaluation. It can be seen from the equation that (Ip/C) and (Ip/It) are different measures of perceptual quantity, and Veryzer Jr. regards them as equally important, each accounting for 50%. We can conduct research based on the nature of the research. And personal needs flexible application.

6.2.2 *How to obtain the user's Kansei demands*

Combined with the application of engineering statistical methods, there are two channels for obtaining user perceptual appeal information:

Provided by the user. By designing more systematic and standardized questionnaires or fixed interview procedures, information about users' perceptual appeals and preferences is obtained.

Data mining of market information. Analyze and predict market trends through the collection of consumption information based on the network platform to guide and improve product development.

User perceptual appeals are characterized by complexity and ambiguity, and they need to present and obtain information in an explicit way. The semantic difference method is an attitude measurement technique developed by the American psychologist Osgood, which is widely used comparative study of differences between individuals and groups. There are about 7 to 11 intervals between each adjective and antonym. You can get the user's feeling information about a certain product.

6.2.3 *Main quantitative methods*

The basic laws of aesthetics are unchanged. Designers use the basic attributes and aesthetic laws of styling characteristics to form a certain shape, and give design genes such as color, material, and man-machine interface to obtain the basic image of the product. This process can be regarded as a process of optimizing various combinations of morphological elements by taking the user's perceptual appeal as the objective function. Transform qualitative analysis into quantitative analysis, and the results are embodied in numerical form, which can provide parameter and threshold basis for generative design.

The modeling quantification method is also known as the shape averaging method. In a group of similar product shapes with large differences, the representative modeling design genes are extracted through the concept of averaging operation and difference compensation, and the concept of weighted average is used. You can derive new shapes.

Generalized Weighted Mean in fuzzy theory is applied to product evaluation, and its purpose is to add subjective measurement factors of human consideration to the weighted average method.

Shape grammar is derived from the concept of linguistic structure. It is considered that a shape is a linear arrangement composed of design genes such as points, lines, and planes. It can be used as a reference point for morphological positioning after conversion through symbolic form. Create a new shape based on the position, direction, reflection or size of the original shape.

6.3 *Establish a system of genetic transformation from em-otional appeal to design*

6.3.1 *Multiple linear regression algorithm*

Applying multiple linear regression algorithm to transform the user's perceptual appeal information into design gene information, and construct the description function of the conversion relationship between the two;The main research of multiple linear regression algorithm explores the relationship between multiple variables, and its purpose is to understand and establish the relationship between a dependent variable and a set of independent variables. The general form of multiple linear regression is:

$$y = \beta_0 + \beta_1 X_1 + \beta_2 X_2 + \ldots\ldots + \beta_P X_p + \varepsilon$$

6.3.2 *Build a description function for the transformation relationship*

From the multiple linear regression algorithm, we can know that if the product design process is regarded as a process of solving the system function, the user's evaluation value of the product shape is used as the dependent variable, the design gene is used as the independent variable, and the statistical analysis can be performed. Genes quantify the physical value of the contribution level of modeling and evaluate the contribution of the product to a certain semantic evaluation.

For a more complex design project, there are many design genes that constitute the product, and this project can be divided into multiple subprojects *Xn*. Extending the definition of multiple linear regression, the functional expression described by its modeling semantics is:

$$\begin{aligned} F_{new} &= \alpha_0 + \alpha_1 X_1 + \alpha_2 X_2 + \cdots\cdots + \alpha_p X_p \\ X_1 &= \beta_{10} + \beta_{11} x_{11} + \beta_{12} x_{12} + \cdots\cdots + \beta_{1p} x_{1p} + \varepsilon_1 \\ X_2 &= \beta_{20} + \beta_{21} x_{21} + \beta_{22} x_{22} + \cdots\cdots + \beta_{2p} x_{2p} + \varepsilon_2 \\ &\vdots \\ X_n &= \beta_{n0} + \beta_{n1} x_{n1} + \beta_{n2} x_{n2} + \cdots\cdots + \beta_{nP} x_{np} + \varepsilon_n \end{aligned}$$

7 GENERATIVE FORM DESIGN OF INDOOR MONITORING PRODUCTS AS AN EXAMPLE

Taking the fall monitoring product for the elderly living alone as an example, the analysis method of design genes and perceptual appeal information and the quantitative correlation model of the two were practically applied. Here are some of the key aspects of generative design:

7.1 *Generation of the main product form*

In order to ensure the G3 continuity of the rounded surface of the product shell, the One line formula of GH is used to generate a high-quality rounded curve, and then the surface is obtained based on the curve, and the natural transition between the rounded surface and the cube and the volume control based on the hardware stack size are completed,as Figure 5.

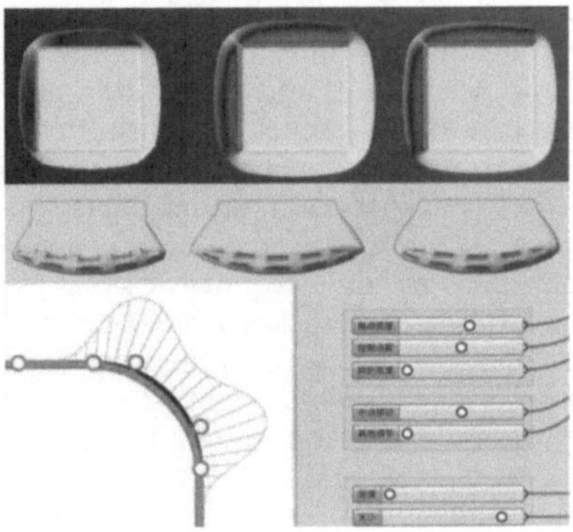

Figure 5. Generation of the main product form.

7.2 *Generation of infrared sensor receiving channel*

Determine the shape of the cellular channel required by the infrared sensor according to the product use environment (height, angle) and monitoring range (area) and other parameters. At the same time, Tyson Segmentation is used to complete the generation of the refractive surface of the sensor lens, as Figure 6.

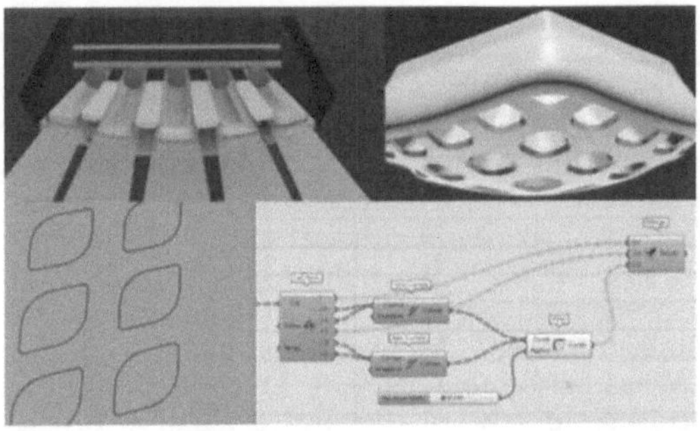

Figure 6. Generation of infrared sensor receiving channel.

7.3 *Lightweight of the core honeycomb components of the product*

Based on the original honeycomb component shape, lightweight design is carried out. Take advantage of the parameter controllable feature of generative design to efficiently test a variety of possibilities, including semi-masks, bases with different orientations, extrusion channels, etc.,and select the best solution with the lightest weight, as Figure 7.

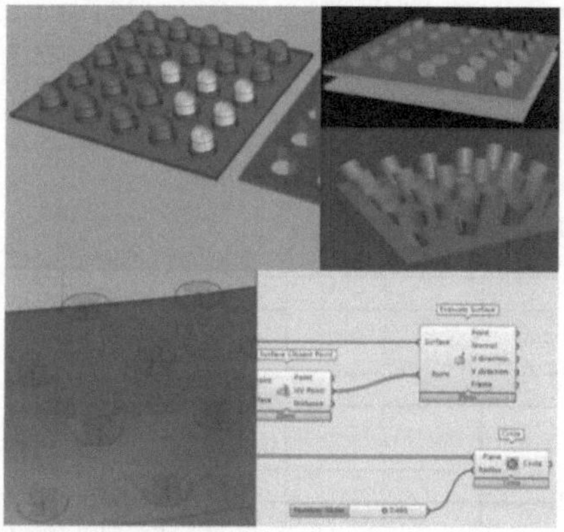

Figure 7. Lightweight of the core honeycomb components.

7.4 *Shell hole array generation based on monitoring function*

Comprehensively consider the user's perceptual requirements for the product (appearance affinity) and the product's own functional requirements (monitoring sensitivity), generate and test the best hole array shape (number, size, distance, Density, etc.), as Figure 8.

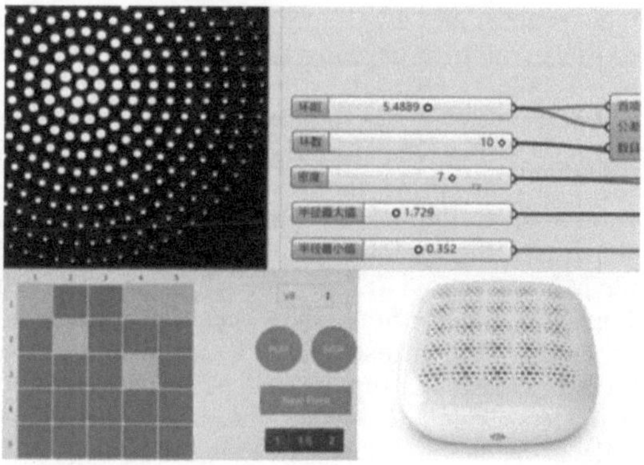

Figure 8. Shell hole array generation based on monitoring function.

The above cases are based on the numerical values and thresholds converted from the quantitative value of perceptual demand information, and the product design work is carried out through parameterization tools, which verifies the feasibility of the process and method of generative product form design based on the PUR model.

8 CONCLUSION AND FUTURE WORK

Design is a cyclical cognitive process, and generative design can help us code, interpret and verify morphological design elements. In order to facilitate designers to further improve product form quality and user experience, this article summarizes the process and method of generative product form design based on PUR model.

At present, products completed by generative design are not common. The problem of development is production technology, how to use existing production technology to support parametric design and generative design. We should combine known things with new things, and combine parametric generation with other processes, such as lost wax casting, inflatable metal, etc. to produce products, not just 3D printing and unique shapes. This is meaningful.

Similarly, the next step of generative design is to integrate traditional design techniques with the latest generative design techniques. The two are not completely opposite, they can be integrated with each other. Combining human design and AI design to obtain truly meaningful products is our expectation for generative design.

REFERENCES

Buxton B., (2012). *Sketching User Experiences*, 1nd Edition. Electronics Industry press.

Liu G., (2012). *Image Set*, 1nd Edition. China Building Industry Press.

Liu L., J. Lin J., (2019). *Design*. 32(7) , 132–133.

Rifkin J., (2010). *The Third Industrial Revolution*, 1nd Edition. CITIC Press.

Shen Y., T. Zhang T., Y. Ke Y., and H. Wang H., (2018). *Journal of Textile Research.*, 39(12), 118–123.

Wu J., L. Sun L., and Q. Liu Q., (2019). *Mechanical Design.*, 36(8), 115–119.

Zhang Y., (2016). *China Information Technology*. 9, 71–73.

Zhang Z., (2011). *PhD thesis*. Shanghai Jiao Tong University.

Zhou Z., Z. Zhang Z., (2021). *Packaging Engineering*. 42 (10) 289–294.

System Innovation for an Artificial Intelligence Era – Lam et al. (Eds)
© 2025 the Author(s), ISBN: 978-1-032-84452-7

Investigating the traditional and modern symbols of Minnan Architecture: A case study of Liu Zai ancient house

Kai-Han Chung & Yun-Ya Huang
Department of Visual Communication Design, Tainan University of Technology

ABSTRACT: This research explores the 'Liu Zai' (劉再) ancient house in preserving cultural heritage in Xigang District, Tainan, Taiwan. As a Minnan architecture, Liu Zai's house faces the challenges of restoration and conservation. This study collects historical and cultural information about the ancestral house through the literature review, field investigation, and semi-structured interviews. Furthermore, it investigates Liu Zai's family culture, the meanings of the couplets, and the transmission of symbols. This study finds that modern-ism may have influenced the ancestral house during its construction process, presenting geometric patterns on its spatial facade that differ from traditional Minnan architecture. In conclusion, the ancestral house exhibits both traditional and modern symbols. It represents the adaptability of Liu Zai's ancient house to modern symbols and material applications during its construction approach.

1 INTRODUCTION: THE LIU FAMILY'S ANCESTRAL HOUSE AND TOMBS IN SIGANG DISTRICT, TAINAN

Sigang District is situated at the southwest end of Tainan City, adjacent to the downstream of the Zengwen River. From the 19th and early 20th centuries, the Zengwen River changed its course multiple times, leading to frequent floods near the ancient Liu family house. The accumulation of silt has caused the coastline to extend outward year by year. In 2001, the Liu family excavated land to build homes and unexpectedly discovered two underground ancient tombs. These two tombs belong to the 'Liu Den-Gkui' (劉登魁) and 'Chen Xiao-Shu' (陳孝淑), who immigrated to Taiwan. The discovery of the tombs has prompted the Liu family to pay more attention to preserving ancestral artifacts. At the same time, the family began editing the genealogy, with a primary focus on documenting the descendants of Liu Den-Gkui. The genealogy comprehensively records the life, marriage, death, burial, and migration data of the Liu family. It also includes information about daughters listed in the family tree, widows remarrying, and daughters remarrying.

According to the Japanese translation document in the Liu family genealogy, a century ago, there were Ting-Kang and Hsia-Kang ports with frequent ship traffic and economic prosperity (Liu, 2018). The Liu family's origin can be traced back over a hundred years to when 'Liu His' (劉喜) and his wife arrived in Taiwan from Dingwei Township, Jinjiang County, Fujian Province, China. They settled in the northern part of Sigang District, engaging in land reclamation work. As time passed, the Liu family gradually migrated southward and eventually settled in the southern part of Sigang District.

There are two Liu family ancestral houses in Sigang District: the 'Liu Jian-Nan' (劉梗楠) and the 'Liu Zai' (劉再) ancient house. The Liu Jian-Nan ancient house was built around 1882 and showcases the architectural style and lifestyle of the Minnan region at that time. Currently, the Liu Jian-Nan ancient house has been renovated and preserved. In contrast, the Liu Zai ancient house faces challenges in repairs and preservation. Built between 1954 and 1964, the ancestral house reflects the changes in social, economic conditions, and residents' lifestyles. Despite being influenced by modernization, the Liu Zai ancient house maintains some traditional elements, revealing its connection to social history.

2 METHODOLOGY AND DATA ANALYSIS: LITERATURE REVIEW, FIELD INVESTIGATION, AND SEMI-STRUCTURED INTERVIEWS

The methodology of this study is based on the literature review, field investigation, and semi-structured interviews to gather relevant information about the Liu Zai ancient house in Sigang District. Firstly, the researcher collected various data related to the Liu Zai ancient house through

DOI: 10.1201/9781003514831-70

a literature review. The data included the history of the Liu family, information about Liu Zai, construction details of the ancient house, and related events. These literary materials will serve as the foundation for the study and provide background information for the upcoming field investigations.

The second stage involves conducting field investigations of the ancient house through on-site exploration to document its structure, spatial layout, and functional usage. This process is complemented by creating the floor plan and elevation plan of the house. At the same time, the researcher observed the surrounding environment of the ancient house to understand the local geography, climate, and socio-cultural background. This exploration also involved delving further into the history and significance of the ancient house.

In the third stage, we conducted semi-structured interviews to explore the memories and impressions of the interviewees regarding the ancient Liu Zai house. Before conducting in-depth interviews, preparing an interview guide enhances the comprehensiveness of the data. By incorporating the interviewees' perspectives and anticipating logical gaps in the data (Patton, 1990), the interview process can be considered a guided conversation (Lofland & Lofland, 2005), ensuring the acquisition of valid oral data. The participants in this study include one local elder and two descendants of Liu Zai. The interview process is primarily recorded using audio, with video as a supplementary tool. The interview content includes Liu Zai's life history, the family's religion, and the meaning of the couplets in the main hall. After the interviews, the audio data will be transcribed into verbatim transcripts to facilitate data summarization and analysis. The comprehensiveness of these methods will contribute to a thorough investigation and analysis of the ancient Liu Zai house.

3 THE SYMBOLS OF LIU ZAI ANCIENT HOUSE

The Liu Zai ancient house is a traditional Minnan architecture in Taiwan, constructed between 1954 and 1964. The architecture is in the 'Dan Shen Shou' (單伸手) style. In the spatial layout of this ancient house, the arrangement of family members' positions follows the hierarchy within the family, corresponding to the relationships within the human body. Apart from meeting the requirements of traditional Chinese design in terms of exterior appearance and spatial functionality, it also embodies a cosmological perspective (Li, 2005). According to interviews with Liu Zai's daughter, the spatial design of the ancient house adopts the form of 'Dan Shen Shou'. The reason was that, at that time, the children in the family had grown up and moved out to work. Hence, the traditional courtyard design of 'San He Yuan' (三合院) was not considered during construction.

Traditional Taiwanese architecture embodies the human form and conveys rich meanings, with both the interior and exterior embellished with symbolic decorations and narrative elements. Through its form, architecture can convey its design principles (Fu, 2005). In addition to symbolism, structures are often adorned with painted motifs depicting folk stories, typically applied to significant areas of the building space. These paintings are often closely connected to the family's well-being and destiny (Lee, 2010). The ridge beam of the Liu Zai ancient house features a painted motif that covers two-thirds of its length. A 'Bagua' (八卦) at the center of the ridge beam symbolizes protection against evil, blessing, and prayer. Flanking the 'Bagua' are the characters 'Tian Ding' (添丁) and 'Jin Cai' (進財), symbolizing the family's aspirations for an increase in male offspring and prosperity. Therefore, the painted decorations on the ridge beam reflect the aspirations and blessings of the homeowner towards the family, as shown in Figure 1.

Figure 1. The painted decorations on the ridge beam of Liu Zai's ancient house consist of 'Bagua', 'Tian Ding' and 'Jin Cai'.

The couplets inside the ancient house were inscribed by Liu Zai using gold paint. The couplets inside the ancient house were inscribed by Liu Zai using gold paint. The content is divided into two parts: family members and blessings. The couplets related to family members correspond to their names. For example, in the couplet 'Li Yu Ren Feng Xi Ri Yung' (里有仁風喜日永), the character 'Li' (里) from Liu Zai's daughter's name, and the entire sentence conveys that their residence is a blessed place. The couplets related to blessings are gifts to the family. For instance, 'Jia Yu De Ze Fu Xing Zhao' (家餘德澤福星照) means that the ancestors of the family are virtuous, and many gods will take care of the family, as shown in Figure 2.

Figure 2. Couplets inside the ancient house.

The exterior wall of the corridor under the eaves used facial-washed terrazzo with decorative symbols. Modernism may have influenced these symbols by avoiding excessive ornamentation and intricate details. It applies new materials and techniques to practical design concepts. Therefore, without altering the symbolic meanings, the ancient homeowner transformed the traditional symbols of blessings into geometric compositions, as shown in Figure 3. Additionally, the hexagonal tiles on the corridor floor resemble turtle shells, symbolizing health and longevity.

Figure 3. The decorative symbols on the exterior wall.

4 CONCLUSION AND DISCUSSION

In the early 20th century, during the Japanese colonial period, Taiwan was influenced by Europe and the United States. Architectural design in Taiwan began to incorporate geometric forms as decorative elements in the Art Deco style. Additionally, Japan started utilizing cement in Taiwan as a construction material. Cement is not only used for architectural structures but also for decorative purposes such as facial-washed terrazzo and monolithic terrazzo. Until the post-war period in Taiwan, Taiwanese residential buildings began to utilize modern building materials and structures. Despite the change in construction materials, people continued to adhere to traditional thinking when building houses.

The Liu Zai ancient house was constructed between 1954 and 1964. It may have been influenced by the modernist and art deco types prevalent during that period. During its construction, the Liu Zai house incorporated elements of modernism and displayed a spatial facade design that differed from traditional Minnan architecture. The architecture also reflects an open attitude towards modern symbols and the adaptability of materials. Therefore, from the perspective of cultural integration, although the symbols of post-war Minnan architecture in Taiwan absorbed influences from modernism and decorative styles, they were not completely replaced or assimilated. They merged traditional elements to create a new expression of cultural integration. This fusion of modernity and tradition reflects the harmony of cultures in architectural vocabulary across different periods.

ACKNOWLEDGEMENTS

We thank the National Science and Technology Council (NSTC) for funding this project.

REFERENCES

Fu, C. C. (2005). Symbolic, architectonic and artistic decorations in structural elements in traditional architecture in Taiwan. *WIT Transactions on The Built Environment*, 83, 35–42.

Lee, Y. X. (2010). Traditional painting on architecture in Taiwan. *Taiwan Historica*, 61(2), 291–318.

Li, C. L. (2005). *20 speeches on the appreciation of ancient Taiwan architecture*. Taipei: Artist Magazine.

Liu, M. F. (2018). *Liushi zupu*. Tainan: Taiwan Liu Family Association of the Descendants of Liu Den-Gkui.

Lofland, J., & Lofland, L. H. (2005). *Analyzing social settings: A guide to qualitative observation and analysis*. CA: Wadsworth.

Patton, M. Q. (1990). *Qualitative evaluation and research methods* (2nd ed.). Thousand Oaks, CA: Sage.

System Innovation for an Artificial Intelligence Era – Lam et al. (Eds)

A service-oriented biological data storage and processing system

Veska Gancheva & Hristo Stoev
Faculty of Computer Systems and Technologies, Technical University of Sofia, Sofia, Bulgaria

ABSTRACT: Given the vast databases of biological data available, the development of service-based systems for accessing computational resources, applications, and data is critically important. The paper presents an extensible storage, management, and analysis system for biological data based on cloud-based Amazon Web Services that enables service modularity. I.e. in addition to searching for sequences and similarity, there is the possibility to add or remove different functionalities (in the form of lambda functions) that can be run at a different stage of processing or completely independently by building workflows. An approach for searching and processing biological data through services is proposed. In order to achieve fast performance in searching for similar biological sequences, a software implementation of the proposed algorithm is used to align a pair of DNA sequences based on the proposed trilateration-based CAT method. A hierarchical structure has been developed for the representation and storage of biological data, so that the data with the best chance of maximally matching the searched sequence are located in the same space. A scheme for recording and searching for a biological sequence in the system has been designed. The proposed system implies facilitating large-scale data integration, enabling repeatable and reproducible analyses, maximizing scope for sharing, and leveraging collective intelligence for knowledge discovery.

1 INTRODUCTION

As new technologies are developed in the biomedical field, a large amount of biomedical data is collected and made available to the medical research community (Raghupathi, W., Raghupathi, V. 2014). These advancements include a broad spectrum of fields, such as medical imaging technologies like nuclear magnetic resonance (MRI) imaging, computed tomography (CT) scans, and multilayer microscopic imaging, which enable advanced cell analysis. Sequencing machines are also producing large amounts of genetic data in the field of genomics.

In recent years, there has been an extraordinary growth in biological databases. High-throughput sequencing research have been made possible by these advancements, which have significantly increased our ability to collect bioinformatics data.

Bioinformatics is a rapidly developing field enabling scientific experiments via computer models and simulations. On the one hand, this abundance of data has the capacity to completely transform how we think about disease mechanisms, intricate biological systems, and patient care using customized medicine techniques. It makes it possible for scientists to identify the underlying genetic causes of many illnesses, create specialized treatments, and raise diagnostic precision.

On the other hand, advanced computational techniques and tools are needed for the management, analysis, and interpretation of this data, in addition to methods to guarantee the confidentiality, integrity, and quality of the data.

However, accurately predicting the course of a disease is one of the most exciting and challenging undertakings for medical practitioners. Three major issues are brought about by these changes for researchers and clinicians: interoperability, security, and scalability (Bahmani, A., Alavi, A., Buergel, T. *et al.* 2021).

Intelligent medicine formulation, tailored therapy, electronic health record mining, and population screening are just a few of the healthcare practices that could be significantly altered by applying contemporary methods to extract meaning from huge amounts of data (Avila K. Sanmartin P., Jabba D., Jimeno M. 2017; Hulsen, *et al.* 2019). The primary concerns are the growing need for collaborative networks with knowledge sharing and the requirement for standardized data content, organization, and clinical criteria.

A major issue with many current healthcare information systems is their inability to scale due to their diverse architecture (Yang T.H., Sun Y.S., Lai F. 2011). Large data sets may be stored in real time by creating integrated data lakes, which can be achieved by combining open source technologies like Hadoop with newly developed data management techniques (McPadden J. *et al.* 2018).

The creation of service-oriented systems for data access, applications, and computational

DOI: 10.1201/9781003514831-71

resources is vital given the huge databases of biological data al-ready available. Developing a cloud-based biological sequence analysis and storage system gives many users access to cutting-edge distributed infrastructure and bioinformatics software tools. Server resources are limited when utilizing a traditional client-server architecture. The system's processing of requests would, at the very least, slow down and eventually fail under stress because of the constraints of the computational resources available to it. By using cloud-based technology, delays and potential faults are reduced to a minimum.

Currently, it is accepted to use Hadoop to build such systems. In this type of computing and data analysis systems, which is largely a time-consuming and resource-intensive process, the task is distributed over multiple nodes (Taylor R. C. 2010). This strategy is good, but it requires software application development skills to configure the relevant workflow in Hadoop and its programming tools, making it practically inaccessible to most biologists or people with no or limited software development experience.

The work presented in this paper is part of a project that suggests a conceptual model for a scientific platform that would enable precision medicine and biomedical scientific research by allowing intelligent administration and analysis of huge data streams. The objective is to develop an integrated open technology platform for intelligent solutions based on novel and efficient techniques, algorithms, and tools for the discovery, integration, storage, and analysis of biomedical data, as well as for knowledge discovery and medical decision-making. Because of this, the platform is anticipated to provide scientists with an integrated and scalable workflow development environment that includes a set of software tools for automating the computational steps involved in carrying out scientific research.

The present research is devoted to the development of an extensible biological data storage and processing system based on Amazon Web Services cloud services. Based on the CAT profiles from the proposed by authors new effective pairwise DNA sequence alignment algorithm (Gancheva V., Stoev H. 2019, 2023, 2024), the sequences can be organized into a hierarchical storage structure to be used as a database for biological data storage in search-optimized systems. A prototype of such a system using cloud technologies to achieve better speed and adaptability to workload is proposed.

The paper is structured as follows: The design of a system for biological fata storage and processing is described in Section 2. Section 3 presents a service-oriented approach for biological data processing. The conclusions are in Section 4.

2 DESIGN A SYSTEM FOR BIOLOGICAL DATA STORAGE AND PROCESSING

2.1 *Purpose and functional requirements*

1. It is necessary to be able to add and remove functionalities to the system in the form of services or modules. Adding and removing functionalities should be done in a standardized way. Not having to recompile existing code. While adding or removing functionality, the system should operate normally without interruption. The system should be highly reliable and accessible at any time from any point.
2. To achieve fast searching with minimal operations for similar sequences.
3. The system should be scalable, i.e. to adapt resources based on the load.
4. To be able to manage users and their access to certain functionality or resource of the system.
5. Individual services/modules must be isolated, and their operation should not affect the data stored in the system. As well as individual services/modules not to affect the operation or data of other modules or services.
6. To provide a user interface for accessing the individual functionalities of the system.
7. The management of system resources should be relatively easy.

To achieve the above goals, it is appropriate to use cloud-based technologies like Amazon Web Services. This allows to design a platform with modularity of the services that make up the system functionality (requirement 1.), it is adaptable and scalable (requirement 3.), there is a ready-made module for managing users, logging into the system, using different login providers (requirement 4.) as well as provides a convenient user interface for resource management (requirement 7.)

2.2 *User interface for accessing individual functionalities*

A React APP to build a user interface with the ability to run tasks, upload data, visualize results, etc. The application is hosted on Amazon Simple Storage Service - S3 (Amazon Simple Storage Service) as a static web page. Cognito (Amason Cognito) and Amplify (AWS Amplify) are used to build the user interface with the ability to manage access to individual services for registered and non-registered users, new user registration, login with Facebook, Google or another platform as needed (AWS Identity and Access Management).

The description of individual service supported by the platform, its configuration and basic startup information is listed in a json file. The user interface is built dynamically based on this json file. (requirement 6. according to requirement 1.)

2.3 *Designing the basis/core of the biological data storage system*

To achieve rapidity in the search for similar biological sequences, it is proposed to use a software implementation of the algorithm proposed by the authors for aligning a pair of DNA sequences based on the proposed trilateration-based CAT method. The idea is to use the generated CAT profile as a key to search for similar sequences, as well as to organize the data. And using the CAT comparison algorithm to select a data set to be input to a more precise comparison algorithm. In this way, we take advantage of the speed of the CAT comparison algorithm and combine the result with the more accurate comparison algorithms.

2.4 *Data storage structure*

The idea behind the proposed CAT method is to build a tree-like structure in the data storage space so that the data with the best chance of maximum matching with the searched sequence are located in the same space. The data structure for storing biological DNA sequences is given in Figure 1.

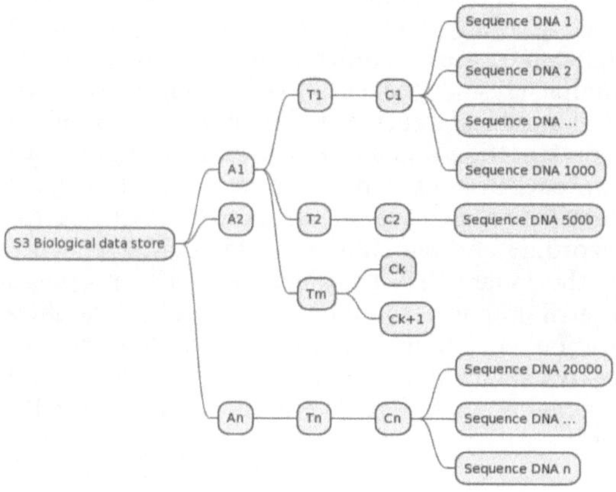

Figure 1. A graphical representation of the data tree structure in S3 for biological data storage. The CAT key corresponds to the path in the tree, i.e. sequence 2 has CAT key A1_T1_C1.

3 A SERVICE-BASED APPROACH TO BIOLOGICAL DATA PROCESSING

With the help of the cloud services provided by AWS, a general approach and operating principle for the functionality of the system is defined, including the processing method, and storing data, a communication graph between individual services and capabilities to define other specialized services for biological data processing (Figure 2).

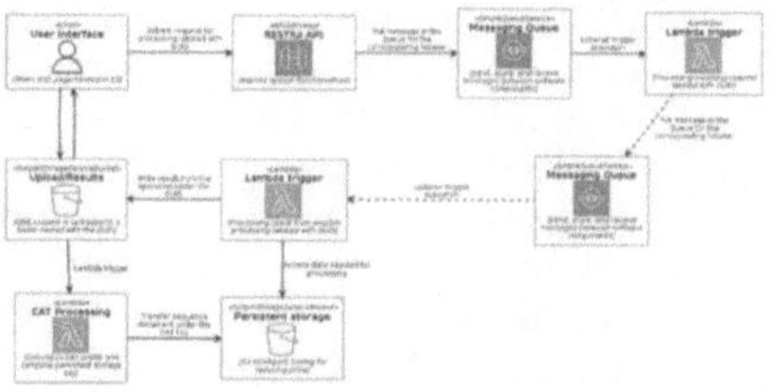

Figure 2. General approach and principle of action. Communication path between individual services.

1. A different S3/s is used for each platform service (requirement 5).
2. Each service organizes the data in S3 designated for it (requirement 5).
   2.1 Two separate S3 spaces are used to process biological sequences.
      2.1.1 S3 for uploading sequences before processing and checking results - Uploads/Results, configuring this S3 to store the data for no more than 24 hours.
      2.1.2 For this S3, a trigger is configured that watches for an uploaded file in Uploads and starts a Lambda to process the CAT sequence - the result is written to another S3, using the CAT-generated result as the file key. The template format for the key looks like: **{A Point}_{T Point}_{C Point}**, where each point is calculated by CAT for the corresponding benchmark and its coordinates x = D and y = h in the form of {D:h}, example {0.49993012144522275: 0.4153371581358836}_{0.50006987 85547772:0.41533715813588357}_ {0.37504599735342037: 0.5308903626512124}. Each benchmark point is a folder/directory from the tree-like folder structure of the storage bucket. In this way, all sequences with the same CAT profile are stored in the same directory. This helps sifting results when searching for sequence similarity or equality and results in a significant reduction in search time (to requirement 2.) – Figure 3.

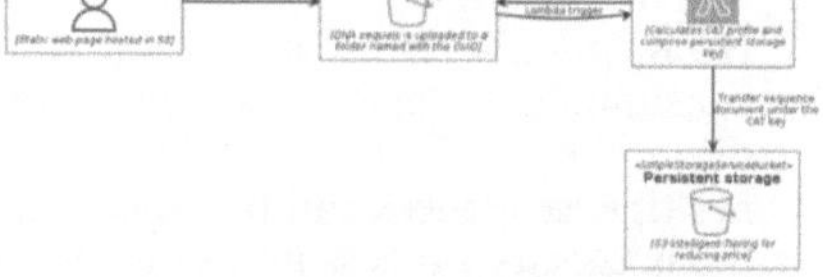

Figure 3. Scheme for recording a biological sequence in the system.

2.1.3 When searching, a GUID identifier is first generated for the search query and folder with this GUID is created in Result bucket. Generated GUID identifier are stored for a period of 24 hours. Access to the folder with name of the generated GUID is managed by Cognito and is only available to the user who generated it. The query identifier is used as key for the returned result. After processing in S3/Results, the user can download the result of the operation. This approach allows the system to evolve better over time, maximizes performance, scalability, and security, and prevents update commands from causing domain-level merge conflicts - Command and Query Responsibility Segregation (CQRS) pattern (to requirement 2 .).

2.1.4 For a search request, an API Gateway (API Gateway) is used to run a Lambda for CAT preprocessing. After the preprocessing is completed, the calculated key is obtained, which is used to select a directory where the similar sequences are stored to be processed by another Lambda according to the Needleman–Wunsch or another precise algorithm. The result of the first Lambda is written to SQS (Amazon Simple Queue Service) along with the query identifier, which starts the Needleman–Wunsch processing Lambda, which in turn, after processing, writes the result to S3/ Results under the query identifier folder – Figure 4.

3. When adding a new service, it is necessary to configure the AWS services required by the service (used S3 storage services, Lambda and others), then update the configuration json file to generate the UI for the added service. Usage of cloud-based technologies allows:

3.1.1 To occupy the necessary resources such as processor time, for a more precise comparison of the results obtained by the CAT method. Processing sequences on the order of 3 million in length using any of the precision methods such as Needleman-Wunch is extremely time consuming and computationally demanding. Using lambda functions, multiple sequences can be processed, and the necessary processing resources used.

3.1.2 The proposed data storage structure implies the creation of multiple directories in the file system. Cloud-based technologies like AWS are designed to work quickly with a large number of directories and have no limit on the amount of data stored.

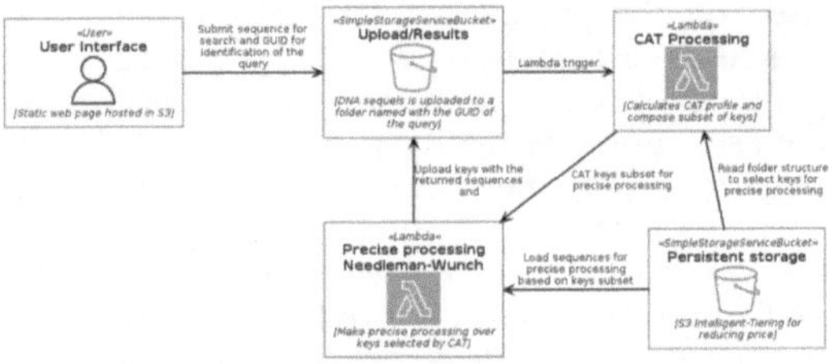

Figure 4. Biological sequence searching scheme.

4 CONCLUSION

The paper presents the architecture of a service-based biological data storage, management, and analysis system that enables service modularity. I.e. in addition to searching for sequences and similarity, there is the possibility to add or remove different functionalities such as serverless compute service (AWS Lambda), streaming data at any scale as a fully managed service (Amason Kinesis) that can be run at a different stage of processing or completely independently by building workflows. A hierarchical structure has been developed for the representation and storage of biological data. An approach for searching and processing biological data through services is proposed. A scheme for recording and searching for a biological sequence in the system has been designed. The proposed system implies facilitating large-scale data integration, enabling repeatable and reproducible analyses, maximizing scope for sharing, and leveraging collective intelligence for knowledge discovery.

ACKNOWLEDGMENT

This research was financially supported by the National Science Fund, Bulgarian Ministry of Education and Science, grant number KP-06-N37/ 24, project "Innovative Platform for Intelligent Management and Analysis of Big Data Streams Supporting Biomedical Scientific Research".

REFERENCES

Amason Cognito https://aws.amazon.com/cognito/

Amason Kinesis https://aws.amazon.com/kinesis/

Amazon Simple Queue Service https://aws.amazon.com/ sqs/

Amazon Simple Storage Service https://aws.amazon.com/s3/
API Gateway https://aws.amazon.com/api-gateway/
Avila K., Sanmartin P., Jabba D., Jimeno M. (2017). Applications based on service-oriented architecture (SOA) in the field of home healthcare. *Sensors.* 17 (8):1703. https://doi.org/10.3390/s17081703
AWS Amplify https://aws.amazon.com/amplify/
AWS Identity and Access Management https://aws.amazon.com/iam/
AWS Lambda https://aws.amazon.com/lambda/
Bahmani, A., Alavi, A., Buergel, T. *et al.* (2021). A scalable, secure, and interoperable platform for deep data-driven health management. *Nat Commun* 12, 5757, https://doi.org/10.1038/s41467-021-26040-1
Gancheva, V., Stoev, H. (2019). DNA sequence alignment method based on trilateration. In: Rojas,I., Valenzuela , O., Rojas,F., Ortuño , F. (eds.) *IWBBIO. LNCS*, vol. 11466, pp. 271–283. Springer, Cham 2019. https://doi.org/10.1007/978-3-030-17935-9_25.
Gancheva, V., Stoev, H. (2023). An algorithm for pairwise DNA sequences alignment. In Bioinformatics and Biomedical Engineering—Proceedings of the 10th International Work-Conference, *IWBBIO*, Meloneras, Gran Canaria, Spain, July 12–14, 2023, Proceedings, Part I; Springer: Cham, Switzerland, 2023; Volume 13919.
Gancheva, V.; Stoev, H. (2024). Optimization and Performance Analysis of CAT Method for DNA Sequence Similarity Searching and Alignment. *Genes*, 15, 341. https://doi.org/10.3390/genes15030341.
Hulsen, T., *et al.* (2019). From big data to precision medicine, *Front. Med.*, https://doi.org/10.3389/fmed.2019.00034.
McPadden J. *et al.* (2018). A scalable data science platform for healthcare and precision medicine research, *arXiv*:1808.04849. https://doi.org/10.48550/arXiv.1808.04849.
Raghupathi, W., Raghupathi, V. (2014). Big data analytics in healthcare: promise and potential, *Health Inf Sci Syst* 2, 3, doi:10.1186/2047-2501-2-3
Taylor R. C., (2010). An overview of the Hadoop/MapReduce/HBase framework and its current applications in bioinformatics. *BMC Bioinformatics*.
Yang T. H., Sun Y. S., Lai F. (2011). A scalable healthcare information system based on a service-oriented architecture. *J Med Syst*. 35(3):391–407. doi:10.1007/s10916-009-9375-5. Epub 2009 Sep 24. PMID: 20703551.

System Innovation for an Artificial Intelligence Era – Lam et al. (Eds)
© 2025 the Author(s), ISBN: 978-1-032-84452-7

A Study on the predicaments and transformation in traditional tea store businesses-take the Hua-Yuan Tea Corp. in Kaohsiung as an example

Chien-Chun Lin
Graduate Student, Department of Cultural and Creative Industries, National Kaohsiung University of Science and Technology

Ying-Jye Lee
Professor, Department of Cultural and Creative Industries, National Kaohsiung University of Science and Technology

ABSTRACT: Tea has played an important part both in Chinese and Taiwan's history and culture. The Tea industry has faced many challenges and difficulties due to the global market, and the competitions from other courtiers. The traditional tea industry has encountered a decrease in exports and a huge decline in sales. This research is anticipated to provide valuable insights for the industrial transformation and marketing strategies of Taiwan's traditional tea businesses. The content completed thus far includes the design of trademarks, product stickers, and the setup and printing of tin cans, as well as the successful introduction of digital payment tools for retailers. These efforts are aimed at assisting traditional tea businesses in maintaining their competitiveness in a fiercely competitive market. It is hoped that the research findings will serve as valuable references for the industrial transformation and marketing of traditional Taiwanese tea businesses.

1 INTRODUCTION

1.1 *Research background*

The advent of the internet and mobile devices has shifted consumer preferences from in-person to online transactions, particularly among younger demographics (Wu, 2022). According to PricewaterhouseCoopers (2021), over three-quarters of the youth prefer to shop via mobile devices. The COVID-19 pandemic has further accelerated this shift, establishing online shopping as the predominant method of consumption. Taiwan's tea industry, a key traditional sector with a rich history, thrived in international markets, especially for Oolong and Pouchong teas since the late Qing dynasty. The early 20th century marked a boom in tea exports, vital to Taiwan's economy. World War II disrupted this, causing lasting impacts on the export market. Post-war, Taiwan shifted focus from exports to domestic sales, improving quality and competitiveness through contests and new techniques. The 21st century brought e-commerce, changing shopping habits and posing challenges for traditional tea businesses, especially during COVID-19, as consumer behavior shifted online. This situation presents both challenges and opportunities for Taiwan's tea industry in adapting to digital trends. This study, focusing on HuaYuan Tea Crop. in Kaohsiung through interviews, explores these challenges, and offers strategies for traditional tea businesses to remain competitive and sustainable.

1.2 *Research purpose*

(1) Investigate the development history and current situation of the tea industry in Taiwan.
(2) Taking Kaohsiung HuaYuan Tea Corp. as an example, explore the operational difficulties faced by traditional tea businesses and the corresponding countermeasures.
(3) Assist Kaohsiung HuaYuan Tea Corp. in Digital Transformation and Brand Identity Design.

2 LITERATURE REVIEW

2.1 *The history of the Taiwan tea*

Ho *et al.* (2021) pointed out that the history of tea can be traced back to the Qin and Han dynasties in China, around 1620 BC, which is before the Tianqi era of the Ming dynasty. By that time, tea had already become a major drink among the Chinese people and gradually evolved into a reflection of culture and lifestyle. People showed significant interest and emphasis on the quality of tea leaves and the brewing process, reflecting not only the status of tea in society at that time but also revealing its profound cultural value. According to "General

 DOI: 10.1201/9781003514831-72

History of Taiwan" by Lian (2015), the history of tea planting in Taiwan is long-standing. Ancient texts mention Shuishaolian tea, known for its unique color similar to pine and rhododendron, and its effectiveness in relieving heat and dispelling storms. Among the teas sold at that time, Wucheng tea, especially Dongding tea, was particularly popular for its excellent quality, despite its limited production. The climate conditions in Taiwan—warm, humid, and with even rainfall—provide an ideal natural environment for the growth of tea trees.

2.2 *The situation of the Taiwan tea*

Since Taiwan's formal accession to the World Trade Organization (WTO) in 2002, the local agriculture sector, particularly the tea industry, began to experience intense market competition from low-priced imported teas. Data extracted from the Taiwan Customs Administration database (2003-2023 changes in Taiwan's tea import and export) reveals that this competitive environment led to tea imports surpassing exports before 2016, reflecting a high dependence on external imports. In response to this challenge, the Taiwanese government has been actively implementing a series of policies since 2017, aimed at strengthening guidance and promotion for the local tea industry. Li *et al.* (2019) pointed out that the four strategies proposed by the government at the "Taiwan Tea Industry R&D Forum"—including "Smart and Eco-Friendly Tea Farming", "Development of New Processing and Value-Added Technologies", "Management and Formulation of Commercial Tea Ingredients" and "Shaping International Tea Industry Fashion and Culture"— are key to establishing a complete tea production and marketing ecosystem.

2.3 *The definition of e-commerce*

Liu (2021) mentioned that e-commerce primarily involves the transformation of traditional industry's offline transactions to online transactions conducted over the Internet, Intranet, and Value-Added Networks (VANs) utilizing electronic funds transfer. E-commerce is not limited to transactions conducted through computers and the Internet; it encompasses transactions made via television programs, radio broadcasts, telephone shopping, fax shopping, and other electronic means. In a broader interpretation, e-commerce includes any transactional activities facilitated by the Internet and various communication protocols and technologies, allowing parties to utilize modern electronic devices such as smartphones, smartwatches, tablets, and computers to complete transactions. Examples include online shopping, electronic ordering, online banking transactions, and mobile payments, all of which fall under the purview of e-commerce.

2.4 *The consumption patterns of e-commerce*

Ryan (2023) categorizes consumer patterns into three main types: B2B, C2C, and B2C. B2B stands for Business to Business, where companies do business with each other through online platforms, with Taiwan Trade Network being a major platform in Taiwan. C2C, or Customer to Customer, is where consumers sell to other consumers, using platforms like Shopee in Taiwan that help with product listings, payments, and shipping. B2C, or Business to Customer, is where companies sell directly to consumers. Platforms like Amazon and Alibaba's Online Mall are popular for B2C transactions, connecting consumers with businesses and their products.

2.5 *The develop of the e-commerce*

Wei (2022) mentioned that in contemporary society, with the widespread popularity of the Internet and the significant impact of the COVID-19 epidemic on the global economic pattern, consumer behavior patterns have experienced fundamental changes. This shift has propelled continuous growth in the performance of the e-commerce sector, gradually becoming the dominant force in contemporary consumer culture, thereby challenging and progressively replacing traditional business models. The traditional tea industry in Taiwan must respond to the changing times by gradually transitioning from traditional physical sales models to integration with e-commerce models. This adaptation is essential to meet the urgent needs of current consumer preferences and consumption channels, thereby enhancing brand visibility and sales performance.

3 RESEARCH METHODOLOGY

The research focuses on Hua Yuan Ming Tea in Kaohsiung as the case study subject. The following sections elaborate on the case study, in-depth interviews, and the scope and limitations of the research.

3.1 *Case study*

McCombes (2019) defines a case study as a detailed examination of a specific topic (such as individuals, groups, locations, events, organizations, or phenomena). It is frequently used in social, educational, clinical, and business research. Case studies usually involve qualitative methods but sometimes also employ quantitative methods. This type of research is suitable for describing, comparing, evaluating, and understanding different aspects of a research problem. Conducting a case study involves following multiple steps to ensure the research's quality and rigor. These steps include defining the research problem, selecting cases, collecting data (using interviews, observations, documents, surveys, and artifacts), analyzing data (using content analysis, thematic analysis, or grounded theory techniques), drawing conclusions, validating findings, and writing the report (Hassan, 2024). For achieving the research objectives, this study selects Hua Yuan Ming Tea in Kaohsiung as the case study subject, hoping to understand the issues currently faced by the traditional tea industry through the process of in-depth interviews.

3.2 *In-depth interviews*

Fan (1994) stated that in-depth interviews are an indispensable method in qualitative research, especially suitable for exploring the multi-dimensional complexity of human experiences and behaviors. By establishing a trust relationship between researchers and interviewees, this method allows for a deep understanding of the interviewees' values, motivations, beliefs, and unexpressed inner thoughts. In-depth interviews emphasize the subjective involvement of the researcher throughout the interview process, where sensitivity and empathy are crucial for uncovering the true thoughts of interviewees. Researchers must possess high levels of creativity, logical thinking, and analytical skills to deeply analyze the collected data and construct a comprehensive interview outline. This method not only provides interpretive descriptions of the research content but is particularly effective in defining unclear situations, offering a unique approach to facilitate the understanding and interpretation of complex social phenomena. In-depth interviews are divided into three parts: pre-interview, during the interview, and post-interview, with each phase elaborated in this study.

3.3 *Research process and steps*

Fan (1994) indicated in her research that the key to in-depth interview studies lies in the depth of the interviews and the breadth of their scope, rather than the representativeness of the sample. This method allows researchers to understand the actual situation of the study subjects more deeply and broadly, which is the main consideration for evaluation. This study conducted in-depth interviews with Mr. Lin Kuan Hung, the second-generation person in charge of Kaohsiung HuaYuan Tea Corp, aiming to obtain information closely related to the research topic. For this purpose, the research team first collected background literature extensively through internet and book resources to enrich the research foundation. Subsequently, detailed communication was conducted with the interviewee, Mr. Lin Kuan Hung, according to the pre-arranged interview outline, to ensure that the content of the interview met the research needs and to obtain his consent for the interview, completing the preparatory work before the interview. In summary, through systematic pre-interview preparation, precise on-site data capture, and rigorous subsequent data processing, this study ensured that the information obtained possessed a high degree of professionalism and practical value, contributing valuable primary data to the research field.

3.4 *Scope and limitations of the study*

3.4.1 *Scope*

This study selects HuaYuan Tea Corp., located in the Qianjin District of Kaohsiung City, which has been in operation for over 40 years, as the research case (Figure 1). HuaYuan Tea Corp, founded in 1982 by Mr. Lin Kunde and his wife Lin Chen Huimei, inherits the family's passion and knowledge for tea. The current second-generation owner, Lin Kuan Hung, abandoned his job in the tech industry at a young age to return to the tea industry, dedicating himself to exploring high-quality tea leaves and deeply studying the characteristics of various tea-producing areas in Taiwan.

Figure 1. The study case: HuaYuan Tea Corp.

3.4.2 *Limitations*

The primary research findings are based on interviews with Mr. Lin Kuan Hung, the second-generation owner of HuaYuan Tea Corp. The analysis is primarily based on Mr. Lin's statements. Whether the proposed management strategies for traditional tea businesses are suitable for all traditional tea businesses in Taiwan remains to be discussed.

4 RESEARCH FINDING AND DISCUSSION

4.1 *In-depth interview record and analysis*

This study employed unstructured, in-depth interviews to conduct a comprehensive discussion with Mr. Lin Kuan Hung, the second-generation proprietor of HuaYuan Tea Corp. (hereinafter referred to as Mr. Lin), on January 20, 2024. The interview, lasting about an hour, took place at Mr. Lin's storefront. During the interview, through Mr. Lin's words and expressions, this research was able to perceive his pride in the tea business inherited from his parents. Mr. Lin detailed the journey of founding "HuaYuan Tea Industry Co., Ltd." in May 1982 by his parents, driven by their profound love for tea, and the operational challenges encountered with the change of eras. Over the past three years, the COVID-19 pandemic and cross-strait political tensions have limited Taiwan's tea exports, reduced public income, and shifted consumption patterns from offline to online. This transition poses a significant challenge for many traditional tea businesses, with some venerable tea houses facing closure due to unfamiliarity with digital transformation. In the contemporary economic wave, traditional tea businesses need substantial financial support whether investing in packaging or undergoing digital marketing transformation. Many faces a financial crunch due to declining business. However, obtaining necessary funding is key to achieving transformation and growth. In summary, for traditional tea businesses, undergoing digital transformation and rebranding, and formulating effective marketing strategies have become crucial. The challenges faced by traditional tea businesses can be summarized into three main issues: packaging design, marketing transformation, and financial needs. In light of this, this study proposes the following strategies for HuaYuan Tea Corp. to address its operational dilemmas:

(1) Packaging Design: Effective communication strategies with consumers at the point of sale should focus on finding a balance, encouraging traditional tea business owners to accept reasonable costs for packaging while not neglecting the quality of their tea products. This is not merely about the aesthetic appeal of the product but a crucial strategy to enhance its competitiveness in the market and meet consumer expectations, essential for the sustainable development of traditional tea businesses.
(2) Marketing Transformation: Traditional tea business operators often display a gap in understanding towards the digital platforms preferred by contemporary young consumers. This gap primarily stems from a lack of understanding of current social media platforms and their mechanisms, leading to communication barriers with the younger consumer demographic. To address this issue, effective communication is necessary to find a common ground acceptable to both parties, thereby narrowing the gap created by differences in digital knowledge and smoothly facilitating the key aspect of digital transformation for traditional tea businesses.
(3) Financial Needs: The government currently offers various subsidy programs to assist with these financial challenges, such as the Taiwan Cloud Marketplace. Furthermore, tea business owners can also seek necessary funds for transformation through legal channels, such as applying for bank loans. Utilizing these financial avenues is essential for helping traditional tea businesses adapt to new era demands and promote business transformation.

4.2 *Trademark design of HuaYuan tea corp.*

The design and strategic application of trademarks are indispensable elements in brand building, playing a crucial role in enhancing corporate image and strengthening market competitiveness. Designing trademarks that are highly recognizable and creative, by deeply analyzing and understanding the needs of target consumers and combining them with the company's brand positioning, is one of the key steps in achieving a successful transformation of traditional brands. Based on the corporate culture, values, and market positioning of HuaYuan Tea Corp., this study assists in designing two trademarks, described as follows:

The first trademark (Figure 2) draws inspiration from tea pots, particularly those symbolizing Chinese tea culture, such as the purple clay pot. This trademark aims to convey the action of brewing tea in a simplistic and abstract form, reflecting a youthful and vibrant image of the brand, which will mainly be used in the bagged tea and tea drink markets. The design concept of this trademark combines traditional and modern elements. The central tea pot pattern is outlined with simple lines to depict the pouring action, presenting the fluidity of tea. This not only symbolizes the heritage and fluidity of tea culture but also conveys a relaxed and joyful sensation associated with the brand. The abstract form of the tea pot gives an impression of a smile, possibly symbolizing the positive emotions and customer satisfaction the brand aims to deliver.

Figure 2. The first trademark.

The second trademark (Figure 3) adopts an octagonal design, incorporating traditional Taiwanese elements and text, with the aim of strengthening brand recognition among existing customers. This trademark will be used for tea products and services. The design of these two trademarks not only reflects the company's deep understanding of market segmentation and target consumer groups but also demonstrates flexibility and foresight in brand strategy. Particularly when the company considers launching new brands targeted at younger consumers, these designed trademarks can be directly applied, thereby saving the costs and time of redesigning.

Figure 3. The second trademark.

4.3 *Design of product packaging*

This study, through in-depth discussions with the owner of the target company, explores how to enhance a product's market competitiveness through packaging design under the consideration of limited costs. Taking cost factors into account, the initial plan adopted sticker-based packaging design and printing, a method applied to vacuum-packed tea and standardized tin can packaging. Moreover, to further improve the visual quality of the packaging and convey the brand's high-end image, the use of hot stamping technology was recommended in this study. However, due to cost constraints, a decision was made to use faux gold printing to simulate the hot stamping effect, aiming to enhance the packaging quality without significantly increasing costs. Additionally, to further enhance brand recognition and product professionalism, this study suggests that the company engage in small-batch production of customized tin cans. This approach not only showcases the uniqueness of the product but also offers a more personalized product experience for consumers. Through this strategy, businesses can effectively enhance their product's market competitiveness and brand value while maintaining cost-effectiveness. The packaging design solutions are detailed in Figures 4 and 5.

Figure 4. Tea product label design production.

Figure 5. Customized tin can design.

4.4 *Marketing transformation and financial assistance goals*

This study employs in-depth interviews to comprehensively understand the intention of the interviewed companies to undergo digital transformation. Specific measures include the introduction of electronic payment systems in physical stores, such as credit card payments, mobile payments, and barcode payments. Throughout the research process, the research team also supported companies in establishing LINE official accounts and managing Meta's social media platforms, such as Facebook fan pages and Instagram business accounts. This is done through social media advertising and the production of short video content to enhance brand awareness. In terms of financial support, this phase of the study has successfully guided companies in applying for the Taiwan Cloud Marketplace Subsidy Program offered by the Small and Medium Enterprise Administration of the Ministry of Economic Affairs, which has played a key role in assisting traditional industries in their digital transformation journey. Currently, the subsidy covers various aspects, including monthly rental fees for equipment used in diverse payment systems. Moving forward, this study will continue to support enterprises in social media advertising and related activities.

5 CONCLUSION AND RECOMMENDATIONS

This study, focusing on HuaYuan Tea Corp. as a case study, delves into the challenges and transformation strategies faced by the traditional tea business in Taiwan. Through literature analysis and in-depth interviews, it has been revealed that digital transformation is crucial for traditional tea businesses to expand their customer base and rebrand. The study indicates that the primary challenges faced by traditional tea businesses include packaging design, marketing transformation, and financial needs, proposing operational strategies accordingly. Traditional tea businesses face obstacles such as a lack of digital skills, uncertainty in consumer behavior, and uncertain return on investment costs in transformation. Therefore, this study recommends that traditional tea businesses adopt a phased digital transformation strategy, from brand reinvention to establishing online platforms, focusing on brand and customer relationship management. In summary, this study underscores the urgency and feasibility of transformation, aiming to provide theoretical and practical guidance for the digital transformation of traditional industries in Taiwan and globally.

REFERENCES

Fan, L. J. (1994). An introduction to in-depth interviews. *Journal of Outdoor Recreation Study*, 7(2), 25–35. https://doi.org/10.6130/jors.1994.7(2)3

Hassan, M. (2024). Case study – methods, examples, and guide. Retrieved from https://researchmethod.net/case-study-research/

Ho, C. F., Lin, H. J., & Ho, C. R. (2021). The transformation of the structure and business model of Taiwan's tea industry. *Journal of Commerce*, (37), 1–25.

Li, H. X., Tsai, H. T., Tsai, W. H., Hong, Z. Y., & Chen, N. H. (2019). The opportunity and discussion for transformation and development towards Taiwan tea industry 3.0. Retrieved from https://agritech-foresight.atri.org.tw/article/contents/1898

Lian, H. (2015). General history of Taiwan. One Person.

Liu, J. P. (2021). Transformation strategies from traditional industry to e-commerce. National Taipei University of Technology, Taipei City.

McCombes, S. (2019, November 20). What is a case study? | Definition, examples & methods. Retrieved from https://www.scribbr.com/methodology/case-study/

PricewaterhouseCoopers (2021). 2021 Global Consumer Insights Survey. Retrieved from https://www.pwc.tw/zh/publications/global-insights/consumer-markets.html

Ryan (2023). What are the b2b, c2c, b2c, c2b models in e-commerce?. Retrieved from https://www.inside.com.tw/article/6947-what-is-b2b-c2c-b2c-c2b

Wei, W. J. (2022). Industry economic statistics brief (No. 398). Department of Statistics, Ministry of Economic Affairs.

Wu, F. Q. (2022). A study on the attitudes of Taiwanese consumers towards the integration of food retail and e-commerce [Unpublished master's thesis]. National Taiwan Normal University, Taipei City.

System Innovation for an Artificial Intelligence Era – Lam et al. (Eds)
© 2025 the Author(s), ISBN: 978-1-032-84452-7

Understanding the political power of water engineering: Factors of Japan, America and China in the development of water engineering in Taiwan

Bin-Yuan Wen
Department of Civil Engineering, National Pingtung University of Science and Technology, Pingtung, Taiwan

Cheh-Shyh Ting
General Research Centre, National Pingtung University of Science and Technology, Pingtung, Taiwan

ABSTRACT: Water engineering has been vital in running Taiwan since the Japanese rule. However, its research has overlooked historical, humanistic, and social contexts. The sustainability of dams has been questioned as they can cause water crises and overexploitation of groundwater. Water engineering can also be a tool for exercising power and a source of power production. This paper examines its development in Taiwan through a technopolitical perspective, investigating how states use it for hidden political goals. The research found that water engineering is an instrument of state power over urban society and nature, and a source of power production.

1 INTRODUCTION

Water engineering has played a crucial role in Taiwan's governance. However, many people think that water engineering is just an application of scientific and technical knowledge, or a profession that creates water-related structures. This traditional view of water engineering is being challenged, and a broader perspective is needed. Shinpei Gotō, the civil administrator of Taiwan during the Japanese rule, believed that Western modern construction, which included water engineering, was a tool for civilization to rule and win over the hearts and minds of the colonies. (Lin, 2014).

The paper discusses the political power of water engineering in Taiwan and the theoretical tools required for water engineering there. It highlights the need for a new perspective and conception of nature and water to address sustainability issues. The traditional understanding of water engineering in natural sciences is not sufficient to understand the political power of water engineering in Taiwan. The paper examines why and how various ruling periods achieved their hidden political goals through the construction of water engineering. Mainstream water engineering in Taiwan still lacks humanistic, historical, and social science perspectives. The challenges to the natural science perspective on water come from scientific study to engineering practice. The climate on the Earth is changing due to significant anthropogenic changes, exposing the past hydrological predictions that relied on stability. Water Alternative was established to break the narrow framing of water issues and highlight the political dimension of water development. Debates on water issues should consider the social and political aspects of water development.

The water cycle is a scientific concept we have long taken for granted as a natural fact. However, it is substantially a recent invention constructed by society and its history. In 2014, an article titled "Modern Water and Its Discontents: A History of Hydrosocial Renewal" was published by Wiley. It sheds light on the fact that the water cycle proposed in 1931, which has dominated hydrological theory for a long time, was created to view hydrology as purely natural science. Consequently, water has been abstracted from its original geographic, social, cultural, religious, and ecological contexts and has been simplified into a substance synthesized by hydroxyl radicals in the natural world (Linton, 2014).

The Gaoping River intake project for hyporheic flow, located on the border of Kaohsiung and Pingtung in Taiwan, is an engineering demonstration that highlights the importance of social communication. According to WRA in 2015, the success of this project was largely due to effective social and political participation, rather than technical or scientific expertise. Despite initial objections from local farmers, the project was eventually able to overcome these challenges through effective communication and collaboration. This project serves as a prime example of the social and political aspects of water engineering, which are increasingly becoming a key focus in this field.

 DOI: 10.1201/9781003514831-73

2 ANALYTICAL FRAMWORK

In recent years, the relationship between water, infrastructure, and political rule has been studied and analyzed by various established schools of thought. The rise of China has further examined and organized this relationship, leading to the emergence of the Chinese Water Machine (CWM). Research into the relationship between technology and politics has been conducted earlier in anthropology and the study of the Cold War. Table 1 below provides a synthesis of theoretical concepts from rich humanistic, historical, and social scientific perspectives to illustrate their epistemological differences from natural science. This table is based on the review works of Bichsel in 2016 and Rogers & Miller in 2017.

Table 1. Approaches to water, water engineering, and political rule.

Theoretical Tools	Political Economy	Political Ecology
Water	Resource	Relationships produced/co-constituted by society
Infrastructure	The economic and symbolic structure	Material mediator between water and politics
Political Rule	State and elite supremacy	State-society relations, governance
W-I-P Connection	Political and economic power	Hydro-social relations
Key Concepts	Hegemony, economic production, state-and-nation-building ideology	Inequality, justice, scale, governance, subjectivity, waterscape
Theoretical Tools	Socio-technical approaches	Socio-material approaches
Water	Multi-purpose/ multi-functional substance	Matter
Infrastructure	Technology	Physical artefact
Political Rule	National and transnational regimes	Relationship between "things" which exert agency
W-I-P Connection	System-building	Assemblage/configuration
Key Concepts	Design, regime, builder, agency	Relationality, contingency, objects, non-humans, agency
Theoretical Tools	Discourse analysis	Governmentality
Water	Text	Political, Bio-politic
Infrastructure	Discursive and non-discursive construction	Manifestation of centralized power
Political Rule	Power-knowledge relationship	The state governs its people and space through water

(continued)

Table 1. Continued

W-I-P Connection	Discourse	Rationality of state governance for securing the health and productivity of the population
Key Concepts	Statements, power, social construction, ideology	Bio-politics, discourse, subjectivity
Theoretical Tools	Infrastructure	Technopolitics
Water	Material	Material
Infrastructure	The material form of water that allows for the possibility of exchange over space	Knowledge, skill, diagram, calculations, and energy that makes its use possible; an assemblage of relationships, practices; knowledge, biophysical dynamics, humans, and things
Political Rule	Select or recognize what is infrastructure and what is not	Enact political goals through water engineering
W-I-P Connection	Infrastructures shape the nature of a network, the speed and direction of movements, its temporalities, and vulnerability to breakdown	Power is not something held by human actors but is seen as a force that comes into effect through the connection between human and non-human actors
Key Concepts	How politics can be constituted in different ways	The role of technology in enacting political goals; technopolitical regime; "Thick" with politics

Source: Rogers & Crow-Miller, 2017; Bichsel, 2016; Larkin, 2013; Edwards & Hecht, 2010. Compiled and adapted by Wen & Ting, 2023.

There are two major theories, technological determinism and social constructionism, that critique modern technological development. Technological determinism sees technology as the cause of all problems and dominates or influences society. Social constructionism argues that technology is the result of interactions between related social groups or actors. Gabrielle Hecht's concept of "technopolitics" highlights the political process of constructing nuclear energy technology in France and the political goals it serves. (Edwards & Hecht, 2012) It also suggests that technology sends specific political messages. (Chang, 2013) Colven (2017) analyzed Jakarta's Great Garuda Sea Wall Project using the concept of technopolitics. The project brought together a network of political and economic interests, world-class urban discourse, engineering expertise, and colonial history. The seawall was considered the best solution to Jakarta's flooding problems by political elites and engineers. The project's techno-political nature was a result of post-colonial relations between Jakarta and the Netherlands.

This paper proposes a theoretical framework (Figure 1) to analyze and explain the development of Taiwan's water engineering technological system. It aims to expand the traditional understanding of hydraulic engineering beyond natural science and into humanistic, historical, and social scientific advancement. This means changing the existing mainstream view of hydraulic engineering in any discipline. As new challenges emerge, it is vital to integrate humanities, history, and social sciences with ecological, social, political, and economic theoretical tools to manage water resources engineering. This will create an integrated science that combines natural and social sciences.

Figure 1. A proposed theoretical framework.

3 THREE FACTS IN THE DEVELOPMENT OF WATER ENGINEERING IN TAIWAN

3.1 *The hydraulic legacy modernized Taiwan under empire's imagination of rule*

During colonial rule, modern water engineering systems, including harbors, water and sewage works, irrigation, and flood control, were constructed in Taiwan. The Japanese mainland and Taiwanese governor's office interfered with the natural water cycle to achieve their political goal of social control. Seaports were initially constructed for military purposes, but the army faced more danger from diseases than from enemies. To avoid epidemics caused by unclean water and soil, the Governor General's Office developed modern sanitation, urban planning, water supply, and sewage works, consolidating the center of Japanese rule. (Li-Ling Liu, 2004, p.2-1;5-1)

The state's implementation of water engineering aimed to dominate Taiwan. Initially, it was for military suppression, and later, for agricultural irrigation, hydroelectric power generation, and wartime tasks. Flood control and water management were utilized to strengthen state rule. These actions demonstrate the use of technology to execute political goals, which is what technopolitical theory refers to as the demonstration of power through technology. (Edwards & Hecht, 2010) The achievement of specific political goals must rely substantially on these materials, artificial technology, and artefacts. (Wen & Ting, 2023)

3.2 *U.S. Aid to Taiwan's water engineering projects after World War II*

The US provided Taiwan with aid in finance and technology. Hydroelectric engineering was a priority in the US Aid's four-year plan for economic and industrial projects. This aid was related to the Cold War geopolitical situation and the US' use of technical assistance for water projects as a diplomatic tool to expand its influence over underdeveloped countries. The US relied on three leading institutes to achieve its political goals, namely the US Army Corps of Engineers (USACE), the Bureau of Reclamation, and the Tennessee Valley Authority (TVA).

After the outbreak of the Korean War in 1950, the United States significantly increased its assistance to Taiwan. This aid was used as a means by the Truman administration to curb the expansion of communism. Additionally, it led to the initiation of several water engineering projects in Taiwan. The J.G. Whites Engineering Corporation and the Bureau of Reclamation provided technical assistance to help Taiwan rehabilitate and restart many of its water projects which were left by the Japanese ruler. American aid was also crucial in building a "model province of a free China" to counter Communist China and Russia after the Nationalist Government exiled Taiwan.

3.3 *Technocrats and practices of engineering water that follow nationalist China's exile to Taiwan*

The story of Yu the Great, who ruled over water in the Xia Dynasty, is considered one of the most legendary tales in China. Irrespective of its actual content, this myth or legend reflects a long-standing political tradition of water governance. The ability to control floods is crucial for gaining the support of the people and establishing legitimacy. By providing the people with the necessary services to avoid disasters, the state or the emperor asserts infrastructural power and gains the legitimacy to rule. This story highlights the power derived from society.

In her 2017 article titled "Reasons to Dam: China's Hydropower Politics and Its Socio-Environmental Consequences," German scholar Habich examines the history of hydropower development in China. This history dates back to Sun Yat-sen and his nationalist followers' efforts to use the scientific method to transform China's natural environment, as they sought to build a robust and prosperous country. Sun Yat-sen's Plan, which was published in 1921, is an excellent example of technopolitics, which refers to using technological initiatives as a political blueprint. Sun Yat-sen's objective was to build a prosperous and powerful nation by constructing engineering projects, including waterways, seaports, railroads, canals, mines, factories, cities, hydroelectricity, and irrigation.

4 FINDING

What political power does water engineering have? Through a historical examination of the development of water engineering in Taiwan during different stages of Japanese rule, U.S. aid, and the exile of the Nationalist government to Taiwan, we find that water engineering is a crucial aspect of the state's technopolitical imagination and blueprint for development and governance. Water engineering has been used as a means to achieve or execute political goals in Taiwan during these different stages of development, regardless of whether the sources were Japanese, American or Chinese. Furthermore, hydraulic engineering has the potential to be an instrument of power to control urban, social and natural objects. Finally, hydraulic engineering has become a wellspring for the production of state power with the characteristics of Tool-power.

In short, this paper aims to explore the technopolitical process of controlling water, water organization and society in Taiwan. It explained the power of technology and politics to shape each other, whereby politics allowed technology to develop into large-scale technological systems, and whereby technology constructed and operated systems based on specific political values. The technological system of hydraulic engineering constructed by these three factors, Japan, America and China regarded water as a material resource to be exploited, but at the same time, limited the possibility of incorporating ecological values into the development of the system, which continued and laid the inevitable fuse for future disputes between development and ecology.

REFERENCES

Bichsel, C. (2016). Water and the (Infra-)Structure of Political Rules: A Synthesis, *Water Alternatives*, 9(2): Pp.356–372.

Chang, Kuo-hui. (2017). A review of the theory of large technological systems: some thoughts about how to study the systems transferred from the West to Taiwan, *Taiwanese Journal for Studies of Science, Technology and Medicine*, 24(April 2017), pp. 091–144. (Published in Chinese)

Chen, K. W. (2014). *The Policy, Implementation, and Development of the Keelung Harbor during the Japanese Colonial Period (1895–1945)*, master's degree, National Chung Cheng University, Chiayi. (Published in Chinese) (Unpublished doctoral dissertation)

Colven, E. (2017). *Understanding the Allure of Big Infrastructure: Jakarta's Great Garuda Sea Wall Project*, *Water Alternatives*, 10(2), pp. 250–264.

Edwards, P. N., Hecht, G. (2010). History and the Technopolitics of Identity: The Case of Apartheid South Africa, *Journal of Southern African Studies 2010*, 36(3): Pp.619–639.

Habich-Sobiegalla, S. (2017). *Reasons to Dam: China's Hydropower Politics and Its Socio-Environmental Consequences*, Research Gate.

Hughes, T. P. (1987). *The Evolution of Large Technological Systems, The Social Construction of Technological Systems*, 4th ed.; W. Bijker, T. Hughes, and T. Pinch, Eds., MIT Press Mass: Cambridge, United States of America, 1987; Pp. 51–82.

Larkin, B. (2013). The politics and poetics of infrastructure, *Annual Review of Anthropology*, volume 43, pp.327–343.

Lin, L. F. (2001). Early electric construction in Taiwan (1877–1919), *Journal of the Department of History* , 18, pp. 243–276. (Published in Chinese)

Linton, J. (2014). Modern water and its discontents: A history of hydrosocial renewal, *WIREs Water*.

Liu, L. L. (2004). *A study on the Water Supply Facilities and Architectures during the Japanese Period in Taiwan*, master's degree, National Cheng Kung University, Tainan. (Published in Chinese) (Unpublished master's thesis)

Meehan, K. M. (2014). *Tool-power: Water Infrastructure as Wellsprings of State Power*, *Geoforum*, volume 57, 2014, pp. 215–224.

Milly, *et al.* (2008). Stationarity is dead: whither water management? *Research Gate*

Rogers, S., Crow-Miller, B. (2017). The politics of water: a review of hydropolitical frameworks and their application in China, *Wires Water*, volume 4 Issue 6.

Winner, L. (1980). Do artifacts have politics? Daedalus, Vol. 109, No. 1, Modern Technology: Problem or Opportunity? Pp.121–136.

Wen, B., Ting, C. (2023). Understanding the political power of water engineering: Colonial modernity and the production of waterscape in taiwan under japanese ruling, *Taiwan Water Conservancy*, Vol. 71, No. 4, pp.14–32.

System Innovation for an Artificial Intelligence Era – Lam et al. (Eds)
© 2025 the Author(s), ISBN: 978-1-032-84452-7

Computer aided platform for healthcare data analytics

Veska Gancheva
Faculty of Computer Systems and Technologies, Technical University of Sofia, Bulgaria

Ivaylo Georgiev
Stephan Angeloff Institute of Microbilogy, Bulgarian Academy of Sciences, Sofia, Bulgaria

ABSTRACT: The research presented in this paper is part of a scientific project aimed to develop an integrated platform for intelligent decisions and automation of efficient algorithms and tools for storing, analyzing and visualizing big biomedical data to support knowledge extraction and decision-making. The platform offers integrated software tools environment for developing workflows to automate computational process and increase efficiency in conducting scientific experiments. The platform architecture is designed, research scenarios and requirements are defined: easy-to-use environment and user interface; supports various data formats; provides interactive tools to perform workflows for data analysis - biological data, medical images, virtual reality visualization; quickly deploy, add and remove functionalities as services; efficiency and speed; scalability and adaptive resource addition. The platform enables quickly and accurately big biomedical data analytics, patterns and trends identification. The features are validated through workflow modeling and software implementation to analyze and search for patterns between data representing part of human health and cardiovascular disease probability. The used dataset is reduced statistical records related to cardiovascular disease analysis. The original database contains 76 attributes, of which 14 are used. The data originates from Cleveland, USA, downloaded from Kaggle web portal. An analysis is performed to detect missing data for the purpose of unification to allow subsequent analysis and processing. A correlation matrix is presented, providing an overview of the level of correlation between pairwise values and the data distribution. The overall analysis provides insight into which data to look for correlation in training a neural network or decision tree.

1 INTRODUCTION

Technological developments have revolutionized biomedical research and treatment, resulting in the massive collection of diverse biomedical data. Biomedical research and healthcare delivery face both opportunities and challenges due to the enormous volume and diversity of data generated on a daily basis. There are many different types of data that are included in this explosion of data: genetic sequences that contain the blueprint for life, clinical reports that describe patient histories and outcomes, reviews and studies that are comprehensive and contained in biomedical literature, detailed medical imaging that offers a window into the human body, and sensor data that continuously monitors physiological parameters. Every one of these data kinds adds to the ever-expanding repository of knowledge that academics and medical professionals have access to. On the one hand, this abundance of data has the capacity to completely transform how we think about disease mechanisms, intricate biological systems, and patient care using customized medicine techniques. It makes it possible for scientists to identify the underlying genetic causes of many illnesses, create specialized treatments, and raise diagnostic precision. On the other hand, advanced computational techniques and tools are needed for the management, analysis, and interpretation of this data, in addition to methods to guarantee the confidentiality, integrity, and quality of the data.

The advent of new tools and technologies for Big Data analysis and visualization has led to a rapid expansion in precision medicine Precision medicine holds that each patient's unique traits should be taken into account when diagnosing an illness, making medical decisions, treating a patient, and recommending medication. This is especially true when the patient has undergone genetic testing (Ginsburg G. S. and Phillips K. A. 2018). In the era of big data and with the development of electronic healthcare records, large and comprehensive databases of genomic, transcriptomics, proteomics, or metabolomics variables, along with traditional clinical patient characteristics and treatment records, have emerged at an increasingly rapid rate.

However, such data are often highly heterogeneous, high dimensional, noisy, and challenging to interpret when employed directly in a clinical setting for accurate diagnosis and illness outcome

 DOI: 10.1201/9781003514831-74

prediction. A limited sample size, poor technology, differences in clinical trials, a range of patient groups, and the structure of the healthcare system can also often limit the quality of the data. The Internet of Medical Things can improve patient outcomes and quality of life by enabling healthcare practitioners to monitor patients' status and access electronic medical records. Hospitalizations that occur frequently are decreased by remote monitoring and consultation (Balasubramanian V., Jolfaei A. 2020).

Big data techniques offer the capability to gather extensive collections of biological samples and efficiently store, manage, and analyse this data (Mallappallil M., *et al.* 2020). Machine learning algorithms have the capacity to generate supplementary data beyond the initial input, thereby deriving knowledge from the available data. Rapid Learning Healthcare (RLHC) models, employing artificial intelligence, can identify data of diverse quality compared to validated datasets. These extracted results are then pro-cessed through decision support systems (DSS) to offer knowledge-based healthcare solutions.

Machine learning techniques have attracted considerable interest from medical re-searchers and clinicians. There is a lot of research going on that addresses the need to develop predictive algorithms, use freely available open source software, and public domain data (Sidey-Gibbons, J., Sidey-Gibbons, C. 2019). The complexity and growth of data in healthcare means that artificial intelligence will be increasingly applied in the field. Key application categories include diagnosis and treatment recommendations, patient engagement and adherence, and administrative activities (Davenport T, Kalakota R. 2019).

Healthcare systems are complex and challenging for all stakeholders, but artificial intelligence has the potential to improve patient care and quality of life. The rapid advancement of artificial intelligence could revolutionize healthcare by integrating it into clinical practice (Alowais, S.A., Alghamdi, S.S., Alsuhebany, N. *et al.* 2023). Artificial intelligence uses machine learning, deep learning and neural networks. In the medical field, it can be used to improve diagnosis, prognosis, treatment, surgery, drug discovery, or for other applications (Hulsen T. 2022).

Mobile health (m-health) is the term for health monitoring using mobile phones and patient monitoring devices, etc. (Khan ZF, Alotaibi SR. 2020). Artificial intelligence and big data analytics are applied within m-health to ensure an efficient healthcare system.

The paper is structured as follows: Section 2 presents the healthcare data sources and analysis. The proposed platform architectute is described in Section 3. An workflow for cardiovascular data analytics is described in Section 4.

2 HEALTHCARE DATA SOURCES AND ANALYSIS

Sources of big data in healthcare include smart devices, genetic databases, government, and more.

- Internet of Things (IoT): Wearable devices; Smartphone applications, Medical devices and sensors.
- Electronic Medical Records/Electronic Health Records (EMR/EHR).
- Other clinical data.

The process of knowledge discovery in databases involves several critical stage (Pareek M., Bhari P. 2020), (Parihar A., Sharma S. 2020), (Borovska P., Gancheva V., Georgiev I. 2019), each designed to refine and focus the analysis toward actionable insights. Given the common imperfections in data, such as inconsistencies and abbreviations, they are often unsuitable for direct application in initiating the data retrieval process. Moreover, the swift increase in data generation rates have to be taken into account when analyzing data across various academic and scientific domains. Most of the collected data necessitate complex analysis mechanisms. Data preprocessing aims to align the data with the requirements of each data retrieval algorithm, thereby enabling the processing of data that would otherwise be unsuitable.

Figure 1 illustrates a conceptual model for data analytics, mapping out the comprehensive steps involved in identifying and interpreting patterns within data. This conceptual model serves as a guide for systematically extracting valuable insights from data, highlighting the iterative nature of the process and the need for careful consideration at each step to ensure the accuracy and relevance of the findings. The outlined stages form a cycle that iteratively improves the understanding and application of data to meet specific goals. This process involves iterating through several steps as described below (Sun S., *et al.* 2023; Sarker I. H. 2021; Mahesh B. 2020):

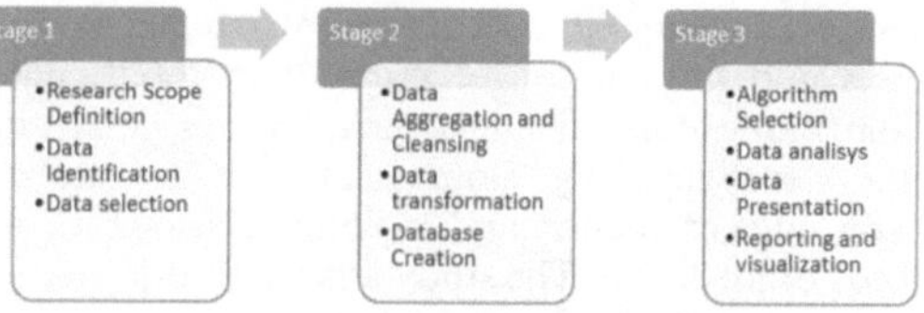

Figure 1. Data-driven knowledge discovery.

1. Goal Definition: This foundational step involves articulating the objectives of the knowledge discovery process. It requires specifying the task at hand, any prior knowledge that exists, and how this knowledge will be applied.

2. Scope Determination: Here, the focus is on understanding the intended end-user's knowledge level and objectives to ensure that the analysis meets their needs.
3. Target Dataset Creation: This involves selecting the specific variables, sets, or instances of data that will form the basis of the analysis.
4. Data Filtering and Preparation: At this stage, the data is cleaned by removing redundant or irrelevant values, gathering the necessary data for modeling, and addressing any gaps in the data.
5. Dataset Simplification: The aim here is to refine the dataset by discarding unnecessary variables and selecting the most appropriate data presentation features for the task.
6. Integration of Discovery Process and Extraction Techniques: This phase combines the overall data discovery process with specific techniques for data extraction, such as classification, clustering, or regression, to clarify the goal of the analysis.
7. Data Extraction Algorithm Selection: This critical step involves choosing the most suitable models and parameters for the task, exploring the data to find these models, and ensuring that the chosen data extraction methods meet established criteria.
8. Data Extraction Execution: The selected algorithms are applied to identify noteworthy patterns or models, such as clusters, regression lines, classification rules, or decision trees.
9. Knowledge Interpretation: The insights derived from the models are interpreted to understand the underlying patterns and relationships within the data.
10. Knowledge Utilization and Integration: Finally, the gleaned knowledge is applied and integrated into another system for further action or decision-making.

3 PLATFORM ARCHITECTURE

The research is part of a scientific project aimed to develop an integrated platform for intelligent decisions and automation of efficient methods, algorithms and tools for integrating, storing, analyzing and visualizing big biomedical data to support knowledge extraction and making decisions for the needs of medicine. The study has led to the creation of an open technological platform to enable the development of automated processes and decision-making based on large-scale biomedical data. Using collection of software tools to automate the computational process in conducting scientific experiments, the platform provides an extensible and integrated environment for the development of workflows to automate the computational process in conducting scientific experiments. Its features include: 1) an intuitive user interface; 2) interactive tools for executing workflows for data analysis, including biological data, medical images, and virtual reality visualization; 3) support for multiple data formats; 4) ability to share and reuse workflows; 5) capability to monitor the outcomes of the workflow execution and the steps involved in creating the workflow; 6) capacity to add, remove, and scale functionalities as services fast; 7) increased speed and efficiency; and 8) scalability and adaptive resource augmentation.

Figure 1 illustrates the SOA based platform architecture, designed to process large-scale biomedical data and derive valuable insights. SOA supports data sharing and exchanges through standardized interfaces, ensuring compatibility between various systems and data sources as well as facilitates the seamless integration and interaction of different applications and data repositories, enabling distributed deployment of services across various platforms and computational environments. These services collaborate to automate complex processes, with each service encapsulating specific data and functionalities. The benefits include adaptability, ease of maintenance, cross-compatibility, support for diverse technologies, and its agility in responding to shifting business requirements. Furthermore, it simplifies the distributed development and management of applications.

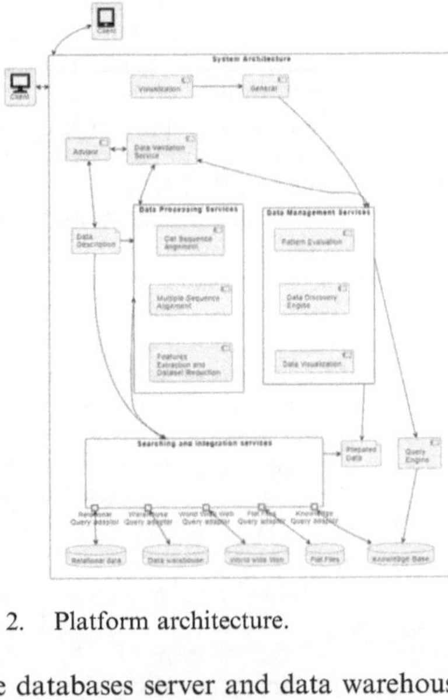

Figure 2. Platform architecture.

The databases server and data warehouse handles essential data management tasks, ensuring that actual data sets are ready for processing and stored appropriately based on their type. Data sources include databases, data warehouses, Internet, text files, and various documents, all of which hold extensive historical information to enable effective knowledge extraction. Data preparation, integration, cleaning, and selection are critical preliminary steps to the knowledge discovery process, addressing issues of incomplete or unreliable data. Given the excess data produced by various sources, it is crucial to filter and select only the data relevant to the investigation, employing various techniques for cleaning and selection. The server's role then involves fetching the pertinent data as per the processing request from the user. The data management system is designed to be platform-independent, user-friendly, and capable of accessing additional databases like PubMed and NCBI.

A key feature is the service advisor role, assisting users with dataset uploads, guiding them through the investigation process, and aiding in the selection of services and data. This module enables effective communication between scientists and the knowledge data discovery system, functioning as an applet or web application. It allows the advisor to interact with the system for managing, analyzing, and understanding diverse data, with results presented through a user-friendly interface.

Integration services facilitate the swift management of large and varied data sets in different formats, such as relational databases, NoSQL, and flat files. These services merge data from numerous sources and translate broad inquiries into specific database queries. The system addresses challenges in handling extensive, varied, and complex datasets found in public databases, exploring continuous access to research updates for comparing findings with existing knowledge.

The pattern evaluation module assesses the relevance of discovered patterns by working with the data discovery engine to focus on identifying significant patterns within the knowledge base. The data discovery engine, a central system component, employs various algorithms for tasks like prediction, clustering, genetic algorithms, and classification, consulting the knowledge base for more accurate and reliable outcomes. A query engine simplifies interaction with complex underlying data sources, while database and data warehouse servers store and manage ready-for-processing datasets. These servers retrieve relevant information from diverse factual sources based on user queries, enabling effective knowledge extraction by preserving substantial historical data.

The knowledge repository stores all identified patterns, models, and rules, supporting the entire knowledge discovery process, particularly in evaluating the significance of outcome models or focusing on specific demands. Preprocessing involves search and optimization techniques for data integration, cleansing, and selection based on relevance and accuracy. Iterative machine learning optimizes feature sets, and post-processing includes verification, validation, visualization, and evaluation of retrieved knowledge through data mining, machine learning, and decision-making methods for high accuracy and precision.

4 CARDIOVASCULAR DATA ANALYTICS

The features of the proposed platform are validated through modeling and software implementation to analyze and search for patterns between data representing part of human health and cardiovascular disease probability. The data used are a reduced set of statistical records of parameters related to cardiovascular disease analysis. The original database contains 76 different attributes, of which 14 are used. The data originates from Cleveland, USA, and can be freely downloaded from the Kaggle (Heart Disease Dataset). The dataset is in .csv format, separated by semicolons. Example of used dataset images are presented in Figure 3.

Data parameters are:

- Age - age in years
- Sex - gender (1 = male; 0 = female)
- CP - type of chest pain
- TRESTBPS – resting blood pressure measured in mm Hg on admission to the hospital
- CHOL – serum cholesterol in mg/dl
- FPS – blood sugar > 120 mg/dl) (1 = yes; 0 = no)
- RESTECH - resting electrocardiographic results
- THALACH – maximum heart rate
- EXANG – angina induced by angina pectoris (1 = yes; 0 = no))
- OLDPEAK - ST depression
- SLOPE – ST slope
- CA – saturation of stained major blood vessels (0-3)
- THAL – defect status (3 = normal; 6 = defect; 7 = reversible defect)
- TARGET – presence of disease (1 = yes; 0 = no)

```
age,sex,cp,trestbps,chol,fbs,restecg,thalach,exang,ol
dpeak,slope,ca,thal,target
52,1,0,125,212,0,1,168,0,1,2,2,3,0
53,1,0,140,203,1,0,155,1,3.1,0,0,3,0
70,1,0,145,174,0,1,125,1,2.6,0,0,3,0
61,1,0,148,203,0,1,161,0,0,2,1,3,0
62,0,0,138,294,1,1,106,0,1.9,1,3,2,0
58,0,0,100,248,0,0,122,0,1,1,0,2,1
```

Figure 3. Example of original dataset.

In the course of the research, experiments are carried out using a specifically designed algorithm and data analysis procedure. A software application is developed to probe and process data concerning human heart disease, aiming to uncover correlations between various health indicators and the likelihood of cardiovascular conditions.

Anaconda, a widely acclaimed platform known for its robust capabilities in data analysis, artificial intelligence, and neural networks, is chosen for this project. Its open nature grants users access to an extensive array of tools, apps, and libraries geared towards data visualization, modification, and analysis. The software tool is implemented using Python programming language, leveraging its versatility and the rich ecosystem of data science libraries.

The process began with loading the heart disease dataset from a heart.csv file, using the pandas library—a staple in Python for data manipulation and analysis. The initial dataset loading and the examination of its first few entries are depicted in Figure 4, showcasing the initial steps in preparing the data for deeper analysis.

	age	sex	cp	trestbps	chol	fbs	restecg	thalach	exang	oldpeak	slope	ca	thal	target
0	63	1	3	145	233	1	0	150	0	2.3	0	0	1	1
1	37	1	2	130	250	0	1	187	0	3.5	0	0	2	1
2	41	0	1	130	204	0	0	172	0	1.4	2	0	2	1
3	56	1	1	120	236	0	1	178	0	0.8	2	0	2	1
4	57	0	0	120	354	0	1	163	1	0.6	2	0	2	1

Figure 4. Input file – cardiovascular data.

Figure 5 illustrates the type information for each column within the dataset, providing a detailed overview that includes the count of elements, the number of null or missing entries, and the specific data type for each column. This aspect of the analysis is crucial for identifying gaps in the dataset, a common challenge encountered in real-world data analysis. The presence of missing data necessitates a series of steps and strategies to address and mitigate its impact, ensuring the dataset's integrity and usability for further analysis. The ultimate aim of this process is to achieve a unified dataset, laying a solid foundation for subsequent analytical and processing tasks. This step is essential for preparing the dataset for in-depth analysis, ensuring that any conclusions drawn are based on complete and accurate information.

```
Data Show Info

<class 'pandas.core.frame.DataFrame'>
RangeIndex: 303 entries, 0 to 302
Data columns (total 14 columns):
age         303 non-null int64
sex         303 non-null int64
cp          303 non-null int64
trestbps    303 non-null int64
chol        303 non-null int64
fbs         303 non-null int64
restecg     303 non-null int64
thalach     303 non-null int64
exang       303 non-null int64
oldpeak     303 non-null float64
slope       303 non-null int64
ca          303 non-null int64
thal        303 non-null int64
target      303 non-null int64
dtypes: float64(1), int64(13)
memory usage: 33.2 KB
```

Figure 5. General type information for cardiovascular data.

4.1 *Distribution by age*

A set of data analysis functions have been implemented. An analysis of the age distribution of the studied patients, showing the total groups with a number of patients greater than 10, is shown in Figure 6.

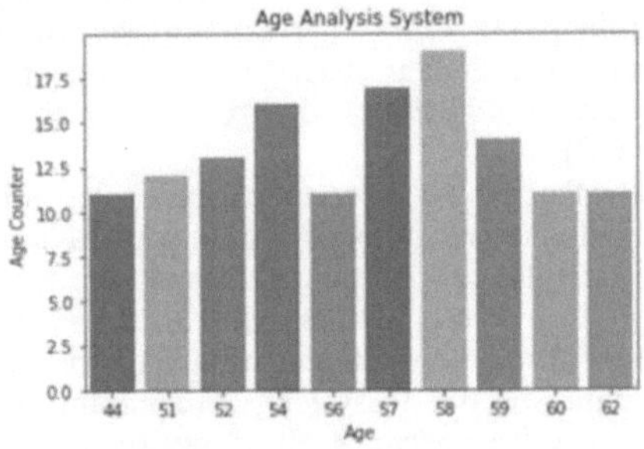

Figure 6. Age distribution of the studied patients.

The correlation matrix gives an overview of the level of correlation between individual columns by visualizing a general map of the data (heatmap). The image in Figure 7 shows a data map of the correlation values calculated in pairs.

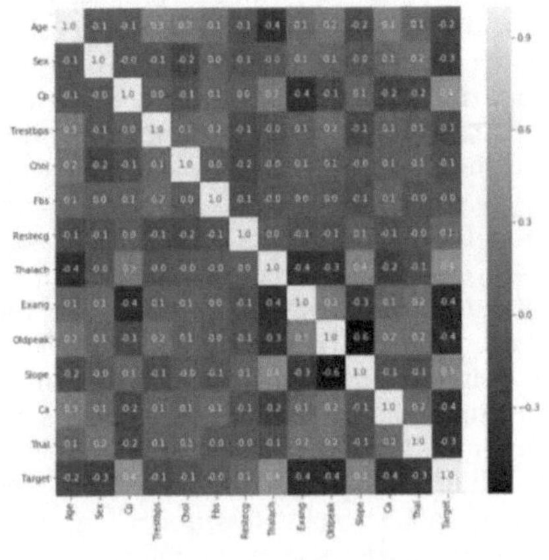

Figure 7. Correlation matrix of cardiovascular data.

4.2 *Distribution by presence of cardiovascular disease*

One of the most revealing analyzes is the one based on the presence of a disease - the 'target' parameter. Initially, the number of unique values of the 'target' parameter is checked. Only two uncial values of the parameter 1 and 0 were confirmed, where 1 is the presence of disease and 0 denotes a healthy patient. Graphical analysis confirms a larger number of sick patients. Figure 8 shows a distribution that separates patients by gender and quantifies the ratio of sick to healthy patients.

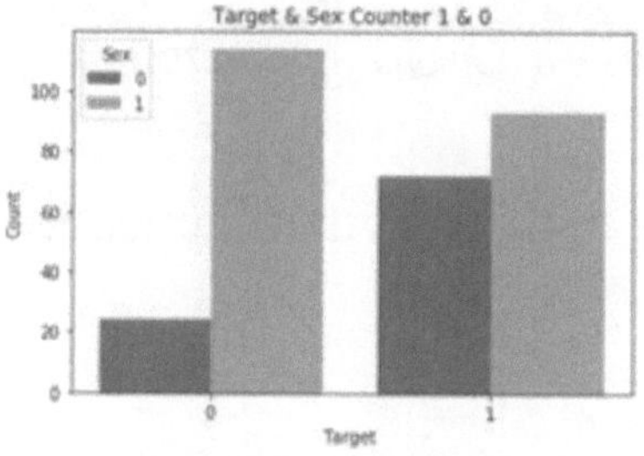

Figure 8. Age distribution of the studied patients.

The visualization of Figure 9 presents the dependence of several parameters - blood pressure on the horizontal axis, cholesterol level on the vertical axis and distribution of the four types of chest pain with blue, orange, green and red dots respectively from patients without chest pain to patients with severe chest pain. The graphical output is done using matplotlib.

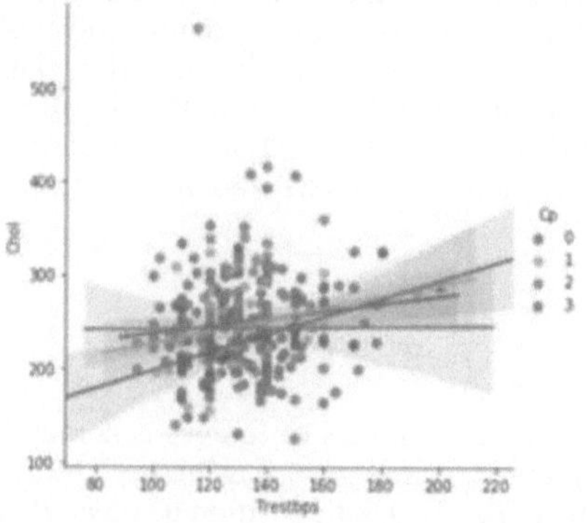

Figure 9. Dependence of parameters blood pressure, cholesterol level and distribution of the four types of chest pain.

The distribution of sick versus healthy patients by age and plotted against the total number in the respective age group requires pre-processing of the data (Figure 10). Two vectors are set - for the healthy and the sick patients, as a cycle goes through the data and distributes the patients in the corresponding vector by age.

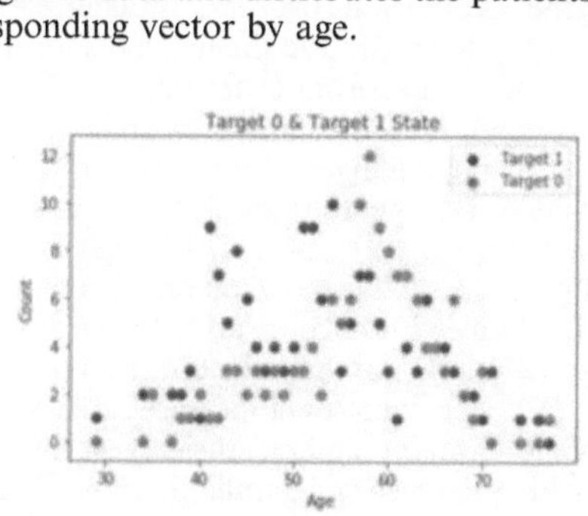

Figure 10. Distribution of sick versus healthy patients.

ACKNOWLEDGMENT

This research is supported by National Science Fund, Bulgarian Ministry of Education and Science, grant KP-06-N37/24 "Innovative Platform for Intelligent Management and Analysis of Big Data Streams Supporting Biomedical Scientific Research".

REFERENCES

Alowais, S. A., Alghamdi, S. S., Alsuhebany, N. *et al.* (2023). Revolutionizing healthcare: the role of artificial intelligence in clinical practice. *BMC Med Educ* 23, 689. https://doi.org/10.1186/s12909-023-04698-z

Balasubramanian V, Jolfaei A. (2020). A scalable framework for healthcare monitoring application using the Internet of Medical Things. *Softw Pract Exp.*, https://doi.org/10.1002/spe.2849.

Borovska P., Gancheva V., Georgiev I. (2019). Platform for adaptive knowledge discovery and decision making based on big genomics data analytics, *Lecture Notes in Computer Science*, vol. 11466, Springer, pp. 297–308

Davenport T, Kalakota R. (2019). The potential for artificial intelligence in healthcare. *Future Healthc J. Jun*;6(2):94–98. doi: 10.7861/futurehosp.6-2-94. PMID: 31363513; PMCID: PMC6616181

Ginsburg G. S. and Phillips K. A. (2018). Precision medicine: from science to value, *Health Aff (Millwood), May*; 37(5): 694–701. doi: 10.1377/hlthaff.2017.1624

Heart Disease Dataset, https://www.kaggle.com/datasets/johnsmith88/heart-disease-dataset, last accessed 2024/03/25

Hulsen T. (2022 Dec). Literature analysis of artificial intelligence in biomedicine. *Ann Transl Med.*;10(23):1284. doi: 10.21037/atm-2022-50. PMID: 36618779.

Khan Z. F, Alotaibi S. R. (2020). Applications of Artificial Intelligence and Big Data Analytics in m-Health: A Healthcare System Perspective. *J Healthc Eng. Aug* 30;2020:8894694. doi: 10.1155/2020/8894694. PMID: 32952992; PMCID: PMC7481991.

Mahesh B. (2020). Machine Learning Algorithms - a Review, *International Journal of Science and Research*, vol. 9, pp. 381–386.

Mallappallil M., *et al.* (2020). A review of big data and medical research, *SAGE Open Med*, vol 8, doi: 10.1177/2050312120934839

Ow G. S. and Kuznetsov V. A. (2013). Big genomics and clinical data analytics strategies for precision cancer prognosis, *Scientific Reports* 6, Article 36493.

Pareek M., Bhari P. (2020). A Review Report on Knowledge Discovery in Databases and Various Techniques of Data Mining, *Open Access International Journal of Science and Engineering*, pp. 79–82

Parihar A., Sharma S. (2020). Knowledge discovery and data mining healthcare, *International Journal of Information Technology Insights & Transformations*, vol. 4, Issue 1.

Sarker I. H. (2021). Machine learning: Algorithms, real-world applications and research directions, *SN Computer Science*, 1–21, https://doi.org/10.1007/s42979-021-00592-x (2021)

Sidey-Gibbons, J., Sidey-Gibbons, C. (2019). Machine learning in medicine: a practical introduction. *BMC Med Res Methodol* 19, 64. https://doi.org/10.1186/s12874-019-0681-4

Sun S., *et al.* (2019). A survey of optimization methods from a machine learning perspective, *IEEE Transactions on Cybernetics*, vol. 50, no. 8, pp. 3668–3681

System Innovation for an Artificial Intelligence Era – Lam et al. (Eds)

Research on the applicability of generative artificial intelligence in image generation

Xin-Zhu Liu & Jiun-Ting Chen
Department of Digital Multimedia Arts, Shih Hsin University, Taipei, Taiwan

ABSTRACT: The maturing Artificial Intelligence technology can produce diverse images through text prompts, useful for scene creation, conceptual modeling, and virtual illustrations. As AI art gains traction, artists increasingly employ AI image generation tools to spark or augment their creative process. This study aims to compare various AI image generation tools' applications across fields, exploring factors like image composition and user evaluations to aid tool selection.Five representative AI image generation tools were chosen: Midjourney, Leonardo.Ai, Deep AI, Stable Diffusion, and Playground AI. Through literature review, factors such as clarity, color, contrast, lighting, and composition were identified as criteria for judging image quality. Differences among these factors across tools influence generated image style, quality, and diversity.A questionnaire survey was conducted to understand image characteristics generated by different AI tools, allowing respondents to evaluate them. Subjects' preferences and evaluations of AI-generated images may vary, influenced by their educational background and personal inclinations.A comprehensive literature review highlights differences and similarities in AI image generation tool applications across fields, delineating their respective advantages and limitations. This study's findings offer insights into AI tool applicability across domains and inform users' tool selection.

1 BACKGROUND

Generative Artificial Intelligence (AI) is a form of AI that can create new content and new ideas by repeatedly using training data to solve complex problems in human language, art, chemistry, and other fields, such as media art creation, product development and design (Amazon, 2023). In art, the mechanism of using tools is to capture and amplify human intent (Joshi, 2023). Technology and art are closely linked, and with the development of AI programs and the increase in AI creative examples, AI programs are gradually involved in various art fields and developing new forms of art (Kurt, 2018). The current era is the era of artificial intelligence, where AI is gradually transitioning from being a tool of humans to being a partner (Kissinger *et al.*, 2021). AI image generation tools can present magnificent visual patterns through textual prompts, and even simulate the traces of human painting, with the rise of AI art creation, many artists have begun to use AI image generation tools to inspire or assist in their creative process, which means non-artists are no longer limited by painting skills to express their creativity (Lyu *et al.*, 2022). Importantly, people also attribute high artistic value to works created by AI (Schwab, 2017).

2 LITERATURE REVIEW

The aesthetic attributes of images generally follow photographic rules, including clarity, tone, color (Aydın *et al.*, 2015), composition (Guo *et al.*, 2013), etc. High-quality images typically have clear and focused subject contours, clear and non-blurry lighting, high contrast, etc. (Ke *et al.*, 2006).

Although the aesthetic properties of images are mentioned in the above literature, the subject's rating of images is still a highly subjective behavior. Tang, Luo and Wang (2013) found in their experiment that professional photographers and amateur enthusiasts view photos differently. Some respondents are influenced by personal preferences during the evaluation process, replacing photography techniques with the theme, considering it as a more important scoring criterion.

According to Sarkar's (2023) research, among the top 50 artificial intelligence tools with the highest traffic between September 2022 and August 2023, seven websites were identified for Image Generator use: Midjourney, Civitai, Craiyon, Leonardo Ai, DeepAI, Stable Diffusion, and Playground AI. Among them, Civitai experienced the fastest net increase in traffic, with a net increase of 23.3 million visits; while Craiyon suffered the largest net loss in traffic, with a net loss of 15 million visits. Therefore, in this study, researchers selected Midjourney,

DOI: 10.1201/9781003514831-75

Leonardo Ai, DeepAI, Stable Diffusion, and Playground AI as experimental websites for AI image generation tools, with Civitai chosen as the original image source website.

3 RESEARCH METHOD

Firstly, this study will adopt an experimental method. The researchers used Civitai as the original image source website and selected 30 images from popular images, including portraits, landscapes, and still life, representing three different domains. For each image, they obtained 30 different textual prompts corresponding to it. Then, they used five different AI image generation tools (Midjourney, Leonardo Ai, DeepAI, Stable Diffusion, and Playground AI) to generate images. Each AI image generation tool generated one image, resulting in a total of 150 generated images. Adding 30 original images, the data set contains 180 items of image data. The five images generated with the same textual prompts were randomly shuffled and added the original images. These images were used as corresponding options for the questions of clarity, color, contrast, lighting, and composition in the questionnaire.

Secondly, this study will adopt a questionnaire survey. The questionnaire respondents for this study were selected 36 respondents, randomly from three first-year to fourth-year students in the Department of Digital Multimedia Design and the first-year to second-year graduate students in non-design and art-related departments. This aimed to explore whether age, gender, and educational background serve as relevant factors influencing the experimental results.

This study chose clarity, color, contrast, lighting, composition, and personal preference as the six aspects for questionnaire design and survey. Respondents evaluated a single aesthetic attribute, and the image source selected the most times was considered to perform better in that particular domain among the five AI image generation tools. To ensure higher accuracy in aesthetic evaluation, researchers aimed to minimize interaction time and frequency during the questionnaire survey, reducing users' aesthetic fatigue (Lv *et al.*, 2023).

Taking the portrait domain as an example, the researchers used Civitai as the original image source website and selected Figure 1 from the "Most Reactions Images". The corresponding prompt obtained was: (score_10, score_9, score_8_up, score_7_up, score_6_up, pretty crazy happy girl). When entering the keywords into Midjourney, Figure 2 was obtained; entering the keywords into Leonardo Ai, Figure 3 was obtained; entering the keywords into DeepAI, Figure 4 was obtained; entering the keywords into Stable Diffusion, Figure 5 was obtained; entering the keywords into Playground AI, Figure 6 was obtained.

Figure 1. The original image.

Figure 2. The image generated by Midjourney.

4 EXPECTED RESULTS

In the survey results, due to the varying performance of the image quality factors of clarity, color, contrast, lighting, and composition in each

Figure 3. The image generated by Leonardo Ai.

Figure 4. The image generated by DeepAI.

Figure 5. The image generated by stable diffusion.

Figure 6. The image generated by Playground AI.

AI image generation tool, it is speculated that respondents may have differences or similarities in their ratings for clarity, color, contrast, lighting, and composition. After the actual questionnaire survey, the researchers will summarize and analyze the questionnaire results. Based on the literature review section above, the researchers predict that due to differences in individual attributes of the respondents, there may also be differences or similarities in preferences and evaluations of images generated by different AI image generation tools.

REFERENCES

Amazon, (2023). What is generative-ai? https://aws.amazon.com/tw/what-is/generative-ai/

Aydın T. O., Smolic A. and Gross M. (2015). Automated Aesthetic Analysis of Photographic Images, *IEEE Transactions on Visualization and Computer Graphics*, vol. 21, no. 1, pp. 31–42.

Göring S., Ramachandra Rao R. R., Merten R. and Raake A. (2023). Analysis of Appeal for Realistic AI-Generated Photos, *IEEE Access*, vol. 11, pp. 38999–39012.

Guo Y. W., Liu M., Gu T. T. and Wang W. P. (2013). Improving Photo Composition Elegantly: Considering Image Similarity During Composition Optimization, *Computer Graphics Forum*, vol. 31, no. 7, pp. 2193–2202.

Joshi B. (2023). *Is AI Going to Replace Creative Professionals?* vol. 30, no. 5, pp. 24–29.

Ke Y., Tang X. and Jing F. (2006). The design of high-level features for photo quality assessment, *IEEE Computer Society Conference on Computer Vision and Pattern Recognition (CVPR'06)*, pp. 419–426.

Kissinger H. A., Schmidt E., and Huttenlocher D., (2021). The age of AI: and our human future. Hachette UK.

Kurt D. E. (2018). *Artistic Creativity in Artificial Intelligence*. URI: http://theses.ubn.ru.nl/handle/123456789/5631

Lv P., Fan J., Nie X., Dong W., Jiang X., Zhou B., Xu M., and Xu C., (2023). User-Guided Personalized Image Aesthetic Assessment Based on Deep Reinforcement Learning, *IEEE Transactions on Multimedia*, vol. 25, pp. 736–749.

Lyu Y., Wang X., Lin R., and Wu J., (2022). Communication in Human-AI Co-Creation: Perceptual Analysis of Paintings Generated by Text-to-Image System, *Applied Sciences*, vol. 12, no. 22: 11312.

Sarkar S., AI Industry Analysis: 50 Most Visited AI Tools and Their 24B+ Traffic Behavior.Schwab K., (2017). The fourth industrial revolution. Crown Currency. https://writerbuddy.ai/blog/ai-industry-analysis

Tang X., Luo W. and Wang X. (2013). Content-Based Photo Quality Assessment, *IEEE Transactions on Multimedia*, vol. 15, no. 8, pp. 1930–1943.

System Innovation for an Artificial Intelligence Era – Lam et al. (Eds)
© 2025 the Author(s), ISBN: 978-1-032-84452-7

A study on the interface usability of ticket booking process for airline websites

Ruan Shihan & Chien-Hsiung Chen
Department of Design, National Taiwan University of Science and Technology, Taipei, Taiwan

ABSTRACT: In the post-epidemic era, there has been an increase in Taiwanese citizens' traveling abroad, leading to the growth in the aviation industry's passenger volume. With the proliferation of smart devices, people are increasingly inclined to book their flights online. To attract customers and maintain profitability, airline companies have begun to focus on the operation and design of their websites. This study examines the interface usability of airline booking websites, and discusses current website features, interface layout, and design. It evaluates user interactions and perceptions of web-based ticket booking systems, proposing interface design suggestions for future airline ticket booking procedures. The experiment adopted one-way analysis of variance (ANOVA) between-subjects design. A total of 30 participants, aged from 19 to 38 (M = 25.07, SD = 4.32) were invited to take part in the experiment via convenience sampling method. The researchers assessed task completion time, gathered subjective responses using the SUS questionnaire, and conducted interviews with the participants. The findings indicated that: (1) A clean, highly recognizable visual interface may improve participants' task performance; (2) Appropriately reducing the amount of information on pages may reduce user stress and frustration; (3) Reducing interface designs with different layouts may help users learn the web-based ticket booking interface; (4) More intuitive and prominent content prompts and easily recognizable icons may help users understand display information.(5)The effectiveness of an interface in terms of performance does not necessarily correlate with its usability. This discrepancy may be related to the visual aesthetic and layout of the interface.

1 INTRODUCTION

This study focuses on the post-pandemic recovery of international travel and the significant improvement in passenger traffic in the aviation industry driven by this recovery. According to the latest data from the Civil Aviation Administration (CAA), as of January to October 2023, the total number of air passengers in Taiwan has reached 42.43 million, showing a gradual growth trend. With the flourishing and maturation of online travel industry, consumers are beginning to change their habits by no longer solely relying on online travel agencies but directly visiting airline websites to book tickets (Yu, 2008). Official websites play a crucial role in attracting customers and maintaining profitability due to the increased reliance of airlines on them (Dominic and Khan, 2014; Powell, 2016). Furthermore, the quality of airline websites has a positive impact on customer satisfaction and purchase intention, playing a key role in attracting new customers and maintaining customer loyalty, with high-quality websites having short-term and long-term effects on airlines' strategic goals (Elkhani, Soltani, and Jamshidi, 2014). Therefore, enhancing the interface usability and user experience of airline web sites have become an important issue that airlines need to address in the current era.

This study investigates the online booking rates on three major Taiwanese airline websites using a single-factor between-subjects experimental design. After the experiment, a System Usability Scale (SUS) survey was adopted to help assess user perceptions, identifying operational issues. Results were analyzed for quantitative statistics, interface operation, design, and web content, leading to recommendations. The research objectives included exploring interface usability across websites, comparing functionalities and layouts, evaluating user interactions and perceptions, and proposing design recommendations. The objectives of this study were as follows: (1) to explore the usability of airline website booking interfaces based on performance measurements across different websites; (2) to examine and compare the strengths and weaknesses of existing website functionalities and interface layouts; (3) to evaluate user interactions and perceptions of web-based booking systems; and (4) to propose interface design recommendations for future web-based booking processes based on the findings.

 DOI: 10.1201/9781003514831-76

Table 1. Descriptions and purposes for 4 tasks.

Task number	Descriptions	Purpose
1	Search for economy class Tickets from Taipei (TPE) to Tokyo (NRT) on Feb 17, 2024	Visual Search for information location
2	Choose the cheapest flight of the day read the information about baggage allowance and seat selection for on that flight	Visual Search for information location and information comparison
3	Fill in the ticket booking information	Visual Search for information location
4	Search for round-trip economy class round-trip tickets from Taipei (TPE) to Tokyo (NRT), selecting March 17, 2024 for the trip and March 24, 2024 for the return.	Visual Search for information location

Table 2. Descriptive statistics and ANOVA results for 4 tasks.

	Sample 1 M(SD)	Sample 2 M(SD)	Sample 3 M(SD)	F	P	Post Hoc
Task 1	29.35(04.11)	49.06(09.96)	37.40(06.81)	18.13	0.000*	Sample 1> Sample 3> Sample 2
Task 2	34.35(07.80)	30.79(05.79)	46.31(07.56)	13.09	0.000*	Sample 1> Sample 2> Sample 3
Task 3	44.00(10.08)	38.72(09.63)	39.92(10.21)	00.77	0.473	
Task 4	41.82(06.69)	49.88(09.92)	55.30(08.53)	06.39	0.005*	Sample 1> Sample 2> Sample 3

Significantly different at the $\alpha = 0.05$ level ($p < 0.05$).

2 EXPERIMENT DESIGN

This study adopted a single-factor between-subjects experimental design. The experimenter explained the purpose, procedures, and tasks to each participant. During the experiment, the focus was on recording the participants' task performance and observing their interaction with the website interface, documenting any issues encountered. After the experiment, participants completed a basic information questionnaire and the "System Usability Scale (SUS)" to assess their experience. A brief interview was conducted based on the recorded issues. The study employed purposive sampling to recruit participants who had no prior experience booking tickets on airline websites. A total of 30 participants were invited, with each of the three sample groups consisting of 10 participants. Participants were aged between 18 and 36 years and had attained a university-level education. All participants had no prior experience in using the user interfaces of the three airline ticket booking websites to prevent learning effects from influencing experimental performance. This study utilized a MacBook Pro laptop with a 14-inch display and iOS operating system as the testing tool. Participants interacted with three airline ticket booking websites using the Google browser. The study designed experimental tasks based on the common ticket booking functionalities across the three websites, as outlined in Table 1.

3 RESULTS AND DISCUSSIONS

The following provides an explanation of the statistical analysis of the timing results for each experimental task. A shorter task completion time by participants indicates higher usability and operational performance of the user interface, while a longer task completion time suggests lower performance. Table 2 presents the descriptive statistics and ANOVA results. For the operation time, the tasks 1, 2, and 4 across the three sample websites, showing significant differences in operational performance.

Figure 1. The booking interface of sample 1,2,3.

Task 1 was a scheduled one-way ticket, and according to its LSD post-test results, there were significant differences between Sample 1 (M = 29.35, SD = 04.11) and Sample 2 (M = 49.06, SD = 09.96) and Sample 3 (M = 37.40, SD = 06.81). Therefore, Sample 1 had the best operational performance, Sample 3 had the second best operational performance, and Sample 2 had the worst operational performance.

Based on the interviews, it was found that the semi-transparent background of the booking interface in Sample 2 (Figure 1) interferes with users' recognition, reducing usability, while Samples 1 and 3 do not have this issue. Additionally, the booking information interface in Sample 2 occupies a relatively small space on the webpage, making it difficult for users to find the booking interface. Some users found the meaning of the icons in Sample 3 unclear. The gray icons placed closer to the inquiry area could not be clicked, leading users to click the red icon after several unsuccessful attempts to access the information. In comparison, Sample 1 performs better in the layout of icons, the readability of text, and the presentation of information.

Task 2 involves searching for flight ticket information. Combining the results of the LSD post-hoc test, there is a significant difference between Sample 1 (M = 34.35, SD = 07.80) and Sample 3(M = 46.31, SD = 07.56), while there is no significant difference between Sample 2 (M = 30.79, SD = 05.79) and Sample 3. However, there is a significant difference between Sample 2 and Sample 3. Therefore, the operational performance of Sample 1 and Sample 2 is better, with Sample 3 exhibiting the poorest operational performance.

In 1971, Paivio proposed the Dual-Coding Theory, suggesting that there is a "referential connection" between language and non-verbal systems. Therefore, using two different coding systems simultaneously to aid memory will be more effective than using a single coding system alone (Paivio, 1990). However, when the meaning of an icon does not match the "referential connection" established by the user, such icons can interfere with the user's own cognitive "conceptual model." Lin and Luo (2023) further validated in their study that including more than four information modes in a message object may have adverse effects, such as attention shifts. As shown in Figure 2, Sample 3 page contains more than four information modes: (1) text, (2) icons, (3) color blocks, (4) graphics, (5) different sizes of text, and (6) text layout. The stacking of too many information modes has a detrimental effect on users, and Sample 3 flight selection interface uses a large number of icons with unclear meanings, significantly increasing the time it takes for participants to find relevant information and reducing operational efficiency. Sample 2 uses an all-text approach to present information (Figure 3), enlarging important information and using different colors to represent it, increasing the contrast between information levels and making information easier to read. Sample 1, on the other hand, uses an "information block" to present information about the three ticket types in a list format (Figure 4), making the information easier to read.

Task 3 involve the completion of passenger information, in conjunction with the post-hoc LSD test results. There were no significant differences observed among the three samples: Sample 1 (M = 44.00, SD = 10.08), Sample 2 (M = 39.91, SD = 10.21), and Sample 3 (M = 38.72, SD = 09.63). The analysis is as follows: The formatting and content of data entry exhibit minimal differences, resulting in negligible variations in performance outcomes.

Task 4 involves booking multi-leg flight tickets. There is a significant difference between sample 1 (M = 41.82, SD = 06.69) and sample 3 (M = 55.30, SD = 08.53), as well as between sample 2 (M = 49.88, SD = 09.92) and sample 3. However, there is no significant difference between sample 2 and sample 3. Therefore, sample 1 demonstrates the best operational performance, while sample 3 and sample 2 exhibit slightly lower operational performance. Analysis is as follows: The layout of the

Figure 2. Interface flight information for Sample 3.

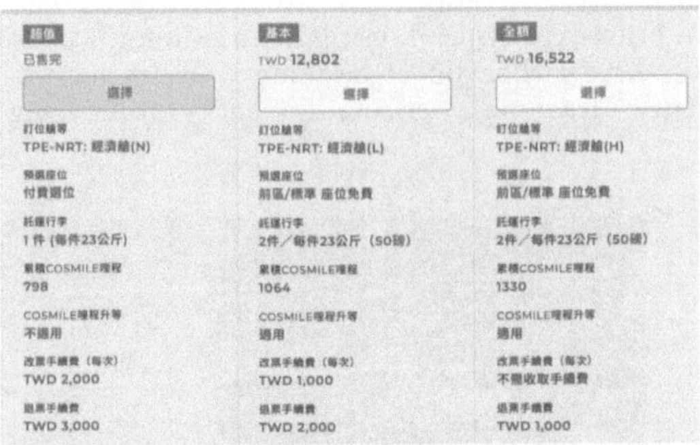

Figure 3. Interface flight information for Sample 2.

Figure 4. Interface flight information for sample.

Table 3. Descriptive statistics and ANOVA Results for SUS.

	Sample 1 M(SD)	Sample 2 M(SD)	Sample 3 M(SD)	F	P	Post Hoc
SUS	65.25(10.24)	79.50(05.37)	68.50(14.35)	04.93	0.015*	Sample 2 > Sample 1 = Sample 3

*Significantly different at the a = 0.05 level (*p < 0.05).

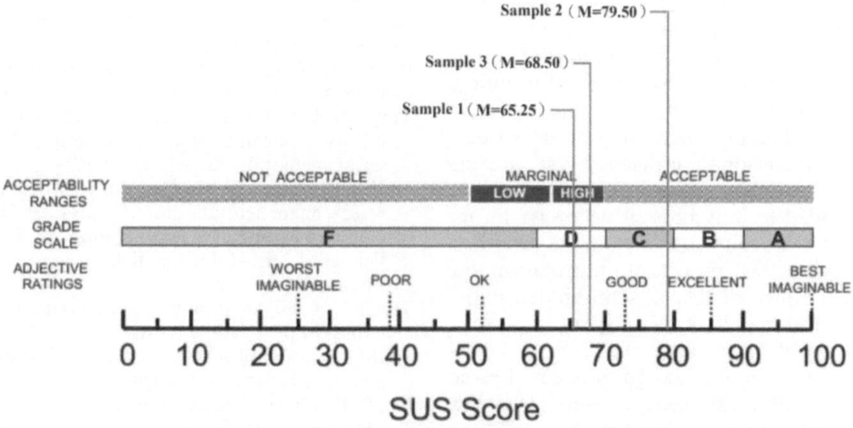

Figure 5. Distribution table of scores for three aviation websites evaluated by the System Usability Scale (SUS).

multi-stop ticket booking interface in Sample 1 is more compact. Its layout is similar to that of the one-way ticket booking interface in Task1, with lower interface learnability. Test subjects only need to follow the visual flow from left to right to operate sequentially.

When navigating the multi-step booking interface of samples 2 and 3, users are redirected to a new page. The layout and content of this interface differ from that of Task 1, requiring participants to familiarize themselves with the new page before proceeding with the booking process.

After the experiment, participants were asked to complete the System Usability Scale (SUS) questionnaire to help assess the usability of the system, using a five-point Likert scale. This was done to evaluate the usability of the samples. Following a one-way analysis of variance, significant differences among the three samples were identified ($F = 04.93$, $P = 0.015 < 0.05$), as shown in Table 3. Based on the LSD post-hoc analysis results presented in Table 3, it is evident that there are significant differences between sample 2 ($M = 79.50$, $SD = 05.37$) and sample 1 ($M = 65.25$, $SD =$

10.24) ($P = 0.006 < 0.05$), as well as between Sample 2 and Sample 3 ($M = 68.50$, $SD = 14.35$) ($P = 0.029 < 0.05$). Conversely, there is no significant difference between Sample 3 and Sample 1 ($P = 0.500 > 0.05$). Therefore, it can be concluded that the usability of the interface in Sample 2 is superior, while the usability of the interfaces in Sample 1 and Sample 3 is comparatively inferior.

As illustrated in Figure 5, Sample 2 obtained a score higher than the mean SUS score of 68 points, falling within the C grade range (70 - 79), while Sample 1 and Sample 3 scores fell within the D grade range (60 - 69). Therefore, it can be inferred that the participants perceive that there is still considerable room for improvement in the system usability of Samples 1 and 2.

4 CONCLUSION

Based on the aforementioned research analysis, this study summarizes the following recommendations for the design of the booking process interface on airline websites:

(1) A visually clean and highly recognizable interface: Many participants found the interface of Sample 1 to have a simple background design with clear primary and secondary prompts, which instilled a sense of reassurance and confidence in users during operation.

(2) Moderately streamline the amount of information on the page: Excessive information may lead users to experience "information overload," thereby increasing user stress, reducing operational efficiency, and even decreasing purchase rates. Therefore, it is recommended to only present necessary information on the main interface in the interface design, place less important information on subpages or present it in a semi-hidden manner, while using a block-style layout to present information of the same level.

(3) Use similar interface design models: Fewer interface models can reduce users' learning curve for booking on web interfaces. Booking flights is highly complex, so the core goal of interface design is to simplify the booking process and prioritize usability in design considerations. It is not advisable to excessively pursue aesthetics at the expense of usability.

(4) Provide more intuitive and prominent content prompts: Accompanied by easily recognizable icons to assist in understanding textual content. Overly complex or non-intuitive icons may disrupt users' operational judgment and consequently reduce overall operational efficiency. The effectiveness of an interface in terms of performance does not necessarily correlate with its usability. This discrepancy may be related to the visual aesthetic and layout of the user interface.

REFERENCES

Dominic, P. D. D. and Khan, H. (2014). Performance measure of air line websites using analytical hierarchy process & fuzzy analytical hierarchy process. November. In: *2014 IEEE International Conference on Control System, Computing and Engineering*. ICCSCE 2014, I

Elkhani, N., Soltani, S., and Jamshidi, M. H. M. (2014). Examining A hybrid model for E- satisfaction and E-loyalty to E-ticketing on airline websites. *J. Air Transport. Manag.* 37, 36–44.

Paivio, A. (1990). *Mental Representations: A Dual Coding Approach.* Oxford University Press.

Powell, C.M. (2016). Twelve years in the life of airline websites. *Sage Open* 6 (2), 2158244016637892.

Tingyi S. Lin, Yue Luo. (2023). Emoji and visual complexity in health information design: A moderated serial mediation model. *Telematics and Informatics.* Volume 85, November 2023,102065.https://www.researchgate.net/publication/372284216_Health_Communication_An_Investigation_on_Emoji_Perceived_Visual_Informativeness_and_Social_Media_Engagement

Traffic of Civil Aviation by International, Domestic and Cross-Strait Lines. (n.d.). *Civil Aviation Administration*,MOTC. https://www.caa.gov.tw/Article.aspx?a=2833&lang=2

Yu, S. F., (2008). Price perception of online airline ticket shoppers. *J. Air Transp. Manag.* 14 (2), 66–69.

System Innovation for an Artificial Intelligence Era – Lam et al. (Eds)

Characteristics of SUS304 stainless steel annealed at different temperatures

Sean Wu & Chien-Chih Chiang
Department of Chemical and Materials Engineering, Lunghwa University of Science and Technology, Guishan, Taoyuan, Taiwan

Yu-Ting Wu & Wen-Jen Lee
Department of Applied Physics, National Pingtung University, Pingtung City, Pingtung, Taiwan

Yee-Shin Chang
Department of Electronic Engineering, National Formosa University, Huwei, Yunlin, Taiwan

ABSTRACT: SUS304 stainless steel annealed at different temperatures from 800 to 1050 °C was performed. The crystal structure, surface morphologies and magnetic properties have been investigated. As the annealing temperature increased, the crystal structures transferred gradually from Fe_2O_3 and Fe_3O_4 phases to Cr_2O_3 and $FeCr_2O_4$ phases, respectively. The SEM morphologies analysis indicated that the surfaces are rougher and the numbers and sizes of pyramidal shape precipitations are increased. All the samples represent a hysteresis behavior, and the hysteresis loop is clearer when the annealing temperature is increased. The maximum saturation magnetization was observed for SUS304 stainless steel annealing at 1050 °C, which is caused by the ferromagnetism $FeCr_2O_4$ phase formation and a greatest amount of oxygen vacancies.

1 INTRODUCTION

Karlsson *et al.* (2004) and Callister *et al.* (1997) point out that stainless steel was first invented in the beginning of the 20th century where the use of conventional carbon steels was restricted to their strong susceptibility to corrosion and oxidation. In August 1913, Harry Brearley in Sheffield (UK) melted stainless steel for the first time which was microstructurally martensitic with a carbon content of 0.24% and chromium content of 12.8% (Honeycombe *et al.*, 1995).

Stainless steel has good strength and good resistance to corrosion, and oxidation at elevated temperatures. Therefore, stainless steel is used to be the main built materials of the centralized receiver (or power tower) which leads a high temperatures solar absorber on stainless steel to be a very important research (López-Marino *et al.*, 2014). These researches include spinel oxide films deposited on the stainless steel for the promise intrinsic solar-selective absorptant properties representation (Ambrosini *et al.*, 2010). Furthermore, the study results by Sean Wu *et al.* (2016) have discovered a relatively low-cost method to grow Fe_2O_3 films using thermal oxidation on SS304 substrates for high-temperature solar absorber. Those Fe_2O_3 films on SUS304 stainless steel substrates at (900–1000 °C) displayed relatively high absorptivity (α) (0.909–0.922) and low emittance values (ε) (0.18–0.38). Oxidation of stainless steel at high-temperatures which applies a simple, low-cost, large-area and effective technique for the high-temperature solar absorber.

Annealing, it means a heat treatment in which a material is exposed to an elevated temperature for a long time and then cooled slowly. During the annealing process, an internal stress releasing and a specific microstructure producing will be carried out, and which is possible to introduce new microstructures and lead the properties to change (M. Ohring, 1992).

In this study, SUS304 stainless steel was annealed at different temperatures from 800 to 1050 °C in air. The effects of annealing temperatures on the crystal structure, surface morphology and specific properties of SUS304 stainless steel were also examined.

2 METHODOLOGY

SUS304 stainless steel were heat treatment in air for 2 h from 800 to 1050 °C in a programmable furnace with a heating rate of 5 °C/min. First, SUS 304 stainless steel was cut to a square shape with a dimension of 5x5 cm, and put into a beaker was full with deionizing water for cleaning using an ultrasonic cleaner. After cleaning, the samples were taken from the ultrasonic cleaner, and dried using a nitrogen gas. After heat treatment, the SUS304 stainless steel were cooling down to room temperature in air for follow up examined.

The crystal structure of the specimens was analyzed by X-ray diffractometer (XRD, D8 Advance

DOI: 10.1201/9781003514831-77

Eco, Bruker, Karlsruhe, Germany), using Cu-Kα radiation with a source power of 30 kV and a current of 20 mA, to identify the possible phases formed in the samples. The surface morphologies were observed using the high-resolution scanning electron microscope (SEM, SU8000, Hitachi, Tokyo, Japan). The Raman spectra of the samples were examined by a Micro Raman Spectrometer (Raman, UniRAM II, UniNanoTech, Yongin, Korea). The binding energy of electrons for the ions in SUS 304 stainless steel annealed at different temperatures were analyzed using an X-ray photoelectron spectroscope (XPS, LAB MK-II, VG Scientific, UK) that was equipped with an Mg Kα X-ray source, and the C 1s line at 285.25 eV was used as a reference for binding energy calibration. A vibrating sample magnetometer (VSM, Lakeshore 7400 series, Laker Shore Cryotronics, Westerville, OH, USA) was used to determine the magnetic properties of these samples.

3 RESEARCH RESULTS

SUS304 stainless steel is the most commonly used stainless steel in the world, containing 18~20 % Chromium (Cr), which is the essential chemical in all stainless steel and it is that which forms the thin passive layer that makes the metal "stainless"; 8~10 % Nikel (Ni) which is added to make the Austenitic structure more stable at normal temperatures and improves high-temperature oxidation resistance makes the steel resistant to stress corrosion cracking. The other elements include Mn, P, S, etc., and the compositions of SUS304 are shown in Table 1.

Figure 1 shows the XRD patterns of SUS304 stainless steel after heat treatment from 800 to 1050 °C in air for 2 h. As can be seen, the diffraction peaks are attributing to the Austenite (γ-Fe) phase in room temperature. When the heat treatment temperature at 800 °C, the diffraction peaks appear in the XRD spectra are composed by the Iron rich compounds including iron (III) oxide (Fe_2O_3) and ferric ferrous oxide (Fe_3O_4). As the heat treatment temperature increased to 1050 °C, the Chromium rich compounds include Chromium (III) oxide (Cr_2O_3) and a spinel group of Chromite ($FeCr_2O_4$) present in the XRD patterns.

Table 1. The compositions of sus304 stainless steel.

SUS304 Composition	Cr	Ni	Mn	P	S	C	Si
wt.%	18-20	8-10	≦2	≦0.05	≦0.03	≦0.08	≦1

Figure 2 shows the diffraction peaks of SUS304 stainless steel after heat treatment from 800 to 1050 °C in air for 2 h. The diffraction peak around 24.29° is verified to the diffraction plane of (012) for Fe_2O_3 phase for SUS304 stainless steel annealing at 800 °C. As the annealing temperature increasing to 1050 °C, the diffraction angle shifts toward a larger angle of 24.48°. According to the JCPD card of Cr_2O_3 (PDF# 821484), the diffraction plane of (012) for Cr_2O_3 phase is positioned at 24.507°. Both of Fe_2O_3 and Cr_2O_3 belong to the same crystal structure (rhombohedral structure with R3c group), and the shift in the diffraction peak position of (012) diffraction plane [Figure 2(a)] between the samples indicating that the partial replacement of the smaller ionic radius of a Fe^{3+} ion (0.645 nm) by the smaller Cr^{3+} ion (0.615 nm) in the Fe_2O_3 system to form the $Fe_{2-x}Cr_xO_3$ phase during the various annealing temperatures, and Cr_2O_3 phase is the final product. The chromium content tended to increase in the outermost surface and the iron content to decrease in the sample heated at a higher temperature proporsed by Nomura and Ujihira (1990).

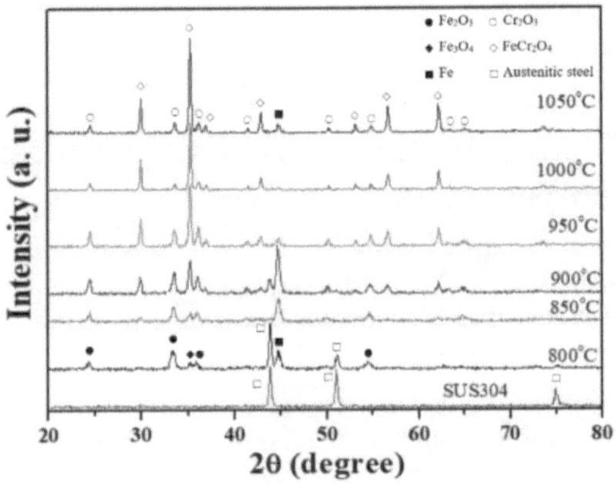

Figure 1. XRD patterns of SUS304 stainless steel annealing at different temperatures in air.

In addition, at the diffraction peak around 62.25° which is identified to the (440) diffraction plane of Fe_3O_4 phase for SUS304 stainless steel annealing in a lower temperature (800 °C), and it shifts to a higher angle as the same as the (012) diffraction plane of Fe_2O_3 phase [shown in Figure 2(b)] when the temperature is increased. Fe_3O_4 is an inverse spinel with half the Fe^{3+} ions in the tetrahedral sites and with the Fe^{2+} ions in the octahedral sites. It is guessed that the Cr^{3+} ion precipitated, and the Fe^{3+} ion substituted by Cr^{3+} ion to form $FeFe_{2-y}CryO_4$ in the Fe_3O_4 system when the heat treatment temperature is increased. According to the results mentioned above, the iron-rich phase of Fe_2O_3 and Fe_3O_4 mixing phases are observed for SUS304 stainless steel annealing at lower temperature region. As the temperature increased, the Cr^{3+} ion precipitate and substitute the Fe^{3+} ion lattice to form $Fe_{2-x}Cr_xO_3$ and $FeFe_{2-y}CryO_4$ phase, respectively.

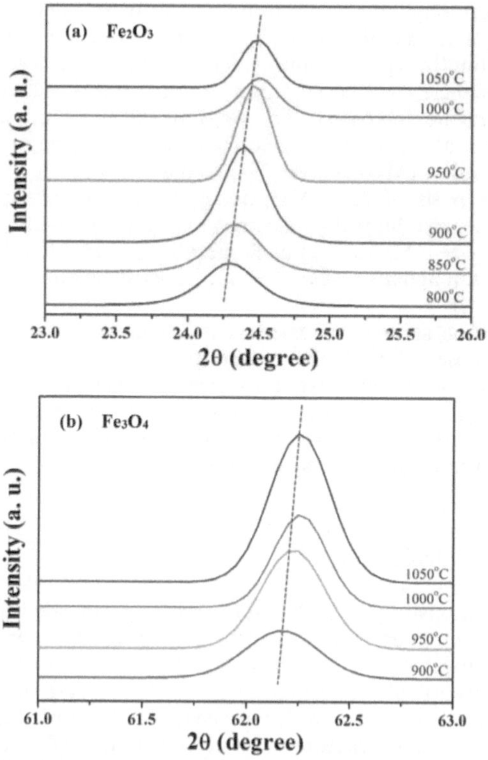

Figure 2. X-ray diffraction peaks position of Fe_2O_3 and Fe_3O_4 for SUS304 stainless steel annealing at different temperatures in air.

Figure 3 is the SEM micrographs of SUS304 stainless steel annealing at different temperatures from 800 to 1050 °C. There are many scratches observed on the surface of SUS304 raw material [Figure 3(a)]. For the SUS304 stainless steel annealed at 800 °C, the surface is flatter and with a dense structure. There are many small particles precipitated, and a few pyramidal shape precipitations [red circled in Figure 3(b)] are observed on the surface. It appears that the surface morphologies of SUS304 stainless steel are significantly affected by the annealing temperatures. As the annealing temperatures increased, the surface morphologies of SUS304 stainless steel become rough more and more, less agglomerated, and the numbers and sizes of precipitations with pyramidal shape are increased. For annealing at 1050 °C, the precipitation is greatest and is widely distributed over the area. Based on the results discussed earlier, the precipitations were guessed to be the spinel structure of $FeCr_2O_4$ phase. In the structure of AB_2O_4 spinel, each cell contains eight AB_2O_4 molecules, that is, thirty-two O^{2-} ions, sixteen B^{3+} ions, and eight A^{2+} ions. The spinel structure can be considered as oxygen ions arranged in a cubic close-packed structure, forming tetrahedral and octahedral vacancies. The A^{2+} and B^{3+} cations are distributed in these two sites (Wang *et al.*, 2021).

Figure 4 shows the Raman spectra of SUS304 stainless steel annealed at different temperatures. According to the study results by McCarty *et al.* (1989), the Raman shift are distributed in the range of 540~560 cm^{-1}, and 670~685 cm^{-1} of the entire $Fe_{3-}xCr_xO_4$ solid solution. In addition, as the x value increased, the Raman shift at about 636 cm^{-1} could be observed for $Fe_{1.8}Cr_{1.2}O_4$ and $Fe_{1.6}Cr_{1.4}O_4$ solid solution. In this study, the Raman shift located at ~550, 675 cm^{-1} are assigned to T_{2g} and A_{1g} symmetries of the $Fe_{3-x}Cr_xO_4$ solid solution, respectively. A shoulder of Raman shift around 640 cm-1 is observed and clearer as the heat treatment temperature increased. Therefore, it is evident that the Cr^{3+} ion precipitate and substitute the Fe^{3+} ion in the $Fe_{2-x}Cr_xO_3$ and $Fe_{3-y}Cr_yO_4$ solid solution when the annealing temperature is increased for SUS304 stainless steel.

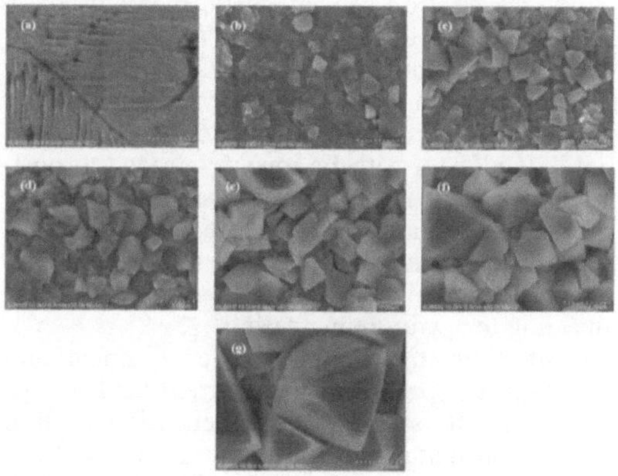

Figure 3. SEM micrographs of SUS304 stainless steel annealed at different temperatures. (a) raw material, (b) 800, (c) 850, (d) 900, (e) 950, (f) 1000 and (g) 1050 °C.

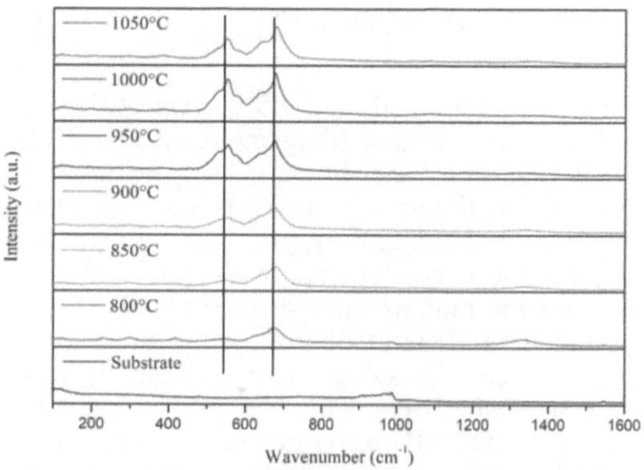

Figure 4. Raman spectra of SUS304 stainless steel annealed at different temperatures in air.

The M–H curves (field dependence of magnetization) for SUS304 stainless steel annealed at different temperatures are shown in Figure 5. The results evident that all samples show hysteresis, and the samples exhibit obvious ferromagnetic properties for the annealing temperature is higher than 900 °C, and there is a maximum saturation magnetization values for SUS304 stainless steel annealed at 1050 °C.

The phase of $FeCr_2O_4$ is observed in the XRD diffraction patterns for SUS304 annealed at a high temperature whereas Fe_3O_4 is observed at a low temperature. Fe_3O_4 has a cubic inverse spinel crystal structure with a space group of Fd3m, and its structure formula can be written as $(Fe^{3+})_A(Fe^{3+}, Fe^{2+})_BO_4$ (Noh *et al.*, 2014). Oxygen atoms are filled in face centered cubic structure, and cations of Fe^{2+} or Fe^{3+} are filled in tetrahedral (A site) or octahedral structure (B site). The magnetic moment direction of ion at the tetrahedral site is downward with a negative value, and the octahedral site one is a positive value with an upward direction. For Fe_3O_4, the electron spin magnetic moments of A and B sites are antiparallel, and it is a ferromagnetic material. Nevertheless, $FeCr_2O_4$ has a normal spinel crystal structure, and belongs to Fd3m for space group. Its structural formula can be written as $(Fe^{2+})_A(Cr^{3+})_BO_4$ with Fe ions in tetrahedral position and Cr ions in octahedral position, and it is a ferromagnetic material. Therefore, the saturation magnetization is increased because the Fe_3O_4 phase transfers gradually to $FeCr_2O_4$ phase as the annealing temperature increasing.

In addition, an increase in oxygen vacancies and the magnetic particles precipitated lead to an increase in the saturation magnetization of thin films annealed at a high temperature. According to the study results developed by R. Moos et al. (1997), at a temperature greater than 1000 °C, an oxygen in the lattice position can emit two electrons into the conduction band and be twofold ionized positively, $V_O^{\bullet\bullet}$. These may be represented by the reaction:

$$O_o^x \leftrightarrow \frac{1}{2}O_2 + V_o^{\bullet\bullet} + 2e^{'} \quad (1)$$

Figure 6 illustrated the XPS spectra of the O1s level and the corresponding curve fitting of the SUS304 annealed at different temperatures. In Figure 6, the fit curve is resolved using two peaks including the binding energy of 530.1 eV (dashed line of curve I) corresponds to oxygen in the lattice, and the binding energy of 532.8 eV (dotted line of curve II) is attributed to the presence of oxygen vacancies (Zou *et al.*, 2002; Ramos-Moorea *et al.*, 2011). The area ratios under the O1s (I) and O1s (II) peaks decreases continuously when the annealing temperature is increased. The minimum ratio is observed when the annealing temperature is 1050 °C. Depending on the thermodynamic conditions, neutral oxygen on regular oxygen sites may leave the lattice, leaving behind oxygen vacancies. The amount of oxygen vacancies will be increased as the annealing temperature increased. The oxygen vacancies could generate a long-range ferromagnetic ordering of the films (Myong *et al.*, 2003; Xue *et al.*, 2019). On the basis of the BMP model, the ferromagnetic order overlapped by electron exchange interaction between Fe ions and oxygen vacancies induces ferromagnetism (Coey *et al.*, 2004; Jaiswal *et al.*, 2015).

The sample annealed at a higher temperature represents a higher saturation magnetization because the increase in oxygen vacancies and the magnetic particles precipitated lead to an increase in the saturation magnetization of SUS304 annealed at a higher temperature.

4 CONCLUSION

In this study, SUS304 stainless steel annealed at different temperatures from 800 to 1050 °C was performed. The crystal structure, surface morphologies and magnetic properties have been investigated. The XRD analysis results show that the crystal structures transferred gradually from Fe_2O_3 and Fe_3O_4 phases to Cr_2O_3 and $FeCr_2O_4$ phases finally as the annealing temperature increased. The SEM morphologies analysis indicated that the surfaces are rougher and the numbers and sizes of pyramidal shape precipitations are increased. All the samples represent a hysteresis behavior, and the hysteresis loop is clearer when the annealing temperature is increased. The mixtures of a Cr-rich phase, $FeCr_2O_4$ formation, and the largest number of oxygen vacancies generated lead the maximum saturation magnetization to be observed for SUS304 stainless steel annealing at 1050 °C.

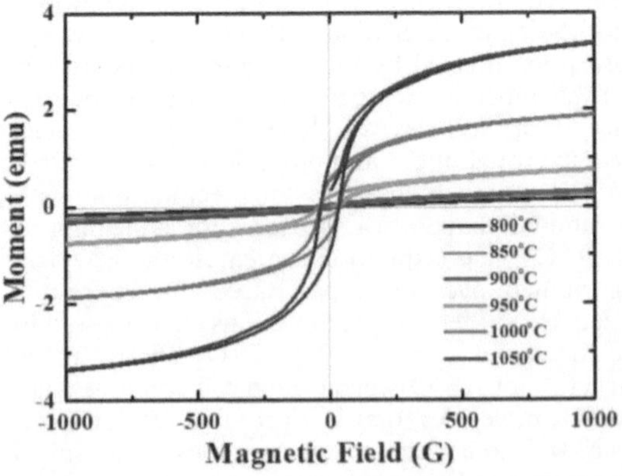

Figure 5. M–H curves for the SUS304 stainless steel annealed at different temperatures in air.

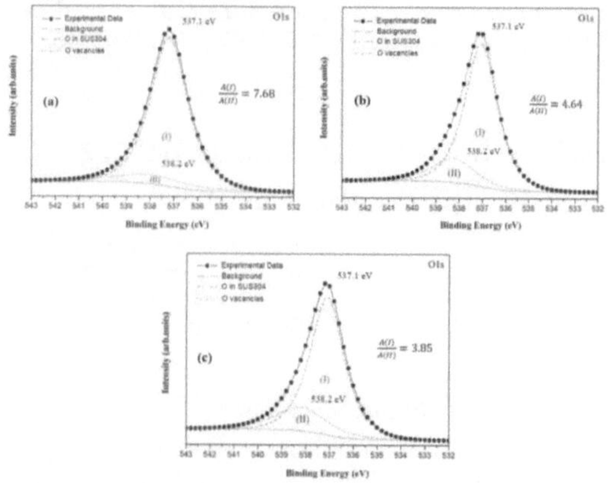

Figure 6. XPS O1S level for SUS304 stainless steel annealing at different temperatures. (a) 800, (b) 950 and (c) 1050 °C.

ACKNOWLEDGMENTS

The authors would like to thank the Ministry Science and Technology of the Republic of China for financially supporting this project under grant MOST 110-2221-E-150-020, and Ms. Hui-Jung Shih (Instrument Center, NCKU, Tainan, Taiwan) for assisting in HR-SEM analysis.

REFERENCES

Ambrosini, A., Lambert, T. N., Staiger, C. L. Hall, A. C. Bencomo, M. Stechel, Sandia, E. B. (2010). National Laboratories, *Sandia Report Sand* 2010–7080.

Callister, W. D., Rethwisch, D. G. (1997). *Materials Science and Engineering: An Introduction*. 4th edition. New York: John Wiley and Sons, Inc.

Coey, J. M. D., Douvalis, A. P., Fitzgerald, C. B., Venkatesan, M. (2004). Ferromagnetism in Fe-doped SnO_2 thin films. *Applied Physics Letters*, 84, 1332.

Honeycombe, R., Bhadeshia, H. (1995). *Steels: Microstructure and Properties"*. 2nd edition. London: Edward Arnold.

Jaiswal, M. K., Kumara, R., Kanjilal, D., Dong, C. L., Chen, C. L., Asokan, K., Ojha, S. (2015). Studies of dense electronic excitation-induced modification in crystalline Fe-doped SnO_2 thin films. *Applied Surface Science*, 332, 726–735.

Karlsson, L., ESAB EB, Gothenburg. (2004). Stainless Steels Past, Present and Future. *Svetsaren*, 1, 47.

López-Marino, S., Neuschitzer, M., Sánchez,Y., Fairbrother, A., Rodriguez, M. E., López-García, J., Placidi, M., Calvo-Barrio, L., Pérez-Rodríguez, A., Saucedo, E. (2014). Earth-abundant absorber based solar cells onto low weight stainless steel substrate. *Solar Energy Materials and Solar Cells*, 130, 347–353.

McCarty, K. F., Boehme, D. R. (1989). A Raman study of the systems $Fe_{3-x}Cr_xO_4$ and $Fe_{2-x}Cr_xO_3$. *Journal of Solid State Chemistry*, 79(1), 19–27.

Moos, R., Hardtl, K. H. (1997). Defect Chemistry of Donor-Doped and Undoped Strontium Titanate Ceramics between 1000° and 1400 °C. *Journal of the American Ceramics Society*, 80, 2549–2562.

Myong, S. Y., Lim, K.S. (2003). Highly stable and textured hydrogenated ZnO thin films. *Applied Physics Letters*, 82, 3026.

Noh, J. H., Osman, O. I., Aziz, S. G., Winget, P., Brédas, J. L. (2014). A density functional theory investigation of the electronic structure and spin moments of magnetite. *Science and Technology of Advanced Materials*, *15*, 044202.

Nomura, K., Ujihira, Y. (1990). Analysis of oxide layers on stainless steel (304, and 316) by conversion electron Mössbauer spectrometry. *Journal of Materials Science*, 25, 1745–1750.

Ohring, M. (1992). *The Materials Science of Thin Films. Academic Press*, San Diego, California, USA

Ramos-Moorea, E., Diaz-Droguett, D. E., Spring, P., Evans, J. T., Cabrera, A. L. (2011). Generation of oxygen vacancies in the surface of ferroelectric Pb(Nb, Zr,Ti)O_3. *Applied Surface Science*, 257, 4695–4698.

Torruella, P., Ruiz-Caridad, A., Walls, M., Roca, A. G., López-Ortega, A., Blanco-Portals, J., López-Conesa, L., Nogués, J., Peiró, F., Estradé, S. (2018). Atomic-Scale Determination of Cation Inversion in Spinel-Based Oxide Nanoparticles. *Nano Letters*, 18, 5854–5861.

Wang, Z. Y., Zhu, H., Ai, L., Ding, J. M., Zhu, P. F., Li, Z. Q., Li, B., Jiang, H. C., Yu, F. P., Duan, X. L., Jiang, H. D. (2021). Synthesis, Electronic Structure, and Electrochemical Properties of the Cubic Mg_2MnO_4 Spinel with Porous-Spongy Structure. *Nanomaterials*, 11(5), 1122.

Wu, S., Cheng, C. H., Hsiao, Y. J., Juang, R. C., Wen, W. F. (2016). Fe_2O_3 films on stainless steel for solar absorbers. *Renewable and Sustainable Energy Reviews*, 58, 574–580.

Xue, F. K., Liu, X. C., Liu, J. W. (2019). Oxygen Vacancy as a Medium-Induced Ferromagnetism in (Fe, Sb) Codoped SnO_2 Films. *The Journal of Physical Chemistry C*, 123(1), 684–690.

Zou, Q., Ruda, H., Yacobi, B. G., Farrell, M. (2002). Microstructural characterization of donor-doped lead zirconate titanate films prepared by sol–gel processing. *Thin Solid Film*, 402, 65–70.

System Innovation for an Artificial Intelligence Era – Lam et al. (Eds)
© 2025 the Author(s), ISBN: 978-1-032-84452-7

The cross cultural difference in University students' mobile learning between Malaysian and Taiwanese by using the A^6 metric

Siu-Tsen Shen
Department of Multi-media Design, National Formosa University, Hu-Wei, Yunlin, Taiwan

Stephen D. Prior
Aeronautics, Astronautics and Computational Engineering, The University of Southampton, Hampshire, Southampton, UK

ABSTRACT: This research used the new A^6 metric created by the authors to compare and contrast the cross-cultural differences between Malaysian and Taiwanese university students' performance and mobile learning experiences. There was a total of 224 Malaysian University students and 416 Taiwanese University students who participated in these surveys during Oct 2023 to Feb 2024. The A^6 metric uses six independent variables of: (i) Accessibility that refers to the capacity of network technology; (ii) Affordability that refers to the cost of ownership of smartphones; (iii) Adoptability that refers to the individual's buying, downloading, installing, and starting to learn and use; (iv) Adaptability that refers to the individual who needs to be able to deal with the unexpected and that which is beyond their experience; (v) Acceptability that refers to individualistic acceptance and use of new technology; (vi) Appropriateness that refers to the course design model (Shen and Prior 2023). The overall results indicated that the Malaysian participants are less positive (67%) in their responses to the questions than their Taiwanese counterparts (54%). Noting that the A^6 metric has an inverse positivity scaling in that a lower overall score denotes higher positivity. It has been postulated that to some degree this relates to the more diverse cultural drivers which exist within the Malaysian society driven by the three large ethnic groups (Malay 70%, Chinese 23% and Indian 7%). In particular, the Malaysian students appear to have a higher level of distrust in mobile phone technology, suffer from high costs issues and relatively low internet speeds. The failure of the Malaysian Government's Wawasan 2020 vision (Tee 2023) to achieve all of its objectives may be a contributory factor in this dissatisfaction. Future research will involve extending the research to other South East Asian countries in order to develop a database of cultural happiness in terms of the A^6 metric.

1 INTRODUCTION

According to Statista (2024) and DataReportal (2024), the number of internet users in Malaysia was 33.6 million with a 97% penetration rate, while the number of internet users in Taiwan was 21.7 million with 91% penetration rate. Malaysian people spend an average screen time of eight hours per day, the same as Taiwanese citizens (Kemp 2024, Statista 2024). Due to the fact that both countries have extremely high internet penetration rates of over 90%, and the average of spending eight hours on the internet per person/day, it is very interesting to conduct the A^6 metric and compare the similarity and differences between Malaysian university students and Taiwanese university students.

It is worth noting that Malaysians were measured as a society with a high-power distance in Hofstede's cultural dimension. Therefore, the emergence of cultural diversity needs to be carefully analyzed. This research aims to identify a set of six criteria of university student performance and feedbacks in relation to the Malaysian and Taiwanese higher education environment. A comparative research is conducted followed by relevant cross-cultural literature reviews. Finally, the A^6 user metric model will be carefully measured, analyzed and constructed for future research.

2 LITERATURE REVIEW

Few researches have addressed cross-cultural issues regarding to mobile learning in higher education (Liñán, Chen *et al.* 2009; Miraz, Ali *et al.* 2018; Zubair Haider and Dilshad 2015). Viberg and Grönlund (2013) used Kearney's pedagogical framework to mobile learning from a socio-cultural perspective to examine students' attitudes (e.g. age, gender and cultural factors) about mobile use for second and foreign language learning in higher education. They compared Yunnan University in China, and Dalarna University in

 DOI: 10.1201/9781003514831-78

Sweden. The result indicated that technology is the most important culture-shaping factor, more important than culture inherited from the physical environment, and more important than age in mobile-assisted language learning (MALL) (Kearney, Schuck *et al.* 2012; Viberg, Grönlund *et al.* 2013).

Mwendia, Wagacha, and Oboko (2014) reported that M-Learning in South Africa, Kenya and Uganda has the highest percentage in Africa, and mobile phones are the major tools for supporting teaching and learning within classrooms or improved students' performance in both formal and informal learning contexts. They further advised researchers to examine the dimensions of cultural variability in the African region because of its multicultural context (Mwendia, Wagacha *et al.* 2014).

June and Lee (2020) explored determinants of open educational resources (OER) adoption by university educators in three cultural contexts i.e. Korea, Japan and the United States, by using the Unified Theory of Acceptance and Use of Technology (UTAUT2). They discovered habit was the strongest determinant of this behavioral intention in each culture. They found that the strongest determinant was performance expectancy in Korea, while it was social influence in Japan, and it was price value in the United States. They concluded that the moderating effect of culture was significant, and the cultural specificity affecting OER adoption was vital (Jung and Lee 2020).

Arpaci, Masrek, Al-Sharafi and Al-Emran (2023) investigated the role of national culture in the relationship between information management and the adoption of cloud computing. They used a multi-group structural equation modeling (SEM) approach to study cross-cultural differences. Their results showed that the relationship between perceived ease of use and usefulness was significant for both Malaysian and Turkish undergraduate students, nevertheless this relationship was stronger for Malaysian students. Furthermore, Malaysian students were better in terms of the relationship between perceived ease of use and behavioral intention, while Turkish students performed better in terms of the relationship between perceived usefulness and behavioral intention (Arpaci, Masrek *et al.* 2023).

3 EXPERIMENTAL

The researchers used the A^6 Metric created by Shen and Prior (2023) to compare the students' attitude in mobile learning in higher education between Malaysia and Taiwan in Oct 2003 - Feb 2024. The A^6 Metric has six categories and includes Accessibility (A^1), Affordability (A^2), Adoptability (A^3), Adaptability (A^4), Acceptability (A^5), and Appropriateness (A^6) (Shen and Prior 2023).

This study consisted of a two-part questionnaire to get a better understanding of university students' preferences on smartphone learning and teaching in a six-month timeframe. A total of 224 Malaysian University students and 416 University students in Taiwan responded. The Malaysian study group included 118 male students (53%) and 106 (47%) female students. The Taiwanese study group comprised 207 male students (50%) and 195 (47%) female students, however, there were 14 students (3%) who preferred not to reveal their gender.

4 RESULTS AND DISCUSSION

The A^6 metric with six ratings, are each graded using a five-point Likert scale, with a score of three being neutral (note that metrics A^1 and A^2 have four questions each (max score = 20; min score = 4) and the others have three questions each (max score = 15; min score = 3)). The maximum overall score is therefore 100, with a mean of 60 and a minimum score of 20. A higher score relates to negativity and a lower score relates to positivity.

In terms of Accessibility (A^1) which represents the capacity of the individuals network technology (broadband speed), including a country's infrastructural development, government initiatives to improve upload and download speeds, and private sector involvement (Telecoms), both study groups had similar results. The Malaysia group scored 13.1 out of 20 points (see Figure 1 – unhappiness with network speed), while the Taiwanese group scored 13.3 out of 20 points, which suggested that these two groups were broadly neutral (score of 12) overall, regarding the accessibility in mobile learning.

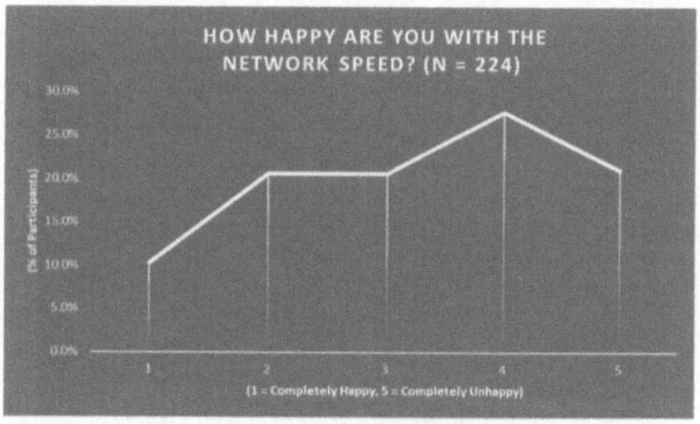

Figure 1. Satisfaction of the Malaysian participants' in terms of network speed.

In terms of Affordability (A^2) represented the cost of ownership of smartphones/tablets, including the consumer spending power, GDP of the country, socio-economic factors, internet penetration rates, and mobile internet subscriber penetration by the country/region, etc., the Malaysian group scored 13.1 out of 20 points (see Figure 2), while Taiwanese group scored 8.5 out of 20 points (neutral score 12). This indicated that the Taiwanese participants tended to be positive with their affordability perspective, whereas the Malaysian group were slightly negative.

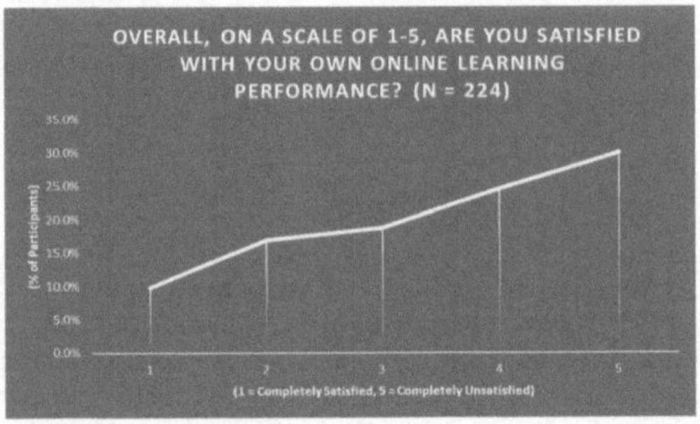

Figure 2. The overall usefulness of smartphones from Malaysian participants perspective.

In terms of Adoptability (A^3) represented the individual's acquiring, downloading, installing, and starting to learn and use, including the mental process of "pre-adoptability" (e.g. user behaviour, individual decision making, etc.) and "post-adoptability" (e.g. commitment to use and being a persistent user), influences of social media, peer pressure, cross-cultural factors, etc., the Malaysian group scored 10.3 out of 15 points (see Figure 3), while the Taiwanese group scored 7.7 out of 15 points (neutral score 9). This suggested that the Taiwanese participants were slightly more positive about their higher adoptability than their Malaysian counterparts.

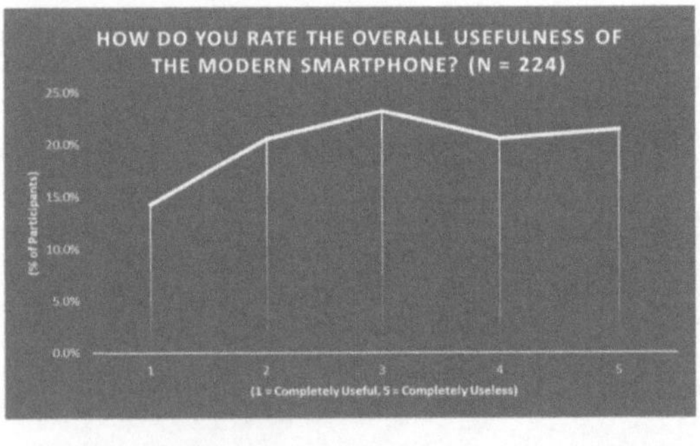

Figure 3. Satisfaction of the Malaysian participants' own learning performance.

In terms of Adaptability (A^4) that represented the individual's ability to deal with the unexpected and that which is beyond their experience, including the different stages of the user learning curve with a new technology, user behaviour/psychology, and user attitude, the Malaysian group scored 10.2 out of 15 points, while Taiwanese group had 8.2 out of 15 points (neutral score 9). This suggested that Taiwanese participants had slightly higher adaptability compared to the Malaysian ones.

In terms of Acceptability (A^5) which represented individualistic acceptance and use of new technology, including the individualistic willingness and eagerness to approach the system and engage with the technology, both groups had similar scores, the Malaysian group scored 10 out of 15 points, while Taiwanese group had 9.9 out of 15 points (neutral score 9). This indicated that both groups had a slightly negative attitude towards embracing up-to-date technology (see Figure 4).

Figure 4. Word Cloud of respondent's feedback on mobile phone user issues and negative aspects.

In terms of Appropriateness (A^6) which represented the course design model, including culturally sensitive issues of the country, religions and taboos, essential requirements of learning in a physical context, etc., both groups had a different opinion.

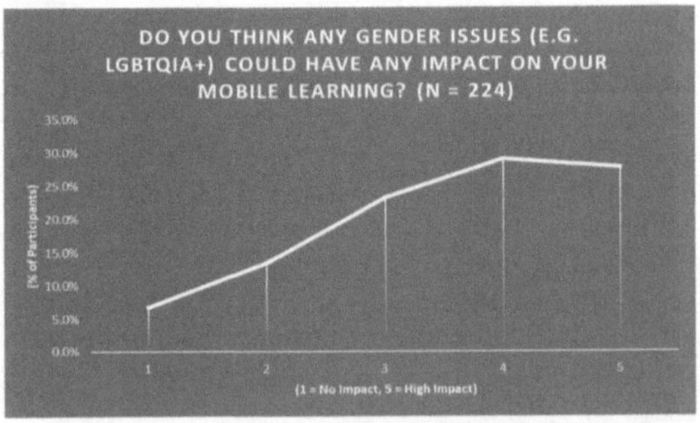

Figure 5. The graph of the Malaysian participants' attitude towards the impact of gender issues.

The Malaysian group scored 10.3 out of 15 points (see Figure 5), whist Taiwanese group scored 6.9 out of 15 points (neutral score 9). This indicated that the Taiwanese group regarded cultural related issues are not that significant compared to the Malaysian group. It could be said that the Taiwanese group originates from a more homogenous society, whereas the Malaysian group have a much more diversified cultural background consisting of three main cultural groups, those being (Malay 70%, Chinese 23% and Indian 7%)(Rahman and Ramli 2018; Statista 2023).

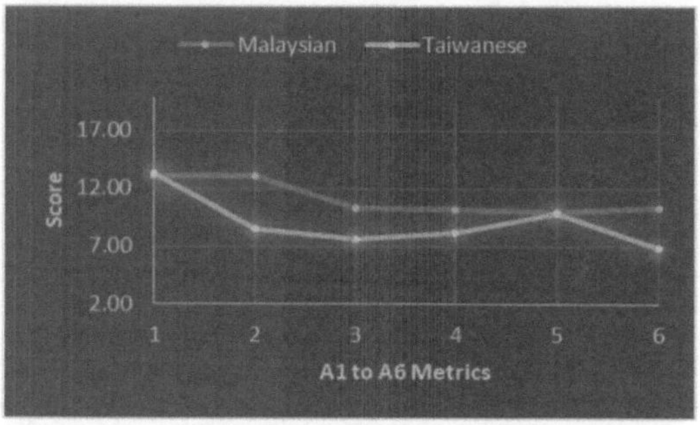

Figure 6. - A^1 to A^6 metric comparison of Taiwanese and Malaysian university students.

The majority of the participants from both countries had the same results from Accessibility (A^1) and Acceptability (A^5) as shown in Figure 1. In regards to Affordability (A^2), the gap of disagreement between Taiwanese and Malaysian students was larger compared to Appropriateness (A^6), Adoptability (A^3), and Adaptability (A^4), respectively (see Figure 6). This could be explained by the economic difference between these two countries. According to Global Finance Magazine (2024), Taiwan was ranked 15 with its USD 72,485 GDP-Adult Per Capita Income (PPP), whist Malaysia was ranked 59 with its USD 37,083 GDP-Adult Per Capita Income (PPP)(Mazzetta 2023, Ventura 2024).

It is interesting to note that the rest of the three categories of differences i.e. Appropriateness (A^6), Adoptability (A^3), and Adaptability (A^4), is correlated to ethnic groups, social status and cultural issues.

According to Hofstede's cultural dimension, Malaysia had a full score of 100 in Power Distance that referred to the attitude of the culture towards inequality in society. Taiwan had a score of 58 on this dimension compared to Malaysia, which suggested Taiwanese society has a less hierarchical order (Factor 2024; Hofstede *et al.* 2010, Minkov and Kaasa 2022).

Furthermore, Malaysia had a low preference of uncertainty avoidance with a score of 36, which referred to people who regard there should be no strict rules and live more flexible. In contrast, Taiwan had a high preference for avoiding uncertainty with a score of 69, which was almost double the Malaysian's score. Taiwan had a high Uncertainty Avoidance that individuals could cope with anxiety by minimizing uncertainty.

Moreover, Malaysian was a short term orientation culture with a socre of 47, which suggested that poeple are focused on the present or past and think them more important than the future. However, Taiwan was a long-term orientation culture with a score of 87. Societies show how flexible a national culture is with regards to change.

5 CONCLUSION

Both Malaysia and Taiwan are fast-moving well-developed countries. Malaysia, in-particular has a wider diverse multi-cultural society. Therefore, the impact of cultural differences in Malaysia would have a great effect on university students' mobile learning styles and performance in higher education.

From the above-mentioned results, the overall A^6 scores from both groups was: Malaysian participants = 67% (slightly negative), and Taiwanese participants = 54% (slightly positive). This result indicated that the A^6 Metric including Accessibility, Affordability, Adoptability, Adaptability, Acceptability, and Appropriateness has a great potential to explore for mobile learning. The future work could be improved with expanding the number of Malaysian and Taiwanese university student participants. The researchers aim to recruit more university students within South East Asian countries in the next phase, and hopefully the A6 Metric could become a well-established pedagogical index for higher education and for future researchers, in the same way that the Hofstede index has defined culrtural aspects.

REFERENCES

Arpaci, I., Masrek M. N., Al-Sharafi M. A., Al-Emran M. J. E. and Technologies I. (2023). Evaluating the actual use of cloud computing in higher education through information management factors: a cross-cultural comparison. *Education Information Technologies*: 1–21.

Factor, T. C. (2024). *Country Comparison Tool.* Retrieved April 14, 2024, from https://www.hofstede-insights.com/country-comparison-tool?countries=malaysia.

Hofstede, G. H., Hofstede G. J. and Minkov M. (2010). *Cultures and Organizations : Software of the Mind : Intercultural Cooperation and Its Importance For Survival.* New York ; London, McGraw-Hill.

Jung, I. and Lee J. (2020). A cross-cultural approach to the adoption of open educational resources in higher education. *British Journal of Educational Technology* **51**(1): 263–280.

Kearney, M., Schuck S., Burden K. and Aubusson P. (2012). Viewing mobile learning from a pedagogical perspective. *Research in Learning Technology is the Journal of the Association for Learning Technology (ALT)* **20**(1).

Kemp, S. (2024). *DIGITAL 2024: MALAYSIA*. Retrieved April 14, 2024, from https://datareportal.com/reports/digital-2024-malaysia.

Liñán, F.,Chen Y. W. (2009). Development and cross–cultural application of a specific instrument to measure entrepreneurial intentions. *Entrepreneurship Theory Practice* **33**(3): 593–617.

Mazzetta, M. (2023). *Taiwan the 14th Richest Country in the World: Global Finance*. Retrieved April 13, 2024, from https://focustaiwan.tw/business/202312250013.

Minkov, M. and Kaasa A. (2022). Do dimensions of culture exist objectively? A validation of the revised Minkov-Hofstede model of culture with World Values Survey items and scores for 102 countries. *Journal of International Management* **28**(4): 100971.

Miraz, M. H., Ali M. and Excell P. S. (2018). Cross-cultural usability issues in e/m-learning. *Annals of Emerging Technologies in Computing (AETiC)* **2**(2): 46–55.

Mwendia, S. N., Wagacha P. W. and Oboko R. (2014). Culture aware M-learning classification framework for African countries. *Cross-Cultural Online Learning in Higher Education and Corporate Training, IGI Global*: 98–111.

Proceedings of the IEEE 9th International Conference on Applied System Innovation (ICASI 2023), April 21-25, 2023, Chiba, Japan. D. K.-T. Lam, S. D. Prior, S.-J. Young, S.-T. Shen and L.-W. Ji. Leiden, The Netherlands, CRC Press, Taylor & Franics Group: 184–187.

Rahman, N. A. A. and Ramli R. (2018). Identifying Important Criteria for Staff Productivity in Relation to Malaysian Cultural Values: A Comparative Review. *The Journal of Social Sciences Research*: 1174–1179: 1176.

Retrieved April 13, 2024, from https://gfmag.com/data/richest-countries-in-the-world/.

Shen, S.-T. and S. D. Prior (2023). Mobile Learning Use of Smartphones in the Post-Pandemic Era. System Innovation for a World in Transition: Applied System Innovation IX.

Statista. (2023, Sep 13, 2023). “Share of population in Malaysia from 2019 to 2023 by ethnicity.” Retrieved April 14, 2024, from https://www.statista.com/statistics/1017372/malaysia-breakdown-of-population-by-ethnicity/.

Statista. (2024). “Number of internet users in Taiwan 2014-2024.” Retrieved April 14, 2024, from https://www.statista.com/statistics/1296415/taiwan-online-population/.

Tee, K. (2023). “PM Anwar: Wawasan 2020 a pipe dream, Malaysia trapped in vicious economic cycle of high costs and low wages.” Retrieved April 14, 2024, from https://www.malaymail.com/news/malaysia/2023/07/27/pm-anwar-wawasan-2020-a-pipe-dream-malaysia-trapped-in-vicious-economic-cycle-of-high-costs-and-low-wages/81967.

Ventura, L. (2024). “Richest Countries in the World 2024.” The Race To Be The World’s Richest Country/Territory (GDP per capita, constant prices)

Viberg, O., Å. J. C. Grönlund and Education (2013). “Cross-cultural analysis of users’ attitudes toward the use of mobile devices in second and foreign language learning in higher education: A case from Sweden and China.” *Computers & Education* **69**: 169–180.

Zubair Haider, S. and M. J. H. E. f. t. f. Dilshad (2015). “Higher education and global development: A cross cultural qualitative study in Pakistan.” *Higher Education for the future* **2**(2): 175–193.

System Innovation for an Artificial Intelligence Era – Lam et al. (Eds)
© 2025 the Author(s), ISBN: 978-1-032-84452-7

A study on the selection of elements and style formation of posters in sugar industry during the Japanese-Occupied Period

Po-Jo Lai & Shyh-Huei Hwang
Graduate School of Design, National Yunlin University of Science and Technology, Yunlin, Taiwan

ABSTRACT: Cross-cultural exchanges across national boundaries have affected all aspects of life under the development of globalization. Imagination, interaction and communication among different cultures has become an inevitable trend. This study took six "Posters in Taiwan Sugar Industry during the Japanese-Occupied Period" under the "Collections Network" of the National Taiwan History Museum as cases, incorporated with qualitative and quantitative analysis. In the first stage, semantic analysis method was used to interpret the symbolic semantics of 6 poster samples from a modern perspective, and the difference among the style images of 6 samples perceived by different genders was analyzed and verified. In the second stage, the author analyzed the poster selection elements and their "style formation" of 6 samples in terms of "text design principles", "presentation of expression techniques", "color application mode", and "arrange structure and composition". After comprehensive analysis, it is found that: (1) Sample S3 has the strongest images (7 items), including Japanese, retro, heavy, restrained, hand-painted, traditional, and Japanese. (2) Subjects of different genders had significant differences among five style images, namely, "cute – mature", "Chinese – Japanese", "thick – light", "hand-painted – photographed" and "Japanese style – Western style". (3) The design expressions imported into Taiwan due to the colonial policy have been imitated and transplanted, combining traditional Chinese, Japanese and Western styles. After integration, they are presented in the visual design of Taiwan's sugar industry posters, showing cross-cultural of multiplicity.

1 INTRODUCTION

The specific research purposes of this study are: (1) to analyze the semantic meaning of symbols conveyed by the posters of the sugar industry in Taiwan during the Japanese-Occupied Period from a modern point of view, and to analyze the comparison and verification of the differences in the style images of the 6 sample samples by subjects of different genders. (2) analyze the selected elements and style formation of the posters of these 6 samples.

This study set the "commercial property" of "artifacts" under the category of the classification retrieval item under "collections network" of the National Taiwan History Museum as the scope of search. The sample was the commercial poster under the category of "commercial property". After searching and summarizing the digital image files, a total of 6 kinds of posters in the "sugar industry" were selected as the research samples, and were sequentially coded as S1 to S6. The list is shown in Table 1.

In the first stage, a semantic survey was conducted on 6 samples, and a 5-scale online questionnaire test was conducted using semantic analysis method, with a hope to interpret the semantics communicated by visual symbols of the posters in sugar industry during the Japanese-occupied period from a modern perspective, while SPSS ® Statistics 19 was used as a test and analysis tool to understand the result in the comparison and verification of the differences behind style images among the 6 samples of the subjects in different

Table 1. List of sugar industry posters during the Japanese-Occupied Period.

coding	name	Historical period	Estimated year	Login ID
S1	Plum house sugar shop poster	1895–1945 (Time of Japan)	unknown	2005.006.0329
S2	Taiwan Sugar Co., Ltd. advertising poster	1895–1945 (Time of Japan)	1900~1945	T2018.002.2737
S3	Advertisement poster for women in kimono of Taiwan Sugar Co., Ltd.	1912–1926 (Taisho prriod)	1918	T2018.001.8468
S4	Advertisement poster of Tai-wan Sugar Co., Ltd	1912–1926 (Taisho period)	1922~1945	T2018.002.1299
S5	Dai Nippon Sugar Co., Ltd. Poster	1926–1945 (Showa period)	1926~1945	T2018.001.3115
S6	Shiomizu Port Sugar Co., Ltd. women's pattern advertising poster	1926–1945 (Showa period)	1935~1945	T2018.002.1624

Source: Collections Network on the Website of National Taiwan Museum of History and compilation of this study.

DOI: 10.1201/9781003514831-79

genders. In the second stage, the 6 samples were analyzed and collected in terms of "text design principles", "presentation of expression techniques", "color application mode", "arrangement and composition" of poster selection elements and investigation of "style formation".

The background as referred in this study was set from 1895 to 1945 chronologically. According to the historical stages of Collections Network of the National Taiwan History Museum, it was staged into three periods: 1895–1912 (Japan time – Meiji period), 1912–1926 (Japan time – Taisei period) and 1926–1945 (Japan time – Showa period). Commercial posters in other years and domains were not included herein.

2 LITERATURE REVIEW

This study combed the meaning of "poster" and "cross culture" through literature review. First of all, it analyzed the definition, category and constituent elements of a "poster", and sorted out the relationship and level among them; it discussed the definition and concept of "cross culture" for future research.

2.1 *Poster*

2.1.1 *Definition of poster*

The term "poster" is translated from the English word "Post", which is translated in-to Traditional Chinese as "poster"; it is a kind of graphic visual design form, which gives a full play of the creativity of a designer in various forms of expression such as painting, photos, text, color, etc., incorporated by the layout and design of concise words or texts, and printed and copied to a large volume, and posted in prominent public places to draw public attention, and then spread a specific theme idea as the purpose of the propaganda media (Hu Ze-min, 1996).

Since ancient times, the word "Lubu" has also been used, "Lubu, a good news document", that is, "announcement". "Encyclopedia Britannica" interprets "Poster" as "poster; a printed picture publicly displayed as a notice or advertisement." "Oxford English Dictionary" interprets it as: "a notice displayed in a public place" (Lin Junliang, 2013).

From the perspective of commercial design, "ocean" means four oceans and "report" means a report, thus are embedded with the meaning of "unveiling and communicating around" (He Yaozong, 1975). Language, characters, and pictures, among which eyes and movements are the ways for human beings to communicate with each other and convey information, in the two-dimensional visual communication. One kind of media which has a shocking visual effect, strong broadcasting feature, while a convenience deeply with thrilling propaganda, that is "Poster" (Fei Guojing, 1999).

2.1.2 *Types of posters*

Generally, a poster is categorized into political posters, commercial posters, social posters and artistic posters. Political posters refer to ones with the theme of political order propaganda, political parties, elections, etc.; commercial posters refer to posters with themes of commercial activities, advertising and marketing, etc.; social posters refer to posters with themes of social issues and public service; Art posters refer to posters with art, performance, drama, music, literature, film, sports, design and other arts and culture as the theme, and can also be called cultural posters (Lin Junliang, 2013). Yuan Manli (1993) proposed four categories of poster generally in her study, that is, commercial posters, which are mainly for commercial services with promotional purpose; cultural posters, targeted to the pro-motion of artistic and cultural activities; cultural posters, for the purpose of publicizing art and cultural activities; social and educational posters, for the purpose of publicity and education ; Artistic posters, mostly art reproductions, both decorative and collectible. In the book "Theory and Practice of Visual Communication Design", In the book "Theory and Practice of Visual Communication Design", Professor Lin Pin-Zhang categorized a poster into commercial posters, sightseeing posters, public welfare posters, political posters, educational posters, cultural posters and activity posters. Professor Jin Taiqiang divided posters into commercial posters, cultural posters, social posters, and artistic posters in "Poster Design-Graphic Designer Series 4" (Cai Xueli, 2018).

2.1.3 *Components of a poster*

The most important thing in design is the idea, because the poster needs to allow the viewers to quickly read and understand the theme in a short time, so the idea conveyed must be eye-catching and fresh and interesting (Jin Taiqiang, 1996).

Nowadays, poster is designed with cultural style and formats unique to all, and design characteristics of "large, beautiful, refined and simple" from the use of textual editing to image techniques (Lin Junliang, 2013). The design of posters is not only the use of editing, the expression of image techniques and the isomorphism of various elements, but also presented through fonts, illustrations, photography, patterns, and computer graphics media (see Lin Peichun, 1993; quoted from Cai Xueli, 2018).

When we confirmed the idea to convey, the course of transforming information into visual images with significance requires the composition of posters. In terms of plane space, composition is a visual pattern formed by arranging and combining different modeling elements (points, lines, surfaces) according to the principle of aesthetic form. The constituent elements of a poster are not only modeling elements composed of dots, lines, and surfaces, but also texts and colors (Chen Yijun, 2011).

There are many experts and scholars in China who put forward their views on the composition of posters, for example: He Yaozong (1975), Lin Pinzhang (1986), Yuan Manli (1993), Lin Peichun (1993), Lin Rongguan (1993) have put forward their own views, as follows the discussions of domestic experts and scholars are summarized in Table 2.

Table 2. List of Poster componentes.

time	author	elements of the poster	book title	Publisher
1975	He Yaozong	Components of print ad design 1. Styling components: illustrations, trademarks, product names, outlines 2. Content composition: title, slogan, explanatory text, company name	Graphic design	lion book co., ltd.
1986	Lin Pinzhang	1. Text: title, description of concept, text, etc. 2. Illustration: special text modeling, painting performance, photographic performance, geometric abstract graphics, etc. 3. Color: text color, picture, background color, etc. 4. Arrangement: visual arrangement, visual space and composition of various plane components	Business Design	artist press
1993	Yuan Manli	Poster Design Elements: 1. Font application. Text is the main element to convey information. Designers often use text to express different languages, fonts, text content, layout methods, levels, spatial ratio and color applications, which produce extinct effects. 2. Combination of pictures and text 3. photography performance 4. illustration performance 5. pattern expression 6. use of color	Introduction to Poster Design: A Brief Introduction to Poster Design	Taipei Fine Arts Museum
1993	Lin Peichun	1. Text: title, subtitle, slogan, copy, company name, address and phone number of the sponsor 2. Modeling: specific signs (such as trademarks or patterns), illustrations extended based on themes, and pictures with special effects of photography 3. Color: Color gives viewers various associations	Introduction to Poster Design: Poster Design Techniques	Taipei Fine Arts Museum
1993	Lin Rongguan	1. Elements belonging to text: headline, subtitle, quotation, slogan, slogan, explanatory text, product name, company name, address, telephone number 2. Elements of graphics: illustration, photography, painting, patterns, illustrations, various models	commercial design	art book company

Source: Li Xiuru (2010) and this study.

2.2 *Cross-culture*

In today's globalized society, "cross-cultural communication" is a necessary ability to adapt to the changes in the globalized world. Living in a different cultural context requires more timely and appropriate cross-cultural communication skills. Having the ability to communicate across cultures, not only in different cultural contexts, but also in the process of interpersonal communication, can conduct appropriate and effective communication (see Spitz-berg & Changnon, 2009; cited from Chen Lishu, 2020). The "cross-cultural" mentioned above refers to the "cross-cultural" observation through translation practice and teaching work based on its own research field.

Experts in the field of intercultural chemistry compare culture to "iceberg". An iceberg floating on the sea usually only exposes 10% of its volume, and the rest 90% are hidden under the sea, which may pose a threat to ships at any time. And "culture" is like an iceberg, like a small corner of an iceberg. The obvious cultural factors such as etiquette, customs and social traditions will not cause serious cross-cultural problems. However, when getting along with people from different cultural backgrounds, cultural factors such as values, beliefs, thinking patterns, and communication styles hidden deep under the cultural surface are like unfathomable oceans. Therefore, cultural differences often result in cross-cultural differences. The main reason for "shipwreck" in cultural relations (Lin Kuanlong, 2019).

"Cross-cultural" is defined as "belonging to or related to two or more different societies, countries, or cultures" (Fox, Manning, Murphy, Urbanm, Mar-wick & O'Shea, 2003, p. 375; quoted from Zhao Yixin, 2013), and many scholars use the root of "cross -", "inter -" and "trans -" to indicate the "cross" of cross-cultural. The meaning of "inter -" and "trans -" are similar, both between two or more groups. How-ever, the meaning of "cross -" is "from one party to the other", "between the two parties and connecting with each other", which is more dynamic than other English roots. Cross-culture includes two or more cultures, which is different from multicultural. Although multicultural refers to "more than one" (Fox *et al.*, 2003, p. 1081; quoted from Zhao Yixin, 2013), multicultural focuses on a country or social group, while cross-cultural refers to the interaction between countries or cultures (Wikipedia, 2012; cited from Zhao Yixin, 2013).

This study attempts to apply the above viewpoints to this article, setting "posters" as the research medium and "sugar posters" as the research object, trying to observe cultural differences across regions, societies, or countries through visual communication, and analyzing each poster The level of expression that is intended to be conveyed through visual symbols.

3 RESEARCH METHOD

This study combines qualitative and quantitative research. Firstly, it collects the definition, category and constituent elements of "poster" and relevant literature on the meaning of "cross culture" through the "literature analysis method". Based on the category search scope of "commercial property" under "Artifacts" of the "collection network" in the website of National Taiwan Museum of History, the "sugar industry" category was selected after searching and summarizing the files using the "conceptual sampling method". Six posters of Taiwan sugar industry during the Japanese-Occupied Period were used as the samples, which were coded sequentially as S1 to S6.

At the preliminary stage, the autor collected 6 samples of adjectives, and added their own antonyms after sorting out and deletion, forming a total of 10 pairs of words, as shown in Table 3.

Secondly, the perceptual cognitive imagery survey was conducted, and the semantic analysis method was used to compile an online questionnaire. 77 subjects (35 males, 41 females, 1 unanswered) were asked to complete the 5-stage scale evaluation. For these 6 samples, the author checked the adjectives that are considered suitable for the image of a poster as a statistical distribution. Then SPSS ® Statistics 19 was used as a test and an analysis tool to understand the comparison and verification of the differences of style images among 6 samples of different genders.

Table 3. 10 sets of adjectives in semantic analysis of this study.

serial number	Opposite Adjective Vocabulary
1	cute – mature
2	Chinese – Japanese
3	retro – modern
4	heavy – lightweight
5	introverted – liberated
6	hand drawn – photographed
7	traditional – avant-garde
8	earthy – ornate
9	Japanese style – Western style
10	smiling – grim-looking

Source: Compiled in this study.

4 RESEARCH RESULT ANALYSIS

4.1 *Descriptive statistics of survey samples*

The formal questionnaire of this study is an online questionnaire survey. A total of 77 questionnaires were collected. After deducting 8 invalid questionnaires with incomplete answers, a total of 69 valid questionnaires were collected. The effective questionnaire recovery rate was 89.6%. For the demographic variables of the entire sample, including gender, age, education level, background, etc., the survey results are explained. The distribution of the data is shown in Table 4.

Table 4. Distribution table of effective sample data in this study.

project	category	number of people	proportion	total
gender	male	32	46.4%	69
	female	37	53.6%	
age	under 20 years old	2	2.9%	69
	21–30 years old	20	29%	
	31–40 years old	31	44.9%	
	41–50 years old	9	13%	
	51–60	6	8.7%	

(*continued*)

Table 4. Continued

project	category	number of people	proportion	total
	years old			
	over 61 years old	1	1.5%	
education level	High school (vocational)	5	7.3%	69
	University	46	66.7%	
	Institute (including above)	18	26%	
background	Students of design related departments	6	8.7%	69
	Design related practitioners	13	18.8%	
	museum practitioners	10	14.5%	
	none of above	40	58%	

Source: Compiled in this study.

4.2 *Analysis of the image communicated by the poster of sugar industry in Taiwan during the Japan-Occupied Period*

In this study, the average and standard deviation of each vocabulary of 6 samples are summarized in Table 5, and a broken line chart of image average is drawn, as shown in Table 6.

Table 6. Average line chart of 6 samples in this study.

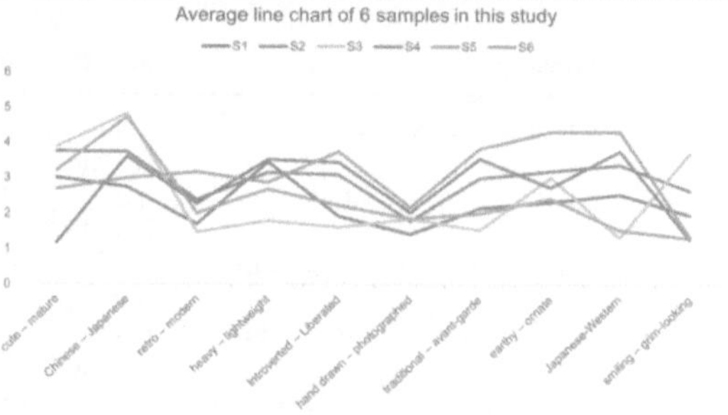

Source: Compiled in this study.

This study adopts a 5-stage evaluation, with a median value of 3, so the scale value higher than 4 or lower than 2 is "strong image". The results of analysis one by one are as follows: S3 is the sample with the strongest images, a total of 7 items; Including Japanese, retro style, thick style, introverted style, hand-painted style, traditional style, and Japanese style. Next are the four items of S5, including Japanese, hand-painted, Japanese style and smiling. The samples with three items are S1 and S6, including retro, introverted, hand-painted and

Table 5. The average and standard deviation of each vocabulary of 6 samples in this study.

	cute/mature		Chinese/ Japanese		retro/modern		heavy/ lightweight		introverted/ liberated	
sample	average	standard deviation	average	standard deviation	average	standard deviation	average	standard deviation	average	standard deviation
S1	3.01	1.231	2.74	1.291	1.71	0.972	3.46	1.158	1.94	1.056
S2	3.75	1.09	3.74	1.133	2.39	1.374	3.16	1.29	3.10	1.308
S3	3.88	1.119	4.83	0.419	1.48	0.72	1.8	0.917	1.62	0.769
S4	1.17	0.419	3.61	1.274	2.28	1.403	3.52	1.208	3.45	1.119
S5	3.2	1.41	4.74	0.585	2.01	1.243	2.68	1.323	2.23	1.165
S6	2.68	1.409	2.99	1.131	3.17	1.424	2.88	1.345	3.75	0.991

	hand drawn/ photographed		traditional/ avant-garde		earthy/ ornate		Japanese Style/ Western Style		smiling/ grim-looking	
sample	average	standard deviation	average	standard deviation	average	standard deviation	average	standard deviation	average	standard deviation
S1	1.43	0.717	2.16	1.146	2.33	1.268	2.54	1.17	1.97	0.891
S2	1.8	1.037	3	1.361	3.2	1.158	3.38	1.273	2.65	1.235
S3	1.86	1.263	1.54	0.833	3.03	1.475	1.33	0.634	3.72	1.069
S4	2.04	1.377	3.55	1.243	2.74	1.171	3.77	1.19	1.28	0.539
S5	1.87	1.248	2.01	1.144	2.42	1.333	1.55	1.037	1.32	0.717
S6	2.17	1.372	3.84	1.093	4.33	0.886	4.33	0.95	1.35	0.819

Source: Compiled in this study.

Table 7. Gender test of subjects with significant differences.

sample	imagery	gender	number	average	standard deviation	t	significance (double tail)
S4	Chinese – Japanese	male	32	4.09	1.146	3.124	.003**
		female	37	3.19	1.244		
S5	heavy – lightweight	male	32	3.03	1.470	2.095	.040*
		female	37	2.38	1.114		
S5	hand drawn – photographed	male	32	2.41	1.542	3.423	.001***
		female	37	1.41	.644		
S6	hand drawn – photographed	male	32	2.56	1.480	2.218	.030*
		female	37	1.84	1.191		
S6	Japanese Style –Western Style	male	32	3.97	1.150	−3.022	.004**
		female	37	4.65	.588		
S6	cute – mature	male	32	2.28	1.250	−2.258	.027*
		female	37	3.03	1.462		

*P<.05 **P<.01 ***P<.001
Source: Compiled in this study.

gorgeous, Western-style and smiling. There are 2 items in S4, including cute ones and smiling ones. S2 has one item, which is hand-painted.

4.3 *Cross analysis of sex attributes of subjects*

In this study, the differences of the poster style image of the sugar industry among the subjects of different genders during the Japan-Occupied Period were statistically tested. After independent sample t-test, the analysis of significant differences among 6 samples herein were summarized and listed as shown in Table 7. The analysis is as follows:

1. The average number of males and females in the "Chinese – Japanese" images of sample S4 (boy dressed as a sailor) was 4.09 and 3.19, respective-ly, with a t value of 3.124 (P = .003<. 05). There was a significant difference between the two. Male person are used to be recognized as Japanese, while female person are used to be recognized as Chinese.
2. The average number of male persons in sample S5 (female dressed in kimono, two-handed push back) in the image of "heavy – lightweight" is 3.03, indicating that male person consider it as "light"; The average number of female persons is 2.38, which means that female person consider it as "heavy"; The t value is 2.095 (P = .040<. 05), and significant difference exist between the two.
3. For sample S5, the average number of males and females in the "hand-painted – photographed" imagery was 2.41, and the t-value was 3.423 (P = .001<.05). There was a significant difference between the two. Both men and women tend to recognize as hand-painted, while women are more strongly.
4. The average number of male person and female person in the "hand-painted – photographed" image of sample S6 is 2.56 and 1.84 respectively, with a t value of 2.218 (P = .030<. 05). There is a significant difference between the two. Both male person and female person tend to recognize hand-painted, while female person are more intense.
5. For sample S6, the average number of men in the image of "Japanese style – Western style" is 3.97, and the average number of women is 4.65, with a t value of −3.022 (P = .004<.05), and there is a significant difference between the two. Both men and women tend to recognize it as foreign style, but women are more strongly.
6. The average number of males in the "cute – mature" image of sample S6 is 2.28; The average number of female people is 3.03; The t value was −2.258 (P = . 027<. 05), and there was a significant difference between the two. Male people tend to think it is "cute", while female person think it is middle.

4.4 *Selection of design elements and style formation of posters in Taiwan Sugar Industry during the Japan-Occupied Period*

This study aims at 6 samples, and deconstructs the "text design principles", "presentation of expression techniques", "color application mode", "arrange structure and composition" and other poster selection elements and "style formation", which are detailed in the following Table 8.

Table 8. Analysis of the selected elements and style formation of the 6 samples in this study.

Icon			
Name		S1–Plum House Sugar Store Poster (Sitting Woman in Tang Shirt)	S2–Advertising poster of Taiwan Sugar Co., Ltd. (half-length woman in dress)
	Text Design Principle	• The name of the store on the lower left side is "Ume House Sugar Shop" in horizontal Chinese characters from right to left. • Only the words "Plum House Sugar Store" can be seen.	• The company name "Taiwan Sugar Co., Ltd." on the lower right side is horizontal Chinese characters from right to left. • Only the words "Taiwan Sugar Co., Ltd." were visible.
Selection of Elements	Presentation of Expression and Method	Illustrator image: • Female person's makeup hair has black thin eyebrows, red lips, and black hair with flow-er hair accessories. She has single eyelids and fair skin. • Female person's clothing consists of a water blue three-quarter sleeve diagonal buttoned Tang shirt with white six-point flowers on the left and right sides of the upper body, a dark gray long skirt on the lower body, white socks on both feet and white short heels with red bow on the toe. • The woman's posture is to look straight ahead, hold a blue cover book with both hands and sit on a white barrel column chair. Extended illustration imagery: • The pink flower tree blooming behind the woman is surrounded by colorful bushes. • The white barrel-style column chair is decorated with patterns and exquisite carving patterns.	Illustrator image: • The woman's makeup and hair are black thick eyebrows, dark red lip gloss, coral blush on both cheeks, black inner eyeliner and thick and slightly curly black wavy hair. • Female person's clothing is a close-fitting low-cut pleated dress with short sleeves on the upper body, revealing a large amount of white skin on the chest and back of the neck, and a bow on the front side of the fake. • The woman is sideways and looking straight ahead, sitting at the table with a small coffee cup in her right hand. The woman's left hand is close to the white porcelain plate on the table. There is a silver spoon and a sugar cube on the plate; The upright table lamp near the woman's left shoulder glowed. Extended illustration imagery: • The upright table lamp near the woman's left shoulder glowed. From this, it is known that "cube sugar" is the main marketing.
	Applied mode of colors	• The company name "Umeya Sugar Shop" is in golden brown, while the illustrator's images, matching objects, and backgrounds are all colorful. • The overall color system is bright and warm, and the left side of the background is a large area of pastel color.	• The company name "Taiwan Sugar Co., Ltd." is bright yellow. • The illustrations, matching objects and back-ground are all colorful. • The overall color system is relatively dark, and the location is set as indoor according to the light and shadow presentation of the light.
	Arranged structure and composition	• The idea of female person's illustration is mainly arranged, supplemented by flowers around and flowering trees behind. • The visual movement starts with the female illustration in the middle of the picture, followed by the flowers around and the flowering trees behind, and finally by the text "Meiwu Sugar Shop".	• The main image of the female illustrator is the main arrangement, and the upright desk lamp with a cover near the left shoulder is the auxiliary. • The visual line starts with the image of the female illustrator in the middle of the screen, followed by the upright table lamp with a cover on the left shoulder, and finally the text "Taiwan Sugar Co., Ltd."
Formation of Style		The woman is wearing a Chinese Tang shirt-style cute three-quarter-sleeve top and sits in the beautiful scenery of the garden. The colorful flowers show the woman's graceful and pleasant demeanor.	Dressed in a low-cut, close-fitting, short-sleeved dress, the makeup woman sits by the dim lamp with a large neck and back exposed. She slightly twists her waist and side-ways to show a sense of movement, presenting a sexy and graceful mature charm.

(*continued*)

Table 8. Continued

Icon			
Name		S3–Advertisement poster for Taiwan Sugar Co., Ltd. woman in kimono (woman in kimono with calendar)	S4–Taiwan Sugar Co., Ltd. Corner Sugar Advertising Poster (Boy in Sailor Suit)
Selection of Elements	Text Design Principle	• The uppermost company name "Taiwan Sugar Co., Ltd." is a horizontal Chinese character from right to left, with a trademark inserted in the center of the company name. On the lower side of the club's name are the horizontal Chinese characters from right to left, such as "capital of 29.8 million yuan" and "founded in the 33rd year of Meiji". • "Taiwan Sugar Co., Ltd. Raw Sugarcane Harvesting Area Map" is written in horizontal Chinese characters from right to left. "Sugar brand" on the upper left of the regional map is the horizontal Chinese characters from right to left and the trademarks of each product. At the bottom right are the straight Chinese characters from top to bottom, such as "Head Office", "Taiwan Sugar Co., Ltd.", "Taiwan Dog Beating". At the bottom left are the straight Chinese characters from top to bottom, such as "contracted sale of Chinese ports" and "Mitsui Foreign Company".	• "CUBE SUGAR" and "TAIWAN SEITO KAISHA LTD." on the uppermost side are horizontal English characters from left to right. • In the middle of the vertical commodity packaging box, there are straight Chinese characters from top to bottom, such as "Fuji Seal" and "angular granulated sugar", and horizontal Chinese characters from right to left, such as "Taiwan Sugar Co., Ltd.". • The company name "Taiwan Sugar Co., Ltd." on the bottom side is horizontal Chinese characters from right to left.
	Presentation of Expression and Method	Illustrator image: • The women's makeup and hair are black thick eyebrows, black eyeliner on double eyelids, light pink lips with small cherry mouth, white powder makeup face symbolizing traditional Japanese culture, and flower hair accessories inserted in black hair bun. • Female person's clothing is a multi-level exquisite and noble kimono style. The outer-most pink lotus color kimono can be seen with detailed pattern patterns and detailed paper crane patterns. The inner lining near the neckline is full of flower patterns. The wide belt is tied to the back waist with a raised belt, which appears heavier. Female person wear rings on the ring fingers of both hands. • The woman's posture is looking straight ahead from a small side, holding a "Seven Years of the Republic of China, Wuwu" calendar in her left hand, and sitting on a sofa-style seat with intricate patterns. Extended illustration imagery: • The upper left corner of the "Taiwan Sugar Co., Ltd. Raw Material Sugarcane Harvesting Area Map" is marked with "Sugar Brand" and various product trademarks. • The location information of the "Head Office" located in "Taiwan Dagou" is marked on the lower right. The map of Taiwan shows the locations of sugar factories established by Taiwan Sugar Corporation. From the words	Illustrator image: • The boy has black thick eyebrows, double eyelids, a toothy smile, rosy cheeks, and a white knitted hat with double-sided red balls on his head. • Boys' clothing is navy blue sailor style, and the wide collar and fold-over cuffs ex-tending to the back are white with three water blue pressing lines. • The boy's posture is to protrude his upper body from the caramel box, support on the box, and hold a silver spoon in his right hand. There is a piece of white sugar in the spoon and hold a white bowl with golden edge in his left hand. Extended illustration imagery: • On the packaging box is "Fuji brand" angular granulated sugar produced by Taiwan Sugar Association. It is known that "sugar" is the main marketing.

(continued)

Table 8. Continued

	"Purchased for various ports in China" and "Mitsui Makoto" marked on the lower left, we can know that this is a promotional calendar issued by Taiwan Sugar Co., Ltd. in China.	
Applied mode of colors	• The company name "Taiwan Sugar Co., Ltd." is golden brown, the trademark in the center of the company name is red, and the "capital of 298 million yuan" and "founded in the 33rd year of Meiji" are blue. • The main image of the illustrator, the map of the sugarcane harvesting area, and the background are all colorful. • The overall color matching is colorful but harmonious.	• "CUBE SUGAR" on the top is orange, and "TAIWAN SEITO KAISHA LTD." is white. • The "Fuji Seal" on the middle upright commodity packaging box is a white character with a golden-brown frame; "Jiao Granose" is a golden-brown letter with a dark gray border; "Taiwan Sugar Co., Ltd." is a white letter placed inside a red block. • The name of the club at the bottom, "Taiwan Sugar Co., Ltd.", is orange.
Arranged structure and composition	• The illustrations of female person dressed in kimono and holding the almanac are mainly arranged, supplemented by the sugar-cane harvest area map and seats on the left. • The visual line starts with the illustrator image of a woman in kimono holding a calendar in the middle of the screen, followed by the area map of the sugarcane harvesting area on the left, and finally the company name "Taiwan Sugar Co., Ltd."	• The capped boy in a sailor suit is the main image of the illustration, and the enlarged version of the horned sugar box is supplemented. • The visual movement route is based on the illustration of the boy holding the sugar cube and bowl in the upper part of the picture, followed by the granulated sugar packaging box in the middle and lower part, and finally the words "CUBE SUGAR", "TAIWAN SEITO KAISHA LTD." and the company name "Taiwan Sugar Co., Ltd." on the top side.
Formation of Style	Female person with formal makeup and hair wears delicate pink lotus color kimonos. The wide belt tied to the back of the waist makes the female person stand upright, showing the beauty of traditional modesty and modesty.	The boys wore sailor suits and white hats. They were shaped by the westernization trend and matched with the Japanese style and the style of the angular sugar packaging box, which was full of a playful and naughty retro atmosphere.
Icon		
Name	S5–Poster of Great Japan Sugar Co., Ltd (Female in kimono, with two hands pushing back)	S6–Shiomizu Port Sugar Co., Ltd. women's pattern advertising poster (female with head and chest in dress)
Selection of Elements — Text Design Principle	• "Dainippon Sugar Co., Ltd." on the right is written in straight Chinese characters from top to bottom. • Only the words "Dainippon Sugar Co., Ltd." can be seen.	• "Shiosui Port Sugar Co., Ltd." on the left is written in straight Chinese characters from top to bottom. • The lower left side is a trian-gular printing company trademark, and "Tokyo Daiichi Printing Co., Ltd." is a horizontal English character from left to right. • There is a horizontal English cursive lettering of "Kita" on the lower side.
Selection of Elements — Presentation of Expression and Method	Illustrator image: • The woman's makeup and hair are black eyebrows, red lips and a toothy smile, and a red hair ornament inserted in the black wavy hair. • The female person's clothing is a kimono style of mustard yellow, with some colorful square patterns visible. The pink lining can be seen near the neckline. The black belt is tied to the back waist with a red belt.	Illustrator image: • The female person's hair is made up of black eyebrows, red lips and grinning teeth, eye black is brushed on both eyes to make it denser, a blue and white thin version of hair band is worn on the black curly hair, and red nail polish is brushed on the nails. • The female person's dress is a water blue sleeved dress style, which is dotted with yellow, white, and blue colors. The cuffs are pendulous flared

(*continued*)

Table 8. Continued

	• The posture of the woman is to raise the side face, looking at the front and upper left, with the backs of the hands crossed and stretched forward.	sleeves, and the cuffs reveal lace fabric. The right hand wears a bracelet with half pearl style, and the left ring finger wears a green ring. • The woman's posture is to raise her face sideways, look forward, and cross her hands with ten fingers on her chin.
Applied mode of colors	• The company name "Dainippon Sugar Co., Ltd." is in white letters with a dark brown border. The background is a dark green gradient color system. • The main image and background of the illustrator are colorful.	• The company name "Shiosui Port Sugar Co., Ltd." is light yellow. • The background is a colorful rendering color scheme. • The main image and background of the illustrator are colorful.
Arranged structure and composition	• Female person's illustrations are mainly arranged as ideas. • The visual movement begins with the female illustration in the middle of the picture, followed by the text "Big Japan Sugar Co., Ltd.".	• The main image of the female illustrator is the main arrangement. • The visual line starts with the image of the female illustrator in the middle of the screen, followed by the text "Shiomizu Port Sugar Co., Ltd."
Formation of Style	The smiling woman is dressed in a mustard yellow kimono, with her arms crossed and stretched forward, presenting an energetic and relaxed atmosphere.	The smiling woman wore a dress with pleated sleeves and pendulous trumpet sleeves, which presented a romantic and confident atmosphere against the colorful fantasy background.

Source: Collection Network of National Taiwan Museum of History and compilation in this study.

5 CONCLUSION

This study focuses on the "poster of sugar industry in Taiwan during the Japan-Occupied Period", and first discusses the meaning of "poster" and "cross culture" according to the literature. Secondly, the semantic analysis method was used to invite the subjects to conduct the perceptual cognitive image evaluation survey of the online questionnaire. Then, the SPSSsystem is used for verification and analysis to understand the comparison and verification of the differences in the style images of the 6 samples among subjects of different genders. Finally, we analyzed the selected elements and style formation of the poster design of the 6 samples, and comprehensive conclusions are as follows.

1. In the symbolic meaning of the posters of Taiwan's sugar industry during the Japan-Occupied Period, S3 (the poster of female person holding the calendar and kimono) has the strongest feeling.

   According to the line chart of the average value of images corresponding to each vocabulary pair in 6 samples, it is known that: the sample with the strongest images is S3, with a total of 7 items; Including Japanese style, retro style, thick style, introverted style, hand-painted style, traditional style, and Japanese style. Followed by the 4 items of S5, including Japanese style, hand-painted, Japanese style and smiling. The samples with three items are S1 and S6, including retro, introverted, hand-painted and gorgeous, Western style and smiling. There are 2 items in S4, including cute ones and smiling ones. S2 has one item, which is hand-painted.
2. The sex attribute analysis results of the subjects in the 6 posters of Taiwan's sugar industry during the Japanese colonial period showed significant differences in S4 (a boy in a sailor suit), S5 (a woman in a kimono who pushes both hands), and S6 (a woman with a head and chest in a western dress)

   After independent sample t-test analysis, subjects of different genders found that there was significant difference in the "Chinese – Japanese style" of sample S4. Sample S5 has a significant difference between "heavy – lightweight". Samples S5 and S6 have significant differences in "hand-painted – photographed". Sample S6 has a significant difference in the "Japanese style – Western style". Sample S6 has significant difference in "cute – mature".
3. Selection of elements and style formation of Posters in the Sugar Industry in Taiwan during the Japanese-occupied Period

   This study focused 6 posters of the sugar industry in Taiwan during the Japan-Occupied period, deconstructing each selection element and style formation, proposing five analysis results. The first point is "Text Design

Principles", the main logo is mainly in Chinese, and the auxiliary origin information is identified in Japanese. The main marketing products and business names are presented in English for the purpose of selling Taiwan sugar around the world. Second, in terms of "presentation of expression techniques", the illustration and drawing techniques are mainly portrait illustrations (female person and boys). Apparel can see obvious differences, such as female person wearing Japanese kimonos that symbolize superior status, dresses influenced by western culture, traditional Chinese Tang shirts and navy feng shui handwear styles with wide collars and fold-over cuffs. The third aspect of "color application mode" is mainly to use hand-painted and converted into color printing for distribution, with the purpose of low price and easy marketing. The fourth point, "Arrangement and composition", mainly uses the main images (women and boys) in the screen as the main visual arrangement, and the relevant business name text is supplemented; The visual moving lines are mostly first contacted by the figures (female person and boys) in the middle of the poster, and then by the names of relevant businesses and other information. Fifth, in terms of "style formation", S1 presents a graceful and pleasant, lovely style, S2 presents a dynamic and sexy mature charm, S3 presents a traditional and respectful and dignified beauty, S4 presents a playful retro atmosphere, S5 presents a vibrant and relaxed atmosphere, and S6 presents a romantic, dreamy and confident atmosphere.

Posters are important medias to convey consciousness, also visual languages. Poster design uses selected elements to transform into visual symbols, giving it the function of transmitting information. The style constructed by the poster has influential power, which attracted the audience, and in fact promoted how they realice it. Today, various industrial posters in Taiwan during the Japan-Occupied Period have been preserved through digitization and become cultural assets of historical significance. The autor also analyzed the selected elements and style formation of these posters, exploring popular poster design style and the flourishing development of the printing industry at that time, feeling the phenomanon of social progress and atmosphere of economic prosperity and development in Taiwan at that time.

REFERENCES

Chen, L. S. (2020). *Teaching Intercultural Communication With an English-Chinese Course Design (Unpublished doctoral dissertation)*. National Taiwan Normal University, Taipei, Taiwan. [in Chinese, semantic translation]

Chen, Y. J. (2011). *Rainbow in Love – The Visual Research of Chinese Homosexuality Film Poster Design (Unpublished Master's Thesis)*. National Taichung University of Science and Technology, Taichung, Taiwan. [in Chinese, semantic translation]

Fei, G. J. (1999). *Research and Application of Surreal Expression Techniques in Poster Design (Unpublished Master's Thesis)*. National Taiwan Normal University, Taipei, Taiwan. [in Chinese, semantic translation]

He, Y. Z. (1975). *Graphic Design*. Taipei City. Taiwan: Lion Art [in Chinese, semantic translation]

Hu, Z. M. (1996). The development of modern poster art. *Journal of Aesthetic Education*. 76, 4. National Taiwan Arts Education Center. [in Chinese, semantic translation]

Kan, T. K. (1996). *Commercial Design Art*. Taipei City. Taiwan: Lion Art [in Chinese, semantic translation]

Lee, H. J. (2010). *Research About Design Form and Image Expression of Taiwan Art Exhibition Propagandistic Poster (Unpublished Master's dissertation)*. National Taiwan University of Arts, Taipei, Taiwan. [in Chinese, semantic translation]

Lin, J. L. (2013). Taiwan art poster development and international design trends. *Newsletter of the National Museum of Taiwan Literature*. 39, 32–41. National Museum of Taiwan Literature. [in Chinese, semantic translation]

Lin, K. L. (2019). *A Study on the Cross-Cultural Management of Service Encounters (Unpublished Master's dissertation)*. Ling Tung University, Taichung, Taiwan. [in Chinese, semantic translation]

Lin, P. C. (1993). *Introduction to Poster Design: The Technique of Poster Design*. Taipei Fine Arts Museum. [in Chinese, semantic translation]

Lin, P. Z. Lin, (1986). *Commercial Design*. Taipei City. Taiwan: Art Book [in Chinese, semantic translation]

Lin, R. G. Lin, (1993). *Commercial Design*. Taipei City. Taiwan: Art Book [in Chinese, semantic translation]

Tsai, H. L. (2018). *Stratification and Symbolization of Poster Design-Using Environment Ecological as a Creative Theme (Unpublished Master's dissertation)*. National University of Kaohsiung, Kaohsiung, Taiwan. [in Chinese, semantic translation]

Tsao, I. C. (2013). *Cross-cultural Experience of International Students in Taiwan (Unpublished Master's dissertation)*. National Chiayi University, Chiayi, Taiwan. [in Chinese, semantic translation]

Yuan, M. L. (1993). *Introduction to Poster Design: A Brief Introduction to Poster Design*. Taipei Fine Arts Museum. [in Chinese, semantic translation]

System Innovation for an Artificial Intelligence Era – Lam et al. (Eds)
© 2025 the Author(s), ISBN: 978-1-032-84452-7

The paradox of universal design's applicability in public spacefrom the church

Xiang-Yuan Zeng
School of Design·Straits Institute of Technology, Fujian University of Technology, Fuzhou, China
Regional Intangible Cultural Heritage Inheritance and Design Research Center and Fujian Institute of Technology, Fuzhou, China

Chen Wang, Bai-Yu Liu & Ya-Ling Yu
School of Design·Straits Institute of Technology, Fujian University of Technology, Fuzhou, China

ABSTRACT: The purpose is to show whether Universal Design can enhance how inclusive public space is. Examine how public space and Universal Design are related. To summarize and evaluate the relationship and present situation between Universal Design and public space, based on the perspective of the unity of opposites, consider the Universal Design in churches as an example. The inclusiveness of general public spaces can be enhanced by universal design, but not the inclusiveness of particular public spaces. By illustrating how space coordinates the relationship between man and nature, this method makes a case for the Universal Design of public space and opens up new territory for the Universal Design.

Keywords: Public space, Architecture, Universal Design, Church, Unity of opposites

1 INTRODUCTION

Architectural design frequently employs the Universal Design (UD) principle, which aims for maximum user acceptance. Church construction should follow the so-called "Universal Design" Principles, according to architects. A church is a limited public area because it is a public building. Public space and UD go hand in hand, as the idea of UD has matured.

We might be able to describe the connection between public space and UD using the method of contradiction and union of opposites. The Public itself carries the notion of universality. From the viewpoint of the church, on the one hand, the church restricts outsider participation; on the other hand, they seem to desire more people to practice their religion, which is perplexing. The position that this paper seeks to challenge is whether the idea of UD can enhance the inclusivity of public space and whether all public space should be inclusive. In the framework of UD, it is necessary to construct public space and present the relevance dilemma.

To uncover the contradictory link between public space and UD, this article seeks to disprove the prevailing wisdom that they work in tandem. Generally speaking, this study comes to the following conclusion: UD can enhance the inclusivity of general public space, but UD cannot enhance the inclusiveness of a particular public space.

This essay attempts to reflect on and clarify its doubts by examining the seemingly contradictory link between public space and UD rather than attempting to resolve it.

2 "SPACE" DISCUSSION

For the discussion of space, Michel Foucault first proposed the term "les heterotopies", compared with "utopia", "heterotopies" is a real, real "utopia" of a kind of real place, this place is the "utopia" that is effectively realized, and "Heterotopies" can be another space created. In architectural planning, the location of shopping malls, schools, hospitals, squares, and streets in the city, as well as walking routes, can be designed through UD, and the designed illusory space can reflect the real space.

In the daily life of human beings, public space is a universal phenomenon. It not only involves the sidewalk, and the space inside and outside the building but also includes the public space with "heterotopia" property. For example, libraries and museums are visible because of the infinite time accumulated in limited space. In public space (churches and houses, etc.), architectural works and sculptures stand. In museums, artworks of different times and peoples are displayed, forming a "heterotopia".

DOI: 10.1201/9781003514831-80

Just as a church is a public space, one can find it according to the route and the exact orientation, it belongs to the physical space; The church is also a ladder connecting with heaven, it is a refuge for human spiritual comfort, it is a psychological space; The church does not exist in isolation but needs to change according to the things in the environment, to deal with social relations and to give social value. It becomes a social space. The church itself not only brought practical space (missionary, prayer, charity, celebration, etc.), but also created mythological space, an invisible space full of fantasy, and at the same time it was a real public space, with "heterophobic" properties.

3 UNIVERSAL DESIGN PRINCIPLES

The Center for Universal Design at North Carolina State University proposes a definition of it: "Universal Design is a product and environment designed to be used by as many people as possible, without the need for adaptation or specialized design. Universal Design aims to simplify the lives of everyone by making products, communications, and the built environment as accessible as possible to as many people as possible at little or no additional cost. Universal Design benefits people of all ages and abilities."

Annett-Hitchcock strives to incorporate the concept of UD into products, where products and space should be designed to meet the needs of people with different physical and intellectual abilities. She gives the example of a shirt with adjustable sleeves that gives amputees the option of showing or not showing a limb.

4 AN EXPERIMENTAL CASE IN THE GROUND

The project of The new St James' Chapel is built on the site of the original church. The new church maintains the height of the original church and maintains the pointed roof of the cross. Considering the height of the cross and wind load, the size and shape of the building frame are determined. The church and outdoor terrace total 940 square meters, and four fulcrums support the entire building, giving a calm, peaceful, tough feeling. The design combines traditional church building elements with modern building forms, combining the church space with other functional space which can also be used alone. The church has a side-by-side layout, which can accommodate up to 400 people at one time.

This design maintains some of the basic elements of the original church. The worship space within the church is presented in a modern way, and a narrow painted window behind the original wooden cross creates a different sense of space. At the same time, through the renovation, the church space can be quickly transformed into another activity space, placing the original church model in a new, modern architecture.

4.1 *Design analysis*

The church's worship space uses lighting and wood trim walls provided by Decor Systems, and has glass stained-flower Windows in recessed angled siding. There are also two self-soundproof rolling walls in the church. When this wall is lowered, the central corridor connects Lober Square to the residential village below; With the rolling walls closed, the church and Dover Hall form a separate space, with a number of bright staircases and ancillary space.

The church is not isolated, but extends into the courtyard of an outdoor public space, which connects the church to a cafe in Lober Square, provides additional housing for retirees and a new cafe, and at the same time, the Lober House building was renovated to achieve a kind of "regenerative" public space. Not only is it a modern place of worship, but it can also house other functional space.

On the outside of the church, there is a lightweight, perforated aluminum-clad ceramic cladding and glass-clad building that allows sunlight to penetrate the pores to create dappled light and shadow, increasing light transmission inside. The interior of the church has excellent audio-visual equipment and a temperature control system, as well as soundproof curtains, allowing for great spatial flexibility. Traditional church building elements are incorporated into modern architecture.

4.2 *Project evaluation*

It is intended to construct a multifunctional space of 400 people like a church, a functional, visual and emotional public building space from formal religious activities to informal social gatherings, and on this basis, a series of space similar to the local community market is planned.

Using the concept of UD, the designer has redefined the church space, giving it more flexibility while retaining the core function of the religious space, providing a practical public building for more people and communities, and reinterpreting the function of the place of worship. In addition, the church completes the renovation of Angley Welfare House's hotel and health area in the Castle Hill area.

Through this case, we can find that public space should have strong variability and flexibility. For example, a church is not only functional but also a multifunctional space like a compound church, so as to meet the needs of different groups, thus achieving the concept of UD. Moreover, UD guides public space, making it more universal, more public, and more in line with people's functional and aesthetic needs.

5 PUBLIC GUIDANCE

Ulm advocates UD, placing people in the public environment at the center of design activities, trying to face all members of society, emphasizing that design is not private or exclusive, and proposing solutions that are universally applicable in production and life, so as to be different from the design centered on indoor everyday objects.

The electronic museum guide system is a good example. When users use it for the first time, try to choose a method that is general, simple, effective, and suitable for everyone. The first museums were inclusive because they were public, but in an era of globalization, where today's museum visitors are admittedly diverse, museums should be invented to anticipate the particularities of visitors and understand their characteristics and needs. For example, according to international convention, museums can be divided into staff, regular visitors, and strangers, and with the aid of text, graphics, speech and touch, visitors should have as many entrances as possible.

In guided public space such as museums and libraries, the public should be quickly familiar with these guiding systems through the concept of UD, so that more people can achieve fair use, simple and intuitive use, so that the guiding information can be perceived, and have the fault-tolerant ability in the face of complex systems. At the same time, it needs to have some flexibility. It is convenient for users to adjust the surrounding space environment and use needs so that the public space is more universal.

6 THE PUBLIC INTERVENTION

The 26th International Association of Museums has revised the definition of "museum": A museum is a long-term, non-profit, permanent institution that serves society and is open to the public... That is accessible and inclusive, that promotes diversity and sustainability... This concept is similar to the idea of UD, which also seeks to be people-oriented and inclusive.

Art galleries are also open to the public like museums, providing the audience with a whole-media experience including physical presence (eyes, ears, nose, tongue, and mind) to acquire knowledge and receive information. The art museum thus becomes an important system for art production, promoting "art into the community". The public naturally interacts with the white box space, thus forming a community system and generating new sequences of works. F.E.I. (Functionality, Elegance, Infinity) Arts serve as a medium connecting people and the city. It does some art projects of community connection and urban publicity in the daily public space. The Art Museum can also recruit influential local enterprises to find suitable patrons and cultivate the public's awareness of participation in public undertakings. The museum naturally intervenes in public space, transcending sociology, anthropology, new media, and other fields to show diversity and sustainability.

In fact, religion has both "public" and "private" attributes. The formation of public space itself needs to be created by various voices including religious groups. Religious groups actively participate in public affairs and let their voices be heard by society and the public. The public is increasingly involved in public space, using the sixth principle of UD: increasing the efficiency of use, making users comfortable, reducing fatigue, and forming a "user-centered" design.

On the one hand, "public" and "citizen" are opposites. They belong to the public and the private. The church separates the churchgoers from other people and is different in size; On the other hand, "public" and "citizen" are unified. The gathering of the "citizen" can lead to the "public", and the "public" can also guide the "citizen". At the same time, the sect hopes more people to join the team of believers, thus achieving the purpose of missionary work.

7 SOCIOLOGICAL CONSIDERATIONS

The concept of UD comes from the concept of architectural design. People with physical and mental disabilities can use barrier-free buildings without any restrictions or discrimination, and the omnidirectional barrier-free space is a space that everyone can use. But UD, even if it is developed on the basis of accessibility, can hardly achieve the goal of accessibility. First of all, it involves the problem of people with physical and mental disabilities themselves. There are places where accessibility is not designed for people with physical and mental disabilities, such as wheelchair ramps, which are also convenient for children and the elderly. In fact, there is no such thing as an ideal body. Everyone's body has certain "obstacles". In public space, we often see the words "old", "infirm", "sick", "disabled", "pregnant", "children" and other priorities. In fact, special rights are not special rights, but obstacles taken for granted in the social structure. From a sociological perspective, UD is based on the values that there is no discrimination, equality, and respect for everyone's rights.

It may be said that some churches have not been designed with the range of religious persons with physical and mental disabilities in mind and that there are few accessible facilities in public areas, making it difficult for these religious persons to reach places of worship. Even in the heart of worship, the dimensions between the seats are often too small to accommodate a wheelchair user, let alone blind paths and ancillary facilities such as handrails. This calls for the seventh principle of UD: to provide appropriate dimensions and Spaces for access, reach, manipulation, and use, regardless of the user's size, posture, or mobility.

However, barriers are a social construct that changes in response to social and cultural circumstances, and people with disabilities are in fact socially diverse, part of a multi-culture. As a social, political, and cultural problem, barriers will become a means of social restructuring.

8 AUXILIARY FUNCTION FOR DISABILITIES

Assistive functions created by Apple for people with limited or inconvenient vision, hearing, physical movement ability and cognitive ability, as well as providing them with helpful accessories and support, are built in from the beginning to achieve the purpose of accessibility.

Accessibility is a fundamental right for everyone. The visual auxiliary function, launched the amplifier and door detection, display setting, and scaling, as well as narration function, taking into account different perspectives; The listening auxiliary function also sets a perceptual reminder, voice recognition, earphone adjustment, and iPhone hearing aid device Made for text, sign language, signal; The auxiliary function of body movement ability introduced Siri to speed up the command and tap the back, and the operation of Apple Watch plus auxiliary touch to trigger various possibilities; In terms of cognitive ability assistance, it launched Safari browser reader, as well as reading content and background sound, in addition, there is voice control, switch control, guided access and iPad OS eye tracking function, which is committed to focusing attention, mining unlimited possibilities, to create a product that everyone can use. Apple announces: Only for you. Built-in features, tailored for you to have fun.

Apple not only built-in accessibility features but also provided users to build their own apps, with design guidelines and sample code, as well as discussions with Apple engineers to create a more inclusive and localized APP experience in the design and development process, taking UD to the extreme, paying timely attention to user dynamics and getting real-time user feedback, Is a very democratizing case.

9 CONCLUSION

Based on the existing theories, this paper makes a new interpretation of the coordination of the church as a public space. The general design concept should make appropriate adjustments and improvements to the special public space and assist with the guidance of anthropological and sociological theories. Perhaps UD can be used to draw new terrain, create a social enlightenment space to improve the public space, rethink how to design complex systems to accommodate multiple types of users and get good feedback from users. In addition, it is necessary to establish the connection between the human heart and the physical space outside, as well as the integration of architecture with the spatial environment. UD also applies to animals, plants, and any other human creation.

REFERENCES

Heideggers Philosophie der Kunst: eine systematische Interpretation der Holzwege-Abhandlung "Der Ursprung des Kunstwerkes". Friedrich-Wilhelm von Herrmann - 1994 - Verlag Vittorio Klostermann.

Maleuvre, Didier. "Must Museums Be Inclusive." *Journal of Educational Media, Memory, and Society* 4 (2012): 112–125.

Michel Foucault. (1966). *Les Mots et les choses : Une Archéologie Des Sciences Humaines*, pressed by Éditions Gallimard.

Moore, A., Lynch, H., & Boyle, B. (2022). Can universal design support outdoor play, social participation, and inclusion in public playgrounds? *A scoping review. Disability and Rehabilitation*, 44(13), 3304–3325.

William Lidwell, Kritina Holden, Jill Butler. (2003). *Universal Principles of Design. Rockport Publishers*, Inc. 224. https://magazine.alumni.ncsu.edu/2022/fashion-forward/

Zhang Heng-Hao. Special Education and Sociology of Disorder: A Theoretical Reflection. *Education and Social Research*, 2007, (13): 71–94. https://www.apple.com.cn/accessibility/.

System Innovation for an Artificial Intelligence Era – Lam et al. (Eds)
© 2025 the Editor(s), ISBN: 978-1-032-84452-7

Author index

Alcaraz, F. 55
Almada, D.G.L. 174
Amiri, A. 23
Armoa, M.A.G. 174
Arteaga, I.L. 23
Azirar, S. 270

Bai, M.-H. 7
Bao, S. 310
Barrail, F.L. 44
Bernijazov, R. 208
Broer, D.J. 23

Caceres, A.C.O. 49, 259
Cao, Y.-T. 316
Chang, C.-W. 140
Chang, H.-W. 134
Chang, Y.-I. 1, 320
Chang, Y.-K. 262
Chang, Y.-S. 387
Chao, L.-T. 332
Chavarnakul, T. 34
Chen 168
Chen, C.-H. 382
Chen, C.-Y. 221
Chen, H.-T. 108, 114
Chen, J.-T. 378
Chen, L.-M. 298
Chen, Y.-H. 221
Chen, Y.-J. 146
Chen, Y.-W. 217
Cheng, C.-Y. 80
Cheng, M.-Y. 97, 103
Chia, T.-L. 229, 235
Chia-fang, H. 197
Chiang, C.-C. 387
Chieh, L.-M. 140
Chien, C.-S. 126
Chien, H.H. 279
Chiu, H.-H. 97
Chiu, W.-C. 168
Chuang, C.-J. 298
Chuang, C.-Y. 194
Chuang, P.-C. 1
Chung, K.-H. 354
Colmán, C. 55
Cooharojananone, N. 34

Dumitrescu, R. 202, 208

Fan, C. 85
Fan, W.-H. 338
Fang, T.-K. 229

Gabbar, H.A. 270
Gancheva, V. 357, 372
Georgiev, I. 372
Gonzalez, R.I.A. 49
Guan, T.-W. 165
Guo, J.-X. 159

Ho, C.-H. 178
Hsiao, K.-H. 29, 92
Hsiao, Y.-Y. 194
Hsieh, T.-Y. 126
Hsu, C.-f. 156, 194, 242
Hsu, H.-C. 140, 146
Hsu, H.-H. 130
Hsu, L.-Y. 108, 114
Hu, N.-Z. 194
Huang, C.-C. 103
Huang, C.-H. 221
Huang, J.-J. 59
Huang, J.-W. 126
Huang, J.-Y. 316
Huang, J. 11
Huang, T.-C. 178
Huang, Y.-Y. 354
Huei, H.S. 279
Hung, Y.-H. 120
Hwang, S.-H. 397

Insfran, K.D.J. 49

Jung-Chin, T. 246, 253, 256

Koldewey, C. 202, 208
Ku, C.-J. 165
Kuan, H.-C. 287
Kuo, L.-Y. 156, 194, 242
Kuo, T.-Y. 126
Kwek, M. 259

Lai, J.-H. 39
Lai, J.-Y. 11
Lai, P.-J. 397
Lai, R.-B. 168
Lai, Y.-H. 97
Lan, C.-Y. 174
Lee, C.-L. 249
Lee, H.-Y. 224
Lee, K.-T. 1
Lee, T.-M. 126
Lee, W.-J. 387
Lee, Y.-J. 362
Leung, K. 197
Li, C. 120
Li, M.-H. 134
Li, Y.-J. 126
Lian, H.-W. 221
Liao, S.-C. 85
Lin, C.-C. 362
Lin, C.-M. 120
Lin, F.-S. 221
Lin, H.-C. 221
Lin, J.-L. 29, 92
Lin, J.-Y. 64
Lin, T.-Y. 190, 214
Lin, W.-T. 126
Lin, Y.-H. 130
Lin, Y.-J. 214
Liou, C.-F. 80
Liu, B.-Y. 408
Liu, H.-H. 108, 114
Liu, M.-K. 44, 49, 55
Liu, M.-R. 150
Liu, P.-C. 221
Liu, X.-Z. 378
Lu, S. 310

Machado, T.N. 259
Martinez, G.D. 49
Miranda, K.J. 168
Mondragon, A.E.C. 305

Pan, C.-Y. 134, 140, 146
Peng, L.-H. 7
Pereira, M.S. 44
Prior, S.D. 392
Putrayudanto, P. 11

Qiu, J.-Z. 320

Rashidi, S. 114
Regoui, C. 270
Rou, L. 85

Samaniego, A.F.G. 44
Sangkittiwan, J. 34
Shen, J.-C. 194
Shen, S.-T. 150, 392
Shen, S.-Y. 59, 262
Shihan, R. 382
Shirahama, K. 55
Smith, E.S.S. 174
Song, P.-P. 11
Stoev, H. 357

Taiphapoon, T. 34
Teng, C.-Y. 70
Ting, C.-S. 368
Tissen, D. 202, 208
Tsai, L.-C. 342
Tsai, Y.-C. 11
Tsai, Y.-T. 235
Tseng, J.-H. 156
Tseng, S.-M. 265
Tz-Ying, L. 190

van de Wouw, N. 23
Velazquez, A.T.T. 259
Vera Gomez, F.D. 44
von Glasenapp, T. 55
Von Glasenapp, T.R. 259

Wang, C.-M. 70, 76, 224, 265
Wang, C.-T. 126
Wang, C. 408
Wang, L.-P. 217
Wang, M.-H. 80
Wang, Q. 11
Wang, R. 347
Wei, K. 168
Wen, B.-Y. 368
Wen, J.-J. 326
Wiederkehr, I. 202, 208
Wu, C.-C. 320
Wu, S. 387
Wu, T.-L. 316
Wu, Y.-H. 130
Wu, Y.-T. 387

Xiao, P.-S. 332
Xu, Y. 347

Yan, W.-M. 108, 114
Yang, H.-S. 59
Yang, Y.-C. 76
Yang, Y.-L. 190
Yang, Y.-Z. 1
Yeh, J.-C. 168
Yeh, Y.-P. 259
Yi-Ling, Y. 246
Yin, K.L. 197
Ying, N.P. 197
You, C. 294
Young, K.-y. 19
Yu, T.-H. 165
Yu, Y.-J. 287
Yu, Y.-L. 408
Yu-Chia, T. 256
Yuan, C.-H. 183
Yuan, Y.-K. 19
Yueh-Ling, H. 253

Zeng, H. 85
Zeng, X.-Y. 408
Zhang, Y. 92
Zheng, M.-C. 64, 159, 183
Zhu, L. 310
Zhuo, X.-J. 92
Zong, Y. 347